T0396460

Springer Proceedings in Physics

Volume 261

Prafulla Kumar Behera · Vipin Bhatnagar ·
Prashant Shukla · Rahul Sinha

Editors

XXIII DAE High Energy Physics Symposium

Select Proceedings

Volume 1

 Springer

Editors
Prafulla Kumar Behera
Department of Physics
Indian Institute of Technology Madras
Chennai, India

Prashant Shukla
Bhabha Atomic Research Center
Mumbai, India

Vipin Bhatnagar
Panjab University
Chandigarh, India

Rahul Sinha
The Institute of Mathematical Sciences
Chennai, India

ISSN 0930-8989   ISSN 1867-4941 (electronic)
Springer Proceedings in Physics
ISBN 978-981-33-4407-5   ISBN 978-981-33-4408-2 (eBook)
https://doi.org/10.1007/978-981-33-4408-2

This Springer imprint is published by the registered company Springer Nature Singapore Pte Ltd.
The registered company address is: 152 Beach Road, #21-01/04 Gateway East, Singapore 189721, Singapore

Preface

Particle Physics has been at the forefront of all Physical sciences since the advent of the electron. Over the years, this field of Particle Physics has given more verities or types of particles than the different types of animals, generally found, in a city zoo! Such things are possible in this field due to the advancement in the theoretical understanding and the technological improvements happening all the time and up to some extent over a similar timescale. Many predictions done by the theoreticians were tested time and again in the experimental labs. Some of these were discovered, inferred or negated. This process is still on, but a bit slowed down due to challenges faced by the technology. For example, the present-day technology allows us to probe a distance of the order of 10-18 m (i.e., a decimal followed by 18 zeros) and at such tiny dimensions the entities that are seen, and are not further resolvable, are Electron and Quarks. These two types of entities, falling under a Generalized category called Fermions, are responsible for all the Matter in the visible Universe!

Moving on to the hunt for the "most sought-after" Higgs Boson ended on 4th July, 2012 by CERN-based Mega-Particle smasher aka LHC, the field appeared to be settling down for a moment for the trust in Standard Model of Particle Physics working extremely well. But, as witnessed with earlier discoveries, as usual it opened up questions on even deeper symmetries to be explored which bring out the various properties of the elementary particles once the "mass" source is accounted for. This in one sense translates to having more and more Center of Mass energy or Even packing more particles in the colliding bunches. These are technological challenges (along with other associated technologies) which attract a large number of manpower in these fields.

From the Particle Physics point of view, which is somewhat synonymous with LHC Physics these days, Super-Symmetry is the next level of improvement in the Standard Model most probed ever. Primordial soup has been extensively probed by many Heavy-ion accelerating instruments with very good outcomes on a new phase of matter—quark–gluon plasma, matter density probes, and hadron freeze-outs. The techniques and tools developed for such physics analyses are the sure shot straight outcomes applicable in many diverse fields in the society.

The "elusive neutrino" appears to be daunting the field heavily with the abrupt rise in the facilities appearing across the globe and opening up of the Astroparticle regime. The beyond SM signature shown by neutrinos has opened up the whole field now with connections being made from Supernovae, Galactic, Solar, Atmospheric, and Geo neutrinos. Interesting, but not many, properties from this sector have far-reaching deep impacts on the cosmological evolution of the Universe, like: Mass hierarchy, Study of CPV in neutrinos, Searches beyond the 3-neutrino framework, and Neutrino cross-sections. Even applications of these weakly interacting neutral leptons are emerging on the horizon having geopolitical and strategic linkage. Experimental setups like LIGO—searching for Gravitational Waves have initiated a complete new revolution in unraveling the mysteries of the Universe with the discovery of the Gravity waves. Thus, the linking of such experiments, with Particle Physics experiment has become stronger. The applications based on the technologies developed in the field were also presented, ranging from Medical Imaging,

The DAE-BRNS symposium, as usual, attracted 450 plus participants and assembled at the sprawling campus of IITM making "Particles" a buzzword for few days! There were invited talks by the experts, followed by sectional talks and parallel talks besides the famous poster sessions for the young minds. The DAE-BRNS-HEP symposium also witnessed participation from the industry and technocrats who have local to global footprints in the field. This led to the development of many Institute–Industry collaborative partnerships for the future of the field and embedded technologies. The proceedings book contains the selected papers covering almost all the aspects of particle physics highlighting the achievements from the past to the present and further illuminating the future path for the field.

Chennai, India   Prafulla Kumar Behera
Chandigarh, India   Vipin Bhatnagar
Mumbai, India   Prashant Shukla
Chennai, India   Rahul Sinha

Contents

Editors and Contributors

About the Editors

Dr. Prafulla Kumar Behera is currently an associate professor at the Department of Physics, Indian Institute of Technology Madras, Chennai. He obtained his Ph.D. from Utkal University, Bhubaneswar. He was a postdoctoral researcher at the University of Pennsylvania, USA and a research scientist at the University of Iowa, USA. His major research interests include experimental particle physics, detector building and data analysis. Dr. Behera was a member of BABAR experiment, Stanford, USA and ATLAS experiment, Geneva, Switzerland. Currently, he is a member of CMS and INO experiments, and he is also a member of the international advisory committee of VERTEX conference. He has published more than 1000 articles in respected international journals. He serves as a referee for several American Physical Society journals.

Dr. Vipin Bhatnagar is currently a professor at the Department of Physics, Panjab University, Chandigarh. He obtained his Ph.D. from Panjab University, Chandigarh, following which he was a postdoctoral fellow at LAL Orsay, France and CERN associate, CERN, Geneva. His major areas of research include experimental particle physics, detector building, computational physics and data analysis. He is a member of NOvA experiment, Fermilab, USA and CMS experiment, Geneva, Switzerland. He has published more than 700 journal articles in international peer-reviewed journals.

Dr. Prashant Shukla is a professor at the Nuclear Physics Division, Bhabha Atomic Research Center (BARC), Mumbai. He obtained his Ph.D. from Mumbai University, India. His research interests include high energy nuclear collisions, quark gluon plasma, and cosmic ray physics. Currently, he is a member of CMS and INO collaboration. He has published more than 150 articles in international journals of repute.

Dr. Rahul Sinha is currently a professor at the Institute of Mathematical Sciences, Chennai. He obtained his M.A and Ph.D from Rochester, USA, which was followed by a postdoctoral fellowship at the University of Alberta, Canada. His research primarily focuses on theoretical particle physics. Dr. Sinha is a member of BELLE and BELLE II experiment at KEK, Japan. He has supervised 11 doctoral students, and has also published more than 100 articles in international peer-reviewed journals.

Contributors

Sandeep Aashish Department of Physics, Indian Institute of Science Education and Research, Bhopal, India

Aman Abhishek Physical Research Laboratory, Navrangpura, Ahmedabad, India

J. Ablinger RISC, Johannes Kepler University, Linz, Austria

Debabrata Adak Department of Physics, Government General Degree College, West Bengal, India

Souvik Priyam Adhya Variable Energy Cyclotron Centre, Kolkata, India

Madan M. Aggarwal Panjab University, Chandigarh, India

Ritu Aggarwal Savitribai Phule Pune University, Pune, India;
Department of Science and Technology, New Delhi, India

Zubayer Ahammed Variable Energy and Cyclotron Centre, Kolkata, India

Asar Ahmed University of Delhi, Delhi, India

Rizwan Ahmed University of Delhi, Delhi, India

C. L. Ahmed Rizwan Department of Physics, National Institute of Technology Karnataka (NITK), Mangaluru, India

K. M. Ajith Department of Physics, National Institute of Technology Karnataka, Mangalore, India

Sampurn Anand Physical Research Laboratory, Ahmedabad, India

Richa Arya Physical Research Laboratory, Ahmedabad, India;
Indian Institute of Technology Gandhinagar, Gandhinagar, India

Abhishek Atreya Center For Astroparticle Physics and Space Sciences, Bose Institute, Kolkata, India

Anjali Attri Panjab University, Chandigarh, India

N. Ayyagiri Electronics Division, Bhabha Atomic Research Centre, Trombay, Mumbai, India

Partha Bagchi Variable Energy Cyclotron Centre, Kolkata, India

Bindu A. Bambah School of Physics, University of Hyderabad, Hyderabad, India

Aritra Bandyopadhyay Departamento de Física, Universidade Federal de Santa Maria, Santa Maria, RS, Brazil

Triparno Bandyopadhyay Department of Theoretical Physics, Tata Institute of Fundamental Research, Mumbai, India

Avik Banerjee Saha Institute of Nuclear Physics, HBNI, Kolkata, India

Pinaki Banerjee ICTS-TIFR, Bengaluru, India

Monika Bansal DAV College, Chandigarh, India

Sunil Bansal UIET, Panjab University, Chandigarh, India

R. C. Baral NISER, HBNI, Jatni, Bhubaneswar, Odisha, India

W. Bari University of Kashmir, Srinagar, India

Mitesh Kumar Behera University of Hyderabad, Hyderabad, India

Nirbhay Kumar Behera Department of Physics, Inha University, Incheon, Republic of Korea

Prafulla Kumar Behera Indian Institute of Technology Madras, Chennai, India

A. Behere Electronics Division, Bhabha Atomic Research Centre, Trombay, Mumbai, India

Rajkumar Bharathi Department of High Energy Physics, Tata Institute of Fundamental Research, Colaba, Mumbai, India

Akanksha Bhardwaj Physical Research Laboratory (PRL), Ahmedabad, Gujarat, India;
IIT Gandhinagar, Gandhinagar, India

Ashutosh Bhardwaj Centre for Detector and Related Software Technology, Department of Physics and Astrophysics, University of Delhi, Delhi, India

Shankita Bhardwaj Department of Physics and Astronomical Science, Central University of Himachal Pradesh, Dharamshala, India

Vishal Bhardwaj Department of Physical Sciences, IISER, Mohali, India

Vipin Bhatnagar Panjab University, Chandigarh, India

A. D. Bhatt Tata Institute of Fundamental Research, Mumbai, India;
Institute of Nuclear Physics Polish Academy of Sciences, Krakow, Poland

Jitesh R. Bhatt Theoretical Physics Division, Physical Research Laboratory, Ahmedabad, India

Sukannya Bhattacharya Theory Divison, Saha Institute of Nuclear Physics, Bidhannagar, Kolkata, India;
Theoretical Physics Division, Physics Research Laboratory, Ahmedabad, India

Gautam Bhattacharyya Saha Institute of Nuclear Physics, HBNI, Kolkata, India

Rik Bhattacharyya School of Physical Sciences, National Institute of Science Education and Research, HBNI, Jatni, India

Debabrata Bhowmik Saha Institute of Nuclear Physics, HBNI, Kolkata, India

P. S. Bhupal Dev Department of Physics and McDonnell Center for the Space Sciences, Washington University, St. Louis, MO, USA

M. Biswal Institute of Physics, Sachivalaya Marg, Bhubaneswar, India

Ambalika Biswas Department of Physics, Vivekananda College, Thakurpukur, India

S. Biswas Department of Physics and CAPSS, Bose Institute, Kolkata, India;
Department of Physics, National Institute of Technology, Durgapur, West Bengal, India

J. Blümlein DESY, Zeuthen, Germany

Debasish Borah Department of Physics, Indian Institute of Technology Guwahati, Guwahati, Assam, India

Ankita Budhraja Indian Institute of Science Education and Research, Bhopal, MP, India

Alexandra Carvalho Estonian Academy of Sciences, Tallinn, Estonia

Aleena Chacko Indian Institute of Technology Madras, Chennai, India

Kaustav Chakraborty Theoretical Physics Division, Physical Research Laboratory, Ahmedabad, India;
Discipline of Physics, Indian Institute of Technology, Gandhinagar, India

S. Chakraborty Department of Physics and CAPSS, Bose Institute, Kolkata, India

H. C. Chandola Department of Physics (UGC-Centre of Advanced Study), Kumaun University, Nainital, India

Sinjini Chandra Variable Energy Cyclotron Centre, Kolkata, India;
Homi Bhabha National Institute, Mumbai, India

Vinod Chandra Indian Institute of Technology Gandhinagar, Gandhinagar, Gujarat, India

V. B. Chandratre Electronics Division, Bhabha Atomic Research Centre, Trombay, Mumbai, India

Akshay Chatla School of Physics, University of Hyderabad, Hyderabad, India

Arindam Chatterjee Indian Statistical Institute, Kolkata, India

Bhaswar Chatterjee Department of Physics, Indian Institute of Technology Roorkee, Roorkee, India

S. Chatterjee Department of Physics and CAPSS, Bose Institute, Kolkata, India

Subhasis Chattopadhyay Variable Energy Cyclotron Centre, HBNI, Kolkata, India

Prakrut Chaubal Physical Research Laboratory, Ahmedabad, India

Ankur Chaubey Department of Physics, Institute of Science, Banaras Hindu University, Varanasi, India

Geetanjali Chaudhary Panjab University, Chandigarh, India

B. C. Chauhan Department of Physics and Astronomical Science, School of Physical and Material Sciences, Central University of Himachal Pradesh (CUHP), Dharamshala, Kangra, HP, India

Bhavesh Chauhan Physical Research Laboratory, Ahmedabad, India; Indian Institute of Technology, Gandhinagar, India

Garv Chauhan Department of Physics and McDonnell Center for the Space Sciences, Washington University, St. Louis, MO, USA

Sushil Singh Chauhan Department of Physics, Panjab University, Chandigarh, India

Sandhya Choubey Harish-Chandra Research Institute, Jhunsi, Allahabad, India; Department of Physics, School of Engineering Sciences, KTH Royal Institute of Technology, AlbaNova University Center, Stockholm, Sweden; Homi Bhabha National Institute, Mumbai, India

Debajyoti Choudhury Department of Physics and Astrophysics, University of Delhi, Delhi, India

S. Choudhury Indian Institute of Technology Hyderabad, Sangareddy, Telangana, India

Maria Agness Ciocci Department of Physics, University of Pisa, Pisa, Italy

Ranjeet Dalal Centre for Detector and Related Software Technology, Department of Physics and Astrophysics, University of Delhi, Delhi, India

Sanskruti Smaranika Dani Institute of Physics, PO: Sainik School, HBNI, Bhubaneswar, India

Arpan Das Theory Division, Physical Research Laboratory, Navrangpura, Ahmedabad, India

Dipankar Das Department of Astronomy and Theoretical Physics, Lund University, Lund, Sweden

Mrinal Kumar Das Department of Physics, Tezpur University, Tezpur, India

Pritam Das Department of Physics, Tezpur University, Tezpur, India

S. Das Department of Physics and CAPSS, Bose Institute, Kolkata, India

Ashutosh Dash School of Physical Sciences, National Institute of Science Education and Research, Jatni, India

V. M. Datar Department of High Energy Physics, Tata Institute of Fundamental Research, Colaba, Mumbai, India

S. De Department of Physics, Indian Institute of Technology Indore, Simrol, Indore, India

K. N. Deepthi Mahindra Ecole Centrale, Hyderabad, India

P. S. Bhupal Dev Department of Physics, McDonnell Center for the Space Sciences, Washington University, St. Louis, MO, USA

Mayuri Devee Department of Physics, University of Science and Technology Meghalaya, Ri-Bhoi, Baridua, India

Ram Krishna Dewanjee Laboratory of High Energy and Computational Physics, KBFI, Tallinn, Estonia

Atri Dey Regional Centre for Accelerator-based Particle Physics, Harish-Chandra Research Institute, HBNI, Jhunsi, Allahabad, India

Ujjal Dey Asia Pacific Center for Theoretical Physics, Pohang, Korea

Lobsang Dhargyal Harish-Chandra Research Institute, HBNI, Jhusi, Allahabad, India

S. Digal The Institute of Mathematical Sciences, Chennai, India

A. K. Dubey Variable Energy Cyclotron Centre, Kolkata, India

Sandeep Dudi Department of Physics, Panjab University, Chandigarh, India

Juhi Dutta Regional Centre for Accelerator-based Particle Physics, Harish-Chandra Research Institute, HBNI, Jhusi, Allahabad, India

Nirupam Dutta National Institute of Science Education and Research Bhubaneswar, Odisha, India

Rupak Dutta National Institute of Technology, Silchar, India

Keval Gandhi Department of Applied Physics, Sardar Vallabhbhai National Institute of Technology, Surat, Gujarat, India

S. Ganesh Department of Physics, Birla Institute of Technology and Science, Pilani, India

Mayukh Raj Gangopadhyay Theory Divison, Saha Institute of Nuclear Physics, Kolkata, India;
Centre for Theoretical Physics, Jamia Millia Islamia, New Delhi, India

Avijit K. Ganguly Department of Physics (MMV), Banaras Hindu University, Varanasi, India

Ila Garg Department of Physics, Indian Institute of Technology Bombay, Powai, Mumbai, India

Renu Garg Department of Physics, Panjab University, Chandigarh, India

V. Gaur Virginia Polytechnic Institute and State University, Blacksburg, Virginia, India

Rajiv V. Gavai Tata Institute of Fundamental Research, Colaba, Mumbai, India

Elizabeth George Department of Physics, Indian Institute of Technology Bombay, Mumbai, India

Pulkit S. Ghoderao Indian Institute of Technology Bombay, Powai, Mumbai, India

C. Ghosh Variable Energy Cyclotron Centre, Kolkata, India;
Homi Bhabha National Institute, Mumbai, India

S. K. Ghosh Department of Physics and CAPSS, Bose Institute, Kolkata, India

Snigdha Ghosh IIT Gandhinagar, Gandhinagar, Gujarat, India

Sovan Ghosh Post Graduate, Department of Physics, Vijaya College, Bangalore, India

Anjan K. Giri Indian Institute of Technology, Hyderabad, Kandi, India

U. Gokhale Department of High Energy Physics, Tata Institute of Fundamental Research, Colaba, Mumbai, India

Mohit Gola University of Delhi, Delhi, India

Srubabati Goswami Theoretical Physics Division, Physical Research Laboratory, Ahmedabad, India

Rajat Gupta Panjab University, Chandigarh, India

Najmul Haque School of Physical Sciences, National Institute of Science Education and Research, HBNI, Jatni, Khurda, India

Nikhil Hatwar Department of Physics, Birla Institute of Technology and Science, Pilani, India

Honey Tata Institute of Fundamental Research, Mumbai, India

D. Indumathi The Institute of Mathematical Sciences, Tharamani, Chennai, Tamil Nadu, India;
Homi Bhabha National Institute, Anushaktinagar, Mumbai, Maharashtra, India

Vijay Iyer School of Physical Sciences, National Institute of Science Education and Research, HBNI, Jatni, India

Ambar Jain Indian Institute of Science Education and Research, Bhopal, MP, India

Chakresh Jain Centre for Detector and Related Software Technology, Department of Physics and Astrophysics, University of Delhi, Delhi, India

Geetika Jain Centre for Detector and Related Software Technology, Department of Physics and Astrophysics, University of Delhi, Delhi, India

Manoj K. Jaiswal Department of Physics, University of Allahabad, Prayagraj, India

Bharti Jarwal Department of Physics and Astrophysics, University of Delhi, Delhi, India

Abhik Jash School of Physical Sciences, National Institute of Science Education and Research, HBNI, Jatni, Odisha, India

Pawan Joshi Department of Physics, Indian Institute of Science Education and Research Bhopal, Bhopal, India

S. R. Joshi Department of High Energy Physics, Tata Institute of Fundamental Research, Colaba, Mumbai, India

Anjan S. Joshipura Theoretical Physics Division, Physical Research Laboratory, Ahmedabad, India

Jan Kalinowski Faculty of Physics, University of Warsaw, Warsaw, Poland

A. B. Kaliyar Indian Institute of Technology Madras, Chennai, India

D. Kalra Department of Physics, Panjab University, Chandigarh, India

Bithika Karmakar Theory Division, Saha Institute of Nuclear Physics, HBNI, Kolkata, India

Saikat Karmakar Tata Institute of Fundamental Research, Mumbai, India

Monal Kashav Department of Physics and Astronomical Science, Central University of Himachal Pradesh, Dharamshala, India

Varchaswi K. S. Kashyap School of Physical Sciences, National Institute of Science Education and Research, HBNI, Jatni, Odisha, India

Daljeet Kaur S.G.T.B. Khalsa College, University of Delhi, New Delhi, India

Manjit Kaur Panjab University, Chandigarh, India

P. K. Kaur Department of High Energy Physics, Tata Institute of Fundamental Research, Colaba, Mumbai, India

Najimuddin Khan Centre for High Energy Physics, Indian Institute of Science, Bangalore, India

Anisa Khatun Department of Physics, Aligarh Muslim University, Aligarh, India

Virendrasinh Kher Applied Physics Department, Polytechnic, The M S University of Baroda, Vadodara, Gujarat, India

Bharti Kindra Physical Research Laboratory, Ahmedabad, India; Indian Institute of Technology, Gandhinagar, India

H. Kolla Electronics Division, Bhabha Atomic Research Centre, Trombay, Mumbai, India

Jyothsna Rani Komaragiri Indian Institute of Science, Bengaluru, India

Partha Konar Theoretical Physics Group, Physical Research Laboratory, Ahmedabad, India

Paweł Kozów Faculty of Physics, University of Warsaw, Warsaw, Poland

A. Kumar Variable Energy Cyclotron Centre, Kolkata, India; Homi Bhabha National Institute, Mumbai, India

Abhass Kumar Physical Research Laboratory, Ahmedabad, India; Harish-Chandra Research Institute, Jhunsi, Allahabad, India

Ajit Kumar Homi Bhabha National Institute, Mumbai, India

Anil Kumar Saha Institute of Nuclear Physics, Kolkata, India; Homi Bhabha National Institute, Trombay, Mumbai, India

Arvind Kumar Dr. B R Ambedkar National Institute of Technology, Jalandhar, Punjab, India

Ashok Kumar Department of Physics and Astrophysics, University of Delhi, Delhi, India

Hemant Kumar Department of Physics & Astrophysics, University of Delhi, Delhi, India

J. Kumar Variable Energy Cyclotron Centre, Kolkata, India

M. Kumar Malaviya National Institute of Technology, Jaipur, India

Nilanjana Kumar Department of Physics and Astrophysics, University of Delhi, Delhi, India

Priyanka Kumar Department of Physics, Cotton University, Guwahati, Assam, India

Rajesh Kumar Dr. B R Ambedkar National Institute of Technology, Jalandhar, Punjab, India

Ramandeep Kumar Akal University Talwandi Sabo, Punjab, India

Sanjeev Kumar Department of Physics and Astrophysics, University of Delhi, New Delhi, India

Sunil Kumar Department of Physics, Panjab University, Chandigarh, India

Utkarsh Kumar Department of Physics, Ariel University, Ariel, Israel

V. Kumar Saha Institute of Nuclear Physics, Kolkata, India; Homi Bhabha National Institute, Trombay, Mumbai, India

Ajay Kumar Rai Department of Applied Physics, Sardar Vallabhbhai National Institute of Technology, Surat, Gujarat, India

Santosh Kumar Rai Regional Centre for Accelerator-based Particle Physics, Harish-Chandra Research Institute, HBNI, Jhusi, Allahabad, India

Priyanka Kumari Panjab University, Chandigarh, India

Suman Kumbhakar Indian Institute of Technology Bombay, Mumbai, India

Sourav Kundu School of Physical Sciences, National Institute of Science Education and Research, HBNI, Jatni, India

Manu Kurian Indian Institute of Technology Gandhinagar, Gandhinagar, Gujarat, India

Apurba Laha Department of Electrical Engineering, Indian Institute of Technology Bombay, Mumbai, India

Amitabha Lahiri SNBNCBS, Kolkata, India

Anirban Lahiri Fakultät für Physik, Universität Bielefeld, Bielefeld, Germany

Jayita Lahiri Regional Centre for Accelerator-based Particle Physics, Harish-Chandra Research Institute, HBNI, Jhunsi, Allahabad, India

S. M. Lakshmi Indian Institute of Technology Madras, Chennai, India

K. Lalwani Malaviya National Institute of Technology, Jaipur, India

Gaetano Lambiase Dipartimento di Fisica "E.R. Caianiello", Universitá di Salerno, Fisciano (SA), Italy

James F. Libby Indian Institute of Technology Madras, Chennai, India

Manisha Lohan IRFU, CEA, Université Paris-Saclay, Gif-sur-Yvette, France

A. Lokapure Department of High Energy Physics, Tata Institute of Fundamental Research, Colaba, Mumbai, India

Kallingalthodi Madhu Department of Physics, BITS-Pilani, Zuarinagar, India

Namit Mahajan Physical Research Laboratory, Ahmedabad, India

Rajesh K. Maiti IISER Mohali, Punjab, India

Snehanshu Maiti Indian Institute of Technology Madras, Chennai, India

Rudra Majhi University of Hyderabad, Hyderabad, India

P. Maji Department of Physics, National Institute of Technology Durgapur, Durgapur, West Bengal, India

Debasish Majumdar Astroparticle Physics and Cosmology Division, Saha Institute of Nuclear Physics, HBNI, Kolkata, India

Nayana Majumdar Saha Institute of Nuclear Physics, HBNI, Kolkata, India

Devdatta Majumder University of Kansas, Lawrence, Kansas, USA

Gobinda Majumder DHEP, Tata Institute of Fundamental Research, Mumbai, India

B. Mallick Institute of Physics, HBNI, Bhubaneswar, India

Debasish Mallick National Institute of Science Education and Research, HBNI, Jatni, India

Rusa Mandal IFIC, Universitat de València-CSIC, València, Spain

A. Manna Electronics Division, Bhabha Atomic Research Centre, Trombay, Mumbai, India

P. Marquard DESY, Zeuthen, Germany

A. Maulik Department of Physics and CAPSS, Bose Institute, Kolkata, India

Arindam Mazumdar Physical Research Laboratory, Ahmedabad, India

Alberto Messineo Department of Physics, University of Pisa, Pisa, Italy

Aditya Nath Mishra Instituto de Ciencias Nucleares, UNAM, CDMX, Mexico

Arvind Kumar Mishra Theoretical Physics Division, Physical Research Laboratory, Ahmedabad, India;
Indian Institute of Technology Gandhinagar, Gandhinagar, India

Dheeraj Kumar Mishra The Institute of Mathematical Sciences, Chennai, India;
Homi Bhabha National Institute, Mumbai, India

Hiranmaya Mishra Theory Division, Physical Research Laboratory, Navrangpura, Ahmedabad, India

M. Mishra Department of Physics, Birla Institute of Technology and Science, Pilani, India

Subhasmita Mishra IIT Hyderabad, Hyderabad, India

Akhila Mohan Department of Physics, BITS-Pilani, Zuarinagar, India

Lakshmi S. Mohan Indian Institute of Technology Madras, Chennai, India

Rukmani Mohanta School of Physics, University of Hyderabad, Hyderabad, Telangana, India

Bedangadas Mohanty School of Physical Sciences, National Institute of Science Education and Research, HBNI, Jatni, Bhubaneswar, Odisha, India; Department of Experimental Physics, CERN, Geneva, Switzerland

G. B. Mohanty Tata Institute of Fundamental Research, Mumbai, India

Subhendra Mohanty Physical Research Laboratory, Ahmedabad, India

Manas K. Mohapatra Indian Institute of Technology, Hyderabad, Kandi, India

R. N. Mohapatra Department of Physics, Maryland Center for Fundamental Physics, University of Maryland, College Park, MD, USA

Ranjita K. Mohapatra Department of Physics, Indian Institute of Technology Bombay, Mumbai, India; Institute of Physics, Bhubaneswar, India; Homi Bhabha National Institute, Mumbai, India

S. Moitra Electronics Division, Bhabha Atomic Research Centre, Trombay, Mumbai, India

Mitali Mondal Variable Energy and Cyclotron Centre, Kolkata, India

N. K. Mondal HENPPD, Saha Institute of Nuclear Physics, Kolkata, India

S. Mondal Tata Institute of Fundamental Research, Mumbai, India

Suryanarayan Mondal NPD, Homi Bhaba National Institute, Mumbai, India; DHEP, Tata Institute of Fundamental Research, Mumbai, India

Sanjib Muhuri Variable Energy Cyclotron Centre, Kolkata, India

Ananya Mukherjee Department of Physics, Tezpur University, Tezpur, India

Arghya Mukherjee Saha Institute of Nuclear Physics, Kolkata, India

Tamal K. Mukherjee Department of Physics, School of Sciences, Adamas university, Kolkata, India

Sourav Mukhopadhyay Bhabha Atomic Research Centre, Mumbai, India

Supratik Mukhopadhyay Saha Institute of Nuclear Physics, HBNI, Kolkata, India

Upala Mukhopadhyay Astroparticle Physics and Cosmology Division, Saha Institute of Nuclear Physics, HBNI, Kolkata, India

Biswarup Mukhopadhyaya Regional Centre for Accelerator-based Particle Physics, Harish-Chandra Research Institute, HBNI, Jhunsi, Allahabad, India

Lakshmi P Murgod Department of Physics, Central University of Karnataka, Kalaburagi, India

M. V. N. Murthy The Institute of Mathematical Sciences, Chennai, India

Munshi G. Mustafa Theory Division, Saha Institute of Nuclear Physics, HBNI, Kolkata, India

P. Nagaraj Department of High Energy Physics, Tata Institute of Fundamental Research, Colaba, Mumbai, India

Srishti Nagu Department of Physics, Lucknow University, Lucknow, India

Bharati Naik Indian Institute of Technology Bombay, Mumbai, India

Md. Naimuddin Department of Physics and Astrophysics, University of Delhi, Delhi, India

Dibyendu Nanda Department of Physics, Indian Institute of Technology Guwahati, Assam, India

Ekata Nandy Variable Energy Cyclotron Centre, HBNI, Kolkata, India; Homi Bhabha National Institute, Mumbai, India

Ashish Narang Physical Research Laboratory, Ahmedabad, India; Indian Institute of Technology Gandhinagar, Gandhinagar, India

Nimmala Narendra Indian Institute of Technology Hyderabad, Kandi, Sangareddy, Telangana, India

Newton Nath Institute of High Energy Physics, Chinese Academy of Sciences, Beijing, China; School of Physical Sciences, University of Chinese Academy of Sciences, Beijing, China

A. Naveena Kumara Department of Physics, National Institute of Technology Karnataka (NITK), Mangaluru, India

Surya Narayan Nayak Jyoti Vihar, Burla, Sambalpur, Odisha, India

T. K. Nayak Variable Energy Cyclotron Centre, Kolkata, India; CERN, Geneva, Switzerland

P. Nayek Department of Physics, National Institute of Technology, Durgapur, West Bengal, India

V. Negi Variable Energy Cyclotron Centre, Kolkata, India

M. Nizam Homi Bhabha National Institute, Mumbai, India; Tata Institute of Fundamental Research, Mumbai, India

Manjunath Omana Kuttan Department of Physics, Central University of Karnataka, Kalaburagi, India

Abhilash Padhy Department of Physics, Indian Institute of Science Education and Research, Bhopal, India

A. Padmini Electronics Division, Bhabha Atomic Research Centre, Trombay, Mumbai, India

Rita Paikaray Department of Physics, Ravenshaw University, Cuttack, India

N. Panchal Homi Bhabha National Institute, Mumbai, India;
Tata Institute of Fundamental Research, Mumbai, India

Sukanta Panda Department of Physics, Indian Institute of Science Education and Research Bhopal, Bhopal, India;
Department of Physics, Ariel University, Ariel, Israel

Susil Kumar Panda Department of Physics, Ravenshaw University, Cuttack, India

H. C. Pandey Department of Physics, Birla Institute of Applied Sciences, Bhimtal, India

Madhurima Pandey Astroparticle Physics and Cosmology Division, Saha Institute of Nuclear Physics, HBNI, Kolkata, India

Sujata Pandey Discipline of Physics, Indian Institute of Technology Indore, Indore, India

Sudhir Pandurang Rode Indian Institute of Technology Indore, Indore, Madhya Pradesh, India

J. N. Pandya Faculty of Technology and Engineering, Applied Physics Department, The Maharaja Sayajirao University of Baroda, Vadodara, Gujarat, India

Lata Panwar Indian Institute of Science, Bengaluru, India

Priyank Parashari Physical Research Laboratory, Ahmedabad, India;
Indian Institute of Technology Gandhinagar, Gandhinagar, India

Bibhuti Parida Tomsk State University, Tomsk, Russia

M. K. Parida CETMS, SOA University, Bhubaneswar, India

Sonia Parmar Panjab University, Chandigarh, India

Avani Patel Indian Institute of Science Education and Research Bhopal, Bhopal, India

Shesha D. Patel Faculty of Science, Physics Department, The Maharaja Sayajirao University of Baroda, Vadodara, Gujarat, India

Vikas Patel Department of Applied Physics, Sardar Vallabhbhai National Institute of Technology, Surat, Gujarat, India

Mahadev Patgiri Department of Physics, Cotton University, Guwahati, Assam, India

Pathaleswar Department of High Energy Physics, Tata Institute of Fundamental Research, Colaba, Mumbai, India

Sourav Patra IISER Mohali, Punjab, India

Avik Paul Astroparticle Physics and Cosmology Division, Saha Institute of Nuclear Physics, HBNI, Kolkata, India

S. Pethuraj NPD, Homi Bhaba National Institute, Mumbai, India; DHEP, Tata Institute of Fundamental Research, Mumbai, India

Aman Phogat Department of Physics and Astrophysics, University of Delhi, Delhi, India

Shailesh Pincha Department of Applied Physics, Sardar Vallabhbhai National Institute of Technology, Surat, Gujarat, India

Soumita Pramanick Department of Physics, University of Calcutta, Kolkata, India; Harish-Chandra Research Institute, Jhunsi, Allahabad, India

S. K. Prasad Department of Physics and CAPSS, Bose Institute, Kolkata, India

Bhabani Prasad Mandal Department of Physics, Institute of Science, Banaras Hindu University, Varanasi, India

Massimiliano Procura Fakultät für Physik, Universität Wien, Wien, Austria; Theoretical Physics Department, CERN, Geneva, Switzerland

M. Punna Electronics Division, Bhabha Atomic Research Centre, Trombay, Mumbai, India

Moh. Rafik Department of Physics and Astrophysics, University of Delhi, Delhi, India

S. Raha Department of Physics and CAPSS, Bose Institute, Kolkata, India

K. V. Rajani Department of Physics, National Institute of Technology Karnataka (NITK), Mangalore, India

N. Rajeev National Institute of Technology, Silchar, India

S. Rajkumarbharathi Tata Institute of Fundamental Research, Mumbai, India

Harishankar Ramachandran Indian Institute of Technology Madras, Chennai, India

P. Ramadevi Indian Institute of Technology Bombay, Powai, Mumbai, India

Arun Rana Department of Physics, Indian Institute of Science Education and Research, Bhopal, India

N. Rana DESY, Zeuthen, Germany; INFN, Milano, Italy

Raghavan Rangarajan School of Arts and Sciences, Ahmedabad University, Ahmedabad, India

Kirti Ranjan Centre for Detector and Related Software Technology, Department of Physics and Astrophysics, University of Delhi, Delhi, India

R. Rath Department of Physics, Indian Institute of Technology Indore, Simrol, Indore, India

Haresh Raval Department of Physics, Institute of Science, Banaras Hindu University, Varanasi, India;
Department of Physics, Indian Institute of Technology Delhi, New Delhi, India

K. C. Ravindran Department of High Energy Physics, Tata Institute of Fundamental Research, Colaba, Mumbai, India

Deependra Singh Rawat Department of Physics (UGC-Centre of Advanced Study), Kumaun University, Nainital, India

Atasi Ray School of Physics, University of Hyderabad, Hyderabad, India

Rajarshi Ray Department of Physics & Center for Astroparticle Physics & Space Science, Bose Institute, Kolkata, India

Amitava Raychaudhuri Department of Physics, University of Calcutta, Kolkata, India

Karaparambil Rajan Rebin Indian Institute of Technology Madras, Chennai, India

P. K. Resmi Indian Institute of Technology Madras, Chennai, India

Niharika Rout Indian Institute of Technology, Madras, India

Promita Roy Saha Institute of Nuclear Physics, HBNI, Kolkata, India

S. Roy Department of Physics and CAPSS, Bose Institute, Kolkata, India

Sahithi Rudrabhatla Department of Physics, University of Illinois at Chicago, Chicago, IL, USA

Samrangy Sadhu Variable Energy Cyclotron Centre, Kolkata, India;
Homi Bhabha National Institute, Mumbai, India

Soumya Sadhukhan Department of Physics and Astrophysics, University of Delhi, Delhi, India

Rajib Saha Indian Institute of Science Education and Research, Bhopal, MP, India

Debashis Sahoo TIFR, Mumbai, India

Pragati Sahoo Discipline of Physics, School of Basic Sciences, Indian Institute of Technology Indore, Indore, India

Raghunath Sahoo Discipline of Physics, School of Basic Sciences, Indian Institute of Technology Indore, Indore, India

S. Sahoo Department of Physics, National Institute of Technology, Durgapur, West Bengal, India

Sarita Sahoo CETMS, SOA University, Bhubaneswar, India; Institute of Physics, HBNI, Bhubaneswar, India

Suchismita Sahoo Theoretical Physics Division, Physical Research Laboratory, Ahmedabad, India

Narendra Sahu Indian Institute of Technology Hyderabad, Kandi, Sangareddy, Telangana, India

P. K. Sahu Institute of Physics, HBNI, Bhubaneswar, India

S. Sahu Institute of Physics, HBNI, Bhubaneswar, India

S. K. Sahu Institute of Physics, HBNI, Bhubaneswar, India

Jogender Saini Variable Energy Cyclotron Centre, Kolkata, India

Jyoti Saini Indian Institute of Technology Jodhpur, Jodhpur, India

Amrutha Samalan Department of Physics, Central University of Karnataka, Kalaburagi, India

Rome Samanta Physics and Astronomy, University of Southampton, Southampton, UK

Subhasis Samanta School of Physical Sciences, National Institute of Science Education and Research, HBNI, Jatni, Bhubaneswar, India

Deepak Samuel Department of Physics, Central University of Karnataka, Kalaburagi, India

Kaur Sandeep Panjab University, Chandigarh, India

S. Sandilya University of Cincinnati, Cincinnati, Ohio, India

M. N. Saraf Department of High Energy Physics, Tata Institute of Fundamental Research, Colaba, Mumbai, India

Pradeep Sarin Department of Physics, Indian Institute of Technology Bombay, Mumbai, India

Sandip Sarkar Saha Institute of Nuclear Physics, HBNI, Kolkata, India

B. Satyanarayana Department of High Energy Physics, Tata Institute of Fundamental Research, Colaba, Mumbai, India

P. S. Saumia Bogoliubov Laboratory of Theoretical Physics, JINR, Dubna, Russia

H. Saveetha Institute of Mathematical Sciences, Chennai, India

S. Sawant Tata Institute of Fundamental Research, Mumbai, India

C. Schneider RISC, Johannes Kepler University, Linz, Austria

J. Selvaganapathy Theoretical Physics Group, Physical Research Laboratory, Ahmedabad, India

Pritam Sen The Institute of Mathematical Sciences, Tharamani, Chennai, Tamil Nadu, India;
Homi Bhabha National Institute, Anushaktinagar, Mumbai, Maharashtra, India

Wadut Shaikh Saha Institute of Nuclear Physics, HBNI, Kolkata, India

Anjali Sharma Panjab University, Chandigarh, India

Ashish Sharma IIT Madras, Chennai, Tamil Nadu, India

Gazal Sharma Department of Physics and Astronomical Science, School of Physical and Material Sciences, Central University of Himachal Pradesh (CUHP), Dharamshala, Kangra, HP, India

Meenakshi Sharma Department of Physics, University of Jammu, Jammu and Kashmir, Jammu, India

Umesh Shas Department of High Energy Physics, Tata Institute of Fundamental Research, Colaba, Mumbai, India

R. R. Shinde Department of High Energy Physics, Tata Institute of Fundamental Research, Colaba, Mumbai, India

Anup Kumar Sikdar Indian Institute of Technology Madras, Chennai, India

G. Sikder University of Calcutta, Kolkata, India

S. Sikder Electronics Division, Bhabha Atomic Research Centre, Trombay, Mumbai, India

D. Sil Department of High Energy Physics, Tata Institute of Fundamental Research, Colaba, Mumbai, India

Dipankar Sil Tata Institute of Fundamental Research, Mumbai, India

R. N. Singaraju Variable Energy Cyclotron Centre, Kolkata, India

Captain R. Singh Department of Physics, Birla Institute of Technology and Science, Pilani, India

J. B. Singh Department of Physics, Panjab University, Chandigarh, India

Jagbir Singh Panjab University, Chandigarh, India

Janardan P. Singh Faculty of Science, Physics Department, The Maharaja Sayajirao University of Baroda, Vadodara, Gujarat, India

Jaydip Singh Department of Physics, Lucknow University, Lucknow, India

Jyotsna Singh Department of Physics, Lucknow University, Lucknow, India

Lakhwinder Singh Institute of Physics, Academia Sinica, Taipei, Taiwan

Ravindra Singh Department of Electrical Engineering, Indian Institute of Technology Bombay, Mumbai, India

S. Somorendro Singh Department of Physics and Astrophysics, University of Delhi, Delhi, India

V. Singhal Variable Energy Cyclotron Centre, Kolkata, India

Shivaramakrishna Singirala Indian Institute of Technology Indore, Simrol, Madhya Pradesh, India;
University of Hyderbad, Hyderabad, India

Roopam Sinha Saha Institute of Nuclear Physics, Kolkata, India

N. R. Soni Faculty of Technology and Engineering, Applied Physics Department, The Maharaja Sayajirao University of Baroda, Vadodara, Gujarat, India

C. Soumya Institute of Physics, Bhubaneswar, India

Vipin Sudevan Indian Institute of Science Education and Research, Bhopal, MP, India

M. Sukhwani Electronics Division, Bhabha Atomic Research Centre, Trombay, Mumbai, India

V. Sunilkumar Department of Physics, UC College, Aluva, Kerala, India

S. Swain Institute of Physics, HBNI, Bhubaneswar, India;
School of Physics, Sambalpur University, Sambalpur, India

Sagarika Swain Institute of Physics, PO: Sainik School, HBNI, Bhubaneswar, India

Michał Szleper National Center for Nuclear Research, High Energy Physics Department, Warsaw, Poland

S. H. Thoker University of Kashmir, Srinagar, India

M. Thomas Electronics Division, Bhabha Atomic Research Centre, Trombay, Mumbai, India

Swatantra Kumar Tiwari Discipline of Physics, School of Basic Sciences, Indian Institute of Technology Indore, Indore, India

Sławomir Tkaczyk Fermi National Accelerator Laboratory, Batavia, IL, USA

K. Trabelsi Laboratory of the Linear Accelerator (LAL), Orsay, France

Jyoti Tripathi Department of Physics, Panjab University, Chandigarh, India

A. Tripathy Institute of Physics, HBNI, Bhubaneswar, India;
Utkal University, Bhubaneswar, India

Sushanta Tripathy Department of Physics, Indian Institute of Technology Indore, Simrol, India

L. Umesh Tata Institute of Fundamental Research, Mumbai, India

S. S. Upadhya Department of High Energy Physics, Tata Institute of Fundamental Research, Colaba, Mumbai, India

Deepak Vaid Department of Physics, National Institute of Technology Karnataka (NITK), Mangalore, India

Ton van den Brink Utrecht University, Utrecht, The Netherlands

P. Verma Tata Institute of Fundamental Research, Mumbai, India

Surender Verma Department of Physics and Astronomical Science, School of Physical and Material Sciences, Central University of Himachal Pradesh (CUHP), Dharamshala, Kangra, HP, India

K. N. Vishnudath Theoretical Physics Division, Physical Research Laboratory, Ahmedabad, India;
Discipline of Physics, Indian Institute of Technology, Gandhinagar, India

A. S. Vytheeswaran Department of Theoretical Physics, University of Madras, Chennai, India

Dinesh Yadav Department of Physics (UGC-Centre of Advanced Study), Kumaun University, Nainital, India

M. Younus Department of Physics, Nelson Mandela University, Port Elizabeth, South Africa

E. Yuvaraj Department of High Energy Physics, Tata Institute of Fundamental Research, Colaba, Mumbai, India

Yongchao Zhang Department of Physics and McDonnell Center for the Space Sciences, Washington University, St. Louis, MO, USA

Chapter 1
Study of Charmless Decays
$B^\pm \to K_S^0 K_S^0 h^\pm$ ($h = K, \pi$) at Belle

A. B. Kaliyar, Prafulla Kumar Behera, G. B. Mohanty, and V. Gaur

Abstract We report a search for charmless hadronic decays of charged B mesons to the final states $K_S^0 K_S^0 K^\pm$ and $K_S^0 K_S^0 \pi^\pm$. The results are based on a $711\,\text{fb}^{-1}$ data sample that contains 772×10^6 $B\bar{B}$ pairs, and was collected at the $\Upsilon(4S)$ resonance with the Belle detector at the KEKB asymmetric-energy e^+e^- collider. For $B^\pm \to K_S^0 K_S^0 K^\pm$ decays, the measured branching fraction and direct CP asymmetry are $[10.42 \pm 0.43(\text{stat}) \pm 0.22(\text{syst})] \times 10^{-6}$ and $[+1.6 \pm 3.9(\text{stat}) \pm 0.9(\text{syst})]\%$, respectively. In the absence of a statistically significant signal for $B^\pm \to K_S^0 K_S^0 \pi^\pm$, we set the 90% confidence level upper limit on its branching fraction at 8.7×10^{-7}.

1.1 Introduction

Charged B-meson decays to three-body charmless hadronic final states $K_S^0 K_S^0 K^\pm$ and $K_S^0 K_S^0 \pi^\pm$ mainly proceed via the $\bar{b} \to \bar{s}$ and $\bar{b} \to \bar{d}$ loop transitions, respectively. Figure 1.1 shows the dominant Feynman diagrams that contribute to the decays. These are flavor changing neutral current transitions, which are suppressed in the standard model (SM) and hence provide a good avenue to search for physics beyond the SM. Further motivation, especially to study the contributions of various quasi-two-body resonances to inclusive CP asymmetry, comes from the recent results on $B^\pm \to K^+ K^- K^\pm$, $K^+ K^- \pi^\pm$ and other such three-body decays [1–3]. LHCb has found large inclusive asymmetries in $B^\pm \to K^+ K^- \pi^\pm$ and $\pi^+ \pi^- \pi^\pm$ decays [2], where the observed phenomena are largely in localized regions of phase space. Recently, Belle has also reported strong evidence for a large CP asymmetry in the low $K^+ K^-$ invariant-mass region of $B^\pm \to K^+ K^- \pi^\pm$ [3].

A. B. Kaliyar (✉) · P. K. Behera
Indian Institute of Technology Madras, Chennai 600036, India
e-mail: basithkaliyar@physics.iitm.ac.in

G. B. Mohanty
Tata Institute of Fundamental Research, Mumbai 400005, India

V. Gaur
Virginia Polytechnic Institute and State University, Blacksburg, Virginia 24061, USA

© Springer Nature Singapore Pte Ltd. 2021
P. K. Behera et al. (eds.), *XXIII DAE High Energy Physics Symposium*,
Springer Proceedings in Physics 261,
https://doi.org/10.1007/978-981-33-4408-2_1

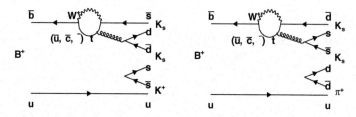

Fig. 1.1 Dominant Feynman diagrams that contribute to the decays $B^{\pm} \rightarrow K_s^0 K_s^0 K^{\pm}$ (left) and $B^{\pm} \rightarrow K_s^0 K_s^0 \pi^{\pm}$ (right)

The three-body decay $B^+ \rightarrow K_s^0 K_s^0 K^+$ [4] has already been observed and subsequently studied by the Belle and BaBar Collaborations [5–7]. Belle measured its branching fraction as $(13.4 \pm 1.9 \pm 1.5) \times 10^{-6}$ based on a small data set of $70\,\text{fb}^{-1}$ [6], while BaBar reported a branching fraction of $(10.6 \pm 0.5 \pm 0.3) \times 10^{-6}$ and an inclusive CP asymmetry of $(4^{+4}_{-5} \pm 2)\%$ using $426\,\text{fb}^{-1}$ of data [5]. The quoted uncertainties are statistical and systematic, respectively. On the other hand, the decay $B^+ \rightarrow K_s^0 K_s^0 \pi^+$ has not yet been observed, with the most restrictive upper limit being available at 90 % confidence level, $\mathcal{B}(B^+ \rightarrow K_s^0 K_s^0 \pi^+) < 5.1 \times 10^{-7}$, from BaBar [7].

We present herein an improved measurement of the branching fraction and direct CP asymmetry of the decay $B^+ \rightarrow K_s^0 K_s^0 K^+$, as well as a search for the decay $B^+ \rightarrow K_s^0 K_s^0 \pi^+$ based on the full $\Upsilon(4S)$ data sample, containing 772×10^6 $B\bar{B}$ pairs, collected with the Belle detector [8] at the KEKB asymmetric-energy e^+e^- (3.5–8.0 GeV) collider [9]. The direct CP asymmetry in the former case is given by

$$\mathcal{A}_{CP} = \frac{N(B^- \rightarrow K_s^0 K_s^0 K^-) - N(B^+ \rightarrow K_s^0 K_s^0 K^+)}{N(B^- \rightarrow K_s^0 K_s^0 K^-) + N(B^+ \rightarrow K_s^0 K_s^0 K^+)}, \quad (1.1)$$

where N is the signal yield obtained for the corresponding mode.

1.2 Event Selection

To reconstruct $B^+ \rightarrow K_s^0 K_s^0 h^+$ decay candidates, we combine a pair of K_s^0 mesons with a charged kaon or pion. Each charged track candidate must have a distance of closest approach with respect to the interaction point (IP) of less than 0.2 cm in the transverse r–ϕ plane and less than 5.0 cm along the z axis. Here, the z axis is the direction opposite the e^+ beam. Charged kaons and pions are identified based on a likelihood ratio $\mathcal{R}_{K/\pi} = \mathcal{L}_K/(\mathcal{L}_K + \mathcal{L}_\pi)$, where $\mathcal{L}_K$ and $\mathcal{L}_\pi$ denote the individual likelihood for kaons and pions, respectively, calculated using specific ionization in the CDC and information from the ACC and the TOF. A requirement, $\mathcal{R}_{K/\pi} > 0.6$, is applied to select the kaon candidates; track candidates failing it are classified as pions. The efficiency for kaon (pion) identification is 86% (91%) with a pion (kaon) misidentification rate of about 14% (9%).

The K_S^0 candidates are reconstructed from pairs of oppositely charged tracks, both treated as pions, and are identified with a neural network (NN) [10]. We require the reconstructed invariant mass of the pion pair to be between 491 and 505 MeV/c^2, corresponding to $\pm 3\sigma$ around the nominal K_S^0 mass [11]. B meson candidates are identified using two kinematic variables: beam-energy constrained mass, $M_{bc} = \sqrt{E_{beam}^2/c^4 - |\sum_i \mathbf{p}_i/c|^2}$, and energy difference, $\Delta E = \sum_i E_i - E_{beam}$, where E_{beam} is the beam energy, and $\mathbf{p}_i$ and E_i are the momentum and energy, respectively, of the i-th daughter of the reconstructed B candidate in the center-of-mass frame. We retain events with $5.271 \,\mathrm{GeV}/c^2 < M_{bc} < 5.287 \,\mathrm{GeV}/c^2$ and $-0.10 \,\mathrm{GeV} < \Delta E < 0.15 \,\mathrm{GeV}$ for further analysis. The M_{bc} requirement corresponds to approximately $\pm 3\sigma$ around the nominal B^+ mass [11]. We apply a looser $(-6\sigma, +9\sigma)$ requirement on ΔE as it is used in the fitter (described below). The average number of B candidates found per event is 1.13 (1.49) for $B^+ \to K_S^0 K_S^0 K^+$ ($K_S^0 K_S^0 \pi^+$). In events with multiple B candidates, we choose the one with the lowest χ^2 value obtained from a B vertex fit. This criterion selects the correct B-meson candidate in 75% (63%) of MC events for $B^+ \to K_S^0 K_S^0 K^+$ ($K_S^0 K_S^0 \pi^+$).

The dominant background is from the $e^+ e^- \to q\bar{q}$ ($q = u, d, s, c$) continuum process. To suppress it, observables based on the event shape topology are utilized. The event shape in the CM frame is expected to be spherical for $B\overline{B}$ events, in contrast to jet-like for continuum events. We employ a NN [10] to combine the event topology variables. The NN training and optimization are performed with signal and $q\bar{q}$ Monte Carlo (MC) simulated events. The signal MC sample is generated with the EvtGen program [12] assuming a three-body phase space. We require the NN output (C_{NB}) to be greater than -0.2 to substantially reduce the continuum background. The relative signal efficiency due to this requirement is approximately 91%, whereas the achieved continuum suppression is close to 84% for both decays. The remainder of the C_{NB} distribution strongly peaks near 1.0 for signal, making it difficult to model it with an analytic function. However, its transformed variable

$$C'_{NB} = \log\left[\frac{C_{NB} - C_{NB,min}}{C_{NB,max} - C_{NB}}\right], \qquad (1.2)$$

where $C_{NB,min} = -0.2$ and $C_{NB,max} \simeq 1.0$, has a Gaussian-like distribution.

The background due to B decays mediated via the dominant $b \to c$ transition is studied with an MC sample comprising such decays. The resulting ΔE and M_{bc} distributions are found to strongly peak in the signal region for both $B^+ \to K_S^0 K_S^0 K^+$ and $K_S^0 K_S^0 \pi^+$ decays. For $B^+ \to K_S^0 K_S^0 K^+$, the peaking background predominantly stems from $B^+ \to D^0 K^+$ with $D^0 \to K_S^0 K_S^0$ and $B^+ \to \chi_{c0}(1P) K^+$ with $\chi_{c0}(1P) \to K_S^0 K_S^0$. To suppress these backgrounds, we exclude candidates for which $M_{K_S^0 K_S^0}$ lies in the ranges of $[1.85, 1.88] \,\mathrm{GeV}/c^2$ and $[3.38, 3.45] \,\mathrm{GeV}/c^2$ corresponding to about $\pm 3\sigma$ window around the nominal D^0 and $\chi_{c0}(1P)$ mass [11], respectively. On the other hand, in case of $B^+ \to K_S^0 K_S^0 \pi^+$, the peaking background largely arises from $B^+ \to D^0 \pi^+$ with $D^0 \to K_S^0 K_S^0$. To suppress this background, we exclude candidates for which $M_{K_S^0 K_S^0}$ lies in the aforementioned D^0 mass window.

There are a few background modes that contribute in the M_{bc} signal region but have the ΔE peak shifted from zero on the positive (negative) side for $B^+ \to K_S^0 K_S^0 K^+$ ($K_S^0 K_S^0 \pi^+$). The so-called "feed-across background" modes, mostly arising due to K–π misidentification, are identified with a $B\bar{B}$ MC sample in which one of the B mesons decays via $b \to u, d, s$ transitions. The feed-across background includes contribution from $B \to K_S^0 K_S^0 \pi$ ($K_S^0 K_S^0 K$) in $B^+ \to K_S^0 K_S^0 K^+$ ($K_S^0 K_S^0 \pi^+$). The events that remain after removing the signal and feed-across components comprise the "combinatorial background." After all selection requirements, the efficiency for correctly reconstructed signal events (ϵ_{rec}) is 24% (28%) for $B^+ \to K_S^0 K_S^0 K^+$ ($K_S^0 K_S^0 \pi^+$). The fraction of misreconstructed signal events (f_{SCF}) is 0.45% (1.05%) for $B^+ \to K_S^0 K_S^0 K^+$ ($K_S^0 K_S^0 \pi^+$). As f_{SCF} represents a small fraction of the signal events for both decays, we consider it as a part of signal.

1.3 Signal Extraction

The signal yield and $\mathcal{A}_{CP}$ are obtained with an unbinned extended maximum likelihood fit to the two-dimensional distributions of ΔE and C'_{NB}. We define a probability density function (PDF) for each event category j (signal, $q\bar{q}$, combinatorial $B\bar{B}$, and feed-across backgrounds) as

$$\mathcal{P}_j^i \equiv \frac{1}{2}(1 - q^i . \mathcal{A}_{CP,j}) \times \mathcal{P}_j(\Delta E^i) \times \mathcal{P}_j(C'^i_{NB}), \tag{1.3}$$

where i denotes the event index, q^i is the charge of the B candidate in the event, $\mathcal{P}_j$ is the PDF corresponding to the component j. Since the correlation between ΔE and C'_{NB} is found to be negligible, the product of two individual PDFs is a good approximation for the total PDF. We apply a tight requirement on M_{bc} instead of including it in the fitter as it exhibits large correlation with ΔE for signal and feed-across components. The extended likelihood function is

$$\mathcal{L} = \frac{e^{-\sum_j n_j}}{N!} \prod_i \left[\sum_j n_j \mathcal{P}_j^i \right], \tag{1.4}$$

where n_j is the yield of the event category j and N is the total number of events. To account for crossfeed between the $B \to K_S^0 K_S^0 K$ and $B \to K_S^0 K_S^0 \pi$ channels, they are simultaneously fitted, with the $B \to K_S^0 K_S^0 K$ signal yield in the correctly reconstructed sample determining the normalization of the crossfeed in the $B \to K_S^0 K_S^0 \pi$ fit region, and vice versa.

Table 1.1 lists the PDF shapes used to model ΔE and C'_{NB} distributions for various event categories for $B \to K_S^0 K_S^0 K$. For $B \to K_S^0 K_S^0 \pi$, we use similar PDF

Table 1.1 List of PDFs used to model the ΔE and C'_{NB} distributions for various event categories for $B \to K_S^0 K_S^0 K$. G, AG, and Poly1 denote Gaussian, asymmetric Gaussian, and first order polynomial, respectively

Event category	ΔE	C'_{NB}
Signal	3 G	G+AG
Continuum $q\bar{q}$	Poly1	2 G
Combinatorial $B\bar{B}$	Poly1	2 G
Feed-across	G+Poly1	G

shapes except for the feed-across background component, where we use a sum of a Gaussian, asymmetric Gaussian, and first order polynomial to parametrize ΔE, and a sum of Gaussian and asymmetric Gaussian functions to parametrize C'_{NB}. For $B \to K_S^0 K_S^0 K$, the yields for all event categories except for that of the combinatorial $B\bar{B}$ background are allowed to vary in the fit. The latter yield is fixed to the MC value as it is found to be correlated with the continuum background yield. For $B \to K_S^0 K_S^0 \pi$, the yields for all event categories are allowed to vary. For both $B \to K_S^0 K_S^0 K$ and $K_S^0 K_S^0 \pi$, the following PDF shape parameters of the continuum background are floated: the slope of the first order polynomial used for ΔE, and one of the means and widths of the Gaussian functions used to model C'_{NB}. The PDF shapes for signal and other background components are fixed to the corresponding MC expectations. We correct the signal ΔE and C'_{NB} PDF shapes for possible data-MC differences, according to the values obtained with a large-statistics control sample of $B \to D^0(K_S^0 \pi^+ \pi^-)\pi$. The same correction factors are also applied for the feed-across background component of $B \to K_S^0 K_S^0 \pi$.

Figure 1.2 shows ΔE and C'_{NB} projections of the fit to B^+ and B^- samples separately for $B \to K_S^0 K_S^0 K$ and overall fit for $B \to K_S^0 K_S^0 \pi$. We determine the branching fraction as

$$\mathcal{B}(B^+ \to K_S^0 K_S^0 h^+) = \frac{N_{\text{sig}}}{\epsilon \times N_{B\bar{B}} \times [\mathcal{B}(K_S^0 \to \pi\pi)]^2} \qquad (1.5)$$

where, N_{sig}, ϵ and $N_{B\bar{B}}$ are the signal yield, corrected reconstruction efficiency and total number of $B\bar{B}$ pairs, respectively. For $B^+ \to K_S^0 K_S^0 \pi^+$, we obtain a signal yield of 69 ± 26, where the error is statistical only. The inclusive branching fraction for $B^+ \to K_S^0 K_S^0 \pi^+$ is $(6.5 \pm 2.6 \pm 0.4) \times 10^{-7}$, where the first uncertainty is statistical and the second is systematic. Its signal significance is estimated as $\sqrt{-2\log(\mathcal{L}_0/\mathcal{L}_{\text{max}})}$, where $\mathcal{L}_0$ and $\mathcal{L}_{\text{max}}$ are the likelihood value with the signal yield set to zero and for the nominal case, respectively. Including systematic uncertainties (described below), we determine the significance to be 2.5 standard deviations (σ). In view of the significance being less than 3σ, we set an upper limit (UL) on the branching fraction of $B \to K_S^0 K_S^0 \pi$. For this purpose, we convolve the likelihood with a Gaussian function of width equal to the systematic error. Assuming a flat prior we set an UL of 8.7×10^{-7} at 90% confidence level.

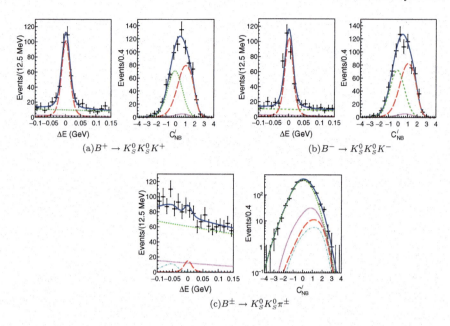

(a)$B^+ \to K_S^0 K_S^0 K^+$ (b)$B^- \to K_S^0 K_S^0 K^-$

(c)$B^\pm \to K_S^0 K_S^0 \pi^\pm$

Fig. 1.2 (color online). Projections of the two-dimensional simultaneous fit to ΔE for $C'_{NB} > 0.0$ and C'_{NB} for $|\Delta E| < 50$ MeV. Black points with error bars are the data, solid blue curves are the total PDF, long dashed red curves are the signal, dashed green curves are the continuum background, dotted magenta curves are the combinatorial $B\bar{B}$ background, and dash-dotted cyan curves are the feed-across background [6]

For $B^+ \to K_S^0 K_S^0 K^+$, we perform the fit in seven bins of $M_{K_S^0 K_S^0}$ to incorporate contributions from possible two-body intermediate resonances. Efficiency, differential branching fraction, and $\mathcal{A}_{CP}$ thus obtained are listed in Table 1.2. Figure 1.3 shows the branching fraction and $\mathcal{A}_{CP}$ plotted as a function of $M_{K_S^0 K_S^0}$. We observe an excess

Table 1.2 Efficiency, differential branching fraction, and $\mathcal{A}_{CP}$ in each $M_{K_S^0 K_S^0}$ bin for $B^+ \to K_S^0 K_S^0 K^+$

$M_{K_S^0 K_S^0}$ (GeV/c^2)	Efficiency (%)	$d\mathcal{B}/dM \times 10^{-6}$ (c^2/GeV)	$\mathcal{A}_{CP}$ (%)
1.0–1.1	24.0 ± 0.4	$10.40 \pm 1.24 \pm 0.38$	$-3.9 \pm 10.9 \pm 0.9$
1.1–1.3	23.4 ± 0.2	$8.60 \pm 0.85 \pm 0.32$	$-0.1 \pm 9.3 \pm 0.9$
1.3–1.6	22.9 ± 0.1	$10.23 \pm 0.73 \pm 0.38$	$+6.6 \pm 6.9 \pm 0.9$
1.6–2.0	21.8 ± 0.1	$3.93 \pm 0.43 \pm 0.15$	$+16.1 \pm 10.3 \pm 0.9$
2.0–2.3	24.1 ± 0.1	$3.90 \pm 0.47 \pm 0.15$	$-3.3 \pm 11.3 \pm 0.9$
2.3–2.7	25.2 ± 0.1	$2.45 \pm 0.33 \pm 0.09$	$-5.7 \pm 12.2 \pm 1.0$
2.7–5.0	26.3 ± 0.0	$0.35 \pm 0.07 \pm 0.01$	$-31.9 \pm 19.7 \pm 1.2$

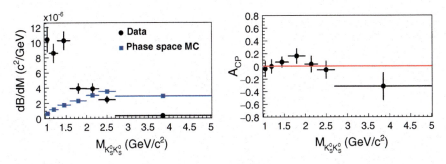

Fig. 1.3 Differential branching fraction (left) and $\mathcal{A}_{CP}$ (right) as functions of $M_{K_S^0 K_S^0}$ for $B^+ \rightarrow K_S^0 K_S^0 K^+$. Black points with error bars are the results from the two-dimensional fits to data and include systematic uncertainties. Blue squares in the left plot show the expectation from a phase space MC sample and the red line in the right plot indicates a zero CP asymmetry [6]

of events around $1.5 \, \text{GeV}/c^2$, whereas no significant evidence for CP asymmetry is found in any of the bins. The inclusive branching fraction obtained by integrating the differential branching fractions over the entire $M_{K_S^0 K_S^0}$ range is

$$\mathcal{B}(B^+ \rightarrow K_S^0 K_S^0 K^+) = (10.42 \pm 0.43 \pm 0.22) \times 10^{-6}, \tag{1.6}$$

where the first uncertainty is statistical and the second is systematic. The $\mathcal{A}_{CP}$ over the full $M_{K_S^0 K_S^0}$ range is

$$\mathcal{A}_{CP}(B^+ \rightarrow K_S^0 K_S^0 K^+) = (+1.6 \pm 3.9 \pm 0.9)\%. \tag{1.7}$$

Table 1.3 Systematic uncertainties in the branching fraction of $B^+ \rightarrow K_S^0 K_S^0 \pi^+$

Source	Relative uncertainty in $\mathcal{B}$ (%)
Tracking	0.35
Particle identification	0.80
Number of $B\bar{B}$ pairs	1.37
Continuum suppression	0.34
Requirement on M_{bc}	0.03
K_S^0 reconstruction	3.22
Fit bias	1.86
Signal PDF	1.30
Combinatorial $B\bar{B}$ PDF	$+1.31, -1.98$
Feed-across PDF	$+3.57, -4.10$
Fixed background yield	$+2.63, -2.27$
Fixed background $\mathcal{A}_{CP}$	0.50
Total	$+6.30, -6.67$

Table 1.4 Systematic uncertainties in the differential branching fraction and $\mathcal{A}_{CP}$ in $M_{K_S^0 K_S^0}$ bins for $B^+ \to K_S^0 K_S^0 K^+$. "†" indicates the uncertainty is independent of $M_{K_S^0 K_S^0}$. An ellipsis indicates a value below 0.05% in $d\mathcal{B}/dM$ and 0.001 in $\mathcal{A}_{CP}$

$M_{K_S^0 K_S^0}$ (GeV/c²)	1.0 – 1.1	1.1 – 1.3	1.3 – 1.6	1.6 – 2.0	2.0 – 2.3	2.3 – 2.7	2.7 – 5.0
Source	Relative uncertainty in $d\mathcal{B}/dM$ (%)						
Tracking†	0.35						
Particle identification†	0.80						
Number of $B\bar{B}$ pairs†	1.37						
Continuum suppression†	0.34						
Requirement on $M_{bc}^†$	0.03						
K_S^0 reconstruction†	3.22						
Fit bias†	0.53						
Signal PDF	$^{+0.33}_{-0.27}$	$^{+0.63}_{-0.48}$	$^{+0.46}_{-0.44}$	$^{+0.22}_{-0.63}$	$^{+0.52}_{-0.38}$	0.67	1.10
Combinatorial $B\bar{B}$ PDF	0.09	$^{+0.08}_{-0.13}$	0.12	$^{+0.17}_{-0.21}$	$^{+0.26}_{-0.34}$	0.40	0.40
Fixed background yield	⋯	0.10	0.10	0.23	⋯	0.11	0.60
Fixed background $\mathcal{A}_{CP}$	⋯	⋯	⋯	0.20	0.10	⋯	0.13
Total	±3.68	±3.72	±3.69	±3.73	±3.72	±3.75	±3.89
$M_{K_S^0 K_S^0}$ (GeV/c²)	1.0 – 1.1	1.1 – 1.3	1.3 – 1.6	1.6 – 2.0	2.0 – 2.3	2.3 – 2.7	2.7 – 5.0
Source	Absolute uncertainty in $\mathcal{A}_{CP}$						
Signal PDF	0.001	0.002	0.001	0.002	0.001	0.001	0.004
Combinatorial $B\bar{B}$ PDF	0.001	0.001	0.001	⋯	0.001	0.002	0.001
Fixed background yield	⋯	⋯	0.001	0.001	0.001	0.001	0.004
Fixed background $\mathcal{A}_{CP}$	⋯	⋯	0.001	0.001	0.001	0.002	0.006
Detector bias†	0.009						
Total	±0.009	±0.009	±0.009	±0.009	±0.009	±0.010	±0.012

This is obtained by weighting the $\mathcal{A}_{CP}$ value in each bin with the fitted yield divided by the detection efficiency in that bin. As the statistical uncertainties are bin-independent, their total contribution is a quadratic sum. On the other hand, for the systematic uncertainties, the total contribution from the bin-correlated sources is taken as a linear sum while that from the bin-uncorrelated sources is determined as a quadratic sum. Major sources of systematic uncertainties are listed along with their contributions in Tables 1.3 and 1.4. The results are in agreement with BaBar [5], where they had reported an overall $\mathcal{A}_{CP}$ consistent with zero, and the presence of intermediate resonances $f_0(1500)$ and $f_2'(1525)$ in the low invariant-mass regions.

1.4 Summary

In summary, we report improved measurements of the suppressed decays $B^+ \to K_S^0 K_S^0 K^+$ and $B^+ \to K_S^0 K_S^0 \pi^+$ using the full $\Upsilon(4S)$ data sample collected with the Belle detector. We perform a two-dimensional simultaneous fit to extract the signal yields of both decays. We report a 90% upper limit on the branching fraction of 8.7×10^{-7} for the decay $B^+ \rightarrow K_S^0 K_S^0 \pi^+$. We also report the branching fraction and $\mathcal{A}_{CP}$ as a function of $M_{K_S^0 K_S^0}$ for $B^+ \rightarrow K_S^0 K_S^0 K^+$. We observe an excess of events at low $M_{K_S^0 K_S^0}$ region, likely caused by the two-body intermediate resonances reported by BaBar [5]. An amplitude analysis with more data is needed to further eluci-date the nature of these resonances. The measured inclusive branching fraction and direct CP asymmetry are $\mathcal{B}(B^+ \rightarrow K_S^0 K_S^0 K^+) = (10.42 \pm 0.43 \pm 0.22) \times 10^{-6}$ and $\mathcal{A}_{CP} = (+1.6 \pm 3.9 \pm 0.9)\%$, respectively. These supersede Belle's earlier mea-surements [6] and constitute the most precise results to date. The results presented in this report have been already published elsewhere [13].

References

1. R. Aaij et al., LHCb collaboration. Phys. Rev. D **90**, 112004 (2014)
2. R. Aaij et al., LHCb collaboration. Phys. Rev. L **112**, 011801 (2014)
3. C.-L. Hsu et al., Belle collaboration. Phys. Rev. D **96**, 031101(R) (2017)
4. Inclusion of charge-conjugate reactions are implicit unless stated otherwise
5. J.P. Lees et al., BaBar collaboration. Phys. Rev. D **85**, 112010 (2012)
6. A. Garmash et al., Belle collaboration. Phys. Rev. D **69**, 012001 (2004)
7. B. Aubert et al., BaBar collaboration. Phys. Rev. D **79**, 051101(R) (2009)
8. A. Abashian et al. (Belle Collaboration), Nucl. Instrum. Methods Phys. Res., Sect. A **479**, 117 (2002)
9. S. Kurokawa, E. Kikutani, Nucl. Instrum. Methods Phys. Res., Sect. A **499**, 1 (2003)
10. M. Feindt, U. Kerzel, Nucl. Instrum. Methods Phys. Res., Sect. A **559**, 190 (2006)
11. M. Tanabashi et al., Particle data group. Phys. Rev. D **98**, 030001 (2018)
12. D.J. Lange, Nucl. Instrum. Methods Phys. Res., Sect. A **462**, 152 (2001)
13. A.B. Kaliyar et al., (Belle Collaboration), "Measurements of branching fraction and direct CP asymmetry in $B^{\pm} \rightarrow K_S^0 K_S^0 K^{\pm}$ and a search for $B^{\pm} \rightarrow K_S^0 K_S^0 \pi^{\pm}$". Phys. Rev. D **99**, 031102(R) (2019)

Chapter 2
Model Independent Analysis
of $\bar{B}^* \to P l \bar{\nu}_l$ Decay Processes

Atasi Ray, Suchismita Sahoo, and Rukmani Mohanta

Abstract A spectacular deviation of Standard Model (SM) value of lepton nonuniversality (LNU) parameters $\left(R_D, R_{D^*}, R_K, R_{K*}, R_{J/\psi} \right)$ from their experimental one, provides a clear hint for the presence of new physics (NP) beyond the SM. In this context, we have studied the effect of NP in the semileptonic $\bar{B}^* \to P l \bar{\nu}_l$ decay process in a model independent way. In this approach, we considered the NP to have scalar, vector, and tensor type of couplings and treated the new Wilson coefficients as complex quantities in our analysis. We constrained the parameter space of new couplings using the measured values of $\mathrm{Br}(B_u^+ \to \tau^+ \nu_\tau)$, $\mathrm{Br}(B \to \pi \tau \bar{\nu}_\tau)$, R_π^l for $b \to u l \bar{\nu}_l$ processes and $\mathrm{Br}(B_c^+ \to \tau^+ \nu_\tau)$, $R_{D^{(*)}}$ and $R_{J/\psi}$ observables for $b \to c l \bar{\nu}_l$ processes. Using the constrained parameters, we analyzed the q^2 variation of branching ratio, forward-backward asymmetry, and LNU parameters of $\bar{B}^* \to P l \bar{\nu}_l$ process, in the context of NP.

2.1 Introduction

Recently B-factory has observed various anomalies associated with several semileptonic B-meson decay processes. Variation of several Lepton nonuniversality parameters is listed in Table 2.1. In this context, we wish to scrutinize the possibility of observing LNU parameters and other asymmetries in $\bar{B}^* \to P l \bar{\nu}_l$ decay process mediated by $b \to (u, c)$ quark level transition.

Various rare B-meson decays have been studied both theoretically and experimentally so far. These play important role in testing SM and probing possible hint

A. Ray (✉) · R. Mohanta
School of Physics, University of Hyderabad, Hyderabad 500046, India
e-mail: atasiray92@gmail.com

R. Mohanta
e-mail: rukmani98@gmail.com

S. Sahoo
Theoretical Physics Division, Physical Research Laboratory, Ahmedabad 380009, India
e-mail: suchismita8792@gmail.com

© Springer Nature Singapore Pte Ltd. 2021
P. K. Behera et al. (eds.), *XXIII DAE High Energy Physics Symposium*,
Springer Proceedings in Physics 261,
https://doi.org/10.1007/978-981-33-4408-2_2

Table 2.1 Lepton nonuniversality Parameters

LNU parameter	Measured value	SM value	Deviation
$R_k = \frac{BR(B^+ \to K^+ \mu^+ \mu^-)}{BR(B^+ \to K^+ e^+ e^-)}$	$0.745^{+0.090}_{-0.074} \pm 0.036$	1.0003 ± 0.0001	2.6σ
$R_{k*} = \frac{BR(B^+ \to K^* \mu^+ \mu^-)}{BR(B^+ \to K^* e^+ e^-)}$	$0.660^{+0.110}_{-0.070} \pm 0.03$	0.92 ± 0.02	2.2σ
$R_D = \frac{BR(B \to D\tau\nu_\tau)}{BR(B \to Dl\nu_l)}$	$0.340 \pm 0.027 \pm 0.013$	0.299 ± 0.003	1.9σ
$R_{D*} = \frac{BR(B \to D^*\tau\nu_\tau)}{BR(B \to D^*l\nu_l)}$	$0.295 \pm 0.011 \pm 0.008$	0.258 ± 0.005	3.3σ
$R_{J/\psi} = \frac{\mathrm{Br}(B_c \to J/\psi\tau\bar{\nu}_\tau)}{\mathrm{Br}(B_c \to J/\psi l\bar{\nu}_l)}$	$0.71 \pm 0.17 \pm 0.184$	0.289 ± 0.01	2σ

of new Physics (NP) beyond it. The vector ground state of $b\bar{q}$ system, B^* meson can also decay through $b \to (u, c)$ quark level transition. With the rapid development of heavy flavor experiments, the B^* weak decays are hopeful to be observed in Belle-II experiment. LHC experiment will provide a lot of experimental information for B^* weak decays due to the larger beauty production cross-section of pp collision. In this aspect, we want to study the rare $B^* \to Pl\bar{\nu}_l$ decay processes in a model independent way.

This paper is organized as follows. Section 2.2 contains the theoretical frame work associated with $B^* \to Pl\bar{\nu}_l$ decay processes. In Sect. 2.3, we present the constraints on the new couplings for decay processes mediated by $b \to c$ and $b \to u$ obtained using χ^2 fit of $R_{D^{(*)}}$, $R_{J/\psi}$, R_π^l, $\mathrm{Br}(B_{u,c} \to \tau\nu)$, $\mathrm{Br}(B \to \pi\tau\bar{\nu})$ observables. Section 4 contains the effect of NP on various observables and we summarize our results in Sec. 2.5.

2.2 Theoretical Framework

The most general effective Lagrangian for $B^* \to Pl\bar{\nu}_l$ processes mediated by $b \to ql^-\bar{\nu}_l$ ($q = u, c$), can be expressed in the effective field theory approach as [1],

$$\mathcal{L}_{eff} = -2\sqrt{2}G_F V_{qb} \Big[(1 + V_L)\, \bar{q}_L \gamma^\mu b_L\, \bar{l}_L \gamma_\mu \nu_L + V_R\, \bar{q}_R \gamma^\mu b_R\, \bar{l}_L \gamma_\mu \nu_L + S_L\, \bar{q}_R b_L\, \bar{l}_R \nu_L$$

$$+ S_R\, \bar{q}_L b_R\, \bar{l}_R \nu_L + T_L\, \bar{q}_R \sigma^{\mu\nu} b_L\, \bar{l}_R \sigma_{\mu\nu} \nu_L + \text{h.c.}\Big], \tag{2.1}$$

where P is any pseudoscalar meson, G_F is the Fermi constant, V_{qb} is the CKM matrix element, $V_{L,R}$, $S_{L,R}$, T_L are the new vector, scalar, and tensor type new physics (NP) couplings. $(q, l)_{L,R} = P_{L,R}(q, l)$, where $P_{L,R} = (1 \mp \gamma_5)/2$ are the chiral projection operators. The NP coefficients are zero in the standard model. In our analysis, we consider all the NP coefficients to be complex and we consider the neutrinos as left-handed. We assume the NP effect is mainly through the third generation leptons and do not consider the effect of tensor operators in our analysis for simplicity.

The double differential decay rate associated with $B^* \to Pl\bar{\nu}_l$ decay process with particular leptonic helicity state $(\lambda = \pm\frac{1}{2})$ are expressed as

$$
\frac{d^2\Gamma(\lambda_l = -\frac{1}{2})}{dq^2 d\cos\theta} = \frac{G_F^2}{768\pi^3} \frac{|\mathbf{p}|}{m_{B^*}^2} |V_{qb}|^2 q^2 \left(1 - \frac{m_l^2}{q^2}\right)^2 \left\{ |1 + V_L|^2 \left[(1 - \cos\theta)^2 H_{-+}^2 + (1 + \cos\theta)^2 H_{+-}^2 \right.\right.
$$
$$
+ 2\sin^2\theta H_{00}^2 \left] + |V_R|^2 \left[(1 - \cos\theta)^2 H_{+-}^2 + (1 + \cos\theta)^2 H_{-+}^2 + 2\sin^2\theta H_{00}^2\right]\right.
$$
$$
\left. - 4\mathcal{R}e\left[(1 + V_L)V_R^*\right]\left[(1 + \cos\theta)^2 H_{+-}H_{-+} + \sin^2\theta H_{00}^2\right]\right\}, \tag{2.2}
$$

$$
\frac{d^2\Gamma(\lambda_l = \frac{1}{2})}{dq^2 d\cos\theta} = \frac{G_F^2}{768\pi^3} \frac{|\mathbf{p}|}{m_{B^*}^2} |V_{qb}|^2 \left(1 - \frac{m_l^2}{q^2}\right)^2 m_l^2 \left\{ (|1 + V_L|^2 + |V_R|^2)\left[\sin^2\theta(H_{-+}^2 + H_{+-}^2)\right.\right.
$$
$$
+ 2(H_{0t} - \cos\theta H_{00})^2\right] - 4\mathcal{R}e\left[(1 + V_L)V_R^*\right]\left[\sin^2\theta H_{-+}H_{+-} + (H_{0t} - \cos\theta H_{00})^2\right]
$$
$$
+ 4\mathcal{R}e[(1 + V_L - V_R)(S_L^* - S_R^*)] \frac{\sqrt{q^2}}{m_l}\left[H_{0t}'(H_{0t} - \cos\theta H_{00})\right]
$$
$$
+ 2|S_L - S_R|^2 \frac{q^2}{m_l^2} H_{0t}'^2 \right\}. \tag{2.3}
$$

Here $H_{\pm,\mp}$, H_{00}, H_{0t} are the helicity amplitudes which are the function of form factors. We used the form factors evaluated in [2]. We used the numerical values of particles masses, CKM elements, and Fermi constant from [3].

From (2.2) and (2.3), the double differential decay rate of $B^* \to Pl\bar{\nu}_l$ decay process can be expressed as

$$
\frac{d\Gamma}{dq^2} = \frac{G_F^2}{288\pi^3} \frac{|\mathbf{p}|}{m_{B^*}^2} |V_{qb}|^2 q^2 \left(1 - \frac{m_\ell^2}{q^2}\right)^2 \left[(|1 + V_L|^2 + |V_R|^2)\right.
$$
$$
\times \left[(H_{-+}^2 + H_{+-}^2 + H_{00}^2)\left(1 + \frac{m_\ell^2}{2q^2}\right) + \frac{3m_\ell^2}{2q^2} H_{0t}^2\right]
$$
$$
- 2\mathcal{R}e[(1 + V_L)V_R^*]\left[(2H_{-+}H_{+-} + H_{00}^2)\left(1 + \frac{m_\ell^2}{2q^2}\right) + \frac{3m_\ell^2}{2q^2} H_{0t}^2\right]
$$
$$
+ 3\frac{m_\ell}{\sqrt{q^2}} \mathcal{R}e\left[(1 + V_L - V_R)(S_L^* - S_R^*)\right] H_{0t}' H_{0t} + \frac{3}{2}|S_L - S_R|^2 H_{0t}'^2\right], \tag{2.4}
$$

Apart from differential decay rate we also considered other observables sensitive towards NP are

– Forward-backward asymmetry:

$$A_{\text{FB}}^P(q^2) = \frac{\int_{-1}^{0} d\cos\theta (d^2\Gamma/dq^2 d\cos\theta) - \int_{0}^{1} d\cos\theta (d^2\Gamma/dq^2 d\cos\theta)}{\int_{-1}^{0} d\cos\theta (d^2\Gamma/dq^2 d\cos\theta) + \int_{0}^{1} d\cos\theta (d^2\Gamma/dq^2 d\cos\theta)} \quad (2.5)$$

– Lepton nonuniversality parameter:

$$R_P^*(q^2) = \frac{d\Gamma(B^* \to P\tau^-\bar{\nu}_\tau)/dq^2}{d\Gamma(B^* \to Pl^-\bar{\nu}_l)/dq^2} . \quad (2.6)$$

– Lepton spin asymmetry:

$$A_\lambda^P(q^2) = \frac{d\Gamma(\lambda_l = -1/2)/dq^2 - d\Gamma(\lambda_l = 1/2)/dq^2}{d\Gamma(\lambda_l = -1/2)/dq^2 + d\Gamma(\lambda_l = 1/2)/dq^2}. \quad (2.7)$$

2.3   Constraints on New Couplings

In this analysis, the new couplings are considered to be complex. Considering the contribution of only one coefficient at a time and all others to be zero, we perform the chi-square fitting for the individual complex couplings. The χ^2 is defined as

$$\chi^2 = \sum_i \frac{(\mathcal{O}_i^{\text{th}} - \mathcal{O}_i^{\text{exp}})^2}{(\Delta\mathcal{O}_i^{\text{exp}})^2}, \quad (2.8)$$

where $\mathcal{O}_i^{\text{th}}$ represents the theoretical prediction of the observables, $\mathcal{O}_i^{\text{exp}}$ symbolizes the measured central value of the observables and $\Delta\mathcal{O}_i^{\text{exp}}$ denotes the corresponding 1σ uncertainty. We constrain the real and imaginary parts of new coefficients related to $b \to cl\bar{\nu}_l$ quark level transitions from the χ^2 fit of $R_{D^{(*)}}$, $R_{J/\psi}$ and $\text{Br}(B_c^+ \to \tau^+\nu_\tau)$ observables and the couplings associated with $b \to u\tau\bar{\nu}_\tau$ processes are constrained from the fit of R_π^l, $\text{Br}(B_u^+ \to \tau^+\nu)$ and $\text{Br}(B^0 \to \pi^+\tau^-\bar{\nu})$ data. The constrained parameter space are presented in Table 2.2.

2.4   Effect of New Physics

Here we considered the contribution of only V_L coefficient in addition to the SM Lagrangian and considered all other coefficients to be zero. In the presence of only V_L coefficient, we showed the q^2 variation of several parameters associated with the considered decay processes. We performed similar analysis in presence of V_R, S_L and S_R coefficient only. Here we present the q^2 variation of the differential decay rate and LNU parameter of $\bar{B}_d^* \rightarrow D^+\tau^-\bar{\nu}_\tau$ and $\bar{B}_d^* \rightarrow \pi^+\tau\bar{\nu}_\tau$ decay processes showing profound deviation of the observables from the SM in presence of NP coupling V_L only. The numerical values of the observables associated with $B^{*0} \rightarrow D^+\tau^-\bar{\nu}_\tau$,

Table 2.2 Best-fit values and corresponding 1σ ranges of new complex coefficients

Decay modes	New coefficients	Best-fit	1σ range	χ^2/d.o.f
$b \rightarrow c\tau\bar{\nu}_\tau$	$(\text{Re}[V_L], \text{Im}[V_L])$	$(-1.1474, 1.1171)$	$([-1.3, -0.7],\ [1.088, 1.148])$	0.988
	$(\text{Re}[V_R], \text{Im}[V_R])$	$(6.57 \times 10^{-3}, -0.5368)$	$([-0.015, 0.025],\ [-0.6, -0.48])$	0.966
	$(\text{Re}[S_L], \text{Im}[S_L])$	$(0.2052, 0)$	$([0.12, 0.28],\ [-0.35, 0.35])$	6.097
	$(\text{Re}[S_R], \text{Im}[S_R])$	$(-1.003, -0.78906)$	$([-1.17, -0.77],\ [-0.89, -0.71])$	3.6
$b \rightarrow u\tau\bar{\nu}_\tau$	$(\text{Re}[V_L], \text{Im}[V_L])$	$(-0.8318, 1.098)$	$([-1.43, -0.43],\ [1.0, 1.2])$	0.265
	$(\text{Re}[V_R], \text{Im}[V_R])$	$(-0.115, 0)$	$([-0.2, -0.025],\ [-0.45, 0.45])$	0.1363
	$(\text{Re}[S_L], \text{Im}[S_L])$	$(-0.0236, 0)$	$([-0.042, -0.006],\ [-0.09, 0.09])$	0.1906
	$(\text{Re}[S_R], \text{Im}[S_R])$	$(-0.439, 0)$	$([-0.46, -0.42],\ [-0.09, 0.09])$	0.1906

Table 2.3 Predicted numerical values of differential decay rate, LNU parameters, lepton spin asymmetry, and forward-backward asymmetry of $\bar{B}^*_{d,(s)} \rightarrow D^+(D_s^+)\tau^-\bar{\nu}_\tau$ and $\bar{B}^*_{d(s)} \rightarrow \pi^+(K^+)\tau\bar{\nu}_\tau$ decay processes in the SM and in the presence of $V_{L,R}$ coefficients

Observables	M Predictions	Values with V_L	Values with V_R
$\text{Br}(B^{*0} \rightarrow D^+\tau^-\bar{\nu}_\tau)$	2.786×10^{-8}	$(3.358 \rightarrow 3.732) \times 10^{-8}$	$(3.394 \rightarrow 3.755) \times 10^{-8}$
R_D^*	0.299	$0.360 \rightarrow 0.40$	$0.364 \rightarrow 0.403$
A_λ^D	0.576	0.576	0.576
A_{FB}^D	-0.054	-0.054	$(0.002 \rightarrow 0.026)$
$\text{Br}(B_s^{*0} \rightarrow D_s^+\tau^-\bar{\nu}_\tau)$	5.074×10^{-8}	$(6.116 \rightarrow 6.797) \times 10^{-8}$	$(6.181 \rightarrow 6.838) \times 10^{-8}$
$R_{D_s}^*$	0.297	$0.358 \rightarrow 0.398$	$0.362 \rightarrow 0.400$
$A_\lambda^{D_s}$	0.573	0.573	0.573
A_{FB}^D	-0.053	-0.053	$0.003 \rightarrow 0.027$
$\text{Br}(B^{*0} \rightarrow \pi^+\tau^-\bar{\nu}_\tau)$	1.008×10^{-9}	$(1.036 \rightarrow 1.479) \times 10^{-9}$	$(1.051 \rightarrow 1.392) \times 10^{-9}$
R_π^*	0.678	$0.697 \rightarrow 0.995$	$0.707 \rightarrow 0.936$
A_λ^π	0.781	0.781	$0.780 \rightarrow 0.781$
A_{FB}^π	-0.209	-0.209	$(-0.198 \rightarrow -0.129)$
$\text{Br}(B_s^{*0} \rightarrow K^+\tau^-\bar{\nu}_\tau)$	1.034×10^{-9}	$(1.063 \rightarrow 1.518) \times 10^{-9}$	$(1.078 \rightarrow 1.421) \times 10^{-9}$
R_K^*	0.639	$0.657 \rightarrow 0.939$	$0.666 \rightarrow 0.878$
A_λ^K	0.747	0.747	$0.745 \rightarrow 0.746$
A_{FB}^K	-0.207	-0.207	$(-0.196 \rightarrow -0.124)$

Table 2.4 Predicted numerical values of differential decay rate, LNU parameters, lepton spin asymmetry, and forward-backward asymmetry of $\bar{B}^*_{d(s)} \to D^+(D^+_s)\tau^-\bar{\nu}_\tau$ and $\bar{B}^*_{d,(s)} \to \pi^+(K^+)\tau\bar{\nu}_\tau$ decay processes in presence of $S_{L,R}$ coefficients

Observables	Values with S_L	Values with S_R
Br($B^{*0} \to D^+\tau^-\bar{\nu}_\tau$)	$(2.731 \to 2.761) \times 10^{-8}$	$(2.670 \to 2.715) \times 10^{-8}$
R^*_D	$0.293 \to 0.296$	$0.289 \to 0.291$
A^D_λ	$0.591 \to 0.608$	$0.617 \to 0.633$
A^D_{FB}	$-0.076 \to -0.064$	$-0.145 \to -0.114$
Br($B^*_s \to D^+_s\tau^-\bar{\nu}_\tau$)	$(4.971 \to 5.027) \times 10^{-8}$	$(4.894 \to 4.941) \times 10^{-8}$
$R^*_{D_s}$	$0.291 \to 0.294$	$0.286 \to 0.289$
$A^{D_s}_\lambda$	$0.588 \to 0.606$	$0.615 \to 0.631$
$A^{D_s}_{FB}$	$-0.075 \to -0.062$	$-0.144 \to -0.113$
Br($B^{*0} \to \pi^+\tau^-\bar{\nu}_\tau$)	$(1.008 \to 1.011) \times 10^{-9}$	$(9.845 \to 9.850) \times 10^{-10}$
R^*_π	$0.678 \to 0.680$	$0.662 \to 0.663$
A^π_λ	$0.774 \to 0.780$	$0.822 \to 0.823$
A^π_{FB}	$-0.208 \to -0.204$	$(-0.254 \to -0.251)$
Br($B^*_s \to K^+\tau^-\bar{\nu}_\tau$)	$(1.034 \to 1.039) \times 10^{-9}$	$(1.002 \to 1.003) \times 10^{-9}$
R^*_K	$0.640 \to 0.642$	$0.619 \to 0.620$
A^K_λ	$0.738 \to 0.745$	$0.800 \to 0.802$
A^K_{FB}	$-0.206 \to -0.202$	$-0.261 \to -0.256$

$B^{*0}_s \to D^+_s\tau^-\bar{\nu}_\tau$, $B^{*0} \to \pi^+, \tau^-\bar{\nu}_\tau$, $B^{*0}_s \to K^+\tau^-\bar{\nu}_\tau$ decay processes in presence of V_L, V_R and S_L, S_R are presented in Tables 2.3 and 2.4 respectively.

2.5 Conclusion

We have studied $B^* \to Pl\bar{\nu}_l$ decay process in a model independent way. We considered the new couplings as Complex. Considering one coefficient at a time the allowed parameter space of the new coefficients were obtained. In the presence of individual complex Wilson coefficients, we have studied their effects on various parameters associated with $B^{*0} \to D^+\tau^-\bar{\nu}_\tau$, $B^{*0}_s \to D^+_s\tau^-\bar{\nu}_\tau$, $B^{*0} \to \pi^+, \tau^-\bar{\nu}_\tau$, $B^{*0}_s \to K^+\tau^-\bar{\nu}_\tau$ decay processes. We show the q^2 variation of differential decay rate and LNU parameter in presence of NP in Fig. 2.1, which shows a significant deviation from the SM.

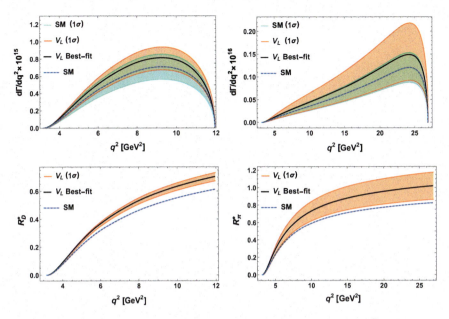

Fig. 2.1 The q^2 variation of differential decay rates and LNU observables of $\bar{B}^*_d \to D^+ \tau^- \bar{\nu}_\tau$ (left panel) and $\bar{B}^*_d \to \pi^+ \tau \bar{\nu}_\tau$ (right panel) in presence of only V_L new coefficient. Here the blue dashed lines represent the standard model predictions. The black solid lines and the orange bands are obtained by using the best-fit values and corresponding 1σ range of V_L coefficient

References

1. M. Tanaka, R. Watanabe, New physics in the weak interaction of $\bar{B} \to D^* \tau^- \bar{\nu}$. Phys. Rev. D **87**, 034028 (2013). https://doi.org/10.1103/PhysRevD.87.034028
2. Wirbel, M., Stch,B., Bauer, M.: Exclusive Semileptonic decays of heavy mesons. https://doi.org/10.1007/BF01560299
3. M. Tanabashi et al., Review of particle physics. Phys. Rev. D **98**, 030001 (2018). https://doi.org/10.1103/PhysRevD.98.030001

Chapter 3
Search for Standard Model Higgs Boson Production in Association with Top Quark Pairs at CMS

Ram Krishna Dewanjee

Abstract After the discovery of a new boson of mass 125 GeV (LHC. Phys. Lett. B716:30, 2012 [1]), one of the main goals of the LHC is to precisely measure its properties. Within the current experimental uncertainties, the properties of this boson are compatible with the expectation for the standard model (SM) Higgs boson. However, due to lack of sufficient data in Run-1, some of the properties of this particle are yet to be measured. The Yukawa coupling between this boson with the top (t) quark is one such crucialy important property. Many beyond SM (BSM) theories predict deviations of this coupling from SM value as evidence for new physics. SM Higgs (H) production in association with top quarks ($t\bar{t}H$) allows a direct measurement of this coupling. In this paper, results of searches for $t\bar{t}H$ process are presented in final states involving bottom (b) quarks, photons (γ), leptons (e/μ), and (hadronically decaying) tau leptons (τ_h) using Run-2 luminosity collected in 2016 (35.9 fb^{-1}) at $\sqrt{s}$ = 13 TeV by the CMS experiment (CERN LHC. J. Instrum. 08: S08004, 2008 [2]).

3.1 Introduction

Within the SM framework, the Higgs boson is responsible for the dynamic generation of masses of all SM particles via the Brout–Englert–Higgs mechanism. In the fermion sector, Higgs interacts via yukawa couplings which are proportional to the fermion masses. Top quark being the heaviest SM fermion, may play a yet unknown role in the electroweak symmetry breaking. This makes the top-Higgs yukawa coupling an important probe for BSM physics. Although indirect constraints on its value (via gluon fusion and the $H \rightarrow \gamma\gamma$ loop contribution) are available, its direct measurement is possible only via studying top quark associated Higgs production ($t\bar{t}H$ and tH). This article overviews all major Run-2 $t\bar{t}H$ searches: $t\bar{t}H(H \rightarrow b\bar{b})$

R. K. Dewanjee (✉)
Laboratory of High Energy and Computational Physics, KBFI, Rävala pst 10,
10143 Tallinn, Estonia
e-mail: ram.krishna.dewanjee@cern.ch; dewanjeeramkrishna5@gmail.com

© Springer Nature Singapore Pte Ltd. 2021
P. K. Behera et al. (eds.), *XXIII DAE High Energy Physics Symposium*,
Springer Proceedings in Physics 261,
https://doi.org/10.1007/978-981-33-4408-2_3

19

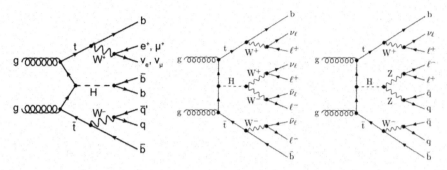

Fig. 3.1 LO Feynman diagrams for top quark associated Higgs boson production ($t\bar{t}H$) with Higgs decaying to pair of b-quarks (left), W-bosons (middle) and Z-bosons (right)

[3, 4], $t\bar{t}H(H \rightarrow \gamma\gamma)$ [5] and $t\bar{t}H$ **Multi-lepton** [6] search (which combines $t\bar{t}H(H \rightarrow VV^*)^1$ [7] and $t\bar{t}H(H \rightarrow \tau\tau)$ [8] searches) (Fig. 3.1).

3.2 Analysis Strategy and Event Categorization

The general analysis strategy of all $t\bar{t}H$ searches involve categorization to seperate events into categories of varying sensitivity. Signal over background discrimination in each category is then enhanced via. Dedicated shape analysis of a usually multi-variate (MVA) discriminator. Finally, a binned maximum likelihood fit on the distribution of this discriminator is performed for signal extraction and limit computation. The major categories employed by the $t\bar{t}H$ searches are as follows:

1. $t\bar{t}H(H \rightarrow b\bar{b})$: This search benefits from the large SM $H \rightarrow b\bar{b}$ branching fraction ($\sim57\%$) but is limited by systematic uncertainity on the (irreducible) $t\bar{t} + b\bar{b}$ background. It is divided into two primary channels (based on the hadronic or leptonic decays of the top quarks):

(a) **Hadronic channel**: Events are first selected using dedicated multi-Jet triggers. They are then divided into six categories depending on jet and b-tagged jet multiplicity. These are (7Jets, 3b-tagged), (7Jets, $\geq$4b-tagged), (8Jets, 3b-tagged), (8Jets, $\geq$4b-tagged), ($\geq$9Jets, 3b-tagged), and ($\geq$9Jets, $\geq$4b-tagged). MEM[2] was employed in all the categories to distinguish signal ($t\bar{t}H$) from background ($t\bar{t} + b\bar{b}$) and was also used for signal extraction.

(b) **Leptonic channel**: Events are first selected using single lepton triggers and divided into two main categories (each of which is further split by lepton flavor).

---

[1]$V = $ W/Z-boson.
[2]Matrix element method.

* **Single Lepton category**: Events are split into (4Jets, $\geq$3b-tagged), (5Jets, $\geq$3b-tagged), and ($\geq$6Jets, $\geq$3b-tagged) categories. DNNs[3] are trained to seperate $t\bar{t}H$ signal from $t\bar{t} + X$ background (where $X = b\bar{b}/c\bar{c}$/2b/b and light flavors).
* **Di-Lepton category**: Events are divided into ($\geq$4Jets, 3b-tagged) and ($\geq$4Jets, $\geq$4b-tagged) categories with BDTs[4] trained to distinguish $t\bar{t}H$ signal from $t\bar{t}$ background in both. While the former uses this BDT itself for signal extraction the latter gives this as input to MEM discrimintaor.

2. $t\bar{t}H(H \rightarrow \gamma\gamma)$: Events are first selected using asymmetric[5] di-photon triggers. This search is itself a channel under the general SM $H \rightarrow \gamma\gamma$ search and utilizes special BDT for tagging $t\bar{t}H$ multi-jet events. This BDT uses the following variables as inputs:

   - Number of jets ($p_T > 25$ GeV).
   - Leading jet p_T.
   - 2 jets with the highest b-tagging score.

Contributions of other SM (non $t\bar{t}H$) $H \rightarrow \gamma\gamma$ processes e.g., Gluon fusion, VBF[6] are treated as background by this BDT. Additional BDTs are used for photon identification and di-photon vertex assignment. This search too is split into Leptonic and Hadronic channels targetting semi-leptonic and hadronic top decays.

(a) **Leptonic channel**: Events enter this channel if they satisfy the following kinematic selections:

* $\geq$1 Lepton ($p_T > 20$ GeV) non-overlapping with any photon.
* $\geq$2 Jets, ($p_T > 25$ GeV, $|\eta| < 2.4$) non-overlapping with any photon or lepton.
* $\geq$1 Medium b-tagged Jet.
* Di-photon BDT score > 0.11.

(b) **Hadronic channel**: This comprises events passing the following kinematic selections:

* $\geq$3 Jets, ($p_T > 25$ GeV, $|\eta| < 2.4$) non-overlapping with any photon.
* $\geq$1 Loose b-tagged Jet.
* No Lepton in the event (passing the lepton selections of the Leptonic channel described above).
* High score (>0.75) on the $t\bar{t}H$ multi-jet tagging BDT (described above).
* Di-photon BDT score > 0.4.

In addition to this, di-photon invariant mass dependent p_T cuts are applied to selected photons in both channels to get a distortion free di-photon mass spectrum.

---

[3] Deep Learning Neural Networks.
[4] Boosted Decision Trees.
[5] $E_T^{\gamma 1} > 30$ GeV, $E_T^{\gamma 2} > 18$ GeV, Loose ECAL based photon identification.
[6] Vector boson fusion.

3. **$t\bar{t}H$ Multi-lepton**: In this search, events are first selected using lepton or lepton $+\tau_h$ triggers. They are then required to have ≥ 2 loose b-tagged jets[7] out of which ≥ 1 is medium b-tagged. A special BDT trained on simulated $t\bar{t}H(t\bar{t})$ events as signal (background) is used to distinguish "prompt" leptons (produced by W/Z/leptonic τ decays) from "non-prompt" leptons (produced in b-hadron decays, decays-in-flight, and photon conversions). Leptons passing (failing) it are called tight (loose) leptons in this analysis. An additional di-lepton invariant mass cut ($m_{ll} > 12$ GeV) is applied to all channels to reject phase space dominated by low mass SM di-lepton resonances (e.g., J/ψ, Υ etc.) which are not well modeled in simulation. The major channels of this search (which are also split by lepton flavor) are

(a) **2 Lepton Same-Sign channel (2/SS)**: Events containing exactly 2 same charge tight leptons ($p_T^{lep1/2} > 25/15$ GeV) and ≥ 4 Jets. Additional selections are applied to reduce backgrounds due to conversions and lepton charge mis-identification.

(b) **3 Lepton channel (3/)**: Events containing exactly 3 leptons ($p_T^{lep1/2/3} > 25/15/15$ GeV) with additional selections for rejection of backgrounds due to $Z \to ll$ events and conversions.

(c) **4 Lepton channel (4/)**: Events passing exactly the same selections as 3/ channel but now having an additional requirement of a fourth lepton ($p_T > 10$ GeV) in the event.

(d) **1 Lepton + $2\tau_h$ channel (1/+$2\tau_h$)**: Events containing exactly 1 tight lepton ($p_T^{e/\mu} > 25/20$ GeV, $|\eta| < 2.1$) and 2 opposite charge tight τ_h leptons ($p_T > 30$ GeV each). Events should also contain ≥ 3 Jets.

(e) **2 Lepton Same-Sign + $1\tau_h$ channel (2/SS+$1\tau_h$)**: Events containing exactly 2 same charge tight leptons ($p_T^{lep1/2} > 25/15$ GeV for electrons, $p_T^{lep1/2} > 25/10$ GeV for muons) and 1 tight τ_h ($p_T > 30$ GeV) with sign opposite to that of the leptons. Events should also contain ≥ 3 Jets. Additional selections were applied to reduce background due to charge mis-identification and $Z \to ll$.

(f) **3 Lepton + $1\tau_h$ channel (3/+$1\tau_h$)**: Events containing ≥ 3 tight leptons ($p_T^{lep1/2/3} > 20/10/10$ GeV) and ≥ 1 tight τ_h ($p_T > 30$ GeV). Sum of the charges of the leptons and τ_h must be zero. Additional selections were applied to reduce background due to charge mis-identification and $Z \to ll$.

Events containing τ_h are vetoed in the pure leptonic channels ((a), (b), and (c)) to keep them orthogonal to the lepton + τ_h channels ((d), (e), and (f)) For the 2/SS and 3/ channels, a pair of BDTs (one each for training against $t\bar{t}V$ and $t\bar{t}$ backgrounds vs the $t\bar{t}H$ signal) was used for enhanced sensitivity. The BDTs for 2/SS included discriminators for "hadronic top tagging". For signal extraction, the BDT pairs were mapped onto one dimension using "Likelihood based clustering". The minimum invariant mass of the opposite sign di-lepton pair was used for signal extraction for the 4/ channel. BDT (trained to distinguish $t\bar{t}H$ from $t\bar{t}$) was used as final discriminator in 1/+$2\tau_h$ channel. MEM (useful in separating $t\bar{t}H$ from $t\bar{t}V$ and $t\bar{t}$ backgrounds)

---

[7] $p_T > 25$ GeV, $|\eta| < 2.4$, non-ovelapping with any lepton/τ_h.

was used for signal extraction in $2lSS+1\tau_h$ channel. BDT pair (one each trained against $t\bar{t}V$ and $t\bar{t}$) mapped into a one-dimensional discriminant was used for signal extraction in $3l+1\tau_h$ channel.

3.3 Backgrounds

Backgrounds in all the $t\bar{t}H$ searches can be classified into three major categories depending on their source/origin.

1. **Reducible/Fake backgrounds**: These backgrounds arise predominantly due to Jets faking the final state particles (γ/leptons/τ_h or b-jets) used in the searches. They are mostly either estimated from data or from simulation (with their yield and shape corrected from background rich sidebands in data). In the $t\bar{t}H$ Multi-Lepton search, the probability of a Jet to pass tight selections is measured in a multi-jet enriched sideband in data (measurement region). This probability is then used to reweight another data sideband having the same selections as the signal region but with relaxed lepton identification requirements (application region). This reweighted sideband is then used to estimate fake background in all channels.
2. **Charge Flip background**: This background is caused due to mis-identification of lepton charge inside the detector (due to inelastic scattering or missing hits in the tracker). It is measured from data via. "Tag and Probe" method in $Z \to ll$ events in bins of lepton p_T and $|\eta|$. It is measured to be $\sim 10^{-3}$ for electrons and was found to be negligible for muons.
3. **Ir-reducible backgrounds**: This background is caused by genuine physical processes having the same final state particles as in the signal process. They are usually estimated via. Simulation and their modeling validated using control regions in data. Examples include $t\bar{t}V$ and $WW/WZ/ZZ$ backgrounds.

3.4 Results

Results of the above mentioned $t\bar{t}H$ searches with the data collected by CMS in 2016 ($35.9\ fb^{-1}$ at $\sqrt{s} = 13$ TeV) were combined with the CMS Run-1 dataset[8] to obtain a "5 σ discovery" of $t\bar{t}H$ process (Fig. 3.2) [9]. For the 2017 dataset,[9] public results were available only for the $t\bar{t}H$ Multi-lepton search at the time of the symposium [10]. A new category ($2l + 2\tau_h$) was added by this search on top of the existing ones and the combined (2016 + 2017) signal strength from the $t\bar{t}H$ Multi-lepton search alone equals $0.96^{+0.34}_{-0.31}$ ($1.00^{+0.30}_{-0.27}$) observed (expected) in units of the SM expectation.

---

[8] $5.1\ (19.7)\ fb^{-1}$ collected at $\sqrt{s} = 7\ (8)$ TeV.
[9] $41.4\ fb^{-1}$ at $\sqrt{s} = 13$ TeV.

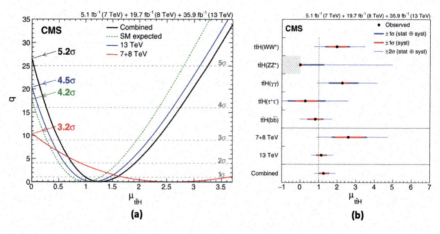

Fig. 3.2 Results of the CMS Run-1+Run-2 $t\bar{t}H$ combination. Likelihood scan of the signal strength (**a**) and Channel-wise signal strength split by era (**b**)

References

1. CMS collaboration, Observation of a new boson at a mass of 125 GeV with the CMS experiment at the LHC. Phys. Lett. B716, 30 (2012)
2. CMS collaboration, The CMS experiment at the CERN LHC. J. Instrum. 3(08), S08004 (2008)
3. CMS collaboration, Search for ttH production in the all-jet final state in proton-proton collisions at $\sqrt{s} = 13$ TeV. J. High Energy Phy. (6) (2018). arXiv:1803.06986
4. CMS collaboration, Search for $t\bar{t}H$ production in the $H \to b\bar{b}$ decay channel with leptonic $t\bar{t}$ decays in proton-protoncollisions at $\sqrt{s} = 13$ TeV. J. High Energy Phys. (3), 26 (2019). arXiv:1804.03682
5. CMS collaboration, Measurements of Higgs boson properties in the diphoton decay channel in proton-proton collisions at $\sqrt{s} = 13$ TeV. J. High Energy Physics. (11), 185 (2018). arXiv:1804.02716
6. CMS collaboration, Evidence for associated production of a Higgs boson with a top quark pair in final states with electrons, muons, and hadronically decaying τ leptons at $\sqrt{s} = 13$ TeV. J. High Energy Phys. (8), 66 (2018). arXiv:1803.05485
7. CMS collaboration, Search for Higgs boson production in association with top quarks in multilepton final states at $\sqrt{s} = 13$ TeV, CMS Physics Analysis Summary CMS-PAS-HIG-17-004
8. CMS collaboration, Search for the associated production of a Higgs boson with a top quark pair in final states with a τ lepton at $\sqrt{s} = 13$ TeV, CMS Physics Analysis Summary CMS-PAS-HIG-17-003
9. CMS collaboration, Observation of $t\bar{t}H$ production. Phys. Rev. Lett. **120**(23), 231801. arXiv:1804.02610
10. CMS collaboration, Measurement of the associated production of a Higgs boson with a top quark pair in final states with electrons, muons and hadronically decaying τ leptons in data recorded in 2017 at $\sqrt{s} = 13$ TeV. CMS Physics Analysis Summary CMS-PAS-HIG-18-019

Chapter 4
Jet Substructure as a Tool to Study Double Parton Scatterings in V + Jets Processes at the LHC

Ramandeep Kumar, Monika Bansal, and Sunil Bansal

Abstract Double parton scatterings (DPS) provide vital information on the parton–parton correlations and parton distributions in a hadron. It also constitute as a background to new physics searches. Measurement of DPS in Vector Boson (V) + jets processes is important because of clean experimental signature and large production cross-section. The available DPS measurements, with V + jets, are dominated by large contamination from V (W or Z) + jets processes produced with single parton scatterings (SPS). In this document, the importance of jet sub-structure in controlling SPS backgrounds for Z + jets DPS processes is discussed.

4.1 Introduction

Two or more than two parton–parton interactions in a single proton–proton (pp) collision are termed as multiple parton interactions (MPI) [1]. The probability of MPI increases with increase in collision energy at the Large Hadron Collider (LHC). MPI may produce particles with small transverse momenta as well as particles with large transverse momenta. Double parton scattering (DPS), a subset of MPI, includes the production of particles with large transverse momenta from at-least two parton–parton interactions. The study of DPS is important to understand parton–parton correlations and parton distributions in a hadron [2]. DPS processes can also contribute as background in the new physics searches [3, 4] as well. The experimental measurements of the DPS processes are usually contaminated by the particles from single

R. Kumar (✉)
Akal University Talwandi Sabo, Punjab 151302, India
e-mail: raman_phy@auts.ac.in; kumardeepraman@gmail.com

M. Bansal
DAV College, Sector 10, Chandigarh 160011, India
e-mail: 83.monu@gmail.com

S. Bansal
UIET, Panjab University, Chandigarh 160014, India
e-mail: sbansal@pu.ac.in

© Springer Nature Singapore Pte Ltd. 2021
P. K. Behera et al. (eds.), *XXIII DAE High Energy Physics Symposium*,
Springer Proceedings in Physics 261,
https://doi.org/10.1007/978-981-33-4408-2_4

parton scatterings (SPS). Usually, the correlation observables are used to disentangle the DPS processes from the SPS ones as in the existing measurements [5, 6]. The jet multiplicity distribution provides other opportunity to enhance the DPS signal contribution [7]. The presented studies [8] demonstrate that the fragmentation properties of a jet can be used to suppress the SPS backgrounds.

This study is performed using Z + jets events which are simulated using MAD-GRAPH [9] and POWHEG [10, 11]. PYTHIA8 [12] is used for the parton showering and hadronization of these events. To investigate effect of hadronization models, events are also simulated using hadronization and parton showering with HERWIG++. The DPS production of Z + 2-jets events is simulated using PYTHIA8, where one parton–parton scattering produces a Z-boson and the second one produces two jets. The following selection criteria, motivated from experimental constraints, is imposed on the simulated events:

– Two muons with transverse momenta larger than 20 GeV/c and absolute pseudo-rapidity less than 2.5.
– The dimuon invariant mass is required to be in range of 60−120 GeV/c^2.
– Two jets with minimum transverse momenta of 20 GeV/c and $|\eta| < 2.5$, which are clustered using anti-k_T algorithm with the radius parameter equal to 0.5.

The dijet production from DPS is dominated by the gluon-initiated jets and most of the jets produced via SPS are supposed to be initiated by quarks as depicted in Fig. 4.1. The jets are identified as initiated by quarks or gluons using jet-parton matching in $\eta \times \phi$ space. It can be seen that the contribution of gluon-initiated jets is ≈75% and ≈ 45% in DPS and SPS processes, respectively. Therefore, the contribution of DPS events can be increased by identifying the flavor of a jet and choosing the events with gluon-initiated jets only.

The fragmentation properties of the quark-initiated jets are different from the gluon-initiated jets [13–15]. The intrinsic properties of jets may be used to construct

Fig. 4.1 The fraction of gluon-initiated jets as a function of the jet p_T in the simulated DPS (blue solid circle markers) and SPS (red hollow circle markers) events

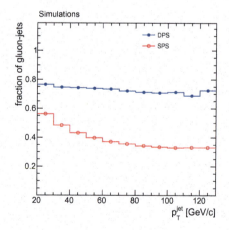

a number of different observables. A certain number of observables, as listed below, are used to construct the quark–gluon discriminator:

– major axis (σ_1^{jet})
– minor axis (σ_2^{jet})
– jet constituents multiplicity (N_p^{jet})
– jet fragmentation function ($p_T^{\text{jet}} D$)

The details of these observables can be found in [15]. Figure 4.2 shows the distributions of these observables for gluon- and quark-initiated jets using the events

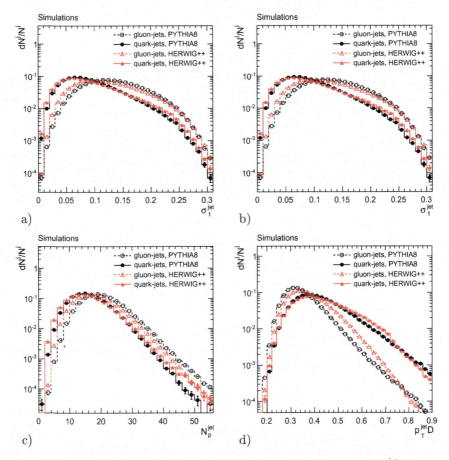

Fig. 4.2 The distributions of discriminating variables: **a** major axis size of jet cone (σ_1^{jet}), **b** minor axis size of jet cone (σ_2^{jet}), **c** jet constituents multiplicity (N_p^{jet}), and (d) jet fragmentation function ($p_T^{\text{jet}} D$) are compared for gluon-initiated jets (hollow markers) and quark-initiated jets (solid markers) events hadronized and parton showered with PYTHIA8 (black colored markers) and HERWIG++ (with red colored markers)

simulated by PYTHIA8 and HERWIG++. As evident from the distributions, jets are broader when initiated by gluons and constitute more number of particles. In addition, jets initiated by quarks constitute harder particles. Therefore, these observables may be used for a clear distinction between two types of jets. A multivariate analysis approach is followed for effective use of these observables along with the optimized cut-based analysis approach.

4.2   Results

It has been observed that in the selected $Z + 2$-jets events, the contribution from DPS processes is about 0.075. A simple cut-based analysis, implementing cuts summarized in Table 4.1 for observables based on the jet fragmentation properties, results in a gain of 41% in the DPS fraction.

The alternate approach for optimized use of the discriminating observables is based on the multivariate analysis, which is based on boosted decision trees (BDT) implemented in the TMVA framework [16]. A clear distinction is observed between two types of jets as depicted by the distribution of BDT discriminant shown in Fig. 4.3. A jet is considered to be initiated by gluon if the value of BDT is more than -0.105, otherwise it is considered to be initiated by quark. By selecting the $Z + 2$-jets events with two jets initiated by gluons, it has been observed that DPS fraction increases to 0.113, which is 51% larger if no jet fragmentation properties are used. A significant gain is observed with use of multivariate analysis approach as compared to cut-based analysis approach.

The effect of different hadronization model is also studied by considering the events hadronized with HERWIG++. The use of HERWIG++ also provides a gain of 43% with DPS fraction equal to 0.107. In addition, the effectiveness of the method is also tested by using event simulated by POWHEG. The gain in the DPS fraction, by using POWHEG, reduces to 36%, which arises due to different treatment at leading order and next-to-leading order for two models. It can be concluded from these studies that fragmentation properties of jets can be used to suppress the SPS background and hence DPS fraction may be enhanced.

Table 4.1 Conditions on observables for selection of gluon-initiated jets in cut-based analysis

Observable	condition
σ_1^{jet}	>0.04
σ_2^{jet}	>0.02
N_p^{jet}	>12.0
$p_T^{\text{jet}} D$	<0.49

Fig. 4.3 BDT output for gluon-initiated (hollow markers) jets and quark-initiated jets (solid markers) in case of dijet events produced with PYTHIA8 (black colored markers) and HERWIG++ (red colored markers)

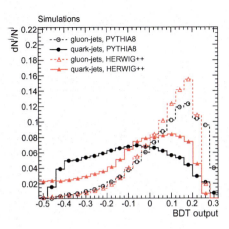

4.3 Summary

This report presents the possibility to explore the jet fragmentation properties for suppression of SPS events using Z + jets events. MADGRAPH and POWHEG Monte Carlo event generators are used to simulate Z + jets events, which are hadronized and parton showered using PYTHIA8. Four different observables are used to discriminate the jets initiated by gluons from those initiated by quarks. By considering the events with jets initiated by gluons, a gain of 40–50% in the DPS fraction is achieved. The presented study may play an important role for DPS studies under actual environmental conditions.

References

1. T. Sjöstrand, M. Van Zijl, A multiple interaction model for the event structure in hadron collisions. Phys. Rev. D **36**, 2019 (1987). https://doi.org/10.1103/PhysRevD.36.2019
2. M. Diehl, D. Ostermeier, A. Schäfer, "Elements of a theory for multiparton interactions in QCD," JHEP **1203**, 089 (2012) Erratum: [JHEP **1603**, 001 (2016)] https://doi.org/10.1007/JHEP03(2012)089, https://doi.org/10.1007/JHEP03(2016)001 arXiv:1111.0910[hep-ph]
3. M.Y. Hussein, A double parton scattering background to associate WH and ZH production at the LHC. Nucl. Phys. Proc. Suppl. **174**, 55 (2007). https://doi.org/10.1016/j.nuclphysbps.2007.08.086
4. C.M.S. Collaboration, Search for new physics with same-sign isolated dilepton events with jets and missing transverse energy at the LHC. JHEP **1106**, 077 (2011). https://doi.org/10.1007/JHEP06(2011)077
5. C.M.S. Collaboration, Study of double parton scattering using W + 2-jet events in proton-proton collisions at $\sqrt{s}$ = 7 TeV. JHEP **1403**, 032 (2014). https://doi.org/10.1007/JHEP03(2014)032
6. C.M.S. Collaboration, Constraints on the double-parton scattering cross section from same-sign W boson pair production in proton-proton collisions at $\sqrt{s}$ = 8 TeV. JHEP **1802**, 032 (2018). https://doi.org/10.1007/JHEP02(2018)032

7. R. Kumar, M. Bansal, S. Bansal, J.B. Singh, New observables for multiple-parton interactions measurements using Z+jets processes at the LHC. Phys. Rev. D **93**, 054019 (2016). https://doi.org/10.1103/PhysRevD.93.054019

8. R. Kumar, M. Bansal, S. Bansal, Jet fragmentation as a tool to explore double parton scattering using Z-boson + jets processes at the LHC. Phys. Rev. D **99**, 094025 (2019). https://doi.org/10.1103/PhysRevD.99.094025

9. J. Alwall et al., MadGraph 5: going beyond. JHEP **1106**, 128 (2011).https://doi.org/10.1007/JHEP06(2011)128

10. S. Frixione, P. Nason, C. Oleari, Matching NLO QCD computations with parton shower simulations: the POWHEG method. JHEP **11**, 070 (2007). https://doi.org/10.1088/1126-6708/2007/11/070

11. J.M. Campbell, R.K. Ellis, P. Nason, G. Zanderighi, W and Z bosons in association with two jets using the POWHEG method. JHEP **1308**, 005 (2013). https://doi.org/10.1007/JHEP08(2013)005

12. T. Sjöstrand, S. Mrenna, P.Z. Skands, A brief introduction to Pythia 8.1. Comput. Phys. Commun. **178**, 852 (2008). https://doi.org/10.1016/j.cpc.2008.01.036

13. J. Gallicchio, M.D. Schwartz, Quark and Gluon Tagging at the LHC. Phys. Rev. Lett. **107**, 172001 (2011). https://doi.org/10.1103/PhysRevLett.107.172001

14. ATLAS Collaboration, "Quark and Gluon Tagging at the LHC," Eur. Phys. J. **C74**, 3023 (2014). https://doi.org/10.1140/epjc/s10052-014-3023-z

15. CMS Collaboration, "Performance of quark/gluon discrimination in 8 TeV pp data," CMS-PAS-JME-13-002 (2013)

16. A. Hocker et al., TMVA—Toolkit for Multivariate Data Analysis. Physics/0703039 [physics.data-an]

Chapter 5
Angularity Distributions at One Loop with Recoil

Ankita Budhraja, Ambar Jain, and Massimiliano Procura

Abstract Angularities are a general class of event shapes which depend on a continuous parameter $b > -1$ that interpolates between recoil-insensitive observables like thrust and observables that are maximally sensitive to recoil effects like jet broadening. We present the first analytic calculations for angularity singular cross section at one-loop order taking into account the recoil effects, irrespective of the exponent b, within the Soft Collinear Effective Theory (SCET) framework. In the differential cross section, these recoil effects contribute to new terms which can have important consequences on resummation of the large logarithms. Our one-loop fixed-order results are checked against numerical results from EVENT2 generator.

5.1 Introduction

Event shapes are inclusive jet observables that quantify the geometrical flow of energy-momentum in a QCD event and probe strong interactions at various energy scales. Over the years, these have been used extensively for tuning parton showers and non-perturbative components of Monte Carlo event generators as well as to gain insight into hadronization effects in QCD. We focus on a class of event shapes known as angularities [1] which are defined as[1]

$$\tau = \frac{1}{Q} \sum_{i \, \epsilon X} |p_{\perp i}| \, e^{-b|\eta_i|},$$

(5.1)

---

[1] The more commonly used convention in the literature uses $a \equiv 1 - b$, for angularity distributions.

A. Budhraja (✉) · A. Jain
Indian Institute of Science Education and Research, Bhopal 462066, MP, India
e-mail: ankitab@iiserb.ac.in

M. Procura
Fakultät für Physik, Universität Wien, Boltzmanngasse 5, Wien 1090, Austria

Theoretical Physics Department, CERN, 1 Esplanade des Particules, Geneva 23, Switzerland

© Springer Nature Singapore Pte Ltd. 2021
P. K. Behera et al. (eds.), *XXIII DAE High Energy Physics Symposium*,
Springer Proceedings in Physics 261,
https://doi.org/10.1007/978-981-33-4408-2_5

where η_i is the rapidity of the ith final-state particle and $p_{\perp i}$ its transverse momentum, with respect to the thrust axis, $\hat{t}$. The thrust axis provides the direction of the net energy-momentum flow in an event. Using the thrust axis, one can divide an event into two hemispheres : particles moving along the direction of the thrust axis, i.e., $\mathbf{p}_i \cdot \hat{t} > 0$ into the right (R) hemisphere, while those moving opposite to the thrust axis, i.e., $\mathbf{p}_i \cdot \hat{t} < 0$ into the left hemisphere (L), and study the angularity of the particles in the given hemisphere. The net angularity τ as given in (5.1) is then obtained by the sum of the angularities of the two hemispheres.

For thrust-like angularities with $b \gtrsim 1$, the direction of the thrust axis is unchanged due to interaction by the soft radiation, but as $b \to 0$, the recoil of the collinear parton due to the soft radiation becomes an $\mathcal{O}(1)$ effect and cannot be ignored. The present analytic results for angularity distributions with respect to thrust axis are only available for $b \gtrsim 0.5$ (by extending the recoil-insensitive analysis applicable for $b \gtrsim 1$ angularities) and $b = 0$ (jet broadening). However, a unified framework applicable to the *whole range of b values* is still lacking. Here we provide novel results applicable for *all* $b > -1$. In addition, we find that compared to the thrust-like analyses of angularities with $0.5 < b < 1$, our results contain an extra integrable singular correction encoding the leading recoil effect. We provide a validation of this new contribution against EVENT2 [7] results.

For $\tau \ll 1$, the angularity distributions for $b \gtrsim 0.5$ and $b = 0$ are known to factorize in terms of a hard function multiplying convolutions of jet functions with a soft function. Here, we use the language of SCET [2], which provides an elegant means to derive factorization theorems. We adopt a factorization theorem similar to the jet broadening factorization presented in [3] for angularities when $b \to 0$, for which SCET$_\mathrm{I}$ factorization theorem is known to break down. The factorization theorem for the double-differential angularity cross section near $b = 0$ is given by the SCET$_\mathrm{II}$-type formula

$$\frac{1}{\sigma_0} \frac{d\sigma}{d\tau_L d\tau_R} = H(Q; \mu) \int d\tau_n \, d\tau_{\bar{n}} \, d\tau_n^s \, d\tau_{\bar{n}}^s \, \delta(\tau_R - \tau_n - \tau_n^s) \, \delta(\tau_L - \tau_{\bar{n}} - \tau_{\bar{n}}^s) \int d\mathbf{p}_\perp^2 d\mathbf{k}_\perp^2$$
$$\mathcal{J}(\tau_n, \mathbf{p}_\perp^2) \, \mathcal{J}(\tau_{\bar{n}}, \mathbf{k}_\perp^2) \, \mathcal{S}(\tau_n^s, \tau_{\bar{n}}^s, \mathbf{p}_\perp^2, \mathbf{k}_\perp^2), \qquad (5.2)$$

where $\tau_{L,R}$ denote the angularity of the particles in the left (L) and right (R) hemisphere, respectively. The hard function $H(Q; \mu)$ is given by the squared amplitude of the matching coefficient obtained by matching the two-jet matrix element of SCET with that of the full theory. The jet function $\mathcal{J}_{n,\bar{n}}$ describes the perturbative evolution of the partons $q, \bar{q}$, produced in the hard scattering into collimated jets of lower energy partons and the soft function S describes the color exchange between the two jets. Both the jet and soft function appearing in (5.2) are given by a generalization of the broadening jet and soft function with two modifications: first, the definition of the observable changes from broadening to that of angularity τ with $b \neq 0$, and secondly, the scaling of gluon fields in the soft Wilson line changes to $A_s \sim Q(\lambda^{1+b}, \lambda^{1+b}, \lambda^{1+b})$ rather than $A_s \sim Q(\lambda, \lambda, \lambda)$ specific to broadening.

5.2 The Jet Angularity Cross Section at $\mathcal{O}(\alpha_s)$

We present the results for angularity cross section for positive-b and negative-b separately. We provide a master formula for the double-differential cross section and discuss the thrust ($b = 1$) and broadening ($b = 0$) limits.

5.2.1 $\mathcal{O}(\alpha_s)$ Cross Section for $b > 0$

Following the factorization theorem defined in (5.2), the one-loop contribution to the double-differential cross section for angularity distributions with $b > 0$ is given as [5]

$$
\left[\frac{1}{\sigma_0}\frac{d^2\sigma^{(+)}}{d\tau_L\,d\tau_R}\right]^{\mathcal{O}(\alpha_s)}_{\mathrm{SCET_{II}}} = \frac{\alpha_s(\mu)\,C_F}{\pi}\,\delta(\tau_L)\left[-\frac{3}{2(1+b)}\left[\frac{1}{\tau_R}\right]_+ - \frac{2}{(1+b)}\left[\frac{\ln\tau_R}{\tau_R}\right]_+\right.
$$
$$
\left. -\frac{2}{1+b}\frac{\ln(1-r)}{\tau_R}+\delta(\tau_R)\left(-\frac{7}{4(1+b)}+\frac{\pi^2}{12\,b}\frac{(b^2+3b-2)}{1+b}-\frac{2}{1+b}i(b)\right)\right]+\{\tau_L \leftrightarrow \tau_R\}, \tag{5.3}
$$

where

$$
i(b) = \int_0^1 dx\left(\frac{x}{2}-1\right)\ln\left(1+\left(\frac{x}{1-x}\right)^b\right) = -\frac{3}{4}\ln 2 + \frac{b}{8} + \mathcal{O}(b^2), \tag{5.4}
$$

and r is given by the solution of the equation

$$
\frac{r}{(1-r)^{1+b}} = \tau_R^b. \tag{5.5}
$$

Integrating (5.3) over hemisphere angularities such that $\tau = \tau_L + \tau_R$, we can obtain the single-differential angularity cross section for $b > 0$ in the small-τ limit as

$$
\left[\frac{1}{\sigma_0}\frac{d\sigma^{(+)}}{d\tau}\right]^{\mathcal{O}(\alpha_s)}_{\mathrm{sing.}} = \frac{\alpha_s(\mu)C_F}{\pi}\left\{-\frac{3}{(1+b)}\left[\frac{1}{\tau}\right]_+ - \frac{4}{1+b}\left[\frac{\ln\tau}{\tau}\right]_+ - \frac{4}{1+b}\frac{\ln(1-r)}{\tau}\right.
$$
$$
\left. +\delta(\tau)\left[-\frac{7}{2(1+b)}+\frac{\pi^2}{6b}\frac{b^2+3b-2}{1+b}-\frac{4}{1+b}i(b)\right]\right\}, \tag{5.6}
$$

where r in the above equation is given by (5.5) with the replacement $\tau_R \to \tau$. The first two terms in the first line of (5.6) as well as the $\delta(\tau)$ piece are present in the result for thrust-like factorization theorem for angularities valid for $b \gtrsim 1$ [4]. Our SCET$_{\mathrm{II}}$ approach, which includes recoil effect, provides a singular correction to the thrust-like factorization result, given by the $\ln(1-r)/\tau$ term in (5.6).

As is evident from (5.5), in the small-τ limit, $r \sim \tau^b$. For $b = 1$, we have $r \sim \tau$, which implies that the recoil term gives only power corrections and the result of

(5.6) reduces to the familiar result of the singular cross section for thrust. When b is small, (5.6) needs to be treated carefully as r is not small even for small τ. Moreover, b regulates the $\tau \to 0$ singularity in the small-b limit. The recoil term in this limit takes the form

$$\frac{\ln(1-r)}{\tau} = \left(-\frac{\pi^2}{12\,b} + \frac{\ln^2 2}{2} - \frac{b}{4}\ln^2 2\right)\delta(\tau) - \ln 2\left[\frac{1}{\tau}\right]_+ - \frac{b}{2}\left[\frac{\ln \tau}{\tau}\right]_+ + \frac{b}{2}\ln 2\left[\frac{1}{\tau}\right]_+ + \mathcal{O}(b^2).$$

(5.7)

Substituting the above result along with the small-b expansion of $i(b)$ (as given in (5.4)) into (5.6), we obtain the single-differential cross section in the small-b limit. Note that the $1/b$ singularity proportional to $\delta(\tau)$ piece cancels out. Dropping the $\mathcal{O}(b)$ terms, we obtain the singular cross section for jet broadening [3]. Thus, we see that (5.6) not only provides the correct thrust or broadening limit but the singular cross section for all $b > 0$ angularities.

5.2.2 $\mathcal{O}(\alpha_s)$ Cross Section for $b < 0$

The master formula for the double-differential $\mathcal{O}(\alpha_s)$ cross section obtained from (5.2) for $b < 0$ has the form [5]

$$\left[\frac{1}{\sigma_0}\frac{d^2\sigma^{(-)}}{d\tau_L\,d\tau_R}\right]^{\mathcal{O}(\alpha_s)}_{\text{SCET}_{II}} = \frac{\alpha_s(\mu)\,C_F}{\pi}\delta(\tau_L)\left[\delta(\tau_R)\left(-\frac{7+2b}{4(1+b)} + \frac{\pi^2}{12\,b}(b+2) - \frac{2}{1+b}i(-b)\right)\right.$$
$$\left. - \frac{3}{2(1+b)}\left[\frac{1}{\tau_R}\right]_+ - \frac{2}{(1+b)^2}\left[\frac{\ln \tau_R}{\tau_R}\right]_+ - \frac{2}{(1+b)^2}\frac{\ln(1-s)}{\tau_R}\right] + \{\tau_L \leftrightarrow \tau_R\},$$

(5.8)

where $i(-b)$ is given by (5.4) with b replaced by $-b$ and s is given by the solution of the equation

$$\frac{s}{(1-s)^{\frac{1}{1+b}}} = \tau_R^{-\frac{b}{1+b}}.$$

(5.9)

Once again, we obtain the single-differential angularity cross section for $b < 0$ in the small-τ limit as,

$$\left[\frac{1}{\sigma_0}\frac{d\sigma^{(-)}}{d\tau}\right]^{\mathcal{O}(\alpha_s)}_{\text{sing.}} = \frac{\alpha_s(\mu)\,C_F}{\pi}\left\{-\frac{3}{1+b}\left[\frac{1}{\tau}\right]_+ - \frac{4}{(1+b)^2}\left[\frac{\ln \tau}{\tau}\right]_+ - \frac{4}{(1+b)^2}\frac{\ln(1-s)}{\tau}\right.$$
$$\left. + \delta(\tau)\left[-\frac{7+2b}{2(1+b)} + \frac{\pi^2(b+2)}{6\,b} - \frac{4}{1+b}i(-b)\right]\right\}.$$

(5.10)

where s in the above equation is given by (5.9) with the replacement $\tau_R \to \tau$. The result in (5.10) is the first analytic result of singular cross section for recoil-sensitive angularities with exponent $b < 0$. This will be validated against EVENT2 in the next section.

The solution of (5.9) in the small-τ limit goes as $s \sim \tau^{-\frac{b}{1+b}}$. When b is close to 0, $\tau^{-\frac{b}{1+b}} \sim 1$ requiring a solution to (5.9) that treats $\tau^{-\frac{b}{1+b}}$ as an $\mathcal{O}(1)$ quantity. Carefully treating the small-b limit, we obtain the following expansion for the $\ln(1 - s)$ term of (5.10)

$$\frac{\ln(1 - s)}{\tau} = \left(\frac{\pi^2}{12b} + \frac{\pi^2}{12} + \frac{\ln^2 2}{2} + \frac{b}{4}\ln^2 2\right)\delta(\tau) - \ln 2\left[\frac{1}{\tau}\right]_+ + \frac{b}{2}\left[\frac{\ln \tau}{\tau}\right]_+ - \frac{b\ln 2}{2}\left[\frac{1}{\tau}\right]_+ + \mathcal{O}(b^2).$$

(5.11)

Substituting the result of (5.11) and (5.4) (with $b \to -b$) into (5.10), we obtain the single-differential angularity cross section in the small-b limit. Taking the limit $b \to 0$ of this result reproduces the singular cross section for jet broadening [3]. Thus we have shown that the $b \to 0^+$ and $b \to 0^-$ limits of the single-differential singular angularity cross section are the same as the well-known broadening result, hence the cross section is a continuous function of the angularity exponent at $b = 0$, as expected.

5.3   Numerical Analysis and Comparison Against EVENT2

We compare both the broadening-like and the thrust-like normalized single-differential angularity distributions against numerical output from the EVENT2 generator [7], and show our results here for three different angularity exponents, $b = \{0.5, 0, -0.2\}$.

We find agreement within numerical uncertainties between EVENT2 and broadening-like SCET$_{\text{II}}$ factorization for sufficiently small values of τ. As an example, the differences between the EVENT2 output and our expressions for $d\sigma/d\log_{10}\tau$ for $b = 0$ (jet broadening case) and $b = -0.2$ are shown in Fig. 5.1 for ranges of τ where no visible cutoff effects are present.

For $b \geq 0.5$, we found agreement within error bars between EVENT2 and both thrust-like and broadening-like factorization for sufficiently small values of τ. Figure 5.2 illustrates the case of $b = 0.5$ for the single-differential cross section $d\sigma/d\log_{10}\tau$. As shown in these plots, the extra terms provided by the SCET$_{\text{II}}$ factorization theorem containing the recoil effects clearly improve the agreement with EVENT2 in the region of intermediate values of τ.

Table 5.1 demonstrates the importance of the recoil effects in comparison to the leading singular contribution. From the table, it is clear that the extra terms given by our SCET$_{\text{II}}$ factorization theorem provide a significant correction in the peak region of the spectrum and are thus expected to effect resummation in this region. For $b = 1$, we obtain a 5–6% correction when $\tau \sim 0.1$, and this is not surprising as this is the typical size of the power correction expected for thrust. For $b = 0$, the extra singular term gives leading contribution which is the largest recoil effect for all values of b.

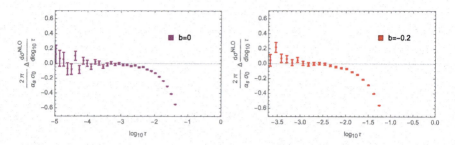

Fig. 5.1 Differences between EVENT2 and our NLO results from SCET$_{\mathrm{II}}$ factorization for $\mathrm{d}\sigma/\mathrm{d}\log_{10}\tau$ in the cases of $b=0$ (jet broadening) and $b=-0.2$

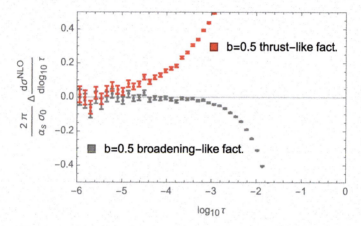

Fig. 5.2 Difference between EVENT2 and thrust-like (SCET$_{\mathrm{I}}$)/broadening-like (SCET$_{\mathrm{II}}$) NLO results for $\mathrm{d}\sigma/\mathrm{d}\log_{10}\tau$ for $b=0.5$

Table 5.1 Relative size of the extra singular contribution compared to the leading singular contribution in the peak region for the τ_b distribution, for various values of b. A 2–6% correction for $b=1$ or -0.5 shows the typical size of the power corrections due to the additional term

b	% correction for $\tau_b = 0.05$	% correction for $\tau_b = 0.1$
1	2	6
0.5	8	16
0.25	16	26
0	31	45
−0.2	15	24
−0.5	2	5

5.4 Conclusion

We have investigated how a theoretical framework based on $SCET_{II}$ factorization allows us to compute fixed-order singular angularity distributions measured with respect to the thrust axis for any value of the exponent b. We have shown that our one-loop results provide the correct thrust and broadening limits, thereby providing a check on our calculations. We have also produced novel one-loop results for the range $b < 1$ that contain recoil as the leading effect. This opens up the possibility to use our formalism to enhance the present analyses of high-precision $e^+ e^-$ studies by extending our results to NLL resummation. All our NLO distributions are found to be in agreement with EVENT2. A detailed study of these fixed-order results is given in [5] while the resummation of large logarithms at NLL accuracy will be presented in a future publication [6].

References

1. C.F. Berger, T. Kucs, G.F. Sterman, Phys. Rev. D **68**, 014012 (2003)
2. C.W. Bauer, S. Fleming, M.E. Luke, Phys. Rev. D63 (2000) 014006, C.W. Bauer, S. Fleming, D. Pirjol, I.W. Stewart, Phys. Rev. D63 (2001) 114020, C.W. Bauer, I.W. Stewart, Phys. Lett. B516 (2001) 134–142
3. J.Y. Chiu, A. Jain, D. Neill, I.Z. Rothstein, JHEP 05 (2012) 084, J.Y. Chiu, A. Jain, D. Neill, I.Z. Rothstein, Phys. Rev. Lett. 108 (2012) 151601
4. A. Hornig, C. Lee, G. Ovanesyan, JHEP **05**, 122 (2009)
5. A. Budhraja, A. Jain, M. Procura, submitted to JHEP
6. A. Budhraja, A. Jain, M. Procura, in preparation
7. S. Catani, M.H. Seymour, Nucl. Phys. B **485**, 291–419 (1997)

Chapter 6
Measurements of Property of Higgs with Mass Near 125 GeV, CMS Collaboration

Ashish Sharma

Abstract During run 2016 at LHC, the total recorded luminosity is 35.9 fb^{-1} by CMS detector [1]. Measurements of properties of Higgs boson can be determined where Higgs decaying to oppositely charged W bosons and W again decaying leptonically. Events are selected on basis of oppositely charged leptons pair, large missing transverse energy and with different number of jets in final state. Higgs producing from vector boson fusion and associated production with W and Z boson are also included on the basis of two jets and three or four leptons in final state. After Combining all these events corresponding to total integrated luminosity 35.9fb^{-1}, total calculated cross-sectional times branching ratio is 1.28 ± 0.17 times the standard model prediction for the Higgs boson with a mass of 125.09 GeV [2].

6.1 Introduction

Electroweak symmetry breaking is achieved through the prediction of a neutral scalar particle known as Higgs boson after the introduction of complex doublet scalar field which also leads masses of W and Z bosons [3]. Higgs was observed by CMS and ATLAS combined at 7 and 8 TeV using Run 1 dataset and the observed mass of Higgs is $m_H = 125.09 \pm 0.21(\text{stat}) \pm 0.11(\text{syst})$GeV.

This paper reports decay of Higgs boson to oppositely charged W boson using 2015 and 2016 dataset at center of mass energy of 13 TeV at total integrated luminosity of 35.9fb^{-1}. The large branching fraction of Higgs to WW, makes this chan- nel to study cross section of Higgs production through gluon–gluon fusion (ggH), vector-boson fusion (VBF) and associated production of Higgs (VH) [4]. In fully leptonic decay of W boson is the cleanest channel despite having presence of neutrino in final state which prevents clear mass peak of Higgs signal.

A. Sharma (✉)
IIT Madras, Chennai, Tamil Nadu, India
e-mail: ashish.sharma@cern.ch

© Springer Nature Singapore Pte Ltd. 2021
P. K. Behera et al. (eds.), *XXIII DAE High Energy Physics Symposium*,
Springer Proceedings in Physics 261,
https://doi.org/10.1007/978-981-33-4408-2_6

39

6.2 CMS Detector

CMS is situated at one collision point of 27 Km large hadron collider (LHC) which accelerated the proton at nearly the speed of light in clockwise and anti-clockwise direc- tion and then collide them. CMS is designed in a compact shape to measure energy and for the tracking of particles. It uses magnetic field of 3.8T for bending of charge particle and to measure momentum accordingly. Charge particle trajectories are measured with silicon pixel and tracker detector, which covers center pseudo-rapidity of $\eta \leq 2.5$. It consists of ECAL, HCAL, and muon system, where electron and photon deposits their energy in ECAL, whereas hadrons in HCAL system. Muon system consists of RPC, drift tube and CSC and Iron yoke.

6.3 Data and Simulated Samples

Events are selected on the basis of one or two electrons and muons. Combination of single or dilepton triggers give total trigger efficiency of more than 98 VBF process which are generated by POWHEG v2 [5].

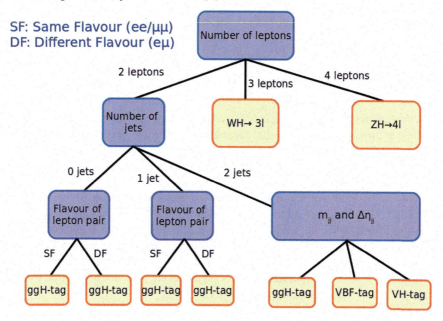

Out of category, one category can be discussed in detail, since ggH is the main pro- duction mode of Higgs. In ggH production mechanism different flavors final state, the main background process are WW, top, DY and W+jets. WW process can be distinguished from signal by different kinematic properties of leptons. To suppress process of three leptons, lepton ≥ 10 GeV is not allowed in final state. Final state transverse mass is defined as $m_T = \sqrt{2p_T^{ll} * E_T^{miss}[1 - \cos \Delta\phi]}$, which should be greater than 30 GeV; where $\Delta\phi$: angle between di lepton momentum and E_T^{miss} .

A shape analysis based on two-dimensional fit of m_{ll} and m_T in different flavors of ggH is done to extract Higgs signal.

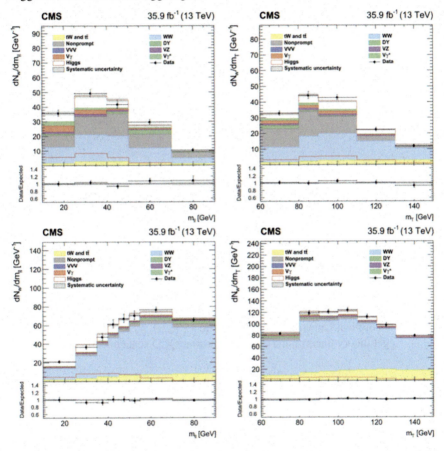

Category	Sub-category	Requirements
Preselection	-	$m_{ll} > 12$ GeV,$p_{T1} > 25$ GeV,$p_{T2} > 13(10)$ GeV for e(μ),$p_T^{miss} > 20$ GeV, $p_T^{ll} > 30$ GeV,no additional lepton with $p_T > 10$ GeV,electron and muon with oposite charge
0 jet ggH-tagged	$e^+\mu^-, e^-\mu^+, \mu^+e^-, \mu^-e^+$ ($p_{T2} > 20$ GeV)	$m_T > 60$ GeV,$p_T^{l2,p_T^{miss}} > 30$ GeV,sub-leading lepton with $p_T > 20$ GeV,no jets with $p_T > 30$ GeV,no b tageed jets with p_T between 20 and 30 GeV
	$e^+\mu^-, e^-\mu^+, \mu^+e^-, \mu^-e^+$ ($p_{T2} < 20$ GeV)	$m_T > 60$ GeV,$p_T^{l2,p_T^{miss}} > 30$ GeV,sub-leading lepton with $p_T < 20$ GeV,no jets with $p_T > 30$ GeV,no b tageed jets with p_T between 20 and 30 GeV
1 jet ggH-tagged	$e^+\mu^-, e^-\mu^+, \mu^+e^-, \mu^-e^+$ ($p_{T2} > 20$ GeV)	$m_T > 60$ GeV,$p_T^{l2,p_T^{miss}} > 30$ GeV,sub-leading lepton with $p_T > 20$ GeV,exactly 1 jets with $p_T > 30$ GeV,no b tageed jets with p_T between 20 and 30 GeV
	$e^+\mu^-, e^-\mu^+, \mu^+e^-, \mu^-e^+$ ($p_{T2} < 20$ GeV)	$m_T > 60$ GeV,$p_T^{l2,p_T^{miss}} > 30$ GeV,sub-leading lepton with $p_T < 20$ GeV,no jets with $p_T > 30$ GeV,no b tageed jets with p_T between 20 and 30 GeV
2 jet ggH-tagged	$e\mu$	at least two jets with $m_T > 30$ GeV,$m_T > 60$ GeV,$p_T^{l2,p_T^{miss}} > 30$ GeV , no b tageed jets with $p_T > 20$ GeV, $m_{jj} > 65$ or 110 GeV $< m_{jj} < 400$ GeV

6.4 Source of Systematical and Statistical Uncertainties

1. Source of systematical and statistical uncertainties

   a. Experimental Uncertainties: sources are—
      luminosity (2.3%), muon momentum (0.2%) and electron energy scale (0.6–1%), jet energy scale uncertainty($<10\%$), trigger efficiency ($<1\%$).

   b. Theoretical Uncertainties: sources are—
      ggH theory uncertainties, Parton Shower, QCD scale uncertainty.

6.5 Results

Signal strength modifier is defined as the ratio of measured and expected signal of Higgs mass of 125.09 GeV including systematical and statistical uncertainty. A summary of signal strength can be seen in Fig. 6.1.

The combined signal strength modifier is $\mu = 1.28^{+0.18}_{-0.17} = 1.28\pm 0.10$(stat)$\pm 0.11(sys)^{+0.10}_{-0.07}$(theo) which means observed significance of Higgs boson is $\sigma = 9.1$ and expected value of significance is $\sigma = 7.1$ (Fig. 6.2).

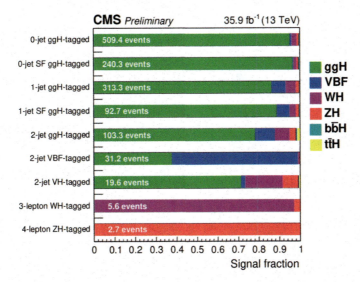

Fig. 6.1 Relative fraction of Higgs signal in measured and expected from SM

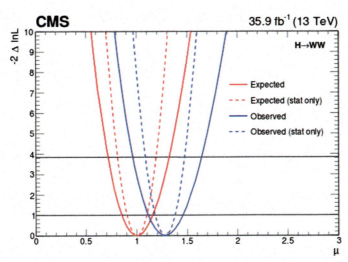

Fig. 6.2 Observed and expected profile likelihood for signal strength modifier

6.6 Summary

Using Run 2 data at total integrated luminosity of 35.9fb^{-1}, W^+, and W^- events are selected on basis of two, three, or four leptons in final state and large missing transverse energy and same flavor(SF) or different flavors(DF) final state. After combining all category, the observed value of significance is $9.1\ \sigma$ and expected (Fig. 6.3).

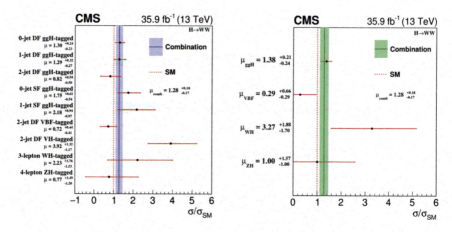

Fig. 6.3 Left plot is the observed signal strength modifier for different category and right plot corre- sponds to Higgs main production mode. Vertical dashed line is SM prediction and vertical line represents combined signal strength best fit value and filled shows 68% confidence interval

Value is 7.1σ. Hence, measured signal strength modifier is $\mu = 1.28^{+0.18}_{-0.17} = 1.28\pm$ $0.10(\text{stat})\pm 0.11(sys)^{+0.10}_{-0.07}(\text{theo})$.

References

1. C.M.S. Collaboration, Measurement of Higgs boson production and properties in the WW decay channel with leptonic final states. JHEP **01**, 096 (2014). https://doi.org/10.1007/JHEP01(2014)096. arXiv:1312.1129
2. Measurements of properties of the Higgs boson decaying to a W boson pair in pp collisions at $\sqrt{s}$=13 TeV. https://doi.org/10.1016/j.physletb.2018.12.073, arXiv.1806.05246
3. CMS Collaboration, Observation of a new boson with mass near 125 GeV in pp collisions at $\sqrt{s} = 7$ and 8 TeV. JHEP 06 (2013) 081, https://doi.org/10.1007/JHEP06(2013)081, arXiv:1303.4571
4. HWW team, Common analysis object definitions and trigger efficiencies for the H $\rightarrow$ WW analysis with 2016 full data. AN-2017/082 (2017)
5. P. Nason, A New method for combining NLO QCD with shower Monte Carlo algorithms. JHEP **11**, 040 (2004). https://doi.org/10.1088/1126-6708/2004/11/040. arXiv:hep-ph/0409146

Chapter 7
Predictions of Angular Observables for $\bar{B}_s \to K^*\ell\ell$ and $B \to \rho\ell\ell$ in Standard Model

Bharti Kindra and Namit Mahajan

Abstract Exclusive semileptonic decays based on $b \to s$ transitions have been attracting a lot of attention as some angular observables deviate significantly from the Standard Model (SM) predictions in specific q^2 bins. B meson decays induced by other Flavor Changing Neutral Current (FCNC), $b \to d$, can also offer a probe to new physics with an additional sensitivity to the weak phase in Cabibbo–Kobayashi–Masakawa (CKM) matrix. We provide predictions for angular observables for $b \to d$ semileptonic transitions, namely $\bar{B}_s \to K^*\ell^+\ell^-$, $\bar{B}^0 \to \rho^0\ell^+\ell^-$, and their CP-conjugated modes including various non-factorizable corrections.

7.1 Introduction

Experimental evidence of new physics has been found in the channels involving FCNC $b \to s\ell^+\ell^-$ and charged current $b \to c\ell\nu$. However, the $b \to d$ counterpart of the weak decay, i.e., $b \to d\ell^+\ell^-$, has not caught much attention perhaps because of low branching ratio. The weak phases incorporate CKM matrix elements $\xi_q^i = V_{qi}^* V_{qb}$, where $q \in \{u, c, t\}$ and $i \in \{s, d\}$. For $b \to s\ell\ell$ transition, $\xi_{c,t}^s \sim \lambda^2$ and $\xi_u^s \sim \lambda^4$ where $\lambda = 0.22$. Since $u\bar{u}$ contribution introduces CKM phase which is negligible for $b \to s\ell\ell$, CP violating quantities are very small in SM. On the other hand, since $\xi_u^d \sim \xi_c^d \sim \xi_t^d \sim \lambda^4$ for $b \to d\ell\ell$, the B decays mediated through this transition allow for large CP violating quantities. Also, leading order contribution in this case is smaller than the leading contribution in $b \to s\ell\ell$ which makes it more sensitive to new particles and interactions. In this work, we focus on two such decay channels, $B_s \to \bar{K}^*\ell^+\ell^-$ and $B \to \rho\ell^+\ell^-$[1].

B. Kindra (✉) · N. Mahajan
Physical Research Laboratory, Ahmedabad, India
e-mail: Bharti.kindra04@gmail.com; bhrt.kndr56@gmail.com

B. Kindra
Indian Institute of Technology, Gandhinagar, India

© Springer Nature Singapore Pte Ltd. 2021
P. K. Behera et al. (eds.), *XXIII DAE High Energy Physics Symposium*,
Springer Proceedings in Physics 261,
https://doi.org/10.1007/978-981-33-4408-2_7

7.2 Decay Amplitude

We follow the effective Hamiltonian approach as used in [2] to write the Hamiltonian and decay amplitude. Th amplitude is written as a product of short-distance contributions through Wilson coefficients and long-distance contribution which is further expressed in terms of form factors,

$$\mathcal{M} = \frac{G_F \alpha}{\sqrt{2}\pi} V_{tb} V_{td}^* \left\{ \left[\langle V | \bar{d}\gamma^\mu (C_9^{\text{eff}} P_L) b | P \rangle - \frac{2m_b}{q^2} \langle V | \bar{d} \, i \, \sigma^{\mu\nu} q_\nu (C_7^{\text{eff}} P_R) b | P \rangle \right] (\bar{\ell}\gamma_\mu \ell) \right.$$

$$\left. + \langle V | \bar{d}\gamma^\mu (C_{10}^{\text{eff}} P_L) b | P \rangle (\bar{\ell}\gamma_\mu \gamma_5 \ell) - 16\pi^2 \frac{\bar{\ell}\gamma^\mu \ell}{q^2} \mathcal{H}_\mu^{\text{non-fac}} \right\}. \tag{7.1}$$

Wilson coefficients $(C_i's)$ are computed upto next-to-next-to leading order (NNLO) [3] and form factors are computed using the method of Light Cone Sum Rules (LCSR) and QCD lattice calculation [4]. $\mathcal{H}_\mu^{\text{non-fac}}$ represents the non-factorizable contribution of non-local hadronic matrix element. This results from four quark and chromomagnetic operators combined with virtual photon emission which then decays to lepton pair through electromagnetic interaction. These corrections are given in terms of hard-scattering kernels $(T_a^q's)$, where $a \in \{\perp, \|\}$ and $q \in \{u, c\}$, which are convoluted with B(B$_S$)-meson and $\rho(\bar{K}^*)$ distribution amplitudes. The non-factorizable corrections included here are spectator scattering $T_a^{q,\text{spec}}$, weak annihilation $T_a^{q,\text{WA}}$, and soft-gluon emission $\Delta C_9^{q,\text{soft}}$. These corrections have been computed in [5–7] except charm loop corrections corresponding to up quark in the loop. For present work, we are assuming that its contribution is less than 10% of C_9: $\Delta C_{9,u}^{\text{soft}} = ae^{i\theta}$; $|a| \in \{0, 0.5\}$, $\theta \in \{0, \pi\}$.

These corrections are then added to transversity amplitudes in the following way:

$$A_{\perp L,R}(q^2) = \sqrt{2\lambda} \, N \left[2\frac{m_b}{q^2} (C_7^{\text{eff}} T_1(q^2) + \Delta T_\perp) + (C_9^{\text{eff}} \mp C_{10} + \Delta C_9^1(q^2)) \frac{V(q^2)}{M_B + M_V} \right] \tag{7.2}$$

$$A_{\| L,R}(q^2) = -\sqrt{2} N (M_B^2 - M_V^2) \left[2\frac{m_b}{q^2} (C_7^{\text{eff}} T_2(q^2) + 2\frac{E(q^2)}{M_B} \Delta T_\perp) + \right.$$

$$\left. (C_9^{\text{eff}} \mp C_{10} + \Delta C_9^2(q^2)) \frac{A_1(q^2)}{M_B - M_V} \right] \tag{7.3}$$

$$A_{0L,R}(q^2) = -\frac{N}{2M_V \sqrt{q^2}} \left[2m_b ((M_B^2 + 3M_V^2 - q^2)(C_7^{\text{eff}} T_2(q^2)) \right.$$

$$- \frac{\lambda}{M_B^2 - M_v^2} (C_7^{\text{eff}} T_3(q^2) + \Delta T_\|)) + (C_9^{\text{eff}} \mp C_{10} + \Delta C_9^3)$$

$$\left. ((M_B^2 + M_V^2 - q^2)(M_B + M_V) A_1(q^2) - \frac{\lambda}{M_B + M_V} A_2(q^2)) \right] \tag{7.4}$$

$$A_t(q^2) = \frac{N}{\sqrt{s}} \sqrt{\lambda} 2 \, C_{10} \, A_0(q^2) \tag{7.5}$$

where,

$$\Delta T_\perp = \frac{\pi^2}{N_c} \frac{f_P f_{V,\perp}}{M_B} \frac{\alpha_s C_F}{4\pi} \int \frac{d\omega}{\omega} \Phi_{P,-}(\omega) \int_0^1 du \, \Phi_{V,\perp}(u)(T_\perp^{c,\text{spec}} + \frac{\xi_u}{\xi_t}(T_\perp^{u,\text{spec}}))$$

(7.6)

$$\Delta T_\| = \frac{\pi^2}{N_c} \frac{f_P f_{V,\|}}{M_B} \frac{M_V}{E} \sum_\pm \int \frac{d\omega}{\omega} \Phi_P(\omega) \int_0^1 du \, \Phi_{V,\|}(u) [T_\|^{c,WA} + \frac{\xi_u}{\xi_t} T_\|^{u,WA}$$

$$\frac{\alpha_s C_F}{4\pi}(T_\|^{c,\text{spec}} + \frac{\xi_u}{\xi_t} T_\|^{,\text{spec}})]$$

(7.7)

$$\Delta C_9^i = \Delta C_{9,c}^{i,soft} + \Delta C_{9,u}^{i,soft}$$

(7.8)

7.3 Observables

The angular decay distribution of $B \to V(\to M_1 M_2)\ell^+\ell^-$ is given in terms of angular functions $(I_i(q^2, \theta_V, \theta_l, \phi)$, the value of which can be obtained by integrating data over specific values of the parameters. We consider an optimized set of observables constricted choosing specific combinations of these angular functions. The observables considered here are

– Form Factor Dependent observables.

$$\frac{d\Gamma}{dq^2} = \frac{1}{4}(3I_1^c + 6I_1^s - I_2^c - 2I_2^s) \qquad A_{FB}(q^2) = \frac{-3I_6^s}{3I_1^c + 6I_1^s - I_2^c - 2I_2^s}$$

$$F_L(q^2) = \frac{3I_1^c - I_2^c}{3I_1^c + 6I_1^s - I_2^c - 2I_2^s}$$

(7.9)

– Form Factor independent observables.

$$P_1 = \frac{I_3}{2I_2^s}, \quad P_2 = \beta_l \frac{I_6^s}{8I_2^s}, \quad P_3 = -\frac{I_9}{4I_2^s}, \quad P_4' = \frac{I_4}{\sqrt{-I_2^c I_2^s}}$$

$$P_5' = \frac{I_5}{2\sqrt{-I_2^c I_2^s}}, \quad P_6' = -\frac{I_7}{2\sqrt{-I_2^c I_2^s}}, \quad P_8' = -\frac{I_8}{2\sqrt{-I_2^c I_2^s}}$$

(7.10)

– Lepton Flavor Universality violating observables.

$$R_{K^*}^{B_s} = \frac{[\mathcal{BR}(B_s \to \bar{K}^*\mu^+\mu^-)]_{q^2 \in \{q_1^2, q_2^2\}}}{[\mathcal{BR}(B_s \to \bar{K}^*e^+e^-)]_{q^2 \in \{q_1^2, q_2^2\}}}$$

(7.11)

These observables are valid for $B_s \to \bar{K}^*\ell\ell$. For the CP-conjugate process, the I_i are replaced by $\bar{I}_i \equiv \xi_i \bar{I}_i$, where $\bar{I}_i$ are I_s only with weak phase conjugated and $\xi_i = 1$ for $i = \{1, 2, 3, 4, 7\}$ and -1 for $i = \{5, 6, 8, 9\}$. For $B \to \rho\ell\ell$, angular functions are replaced with time-dependent angular functions, since the final state in this case is self conjugate [1]. Thus, observables are sensitive to $B^0 - \bar{B}^0$ oscillations in this case and the $I_i's$ are replaced by $J_i's$ in the definition of observables, where $J_i's$ are given as [9],

$$J_i(t) + \tilde{J}_i(t) = e^{-\Gamma t}[(I_i + \bar{I}_i)\cosh(y\Gamma t) - h_i\sinh(y\Gamma t)] \tag{7.12}$$

$$J_i(t) - \tilde{J}_i(t) = e^{-\Gamma t}[(I_i - \bar{I}_i)\cosh(y\Gamma t) - s_i\sinh(y\Gamma t)] \tag{7.13}$$

where $x = \Delta m/\Gamma$, $y = \Delta\Gamma/\Gamma$, and $\tilde{J}_i \equiv \xi_i \bar{J}_i$. The extra terms h_i and s_i are the cross terms because of meson mixing [9]. These are time-dependent angular functions. To construct time-independent observables, these are integrated over a range of time which is $t \in \{-\infty, \infty\}$ in the case of LHCb and $t \in [0, \infty\}$ in case of Belle. Because of this difference, the integrated angular functions are slightly different for Belle and LHCb. We have taken this into account and given the prediction of angular observables separately.

7.4 Results

The binned values for the decay modes in study are listed in Tables 7.1 and 7.2, where the first uncertainty is due to form factors and scond uncertainty is due to soft-

Table 7.1 Observables for $\bar{B}_s \rightarrow K^*\mu^+\mu^-$ and $B_s \rightarrow \bar{K}^*\mu^+\mu^-$ using form factors based on LCSR and QCD lattice calculation

Observable	$\bar{B}_s \rightarrow K^*\mu^+\mu^-$		$B_s \rightarrow \bar{K}^*\mu^+\mu^-$	
	[0.1-1] GeV^2	[1-6] GeV^2	[0.1-1] GeV^2	[1-6] GeV^2
P_1	$0.017 \pm 0.132 \pm 0.001$	$-0.096 \pm 0.128 \pm 0.005$	$0.015 \pm 0.135 \pm 0.001$	$-0.087 \pm 0.118 \pm 0.005$
P_2	$0.122 \pm 0.013 \pm 0.001$	$0.026 \pm 0.081 \pm 0.036$	$0.114 \pm 0.012 \pm 0.001$	$0.054 \pm 0.081 \pm 0.034$
P_3	$0.001 \pm 0.003 \pm 0.0$	$0.004 \pm 0.009 \pm 0.002$	$0.001 \pm 0.006 \pm 0.0$	$0.004 \pm 0.009 \pm 0.002$
P_4'	$-0.704 \pm 0.063 \pm 0.009$	$0.543 \pm 0.167 \pm 0.014$	$-0.736 \pm 0.064 \pm 0.008$	$0.453 \pm 0.176 \pm 0.016$
P_5'	$0.437 \pm 0.044 \pm 0.016$	$-0.422 \pm 0.124 \pm 0.046$	$0.445 \pm 0.045 \pm 0.016$	$-0.377 \pm 0.130 \pm 0.047$
P_6'	$-0.091 \pm 0.005 \pm 0.016$	$-0.087 \pm 0.010 \pm 0.002$	$-0.048 \pm 0.004 \pm 0.001$	$-0.064 \pm 0.004 \pm 0.002$
P_8'	$0.027 \pm 0.007 \pm 0.016$	$0.042 \pm 0.010 \pm 0.017$	$0.048 \pm 0.009 \pm 0.016$	$0.036 \pm 0.008 \pm 0.019$
$R_{K^*}^{B_s}$	$0.945 \pm 0.008 \pm 0.001$	$0.998 \pm 0.004 \pm 0.0$	$0.944 \pm 0.007 \pm 0.001$	$0.998 \pm 0.004 \pm 0.0$
$BR \times 10^9$	$4.439 \pm 0.648 \pm 0.086$	$8.251 \pm 1.872 \pm 0.357$	$5.082 \pm 0.699 \pm 0.101$	$8.763 \pm 1.959 \pm 0.375$
A_{FB}	$-0.048 \pm 0.008 \pm 0.001$	$0.001 \pm 0.021 \pm 0.009$	$-0.047 \pm 0.007 \pm 0.001$	$-0.012 \pm 0.020 \pm 0.009$
F_L	$0.576 \pm 0.066 \pm 0.014$	$0.872 \pm 0.035 \pm 0.007$	$0.553 \pm 0.065 \pm 0.014$	$0.862 \pm 0.035 \pm 0.007$

Table 7.2 Binned values of observables for the process $B \to \rho\mu^+\mu^-$ and $\bar{B} \to \rho\mu^+\mu^-$ for tagged events to be measured at Belle. Form factors are based on LCSR form factors

	$B \to \rho\mu^+\mu^-$		$\bar{B} \to \rho\mu^+\mu^-$	
	[0.1-1] GeV^2	[1-6] GeV^2	[0.1-1] GeV^2	[1-6] GeV^2
$\langle P_1 \rangle$	$0.009 \pm 0.177 \pm 0.001$	$-0.065 \pm 0.116 \pm 0.003$	$0.010 \pm 0.175 \pm 0.001$	$-0.069 \pm 0.120 \pm 0.003$
$\langle P_2 \rangle$	$0.082 \pm 0.0 \pm 0.001$	$0.021 \pm 0.056 \pm 0.023$	$0.076 \pm 0.008 \pm 0.0$	$-0.042 \pm 0.050 \pm 0.024$
$\langle P_3 \rangle$	$0 \pm 0.005 \pm 0.0$	$0.001 \pm 0.005 \pm 0.002$	$0.001 \pm 0.001 \pm 0.0$	$0.002 \pm 0.005 \pm 0.002$
$\langle P_4' \rangle$	$-0.724 \pm 0.081 \pm 0.047$	$0.508 \pm 0.161 \pm 0.029$	$-0.703 \pm 0.080 \pm 0.046$	$0.569 \pm 0.154 \pm 0.017$
$\langle P_5' \rangle$	$0.276 \pm 0.004 \pm 0.027$	$-0.270 \pm 0.083 \pm 0.085$	$0.246 \pm 0.003 \pm 0.030$	$-0.321 \pm 0.074 \pm 0.098$
$\langle P_6' \rangle$	$-0.043 \pm 0.003 \pm 0.001$	$-0.061 \pm 0.004 \pm 0.002$	$-0.075 \pm 0.005 \pm 0.001$	$-0.073 \pm 0.010 \pm 0.002$
$\langle P_8' \rangle$	$0.025 \pm 0.005 \pm 0.016$	$0.025 \pm 0.005 \pm 0.018$	$0.031 \pm 0.005 \pm 0.007$	$0.030 \pm 0.006 \pm 0.017$
$\langle R_\rho \rangle$	$0.936 \pm 0.008 \pm 0.001$	$0.997 \pm 0.003 \pm 0.0$	$0.950 \pm 0.167 \pm 0.002$	$1.064 \pm 0.392 \pm 0.0$
$\langle BR \rangle \times 10^9$	$5.233 \pm 0.711 \pm 0.080$	$8.714 \pm 1.668 \pm 0.366$	$4.736 \pm 0.656 \pm 0.077$	$8.414 \pm 1.649 \pm 0.365$
$\langle A_{FB} \rangle$	$-0.038 \pm 0.005 \pm 0.001$	$-0.007 \pm 0.019 \pm 0.007$	$-0.034 \pm 0.005 \pm 0.001$	$0.014 \pm 0.022 \pm 0.006$
$\langle F_L \rangle$	$0.495 \pm 0.067 \pm 0.014$	$0.813 \pm 0.037 \pm 0.007$	$0.514 \pm 0.072 \pm 0.014$	$0.838 \pm 0.046 \pm 0.006$

gluon emission from up quark. Moreover, the full branching ratio for $B_s \to \bar{K}^*\ell\ell$ is $(3.356 \pm 0.814) \times 10^{-8}$ which is consistent with the recent measurement [8].

References

1. B. Kindra, N. Mahajan. arXiv:1803.05876 [hep-ph]
2. W. Altmannshofer, P. Ball, A. Bharucha, A.J. Buras, D.M. Straub, M. Wick, JHEP **0901**, 019 (2009). https://doi.org/10.1088/1126-6708/2009/01/019. arXiv:0811.1214 [hep-ph]
3. H.M. Asatrian, K. Bieri, C. Greub, M. Walker, Phys. Rev. D **69**, 074007 (2004). https://doi.org/10.1103/PhysRevD.69.074007 [hep-ph/0312063]
4. A. Bharucha, D.M. Straub, R. Zwicky, JHEP **1608**, 098 (2016). https://doi.org/10.1007/JHEP08(2016)098. [arXiv:1503.05534 [hep-ph]]
5. M. Beneke, T. Feldmann, D. Seidel, Nucl. Phys. B **612**, 25 (2001). https://doi.org/10.1016/S0550-3213(01)00366-2 [hep-ph/0106067]
6. M. Beneke, T. Feldmann, D. Seidel, Eur. Phys. J. C **41**, 173 (2005). https://doi.org/10.1140/epjc/s2005-02181-5 [hep-ph/0412400]
7. A. Khodjamirian, T. Mannel, A.A. Pivovarov, Y.-M. Wang, JHEP **1009**, 089 (2010). https://doi.org/10.1007/JHEP09(2010)089. [arXiv:1006.4945 [hep-ph]]

8. R. Aaij et al., [LHCb Collaboration], JHEP **1807**, 020 (2018). https://doi.org/10.1007/JHEP07(2018)020 arXiv:1804.07167 [hep-ex]
9. S. Descotes-Genon, J. Virto, JHEP **1504**, 045 (2015) Erratum: JHEP **1507**, 049 (2015). https://doi.org/10.1007/JHEP04(2015)045, https://doi.org/10.1007/JHEP07(2015)049 arXiv:1502.05509 [hep-ph]

Chapter 8
Search for Lepton Flavor, Lepton Number, and Baryon Number Violating Tau Decay $\tau \to p\mu\mu$ at Belle

Debashis Sahoo

8.1 Introduction

To explain the matter–antimatter asymmetry observed in the universe, three conditions formulated by Sakharov, must be satisfied [1].

- Baryon number violation (BNV): don't have any experimental confirmation yet.
- C-symmetry and CP-symmetry violation: observed experimentally.
- Interaction out of thermal equilibrium.

Any observation of BNV would be a clear signal of new physics. This phenomenon is presumed to have happened in the early universe. There is an indirect way of looking into BNV by means of collider experiments. For instance, BNV in charged lepton decays would imply lepton number and lepton flavor violation; with angular momentum conservation it would require the change $\Delta(B - L) = 0$ or 2, where B and L are the net baryon and lepton numbers. We report herein the expected upper limit on the branching fraction of $\tau^- \to p\mu^-\mu^-$ based on a Monte Carlo (MC) study. A study of low-momentum muons that are unable to reach the dedicated muon detector using $J/\psi \to \mu^+\mu^-$ events is also presented in the report.

8.2 Reconstruction of $\tau^- \to p\mu^-\mu^-$

The Belle detector is placed at the interaction point (IP) of the KEKB asymmetric e^+e^- collider. The detector is a large-solid-angle magnetic spectrometer consisting of a silicon vertex detector, a central drift chamber (CDC), an array of aerogel threshold Cherenkov counters (ACC), time-of-flight scintillation counters (TOF),

D. Sahoo (✉)
TIFR, Mumbai, India
e-mail: debashis.sahoo@tifr.res.in; sahoodev1994@gmail.com

© Springer Nature Singapore Pte Ltd. 2021   51
P. K. Behera et al. (eds.), *XXIII DAE High Energy Physics Symposium*,
Springer Proceedings in Physics 261,
https://doi.org/10.1007/978-981-33-4408-2_8

and a CsI(Tl) crystal electromagnetic calorimeter (ECL); all located inside a super-conducting solenoid providing an axial magnetic field of 1.5 T. An iron flux-return located outside the coil is instrumented with resistive plate chambers to detect K^0_L mesons and muons (KLM). A more detailed description of the Belle detector can be found in [2].

In this analysis, we search for $e^+e^- \rightarrow \tau^+\tau^-$ events where one τ (called the signal τ) decays to one proton or antiproton and two muons while other τ (denoted as the tag τ) decays to one charged track, neutrinos, and single/multiple neutrals. We select tau pair events within a fiducial volume of $150° < \theta < 17°$, where θ is the polar angle relative to the direction opposite the e^+ beam in the laboratory frame. The transverse momentum (p_T) of each charged track is required to be greater than $0.1\text{GeV}/c$ and energy of each photon (E_γ) greater than 0.1 GeV. Each track must have a distance of closest approach with respect to the IP within ±0.5 cm in the transverse plane and within ±3.0 cm along the beam direction.

Backgrounds to this analysis arise from $e^+e^- \rightarrow \tau^+\tau^-$ (generic), $B\bar{B}$(charged and mixed), $c\bar{c}$(charm), $q\bar{q}$(uds), $\mu\mu$ and two-photon events. To suppress high-multiplicity events from $B\bar{B}$, $c\bar{c}$ and $q\bar{q}$, we require the total number of tracks to be within 2 and 8. Such events are further suppressed by applying a 3-1 event topology. This classification is done by means of the thrust axis which is calculated from the observed tracks and neutral candidates. The [Thrust.$p_i^{\text{sig}} > 0$ and Thrust.$p_i^{\text{tag}} < 0$] or [Thrust.$p_i^{\text{sig}} < 0$ and Thrust.$p_i^{\text{tag}} > 0$] criteria separate the events into two hemi-spheres, called as the signal and tag side; the signal side contains three charged tracks while the tag side contains one charged track. In addition, we require the charge sum of all the tracks in an event be zero. The absolute missing momentum must exceed 0.4 GeV/c in order to ensure that the missing particles are neutrinos instead of being photons or charged particles that lie outside the detector acceptance. To reject $e^+e^- \rightarrow \mu\mu$, two-photon and more continuum backgrounds, we require the magnitude of thrust, $0.9 < |\text{Thrust}| < 0.99$ (Fig. 8.1). Since neutrinos are emitted only on the tag side, the missing momentum direction should also lie within the same side. The cosine of the angle between the missing momentum and the charged track on the tag side in the center-of-mass (CM) system should lie in the range $[0.0, 0.98]$. To reject the surviving two-photon and $\mu\mu$ backgrounds, we apply a requirement on the total visible energy in the CM frame, $5.29 \text{ GeV} \leq E_{\text{vis}}^{\text{CM}} \leq 9.50 \text{ GeV}$ (Fig. 8.2). The reconstructed mass on the tag side calculated using the charged track (with a pion mass assumed) and photons, m_{tag}, is required to be less than 1.78 GeV/c^2.

We require one of the charged tracks to be identified as proton in the signal side. The track is selected as proton if $P(p/K) > 0.6$ and $P(p/\pi) > 0.6$ where $P(i/j) = L(i)/(L(i) + L(j))$ with $L(i)$ and $L(j)$ being the likelihood for a track to be identified as i and j, respectively; these are obtained with the information from the ACC, TOF and CDC.

At Belle muons are identified based on the information from KLM which is the outermost subdetector. As we are dealing with a three-body decay of the tau, there is a good possibility of one of tracks having low momentum. From the generator level information, we verify that one muon has indeed low momentum and hence

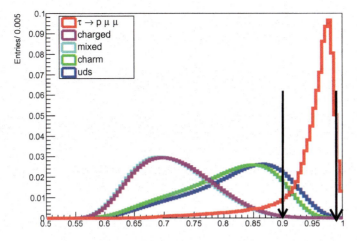

Fig. 8.1 |Thrust|

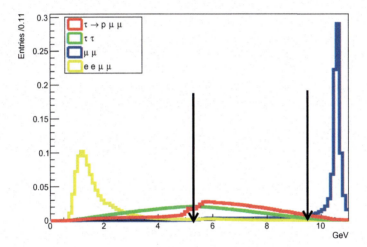

Fig. 8.2 E_{vis}^{CM}

is unable to reach the KLM. For the signal MC study, one million signal events are generated using KKMC and Tauola and are simulated by the GEANT3 package, which includes all subdetectors. The identification of low-momentum muons become difficult as it mostly relies on only the KLM subdetector. So we relax the criteria on $P(\mu)$ and calculate the signal reconstruction efficiency. The obtained efficiency (ϵ) is 19% with atleast one muon satisfying $P(\mu) > 0.9$ in the signal region which is defined as $1.745 \leq M_{rec} \leq 1.815$ GeV and $-125 \leq \Delta E \leq 100$ MeV. The background shape in the signal region is determined by taking background samples that have same luminosity as the Belle data sample. There is no peaking structure from the backgrounds and about 50 background candidates found in the signal region with the most dominant contribution coming from $\tau^- \to \pi^-\pi^+\pi^-\nu_\tau$. The shape parameters of signal and background probability density functions (PDFs) are extracted from a two-dimensional M_{rec} versus ΔE fitting to the respective MC samples. Then a maximum likelihood fit is done to determine the signal and background yields by using the combined PDFs extracted earlier. No signal candidate is found from the fit, so an upper limit is set on the signal yield by the Frequentist method. The expected signal yield at 90% confidence level ($N_{\text{sig}}^{\text{UL}}$) is 5. The branching fraction is given by

$B(\tau^- \to p\mu^-\mu^-) < \dfrac{N_{\text{sig}}^{\text{UL}}}{2N_{\tau\tau}\epsilon}$ where $\epsilon = 19\%$, $N_{\tau\tau} = 7.1 \times 10^8$ corresponding to the 770 fb^{-1} of MC data. So the expected 90% confidence-level upper limit from the MC study is 1.8×10^{-8}, compared to the LHCb's observed limit of 4.4×10^{-7} [3].

8.3 Low-Momentum Muon Identification

In this section, we study the behavior of muons with momentum less than 1.2 GeV in the non-KLM subdetectors and use a neural network (NN) [4] to combine the obtained information. Muons from $J/\psi \to \mu^+\mu^-$ sample are taken as signal muons, whereas charged pions, kaons, and electrons in the same momentum range are considered to be backgrounds. The ratio of measured and expected muon energy losses in the CDC, the likelihood for the muon hypothesis based on the ACC and the TOF, the ratio of the matched cluster energy in the ECL to the track momentum, the ratio of energy deposited in 3×3 to 5×5 crystal array around the central crystal, and the electromagnetic shower width are taken as the discriminating variables. The combined output results are shown in Fig. 8.3.

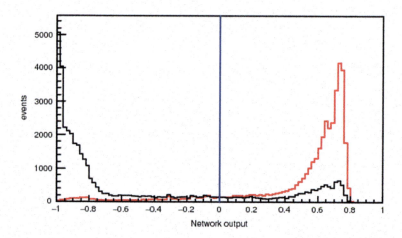

Fig. 8.3 NN output, the red histogram corresponds to muon and the black histogram is for background

Clearly, there is a difference between muons and the background particles based upon the non-KLM information. We plan to use this method for low-momentum muon identification in the $\tau^- \to p\mu^-\mu^-$ analysis.

In summary, the first study of baryon number, lepton number, and lepton flavor violating tau decay at Belle is reported. The signal reconstruction efficiency and background yield are obtained by applying robust selection criteria. We expect a substantially better result compared to LHCb.

References

1. A.D. Sakharov, JETP Lett. **5**, 24–27 (1967)
2. A. Abashian et al. (Belle Collaboration), Nucl. Instrum. Methods Phys. Res., Sec. A **479**, 117 (2002)
3. R. Aaij et al., LHCb collaboration. Phys. Lett. B **724** (2013)
4. M. Feindt, U. Kerzel, Nucl. Instrum. Methods Phys. Res., Sect. A 559, 190 (2006)

Chapter 9
Probing Anomalous *tcZ* Couplings with Rare *B* and *K* Decays

Jyoti Saini and Suman Kumbhakar

Abstract In this work, we study the effects of anomalous tcZ couplings. Such couplings would potentially affect several neutral current decays of K and B mesons via Z-penguin diagrams. Using constraints from relevant observables in K and B sectors, we calculate $\mathscr{B}(t \to cZ)$ and $\mathscr{B}(K_L \to \pi^0 \nu \bar{\nu})$ in the presence of anomalous tcZ coupling. Further, we find that the complex tcZ coupling can also provide large enhancements in many CP violating angular observable in $B \to K^* \mu^+ \mu^-$ decay.

9.1 Introduction

The measurement of several observables in B meson decays do not agree with their Standard Model (SM) predictions. These observables include the measurement of $R_{K^{(*)}}$, angular observables in $B \to K^* \mu^+ \mu^-$ (in particular P_5'), $\mathscr{B}(B_s \to \phi \mu^+ \mu^-)$ in the neutral current sector and $R_{D^{(*)}, J/\psi}$ in the charged current sector. These measurements can be considered as hints of physics beyond the SM.

Apart from the decays of B meson, the top quark decays are particularly important for hunting physics beyond the SM. As it is the heaviest of all the SM particles, it is expected to feel the effect of new physics (NP) most. Also, LHC is primarily a top factory producing abundant top quark events. Hence one expects the observation of possible anomalous couplings in the top sector at the LHC. The SM predictions for the branching ratios of the flavor changing neutral current (FCNC) top quark decays, such as $t \to uZ$ and $t \to cZ$ decays are $\sim 10^{-17}$ and 10^{-14}, respectively [1, 2], and are probably immeasurable at the LHC until NP enhances their branching ratios up to the detection level of LHC.

J. Saini (✉)
Indian Institute of Technology Jodhpur, Jodhpur 342037, India
e-mail: saini.1@iitj.ac.in

S. Kumbhakar
Indian Institute of Technology Bombay, Mumbai 400076, India
e-mail: suman@phy.iitb.ac.in

© Springer Nature Singapore Pte Ltd. 2021
P. K. Behera et al. (eds.), *XXIII DAE High Energy Physics Symposium*,
Springer Proceedings in Physics 261,
https://doi.org/10.1007/978-981-33-4408-2_9

57

In this work, we study the effects of anomalous tcZ couplings on rare B and K meson decays. Using these decays, we obtain constraints on anomalous $t \to cZ$ coupling. We then look for flavor signatures of anomalous tcZ coupling. In particular, we examine $\mathscr{B}(K_L \to \pi \nu \bar{\nu})$ and various CP violating angular observables in $B \to K^* \mu^+ \mu^-$. We find that the complex tcZ coupling can give rise to large new physics effects in these CP violating observables.

9.2 Effect of Anomalous $t \to cZ$ Couplings on Rare B and K Decays

The effective tcZ Lagrangian can be written as [3]

$$
\mathscr{L}_{tcZ} = \frac{g}{2\cos\theta_W} \bar{c}\gamma^\mu \left(g_{ct}^L P_L + g_{ct}^R P_R \right) t Z_\mu
$$
$$
+ \frac{g}{2\cos\theta_W} \bar{c} \frac{i\sigma^{\mu\nu}p_\nu}{M_Z} \left(\kappa_{ct}^L P_L + \kappa_{ct}^R P_R \right) t Z_\mu + \text{h.c.}, \tag{9.1}
$$

where $P_{L,R} \equiv (1 \mp \gamma_5)/2$ and $g_{ct}^{L,R}$ and $\kappa_{ct}^{L,R}$ are NP couplings. The anomalous tcZ couplings can provide additional contributions to $b \to s\, l^+ l^-$, $b \to d\, l^+ l^-$ and $s \to d\nu\bar{\nu}$ decays via Z penguin diagrams and hence have the potential to affect the decays of several B and K mesons.

Let us now consider the contribution of anomalous tcZ couplings to the rare B decays induced by the quark-level transition $b \to s\, \mu^+ \mu^-$. The effective Hamiltonian for the quark-level transition $b \to s\, \mu^+ \mu^-$ in the SM can be written as

$$
\mathscr{H}_{eff} = -\frac{4G_F}{\sqrt{2}} V_{ts}^* V_{tb} \sum_{i=1}^{10} C_i(\mu)\, O_i(\mu), \tag{9.2}
$$

where the form of the operators O_i are given in [4]. The effective tcZ vertices, given in (9.1), affect $b \to s\, \mu^+ \mu^-$ transition. This contribution modifies the Wilson coefficients (WCs) C_9 and C_{10}. The NP contributions to these WCs are [5]

$$
C_9^{s,NP} = -C_{10}^{s,NP} = -\frac{1}{8\sin^2\theta_W} \frac{V_{cs}^*}{V_{ts}^*} \left[\left(-x_t \ln \frac{M_W^2}{\mu^2} + \frac{3}{2} + x_t - x_t \ln x_t \right) g_{ct}^L \right], \tag{9.3}
$$

with $x_t = \bar{m}_t^2 / M_W^2$. Here the right-handed coupling, g_{ct}^R, is neglected as it is suppressed by a factor of $\bar{m}_c / M_W$. Here we have also neglected the contributions from CKM suppressed Feynman diagrams. The NP contributions to $C_{9,10}$ have been calculated in the unitary gauge with the modified minimal subtraction ($\overline{\text{MS}}$) scheme [5]. The effective Hamiltonian and the NP contributions to the WCs C_9 and C_{10} for the process $b \to d\, \mu^+ \mu^-$ can be obtained from (9.2) to (9.3), respectively, by replacing s by d.

We now consider NP contribution to $s \to d \, v\bar{v}$ transition. The $K^+ \to \pi^+ v\bar{v}$ decay is the only observed decay in this sector. The effective Hamiltonian for $K^+ \to \pi^+ v\bar{v}$ in the SM can be written as

$$\mathcal{H}_{eff} = \frac{G_F}{\sqrt{2}} \frac{\alpha}{2\pi \sin^2 \theta_W} \sum_{l=e,\mu,\tau} \left[V_{cs}^* V_{cd} X_{NL}^l + V_{ts}^* V_{td} X(x_t) \right] \times (\bar{s}d)_{V-A} (\bar{v}_l v_l)_{V-A} \, ,$$

$$(9.4)$$

where X_{NL}^l and $X(x_t)$ are the structure functions corresponding to charm and top sector, respectively [4, 6, 7]. The contribution of anomalous tcZ coupling to $\bar{s} \to \bar{d} \, v\bar{v}$ transition then modifies the structure function $X(x_t)$ in the following way:

$$X(x_t) \to X^{tot}(x_t) = X(x_t) + X^{NP}, \qquad (9.5)$$

where

$$X(x_t) = \eta_X \frac{x_t}{8} \left[\frac{2 + x_t}{x_t - 1} + \frac{3x_t - 6}{(1 - x_t)^2} \ln x_t \right], \qquad (9.6)$$

$$X^{NP} = -\frac{1}{8} \left(\frac{V_{cd} V_{ts}^* + V_{td} V_{cs}^*}{V_{td} V_{ts}^*} \right) \left(-x_t \ln \frac{M_W^2}{\mu^2} + \frac{3}{2} + x_t - x_t \ln x_t \right) (g_{ct}^L)^* . \quad (9.7)$$

Here $\eta_X = 0.994$ is the NLO QCD correction factor.

9.3 Constraints on the Anomalous tcZ Couplings

In order to obtain the constraints on the anomalous tcZ coupling g_{ct}^L, we perform a χ^2 fit using all measured observables in B and K sectors. The total χ^2 is written as a function of two parameters: $\text{Re}(g_{ct}^L)$ and $\text{Im}(g_{ct}^L)$. The χ^2 function is defined as

$$\chi^2_{total} = \chi^2_{b \to s \, \mu^+ \mu^-} + \chi^2_{b \to d \, \mu^+ \mu^-} + \chi^2_{s \to d v\bar{v}} \, . \qquad (9.8)$$

In our analysis, we include all recent CP conserving data from $b \to s\mu^+\mu^-$ to obtain constraints on $C_{9,10}^{s,NP}$. Assuming the WCs C_i to be real, we obtain $C_9^{NP} = -C_{10}^{NP} = -0.51 \pm 0.09$ [8]. This is consistent with several global fit results such as [9–11]. For complex WCs, we get $C_9^{NP} = -C_{10}^{NP} = (-0.56 \pm 0.26) + i(0.55 \pm 1.36)$. The fit values thus obtained can be used to constrain g_{ct}^L. For real g_{ct}^L coupling, we have

$$\chi^2_{b \to s\mu^+\mu^-} = \left(\frac{C_9^{s,NP} + 0.51}{0.09} \right)^2 . \qquad (9.9)$$

For complex g_{ct}^L couplings, the χ^2 function can be written as

$$\chi^2_{b \to s\mu^+\mu^-} = \left(\frac{Re(C_9^{s,NP}) + 0.56}{0.26} \right)^2 + \left(\frac{Im(C_9^{s,NP}) - 0.55}{1.36} \right)^2 . \qquad (9.10)$$

From $b \to d\mu^+\mu^-$ sector the branching ratio of $B^+ \to \pi^+\mu^+\mu^-$ and $B_d \to \mu^+\mu^-$ decay are included in our analysis:

$$\chi^2_{b \to d\mu^+\mu^-} = \chi^2_{B^+ \to \pi^+\mu^+\mu^-} + \chi^2_{B_d \to \mu^+\mu^-}. \tag{9.11}$$

For $B^+ \to \pi^+\mu^+\mu^-$ decay,

$$\chi^2_{B^+ \to \pi^+\mu^+\mu^-} = \left(\frac{\mathscr{B}(B^+ \to \pi^+\mu^+\mu^-) - 2.3 \times 10^{-8}}{0.66 \times 10^{-8}} \right)^2, \tag{9.12}$$

where, following [12], a theoretical error of 15% is included in $\mathscr{B}(B^+ \to \pi^+\mu^+\mu^-)$. For $\mathscr{B}(B_d \to \mu^+\mu^-)$ decay,

$$\chi^2_{B_d \to \mu^+\mu^-} = \left(\frac{\mathscr{B}(B_d \to \mu^+\mu^-) - 3.9 \times 10^{-10}}{1.6 \times 10^{-10}} \right)^2. \tag{9.13}$$

The branching ratio of $B_d \to \mu^+\mu^-$ in the presence of anomalous tcZ coupling is given by

$$\mathscr{B}(B_d \to \mu^+\mu^-) = \frac{G_F^2 \alpha^2 M_{B_d} m_\mu^2 f_{B_d}^2 \tau_{B_d}}{16\pi^3} |V_{td} V_{tb}^*|^2 \sqrt{1 - 4(m_\mu^2/M_{B_d}^2)} \left| C_{10} + C_{10}^{d,NP} \right|^2. \tag{9.14}$$

The branching ratio of $K^+ \to \pi^+ \nu\bar{\nu}$, the only measurement in $s \to d \nu\bar{\nu}$ sector, in the presence of anomalous tcZ coupling is given by

$$\frac{\mathscr{B}(K^+ \to \pi^+\nu\bar{\nu})}{\kappa_+} = \left(\frac{\text{Re}(V_{cd} V_{cs}^*)}{\lambda} P_c(X) + \frac{\text{Re}(V_{td} V_{ts}^*)}{\lambda^5} X^{\text{tot}}(x_t) \right)^2$$
$$+ \left(\frac{\text{Im}(V_{td} V_{ts}^*)}{\lambda^5} X^{\text{tot}}(x_t) \right)^2, \tag{9.15}$$

where $P_c(X) = 0.38 \pm 0.04$ [13] is the NNLO QCD-corrected structure function in the charm sector and

$$\kappa_+ = r_{K^+} \frac{3\alpha^2 \mathscr{B}(K^+ \to \pi^0 e^+ \nu)}{2\pi^2 \sin^4 \theta_W} \lambda^8. \tag{9.16}$$

Using $r_{K^+} = 0.901$, we estimate

$$\frac{\mathscr{B}(K^+ \to \pi^+\nu\bar{\nu})}{\kappa_+} = 3.17 \pm 2.05. \tag{9.17}$$

Table 9.1 Values of anomalous tcZ couplings

Real coupling	Complex coupling
$g_L^{ct} = (-7.04 \pm 1.28) \times 10^{-3}$	$\mathrm{Re}(g_L^{ct}) = (-7.63 \pm 3.69) \times 10^{-3}; \mathrm{Im}(g_L^{ct}) = (1.87 \pm 1.02) \times 10^{-2}$

In order to include $\mathcal{B}(K^+ \to \pi^+ \nu \bar{\nu})$ in the fit, we define

$$\chi^2_{K^+ \to \pi^+ \nu \bar{\nu}} = \left(\frac{\mathcal{B}(K^+ \to \pi^+ \nu \bar{\nu})/\kappa_+ - 3.17}{2.05} \right)^2 + \left(\frac{P_c(X) - 0.38}{0.04} \right)^2. \quad (9.18)$$

Thus, the error on $P_c(X)$ has been taken into account by considering it to be a parameter and adding a contribution to χ^2_{total}.

The $\mathcal{B}(t \to cZ)$ in the presence of tcZ coupling is given as [14–16]

$$\mathcal{B}(t \to c Z) = \frac{\beta_Z^4(3 - 2\beta_Z^2)}{2\beta_W^4(3 - 2\beta_W^2)} \frac{|g_{ct}^L|^2 + |g_{ct}^R|^2}{|V_{tb}|^2}, \quad (9.19)$$

with $\beta_x = (1 - m_x^2/m_t^2)^{1/2}$, being the velocity of the $x = W, Z$ boson in the top quark rest frame.

The fit results for real and complex tcZ couplings are presented in Table 9.1. Using the fit results, we find that for real tcZ coupling, $\mathcal{B}(t \to c Z) = (0.90 \pm 0.33) \times 10^{-5}$. For complex tcZ coupling, 2σ upper bound on the branching ratio is 2.14×10^{-4}. Hence, any future measurement of this branching ratio at the level of 10^{-4} would imply the coupling to be complex.

9.4 Predictions for Various *CP* Violating Observables

We now see whether large deviation is possible in some of the flavor physics observables due to the anomalous tcZ coupling.

$\mathcal{B}(\mathbf{K_L} \to \pi^0 \nu \bar{\nu})$: The preset upper bound on $\mathcal{B}(K_L \to \pi^0 \nu \bar{\nu})$ is 2.6×10^{-8} [17] at 90% C.L. which is about three orders of magnitude above the SM prediction. The branching ratio of $K_L \to \pi^0 \nu \bar{\nu}$ is a purely CP violating quantity. The branching ratio of $K_L \to \pi^0 \nu \bar{\nu}$ in the presence of tcZ coupling is given by

$$\mathcal{B}(K_L \to \pi^0 \nu \bar{\nu}) = \kappa_L \left[\frac{\mathrm{Im}\left(V_{ts}^* V_{td} X^{\text{tot}}(x_t) \right)}{\lambda^5} \right]^2, \quad (9.20)$$

where $X^{\text{tot}}(x_t)$ is given in (9.5).

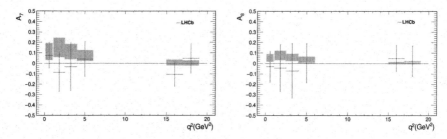

Fig. 9.1 (Color Online) The plots depicts various CP violating observables in $B \to K^* \mu^+ \mu^-$ decays

Using fit result for the complex tcZ coupling, we get $Br(K_L \to \pi^0 \nu \bar{\nu}) = (9.88 \pm 5.96) \times 10^{-11}$. The 2σ upper bound on $\mathscr{B}(K_L \to \pi^0 \nu \bar{\nu})$ is obtained to be $\leq 2.18 \times 10^{-10}$, an order of magnitude higher than its SM prediction.

CP violating observables in B $\to$ K$^* \mu^+ \mu^-$: We study various CP violating observables in $B \to K^* \mu^+ \mu^-$ decays in the presence of complex anomalous tcZ couplings.The CP-violating observables for these decays are defined as [18]

$$A_i = \frac{I_i - \bar{I}_i}{d(\Gamma + \bar{\Gamma})/dq^2}, \tag{9.21}$$

where $I_i s$ are given in [18]. These asymmetries are largely suppressed in SM because of the small weak phase of CKM and hence they are sensitive to complex NP couplings. These symmetries can get significant contribution from the NP in the presence of CP-violating phase [19–21].

The predictions for CP-violating asymmetries A_7 and A_8 in the presence of complex anomalous tcZ couplings are shown in Fig. 9.1. It can be seen from our results that the asymmetry A_7 can be enhanced up to 20% whereas enhancement in A_8 can be up to 10% in the low-q^2 region. For all other asymmetries, large enhancement is not possible.

References

1. G. Eilam, J.L. Hewett and A. Soni, Phys. Rev. D **44**, 1473 (1991) [Erratum-ibid. D **59**, 039901 (1999)]
2. J.A. Aguilar-Saavedra, Acta Phys. Polon. B **35**, 2695 (2004)
3. J.A. Aguilar-Saavedra, Nucl. Phys. B **812**, 181 (2009)
4. G. Buchalla, A.J. Buras, M.E. Lautenbacher, Rev. Mod. Phys. **68**, 1125 (1996)
5. X.Q. Li, Y.D. Yang, X.B. Yuan, JHEP **1203**, 018 (2012)
6. G. Buchalla, A.J. Buras, Nucl. Phys. B **412**, 106 (1994)
7. G. Buchalla, A.J. Buras, Nucl. Phys. B **548**, 309 (1999)
8. S. Kumbhakar, J. Saini, arXiv:1905.07690 [hep-ph]
9. A.K. Alok, B. Bhattacharya, A. Datta, D. Kumar, J. Kumar, D. London, Phys. Rev. D **96**(9), 095009 (2017)

10. A.K. Alok, A. Dighe, S. Gangal, D. Kumar, arXiv:1903.09617 [hep-ph]
11. M. Algueró, B. Capdevila, A. Crivellin, S. Descotes-Genon, P. Masjuan, J. Matias, J. Virto, arXiv:1903.09578 [hep-ph]
12. J.J. Wang, R.M. Wang, Y.G. Xu, Y.D. Yang, Phys. Rev. D **77**, 014017 (2008)
13. A.J. Buras, M. Gorbahn, U. Haisch, U. Nierste, JHEP **0611**, 002 (2006) [Erratum-ibid. **1211**, 167 (2012)]
14. T. Han, R.D. Peccei, X. Zhang, Nucl. Phys. B **454**, 527 (1995)
15. M. Beneke et al., hep-ph/0003033
16. W. Bernreuther, J. Phys. G **35**, 083001 (2008)
17. J.K. Ahn et al., E391a collaboration. Phys. Rev. D **81**, 072004 (2010)
18. W. Altmannshofer, P. Ball, A. Bharucha, A.J. Buras, D.M. Straub, M. Wick, JHEP **0901**, 019 (2009)
19. A.K. Alok, A. Dighe, S. Ray, Phys. Rev. D **79**, 034017 (2009)
20. A.K. Alok, A. Datta, A. Dighe, M. Duraisamy, D. Ghosh, D. London, JHEP **1111**, 122 (2011)
21. A.K. Alok, B. Bhattacharya, D. Kumar, J. Kumar, D. London, S.U. Sankar, Phys. Rev. D **96**(1), 015034 (2017)

Chapter 10
Impact of Nonleptonic $\bar{B}_{d,s}$ Decay Modes on $\bar{B} \to \bar{K}^*\mu^+\mu^-$ Process

Manas K. Mohapatra, Suchismita Sahoo, and Anjan K. Giri

Abstract We scrutinize the effect of nonleptonic B decay modes on the branching ratio and angular observables of $\bar{B} \to \bar{K}^*\mu^+\mu^-$ process involving $b \to s$ quark level transition in the non-universal Z' model. The new couplings are constrained by using the experimental limits on the branching ratios of $B_d \to \pi K$, $B_d \to \rho K$, and $B_s \to \eta'\eta'$, K^*K^* nonleptonic processes. Using the allowed parameter space, we perform an angular analysis of the $\bar{B} \to \bar{K}^*\mu^+\mu^-$ process. We observe significant impact of nonleptonic decay modes on $\bar{B} \to \bar{K}^*\mu^+\mu^-$ observables.

10.1 Introduction

Although Standard Model (SM) is a successfully fundamental theory, it fails to explain the open puzzles such as matter–antimatter asymmetry, hierarchy problem, neutrino mass, dark matter, and dark energy. Thus, it implies the existence of new physics (NP) beyond it. In this regard, the study of rare B decays, which provide not only deep understanding on CP violation but also different anomalies both in nonleptonic as well as semileptonic sectors, is quite interesting. The decay rate and P'_5 observable of $\bar{B} \to \bar{K}^*\mu^+\mu^-$ process have 3σ [1] deviation from their SM results. The decay distribution of $B_s \to \phi\mu^+\mu^-$ also has tension [2]. Furthermore the lepton universality violating ratio, $R_K = \mathrm{Br}(B^+ \to K^+\mu^+\mu^-)/\mathrm{Br}(B^+ \to K^+e^+e^-)$ disagrees with SM prediction at the level of 2.5σ [3]. Discrepancy of $2.2\sigma(2.4\sigma)$ has been observed in R_{K^*} measurement by LHCb experiment [4]

M. K. Mohapatra (✉) · A. K. Giri
Indian Institute of Technology, Hyderabad 502285, Kandi, India
e-mail: manasmohapatra12@gmail.com

A. K. Giri
e-mail: giria@iith.ac.in

S. Sahoo
Physical Research Laboratory, Ahmedabad 380009, India
e-mail: suchismita8792@gmail.com

© Springer Nature Singapore Pte Ltd. 2021
P. K. Behera et al. (eds.), *XXIII DAE High Energy Physics Symposium*,
Springer Proceedings in Physics 261,
https://doi.org/10.1007/978-981-33-4408-2_10

$$R_{K^*}^{\text{Expt}} = \frac{\text{Br}(B^0 \to K^{*0}\mu^+\mu^-)}{\text{Br}(B^0 \to K^{*0}e^+e^-)} = 0.66^{+0.11}_{-0.07} \pm 0.03, \quad q^2 \in [0.045, 1.1]\,\text{GeV}^2,$$

$$= 0.69^{+0.11}_{-0.07} \pm 0.05, \quad q^2 \in [1.1, 6]\,\text{GeV}^2, \quad (10.1)$$

from their SM predictions [5]. Though the measurements on R_{K^*} by Belle Collaboration [6] is toward the SM results, the error values are comparatively higher than the previous LHCb result. Additionally, the mismatch between the measured data and the SM results are also observed in the two body hadronic decay processes like $B \to PP, PV, VV$, where $P = \pi, K, \eta^{()}$ are the pseudoscalar mesons and $V = K^*, \phi, \rho$ are the vector mesons. Inspired by these anomalies, we would like to see whether the new physics (arising due to an additional Z' boson) influencing the nonleptonic B decays also have significant impact on rare semileptonic B decay processes.

The paper is organized as follows. In Sect. 10.2, we discuss the effective Hamiltonian of $b \to sll(q\bar{q})$ processes in both SM and in Z' model. We also present the new physics contribution in this section. Section 10.3 describes the constraints on new parameters from the nonleptonic B modes. The impact of new couplings on $\bar{B} \to \bar{K}^*\mu\mu$ is presented in Sects. 10.4 and 10.5 summarize our results.

10.2 Effective Hamiltonian

The generalized effective Hamiltonian for $b \to sq\bar{q}$ process, where q is any light quark, is given as [7]

$$\mathcal{H}_{\text{eff}} = \frac{G_F}{\sqrt{2}}\left[\sum_{p=u,c}\lambda_p(C_1\mathcal{O}_1{}^p + C_2\mathcal{O}_2{}^p) - \lambda_t\sum_{i=3}^{10}(C_i\mathcal{O}_i + C_{7\gamma}\mathcal{O}_{7\gamma} + C_{8\gamma}\mathcal{O}_{8\gamma})\right] + h.c,$$

(10.2)

where G_F is the Fermi constant, $\lambda_p = V_{pb}V_{ps}^*$, $\lambda_t = V_{tb}V_{ts}^*$ are the product of CKM matrix elements. Here $\mathcal{O}_{1,2}^p$ are left-handed current–current operators; $\mathcal{O}_{3,...6}$ and $\mathcal{O}_{7,...,10}$ are QCD and electroweak penguin operators; and $\mathcal{O}_{7\gamma}$, $\mathcal{O}_{8g}$ are the electromagnetic and chromomagnetic dipole operators. The relevant $\mathcal{O}_{7,...,10}$ operators are defined as

$$\mathcal{O}_{7(9)} = (\bar{s}b)_{V-A}\sum_q e_q(\bar{q}q)_{V+A(V-A)}, \quad \mathcal{O}_{8(10)} = (\bar{s}_\alpha b_\beta)_{V-A}\sum_q e_q(\bar{q}_\beta q_\alpha)_{V+A(V-A)},$$

where $V \mp A$ denotes $\gamma^\mu P_{L(R)}$ with $P_{L(R)} = (1 \mp \gamma_5)/2$ are the projection operators and e_q stand for the charge of q quark. The effective Hamiltonian for $b \to sq\bar{q}$ transition in the Z' model is given by [8]

$$\mathcal{H}_{\text{eff}}^{Z'} = \frac{2G_F}{\sqrt{2}}\left(\frac{g'M_Z}{g_1M_{Z'}}\right)^2 B_{sb}^L (\bar{s}b)_{V-A}\sum_q\left[(B_{qq}^L (\bar{q}q)_{V-A} + B_{qq}^R (\bar{q}q)_{V+A}\right], (10.3)$$

where $g_1(g')$ are the coupling constants of $Z^{(\prime)}$ boson and $B_{bs}^{L(R)}$, $B_{qq}^{L(R)}$ are the new couplings. Now, assuming $B_{uu}^{L(R)} \simeq -2B_{dd}^{L(R)}$ and comparing the Hamiltonian of Z' (10.3) with SM (10.2), we find an extra contribution to the electroweak penguin sector of nonleptonic decay modes as

$$\Delta C_9^{Z'} = \left(\frac{g' M_Z}{g_1 M_{Z'}}\right)^2 \left(\frac{B_{sb}^L B_{dd}^L}{V_{tb} V_{ts}^*}\right) \,, \quad \Delta C_7^{Z'} = \left(\frac{g' M_Z}{g_1 M_{Z'}}\right)^2 \left(\frac{B_{sb}^L B_{dd}^R}{V_{tb} V_{ts}^*}\right). \quad (10.4)$$

The most general effective Hamiltonian describing $b \to sl^+l^-$ processes in the SM is given by [9]

$$\mathcal{H}_{\text{eff}} = -\frac{4G_F}{\sqrt{2}} V_{tb} V_{ts}^* \left(\sum_{i=1,\cdots 10,S,P} C_i \mathcal{O}_i + \sum_{i=7,\cdots 10,S,P} C_i' \mathcal{O}_i' \right), \quad (10.5)$$

where $V_{qq'}$ are the CKM matrix elements, $\mathcal{O}_i$'s are the effective operators and C_i's are the corresponding Wilson coefficients. Though only $\mathcal{O}_7$ and $\mathcal{O}_{9,10}$ operators have contributions to the SM, additional $\mathcal{O}_{9,10}^{(\prime)}$ can be generated due to the presence of Z' gauge boson, defined as

$$\mathcal{O}_7^{(\prime)} = \frac{e}{16\pi^2} \left[\bar{s} \sigma_{\mu\nu}(m_s P_{L(R)} + m_b P_{R(L)})b \right] F^{\mu\nu} \,,$$
$$\mathcal{O}_9^{(\prime)} = \frac{\alpha_{\text{em}}}{4\pi} \left(\bar{s}\gamma^\mu P_{L(R)}b \right) \left(\bar{l}\gamma_\mu l \right), \quad \mathcal{O}_{10}^{(\prime)} = \frac{\alpha_{\text{em}}}{4\pi} \left(\bar{s}\gamma^\mu P_{L(R)}b \right) \left(\bar{l}\gamma_\mu \gamma_5 l \right),$$

where α_{em} denotes the fine structure. The effective Hamiltonian of $b \to sl^+l^-$ in the Z' model can be written as [10]

$$\mathcal{H}_{\text{eff}}^{Z'}(b \to sl^+l^-) = -\frac{2G_F}{\sqrt{2}} V_{tb} V_{tq}^* \left(\frac{g_2 M_Z}{g_1 M_{Z'}}\right)^2 \left[-\frac{B_{sb}^L B_{ll}^L}{V_{tb} V_{tq}^*} (\bar{q}b)_{V-A}(\bar{l}l)_{V-A} \right.$$
$$\left. -\frac{B_{qb}^L B_{ll}^R}{V_{tb} V_{tq}^*} (\bar{s}b)_{V-A}(\bar{l}l)_{V+A} \right] + \text{h.c.} \,,$$

which after comparing with (10.5) gives additional coefficients as well as new contributions to the SM Wilson coefficients ($C_{9,10}^{Z'(\prime)}$) as

$$C_9^{Z'}(M_W) = -2 \left(\frac{g_2 M_Z}{g_1 M_{Z'}}\right)^2 \frac{B_{sb}^L}{V_{tb} V_{ts}^*} (B_{ll}^L + B_{ll}^R) \,, \quad (10.6)$$

$$C_{10}^{Z'}(M_W) = 2 \left(\frac{g_2 M_Z}{g_1 M_{Z'}}\right)^2 \frac{B_{sb}^L}{V_{tb} V_{ts}^*} (B_{ll}^L - B_{ll}^R) \,. \quad (10.7)$$

Table 10.1 The experimental values and SM predictions on the branching ratio of nonleptonic $B_{d,s}$ decay modes

Decay processes	SM values	Experimental values [11]
$\bar{B}_d \to \pi^- K^+$	20.11×10^{-6}	$(1.96 \pm .05) \times 10^{-5}$
$\bar{B}_d \to \pi^0 K^0$	6.57×10^{-6}	$(9.9 \pm .5) \times 10^{-6}$
$\bar{B}_d \to \rho^0 K^0$	2.80×10^{-6}	$(4.7 \pm .6) \times 10^{-6}$
$\bar{B}_d \to \rho^- K^+$	2.77×10^{-6}	$(7 \pm .9) \times 10^{-6}$
$\bar{B}_s \to \eta'\eta'$	57.53×10^{-6}	$(3.3 \pm .7) \times 10^{-5}$
$\bar{B}_s \to K^{0*}\bar{K}^{0*}$	3.72×10^{-6}	$(1.11 \pm .27) \times 10^{-5}$

10.3 Constraints on New Couplings

After getting an idea on new coefficients, we now proceed to constrain the coefficients by using the branching ratios of nonleptonic B decay modes. Using the CKM matrix elements, particles masses, life time of $B_{d,s}$ meson from [11], the form factors, decay constants except $f_\pi = .131$, $f_K = .160$ from [12], the predicted SM branching ratios of $B_d \to (\pi, \rho)K$, $B_s \to \eta'\eta'$, K^*K^* decay modes, and their respective measured values are presented in Table 10.1.

We consider two cases, (a) $B_{dd}^R = 0$, which implies $\Delta C_7^{Z'} = 0$ (b) $B_{dd}^R = B_{dd}^L$, which implies $\Delta C_7^{Z'} = \Delta C_9^{Z'}$ in order to constrain the new parameters. In this manuscript, we will only discuss the first case. Comparing the theoretical predictions from Table 10.1 with their experimental results, the constraints on $B_{sb}^L - \phi_s^L$ (left panel) and $B_{sb}^L - B_{dd}^L$ (right panel) planes for first case are shown in Fig. 10.1.

10.4 Impact on $\bar{B} \to \bar{K}^* \mu^+ \mu^-$ Decay Mode

In this section, we present the impact of new parameters constrained from the nonleptonic B modes on the $\bar{B} \to \bar{K}^* \mu^+ \mu^-$ process, which can be completely described in terms of only four kinematical variables; the lepton invariant mass squared (q^2) and three angles θ_l, θ_V and ϕ, where θ_l is the angle between l^- and $B_{(s)}$ in the dilepton frame, θ_V is defined as the angle between K^- and $B_{(s)}$ in the $K^-\pi^+$ (K^-K^+) frame, the angle between the normal of the $K^-\pi^+$ (K^-K^+) and the dilepton plane is given by ϕ.

The decay rate, forward–backward (A_{FB}) asymmetry and $P_{4,5}'$ observables are defined as [13]

$$\frac{d\Gamma}{dq^2} = \frac{3}{4}\left(J_1 - \frac{J_2}{3}\right), \quad A_{FB}\left(q^2\right) = -\frac{3}{8}\frac{J_6}{d\Gamma/dq^2},$$
$$P_4' = \frac{J_4}{\sqrt{-J_2^c J_2^s}}, \quad P_5' = \frac{J_5}{2\sqrt{-J_2^c J_2^s}}, \tag{10.8}$$

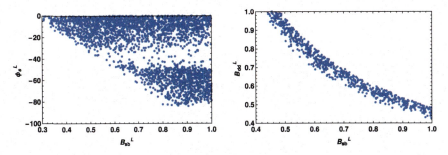

Fig. 10.1 Constraints on new parameters from the branching ratios of nonleptonic B processes for $B_{qq}^R = 0$ case

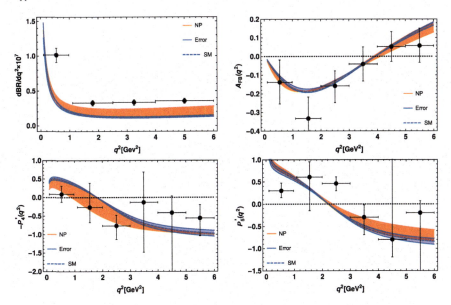

Fig. 10.2 The q^2 variation of branching ratio (top-left), forward–backward asymmetry (top-right), P_4' (bottom-left) and P_5' (bottom-right) observables of $\bar{B} \to \bar{K}^* \mu\mu$ process. Here $P_4'|^{\text{LHCb}} = -P_4'$

where $J_i = 2J_i^s + J_i^c$ contain the transversity amplitudes which are the functions form factors and Wilson coefficients. All the input parameters are taken from [11] and the form factors from [14].

Using the allowed parameter space from Fig. 10.1, we show the variation of branching ratio (top-left), A_{FB} (top-right), P_4' (bottom-left) and P_5' (bottom-right) of $\bar{B} \to \bar{K}^* \mu\mu$ with respect to q^2 in Fig. 10.2. Here the dashed blue lines (light blue bands) represent the SM predictions (uncertainties arising due to the input parameters) and orange bands stand for the NP contributions. The experimental results are shown in black color [1]. We observe that NP contribution provide significant deviation from their SM results and can accommodate experimental data.

10.5 Conclusion

We have studied the rare semileptonic $\bar{B} \to \bar{K}^*\mu\mu$ in a non-universal Z' model. We constrain the new parameters from the branchings ratios of nonleptonic B decay modes. We mainly check whether the new physics couplings influencing the nonleptonic modes also have impact on semileptonic processes. We found that the constraint from nonleptonic decays significantly affect the branching ratios and angular observables of $\bar{B} \to \bar{K}^*\mu\mu$ process.

Acknowledgments MM would like to thank DST, Government of India for the financial support through Inspire Fellowship.

References

1. LHCb, R. Aaij et al., JHEP **6**, 133 (2014); LHCb, R. Aaij et al., Phys. Rev. Lett. **111**, 191801 (2013)
2. LHCb, R. Aaij et al., JHEP **9**, 179 (2015)
3. LHCb, R. Aaij et al., Phys. Rev. Lett. **113**, 151601 (2014); LHCb, R. Aaij et al., arXiv:1903.09252
4. LHCb, R. Aaij et al., JHEP **8**, 055 (2017)
5. B. Capdevila, A. Crivellin, S. Descotes-Genon, J. Matias, J. Virto, JHEP **01**, 093 (2018)
6. Belle, M. Prim, talk given at Moriond, March 22 2019
7. G. Buchalla, A.J. Buras, M.E. Lauteubacher, Rev. Mod. Phys. **68**, 1125 (1996)
8. V. Barger, C.W. Chiang, P. Langacker, H.S. Lee, Phys. Lett. B **598**, 218 (2004)
9. C. Bobeth, M. Misiak, J. Urban, Nucl. Phys. B **574**, 291 (2000)
10. Q. Chang, Xin-Qiang Li, Ya-Dong Yang, JHEP **1002**, 082 (2010); V. Barger, L. Everett, J. Jiang, P. Langacker, T. Liu and C. Wagner, Phys. Rev. D **80**, 055008 (2009); JHEP **0912**, 048 (2009)
11. M. Tanabashi et al., Particle data group. Phys. Rev. D **98**, 030001 (2018)
12. M. Beneke, N. Matthias, Nucl. Phys. B **675**, 333 (2003); X. Li, G. Lu and Y. Yang, Phys. Rev. D **68**, 114015 (2003), [Erratum: D **71**,019902 (2005)]; C.-D. Lü, Y.-L. Shen, Y.-M. Wang and Y.-B. Wei, JHEP **01**, 024 (2019); N. Gubernari, A. Kokulu and D. V. Dyk, JHEP **01**, 150 (2019); A. Bharucha, D. M. Straub and R. Zwicky, JHEP **08**, 098 (2016); G. Duplancic and B. Melic, JHEP **11**, 138 (2015); H.Y. Cheng and C. K. Chua, Phys. Rev. **D80**, 114026(2009); A.Bharucha, D. M. Straub, R. Zwicky, JHEP **08**, 098(2016)
13. C. Bobeth, G. Hiller, G. Piranishivili, JHEP **7**, 106 (2008); U. Egede, T. Hurth, J. Matias, M. Ramon, W. Reece, JHEP **10**, 056 (2010); U. Egede, T. Hurth, J. Matias, M. Ramon, W. Reece, JHEP **11**, 032 (2008); J. Matias, F. Mescia, M. Ramon and J. Virto, JHEP **4**, 104 (2012)
14. M. Beneke, T. Feldmann, D. Seidel, Eur. Phys. J. C **41**, 173 (2005)

Chapter 11
Underlying Event Measurements Using CMS Detector at LHC

Manisha Lohan

Abstract Results of recent underlying event (UE) measurements using data collected by CMS detector are presented. UE measurements are done using events containing a leading charged particle jet, a leading charged particle, a Drell-Yan (DY) lepton pair, and a $t\bar{t}$ pair. Measurements are corrected to remove detector effect and are compared to various Monte Carlo (MC) predictions.

11.1 Introduction

The Underlying Event (UE) is composed of everything which is not originated from hard scatter outgoing partons (quarks and gluons). All QCD interactions except hard interactions constitute UE. The main components of UE are initial state radiations (ISR), final state radiations (FSR), multiple parton interactions (MPI) from semi-hard interactions and beam–beam remnants (BBR) concentrated along the beam direction. The UE measurements help in probing the hadron production in high energy p-p collisions. It is also an important background for precision measurements of new physics searches at LHC and MC modeling. In the present article, the results of UE measurements using leading charged particle & jet, Drell-Yan (DY) events, and $t\bar{t}$ pair are reported. Presented UE measurements are performed using 13 TeV p-p collisions data collected by CMS detector [1] at LHC [2].

Manisha Lohan—On behalf of the CMS Collaboration.

M. Lohan (✉)
IRFU, CEA, Université Paris-Saclay, Gif-sur-Yvette 91191, France
e-mail: manisha1.lohan@gmail.com

© Springer Nature Singapore Pte Ltd. 2021
P. K. Behera et al. (eds.), *XXIII DAE High Energy Physics Symposium*,
Springer Proceedings in Physics 261,
https://doi.org/10.1007/978-981-33-4408-2_11

11.2 UE Measurements Using Leading Charged Particles and Jets

UE measurements are done using leading charged particles (jets) [3], having $p_T > 0.5$ GeV ($p_T > 1$ GeV) and $|\eta| < 2$ to ensure good reconstruction efficiency. The leading charged particles having $p_T > 0.5$ GeV and $|\eta| < 2.5$ are used to construct jets. The leading charged particle jets are constructed using the Seedless Infrared-Safe Cone (SIS Cone) jet algorithm. ZeroBias datasets are used for the present measurements, corresponding to an integrated luminosity of 281 nb^{-1}. Events with exactly one primary vertex are selected for the analysis. UE activity is measured in terms of the average charged particle density and average energy density of the leading charged particles (jets). The average charged particle density is defined as the number of tracks (N_{ch}) in any region divided by the area of the same region in η–ϕ space, and average energy density is the scalar sum of tracks transverse momenta (Σp_T) in any region divided by its area in η–ϕ space.

The regions of measurements are defined using the ϕ direction of leading charged particle (jet) as the reference direction:

- $\Delta\phi < 60°$: Towards
- $\Delta\phi > 120°$: Away
- $60° < \Delta\phi < 120°$: Transverse 1
- $-60° < \Delta\phi < -120°$: Transverse 2

$\Delta\phi$ is the difference between the ϕ direction of leading charged particle and any other charged particle. Among the Transverse 1 and Transverse 2 regions, the region having higher value of UE observables, i.e., the average charged particle density and average energy density is defined as TransMax region and another one as TransMin region.

In order to compare the data with various theory predictions, data measurements are corrected using RooUnfold package. Corrected distributions are compared with predictions of PYTHIA8, EPOS & Herwig++ MC samples. The predictions of Herwig++ fail in the low p_T region and EPOS fail in the plateau region. Overall Monash tune of PYTHIA8 provides the best description of data as shown in Fig. 11.1 (left). UE observables at $\sqrt{s} = 2.76$ TeV are compared with the measurements at $\sqrt{s} = 13$ TeV, an increase of 60–70% is observed in UE activity with increase in $\sqrt{s}$ from 2.76 TeV to 13 TeV (shown in Fig. 11.1 (middle & right)). Increase is observed in the UE activity with increase in $\sqrt{s}$ since contribution from MPI increases with increasing $\sqrt{s}$.

11.3 UE Measurements Using DY Lepton Pair

UE measurements are performed using DY events with dimuon final state ($q\bar{q} \rightarrow \mu^+\mu^-$) [4]. These events have a clean experimental signature and also theoretically well understood, which implies precise and accurate measurements of the UE activ-

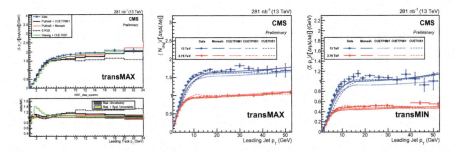

Fig. 11.1 Unfolded distribution of energy density as a function of leading track p_T is compared with MCs, i.e., PYTHIA8, EPOS, Herwig++, etc. predictions. Bottom panel shows the ratio of the corrected measurements to the MCs predictions (left). UE densities at 2.76 TeV are compared with the same at 13 TeV (middle & right). UE activity grows strongly (by 60–70%) with increase in centre-of-mass energy [3]

ity. UE activity is studied as a function of resultant transverse momentum ($p_T^{\mu\mu}$), using p-p collisions data corresponding to an integrated luminosity of 2.1 fb^{-1} and pileup (PU) $\sim$20. Events having at least two isolated muons with $p_T > 17$ GeV/c (8 GeV/c) for leading (subleading) muon are selected. Also selected events lie within a narrow mass window, $81 < M_{\mu\mu} < 101$ GeV/c^2 are having at least one well reconstructed vertex. Charged particles having $p_T > 0.5$ GeV and $|\eta| < 2$ are used for the analysis. Mis-reconstructed tracks are removed using high-purity reconstruction algorithm. Muons are reconstructed using particle-flow algorithm. Both muons are required to lie in the region, $|\eta| < 2.4$. Corrected data measurements are compared with different MC predictions, i.e., MADGRAPH + PYTHIA8, POWHEG + Herwig++, and POWHEG + PYTHIA8 MC combinations. Corrected distribution of particle density as a function of $p_T^{\mu\mu}$ in Away region is compared with different MC predictions as shown in Fig. 11.2. POWHEG in combination with Herwig++ overestimates the data by 10–15% whereas PYTHIA8 in combination with POWHEG and MADGRAPH show good agreement with data measurements having difference up to 5%. Charged particle density is also compared in Towards, Away, and Transverse regions (shown in Fig. 11.3 (left)). In Away region, fast rise is observed due to recoiling hadronic activity but in Towards and Transverse regions, growth is comparatively slow due to large spatial separation. UE measurements at $\sqrt{s} = 13$ TeV are compared with the previous measurements at $\sqrt{s} = 7, 1.96$ TeV presented in Fig. 11.3 (right). UE activity shows 25–30% rise on moving from 7 to 13 TeV and 60–80% on moving from 1.96 TeV to 7 TeV. For lower values of $p_T^{\mu\mu}$, POWHEG + PYTHIA8, POWHEG + HERWIG++ predictions show bit slow increase as compared to data measurements, but data-MC agreement improves for the higher values of $p_T^{\mu\mu}$.

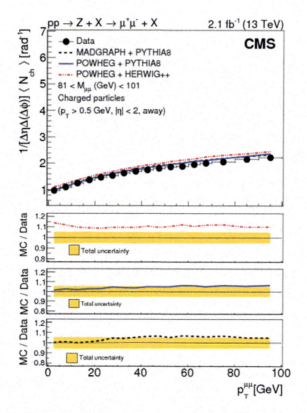

Fig. 11.2 Unfolded distribution of particle density as a function of $p_T^{\mu\mu}$ (Away region) is compared with MC predictions. POWHEG and MADGRAPH in combination with PYTHIA8 gives the best description of data (within 5%). Bottom panel shows the ratio of the corrected measurements to the MCs predictions [4]

11.4   UE Measurements Using $t\bar{t}$ Pair

UE measurements in $t\bar{t}$ pair production channel are performed using data collected at $\sqrt{s} = 13$ TeV [5], corresponding to an integrated luminosity of 35.9 fb^{-1}. Events having one electron, one muon (having $p_T > 20$ GeV and $|\eta| < 2.4$) with opposite charge sign and two jets (originated from hadronization and fragmentation of b quarks) in final state are used for the measurements as this final state has high purity. Also the decay products of hard processes can be easily distinguished. Jets are reconstructed using the infrared and collinear safe anti-k_T algorithm. The jets having $p_T > 15$ GeV are selected for the measurements. Dilepton mass (m(ll)) distribution is compared to the sum of expectations of signal and background as shown in Fig. 11.4 (left). The m(ll) variable is estimated with a resolution >2%. Good resolution of m(ll) variable estimation implies precise measurements of UE dependence on the energy scale of hard process which is correlated with m(ll) variable. The

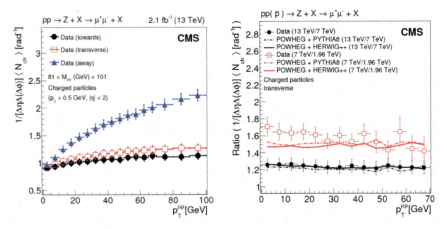

Fig. 11.3 The average charged particle density as a function of $p_T^{\mu\mu}$ is compared in the Away, Towards, and Transverse regions (left). The average charged particle at 13 TeV is compared with the previous measurements at 7 and 1.96 TeV (right) [4]

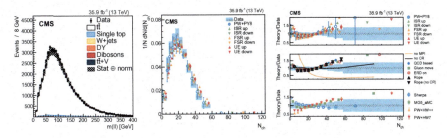

Fig. 11.4 Dilepton mass, m(ll) variable distribution is compared to the sum of expectations of signal and backgrounds (left). Shaded region represents the total uncertainty, i.e., systematic plus statistical. The differential cross section as a function of N_{ch} is compared to POWHEG + PYTHIA8 predictions as well as to different setups (middle). Corrected data is compared to different models (right). POWHEG + PYTHIA8 setup shows overall nice agreement with the data [5]

differential cross-sectional distribution as a funcion of N_{ch} (shown in Fig. 11.4 (middle)) is compared to the predictions of POWHEG + PYTHIA8 and different setups (in Fig. 11.4 (right)), obtained by varying the parameters of CUETP8M2T4 tune of PYTHIA8. POWHEG + Herwig++ and POWHEG + Herwig7-based setups show different behavior from PYTHIA8-based setups. POWHEG + PYTHIA8 combination provides the best description of data measurements.

11.5 Summary

The evolution of UE activity is studied as a function of centre-of-mass energy, and strong dependency is observed. UE measurements in top quark pair production show no deviation from universality hypothesis even at higher energies. The overall good description of the UE activity by MC predictions confirms the universality of the physical processes producing the underlying event in p-p collisions at high energies. Results obtained from UE measurements via different channels are valuable feedback to further constrain phenomenological models, useful for the understanding of particle production at low p_T.

References

1. CMS Collaboration, The CMS experiment at the CERN LHC. JINST **3**, S08004 (2008). https://doi.org/10.1088/1748-0221/3/08/S08004
2. L. Evans et al., LHC machine. JINST **3**, S08001 (2008). https://doi.org/10.1088/1748-0221/3/08/S08001
3. CMS Collaboration, Underlying Event Measurements with Leading Particles and Jets in pp collisions at $\sqrt{s}$ = 13 TeV. CMS-PAS-FSQ-15-007 (2015), https://cds.cern.ch/record/2104473
4. CMS Collaboration, Measurement of the underlying event activity in inclusive Z boson production in proton-proton collisions at $\sqrt{s}$ = 13 TeV. JHEP **7**, 032 (2018). https://doi.org/10.1007/JHEP07(2018)032
5. CMS Collaboration, Study of the underlying event in top quark pair production in pp collisions at 13 TeV. Eur. Phys. J. C **79**, 123 (2019). https://doi.org/10.1140/epjc/s10052-019-6620-z

Chapter 12
Probing New Physics in
$B_s \to (K, K^*)\tau\nu$ and $B \to \pi\tau\nu$ Decays

N. Rajeev and Rupak Dutta

Abstract Motivated by the anomalies present in $b \to u$ and $b \to c$ semileptonic decays, we study the corresponding $B_s \to (K, K^*)\tau\nu$ and $B \to \pi\tau\nu$ decays within an effective field theory formalism. Our analysis is based on a strict model-dependent assumption, i.e., we assume that $b \to u$ and $b \to c$ transition decays exhibit similar new physics pattern. We give a prediction of various observables such as the branching fraction, ratio of branching ratio, lepton side forward-backward asymmetry, longitudinal polarization fraction of the charged lepton, and convexity parameter in the standard model and in the presence of vector type new physics couplings.

12.1 Introduction

Study of lepton flavor non-universality in the B meson systems have been the center of interest both theoretically and experimentally over the last decade. Disagreement between the SM expectations and the experimental measurements (BaBar, Belle, and LHCb) in $B \to D^{(*)}l\nu$ and $B_c \to J/\Psi l\nu$ undergoing $b \to (c, u)l\nu$ quark level transitions are well reflected in the flavor ratios R_D, R_{D^*} and $R_{J/\psi}$ defined as

$$R_{D^{(*)}} = \frac{\mathcal{B}(B \to D^{(*)}\tau\nu)}{\mathcal{B}(B \to D^{(*)}l\nu)}, \qquad R_{J/\psi} = \frac{\mathcal{B}(B_c \to J/\Psi\tau\nu)}{\mathcal{B}(B_c \to J/\Psi l\nu)}$$

In Table 12.1, we report the precise SM predictions and the experimental measurements of the various decay modes. The combined deviation of 3.78σ in R_D and R_{D^*} and around 1.3σ in $R_{J/\psi}$ from SM expectation is observed. Similarly, the average value of the branching ratio $\mathcal{B}(B \to \tau\nu)$ reported by BaBar and Belle

N. Rajeev (✉) · R. Dutta
National Institute of Technology, Silchar 788010, India
e-mail: rajeev_rs@phy.nits.ac.in; rajeevneutrino@gmail.com

R. Dutta
e-mail: rupak@phy.nits.ac.in

© Springer Nature Singapore Pte Ltd. 2021
P. K. Behera et al. (eds.), *XXIII DAE High Energy Physics Symposium*,
Springer Proceedings in Physics 261,
https://doi.org/10.1007/978-981-33-4408-2_12

Table 12.1 The SM prediction and the world averages of the ratio of branching ratios for various decay modes

Ratio of branching ratio	SM prediction	Experimental prediction
R_D	0.300 ± 0.008 [1–4]	$0.407 \pm 0.039 \pm 0.024$ [12–16]
R_{D^*}	0.258 ± 0.005 [5–8]	$0.304 \pm 0.013 \pm 0.007$ [12–16]
$R_{J/\psi}$	[0.20, 0.39] [9]	$0.71 \pm 0.17 \pm 0.18$ [17]
$\mathcal{B}(B \to \tau\nu)$	$(0.84 \pm 0.11) \times 10^{-4}$ [10]	$(1.09 \pm 2.4) \times 10^{-4}$ [18]
R_π^l	0.566	0.698 ± 0.155 [11]
R_π	0.641 [11]	<1.784 [18]

experiments is not in good agreement with the SM expectations. Although, the $\mathcal{B}(B \to \pi l\nu)$ is consistent with the SM, the ratio $R_\pi^l = (\tau_{B^0}/\tau_{B^-}) \mathcal{B}(B \to \tau\nu)/ \mathcal{B}(B \to \pi l\nu)$ shows mild deviation. Similar deviations are also observed in the ratio $R_\pi = \mathcal{B}(B \to \pi\tau\nu)/\mathcal{B}(B \to \pi l\nu)$ as well. Motivated by these anomalies, we study the implications of R_D, R_{D^*}, $R_{J/\psi}$, and R_π^l anomalies on $B_s \to (K, K^*)\tau\nu$ and $B \to \pi\tau\nu$ semileptonic decays in a model dependent way.

12.2 Theory

12.2.1 Effective Lagrangian

The effective Lagrangian for $b \to u l \nu$ transition that decays in the presence of vector type NP couplings is of the form [19]

$$
\mathcal{L}_{\text{eff}} = -\frac{4 G_F}{\sqrt{2}} V_{ub} \left\{ (1 + V_L) \bar{l}_L \gamma_\mu \nu_L \bar{c}_L \gamma^\mu b_L + V_R \bar{l}_L \gamma_\mu \nu_L \bar{c}_R \gamma^\mu b_R \right.
$$

$$
\left. + \tilde{V}_L \bar{l}_R \gamma_\mu \nu_R \bar{c}_L \gamma^\mu b_L + \tilde{V}_R \bar{l}_R \gamma_\mu \nu_R \bar{c}_R \gamma^\mu b_R \right\} + \text{h.c.}, \qquad (12.1)
$$

where G_F is the Fermi coupling constant and $|V_{ub}|$ is the CKM matrix element. V_L, V_R are the NP Wilson coefficients (WCs) involving left-handed neutrinos, and the WCs referring to tilde terms involve right-handed neutrinos.

Using the effective Lagrangian, we calculate the three-body differential decay distribution for the $B \to (P, V) l \nu$ decays. The final expressions pertaining to the pseudoscalar and vector differential decay rates can be found in [20].

In general, we define the ratio of branching ratio as

$$
R = \frac{\mathcal{B}(B_q \to M \tau \nu)}{\mathcal{B}(B_q \to M l \nu)}, \qquad (12.2)
$$

where $M = K$, K^*, π and $l = \mu$. We also define various q^2 dependent observables such as differential branching ratio $DBR(q^2)$, ratio of branching ratio $R(q^2)$, forward-backward asymmetry $A^l_{FB}(q^2)$, polarization fraction of the charged lepton $P^l(q^2)$, and convexity parameter $C^l_F(q^2)$ for the decay modes. For details one can refer to [20].

12.3 Results and Discussion

12.3.1 Standard Model Predictions

The SM central values are reported in Table 12.2. We calculate the central values by considering the central values of the input parameters. For the 1σ ranges, we perform a random scan over the theoretical inputs such as CKM matrix elements and the form factor inputs within 1σ of their central values. The significant difference in the μ mode and the τ mode are observed. The branching ratio of the order of 10^{-4} is observed in all the decay modes. The results pertaining $\langle P^l \rangle$ and $\langle C^l_F \rangle$ are calculated for the first time for these decay modes. In Fig. 12.1, we show the q^2 dependency of all the observables for the μ mode and the τ mode.

12.3.2 Beyond the SM Predictions

We discuss the NP contributions coming from V_L and $\widetilde{V}_L$ NP couplings. To get the allowed NP parameter space, we impose 2σ constraint coming from the measured values of R_D, R_{D^*}, $R_{J/\psi}$, and R^l_π. In the left panel of Fig. 12.2, we show the allowed range of V_L and $\widetilde{V}_L$ NP couplings once the 2σ constraints are imposed. Similarly, in the right panel the corresponding ranges in $\mathcal{B}(B \to \pi\tau\nu)$ and R_π using the allowed ranges of V_L and $\widetilde{V}_L$ NP couplings are shown. In Table 12.3 we display the allowed ranges of each observable in the presence of V_L and $\widetilde{V}_L$ NP couplings. Also, in Figs. 12.3 and 12.4, we display the q^2 dependency of the various observables in the presence of V_L and $\widetilde{V}_L$ NP couplings for the $B_s \to K\tau\nu$, $B_s \to K^*\tau\nu$, and $B \to \pi\tau\nu$ decays. The detailed observations are as follows:

- For the V_L NP coupling, we notice a significant deviation from the SM prediction in DBR(q^2) and $R(q^2)$ for all the decay modes. In addition, in the presence of $\widetilde{V}_L$ NP coupling the τ polarization fraction show deviation along with $R(q^2)$ and DBR(q^2). So the measurement of $P^\tau(q^2)$ can easily differentiate V_L and $\widetilde{V}_L$ NP contributions.
- The other observable such as $A^\tau_{FB}(q^2)$, $P^\tau(q^2)$, and $C^\tau_F(q^2)$ are not affected by V_L NP coupling. Similarly, $A^\tau_{FB}(q^2)$ and $C^\tau_F(q^2)$ are not affected by $\widetilde{V}_L$ NP coupling.

Table 12.2 The central values and 1σ ranges of each observable for both μ and τ modes in SM are reported for $B_s \to Kl\nu$, $B_s \to K^*l\nu$ and $B \to \pi l\nu$ decays

$B_s \to Kl\nu$		$BR \times 10^{-4}$	$\langle A^l_{FB} \rangle$	$\langle P^l \rangle$	$\langle C^l_F \rangle$	R_{B_sK}
μ mode	Central value	1.520	6.647×10^{-3}	0.982	−1.479	
	1σ range	[1.098, 2.053]	[0.006, 0.007]	[0.979, 0.984]	[−1.482, −1.478]	0.636
τ mode	Central value	0.966	0.284	0.105	-0.607	
	1σ range	[0.649, 1.392]	[0.262, 0.291]	[−0.035, 0.279]	[−0.711, −0.525]	[0.586, 0.688]
$B_s \to K^*l\nu$		$BR \times 10^{-4}$	$\langle A^l_{FB} \rangle$	$\langle P^l \rangle$	$\langle C^l_F \rangle$	$R_{B_sK^*}$
μ mode	Central value	3.259	−0.281	0.993	−0.417	
	1σ range	[2.501, 4.179]	[−0.342, −0.222]	[0.989, 0.995]	[−0.575, −0.247]	0.578
τ mode	Central value	1.884	−0.132	0.539	−0.105	
	1σ range	[1.449, 2.419]	[−0.203, −0.061]	[0.458, 0.603]	[−0.208, −0.007]	[0.539, 0.623]
$B \to \pi l\nu$		$BR \times 10^{-4}$	$\langle A^l_{FB} \rangle$	$\langle P^l \rangle$	$\langle C^l_F \rangle$	R_π
μ mode	Central value	1.369	4.678×10^{-3}	0.988	−1.486	
	1σ range	[1.030, 1.786]	[0.004, 0.006]	[0.981, 0.991]	[−1.489, −1.481]	0.641
τ mode	Central value	0.878	0.246	0.298	−0.737	
	1σ range	[0.690, 1.092]	[0.227, 0.262]	[0.195, 0.385]	[−0.781, −0.682]	[0.576, 0.725]

Table 12.3 Allowed ranges of each observable in the presence of V_L and $\widetilde{V}_L$ NP coupling

	V_L		$\widetilde{V}_L$		
	$\langle R \rangle$	$\langle BR \rangle \times 10^{-4}$	$\langle R \rangle$	$\langle BR \rangle \times 10^{-4}$	$\langle P^\tau \rangle$
$B_s \to K\tau\nu$	[0.644, 0.891]	[0.735, 1.746]	[0.638, 0.898]	[0.731, 1.774]	[−0.026, 0.217]
$B_s \to K^*\tau\nu$	[0.593, 0.804]	[1.684, 2.993]	[0.582, 0.802]	[1.579, 3.098]	[0.249, 0.513]
$B \to \pi\tau\nu$	[0.630, 0.915]	[0.793, 1.368]	[0.631, 0.926]	[0.765, 1.391]	[0.117, 0.315]

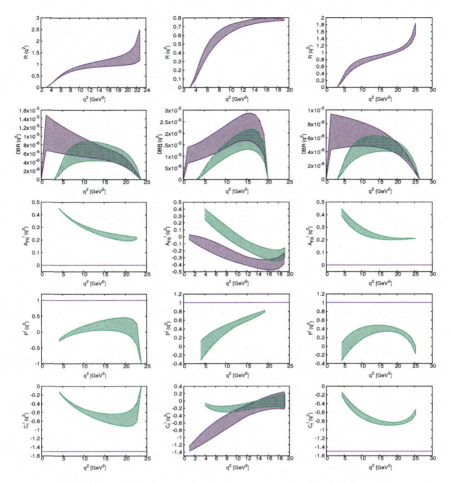

Fig. 12.1 q^2 dependent observables of $B_s \to K\,l\,\nu$ (first column), $B_s \to K^*\,l\,\nu$ (second column), and $B \to \pi\,l\,\nu$ (third column) decays in the SM for the μ (violet) and τ (green) modes

12.4 Conclusion

We study $B_s \to (K, K^*)\tau\nu$ and $B \to \pi\tau\nu$ decay modes within the SM and within the various NP scenarios. Although, there are hints of NP in various B meson decays, the NP is not yet established. Studying $B_s \to (K, K^*)\tau\nu$ and $B \to \pi\tau\nu$ decay modes theoretically as well as experimentally are well motivated as these can provide complementary information regarding NP.

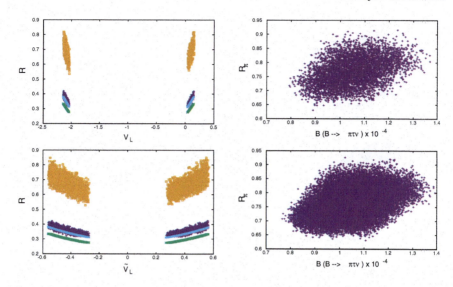

Fig. 12.2 In the left panel we show the allowed ranges in V_L (above) and $\widetilde{V}_L$ (below) NP coupling and the corresponding ranges in R_D (violet), R_{D^*} (green), $R_{J/\psi}$ (blue), and R_π^l (yellow) once 2σ experimental constraint is imposed. The corresponding ranges in $\mathcal{B}(B \rightarrow \pi\tau\nu)$ and R_π are shown in the right panel

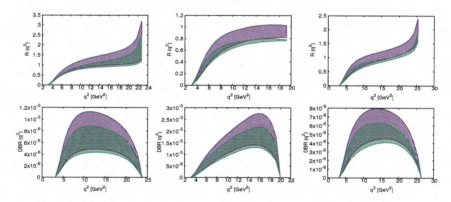

Fig. 12.3 $R(q^2)$ and DBR(q^2) for $B_s \rightarrow K\tau\nu$ (first column), $B_s \rightarrow K^*\tau\nu$ (second column), and $B \rightarrow \pi\tau\nu$ (third column) decays using the V_L NP coupling of Fig. 12.2 are shown with violet band. The corresponding SM ranges are shown with green band

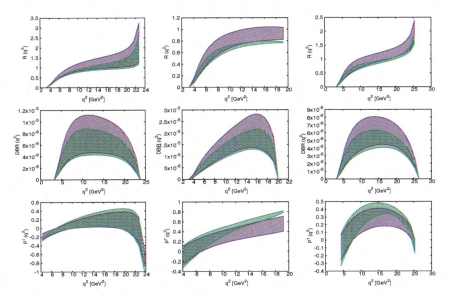

Fig. 12.4 $R(q^2)$, DBR(q^2) and $P^\tau(q^2)$ for $B_s \rightarrow K\tau\nu$ (first column), $B_s \rightarrow K^*\tau\nu$ (second column), and $B \rightarrow \pi\tau\nu$ (third column) decays using the $\widetilde{V}_L$ NP coupling is shown in violet band. The corresponding 1σ SM band is shown in green color

References

1. J.A. Bailey et al., [MILC Collaboration]. Phys. Rev. D **92**(3), 034506 (2015)
2. H. Na et al., [HPQCD Collaboration]. Phys. Rev. D **92**(5), 054510 (2015). Erratum: [Phys. Rev. D **93**(11), 119906 (2016)]
3. S. Aoki et al., Eur. Phys. J. C **77**(2), 112 (2017)
4. D. Bigi, P. Gambino, Phys. Rev. D **94**(9), 094008 (2016)
5. S. Fajfer, J.F. Kamenik, I. Nisandzic, Phys. Rev. D **85**, 094025 (2012)
6. F.U. Bernlochner, Z. Ligeti, M. Papucci, D. Robinson, Phys. Rev. D **95**(11), 115008 (2017). Erratum: [Phys. Rev. D **97**(5), 059902 (2018)]
7. D. Bigi, P. Gambino, S. Schacht, JHEP **1711**, 061 (2017)
8. S. Jaiswal, S. Nandi, S.K. Patra, JHEP **1712**, 060 (2017)
9. T.D. Cohen, H. Lamm, R.F. Lebed, arXiv:1807.02730 [hep-ph]
10. M. Bona et al., UTfit collaboration. Phys. Lett. B **687**, 61 (2010)
11. C. Patrignani et al., [Particle Data Group]. Chin. Phys. C **40**(10), 100001 (2016)
12. J.P. Lees et al., [BaBar Collaboration]. Phys. Rev. D **88**(7), 072012 (2013)
13. M. Huschle et al., [Belle Collaboration]. Phys. Rev. D **92**(7), 072014 (2015)
14. Y. Sato et al., [Belle Collaboration]. Phys. Rev. D **94**(7), 072007 (2016)
15. S. Hirose et al., [Belle Collaboration]. Phys. Rev. Lett. **118**(21), 211801 (2017)
16. R. Aaij et al., [LHCb Collaboration]. Phys. Rev. Lett. **115**(11), 111803 (2015). Erratum: [Phys. Rev. Lett. **115**(15), 159901 (2015)]
17. R. Aaij et al., [LHCb Collaboration]. arXiv:1711.05623 [hep-ex]
18. F.U. Bernlochner, Phys. Rev. D **92**(11), 115019 (2015)
19. R. Dutta, A. Bhol, A.K. Giri, Phys. Rev. D **88**(11), 114023 (2013)
20. N. Rajeev, R. Dutta, Phys. Rev. D **98**(5), 055024 (2018)

Chapter 13
Semileptonic Decays of Charmed Meson

N. R. Soni and J. N. Pandya

Abstract In this work, we present the charmed meson semileptonic decays in the covariant confined quark model. The necessary transition form factors for the channels $D \to (K, \pi)$ are computed in the whole physical range of momentum transfer. These form factors are then utilized for computation of semileptonic branching fractions. We also compare our results with the recent BESIII data and CLEO data.

13.1 Introduction

Semileptonic decays of charmed meson provide the key window to understand the decay of heavy quark dynamics because of the involvement of strong as well as weak interaction. The CKM matrix elements $|V_{cd}|$ and $|V_{cs}|$ can be extracted from semileptonic decays of $D_{(s)}$ mesons as they are parameterized by the form factor calculations. Experimentally, the data on form factors and branching fractions are reported by BESIII [1–4], *BABAR* [5], Belle [6], and CLEO collaborations [7]. The form factors for the channels $D \to (K, \pi)$ have also been reported using lattice quantum charmodynamics (LQCD) by ETM Collaborations [8, 9] and light cone sum rules (LCSR) [10]. The form factors and branching fractions are also computed using light front quark model [11], heavy meson chiral theory [12], constituent quark model [13], and chiral unitary approach [14]. The charmed meson decay properties are also studied in the potential model formalism [15–18].

In this article, we employ the Covariant Confined Quark Model for computation of semileptonic transition form factors and branching fractions of D mesons. We compare our findings with the experimental data from the BESIII, *BABAR*, CLEO, and Belle collaborations.

N. R. Soni (✉) · J. N. Pandya
Faculty of Technology and Engineering, Applied Physics Department, The Maharaja Sayajirao University of Baroda, Vadodara 390001, Gujarat, India
e-mail: nrsoni-apphy@msubaroda.ac.in; nakulphy@gmail.com

J. N. Pandya
e-mail: jnpandya-apphy@msubaroda.ac.in

© Springer Nature Singapore Pte Ltd. 2021
P. K. Behera et al. (eds.), *XXIII DAE High Energy Physics Symposium*,
Springer Proceedings in Physics 261,
https://doi.org/10.1007/978-981-33-4408-2_13

13.2 Form Factors and Branching Fractions in Covariant Confined Quark Model

The Covariant Confined Quark Model (CCQM), developed by Efimov and Ivanov [19–22] is a quantum field theoretical approach for hadronic interaction with the constituent quark via exchange of quark only. The effective interaction Lagrangian for meson $M(q_1, \bar{q}_2)$ corresponding to the constituent quarks q_1 and $\bar{q}_2$ is given by

$$L_{\text{int}} = g_M M(x) \int dx_1 \int dx_2 F_M(x; x_1, x_2) \cdot \bar{q}_{f_1}^a(x_1) \Gamma_M q_{f_2}^a(x_2) + H.c. \quad (13.1)$$

with $F_M(x; x_1, x_2)$ as the vertex function that characterizes the quark distribution within the mesons. For simplicity, we choose the vertex function to be of the Gaussian form. Here, Γ_M is the Dirac matrix corresponding to spin of the respective mesonic field $M(x)$ and g_M is the meson coupling constant computed from the meson self-energy diagram. The model parameters, namely quark masses and size parameters are given in Tab. 13.1.

For computation of semileptonic branching fractions, the invariant matrix element can be written as

$$M(D \to (P, V)\ell^+ \nu_\ell) = \frac{G_F}{\sqrt{2}} V_{cq} \langle P, V | \bar{q} \gamma^\mu (1 - \gamma_5) c | D \rangle \, \ell^+ \gamma^\mu (1 - \gamma_5) \nu_\ell, \quad (13.2)$$

where G_F is the Fermi coupling constant, P, V correspond to the pseudoscalar or vector mesons in the final state, respectively. The matrix element for the semileptonic decays can be written in terms of transition form factors as

$$\langle P(p_2) | \bar{s} O^\mu c | D(p_1) \rangle = F_+(q^2) P^\mu + F_-(q^2) q^\mu \quad (13.3)$$

$$\langle V(p_2, \epsilon_\nu) | \bar{s} O^\mu c | D(p_1) \rangle = \frac{\epsilon_\nu^\dagger}{m_1 + m_2} \left[-g^{\mu\nu} P \cdot q A_0(q^2) + P^\mu P^\nu A_+(q^2) \right.$$
$$\left. + q^\mu P^\nu A_-(q^2) + i \varepsilon^{\mu\nu\alpha\beta} P_\alpha q_\beta V(q^2) \right], \quad (13.4)$$

where $P = p_1 + p_2, q = p_1 - p_2$ and p_1, p_2 are the momenta of parent and daughter mesons, respectively. We present our form factors in Fig. 13.1. The form factors in the double pole approximation are given as

Table 13.1 Constituent quark masses and size parameters (in GeV)

m_u	m_s	m_c	m_b	Λ_K	Λ_{K^*}	Λ_π	λ	
0.241	0.428	1.67	5.05	1.04	0.72	0.87	0.181	GeV

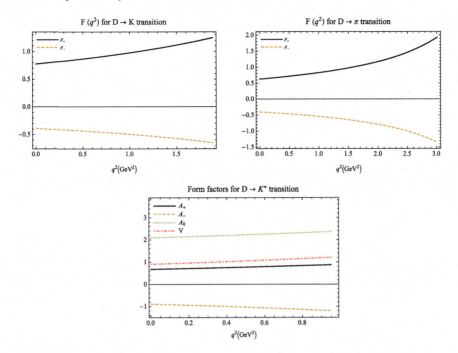

Fig. 13.1 Form factors

Table 13.2 Double pole parameters

	$D \to K$		$D \to \pi$		$D \to K^*$			
	F_+	F_-	F_+	F_-	A_0	A_+	A_-	V
$F(0)$	0.77	−0.39	0.63	−0.41	2.08	0.67	−0.90	0.89
a	0.72	0.78	0.86	0.93	0.39	0.86	0.96	0.97
b	0.047	0.070	0.096	0.13	−0.10	0.091	0.14	0.14

$$F(q^2) = \frac{F(0)}{1 - a\frac{q^2}{m_1^2} + b(\frac{q^2}{m_1^2})^2}. \tag{13.5}$$

The form factors and associated double pole parameters are given in Table 13.2.

After defining the form factors, we compute the semileptonic branching fractions using the relations [23, 24]

$$\frac{d\Gamma(D \to (P, V)\ell^+\nu_\ell)}{dq^2} = \frac{G_F^2|V_{cq}|^2|\mathbf{p_2}|q^2}{12(2\pi)^3 m_D^2}\left(1 - \frac{m_\ell^2}{q^2}\right)^2$$
$$\times \left[\left(1 + \frac{m_\ell^2}{2q^2}\right)\sum_{n=0,+,-}|H_n|^2 + \frac{3m_\ell^2}{2q^2}|H_t|^2\right]. \tag{13.6}$$

Table 13.3 Charmed semileptonic branching fractions (in %)

Channel	Present	Data	References
$D^+ \to \pi^0 e^+ \nu_e$	0.291	$0.350 \pm 0.011 \pm 0.010$	BESIII [2]
$D^+ \to \pi^0 \mu^+ \nu_\mu$	0.285		
$D^+ \to \bar{K}^0 e^+ \nu_e$	9.287	$8.60 \pm 0.06 \pm 0.15$	BESIII [3]
		$8.83 \pm 0.10 \pm 0.20$	CLEO [7]
$D^+ \to \bar{K}^0 \mu^+ \nu_\mu$	9.022	$8.72 \pm 0.07 \pm 0.18$	BESIII [29]
$D^+ \to \bar{K}^{*0} e^+ \nu_e$	7.613	5.40 ± 0.10	PDG [30]
$D^+ \to \bar{K}^{*0} \mu^+ \nu_\mu$	7.207	5.27 ± 0.16	CLEO [31]

Here $|\mathbf{p_2}| = \lambda^2(m_D^2, m_{P/V}^2, q^2)/2m_D^2$ is the momentum of daughter meson in the rest frame of the D meson with λ is the Källen function. $H_\pm$, H_0, and H_t are the helicity amplitudes expressed as

For $D \to P$ channel:

$$H_t = \frac{1}{\sqrt{q^2}}(Pq F_+ + q^2 F_-),$$

$$H_\pm = 0 \quad \text{and} \quad H_0 = \frac{2m_1|\mathbf{p_2}|}{\sqrt{q^2}} F_+ \tag{13.7}$$

For $D \to V$ channel:

$$H_t = \frac{1}{m_1 + m_2} \frac{m_1|\mathbf{p_2}|}{m_2\sqrt{q^2}}\left((m_1^2 - m_2^2)(A_+ - A_-) + q^2 A_-\right)$$

$$H_\pm = \frac{1}{m_1 + m_2}(-(m_1^2 - m_2^2)A_0 \pm 2m_1|\mathbf{p_2}|V)$$

$$H_0 = \frac{1}{m_1 + m_2} \frac{1}{2m_2\sqrt{q^2}}(-(m_1^2 - m_2^2)(m_1^2 - m_2^2 - q^2)A_0 + 4m_1^2|\mathbf{p_2}|^2 A_+).$$

$$\tag{13.8}$$

The computed semileptonic branching fractions for D^+ meson are given in Table 13.3 in comparison with experimental observations.

13.3 Results and Discussion

Having defined the model parameters and form factors, we compute the semileptonic branching fractions within the framework of the Covariant Confined Quark Model. The numerical results of semileptonic branching fractions are tabulated in Table 13.3. Our results for $D^+ \to \pi^0 e^+ \nu_e$ are found to be nearer to the BESIII data and for muon channel the experimental results are still not available. For

$D^+ \to \bar{K}^0 \ell^+ \nu_\ell$ channel, our results are in good agreement with the CLEO and BESIII data. For $D^+ \to \bar{K}^*(892)^0 \ell^+ \nu_\ell$ channel our results overestimate the experimental data. For detailed description of the model and computation technique we suggest the readers to refer our papers [25–27] as well as our recent review article [28]. In these papers, we have extensively studied the leptonic and semileptonic decay of D and D_s mesons. We study $D^{+(0)} \to (K, K^*(892), \pi, \rho, \omega, \eta, \eta', D^0) \ell^+ \nu_\ell$, $D_s^+ \to (K, K^*(892)^0, \phi, \eta, \eta', D^0) \ell^+ \nu_\ell$. We also study the other physical observables such as forward-backward asymmetry, longitudinal and transverse polarizations, lepton-side, and hadron-side convexity parameters.

The study of hadronic properties in Covariant Confined Quark Model is very general and this formalism is applicable to any number of quarks with any number of loops. In the last few years this formalism is successfully employed for computation of decay properties of $B_{(s)}$ mesons [32–38], B_c mesons [39–41], charmed and beauty baryons [42–46] and exotic states [47–49].

Acknowledgments We thank our collaborators Prof. M. A. Ivanov, Prof. J. G. Körner, Prof. P. Santorelli and Dr. C. T. Tran for their support. J. N. P. acknowledges the financial support from the University Grants Commission of India under Major Research Project F.No.42-775/2013(SR).

References

1. M. Ablikim et al., BESIII collaboration. Phys. Rev. Lett. **122**, 011804 (2019)
2. M. Ablikim et al., BESIII collaboration. Phys. Rev. Lett. **121**, 171803 (2018)
3. M. Ablikim et al., BESIII collaboration. Phys. Rev. D **96**, 012002 (2017)
4. M. Ablikim et al., BESIII collaboration. Phys. Rev. D **92**, 072012 (2015)
5. J.P. Lees et al., BABAR collaboration. Phys. Rev. D **91**, 052022 (2015)
6. L. Widhalm et al., Belle collaboration. Phys. Rev. Lett. **97**, 061804 (2006)
7. D. Besson et al., CLEO collaboration. Phys. Rev. D **80**, 032005 (2009)
8. V. Lubicz et al., ETM collaboration. Phys. Rev. D **96**, 054514 (2017)
9. V. Lubicz et al., ETM collaboration. Phys. Rev. D **98**, 014516 (2018)
10. Y.L. Wu, M. Zhong, Y.B. Zuo, Int. J. Mod. Phys. A **21**, 6125 (2006)
11. H.-Y. Cheng, X.-W. Kang, Eur. Phys. J. C **77**, 587 (2017)
12. S. Fajfer, J.F. Kamenik, Phys. Rev. D **72**, 034029 (2005)
13. D. Melikhov, B. Stech, Phys. Rev. D **62**, 014006 (2000)
14. T. Sekihara, E. Oset, Phys. Rev. D **92**, 054038 (2015)
15. V. Kher, N. Devlani, A.K. Rai, Chin. Phys. C **41**, 073101 (2017)
16. N. Devlani, A.K. Rai, Int. J. Theor. Phys. **52**, 2196 (2013)
17. N. Devlani, A.K. Rai, Eur. Phys. J. A **48**, 104 (2012)
18. N. Devlani, A.K. Rai, Phys. Rev. D **84**, 074030 (2011)
19. G.V. Efimov, M.A. Ivanov, Int. J. Mod. Phys. A **4**, 2031 (1989)
20. G.V. Efimov, M.A. Ivanov, *The Quark Confinement Model of Hadrons* (IOP, Bristol, 1993)
21. A. Faessler, T. Gutsche, M.A. Ivanov, J.G. Körner, V.E. Lyubovitskij, Eur. Phys. J. Direct C **4**, 1 (2002)
22. T. Branz, A. Faessler, T. Gutsche, M.A. Ivanov, J.G. Körner, V.E. Lyubovitskij, Phys. Rev. D **81**, 034010 (2010)
23. M.A. Ivanov, J.G. Körner, C.T. Tran, Phys. Rev. D **92**, 114022 (2015)
24. T. Gutsche, M.A. Ivanov, J.G. Körner, V.E. Lyubovitskij, P. Santorelli, N. Habyl, Phys. Rev. D **91**, 074001 (2015) [Erratum: Phys. Rev. D **91**, 119907 (2015)]

25. N.R. Soni, J.N. Pandya, Phys. Rev. D **96**, 016017 (2017) [Erratum: Phys. Rev. D **99**, 059901 (2019)]
26. N.R. Soni, M.A. Ivanov, J.G. Körner, J.N. Pandya, P. Santroelli, C.T. Tran, Phys. Rev. D **98**, 114031 (2018)
27. N.R. Soni, J.N. Pandya, EPJ Web Conf. **202**, 06010 (2019)
28. M.A. Ivanov, J.G. Korner, J.N. Pandya, P. Santroelli, N.R. Soni, J.N. Pandya, Front. Phys. **14**, 64401 (2019)
29. M. Ablikim et al., BESIII collaboration. Eur. Phys. J C **76**, 369 (2016)
30. M. Tanabashi et al., Particle data group. Phys. Rev. D **98**, 030001 (2018)
31. R.A. Briere et al., CLEO collaboration. Phys. Rev D **81**, 112001 (2010)
32. S. Dubnicka, A.Z. Dubnickova, M.A. Ivanov, A. Liptaj, P. Santorelli, C.T. Tran, Phys. Rev. D **99**, 014042 (2019)
33. M.A. Ivanov, J.G. Körner, C.T. Tran, Phys. Part. Nucl. Lett. **14**, 669 (2017)
34. M.A. Ivanov, J.G. Körner, C.T. Tran, Phys. Rev. D **95**, 036021 (2017)
35. M.A. Ivanov, J.G. Körner, C.T. Tran, Phys. Rev. D **94**, 094028 (2016)
36. S. Dubniccka, A.Z. Dubnickova, A. Issadykov, M.A. Ivanov, A. Liptaj, S.K. Sakhiyev, Phys. Rev. D **93**, 094022 (2016)
37. A. Issadykov, M.A. Ivanov, S.K. Sakhiyev, Phys. Rev. D **91**, 074007 (2015)
38. M.A. Ivanov, J.G. Körner, S.G. Kovalenko, P. Santorelli, G.G. Saidullaeva. Phys. Rev. D **85**, 034004 (2012)
39. A. Issadykov, M.A. Ivanov, Phys. Lett. B **783**, 178 (2018)
40. C.T. Tran, M.A. Ivanov, J.G. Körner, P. Santorelli, Phys. Rev. D **97**, 054014 (2018)
41. S. Dubnicka, A.Z. Dubnickova, A. Issadykov, M.A. Ivanov, A. Liptaj, Phys. Rev. D **96**, 076017 (2017)
42. T. Gutsche, M.A. Ivanov, J.G. Körner, V.E. Lyubovitskij, arXiv:1905.06219
43. T. Gutsche, M.A. Ivanov, J.G. Körner, V.E. Lyubovitskij, Z. Tyulemissov, Phys. Rev. D **99**, 056013 (2019)
44. T. Gutsche, M.A. Ivanov, J.G. Körner, V.E. Lyubovitskij, P. Santorelli, C.T. Tran, Phys. Rev. D **98**, 053003 (2018)
45. T. Gutsche, M.A. Ivanov, J.G. Körner, V.E. Lyubovitskij, Phys. Rev. D **98**, 074011 (2018)
46. T. Gutsche, M.A. Ivanov, J.G. Körner, V.E. Lyubovitskij, Phys. Rev. D **96**, 054013 (2017)
47. F. Goerke, T. Gutsche, M.A. Ivanov, J.G. Körner, V.E. Lyubovitskij, Phys. Rev. D **96**, 054028 (2017)
48. F. Goerke, T. Gutsche, M.A. Ivanov, J.G. Körner, V.E. Lyubovitskij, P. Santorelli, Phys. Rev. D **94**, 094017 (2016)
49. T. Gutsche, M.A. Ivanov, J.G. Körner, V.E. Lyubovitskij, K. Xu, Phys. Rev. D **96**, 114004 (2017)

Chapter 14
Three-Loop Heavy Quark Form Factors and Their Asymptotic Behavior

J. Ablinger, J. Blümlein, P. Marquard, N. Rana, and C. Schneider

Abstract A summary of the calculation of the color-planar and complete light quark contributions to the massive three-loop form factors is presented. Here a novel calculation method for the Feynman integrals is used, solving general uni-variate first-order factorizable systems of differential equations. We also present predictions for the asymptotic structure of these form factors.

14.1 Introduction

The detailed description of top quark pair production to high perturbative order is of importance in various respects, including precision studies of QCD, the measurement of the top-quark mass, and its other properties, and in the search for effects from potential physics beyond the Standard Model. The heavy quark form factors act as the basic building block of the related observables. In a series of publications [1–4], two-loop QCD contributions of these form factors for vector, axial-vector, scalar, and pseudo-scalar currents were first computed. In an independent calculation in [5], the $\mathcal{O}(\varepsilon)$ terms were included for the vector form factors, where ε is the dimensional regularization parameter in $D = 4 - 2\varepsilon$ space–time dimensions. Later in [6], two-loop QCD contributions up to $\mathcal{O}(\varepsilon^2)$ for all these form factors were obtained.

At the three-loop level, the color-planar contributions to the vector form factors were obtained in [7, 8] and the complete light quark contributions in [9]. We have computed both the color-planar and complete light quark contributions to the three-loop form factors for the axial-vector, scalar, and pseudo-scalar currents in [10] and for the vector current in [11], which are the subject of the first part of this article.

J. Ablinger · C. Schneider
RISC, Johannes Kepler University, Altenbergerstraße 69, 4040 Linz, Austria

J. Blümlein · P. Marquard · N. Rana (✉)
DESY, Platanenallee 6, 15738 Zeuthen, Germany
e-mail: Narayan.Rana@mi.infn.it; narayan.rana@desy.de

N. Rana
INFN, Sezione di Milano, Via Celoria 16, 20133 Milano, Italy

© Springer Nature Singapore Pte Ltd. 2021
P. K. Behera et al. (eds.), *XXIII DAE High Energy Physics Symposium*,
Springer Proceedings in Physics 261,
https://doi.org/10.1007/978-981-33-4408-2_14

In [11], we have presented a detailed description of the method which we have used to obtain the master integrals in this case. The method is generic to compute any first-order factorizable and uni-variate system of differential equations. In a parallel calculation, the same results have been obtained in [12].

Amplitudes for hard scattering processes in QCD provide a clear insight into underlying principles such as factorization or the universality of infrared (IR) singularities. In the case of massless QCD amplitudes, a plethora of work [13–16] has been performed to understand the structure of IR divergences which is due to the interplay of the soft- and collinear dynamics. In the case of two parton amplitudes, i.e., the form factors, the IR structure is more prominent. The interplay of the soft and collinear anomalous dimensions building up the singular structure of the massless form factors has first been noticed in [17] at two-loop order and has been later established at the three-loop order in [18]. The generalization of this universal structure to the case of massive form factors is also of interest. First steps were taken in [19] in the asymptotic limit, i.e., in the limit where the quark mass is small compared to the center of mass energy, followed by the proposition of a factorization theorem [20–22] in the asymptotic limit. Finally, in [23], a general solution was presented following a soft-collinear effective theory approach. While, the solution in [23] provides the structure of IR poles for the exact computation, the study of the Sudakov behavior in the asymptotic limit also elucidates the logarithmic behavior for the finite contributions. Following the method proposed for massless form factors in [24, 25], we have performed a rigorous study in [26] in the asymptotic limit to obtain all the poles and also all logarithmic contributions to finite pieces of the three-loop heavy quark form factors for vector, axial-vector, scalar, and pseudo-scalar currents. A similar study has been performed in [27] obtaining the poles for the vector form factor. In the second part of this article, we summarize the contents of [26].

14.2 Heavy Quark Form Factors

We consider a virtual massive boson of momentum q, which can be a vector (V), an axial-vector (A), a scalar (S), or a pseudo-scalar (P), decaying into a pair of heavy quarks of mass m, color c and d, and momenta q_1 and q_2, at a vertex X_{cd}, where $X_{cd} = \Gamma^\mu_{V,cd}, \Gamma^\mu_{A,cd}, \Gamma_{S,cd}$ and $\Gamma_{P,cd}$. The general forms of the amplitudes are

$$\bar{u}_c(q_1)\Gamma^\mu_{V,cd}v_d(q_2) \equiv -i\bar{u}_c(q_1)\left[\delta_{cd}v_Q\left(\gamma^\mu\, F_{V,1} + \frac{i}{2\,m}\sigma^{\mu\nu}q_\nu\, F_{V,2}\right)\right]v_d(q_2),$$

$$\bar{u}_c(q_1)\Gamma^\mu_{A,cd}v_d(q_2) \equiv -i\bar{u}_c(q_1)\left[\delta_{cd}a_Q\left(\gamma^\mu\gamma_5\, F_{A,1} + \frac{1}{2\,m}q_\mu\gamma_5\, F_{A,2}\right)\right]v_d(q_2),$$

$$\bar{u}_c(q_1)\Gamma_{S,cd}v_d(q_2) \equiv -i\bar{u}_c(q_1)\left[\delta_{cd}s_Q\left(\frac{m}{v}(-i)\, F_S\right)\right]v_d(q_2),$$

$$\bar{u}_c(q_1)\Gamma_{P,cd}v_d(q_2) \equiv -i\bar{u}_c(q_1)\left[\delta_{cd}p_Q\left(\frac{m}{v}(\gamma_5)\, F_P\right)\right]v_d(q_2). \tag{14.1}$$

Here $\bar{u}_c(q_1)$ and $v_d(q_2)$ are the bi-spinors of the quark and the anti-quark, respectively, with $\sigma^{\mu\nu} = \frac{i}{2}[\gamma^{\mu}, \gamma^{\nu}]$. v_Q, a_Q, s_Q, and p_Q are the Standard Model (SM) coupling constants for the vector, axial-vector, scalar, and pseudo-scalar, respectively. $v = (\sqrt{2}G_F)^{-1/2}$ denotes the SM vacuum expectation value of the Higgs field, with the Fermi constant G_F. For more details, see [6]. The form factors can be obtained from the amplitudes by multiplying appropriate projectors [6] and performing the trace over the color and spinor indices.

14.2.1 Details of the Computation

The computational procedure is described in detail in [6]. The Feynman diagrams are generated using QGRAF [28]. The packages Q2e/Exp [29, 30], FORM [31, 32], and Color [33] are used in the computation. By decomposing the dot products among the loop and external momenta into the combination of inverse propagators, each Feynman diagram can be expressed in terms of a linear combination of a large set of scalar integrals. These integrals are related to each other through integration-by-parts identities (IBPs) [34, 35], and are reduced to 109 master integrals (MIs) by using the package Crusher [36].

We apply the method of differential equations [37–40] to calculate the master integrals. The method and the corresponding algorithm is presented in detail in [11].[1] The principal idea of this method is to obtain a set of differential equations of the MIs by performing differentiation with respect to the variable x, with $q^2/m^2 = -(1 - x)^2/x$ and then to use the IBP relations on the output to obtain a linear combination of MIs for each differentiated integral for general bases. One obtains a $n \times n$ system of coupled linear differential equations for n master integrals $\mathcal{I}$

$$\frac{d}{dx}\mathcal{I} = \mathcal{M}\mathcal{I} + \mathcal{R}. \tag{14.2}$$

Here the $n \times n$ matrix $\mathcal{M}$ consists out of entries from the rational function field $\mathbb{K}(D, x)$ (or equivalently from $\mathbb{K}(\varepsilon, x)$), where $\mathbb{K}$ is a field of characteristic 0. The inhomogeneous part $\mathcal{R}$ contains MIs which are already known. In simple cases, $\mathcal{R}$ turns out to be just the null vector. The first step to solve such a coupled system of differential equations is to find out whether the system factorizes to first order or not. Using the package Oresys [42], based on Zürcher's algorithm [43] and applying a corresponding solver [11, 44] we have first confirmed that the present system is indeed first-order factorizable in x-space.

Without the need to choose a special basis, we solve the system in terms of iterated integrals over whatsoever alphabet, cf. [11] for details. To proceed, we first arrange the differential equations in such a manner that it appears in upper block-triangular form. Then, we compute the integrals block-by-block starting from the last in the

---

[1] For a review on the computational methods of loop integrals in quantum field theory, see [41].

arrangement. While solving for each block, say of order $m \times m$, the differential equations are solved order by order in ε successively, starting at the leading pole terms, $\propto 1/\varepsilon^3$ for our case. The successive solutions in ε also contribute to the inhomogeneities in the next order. We use the package Oresys [42], based on Zürcher's algorithm [43] to uncouple the differential equations. At each order in ε, l inhomogeneous ordinary differential equations are obtained, where $1 \leq l \leq m$. The orders of these differential equations are $m_1, \dots, m_l$ such that $m_1 + \cdots + m_l = m$. We have solved these differential equations using the method of variation of constant. In our case, the spanning alphabet is

$$\frac{1}{x}, \; \frac{1}{1-x}, \; \frac{1}{1+x}, \; \frac{1}{1-x+x^2}, \; \frac{x}{1-x+x^2}, \tag{14.3}$$

i.e., the usual harmonic polylogarithms (HPLs) [45] and their cyclotomic extension (CHPL) [46]. While integration over a letter is a straightforward algebraic manipulation, often k-th power of a letter, $k \in \mathbb{N}$, appears which needs to be transformed to the letters of (14.3) by partial integration. The other $m - l$ solutions are immediately obtained from the former solutions. The constants of integration are determined using boundary conditions in the low energy limit, i.e., at $x = 1$. The boundary values for the HPLs and CHPLs give rise to the respective constants in the limit $x \to 1$, i.e., the multiple zeta values (MZVs) [47] and the cyclotomic constants [46]. The computation is performed by intense use of HarmonicSums [46, 48–54], which uses the package Sigma [55, 56]. Finally, all the MIs have been checked numerically using FIESTA [57–59].

14.2.2 Ultraviolet Renormalization and Universal Infrared Structure

We perform the ultraviolet (UV) renormalization of the form factors in a mixed scheme. The heavy quark mass and wave function have been renormalized in the on-shell (OS) renormalization scheme. The strong coupling constant has been renormalized using the $\overline{\text{MS}}$ scheme, by setting the universal factor $S_\varepsilon = \exp(-\varepsilon(\gamma_E - \ln(4\pi)))$ for each loop order to one at the end of the calculation.

The required renormalization constants are already well-known and are denoted by $Z_{m,\text{OS}}$ [60–64], $Z_{2,\text{OS}}$ [60–62, 65], and Z_{a_s} [66, 67], with $a_s = \alpha_s/(4\pi)$, for the heavy quark mass, wave function, and strong coupling constant, respectively. $Z_{2,\text{OS}}$ and Z_{a_s} are multiplicative, while the renormalization of massive fermion lines has been taken care of by properly considering the counter terms. For the scalar and pseudo-scalar currents, the presence of the heavy quark mass in the Yukawa coupling employs another overall mass renormalization constant, which also has been performed in the OS renormalization scheme.

The universal behavior of IR singularities of the massive form factors was first investigated in [21] considering the high energy limit. Later in [23], a general argument was provided to factorize the IR singularities as a multiplicative renormalization constant as

$$F_I = Z(\mu)\, F_I^{\text{fin}}(\mu)\,,\tag{14.4}$$

where F_I^{fin} is finite as $\varepsilon \to 0$. The renormalization group equation (RGE) for $Z(\mu)$ is constrained by the massive cusp anomalous dimension [68, 69].

14.2.3 Checks of the Results

To perform checks, we have maintained the gauge parameter ξ to first order and have thus obtained a partial check on gauge invariance. Fulfillment of the chiral Ward identity gives another strong check on our calculation.

Considering α_s-decoupling appropriately, we obtain the universal IR structure for all the UV renormalized results, confirming again the universality of IR poles. Also, in the low energy limit, the magnetic vector form factor produces the anomalous magnetic moment of a heavy quark which we cross check with [70] in this limit. Finally, we have compared our results with those of [7, 9, 12], which have been computed using partly different methods. Both results agree.

14.3 Asymptotic Behavior of Massive Form Factors

We consider from now on only the renormalized electric form factor (F_V) for the vector current and the renormalized scalar form factor (F_S), in the asymptotic limit. All other massive form factors either agree to one of them or vanish in this limit. To start with, we write down a Sudakov type integro-differential equation [71, 78] for a function $\hat{F}_I\left(a_s(\mu), \frac{Q^2}{\mu^2}, \frac{m^2}{\mu^2}, \varepsilon\right)$ in the asymptotic limit as follows:

$$\mu^2 \frac{\partial}{\partial \mu^2} \ln \hat{F}_I\left(\frac{Q^2}{\mu^2}, \frac{m^2}{\mu^2}, a_s, \varepsilon\right) = \frac{1}{2}\left[K_I\left(\frac{m^2}{\mu^2}, a_s, \varepsilon\right) + G_I\left(\frac{Q^2}{\mu^2}, a_s, \varepsilon\right)\right],\tag{14.5}$$

where $I = V, S$ only. Here $\hat{F}_I$ contains all logarithmic behavior and singular contributions of the respective form factor. As evident from the functional dependence, K_I incorporates the contributions from the quark mass m and does not depend on the kinematic invariants, while G_I contains the information of the process. Along with the evolution of the strong coupling constant, (14.5), and the renormalization group (RG) invariance of $\hat{F}$, individual solutions for K_I and G_I are provided as follows:

$$K_I = K_I\left(a_s(m^2), 1, \varepsilon\right) - \int_{\frac{m^2}{\mu^2}}^{1} \frac{d\lambda}{\lambda} A_q\left(a_s(\lambda\mu^2)\right),$$

$$G_I = G_I\left(a_s(Q^2), 1, \varepsilon\right) + \int_{\frac{Q^2}{\mu^2}}^{1} \frac{d\lambda}{\lambda} A_q\left(a_s(\lambda\mu^2)\right). \tag{14.6}$$

Here A_q denotes the quark cusp anomalous dimension. $K_I(a_s(m^2), 1, \varepsilon)$ and $G_I(a_s(Q^2), 1, \varepsilon)$ are initial conditions arising while solving the RG equations. Using (14.6), one can solve (14.5) to obtain $\hat{F}_I$, from which the form factors can be obtained through the following matching relation

$$F_I\left(a_s, \frac{Q^2}{\mu^2}, \frac{m^2}{\mu^2}, \varepsilon\right) = C_I(a_s, \varepsilon)\hat{F}_I\left(a_s, \frac{Q^2}{\mu^2}, \frac{m^2}{\mu^2}, \varepsilon\right). \tag{14.7}$$

The solutions for $\hat{F}_I$ up to four-loop are presented in [26]. At each order in a_s, say n, the solution consists of $A_I^{(n)}$, $K_I^{(n)}$, and $G_I^{(n)}$, the expansion coefficients of A_I, $K_I(a_s(m^2), 1, \varepsilon)$, and $G_I(a_s(Q^2), 1, \varepsilon)$, respectively, and lower order terms.

In the massless quark form factor, the soft (f_q) and collinear (B_q) anomalous dimensions govern the infrared structure in the form $\gamma_q = B_q + \frac{f_q}{2}$. Intuitively, in the massive case, γ_q, along with similar contributions (γ_Q) from the heavy quark anomalous dimension, will control the singular structure. Hence, it is suggestive to write

$$K_I^{(n)} = -2(\gamma_q^{(n)} + \gamma_Q^{(n)} - \gamma_I^{(n-1)}). \tag{14.8}$$

The anomalous dimension $\gamma_I^{(n-1)}$ [60–62, 72–75] arises due to renormalization of the current. Note that the power of each term γ^n indicates the series expansion in a_s. For γ_I, the contribution is of the same order also, however, we denote it by $(n-1)$ to match with general notation of [72]. The other finite functions $G_I^{(n)}$ contain the information on the process through its dependence on Q^2. Hence, it is similar to the one in case of massless form factors [76, 77]

$$G_I^{(n)} = 2(B_q^{(n)} - \gamma_I^{(n-1)}) + f_q^{(n)} + C_I^{(n)} + \sum_{k=1}^{\infty} \varepsilon^k g_I^{n,k}. \tag{14.9}$$

Given the structural similarities, $C_I^{(n)}$ and $g_I^{n,k}$ are the same as in the massless cases. All the required anomalous dimensions, except γ_Q, are known from different computations. On the other hand, γ_Q can be obtained from the non-logarithmic contribution of the massive cusp anomalous dimension in the asymptotic limit. With all the ingredients, we obtain the full singular contributions and all logarithmic contributions to the finite part for vector and scalar form factors in the asymptotic limit. The non-logarithmic part of the finite piece gets contributions from the matching function C_I which can only be obtained by an exact computation. Using our results of [11], we obtain the color-planar and complete light quark contributions for $C_I^{3,0}$.

14.4 Conclusion

In the first part, we have summarized the computational details to obtain the color-planar and complete light quark contributions to the three-loop heavy quark form factors along with a new method to solve uni-variate first-order factorizable systems of differential equations. The system is solved in terms of iterative integrals over a finite alphabet of letters. Finally, we have computed all the corresponding contributions to the massive three-loop form factors for vector, axial-vector, scalar, and pseudo-scalar currents, which play an important role in the phenomenological study of the top quark. We then have studied the asymptotic behavior of these form factors. A Sudakov type integro-differential equation can be written down for the massive form factors and along with the study of RGE, we have obtained all the logarithmic contributions of the finite part of the vector and scalar form factors.

Acknowledgments This work was supported in part by the Austrian Science Fund (FWF) grant SFB F50 (F5009-N15), by the bilateral project DNTS-Austria 01/3/2017 (WTZ BG03/2017), funded by the Bulgarian National Science Fund and OeAD (Austria), by the EU TMR network SAGEX Marie Sklodowska-Curie grant agreement No. 764850 and COST action CA16201: Unraveling new physics at the LHC through the precision frontier.

References

1. W. Bernreuther, R. Bonciani, T. Gehrmann, R. Heinesch, T. Leineweber, P. Mastrolia, E. Remiddi, Nucl. Phys. B **706**, 245–324 (2005). https://doi.org/10.1016/j.nuclphysb.2004.10.059
2. W. Bernreuther, R. Bonciani, T. Gehrmann, R. Heinesch, T. Leineweber, P. Mastrolia, E. Remiddi, Nucl. Phys. B **712**, 229–286 (2005). https://doi.org/10.1016/j.nuclphysb.2005.01.035
3. W. Bernreuther, R. Bonciani, T. Gehrmann, R. Heinesch, T. Leineweber, E. Remiddi, Nucl. Phys. B **723**, 91–116 (2005). https://doi.org/10.1016/j.nuclphysb.2005.06.025
4. W. Bernreuther, R. Bonciani, T. Gehrmann, R. Heinesch, P. Mastrolia, E. Remiddi, Phys. Rev. D **72**, 096002 (2005). https://doi.org/10.1103/PhysRevD.72.096002
5. J. Gluza, A. Mitov, S. Moch, T. Riemann, JHEP **7**, 001 (2009). https://doi.org/10.1088/1126-6708/2009/07/001
6. J. Ablinger, A. Behring, J. Blümlein, G. Falcioni, A. De Freitas, P. Marquard, N. Rana, C. Schneider, Phys. Rev. D **97**(9), 094022 (2018). https://doi.org/10.1103/PhysRevD.97.094022
7. J. Henn, A.V. Smirnov, V.A. Smirnov, M. Steinhauser, JHEP **1**, 074 (2017). https://doi.org/10.1007/JHEP01(2017)074
8. J.M. Henn, A.V. Smirnov, V.A. Smirnov, JHEP **12**, 144 (2016). https://doi.org/10.1007/JHEP12(2016)144
9. R.N. Lee, A.V. Smirnov, V.A. Smirnov, M. Steinhauser, JHEP **3**, 136 (2018). https://doi.org/10.1007/JHEP03(2018)136
10. J. Ablinger, J. Blümlein, P. Marquard, N. Rana, C. Schneider, Phys. Lett. B **782**, 528–232 (2018). https://doi.org/10.1016/j.physletb.2018.05.077
11. J. Ablinger, J. Blümlein, P. Marquard, N. Rana, C. Schneider, Nucl. Phys. B **939**, 253–291 (2019). https://doi.org/10.1016/j.nuclphysb.2018.12.010

12. R.N. Lee, A.V. Smirnov, V.A. Smirnov, M. Steinhauser, JHEP **5**, 187 (2018). https://doi.org/10.1007/JHEP05(2018)187

13. S. Catani, Phys. Lett. B **427**, 161–171 (1998). https://doi.org/10.1016/S0370-2693(98)00332-3

14. G.F. Sterman, M.E. Tejeda-Yeomans, Phys. Lett. B **552**, 48–56 (2003). https://doi.org/10.1016/S0370-2693(02)03100-3

15. T. Becher, M. Neubert, Phys. Rev. Lett. **102**, 162001 (2009). https://doi.org/10.1103/PhysRevLett.102.162001 [Erratum: Phys. Rev. Lett. 111(19) 199905(2013). https://doi.org/10.1103/PhysRevLett.111.199905]

16. E. Gardi, L. Magnea, JHEP **3**, 079 (2009). https://doi.org/10.1088/1126-6708/2009/03/079

17. V. Ravindran, J. Smith, W.L. van Neerven, Nucl. Phys. B **704**, 332–348 (2005). https://doi.org/10.1016/j.nuclphysb.2004.10.039

18. S. Moch, J.A.M. Vermaseren, A. Vogt, Phys. Lett. B **625**, 245–252 (2005). https://doi.org/10.1016/j.physletb.2005.08.067

19. A. Mitov, G.F. Sterman, I. Sung, Phys. Rev. D **79**, 094015 (2009). https://doi.org/10.1103/PhysRevD.79.094015

20. A.A. Penin, Nucl. Phys. B **734**, 185–202 (2006). https://doi.org/10.1016/j.nuclphysb.2005.11.016

21. A. Mitov, S. Moch, JHEP **5**, 001 (2007). https://doi.org/10.1016/j.nuclphysb.2005.11.016

22. T. Becher, K. Melnikov, JHEP **6**, 084 (2007). https://doi.org/10.1088/1126-6708/2007/06/084

23. T. Becher, M. Neubert, Phys. Rev. D **79**, 125004 (2009). https://doi.org/10.1103/PhysRevD.79.125004 [Erratum: Phys. Rev. D **80**, 109901 (2009)]

24. V. Ravindran, Nucl. Phys. B **746**, 58–76 (2006). https://doi.org/10.1016/j.nuclphysb.2006.04.008

25. V. Ravindran, Nucl. Phys. B **752**, 173–196 (2006). https://doi.org/10.1016/j.nuclphysb.2006.06.025

26. J. Blümlein, P. Marquard, N. Rana, Phys. Rev. D **99**(1), 016013 (2019). https://doi.org/10.1103/PhysRevD.99.016013

27. T. Ahmed, J.M. Henn, M. Steinhauser, JHEP **6**, 125 (2017). https://doi.org/10.1007/JHEP06(2017)125

28. P. Nogueira, J. Comput. Phys. **105**, 279–289 (1993). https://doi.org/10.1006/jcph.1993.1074

29. R. Harlander, T. Seidensticker, M. Steinhauser, Phys. Lett. B **426**, 125–132 (1998). https://doi.org/10.1016/S0370-2693(98)00220-2

30. T. Seidensticker (1999), arXiv:hep-ph/9905298 [hep-ph]

31. J.A.M. Vermaseren, arXiv:math-ph/0010025 [math-ph]

32. M. Tentyukov, J.A.M. Vermaseren, Comput. Phys. Commun. **181**, 1419–1427 (2010). https://doi.org/10.1016/j.cpc.2010.04.009

33. T. van Ritbergen, A.N. Schellekens, J.A.M. Vermaseren, Int. J. Mod. Phys. A **14**, 41–96 (1999). https://doi.org/10.1142/S0217751X99000038

34. K.G. Chetyrkin, F. V. Tkachov, Nucl. Phys. **B192**(159–204) (1981). https://doi.org/10.1016/0550-3213(81)90199-1

35. S. Laporta, Int. J. Mod. Phys. A **15**, 5087–5159 (2000). https://doi.org/10.1016/S0217-751X(00)00215-7

36. P. Marquard, D. Seidel. (unpublished)

37. A.V. Kotikov, Phys. Lett. B **254**, 158–164 (1991). https://doi.org/10.1016/0370-2693(91)90413-K

38. E. Remiddi, Nuovo Cim. A **110**, 1435–1452 (1997)

39. J.M. Henn, Phys. Rev. Lett. **110**, 251601 (2013). https://doi.org/10.1103/PhysRevLett.110.251601

40. J. Ablinger, A. Behring, J. Blümlein, A. De Freitas, A. von Manteuffel, C. Schneider, Comput. Phys. Commun. **202**, 33–112 (2016). https://doi.org/10.1016/j.cpc.2016.01.002

41. J. Blümlein, C. Schneider, Int. J. Mod. Phys. A **33**(17), 1830015 (2018). https://doi.org/10.1142/S0217751X18300156

42. S. Gerhold, Uncoupling systems of linear Ore operator equations. Master's thesis, RISC (J. Kepler University, Linz, 2002)
43. B. Zürcher, Rationale Normalformen von pseudo-linearen Abbildungen. Master's thesis, Mathematik, ETH Zürich (1994)
44. J. Ablinger, PoS (RADCOR2017) 69, arXiv:1801.01039 [cs.SC]
45. E. Remiddi, J.A.M. Vermaseren, Int. J. Mod. Phys. A **15**, 725–754 (2000). https://doi.org/10.1142/S0217751X00000367
46. J. Ablinger, J. Blümlein, C. Schneider, J. Math. Phys. **52**, 102301 (2011). https://doi.org/10.1063/1.3629472
47. J. Blümlein, D.J. Broadhurst, J.A.M. Vermaseren, Comput. Phys. Commun. **181**, 582–625 (2010). https://doi.org/10.1016/j.cpc.2009.11.007
48. J.A.M. Vermaseren, Int. J. Mod. Phys **A14**, 2037–2076 (1999). https://doi.org/10.1142/S0217751X99001032
49. J. Blumlein, S. Kurth, Phys. Rev. D **60**, 014018 (1999). https://doi.org/10.1103/PhysRevD.60.014018
50. J. Ablinger, PoS **LL2014**, 019 (2014)
51. J. Ablinger (2009), arXiv:1011.1176 [math-ph]
52. J. Ablinger (2012-04), arXiv:1305.0687 [math-ph]
53. J. Ablinger, J. Blümlein, C. Schneider, Math. Phys. **54**, 082301 (2013). https://doi.org/10.1063/1.4811117
54. J. Ablinger, J. Blümlein, C.G. Raab, C. Schneider, J. Math. Phys. **55**, 112301 (2014). https://doi.org/10.1063/1.4900836
55. C. Schneider, Sém. Lothar. Combin. **56**, 1 (2007), article B56b
56. C. Schneider, arXiv:1304.4134 [cs.SC]
57. A.V. Smirnov, M.N. Tentyukov, Comput. Phys. Commun. **180**, 735–746 (2009). https://doi.org/10.1016/j.cpc.2008.11.006
58. A.V. Smirnov, V.A. Smirnov, M. Tentyukov, Comput. Phys. Commun. **182**, 790–803 (2011). https://doi.org/10.1016/j.cpc.2010.11.025
59. A.V. Smirnov, Comput. Phys. Commun. **204**, 189–1999 (2016). https://doi.org/10.1016/j.cpc.2016.03.013
60. D.J. Broadhurst, N. Gray, K. Schilcher, Z. Phys, C **52**, 111–122 (1991). https://doi.org/10.1007/BF01412333
61. K. Melnikov, T. van Ritbergen, Nucl. Phys. B **591**, 515–546 (2000). https://doi.org/10.1016/S0550-3213(00)00526-5
62. P. Marquard, L. Mihaila, J.H. Piclum, M. Steinhauser, Nucl. Phys. B **773**, 1–18 (2007). https://doi.org/10.1016/j.nuclphysb.2007.03.010
63. P. Marquard, A.V. Smirnov, V.A. Smirnov, M. Steinhauser, Phys. Rev. Lett. **114**(14), 142002 (2015). https://doi.org/10.1103/PhysRevLett.114.142002
64. P. Marquard, A.V. Smirnov, V.A. Smirnov, M. Steinhauser, D. Wellmann, Phys. Rev. D **94**(7), 074025 (2016). https://doi.org/10.1103/PhysRevD.94.074025
65. P. Marquard, A.V. Smirnov, V.A. Smirnov, M. Steinhauser, Phys. Rev. D **97**(5), 054032 (2018). https://doi.org/10.1103/PhysRevD.97.054032
66. O.V. Tarasov, A.A. Vladimirov, A.Y. Zharkov, Phys. Lett. B **93**, 429–432 (1980). https://doi.org/10.1016/0370-2693(80)90358-5
67. S.A. Larin, J.A.M. Vermaseren, Phys. Lett. B **303**, 334–336 (1993). https://doi.org/10.1016/0370-2693(93)91441-O
68. A. Grozin, J.M. Henn, G.P. Korchemsky, P. Marquard, Phys. Rev. Lett. **114**(6), 062006 (2015). https://doi.org/10.1103/PhysRevLett.114.062006
69. A. Grozin, J.M. Henn, G.P. Korchemsky, P. Marquard, JHEP **1**, 140 (2016). https://doi.org/10.1007/JHEP01(2016)140
70. A.G. Grozin, P. Marquard, J.H. Piclum, M. Steinhauser, Nucl. Phys. B **789**, 277–293 (2008). https://doi.org/10.1016/j.nuclphysb.2007.08.012
71. J.C. Collins, Phys. Rev. D **22**, 1478 (1980). https://doi.org/10.1103/PhysRevD.22.1478

72. J.A.M. Vermaseren, S.A. Larin, T. van Ritbergen, Phys. Lett. B **405**, 327–333 (1997). https://doi.org/10.1016/S0370-2693(97)00660-6
73. J.A. Gracey, Phys. Lett. B **488**, 175–181 (2000). https://doi.org/10.1016/S0370-2693(00)00859-5
74. T. Luthe, A. Maier, P. Marquard, Y. Schröder, JHEP **1**, 081 (2017). https://doi.org/10.1007/JHEP01(2017)081
75. P.A. Baikov, K.G. Chetyrkin, J.H. Kühn, JHEP **4**, 119 (2017). https://doi.org/10.1007/JHEP04(2017)119
76. T. Ahmed, M. Mahakhud, N. Rana, V. Ravindran, Phys. Rev. Lett. **113**(11), 112002 (2014). https://doi.org/10.1103/PhysRevLett.113.112002
77. T. Ahmed, N. Rana, V. Ravindran, JHEP **10**, 139 (2014). https://doi.org/10.1007/JHEP10(2014)139
78. J.C. Collins, *Perturbative Quantim Chromodynamics*, Advanced Series on Directions in High Energy Physics, ed. by A.H. Mueller, vol. 5 (World Scientific, Singapore, 1989), pp. 573–614

Chapter 15
Rediscoveries from the First Data of Belle II

Niharika Rout

Abstract The Belle II experiment at the SuperKEKB asymmetric e^+e^- collider in KEK, Japan, will accumulate a data sample corresponding to an integrated luminosity of 50 ab^{-1}. The high luminosity will probe for new physics beyond the standard model in rare decays and make high-precision measurements of the CKM matrix parameters. The accelerator commissioning of the Belle II experiment, also known as Phase I, was completed in 2016. The detector entered the second commissioning (Phase II) in February 2018, with the first collisions taking place on April 25, 2018. A data sample was collected corresponding to an integrated luminosity of 0.5 fb^{-1} of data. Here, we present the first results in which D and B mesons are reconstructed. We report measurements of D^{*+} and D^{*0} decay modes into various final states containing both charged and neutral particles. In addition, we report the observation of 245 events consistent with B meson decay to final states containing charmed mesons.

15.1 Introduction

The Standard Model (SM) is quite successful in explaining the fundamental particles of nature and their interactions. Despite the tremendous success, there are still a few unanswered questions such as the matter-antimatter asymmetry, mass and flavor hierarchy of the quarks and leptons, existence of too many parameters in SM, etc. Many New Physics (NP) scenarios have been proposed to explain such blind-spots of the SM. One of the approaches to search for NP is making the measurements of the parameters in the flavor sector to see if they deviate from the SM predictions. Belle II is a unique opportunity to constrain and search for the NP at the intensity frontier level.

N. Rout—On behalf of the Belle II Collaboration.

N. Rout (✉)
Indian Institute of Technology, Madras, India
e-mail: niharikarout@physics.iitm.ac.in; niharikarout027@gmail.com

© Springer Nature Singapore Pte Ltd. 2021
P. K. Behera et al. (eds.), *XXIII DAE High Energy Physics Symposium*,
Springer Proceedings in Physics 261,
https://doi.org/10.1007/978-981-33-4408-2_15

The Belle II experiment, a second-generation B-factory experiment, was designed to search for NP using precision measurements and rare decays. It has already started collecting data and will accumulate total integrated luminosity of 50 ab^{-1} by 2027. With this large data set, we can perform precision measurements of CKM parameters, such as ϕ_3 and V_{ub} [1], and search for New Physics (NP), such as CP violation in charm mesons, lepton flavor violations in τ decays, new particles affecting rare flavor-changing neutral current processes and search for light dark matter candidates [2].

15.2 SuperKEKB and Belle II Detector

The SuperKEKB colliding-beam accelerator provides e^+e^- collisions at an energy corresponding to the mass of the $\Upsilon(4S)$ resonance, which are being recorded by the Belle II detector. It is consisting of two storage rings of 3.012 km length each, one for the 7 GeV electrons (High Energy Ring, HER) and one for the 4 GeV positrons (Low Energy Ring, LER). The design peak instantaneous luminosity of SuperKEKB is 8×10^{35} cm^{-2}s^{-1}, approximately forty times higher than what has been achieved at the KEKB accelerator [3]. This will allow a data sample to be accumulated that corresponds to an integrated luminosity of 50 ab^{-1}.

Belle II is the upgraded version of the Belle detector [4]. It has better performance and can tolerate the much higher level of beam-related background that arises from the increase in instantaneous luminosity [5]. The different sub-detectors are shown in Fig. 15.1.

15.2.1 Data Taking and On-Resonance Test

The accelerator commissioning of the Belle II experiment, also known as Phase I, was completed in 2016. The detector entered to the second commissioning period (Phase II) in February 2018, with the first collisions taking place on April 25, 2018.

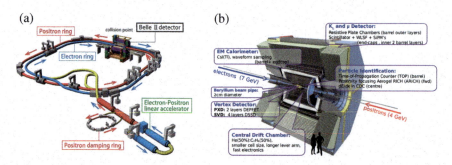

Fig. 15.1 **a** SuperKEKB accelerator and **b** Belle II detector

Fig. 15.2 R_2 distribution from phase II data

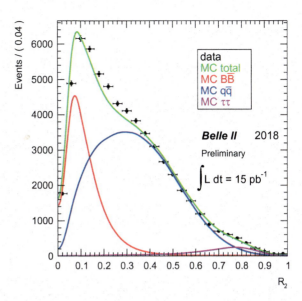

A data sample was collected corresponding to an integrated luminosity of 0.5 fb^{-1}. Only one ladder of each layer of the vertex detector was present during the data taking, which corresponds to 1/8th of the full detector.

A test was performed to validate the $\Upsilon(4S)$ resonance in data using the event-shape variable R_2, which is the ratio of second and zeroth Fox-Wolfram moments [6]. The distribution for phase II data is shown in Fig. 15.2. The value is close to zero for $e^+e^- \rightarrow \Upsilon(4S) \rightarrow B\overline{B}$ events and close to one for $e^+e^- \rightarrow q\overline{q}$ where, $q = u, d, s, c$, also known as continuum events. The distribution agrees reasonably with the MC expectations, hence proving the existence of $B\overline{B}$ pair production in data.

15.3 Rediscoveries of the Charm and B Modes

This section describes $D^{*\pm}$ and D^{*0} reconstruction in phase II data and provides information required to obtain those plots. In the end, the reconstruction of $B^+ \rightarrow D^{(*)}\pi^+$ and $B^+ \rightarrow D^{(*)}\rho$ final state is also discussed; this is the principal control channel for the analysis of $B^+ \rightarrow D^{(*)}K^+$, which determines the unitary triangle angle ϕ_3 [8].

15.3.1 Selection Criteria

We use events that are selected by the High-Level Trigger (HLT) stream and are required to have at least three charged tracks to be inconsistent with a Bhabha $e^+e^- \to e^+e^-$ event. Charged tracks are selected with absolute values of the impact parameters parallel and perpendicular to the beam direction <0.5 cm and 3 cm, respectively. In addition, PID criteria are applied such that the likelihood of the charged track for the pion or kaon hypothesis being >0.6. PID criteria are only applied to the daughters of the D meson. An additional charged track, which is assumed to be a pion, is combined with the D candidate to form a $D^{*\pm}$ candidate and the momentum in the center-of-mass frame of this candidate is required to be more than 2.5 GeV/c to select D^* mesons from $c\bar{c}$ events, which tend to have less combinatorial background than D^* mesons produced in decays of the $\Upsilon(4S)$. The following selection criteria are applied for $D^{*\pm}$ candidates:

- $1.7 < M_{D^0} < 2.1$ GeV/c^2;
- $0.14 < \Delta M < 0.16$ GeV/c^2, where ΔM is defined as $M(D^*) - M(D^0)$, it has a much better resolution and discriminates more effectively between signal D^{*+} and background than the D^{*+} invariant mass;

The B candidates selection criteria are the same as those of the D^* candidates with an additional $R_2 < 0.3$ cut for rejection of the background events coming from the continuum.

15.3.2 Charm Modes

Figures 15.3 and 15.4 show the ΔM and M_D distributions for the Cabibbo-favored final state $K\pi$ coming from $D^{*\pm}$ and D^{*0}, which has branching fraction of $(3.88 \pm 0.05)\%$ [7]. A two-dimensional maximum likelihood fit is performed between ΔM and M_D. There are three components: signal, combinatorial background, and events with correct D but fake D^* due to the presence of a random π. The signal yields for $D^{*\pm}$ and D^{*0} are 1188 ± 37 and 523 ± 31, respectively.

$D \to K^+K^-$ is a CP even final state and also a singly Cabibbo-suppressed mode with a branching fraction is $(3.96 \pm 0.08) \times 10^{-3}$ [7]. A PID cut >0.5 is applied on both the K. $D \to K_S^0 \pi^0$ is a CP odd final state and contains two neutral final state particles; branching fraction for this mode is $(1.19 \pm 0.04)\%$ [7]. Figures 15.5 and 15.6 shows the ΔM and M_D distributions for these two modes. The signal yields for K^+K^- and $K_S^0\pi^0$ from the maximum likelihood fit are found to be 58 ± 9 and 91 ± 11, respectively.

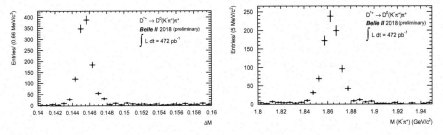

Fig. 15.3 Signal-enhanced projections: ΔM for $1.845 < M(K\pi) < 1.885$ GeV/c^2 (left) and $M(K\pi)$ for $0.144 < \Delta M < 0.146$ GeV/c^2 (right) for the mode $D^{*\pm} \to D^0(K^-\pi^+)\pi^\pm$

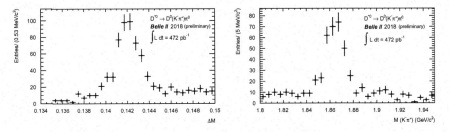

Fig. 15.4 Signal-enhanced projections: ΔM for $1.845 < M(K\pi) < 1.885$ GeV/c^2 (left) and $M(K\pi)$ for $0.144 < \Delta M < 0.146$ GeV/c^2 (right) for the mode $D^{*0} \to D^0(K^-\pi^+)\pi^0$

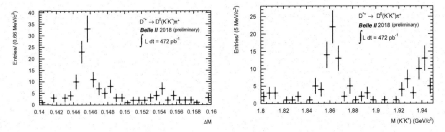

Fig. 15.5 Signal-enhanced projections: ΔM for $1.845 < M(KK) < 1.885$ GeV/c^2 (left) and $M(KK)$ for $0.144 < \Delta M < 0.146$ GeV/c^2 (right) for the mode $D^{*\pm} \to D^0(K^-K^+)\pi^\pm$

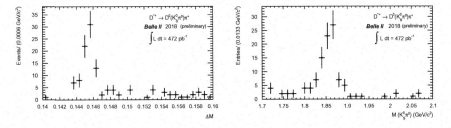

Fig. 15.6 Signal-enhanced projections: ΔM for $1.845 < M(K_S^0\pi^0) < 1.885$ GeV/c^2 (left) and $M(K_S^0\pi^0)$ for $0.144 < \Delta M < 0.146$ GeV/c^2 (right) for the mode $D^{*\pm} \to D^0(K_S^0\pi^0)\pi^\pm$

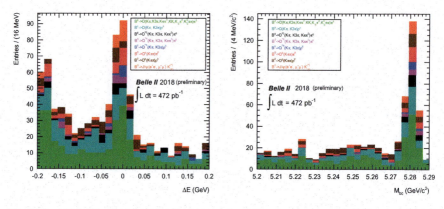

Fig. 15.7 ΔE (left) and M_{bc} (right) distributions for the B candidates selected in the Phase II data

15.3.3 B Rediscovery

A total of 245 B signal candidates were obtained from various hadronic channels, mostly from the reconstruction of $B^{\pm} \to D\pi^{\pm}$ and $B^{\pm} \to D\rho^{\pm}$. The signal-enhanced ΔE and M_{bc} distributions are shown in Fig. 15.7, which are defined like

$$M_{bc} = \sqrt{E_{beam}^2 - (\Sigma \overrightarrow{p_i})^2}, \quad \Delta E = \Sigma E_i - E_{beam}, \qquad (15.1)$$

where E_{beam} is the beam energy in the center-of-mass frame and E_i and $\overrightarrow{p_i}$ are the energy and momenta of B daughter particles in the center-of-mass frame. For signal, the value of M_{bc} peaks at the nominal B meson mass and ΔE at zero. The signal region chosen was $|\Delta E| < 0.05$ GeV and $M_{bc} > 5.27$ GeV/c^2, respectively.

15.4 Summary

The phase II run of Belle II was successful despite the very small data sample collected. We observed various charm modes validating the potential for charm physics at Belle II. Possible charged and neutral B meson candidates are reconstructed from different charmed mesons and a total of 245 signal events are obtained, which is consistent with the yields from the data samples of the ARGUS/CLEO experiment [9]. Many more particles like J/ψ, ϕ, Λ, semi-leptonic B modes are also confirmed in the early data, giving the green light for the rich physics program ahead at Belle II. For example, the aim is to reach a precision level of 1° for the ϕ_3 measurements using the 50 ab^{-1} data sets of Belle II in near future [2].

References

1. M. Kobayashi, T. Maskawa, Progr. Theor. Phys. **49**, 2 (1973)
2. E. Kou, P. Urquijo (eds.), arXiv:1808.10567 [hep-ex]
3. S. Kurokawa, E. Kikutani, Nucl. Instr. Meth. A **499** (2003)
4. A. Abashian et al., Belle collaboration. Nucl. Inst. Methods Phys. Res. A **479**, 117232 (2002)
5. T. Abe et al., Belle II collaboration, arXiv:1011.0352 [physics.ins-det]
6. G.C. Fox, S. Wolfram, Phys. Rev. Lett. **41**, 1581 (1978)
7. M. Tanabashi et al., Particle data group. Phys. Rev. D **98**, 030001 (2018)
8. H. Aihari et al., Belle collaboration. Phys. Rev. D **85**, 112014 (2012)
9. H. Albrecht et al., ARGUS. Phys. Lett. B **192**, 245 (1987)

Chapter 16
Double Parton Scattering Measurements at CMS

Rajat Gupta

Abstract Recent results on double parton scattering (DPS) studies using data collected during Run 1 and Run 2 of the LHC with the CMS experiment are presented. Double parton scattering is investigated in several final states including vector bosons and multi-jets. Measurements of observables designed to highlight the DPS contribution are shown and compared to MC predictions from models based on multiple partonic interactions (MPIs) phenomenology.

16.1 Introduction

Production of particles in a hadron-hadron collision involves parton-parton scatterings, initial-state radiation (ISR), final-state radiation (FSR), and beam-beam remnants (BBR) interactions. The large parton densities available in the proton-proton (pp) collisions at the CERN LHC result in a significant probability of more than one parton-parton scattering in the same pp collision, a phenomenon known as multiple parton interactions (MPIs) [1]. In general, MPI produces mostly low p_T particles, and there is small probability of the production of high p_T particles from MPI. Double parton scattering (DPS) corresponds to events where two hard parton-parton interactions occur in a single proton-proton collisions.

16.2 DPS Measurements at CMS

The study of DPS processes provides valuable information on the transverse distribution of partons in the proton [2] and on the parton correlations in the hadronic wave function [3, 4]. Under the assumption of transverse and longitudinal factorization

R. Gupta—On behalf of the CMS collaboration.

R. Gupta (✉)
Panjab University, Chandigarh, India
e-mail: rajat.gupta@cern.ch; rajatgupta116@gmail.com

© Springer Nature Singapore Pte Ltd. 2021
P. K. Behera et al. (eds.), *XXIII DAE High Energy Physics Symposium*,
Springer Proceedings in Physics 261,
https://doi.org/10.1007/978-981-33-4408-2_16

of the two single parton interactions, the cross section of a double parton scattering (DPS) process can be written as

$$\sigma_{AB}^{DPS} = \frac{n}{2} \frac{\sigma_A \times \sigma_B}{\sigma_{\text{eff}}} \tag{16.1}$$

where A and B denote the single parton scattering (SPS) processes, and σ_A and σ_B their respective SPS cross sections. The factor 'n' is unity if processes A and B are the same, and $n = 2$ if $A \neq B$. The parameter σ_{eff} is related to the extent of the parton distribution in the plane orthogonal to the direction of motion of the protons.

The production of same-sign ww production via DPS, from pp collisions of $\sqrt{s} = 8$ [5] and 13 TeV [6] at integrated luminosity of 19.7fb^{-1} and 35.9fb^{-1} respectively, is studied. A multivariate analysis has been performed in order to enhance the signal sensitivity; a limit on the DPS yield, along with corresponding σ_{eff}, has been estimated.

The first step is involved in selecting an inclusive region of phase space with minimal cuts for trigger and some QCD suppression. After the event selection, the remaining background contributions include WZ production, backgrounds in which one of the two leptons is fake, as well as minor contributions of opposite-sign di-lepton events in which the charge of the electron is mismeasured, and rare processes such as tri-boson production, or ZZ production. Since the most important background is WZ production in which both bosons decay leptonically and one of the leptons from the Z boson is subsequently out of acceptance or not reconstructed, therefore, a multivariate discriminator is trained with a boosted decision tree (BDT) algorithm [7] in order to optimize the discrimination between the signal process and the WZ process.

16.2.1 Constraints on the Double Parton Scattering Cross Section from Same-Sign W Boson Pair Production in Proton-Proton Collisions at $\sqrt{s} = 8$ TeV

A first search for same-sign W boson pair production via DPS in pp collisions at a center-of-mass energy of 8 TeV is performed. The results presented are based on the analysis of events containing two same-sign W bosons decaying into either same-sign muon-muon or electron-muon pairs. The analyzed data were collected by the CMS detector at the LHC during 2012 and correspond to an integrated luminosity of 19.7 fb^{-1}. Table 16.1 shows the list of same-sign WW selection criteria chosen to reduce various background processes. The majority of background events originate from processes in which one or both of the leptons, coming from leptonic decays of heavy quarks or in-flight decays of light mesons, pass the event selection criteria.

Thirteen input variables were selected, which exhibit differences in the signal and the WZ background. Overall, the data and simulation are found to be consistent within the uncertainties for all input variables. The BDT discriminant after the full

Table 16.1 Event selection criteria for same-sign W boson pair production in dimuon and electron-muon channels [5]

Dimuon channel	Electron-muon channel
Pair of same-sign leptons	
Leading lepton $p_T > 20$ GeV	
Subleading lepton $p_T > 10$ GeV	
No third isolated and identified lepton with $p_T > 10$ GeV	
$p_t^{\text{miss}} > 20$ GeV	
$m_{ll} > 20$ GeV	
$m_{ll} \notin [75, 105]$ GeV	–
$\lvert p_{T_{\mu_1}} \rvert + \lvert p_{T_{\mu_2}} \rvert > 45$ GeV	–
	No b-tagged jet with $p_T > 30$ GeV and $\lvert \eta \rvert < 2.1$

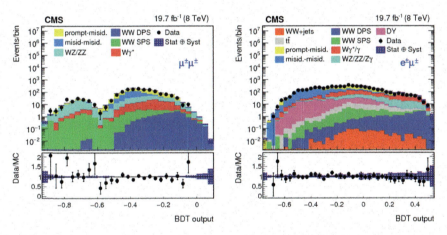

Fig. 16.1 Distribution of the BDT discriminant, for the dimuon channel (left) and for the electron-muon channel (right). The data are represented by the black dots and the shaded histograms represent the pre-fit signal and post-fit background processes. The bottom panels show the ratio of data to the sum of all signal and background contributions [5]

event selection has been applied, is used to extract the limits on the DPS cross section and σ_{eff} using statistical analysis techniques.

The expected and observed upper limits at 95% confidence level (CL) on the cross section for inclusive same-sign WW production via DPS have been extracted. Figure 16.1 shows the distributions of the BDT discriminant having post-fit contributions for the backgrounds and pre-fit ones for the signal, for the dimuon and electron-muon final states with the corresponding uncertainty bands (shown as hatched bands). The expected and observed 95% CL limits on the cross section for same-sign WW production via DPS ($\sigma_{W^{\pm}W^{\pm}}^{\text{DPS}}$) are summarized in Table 16.2.

Table 16.2 Expected and observed 95% CL limits on the cross section for inclusive same-sign WW production via DPS for the dimuon and electron-muon channels along with their combination [5]

95% CL	Dimuon	Electron-muon	Combined
Expected	0.67 pb	0.78 pb	0.48 pb
Expected $\pm 1\sigma$	[0.46, 1.00] pb	[0.52, 1.16] pb	[0.33, 0.72] pb
Expected $\pm 2\sigma$	[0.34, 1.45] pb	[0.37, 1.71] pb	[0.24, 1.04] pb
Observed	0.72 pb	0.64 pb	0.32 pb

Assuming the two scatterings to be independent, a limit can be placed on σ_{eff}. A lower 95% CL limit on σ_{eff} can be calculated as

$$\sigma_{\text{eff}} > \frac{\sigma_{W^+}^2 + \sigma_{W^-}^2}{2\,\sigma_{W^\pm W^\pm}^{\text{DPS}}} = 12.2 \text{mb}.$$

The obtained lower limit on σ_{eff} is compatible with the values of $\sigma_{\text{eff}} \approx 10\text{--}20 \text{mb}$ obtained from measurements at different center-of-mass energies using a variety of processes [8].

16.2.2 Measurement of Double Parton Scattering in Same-Sign WW Production in p-p Collisions at $\sqrt{s} = 13$ TeV with the CMS Experiment

Same-sign $W^\pm W^\pm$ production in which the bosons originate from two distinct parton-parton interactions within the same proton-proton collision is studied in the $\mu^\pm \mu\pm$ and $e^\pm \mu\pm$ final states. A data set of 35.9 fb^{-1} of proton-proton collisions at $\sqrt{s} = 13$ TeV, recorded with the CMS detector at the LHC in 2016, is used. The summary of all kinematic selection imposed at 13 TeV is given in Table 16.3.

As compared to 8 TeV DPS analysis, three new variables were included at 13 TeV measurement, namely the product of the two lepton-η's, the absolute sum of two lepton-η's, as well as the M_{T2}^{ll} of the two-lepton system and the E_T^{miss}.

Table 16.3 Event selection criteria for same-sign W boson pair production in dimuon and electron-muon channels [6]

Two leptons $e^\pm \mu^\pm$ or $\mu^\pm \mu^\pm$
$p_{T1\,2} > 25\ 20$ GeV
$|\eta_e| < 2.5,\ |\eta_\mu| < 2.4$
MET > 15 GeV
nj < 2
nb == 0
Veto on additional leptons
Veto on hadronic τ leptons

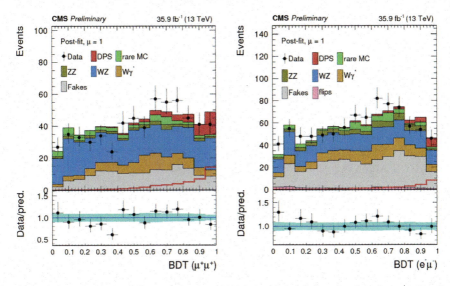

Fig. 16.2 Final BDT classifier output with all background estimations in place for $\mu^+\mu^-$ and $e^-\mu^-$ channel. Observed data are shown in black markers with the signal pre-fit expectation as a red histogram and separately imposed as a red line to show the behavior of the signal in the BDT classifier [6]

Table 16.4 Results obtained from a constrained fit to the BDT classifier [6]

	Expected	Observed
σ_{DPSWW}^{pythia}	1.64 pb	$1.09^{+0.50}_{-0.49}$ pb
$\sigma_{DPSWW}^{factorized}$	0.87 pb	
Significance for σ_{DPSWW}^{pythia}	3.27 σ	2.23 σ
Significance for $\sigma_{DPSWW}^{factorized}$	1.81 σ	
UL in the absence of signal	<0.97 pb	<1.94 pb

Figure 16.2 shows the distribution of the BDT classifier in one of the $\mu^+\mu+$ and $e^-\mu^-$ channel. Overall good agreement between the background predictions is observed in the low-BDT classifier region. Indicative from Fig. 16.2, and evident from Table 16.4, the observed yield of the DPS WW signal process is lower than the expectation. Therefore, although from the PYTHIA8 cross section of 1.64 pb a significance of 3.27 is expected, the measured cross section is below that value at $1.09^{+0.50}_{-0.49}$ pb with a significance of 2.23 σ. Conversely, applying the factorization approach with an expected cross section of 0.87 pb, and an expected significance of 1.81 σ results in a larger than expected cross section. The upper limit on the cross section in the absence of signal is expected to be <0.97 pb and measured to be <1.94 pb.

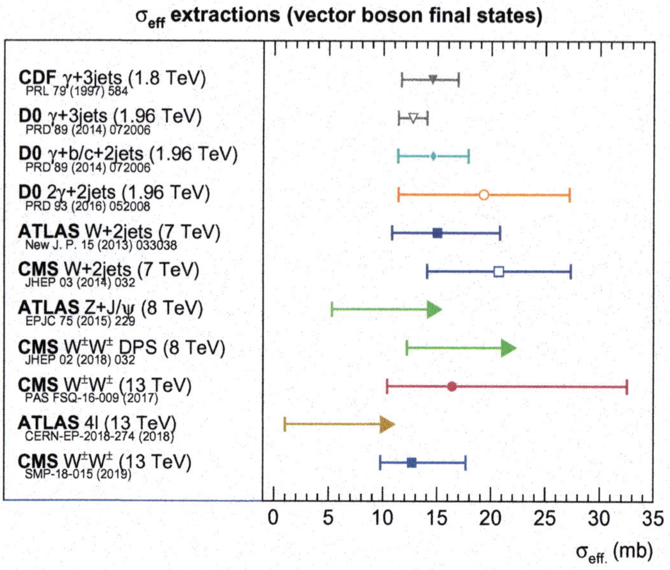

Fig. 16.3 The effective cross section of DPS measured at various energies and final states by different experiments

16.3 Summary

DPS measurements using same-sign WW process at 8 and 13 TeV are presented. DPS studies are important for better understanding of new physics searches and partonic structure of hadrons. In Fig. 16.3, the CMS results for effective sigma at 8 and 13 TeV are compared to measurements done at different energies and final states by various experiments. Generally, all measurements of σ_{eff} are consistent between each other.

References

1. T. Sjostrand, M. van Zijl, A multiple interaction model for the event structure in Hadron collisions. Phys. Rev. D **36**, 2019 (1987). https://doi.org/10.1103/PhysRevD.36.2019
2. M. Diehl, D. Ostermeier, A. Schafer, Elements of a theory for multiparton interactions in QCD. JHEP **1203**, 089 (2012). Erratum: [JHEP **1603**, 001 (2016)]. https://doi.org/10.1007/JHEP03(2012)089, https://doi.org/10.1007/JHEP03(2016)001
3. D. Treleani, G. Calucci, Inclusive and exclusive cross-sections, sum rules. Adv. Ser. Direct. High Energy Phys. **29**, 29 (2018). https://doi.org/10.1142/9789813227767_0003, arXiv:1707.00271 [hep-ph]
4. M. Rinaldi, S. Scopetta, V. Vento, Double parton correlations in constituent quark models. Phys. Rev. D **87**, 114021 (2013). https://doi.org/10.1103/PhysRevD.87.114021, arXiv:1302.6462 [hep-ph]

5. A.M. Sirunyan et al., [CMS Collaboration], Constraints on the double-parton scattering cross section from same-sign W boson pair production in proton-proton collisions at $\sqrt{s} = 8$ TeV. JHEP **1802**, 032 (2018). https://doi.org/10.1007/JHEP02(2018)032, arXiv:1712.02280 [hep-ex]
6. CMS Collaboration, Measurement of double parton scattering in same-sign WW production in p-p collisions at $\sqrt{s} = 13$ TeV with the CMS experiment, CMS-PAS-FSQ-16-009 (2017)
7. A. Hocker et al., TMVA—Toolkit for multivariate data analysis. physics/0703039 [physics.data-an]
8. S. Chatrchyan et al., [CMS Collaboration], Study of double parton scattering using W + 2-jet events in proton-proton collisions at $\sqrt{s} = 7$ TeV. JHEP **1403**, 032 (2014). https://doi.org/10.1007/JHEP03(2014)032, arXiv:1312.5729 [hep-ex]

Chapter 17
Search for $Y(4260)$ in $B \rightarrow Y(4260)K$ Decay Mode at Belle

Renu Garg, Vishal Bhardwaj, and J. B. Singh

Abstract $Y(4260)$ is an exotic charmonium-like state with 4230 ± 8 MeV/c^2 mass and 55 ± 19 MeV width. The $Y(4260)'$s decay to $J/\psi\pi\pi$ suggests, it to be a charmonium ($c\bar{c}$) meson. But its mass is not consistent with any of 1^{--} $c\bar{c}$ state. Several models have been proposed to explain the nature of $Y(4260)$ including $c\bar{c}g$ hybrid model, tetraquark, D_1D, D^0D^* molecule, $J/\psi f_0(980)$ molecule, and so on. This state is so interesting that a charged state $Z_c(3900)$ is observed in its decay mode and $Z_c(3900)$ is a tetraquark state. Some recent studies suggest $Y(4260)$ be an admixture of tetraquark and charmonium state. It has been suggested that the structure of $Y(4260)$ can be estimated if one measured branching fraction of $B \rightarrow Y(4260)K$. Till now, this state has been only produced by ISR or e^+e^- annihilation. We search for $B \rightarrow Y(4260)K$, where $Y(4260) \rightarrow J/\psi\pi\pi$ decay mode using the full $\Upsilon(4S)$ data collected by the Belle detector at the asymmetric KEKB e^+e^- collider.

17.1 Introduction

The $Y(4260)$ state was first observed in the initial state radiation (ISR) process $e^+e^- \rightarrow \gamma_{ISR}J/\psi\pi^+\pi^-$ by the BABAR collaboration [1]. It has been confirmed by the Belle [2] and CLEO [3] collaborations in the same process. J^{PC} of $Y(4260)$ is expected to be 1^{--} as it is produced in ISR and its decay to J/ψ modes indicate the presence of $c\bar{c}$ in its contents. However, its mass and properties are not consistent with any of the $c\bar{c}$ states in the charmonium spectrum as low lying $\psi(3S)$, $\psi(2D)$ and $\psi(4S)$ $c\bar{c}$ states have been assigned to well established states $\psi(4040)$, $\psi(4160)$,

R. Garg (✉) · J. B. Singh
Department of Physics, Panjab University, Chandigarh, India
e-mail: renu92garg@gmail.com

J. B. Singh
e-mail: singhjb@pu.ac.in

V. Bhardwaj
Department of Physical Sciences, IISER, Mohali, India
e-mail: vishstar@gmail.com

© Springer Nature Singapore Pte Ltd. 2021
P. K. Behera et al. (eds.), *XXIII DAE High Energy Physics Symposium*,
Springer Proceedings in Physics 261,
https://doi.org/10.1007/978-981-33-4408-2_17

and $\psi(4415)$, respectively, and $\psi(3D)$ has a higher mass (4.520 GeV/c^2) [4]. This results in difficulty to assign $Y(4260)$ as one of the conventional states. Several models have been proposed to explain the nature of $Y(4260)$ including tetraquark [5], hybrid [6], molecule [7, 8] or even charmonium baryonium [9]. Observation of charged charmonium candidate $Z_c(3900)^{\pm}$ by BESIII [10] and Belle [11] collaborations in the $J/\psi\pi^{\pm}$ invariant mass spectrum of $Y(4260) \rightarrow J/\psi\pi^+\pi^-$ decay provides the strong evidence for the $Y(4260)$ being a exotic state.

Measurement of the branching fraction $\mathcal{B}(B \rightarrow Y(4260)K)$ and its decay property can help us to understand the structure of $Y(4260)$. It has been suggested [12] on the basis of QCD sum rules that branching fractions of $\mathcal{B}(B \rightarrow Y(4260)K)$, $\mathcal{B}(Y(4260) \rightarrow J/\psi\pi\pi)$ to be in the range $3.0 \times 10^{-8} - 1.8 \times 10^{-6}$. Till now, only the BABAR collaboration has provided the limit [13] on $\mathcal{B}(B^- \rightarrow Y(4260)K^-)$ with statistical significance of 3.1σ, $1.2 \times 10^{-5} < \mathcal{B}(B^- \rightarrow K^-Y(4260)) \times \mathcal{B}(Y(4260) \rightarrow J/\psi\pi\pi) < 2.9 \times 10^{-5}$ based on $211 fb^{-1}$ data that contains $(232 \pm 3) \times 10^6 B\bar{B}$ pairs. Due to limited statistics, it is not sufficient to conclude. Our aim is to provide precise measurements of the branching fraction. Recently, BESIII [14] observed two resonances in a fit to the cross section of $e^+e^- \rightarrow J/\psi\pi^+\pi^-$ process, one at $Y(4260)$ resonance and other at $Y(4360)$ resonance. $Y(4360)$ has not been confirmed yet. In the present analysis, we assume $Y(4260)$ to be a single resonance as measured by Belle and BABAR.

We report on the MC study for $B \rightarrow Y(4260)K$ decay. Monte Carlo (MC) sample for each decay mode is generated using EvtGen [15] and radiative effects are taken into account using PHOTOS [16]. Detector response is added by detector simulation software based on GEANT3.4 [17] software tool.

17.2 Particle Selection and Reconstruction

The charged tracks like kaons, pions, and protons are required to originate from the interaction point (IP). The closest approach w.r.t IP is required to be within 3.5 cm in the beam direction (z) and 1.0 cm in the transverse plane (xy-plane). Charged kaon and pion selections are based on the information from aerogel Cerenkov counters (number of Cherenkov photons), time-of-flight, and central drift chamber (dE/dx measurement) detectors.

The J/ψ is reconstructed via its decay mode $J/\psi \rightarrow \ell^+\ell^-$, where ℓ stands for e or μ. There is a loss of energy from a electron in the form of emission of bremsstrahlung photons. In $J/\psi \rightarrow e^+e^-$, the four momenta of the photons within 0.05 radian of e^+ or e^- direction are included in the invariant mass calculation [hereinafter denoted as $e^+e^-(\gamma)$]. The invariant mass of the J/ψ is required to be within 3.05 GeV/$c^2 \leq M_{ee(\gamma)} \leq 3.13$ GeV/c^2 or 3.07 GeV/$c^2 \leq M_{\mu\mu} \leq 3.13$ GeV/c^2 as shown in Fig. 17.1. The asymmetric interval is taken for $e^+e^-(\gamma)$ to include the radiative tail. The vertex- and mass-constrained fit is performed to the selected J/ψ candidates.

ψ', $X(3872)$, and $Y(4260)$ candidates are formed by combining the selected J/ψ candidate with a $\pi^+\pi^-$ pair. The invariant mass of ψ', $X(3872)$, and $Y(4260)$ is

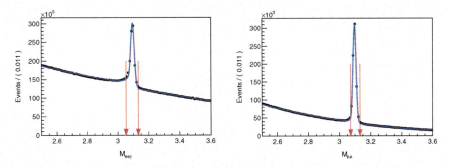

Fig. 17.1 Fit to $M_{\ell\ell}$ invariant mass of J/ψ, $J/\psi \rightarrow e^+e^-$ [left] and $J/\psi \rightarrow \mu^+\mu^-$ [right]

Fig. 17.2 Comparison of ψ', $X(3872)$, and $Y(4260)$ signal

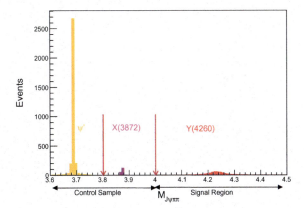

required to be in the range $3.67 \text{ GeV}/c^2 \leq M_{J/\psi\pi\pi} \leq 3.70 \text{ GeV}/c^2$, $3.835 \text{ GeV}/c^2 \leq M_{J/\psi\pi\pi} \leq 3.910 \text{ GeV}/c^2$, and $4.0 \text{ GeV}/c^2 \leq M_{J/\psi\pi\pi} \leq 4.6 \text{ GeV}/c^2$, respectively. Then, finally B candidates are formed by combining ψ', $X(3872)$ and $Y(4260)$ candidate with K candidates. As $B \rightarrow \psi'K$ and $B \rightarrow X(3872)K$ decay modes have the same topology and are well established, we used them as our control sample to validate and calibrate our MC simulations. Comparison of ψ', $X(3872)$, and $Y(4260)$ signal is shown in Fig. 17.2.

We use two kinematical variables to identify the B meson: the beam constrained mass $(M_{\text{bc}} = \sqrt{E_{\text{beam}}^2 - \sum_i p_i^{*2}})$ and the energy difference $(\Delta E = \sum_i E_i^* - E_{\text{beam}})$. Here, E_{beam} is the beam energy in the center of mass (CM) frame and p_i^* (E_i^*) is the momentum (energy) of the ith particle in the CM frame of the $\Upsilon(4S)$. There are multiple reconstructed B candidates in an event due to wrong combination of selected particles. We have to select best B meson in an event which has least χ^2

$$\chi^2 = \chi_{\text{vtx}}^2 + \left(\frac{M_{J/\psi} - m_{PDG}^{J/\psi}}{\sigma_{J/\psi}}\right)^2 + \left(\frac{M_{\text{bc}} - m_B^{PDG}}{\sigma_{M_{\text{bc}}}}\right)^2 \qquad (17.1)$$

where $\sigma_{J/\psi}$ and $\sigma_{M_{bc}}$ represent the $M_{J/\psi}$ and M_{bc} resolutions, respectively, and are taken to be 9.835 MeV/c^2 and 2.59 MeV/c^2, respectively, from a fit to the $B \to \psi' K$ events. χ^2_{vtx} represents vertex fit for all the charged particles. This procedure to select most probable B candidate is called the best candidate selection. The best candidate selection chooses true candidate 76% of the time for $B^+ \to Y(4260)K^+$.

17.3 Background Study

Continuum events $e^+e^- \to q\bar{q}$ (where $q = u, d, s$ or c) are suppressed by requiring $R_2 = H_2/H_0 < 0.5$, where R_2 is the ratio of the second- to zeroth-order Fox-Wolfram moments [18]. The main background contribution is expected to be arise from inclusive B decays to J/ψ. With the above selection criteria, we didn't find any peaking structure in the signal region for $B \to \psi' K$, $B \to X(3872)K$, and $B \to Y(4260)K$ decay modes. We expect negligible contribution from J/ψ mass sidebands (2.54 GeV/c^2 < $M_{J/\psi}$ < 2.72 GeV/c^2 and 3.32 GeV/c^2 < $M_{J/\psi}$ < 3.5 GeV/c^2).

17.4 Signal Extraction

We performed unbinned extended maximum likelihood (UML) fit to the ΔE variable for each mode and get the background subtracted $_s\mathcal{P}$lot [19] distribution of $M_{J/\psi\pi\pi}$. The likelihood function used is

$$\mathcal{L}(N_S, N_B) = \frac{e^{-(N_S+N_B)}}{N!} \prod_{i=1}^{N} (N_S \times P_S + N_B \times P_B) \tag{17.2}$$

where N is the total number of events. N_S and N_B are the signal events and background events, respectively. P_S (P_B) is the signal (background) probability density function (PDF) model. Signal PDF (P_S) for ΔE is modeled by a sum of two Gaussians and bifurcated Gaussian for the $B \to \psi' K$ and $B \to X(3872)K$ decay mode. Background PDF (P_B) is modeled by the first order polynomial for $\psi' K$, while second order polynomial is used for $X(3872)K$. We extract the signal yield from a UML fit to the $_s\mathcal{P}$lot distribution of $M_{J/\psi\pi\pi}$. Here also, the PDF comprises of signal (P_S) and a flat background (P_B). P_S for $M_{J/\psi\pi\pi}$ is a sum of two Gaussians for $\psi' K$, and a sum of two Gaussians plus bifurcated Gaussian for $X(3872)K$. P_B for $M_{J/\psi\pi\pi}$ is second order polynomial for $\psi' K$, while the first order polynomial is used for $X(3872)K$. The mean and width of the core Gaussian are varied and remaining parameters are fixed according to the MC. The results of the fit for the control samples $B \to \psi' K$ and $B \to X(3872)K$ are shown in Fig. 17.3 and Fig. 17.4, respectively.

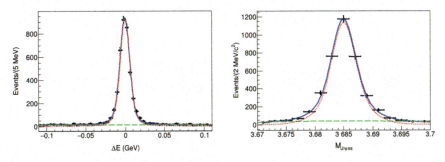

Fig. 17.3 Fit to the ΔE and $_sPlot$ of $M_{J/\psi\pi\pi}$ distributions for $B^+ \rightarrow \psi(2S)(\rightarrow J/\psi\pi^+\pi^-)K^+$

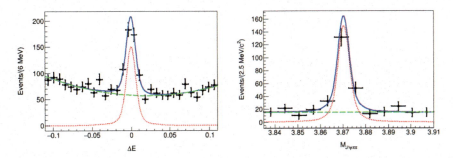

Fig. 17.4 Fit to the ΔE and $_sPlot$ of $M_{J/\psi\pi\pi}$ distributions for $B^+ \rightarrow X(3872)(\rightarrow J/\psi\pi^+\pi^-)K^+$

17.5   Branching Fraction

We determine the branching fraction, $\mathcal{B}(B \rightarrow \psi'K)$ and $\mathcal{B}(B \rightarrow X(3872)K) \times \mathcal{B}(X(3872) \rightarrow J/\psi\pi\pi)$ via relation

$$\mathcal{B} = \frac{N_{\text{event}}}{N_{B\bar{B}} \times \epsilon \times \mathcal{B}_{\text{secondary}}} \quad (17.3)$$

where N_{event} is the number of events for a particular mode, $N_{B\bar{B}} = (772 \pm 11) \times 10^6$ is the number of $B\bar{B}$ events in the data, $\mathcal{B}_{\text{secondary}}$ is the secondary branching fractions ($\mathcal{B}(\psi' \rightarrow J/\psi\pi\pi) = 0.3449 \pm 0.0030$ [20] and $\mathcal{B}(J/\psi \rightarrow \ell\ell) = 0.119 \pm 0.001$ [20]) based on the mode, ϵ is the efficiency estimated from the signal MC after MC/data correction. Results are summarized in Table 17.1 and agrees well with previous results [21].

Table 17.1 Summary of the reconstruction efficiency (ϵ), signal yield (N_S), branching fraction ($\mathcal{B}$) measured and PDG branching fraction ($\mathcal{B}_{PDG}$) for the $B \to \psi(2S)K$ and $B \to X(3872)K$, $X(3872) \to J/\psi\pi^+\pi^-$ decays

Decay	$B^+ \to \psi(2S)K^+$	$B^+ \to X(3872)K^+$, $X(3872) \to J/\psi\pi^+\pi^-$
ϵ (%)	16.8	22.2
N_S	3481 ± 95	185 ± 13
$\mathcal{B}$	$(6.54 \pm 0.18) \times 10^{-4}$	$(9.07 \pm 0.64) \times 10^{-6}$
$\mathcal{B}_{PDG}$	$(6.21 \pm 0.23) \times 10^{-4}$	$(8.6 \pm 0.8) \times 10^{-6}$

17.6 Summary

In summary, a search for $B \to Y(4260)K$ is crucial to understand the structure of $Y(4260)$. This study is performed using $B\bar{B}$ pairs collected at $\Upsilon(4S)$ resonance by the Belle at KEKB. The branching fractions obtained for $B \to \psi'K$ and $B \to X(3872)K$ are $(6.54 \pm 0.18) \times 10^{-4}$ and $(9.07 \pm 0.64) \times 10^{-6}$, respectively, and found to be consistent with the previous results [21]. Final results for $B \to Y(4260)K$ are obtained and published [22].

References

1. B. Aubert et al., BaBar collaboration. Phys. Rev. Lett. **95**, 142001 (2005)
2. C.Z. Yuan et al., Belle collaboration. Phys. Rev. Lett. **99**, 182004 (2007)
3. Q. He et al., CLEO collaboration. Phys. Rev. D **74**, 091104(R) (2006)
4. S. Godfrey, N. Isgur, Phys. Rev. D **32**, 189 (1985)
5. L. Maiani et al., Phys. Rev. D **72**, 031502 (2005)
6. S.L. Zhu et al., Phys. Lett. B **625**, 212 (2005)
7. X. Liu, X.Q. Zeng, X.Q. Li et al., Phys. Rev. D **72**, 054023 (2005)
8. G.J. Ding et al., Phys. Rev. D **79**, 014001 (2009)
9. C.F. Qiao et al., Phys. Lett. B **639**, 263 (2006)
10. M. Ablikim et al., BESIII collaboration. Phys. Rev. Lett. **110**, 252001 (2013)
11. Z.Q. Liu et al., Belle collaboration. Phys. Rev. Lett. **110**, 252002 (2013)
12. R.M. Albuquerque, M. Nielsen, C.M. Zanetti et al., Phys. Lett. B **747**, 83 (2015)
13. B. Aubert et al., BaBar collaboration. Phys. Rev. D **73**, 011101(R) (2006)
14. M. Ablikim et al., BESIII collaboration, Phys. Rev. Lett. **118**, 092002 (2017) and following articles up to 03A011
15. D. J. Lange, Nucl. Instrum. Methods, Phys. Res., Sect. A **462**, 152 (2001)
16. E. Barberio, Z. Wąs, Comput. Phys. Commun. **79**, 291 (1994); P. Golonka and Z. Wąs, Eur. Phys. J. C **45**, 97 (2006); **50**, 53 (2007)
17. R. Brunet al., GEANT3.21, CERN Report No. DD/EE/84-1 (1984)
18. G.C. Fox, S. Wolfram, Phys. Rev. Lett. **41**, 1581 (1978)
19. M. Pivk, F. R. Le Diberder, Nucl. Instrum. Methods, Phys. Res. Sect. A **555**, 356 (2005)
20. M. Tanabashi et al., Particle data group. Phys. Rev. D **98**, 030001 (2018)
21. S.-K. Choi et al., Belle collaboration. Phys. Rev. D **84**, 052004(R) (2011)
22. R. Garg et al., Belle collaboration. Phys. Rev. D **99**, 071102(R) (2019)

Chapter 18
Measurement of CKM Angle ϕ_3 Using $B^\pm \to D(K^0_S \pi^+ \pi^- \pi^0)K^\pm$ Decays at Belle

P. K. Resmi, James F. Libby, and K. Trabelsi

Abstract The CKM angle ϕ_3 can be determined in a theoretically clean way as it is accessible via the tree-level decays, $B^\pm \to DK^\pm$. The current uncertainty on ϕ_3 is significantly larger than that of the standard model (SM) prediction. A more precise measurement of ϕ_3 is crucial for testing the SM description of CP violation and probing for new physics effects. The statistical uncertainty can be reduced if information from additional D meson final states is included, which in practice means new three and four-body decay modes. Here, we measure ϕ_3 with $B^\pm \to D(K^0_S \pi^+ \pi^- \pi^0)K^\pm$ decays using a data sample corresponding to an integrated luminosity of 711 fb^{-1} collected with the Belle detector at KEKB asymmetric e^+e^- collider. This four-body D final state has a large branching fraction of 5.2% and the phase space is rich with different resonance substructures. We adopt a model-independent method to estimate ϕ_3 by studying various regions of the phase space. The D decay strong-phase information is obtained from the quantum-correlated D meson pairs produced at CLEO-c.

18.1 Introduction

The current best measurement of the CKM [1, 2] angle ϕ_3, combining all the results from different experiments, is $(73.5^{+4.2}_{-5.1})°$ [3]. This large uncertainty is due to the small branching fractions of the decays sensitive to ϕ_3. The value of ϕ_3 estimated indirectly from other parameters of the unitarity triangle is $(65.3^{+1.0}_{-2.5})°$ [3]. Any disagreement between these results could imply that there is new physics beyond the standard model (SM). But a comparison would be meaningful only if the associated uncertainties are

P. K. Resmi, J. F. Libby, and K. Trabelsi on behalf of Belle Collaboration.

P. K. Resmi (✉) · J. F. Libby
Indian Institute of Technology Madras, Chennai, India
e-mail: resmipk@physics.iitm.ac.in

K. Trabelsi
Laboratoire de L'accélérateur Linéaire, Orsay, France

© Springer Nature Singapore Pte Ltd. 2021
P. K. Behera et al. (eds.), *XXIII DAE High Energy Physics Symposium*,
Springer Proceedings in Physics 261,
https://doi.org/10.1007/978-981-33-4408-2_18

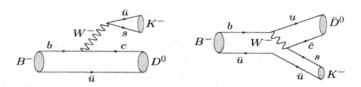

Fig. 18.1 Color-favored (left) and color-suppressed (right) $B^- \to DK^-$ processes

comparable. Thus, an improved measurement of ϕ_3 is essential for testing the SM description of CP violation. The color-favored $B^- \to D^0 K^-$ and color-suppressed $B^- \to \overline{D^0} K^-$ decays, where D indicates a neutral charm meson reconstructed in a final state common to both D^0 and $\overline{D^0}$, provide CP-violating observables, that are sensitive to ϕ_3. Here and elsewhere in this paper, charge conjugation of final states is implied unless explicitly stated otherwise. The Feynman diagrams are shown in Fig. 18.1. These are tree-level decays and hence the theoretical uncertainty is negligible $(\mathcal{O}(10^{-7}))$ [4].

If the amplitude for the color-favored decay is $A_{\text{fav}} = A$, then the color-suppressed one can be written as $A_{\text{sup}} = A r_B e^{i(\delta_B - \phi_3)}$, where δ_B is the strong-phase difference between the decay processes, and

$$r_B = \frac{|A_{\text{sup}}|}{|A_{\text{fav}}|}. \tag{18.1}$$

The statistical uncertainty on ϕ_3 is proportional to r_B. For $B^+ \to DK^+$ decays, $r_B \sim 0.1$, whereas for $B^+ \to D\pi^+$, it is 0.005. Though $B^+ \to D\pi^+$ decays are not very sensitive to r_B and ϕ_3, they serve as excellent control sample modes for signal extraction procedure in $B^+ \to DK^+$ due to their similar kinematics. This also helps in determining the cross-feed background due to the misidentification of kaons and pions from data.

The limitations on the current ϕ_3 measurements due to statistical precision can be reduced by exploring more and more D final states. Here, we study the four-body self-conjugate state, $D \to K_S^0 \pi^+ \pi^- \pi^0$. This decay mode has a branching fraction of 5.2% [5], which is almost twice that of $D \to K_S^0 \pi^+ \pi^-$, the dominant multi-body D final state used to determine ϕ_3 [6, 7]. This decay proceeds via interesting resonance substructures like $K_S^0 \omega$, $K^* \rho$, etc., thus facilitating a model-independent extraction of ϕ_3 by studying the D phase space regions. We present the expected results from $B^+ \to DK^+$ decays by analyzing simulated samples and preliminary results obtained from the $B^+ \to D\pi^+$ data sample, i.e., the calibration mode.

18.2 Formalism to Measure ϕ_3 Sensitive Parameters

The methods to determine ϕ_3 vary according to the D meson final state under consideration. When it is a multibody self-conjugate state, there are two methods: model-dependent and model-independent. In the model-dependent method, the D amplitudes are fitted to a model corresponding to the intermediate resonances. The model assumptions cause large uncertainties that could limit the precision of the ϕ_3 measurement. The model-independent approach provides measurements of CP violating asymmetries made in independent regions of the D phase space [8, 9]. This binning reduces the statistical precision, but the uncertainty due to model assumptions are no longer present as the average strong-phase measurements are used. This analysis follows the model-independent method.

The D phase space is binned into regions with differing strong phases, which allows ϕ_3 to be determined from a single channel in a model-independent manner. The signal yield for $B^{\pm} \to DK^{\pm}$ decays in each bin is given as

$$\Gamma_i^{\pm} \propto K_i + r_B^2 \overline{K_i} + 2\sqrt{K_i \overline{K_i}}(c_i x_{\pm} \mp s_i y_{\pm}), \tag{18.2}$$

where $x_{\pm} = r_B \cos(\delta_B \pm \phi_3)$ and $y_{\pm} = r_B \sin(\delta_B \pm \phi_3)$. The $x_{\pm}$ and $y_{\pm}$ parameters, that are sensitive to ϕ_3, can be obtained when the phase space is divided into three or more bins. Here, K_i and $\overline{K_i}$ are the fraction of flavor-tagged D^0 and $\overline{D^0}$ events in the ith bin, respectively, which can be estimated from $D^{*+} \to D^0 \pi^+$ decays with good precision due to their large sample size. The parameters c_i and s_i are the amplitude-weighted average of the cosine and sine of the strong-phase difference between D^0 and $\overline{D^0}$ over the ith bin; these parameters need to be determined at a charm factory experiment like CLEO-c or BESIII, where the quantum-entangled $D^0 \overline{D^0}$ pairs are produced via $e^+ e^- \to \psi(3770) \to D^0 \overline{D^0}$ [10]. The values of c_i and s_i parameters for $D \to K_S^0 \pi^+ \pi^- \pi^0$ decays as well as the binning scheme to divide the D phase space reported in [11] are used in this analysis.

18.3 Data Samples and Event Selection

The $e^+ e^-$ collision data sample at a center-of-mass energy corresponding to the pole of the $\Upsilon(4S)$ resonance collected by the Belle detector [12, 13] is used in this analysis. It corresponds to an integrated luminosity of 711 fb^{-1} and contains 772×10^6 $B\overline{B}$ pairs. The Belle detector is located at the interaction point (IP) of KEKB asymmetric $e^+ e^-$ collider [14]. A detailed description of the Belle detector is given in [12, 13]. Monte Carlo (MC) samples are used to optimize the selection criteria, determine the efficiencies, and identify various sources of background.

We reconstruct $B^+ \to DK^+$ and $B^+ \to D\pi^+$ decays in which the D decays to the four-body final state of $K_S^0 \pi^+ \pi^- \pi^0$. The decays $D^{*+} \to D\pi^+$ produced via the $e^+ e^- \to c\bar{c}$ continuum process are also selected to measure the K_i and $\overline{K_i}$ parameters. We select the charged particle candidates produced within 0.5 cm and ± 3.0 cm of the IP in perpendicular and parallel directions to the z-axis, respectively, where the z-axis is defined to be opposite to the e^+ beam direction. These tracks are then identified as kaons or pions with the help of the particle identification system at Belle [12]. We reconstruct the K_S^0 candidates from two oppositely charged pion tracks. The invariant mass of these pion candidates is required to be within ±3σ of the nominal K_S^0 mass [5], where σ is the mass resolution. The background due to random combinations of pions is reduced with the help of a neural network [15] based selection with 87% efficiency [16].

We reconstruct π^0 candidates from a pair of photons detected in the electromagnetic calorimeter (ECL). The π^0 candidates within the diphoton invariant mass range 0.119–0.148 GeV/c^2 are retained. The photon energy thresholds are optimized separately for candidates detected in the barrel, forward endcap, and backward endcap regions of the ECL. Furthermore, kinematic constraints are applied to K_S^0, π^0, and D invariant masses and decay vertices. This improves the energy and momentum resolution of the B candidates and the invariant masses used to divide the D phase space into bins.

While reconstructing $D^{*+} \rightarrow D\pi^+$ decays, it is required that the accompanying pion has at least one hit in the silicon vertex detector. This pion carries a small fraction of the momentum due to the limited phase space of the decay and hence is known as a slow pion. The D meson momentum in the laboratory frame is chosen to be between 1–4 GeV/c so that it matches to that in $B^+ \rightarrow Dh^+ (h = K, \pi)$ sample. The signal candidates are identified by the kinematic variables M_D, the invariant mass of D candidate and ΔM, the difference in the invariant masses of D^* and D candidates. We retain events that satisfy the criteria, $1.80 < M_D < 1.95$ GeV/c^2 and $\Delta M < 0.15$ GeV/c^2. A kinematic constraint is applied so that the D and π candidates come from the common vertex position. When there are more than one candidate in an event, the one with the smallest χ^2 value from the D^* vertex fit is retained for further analysis. The overall selection efficiency is 3.7%.

The B meson candidates are reconstructed by combining a D candidate with a charged kaon or pion track. Events with D meson invariant mass in the range 1.835–1.890 GeV/c^2 are selected. The kinematic variables energy difference ΔE and beam constrained mass M_{bc} are used to identify the signal candidates. They are defined as $\Delta E = E_B - E_{beam}$ and $M_{bc} = c^{-2}\sqrt{E_{beam}^2 - |\mathbf{p}_B|^2 c^2}$, where E_B and $\mathbf{p}_B$ are the energy and momentum of the B candidate and E_{beam} is the beam energy in the center-of-mass frame. The candidates that satisfy the criteria $M_{bc} > 5.27$ GeV/c^2 and $-0.13 < \Delta E < 0.30$ GeV are selected. In events with more than one candidate, the candidate with the smallest value of $(\frac{M_{bc} - M_B^{PDG}}{\sigma_{M_{bc}}})^2 + (\frac{M_D - M_D^{PDG}}{\sigma_{M_D}})^2 + (\frac{M_{\pi^0} - M_{\pi^0}^{PDG}}{\sigma_{M_{\pi^0}}})^2$ is retained. Here, the masses M_i^{PDG} are those reported by the Particle Data Group in [5] and the resolutions $\sigma_{M_{bc}}$, σ_{M_D} and $\sigma_{M_{\pi^0}}$ are obtained from MC simulated samples of signal events.

The main source of background is from $e^+e^- \rightarrow q\bar{q}$, $q = u, d, s, c$ continuum processes, and these are suppressed by exploiting the difference in their event topology to that of $B\overline{B}$ events. The continuum events are jet-like in nature and $B\overline{B}$ events have a spherical topology. These events are separated with the help of a neural-network-based algorithm [15]. We require the neural network output to be greater than -0.6, which reduces the continuum background by 67% at the cost of 5% signal loss. The overall selection efficiency is 4.7 and 5.3% for $B^+ \rightarrow DK^+$ and $B^+ \rightarrow D\pi^+$ modes, respectively.

18.4 Determination of K_i and $\overline{K_i}$

The K_i and $\overline{K_i}$ parameters indicate the fraction of D^0 and $\overline{D^0}$ events in each D phase space bin. They are measured from the $D^{*+} \rightarrow D\pi^+$ sample; the charge of the pion determines the flavor of the D meson. The signal yield is obtained from a two-dimensional extended maximum-likelihood fit to M_D and ΔM distributions independently in each bin. Appropriate probability density functions (PDF) are used to model the distributions. A quadratic correlation between M_D and ΔM is taken into account for the signal component. The yields along with K_i and $\overline{K_i}$ values are given in Table 18.1.

18.5 Signal Extraction in $B^+ \rightarrow Dh^+$ Sample

The signal yield in each D phase space bin is determined from a two-dimensional extended maximum-likelihood fit to ΔE and neural network output (NB). The latter is transformed as

$$NB' = \log\left(\frac{NB - NB_{\text{low}}}{NB_{\text{high}} - NB}\right), \tag{18.3}$$

Table 18.1 D^0 and $\overline{D^0}$ yield in each bin of D phase space along with K_i and $\overline{K_i}$ values measured in D^* tagged data sample

Bin no.	N_{D^0}	$N_{\overline{D^0}}$	K_i	$\overline{K_i}$
1	51048±282	50254±280	0.2229±0.0008	0.2249±0.0008
2	137245±535	58222±382	0.4410±0.0009	0.1871±0.0007
3	31027±297	105147±476	0.0954±0.0005	0.3481±0.0009
4	24203±280	16718±246	0.0726±0.0005	0.0478±0.0004
5	13517±220	20023±255	0.0371±0.0003	0.0611±0.0004
6	21278±269	20721±267	0.0672±0.0005	0.0679±0.0005
7	15784±221	13839±209	0.0403±0.0004	0.0394±0.0004
8	6270±148	7744±164	0.0165±0.0002	0.0183±0.0002
9	6849±193	6698±192	0.0070±0.0002	0.0054±0.0001

where $NB_{low} = -0.6$ and $NB_{high} \approx 1.0$ are the minimum and maximum values of NB in the sample, respectively. The three background components are continuum background, combinatorial $B\bar{B}$ background due to final state particles from both the B mesons and cross-feed peaking background due to the misidentification of a kaon as a pion or *vice versa*.

The sum of a Crystal Ball (CB) [17] function and two Gaussian functions with a common mean is used as the PDF to model the ΔE signal component in both the B samples. The sum of a Gaussian and an asymmetric Gaussian with different mean values is used to parametrize the PDF that describes the NB' signal component. The continuum background distribution in ΔE is modeled with a first-order Chebyshev polynomial and that in NB' is described by the sum of two Gaussian PDFs with different mean values. The ΔE distribution of random $B\bar{B}$ background in $B^+ \rightarrow D\pi^+$ is described by an exponential function. There is a small peaking structure due to misreconstructed π^0 events and this is modeled by a CB function. A first-order Chebyshev polynomial is added to the above two PDFs in the case of $B^+ \rightarrow DK^+$ decays. The NB' distribution for both the samples are modeled by an asymmetric Gaussian function. The cross-feed peaking background in ΔE is modeled with the sum of three Gaussian functions, whereas the signal PDF itself is used for the NB' distribution. The fit projections in $B^+ \rightarrow DK^+$ MC sample are shown in Fig. 18.2. These are signal-enhanced projections with events in the signal region of the other variable, where the signal regions are defined as $|\Delta E| < 0.05$ GeV and $0 < NB' < 12$.

The ϕ_3 sensitive parameters are determined directly from the fit by expressing the signal yield as in (18.2). The K_i and $\overline{K}_i$ values along with the c_i and s_i measurements

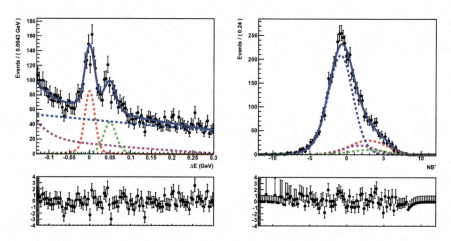

Fig. 18.2 Signal-enhanced fit projections of ΔE (left) and NB' (right) for $B^\pm \rightarrow DK^\pm$ MC sample having equivalent luminosity as that of full data sample collected by Belle. The black points with the error bar are the data and the solid blue curve is the total fit. The dotted red, blue, magenta, and green curves represent the signal, continuum, random $B\bar{B}$ backgrounds, and cross-feed peaking background components, respectively. The pull between the data and the fit are shown for both the projections

Table 18.2 Preliminary results of $x_\pm$ and $y_\pm$ parameters from $B^\pm \to D\pi^\pm$ data sample. The first uncertainty is statistical, second systematic and the third one is due to the uncertainty on the c_i, s_i measurements

Parameter	Result
$x_+^{D\pi}$	$0.04 \pm 0.03 \pm 0.03 \pm 0.01$
$y_+^{D\pi}$	$0.06_{-0.20}^{+0.08} \pm 0.10_{-0.03}^{+0.07}$
$x_-^{D\pi}$	$0.01 \pm 0.02_{-0.03}^{+0.02} \pm 0.02$
$y_-^{D\pi}$	$-0.02 \pm 0.06_{-0.04}^{+0.03} \pm 0.06$

Fig. 18.3 One (solid line), two (dashed line), and three (dotted line) standard deviation likelihood contours for the $(x_\pm, y_\pm)$ parameters for $B^\pm \to D\pi^\pm$ data sample

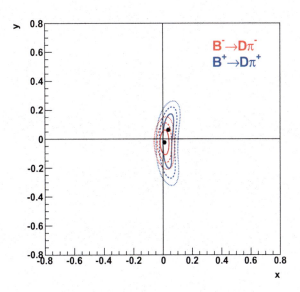

reported in [11] are used as input parameters. Efficiency corrections are applied and the effect of migration of events between the bins due to finite momentum resolution is also taken into account. The preliminary results obtained from $B^+ \to D\pi^+$ data sample are summarized in Table 18.2. The dominant source of systematic uncertainty is the size of the signal MC sample used for estimating the efficiency and the extent of migration between the bins. The statistical likelihood contour is given in Fig. 18.3.

18.6 Summary

A precise measurement of the CKM angle ϕ_3 is essential to establish the SM description of CP violation. Here, we present the feasibility of $D \to K_S^0 \pi^+ \pi^- \pi^0$ final state to do so in $B^+ \to DK^+$ decays. This is the first attempt to analyze this particular decay mode. The signal extraction procedure is established in an MC sample, as well as $B^+ \to D\pi^+$ data sample, the calibration mode. MC predictions estimate the statistical uncertainty on $x_\pm$, $y_\pm$ in $B^+ \to DK^+$ to be 0.08 and 0.17, respectively.

An improved measurement is possible once an amplitude model for $D^0 \to K_S^0\pi^+\pi^-\pi^0$ is available to guide the binning of the phase space such that maximum sensitivity to ϕ_3 is obtained. Furthermore, a more precise measurement of c_i, s_i parameters could be performed with a larger sample of $e^+e^- \to \psi(3770)$ data that has been collected by BESIII, thus reducing the systematic uncertainty. The Belle II detector is expected to collect about 50 times larger B sample. Thus, the improved binning combined with the larger B sample make $B^+ \to D(K_S^0\pi^+\pi^-\pi^0)K^+$ a promising addition to the set of modes to be used to determine ϕ_3 to a precision of 1–2° [18].

References

1. N. Cabibbo, Unitary symmetry and leptonic decays. Phys. Rev. Lett. **10**, 531 (1963)
2. M. Kobayashi, T. Maskawa, CP violation in the renormalizable theory of weak interaction. Prog. Theor. Phys. **49**, 652 (1973)
3. Y. Amhis et al., (Heavy Flavor Averaging Group Collaboration): averages of b-hadron, c-hadron and τ-lepton properties as of November 2016. Eur. Phys. J **C77**, 895 (2017)
4. J. Brod, J. Zupan, The ultimate theoretical error on gamma from $B \to DK$ decays. J. High. Energ. Phys. **01**, 051 (2014)
5. M. Tanabashi et al., (Particle Data Group Collaboration): review of particle physics. Phys. Rev. D. **98**, 030001 (2018)
6. H. Aihara et al., (Belle Collaboration): first measurement of ϕ_3 with a model-independent Dalitz plot analysis of $B^\pm \to DK^\pm$, $D \to K_S^0\pi^+\pi^-$ decay. Phys. Rev. D **85**, 112014 (2012)
7. A. Poluektov et al., (Belle Collaboration): evidence for direct CP violation in the decay $B^\pm \to D^{(*)}K^\pm$, $D \to K_S^0\pi^+\pi^-$ and measurement of the CKM phase ϕ_3. Phys. Rev. D **81**, 112002 (2010)
8. A. Giri, Y. Grossman, A. Soffer, J. Zupan, Determining γ using $B^\pm \to DK^\pm$ with multibody D decays. Phys. Rev. D **63**, 054018 (2003)
9. A. Bondar, Proceedings of BINP special analysis meeting on Dalitz analysis, 2002 (unpublished). A. Giri, Y. Grossman, A. Soffer, J. Zupan, Determining γ using $B^\pm \to DK^\pm$ with multibody D decays. Phys. Rev. D **63**, 054018 (2003)
10. P.K. Resmi, Input from the charm threshold for the measurement of γ. arXiv:1810.00836 [hep-ex]
11. P.K. Resmi et al., Quantum-correlated measurements of $D \to K_S^0\pi^+\pi^-\pi^0$ decays and consequences for the determination of the CKM angle γ. J. High. Energ. Phys. **01**, 82 (2018)
12. A. Abashian et al., (Belle Collaboration): The Belle Detector. Nucl. Instrum. Methods Phys. Res. A **479**, 117 (2002)
13. J. Brodzicka et al., (Belle Collaboration): physics achievements from the belle experiment. Prog. Theor. Exp. Phys. **2012**, 04D001 (2012)
14. S. Kurokawa, E. Kikutani, Overview of the KEKB accelerators. Nucl. Instr. Meth. A **499**, 1 (2003). and other papers included in this volume
15. M. Feindt, U. Kerzel, The NeuroBayes neural network package. Nucl. Instrum. Methods Phys. Res. A **559**, 190 (2006)
16. H. Nakano, Search for new physics by a time-dependent CP violation analysis of the decay $B \to K_S\eta\gamma$ using the Belle detector, Ph.D. Thesis, Tohoku University, 2014, Chap. 4 (unpublished)

17. Skwarnicki, T.: A study of the radiative cascade transitions between the Υ and Υ' resonances, Ph.D thesis (Appendix E), DESY F31-86-02 (1986)
18. E. Kou, P. Urquijo, The Belle II collaboration, and The B2TiP theory community. *The Belle II Physics Book*. arXiv:1808.10567

Chapter 19
$b \rightarrow s\ell\ell$ Decays at Belle

S. Choudhury, S. Sandilya, K. Trabelsi, and Anjan K. Giri

Abstract The observable R_K which is the ratio of branching fractions for $B \rightarrow K\mu\mu$ to $B \rightarrow Kee$, tests lepton flavor universality (LFU) in the standard model (SM), and hence constitutes an important probe for new physics (NP). We report herein a sensitivity study of R_K in $B \rightarrow K\ell\ell$ and of the equivalent $R_K(J/\psi)$ in $B \rightarrow KJ/\psi(\rightarrow \ell\ell)$. The latter is measured with Belle's full data sample of 772×10^6 $B\bar{B}$ pairs and the result is consistent with unity. In a variety of NP models, lepton flavor violation (LFV) comes together with LFU violation. We also report on searches for LFV in $B \rightarrow K\mu e$ and $B \rightarrow Ke\mu$ modes. Belle has recently measured LFV $B^0 \rightarrow K^{*0}\ell\ell'$ and the most stringent upper limit is found.

19.1 Introduction

The flavor changing neutral current (FCNC) decays $B \rightarrow K\mu\mu$ and $B \rightarrow Kee$ involve the $b \rightarrow s$ quark-level transition and are forbidden at tree level in the SM. These type of reactions are mediated through electroweak penguin and box diagrams, shown in Fig. 19.1. These processes are highly suppressed, have very small branching ratio ($\mathscr{B}$), and are very sensitive to NP. NP can either enhance or suppress the amplitude of the decay or may modify the angular distribution of the final state particles. The variable R_K is theoretically very clean as most of the hadronic uncertainties cancel out in the ratio. This observable is measured by LHCb [1] and the result shows a deviation of 2.6 standard deviation in the bin of $1 < q^2 < 6 \, \text{GeV}^2/c^4$ (q^2 = invariant-mass square of two leptons), measured for a data sample of $1 \, fb^{-1}$. The R_K is again measured by LHCb [2] for a data sample of $3 \, fb^{-1}$ for a bin of

S. Choudhury (✉) · A. K. Giri
Indian Institute of Technology Hyderabad, Sangareddy 502285, Telangana, India
e-mail: ph16resch11007@iith.ac.in

S. Sandilya
University of Cincinnati, Cincinnati, Ohio 45221, India

K. Trabelsi
Laboratory of the Linear Accelerator (LAL), 91440 Orsay, France

© Springer Nature Singapore Pte Ltd. 2021
P. K. Behera et al. (eds.), *XXIII DAE High Energy Physics Symposium*,
Springer Proceedings in Physics 261,
https://doi.org/10.1007/978-981-33-4408-2_19

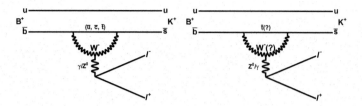

Fig. 19.1 Penguine diagram of $B \rightarrow K\ell\ell$ in SM (left) and Beyond SM (right) scenario

$1.1 < q^2 < 6\,\text{GeV}^2/c^4$ having $2.5\,\sigma$ deviation. Earlier Belle [3] had also measured R_K for the whole q^2 region using a data sample of 657×10^6 $B\bar{B}$ pairs and the result was consistent with unity having very high uncertainty. The deviation from SM expectation in R_K or R_{K*} from LHCb result may possibly show LFU violation. LFV is also an important probe to search for NP, where, LFV and LFU violation are complimentary of each other.

19.2 Monte Carlo (MC) Simulation

Our selection is based on, and optimized with an MC simulation study. One million signal events are generated using the BTOSLLBALL decay model [4] for LFU modes and phase-space for LFV modes with the EvtGen package [5]. The detector simulation is subsequently performed with GEANT3 [6].

19.3 Event Selection

We reconstruct $B \rightarrow K\ell\ell(')$ by combining a kaon (charged or neutral) with two oppositely charged leptons. Here, ℓ can be either electron or muon. The impact parameter criteria for the charged particle tracks are, along the z-axis $|dz| < 4$ cm and in the transverse plane $|dr| < 1$ cm. Charged kaon are selected based on a ratio $L_{(K/\pi)} = L_K/(L_K + L_\pi)$, where L_K and L_π are the individual likelihood of kaon and pion, respectively. For our selection, we require $L_{K/\pi} > 0.6$, which corresponds to an efficiency of above 92% with a pion fake rate below 10%. Similarly, electrons (muons) are selected with $L_e > 0.9$ ($L_\mu > 0.9$), and these correspond to an efficiency of >92% (90%) and a pion fake rate of <0.3% (<1.4%). The bremsstrahlung photon emitted by high energy electrons are recovered by considering energy deposit in a cone of 50 mrad around the initial direction of the electron track. The K_S^0 candidates are reconstructed from pairs of oppositely charged tracks, both treated as pions, and are identified with a neural network (NN). The kinematic variables that distinguish signal from background are the beam-energy constrained mass

$M_{bc} = \sqrt{(E_{beam}/c^2)^2 - p_B^2/c^2}$ and energy difference $\Delta E = E_B - E_{beam}$, where, E_B and p_B are the energy and momentum of B candidate, respectively, and E_{beam} is the beam energy. Events are selected within the range of $5.20 < M_{bc} < 5.29$ GeV/c^2 and $-0.1 < \Delta E < 0.25$ GeV.

19.4 Background Rejection

The main sources of background are continuum ($e^+e^- \rightarrow q\bar{q}$) and $B\bar{B}$ events. We find that some event shape and vertex quality variables can well separate signal from background. An artificial neural network (*NN*) is developed using an equal number of signal and background events, where the latter is taken from continuum as well as $B\bar{B}$ samples according to their luminosity. The NN output (*NN*) is translated to *NN'* using the following transformation

$$NN' = \frac{(NN - NN_{min})}{(NN_{max} - NN)}$$

Here, NN_{min} is the minimum NN value, chosen to be -0.6. This criterion reduces 75% of the background with only 5–6% loss in the signal efficiency. NN_{max} is the maximum NN value, found from signal MC. The *NN'* distributions Fig. 19.2a, integrated as well as for different q^2 bins, are shown in Fig. 19.2b. It has similar shape for different q^2 regions in signal and backgrounds.

The peaking backgrounds which pass these criteria are mainly coming from $B \rightarrow J/\psi K$ because of misidentification and swapping between the leptons or lepton and kaon. These backgrounds are removed by applying invariant mass cut around J/ψ mass region. The backgrounds coming from $B^+ \rightarrow D^0(\rightarrow K^+\pi^-)\pi^+$ due to lepton

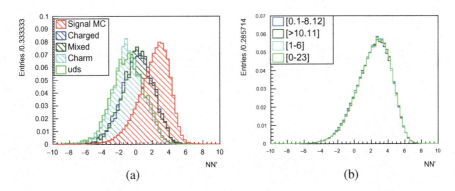

Fig. 19.2 (a) *NN'* distribution, where the red histogram represents signal MC, deep green and blue histograms are continuum and $B\bar{B}$ background, respectively. (b) *NN'* shape for different q^2 regions in signal MC events

candidates are faked by pions, and are removed by applying invariant mass cut in D^0 mass region.

19.5 Signal Yield Extraction

We perform a three-dimensional (3D) fit with M_{bc}, ΔE, and NN'. The signal of ΔE is modeled with Crystal Ball (CB) and a gaussian function. Similarly, M_{bc} and NN' of signal are modeled with Gaussian and bifurcated gaussian-gaussian, respectively. For continuum background, the ΔE, M_{bc}, and NN' are modeled with chebychev polynomial, argus function, and Gaussian, respectively. Similarly, the $B\bar{B}$ background is fitted with exponential, argus fuction, and gaussian for ΔE, M_{bc}, and NN', respectively. From the 3D fit, the $R_K(J/\psi)$ is found to be consistent with unity and $B \rightarrow KJ/\psi(\rightarrow \ell\ell)$ is used as a control sample for $B \rightarrow K\ell\ell$. The fit result for $B^+ \rightarrow K^+\mu^+\mu^-$ is shown in Fig. 19.3. The signal enhanced projections are shown in Fig. 19.4. Candidate events with $M_{bc} > 5.27$ GeV/c^2, $|\Delta E| < 0.05$ GeV and $NN' > 0.5$ are considered to be part of the signal region.

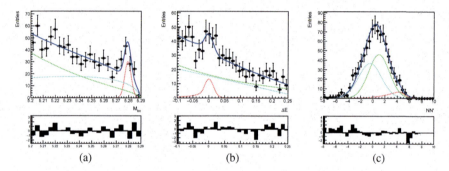

Fig. 19.3 3D fit result for a bin of $1 < q^2 < 6$ GeV^2/c^4 in case of $B^+ \rightarrow K^+\mu^+\mu^-$ mode. **a** M_{bc}, **b** ΔE and **c** NN'

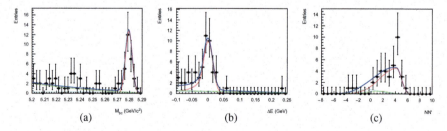

Fig. 19.4 **a** M_{bc} projection in the ΔE and NN' signal region **b** ΔE projection in the M_{bc} and NN' signal region, and **c** NN' projection in the M_{bc} and ΔE signal region for the bin of $1 < q^2 < 6$ GeV^2/c^4 for $B^+ \rightarrow K^+\mu^+\mu^-$

19.6 Results

19.6.1 LFU Test

The statistical uncertainty of Belle for the whole q^2 region measured for a data sample of 605 fb^{-1} was 0.19 [3]. For this analysis, the expected uncertainty for the bin of $1 < q^2 < 6 \, \text{GeV}^2/c^4$ is 20%, which is represented by a violet box in Fig. 19.5, here we have considered LHCb result as central value. The expected statistical uncertainty of R_K for the whole q^2 bin is 10%.

19.6.2 Search for LFV

The modes that we are studying to search LFV are $B^+ \to K^+\mu^+e^-$ and $B^+ \to K^+\mu^-e^+$. We extracted the signal from these modes by performing 3D extended maximum likelihood fit as that of LFU modes. The signal enhanced projection plots for $B^+ \to K^+\mu^+e^-$ is shown in Fig. 19.6. The upper limit is estimated from $N_{sig}^{(UL)}$, efficiency (ε) of particular mode and number of $B\bar{B}$ pairs ($N_{B\bar{B}}$), which is represented by a formula

$$\mathscr{B}^{(UL)} = \frac{N_{sig}^{(UL)}}{N_{B\bar{B}} \times \varepsilon}$$

Our estimated upper limit for LFV $B^+ \to K^+\mu^+e^-$ and $B^+ \to K^+\mu^-e^+$ are $<2.0 \times 10^{-8}$ and $<2.1 \times 10^{-8}$, respectively, as tabulated in Table 19.1, and these results are one order of magnitude better than that of the PDG values.

Belle [7] has recently searched LFV $B^0 \to K^{*0}\ell\ell'$ decays, where $\ell = \mu$ or e with full data sample. In this analysis, strong contribution from contiunnm and $B\bar{B}$

Fig. 19.5 Expected sensitivity of R_K for a bin of $1 < q^2 < 6 \, \text{GeV}^2/c^4$. Here, we have considered the LHCb result as central value and the violet box represent our expected uncertainty

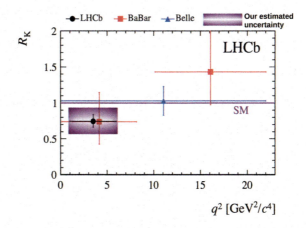

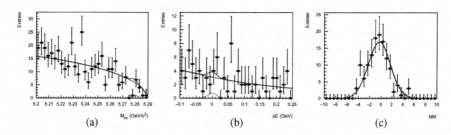

Fig. 19.6 Signal enhanced projection plots for $B^+ \to K^+\mu^+e^-$ mode. **a** M_{bc} projection in the ΔE and NN' signal region **b** ΔE projection in the M_{bc} and NN' signal region, and **c** NN' projection in the M_{bc} and ΔE signal region

Table 19.1 Upper limit estimation in MC for LFV $B \to K\ell\ell'$ modes

Mode	ε (%)	$N_{sig}^{(UL)}$	$\mathscr{B}^{(UL)}$ (10^{-8})	PDG $\mathscr{B}$ (10^{-7})
$B^+ \to K^+\mu^+e^-$	29.3	4.4	**2.0**	< 1.3
$B^+ \to K^+\mu^-e^+$	30.0	4.9	**2.1**	< 0.9

background is found. So, we have used two stage NN to suppress the backgrounds. The signal is extracted by performing extended maximum likelihood fit to M_{bc} but no evidence of signal is found and upper limit is estimated. The upper limits are $<1.2 \times 10^{-7}$, $<1.6 \times 10^{-7}$, and $<1.8 \times 10^{-7}$ for $B^0 \to K^{*0}\mu^+e^-$, $B^0 \to K^{*0}\mu^-e^+$, and $B^0 \to K^{*0}\mu^{\pm}e^{\mp}$, respectively. These observed limits are most stringent to date.

19.7   Conclusion

Several anomalies in B decays indicates lepton non-universal interaction. The LFU test is an extremely clean probe to search for NP as most of the hadronic uncertainties cancel out in the ratio of R_K. Many theoretical models predict LFV in presence of LFU violation. Belle has recently search LFV $B^0 \to K^{*0}\mu^{\pm}e^{\mp}$ and most stringent limit is found. Belle [8] will publish soon the result of R_K and LFV $B^{\pm} \to K^{\pm}\mu^{\pm}e^{\mp}$ for full data sample of $711\ fb^{-1}$.

Acknowledgements We thank the KEKB group for excellent operation of the accelerator; the KEK cryogenics group for efficient solenoid operations; and the KEK computer group the NII, and PNNL/EMSL for valuable computing and SINET5 network support. We acknowledge support from MEXT, JSPS and Nagoya's TLPRC (Japan); ARC (Australia); FWF (Austria); NSFC and CCEPP (China), MSMT (Czechia); CZF, DFG, EXC153, and versus (Germany); DST (India); INFN (Italy); MOE, MSIP, NRF, RSRI, FLRFAS project and GSDC of KISTI (Korea); MNiSW and NCN (Poland); MES and RFAAE (Russia); ARRS (Slovenia); IKERBASQUE and MINECO (Spain); SNSF (Switzerland); MOE and MOST (Taiwan); and DOE and NSF (USA).

References

1. R. Aaij et al., (LHCb Collaboration). JHEP **02**, 104 (2016)
2. R. Aaij et al., (LHCb Collaboration). Phys. Rev. Lett. **112**, 191801 (2019)
3. S. Wehle et al., (Belle Collaboration). Phys. Rev. Lett. **118**, 111801 (2017)
4. A. Ali, P. Ball, L.T. Handoko, G. Hiller, Phys. Rev. D **61**, 074024 (2000)
5. D.J. Lange et al., Nucl. Instrum. Meth. **A462**, 152 (2001)
6. R. Brun et al., CERN Report No. **DD/EE**, 84-1 (1984)
7. S. Sandilya et al., (Belle Collaboration). Phys. Rev. D **98**, 071101 (2018)
8. S. Choudhury et al., (Belle Collaboration), arXiv:1908.01848

Chapter 20
Magnetic Moments and the Decay Properties of the D and D_s Mesons

Keval Gandhi, Vikas Patel, Shailesh Pincha, Virendrasinh Kher, and Ajay Kumar Rai

Abstract Using the semi-relativistic approach of quark-antiquark Coulomb plus linear confinement potential with a Gaussian wave function the masses of the radial states of D and D_s mesons are used. In this article, their ground state masses, the magnetic moments, and magnetic dipole (M1) transitions of D and D_s mesons of $J^P = 1^-$ are calculated in the constituent quark model. Moreover, their M1 decay rates are also analyzed. We compared our results with other theoretical predictions.

20.1 Introduction

Experimentally, the ground state masses of the charm and strange-charm mesons are well established and recently the LHCb Collaboration [1–3] observed many new excited states. Still the J^P value of $D_1(2420)^{\pm}$, $D(2550)^0$, $D_J^*(2600)$, $D^*(2640)^{\pm}$, $D(2740)^0$, and $D(3000)^0$ mesons are unknown from the known experimental resonances [4]. Upcoming experimental facilities J-PARC, $\overline{P}$ANDA [5] will be expected to provides more information in the low-energy regime of quantum chromodynamics (QCD). The J^P value assignments from the heavy-light hadrons mass spectra are crucial. It helps to extract the experimental information like decay widths, branching fractions, the hyperfine mass splitting, etc. This makes it more interesting to look back to the theory and the phenomenology study for the predictions of their spin-parity. In the past few years, the mass spectra of heavy-light mesons have been studied in various potential models using different approaches: two-loop static potential in a

K. Gandhi (✉) · V. Patel · S. Pincha · A. Kumar Rai
Department of Applied Physics, Sardar Vallabhbhai National Institute of Technology, Surat 395007, Gujarat, India
e-mail: keval.physics@yahoo.com

A. Kumar Rai
e-mail: raiajayk@gmail.com

V. Kher
Applied Physics Department, Polytechnic, The M S University of Baroda, Vadodara 390002, Gujarat, India

© Springer Nature Singapore Pte Ltd. 2021
P. K. Behera et al. (eds.), *XXIII DAE High Energy Physics Symposium*, Springer Proceedings in Physics 261, https://doi.org/10.1007/978-981-33-4408-2_20

variational approach [6], perturbative approach [7], Regge-like mass relations [8], Regge trajectories from the quadratic form of the spinless Salpeter-type equation (QSSE) [9], semi-relativistic approach using Gaussian wave function [10], relativistic Dirac formalism [11], relativistic quark model [12–14], non-relativistic approach using hydrogenic and Gaussian wave function [15], lattice QCD [16], non-relativistic constituent quark model [17], etc.

The D-meson contains one light antiquark, antiup ($\bar{u}$) or antidown ($\bar{d}$), and one charm quark (c) as a heavy quark. And the D_s meson has antistrange ($\bar{u}$) quark as a light quark and the charm (c) as a heavy quark. In the present study, we mainly concentrate on the ground state of D and D_s mesons of $J^P = 1^-$. We calculate their magnetic moments, magnetic dipole transitions (say M1 transitions), and analyzed their M1 decay widths. There are several approaches available for the study of electromagnetic properties of heavy-light mesons and they are bag model [18, 19], Nambu-Jona-Lasinio model [20], Blankenbecler-Sugar equation with a scalar confining interaction [21], relativistic quark model [22], relativistic potential model [23], chiral effective theory [24], etc. Recently, the [25, 26] used the covariant constituent quark model (CCQM) for the study of semileptonic decays of D and D_s mesons.

This article is organized as follows: the semi-relativistic potential model is described in brief in Sect. 20.2, the magnetic properties of the ground state D and D_s mesons of $J^P = 1^-$ are calculated in Sect. 20.3. At last, we summarize our work in Sect. 20.3.

20.2 Methodology

The hadron spectroscopy is usually studied in the non-relativistic and the relativistic frame of quantum mechanics. The study of hadron spectroscopy helps us to determine how the strong interaction binds the quarks and gluons inside the hadrons. For the study of heavy and light quark mesons, we use the Hamiltonian as [10, 27, 28]

$$H = \sqrt{\mathbf{p}^2 + m_Q^2} + \sqrt{\mathbf{p}^2 + m_{\bar{q}}^2} + V(\mathbf{r}), \tag{20.1}$$

where $\mathbf{p}$ is the relative momentum, $V(\mathbf{r})$ is the potential of quark-antiquark, and m_Q and $m_{\bar{q}}$ give masses of the heavy quark and the light antiquark, respectively. Here, we use the constitute quark masses $m_{\bar{u}/\bar{d}} = 460$ MeV, $m_{\bar{s}} = 586$ MeV, and $m_c = 1400$ MeV for the study of D and D_s mesons spectroscopy [10]. An expression of quark-antiquark potential ($V(\mathbf{r})$) with $\mathcal{O}(\frac{1}{m})$ is [29]

$$V(\mathbf{r}) = V^{(0)}(\mathbf{r}) + \left(\frac{1}{m_Q} + \frac{1}{m_{\bar{q}}}\right) V^{(1)}(\mathbf{r}) + \mathcal{O}\left(\frac{1}{m^2}\right), \tag{20.2}$$

where $V^{(0)}(\mathbf{r})$ is the Coulomb (considering color wave function) plus linear potential given by [30]

Table 20.1 The ground state masses of D and D_s mesons (in MeV)

State	J^P	Meson	Present [10]	[8]	[9]	[11]	[12]	[13]	[14]	PDG [4]
$1^3 S_0$	0^-	D^0	1865	1869.7	1884	1869.57	1877	1874	1871	1864.83±0.05
		$D^\pm$								1869.58±0.09
$1^3 S_1$	1^-	$D^*(2007)^0$	2003	2010.3	2010	2009.54	2041	2038	2010	2006.85±0.05
		$D^*(2010)^\pm$								2010.26±0.05
$1^3 S_0$	0^-	$D_s^\pm$	1953	1968.3	–	–	1979	–	1969	1968.34±0.07
$1^3 S_1$	1^-	$D_s^{*\pm}$	2112	2112.2	–	–	2129	–	2111	2112.2±0.4

$$V^{(0)}(\mathbf{r}) = -\frac{4}{3}\frac{\alpha_s}{r} + Ar + V_0, \tag{20.3}$$

and from the leading order perturbation theory,

$$V^{(1)}(\mathbf{r}) = -\frac{C_F C_A \alpha_s^2}{4r^2}. \tag{20.4}$$

Here, A is the potential parameter, V_0 is the constant, α_s is the strong running coupling constant, and $C_F = 4/3$ and $C_A = 3$ are the Casimir charges of the fundamental and the adjoint representation, respectively [29].

The heavy-light mesons are needed to be treated relativistically. So we expand kinetic energy term appearing in the (20.1), retaining up to $\mathcal{O}(\mathbf{p}^{10})$ power, which gives the relativistic effect to the kinetic energy of the heavy-light systems. The more details can be found from [10]. We will use the masses of the ground state with $J^P = 0^-$ and $J^P = 1^-$ (see in Table 20.1) and calculate the magnetic moments, M1 dipole transitions, and the M1 decay widths of the ground state charm and strange-charm mesons of $J^P = 1^-$ in the next section.

20.3 The Magnetic Properties

The magnetic moments of the particles determines their structural properties and provides an important role in the study of their internal dynamics. The radiative decay rates depending upon the magnetic dipole transition can also probe the charge structure of the hadrons. The radiative decay is done by an exchange of massless photon among the participating hadrons without contained phase space restriction. Therefore, some of the radiative decay rates, especially the low-lying states, contribute strongly to their total branching fraction.

20.3.1 Magnetic Moments

The magnetic moment of the hadron is purely the function of flavor, spin, and charge of the constituent quarks. An expression of the magnetic moment can be written in the form of expectation value as [31–35]

$$\mu_M = \sum_q \langle \Phi_{sf} | \hat{\mu}_{q_z} | \Phi_{sf} \rangle; \qquad q = u, d, c, \tag{20.5}$$

where μ_M denotes the magnetic moment of the particular mesonic state, Φ_{sf} represents the spin-flavor wave function of a participating meson, and the $\hat{\mu}_{q_z}$ is the z-component of the magnetic moment of the individual quark given by

$$\hat{\mu}_{q_z} = Q \frac{e}{2 m_q^{eff}} \sigma_{q_z}. \tag{20.6}$$

Here Q is the charge and σ_{q_z} is the z-component of the constitute quark spin, and m_q^{eff} is the effective quark mass gives the mass of the bound quark inside the meson given by [17, 31–35, 37]

$$m_q^{eff} = m_q \left(1 + \frac{\langle H \rangle}{\sum_q m_q} \right). \tag{20.7}$$

The Hamiltonian $\langle H \rangle$ is $\langle H \rangle = M - \sum_q m_q$, where M is the measured or predicted meson mass and $\sum_q m_q$ is the sum of the masses of three constituent quarks. To determine the magnetic moment of the ground state D^{*0} meson, the (20.5) will be

$$\mu_{D^{*0}} = \sum_q \langle \Phi_{sf_{D^{*0}}} | \hat{\mu}_{q_z} | \Phi_{sf_{D^{*0}}} \rangle, \tag{20.8}$$

and we write its spin-flavor wave function as

$$\left| \Phi_{sf_{D^{*0}}} \right\rangle = c \uparrow \bar{u} \uparrow . \tag{20.9}$$

Hence, the (20.8) becomes

$$\mu_{D^{*0}} = \sum_q \langle c \uparrow \bar{u} \uparrow | \hat{\mu}_{q_z} | c \uparrow \bar{u} \uparrow \rangle. \tag{20.10}$$

An expectation values of the z-component of the magnetic moment of the up and charm quarks are $\langle u \uparrow | \hat{\mu}_z | u \uparrow \rangle = +\mu_u$ and $\langle c \uparrow | \hat{\mu}_z | c \uparrow \rangle = +\mu_c$ & applying the quark charge(Q), so we will have

Table 20.2 Magnetic moments of D^{*0}, D^{*+} and D_s^{*+} mesons of $J^P = 1^-$ (in μ_N)

Meson	Magnetic moment	Present	[18]	[20]	[19]
D^{*0}	$\frac{2}{3}\mu_c - \frac{2}{3}\mu_{\bar{u}}$	−0.845	−1.21	–	−0.89
$D^{*\pm}$	$\frac{2}{3}\mu_c + \frac{1}{3}\mu_{\bar{d}}$	1.042	1.06	1.16	1.17
D_s^{*+}	$\frac{2}{3}\mu_c + \frac{1}{3}\mu_{\bar{s}}$	0.922	0.87	0.98	1.03

$$\mu_{D^{*0}} = \frac{2}{3}\mu_c - \frac{2}{3}\bar{u}, \tag{20.11}$$

which is the quark model prediction for the magnetic moment of D^{*0} meson. In the similar manner, we will get an expression of the magnetic moment of $D^{*\pm}$ meson as $\mu_{D^{*\pm}} = \frac{2}{3}\mu_c + \frac{2}{3}\bar{d}$ & $D_s^{*+} = \frac{2}{3}\mu_c + \frac{1}{3}\mu_{\bar{s}}$. Our results are listed in Table 20.2 with other theoretical predictions.

20.3.2 M1 Transitions and Decay Rates

The M1 transitions flip the quark spin in the same orbital state which results in the transition from vector (spin $= 1$) to pseudoscalar (spin $= 0$) meson. Specially, it is necessary for the study of radiative decay rates of the low-lying state which contributes more to the total decay rates. An expression of M1 decay rate is [18, 35, 36],

$$\Gamma_{M_V \to M_P + \gamma} = \frac{4\alpha k^3}{3m_p} \frac{2}{2J+1} (\mu_{M_V \to M_P})^2, \tag{20.12}$$

where M_V and M_P are initial and final state vector and pseudoscalar meson, respectively. $\alpha \sim \frac{1}{137}$ is the fine structure constant for the electromagnetic transitions, k is the photon momentum, m_p is the proton mass, and J represents the total angular momentum of the vector meson (M_V). For the transition of $\mu_{D^{*+} \to D^+}$, the spin-flavor combinations of D^{*+} and D^+ are

$$\left| \Phi_{sf_{D^{*+}}} \right\rangle = (c\bar{d}) \cdot \left(\frac{1}{\sqrt{2}} (\uparrow\downarrow + \downarrow\uparrow) \right) \quad and \quad \left| \Phi_{sf_{D^+}} \right\rangle = (c\bar{d}) \cdot \left(\frac{1}{\sqrt{2}} (\uparrow\downarrow - \downarrow\uparrow) \right), \tag{20.13}$$

respectively. Using the appropriate combination of the spin-flavor wave function for the transitions $\mu_{D^{*0} \to D^0}$ and $\mu_{D_s^{*+} \to D_s^+}$, we will get an expression of their magnetic dipole transition by following the process as we discussed in the above section and the orthogonality of quark flavor and spin states, for example, $\langle c \uparrow \bar{d} \downarrow | c \downarrow \bar{d} \uparrow \rangle = 0$. So we have M1 dipole transitions (in μ_N) as

Table 20.3 M1 decay widths of D^{*0}, D^{*+}, and D_s^{*+} mesons of $J^P = 1^-$ (in keV)

Radiative decay	Present	[18]	[21]	[22]	[23]	[24]
$D^{*0} \to D^0 + \gamma$	37.45	19.7	1.25	32±1	11.5	33.5
$D^{*+} \to D^+ + \gamma$	0.852	1.10	1.10	1.5	1.04	1.63
$D_s^{*+} \to D_s^+ + \gamma$	0.176	0.40	0.337	0.32±0.01	0.19	0.43

- $\mu_{D^{*0} \to D^0}$ $\quad \frac{2}{3}\mu_c + \frac{2}{3}\mu_{\bar{u}}$ $\quad$ 1.671
- $\mu_{D^{*+} \to D^+}$ $\quad \frac{2}{3}\mu_c - \frac{1}{3}\mu_{\bar{d}}$ $\quad$ −0.216
- $\mu_{D_s^{*+} \to D_s^+}$ $\quad \frac{2}{3}\mu_c - \frac{1}{3}\mu_{\bar{s}}$ $\quad$ −0.082.

Using these transition magnetic moments, the M1 decay widths are calculated. Our results are listed in Table 20.3 and compared with other theoretical predictions.

20.4 Summary

In this work, the masses of D and D_s mesons are calculated in the semi-relativistic framework of potential model. Their ground state masses are used to calculate the magnetic properties: magnetic moments, M1 transitions, and decay widths, which are listed in Tables 20.2 and 20.3. Our results of the ground state magnetic properties of D and D_s mesons of $J^P = 1^-$ are in accordance with other theoretical predictions. So we would like to extend our work for the ground state magnetic properties of the light-light, heavy-light, and the heavy-heavy flavored mesons.

References

1. R. Aaij et al., LHCb collaboration. Phys. Rev. D **94**, 072001 (2016)
2. R. Aaij et al., LHCb collaboration. Phys. Rev. D **92**, 032002 (2015)
3. R. Aaij et al., LHCb collaboration. JHEP **09**, 145 (2013)
4. M. Tanabashi et al., Particle data group. Phys. Rev. D **98**, 030001 (2018)
5. B. Singh et al., $\bar{P}$ANDA collaboration. Phys. Rev. D **95**, 032003 (2017); Eur. Phys. J. A 52, 325 (2016); Nucl. Phys. A 954, 323 (2016); Eur. Phys. J. A 51, 107 (2015); arxiv: 1704.02713 (2017); G. Barucca et al., Eur. Phys. J. A **55**, 42 (2019)
6. J. Lahkar, R. Hoque, D.K. Choudhury, Mod. Phys. Lett. A **34**, 1950106 (2019)
7. J. Lahkar, D.K. Choudhury, B.J. Hazarika (2019). arxiv: 1902.02079
8. D. Jia, W.-C. Dong, Eur. Phys. J. Plus **134**, 123 (2019)
9. J.-K. Chen, Eur. Phys. J. C **78**, 648 (2018)
10. V. Kher, N. Devlani, A.K. Rai, Chin. Phys. C **41**, 073101 (2017)
11. M. Shah, B. Patel, P.C. Vinodkumar, Eur. Phys. J. C **76**, 36 (2016)
12. S. Godfrey, K. Moats, Phys. Rev. D **93**, 034035 (2016)
13. Y. Sun, X. Liu, T. Matsuki, Phys. Rev. D **88**, 094020 (2013)
14. D. Ebert, R. Faustov, V. Galkin, Eur. Phys. J. C **66**, 197 (2010)
15. N. Devlani, A.K. Rai, Int. J. Theor. Phys. **52**, 2196 (2013)
16. G. Moir et al., JHEP **05**, 021 (2013)

17. A.K. Rai, R.H. Parmar, P.C. Vinodkumar, J. Phys. G **28**, 2275 (2002)
18. V. Simonis, Eur. Phys. J. A **52**, 90 (2016)
19. S.K. Bose, L.P. Singh, Phys. Rev. D **22**, 773 (1980)
20. Y.-L. Luan, X.-L. Chen, W.-Z. Deng, Chin. Phys. C **39**, 113103 (2015)
21. T.A. Löahde, C.J. Nyfält, D.O. Riska, Nucl. Phys. A **674**, 141 (2000)
22. J.L. Goity, W. Roberts, Phys. Rev. D **64**, 094007 (2001)
23. D. Ebert, R.N. Faustov, V.O. Galkin, Phys. Lett. B **537**, 241 (2002)
24. W.A. Bardeen, E.J. Eichten, C.T. Hill, Phys. Rev. D **68**, 054024 (2003)
25. N.R. Soni, J.N. Pandya, Phys. Rev. D **96**, 016017 (2017)
26. N.R. Soni et al., Phys. Rev. D **98**, 114031 (2018)
27. N. Devlani, A.K. Rai, Phys. Rev. D **84**, 074030 (2011)
28. V. Kher, N. Devlani, A.K. Rai, Chin. Phys. C **41**, 093101 (2017)
29. Y. Koma, M. Koma, H. Wittig, Phys. Rev. Lett. **97**, 122003 (2006)
30. E. Eichten et al., Phys. Rev. D **17**, 3090 (1978)
31. B. Patel, A.K. Rai, P.C. Vinodkumar, J. Phys. G **35**, 065001 (2008)
32. Z. Shah et al., Chin. Phys. C **40**, 123102 (2016); Eur. Phys. J. A **52**, 313 (2016)
33. K. Gandhi, Z. Shah, A.K. Rai, Eur. Phys. J. Plus **133**, 512 (2018)
34. Z. Shah, K. Gandhi, A.K. Rai, Chin. Phys. C **43**, 034102 (2019)
35. K. Gandhi et al., DAE Symp. Nucl. Phys. **63**, 824 (2018)
36. V. Simonis, (2018). arxiv: 1804.04872
37. A.K. Rai, J.N. Pandya, P.C. Vinodkumar, J. Phys. G **31**, 1453 (2005)

Chapter 21
Search for Resonant Higgs Boson Pair Production in the $4W$ Channel with $3l + 2j$ Final State at $\sqrt{s} = 13$ TeV with CMS Detector

S. Sawant

21.1 Introduction

The discovery of the Standard Model (SM) like Higgs boson (H) at 125 GeV/c^2 mass by the CMS and the ATLAS experiments in 2012 [1, 2] corroborates the mechanism of electroweak symmetry breaking (EWSB). Precise measurements of coupling of this discovered H with itself and other elementary particles in SM are crucial in determining its exact role in EWSB mechanism. Pair production of Higgs boson (HH) provides a direct probe to the Higgs boson self-coupling. The SM cross section of HH production in proton-proton collisions at $\sqrt{s} = 13$ TeV is 33.5 fb at next-to-next-to-leading order in quantum chromodynamics for the gluon-gluon fusion process [3]. Many beyond SM studies predict enhancement to this production rate which is attributed to either anomalous Higgs coupling [4–6] or resonant HH production [7, 8].

This article presents an update on a search for resonant HH production in $X \rightarrow$ HH $\rightarrow 4W \rightarrow 3(\ell\,\nu)\,2j$² channel, where resonance particle X is assumed to be a spin-0 radion with mass 400, 700 GeV/c^2. Here ℓ refers to a lepton (electron or muon) and j refers to a jet.

---

[1] Branching fraction (BF) of HH$\rightarrow 4W \rightarrow 3(\ell\,\nu)\,2j$ is 2.1×10^{-3}, following BR of H and W from [9].

S. Sawant for the CMS Collaboration

S. Sawant (✉)
Tata Institute of Fundamental Research, Mumbai, India
e-mail: siddhesh.sawant@tifr.res.in; sawantsiddhesh08@gmail.com

© Springer Nature Singapore Pte Ltd. 2021
P. K. Behera et al. (eds.), *XXIII DAE High Energy Physics Symposium*,
Springer Proceedings in Physics 261,
https://doi.org/10.1007/978-981-33-4408-2_21

21.2   CMS Experiment

The Compact Muon Solenoid (CMS) experiment is one of the four particle physics experiments situated at the Large Hadron Collider (LHC). It has a multipurpose detector system consisting of a silicon tracker, superconducting solenoid providing 3.8 T magnetic field, electromagnetic and hadron calorimeters, and a muon tracking system [10]. The search for resonant HH production reported here is performed with proton-proton collision data collected at $\sqrt{s} = 13$ TeV in 2017, amounting to an integrated luminosity of 41.5 fb^{-1}.

21.3   Event Selection

The analyzed data are collected using a collection of single, double or triple lepton triggers. The p_T threshold for the triggers are 32, 23, and 9 GeV for leading electrons and 24, 17, and 12 GeV for leading muons, respectively.

A particle flow (PF) algorithm [11] is utilized to combine the information from all the CMS detectors to identify and reconstruct particles in an event, namely electrons, muons, photons, neutral, and charged hadrons. These particles are then used to reconstruct jets, hadronically decaying τ leptons and the missing transverse momentum (p_T^{miss}) vector. Jets are reconstructed using the anti-kT algorithm with 0.4 cone radius parameter. Transverse component of a negative vector sum of momenta of all PF candidates is defined as p_T^{miss} to account for the contribution of neutrinos from the event.

Events with at least three leptons with 25, 15, 10 GeV p_T thresholds and at least two jets with p_T more than 25 GeV are selected. The first three leading p_T leptons and the first two leading p_T jets are considered for the analysis. Events with b-tagged (with "combined secondary vertex" algorithm [12]) jets, a lepton pair with its mass close to the Z-boson mass in a window of 10 GeV/c^2 and four leptons with their invariant mass less than 140 GeV/v^2 are vetoed. This veto condition helps to reduce background contribution from the SM Higgs boson and Drell-Yan+jets reactions.

For the selected events, a distribution of the invariant mass of three leptons, two jets, and p_T^{miss} is shown in Figure 21.1 left panel, and the right panel displays the histogram of the number of same flavor opposite sign lepton pairs in an event. Major background contribution comes from events containing VV+jets, Drell-Yan+jets, $t\bar{t}$+jets, VH+jets, tH+jets, $t\bar{t}H$+jets, and $t\bar{t}V$+jets, where V refers to W, Z bosons. These background contributions are estimated from simulation. A background contribution from non-prompt leptons or jets faking leptons of interest is estimated from data using the fake factor method [13].

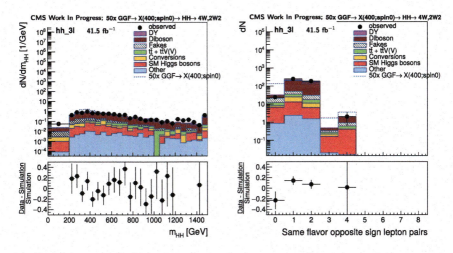

Fig. 21.1 **Left**: The invariant mass of three leptons, two jets and $p_{\mathrm{T}}^{\mathrm{miss}}$ system. **Right**: Histogram of multiplicity of same flavor opposite sign lepton pairs. The cross section of resonant HH production is assumed to be 1 pb for plotting purpose

21.4 Signal Extraction

A multivariate technique called Boosted Decision Tree (BDT) is utilized to achieve better signal over background classification by combining information from many kinematics variables. A BDT is trained on MC samples of HH signal and the major contributing background process, WZ+jets, Drell-Yan+jets, and $t\bar{t}$+jets. Variables used as input to the BDT are: $m_{\ell\ell}$, $\Delta R(\ell\ell\,\mathrm{OS})$, $\Delta R(\ell\ell\,\mathrm{SS})$, $\Delta R(\ell, j)$, $p_{\mathrm{T}}(\ell)$, $\eta(\ell)$, $\Delta R(jj)$, $p_{\mathrm{T}}^{\mathrm{miss}}$ and m_{jj}. Here OS and SS refer to opposite and same sign pair, respectively, and $\Delta R = \sqrt{(\Delta\eta)^2 + (\Delta\phi)^2}$. The receiver operating characteristics (ROC) curve of the BDT is shown in the left panel of Figure 21.2. The right panel of the figure shows a distribution of the BDT discriminating score.

The best fit signal is extracted by performing a binned maximum likelihood fit to the BDT discriminating score distribution. The likelihood function is the product of the Poisson likelihoods over all bins of the distribution and is given by

$$L(\beta_{\mathrm{signal}}, \beta_{\mathrm{background}} \mid \mathrm{data}) = \prod_{i=1}^{N_{\mathrm{bins}}} \frac{\mu_i^{n_i} e^{-\mu_i}}{n_i!} , \qquad (21.1)$$

where n_i is the number of observed events in bin i and the Poisson mean for bin i is given by

$$\mu_i = \beta_{\mathrm{signal}} S_i + \sum_k \beta_k T_{k,i} , \qquad (21.2)$$

where k denotes all of the considered background processes, $T_{k,i}$ is the bin content of bin i of the distribution for process k, and S_i is the bin content of bin i of the signal

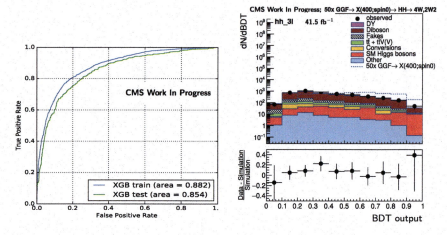

Fig. 21.2 **Left**: ROC curve of BDT. **Right**: A distribution of BDT discriminator score. The cross section of resonant HH production is assumed to be 1 pb for plotting purpose

distribution. The parameter β_k is the nuisance parameter for the normalization of the process k, constrained by theoretical uncertainty with a log-normal prior, and β_{signal} is the unconstrained signal strength.

Preliminary expected upper limits at 95% confidence level (CL) on the product of production cross section for $pp \rightarrow X \rightarrow$ HH using the asymptotic modified frequentist method (asymptotic CLs) [14] is 6.1 pb and 1.1 pb limit for $m_X = 400, 700$ GeV/c^2, respectively.

21.5 Outlook

A next step would be to include reconstruction of boosted W jets using the anti-kT algorithm with 0.8 cone radius parameter, followed by estimation of the effect of possible systematics sources on the final result and calculation of systematic uncertainties. In future, the analysis will include more mass points of resonance X in a range from 250 to 1000 GeV/c^2 and also its spin-2 nature.

References

1. ATLAS Collaboration, Observation of a new particle in the search for the Standard Model Higgs boson with the ATLAS detector at the LHC. Phys. Lett. B **716**, 1 (2012)
2. C.M.S. Collaboration, Observation of a new boson at a mass of 125 GeV with the CMS experiment at the LHC. Phys. Lett. B **716**, 30 (2012)

3. D. de Florian et al., Handbook of LHC Higgs cross sections: 4. deciphering the nature of the Higgs sector. CERN Yellow Report CERN-2017-002-M (2016)
4. A. Azatov, R. Contino, G. Panico, M. Son, Effective field theory analysis of double Higgs boson production via gluon fusion. Phys. Rev. D **92**, 035001 (2015)
5. F. Goertz, A. Papaefstathiou, L.L. Yang, J. Zurita, Higgs boson pair production in the D = 6 extension of the SM. JHEP **04**, 167 (2015)
6. B. Hespel, D. Lopez-Val, E. Vryonidou, Higgs pair production via gluon fusion in the Two-Higgs-Doublet Model. JHEP **09**, 124 (2014)
7. G.C. Branco et al., Theory and phenomenology of two-Higgs-doublet models. Phys. Rept. **516**, 1 (2012)
8. L. Randall, R. Sundrum, A large mass hierarchy from a small extra dimension. Phys. Rev. Lett. **83**, 3370 (1999)
9. C. Patrignani et al., Particle data group. Chin. Phys. C **40**, 100001 (2016)
10. CMS collaboration, The CMS experiment at the CERN LHC, 2008 JINST 3 S08004 (2008)
11. C.M.S. Collaboration, Particle-flow reconstruction and global event description with the CMS detector. JINST **12**, P10003 (2017)
12. CMS Collaboration, Identification of heavy-flavour jets with the CMS detector in pp 731 collisions at 13 Tev. J. Instrum. **13**(05), P05011 (2018)
13. CMS Collaboration, Search for the associated production of a Higgs boson with a top quark pair in final states with electrons, muons and hadronically decaying τ leptons at $\sqrt{s}$ = 13 TeV (2017 data), PAS HIG-18-019
14. G. Cowan, K. Cranmer, E. Gross, O. Vitells, Asymptotic formulae for 579 likelihood-based tests of new physics. Eur. Phys. J. C **71**, 1554 (2011)

Chapter 22
Study of $X(3872)$ and $X(3915)$ in $B \rightarrow (J/\psi\omega)K$ at Belle

Sourav Patra, Rajesh K. Maiti, and Vishal Bhardwaj

Abstract We present a preliminary study of $X(3872)$ and $X(3915)$ in the $B \rightarrow (J/\psi\omega)K$ decay at Belle. This study is based on MC simulated events on the Belle detector at the KEK asymmetric-energy $e + e-$ collider.

22.1 Motivation

The $X(3872)$ was discovered by the Belle collaboration in the $B \rightarrow (J/\psi\pi^+\pi^-)K$ decay mode [1]. It is difficult to assign $X(3872)$ as a conventional state due to its mass near DD^* threshold and narrow width (<1.2 MeV) [2]. As per the current scenario, it is expected to be an admixture of DD^* molecular state and $c\bar{c}$ state. There is no signature for the charge partner in $J/\psi\pi^+\pi^0$ and no signature of odd charge conjugate ($C = -1$) partner in the $\eta_c\omega$ and $\eta_c\pi^+\pi^-$ decay [3]. So, $X(3872)$ is suggested to be an iso-singlet state. In that scenario, $X(3872) \rightarrow J/\psi\pi^+\pi^-$ is isospin violating decay. On the other hand, $X(3872)$ decays to $J/\psi\omega$ which is an isospin allowed decay. It has been suggested that the ratio of $\mathcal{B}[X(3872) \rightarrow J/\psi\pi^+\pi^0\pi^-]$ to $\mathcal{B}[X(3872) \rightarrow J/\psi\pi^+\pi^-]$ should be 30. However, BaBar collaboration has measured this ratio to be 0.8 ± 0.3 [4]. Measuring this ratio with precision will be very useful in understanding the nature of the $X(3872)$.

S. Patra (✉) · R. K. Maiti · V. Bhardwaj
IISER Mohali, Punjab 140306, India
e-mail: souravpatra3012@gmail.com

R. K. Maiti
e-mail: rkumar30795@gmail.com

V. Bhardwaj
e-mail: vishstar@gmail.com

© Springer Nature Singapore Pte Ltd. 2021
P. K. Behera et al. (eds.), *XXIII DAE High Energy Physics Symposium*,
Springer Proceedings in Physics 261,
https://doi.org/10.1007/978-981-33-4408-2_22

155

22.2 Analysis Strategy

We generated events for each of $X(3872)$ and $X(3915)$ decay using EvtGen package. Those events were simulated according to the Belle detector using GSIM [5]. Generated MC was used to optimize and validate our study. For this analysis, we reconstruct $B^{\pm}(B^0)$ from $J/\psi\omega K^{\pm}(K_S^0)$, where we further reconstruct J/ψ from ee, $\mu\mu$, and ω from $\pi^+\pi^0\pi^-$. We identify K_s in $\pi^+\pi^-$ decay and π^0 in $\gamma\gamma$ decay. Maximum unbinned likelihood fit is performed for $J/\psi\omega$ invariant mass to measure the yield for corresponding signal and backgrounds.

22.2.1 Particle Identification and Basic Selection

The distance of closest approach from the IP in azimuthal direction ($|dr|$) is less than 1 cm and that in horizontal direction ($|dz|$) is less than 3.5 cm. Fox-Wolfram moment (R_2) less than 0.5 is used to suppress continuum background events. We select the $K^{\pm}$ with kaon vs pion likelihood, $\mathcal{R}_K/(\mathcal{R}_K + \mathcal{R}_\pi)$ greater than 0.6 and that for $\pi^{\pm}$ is less than 0.4. All gamma candidates having energy more than 60 MeV and E_9/E_{25} in ECL crystal > 0.85 are selected. π^0 candidates having mass from 123 to 147 MeV/c^2 are kept for future combination. We select K_s^0 having mass within [482,524] MeV/c^2. We choose the mass window for selected omega from 0.7 GeV/c^2 to 0.85 GeV/c^2. J/ψ candidates having mass from 3.07 to 3.13 GeV/c^2 for $\mu\mu$ events and from 3.05 to 3.13 GeV/c^2 for ee events are selected. Photons within 50 mrad of each $e^{\pm}$ track are selected as bremsstrahlung photon to get the corrected mass and momentum for J/ψ. We use two parameters: beam constrained mass (M_{bc}) and ΔE where $M_{bc} = \sqrt{E_{cm}^{*2} - \mathbf{p}_B^{*2}}$ and $\Delta E = E_{beam}^* - E_B^*$ to set the proper signal window. One should expect ΔE to peak at 0 and M_{bc} to peak at nominal B mass. Events within $|\Delta E| < 0.2$ and $M_{bc} > 5.27$ GeV/c^2 are selected as reconstructed events for further study.

22.2.2 Omega Selection with Dalitz Method

We reconstruct ω from $\pi^+\pi^-\pi^0$. Due to its broad width and poor efficiency in π^0 reconstruction, large number of fake combinations for ω are selected. In order to avoid those fake combinations, we use Dalitz cuts. Kinematics of $\omega \to \pi^+\pi^0\pi^-$ decay can be represented in XY plane, where $X = \sqrt{3}(T_{\pi^+} - T_{\pi^-})/Q$ and $Y = (2T_{\pi^0} - T_{\pi^+} - T_{\pi^-})/Q$. Here, T is the kinetic energy of the corresponding particle and Q implies the total kinetic energy of all three particles. We apply two concentric circular cuts centered at (0,3) in XY plane, $1.5 < |\sqrt{X^2 + (Y - 3)^2}| < 3.8$, which give the maximum fake events rejection (28.61%) and minimum true events rejection (7.18%).

22.2.3 Best Candidate Selection

Multiple B candidates are reconstructed for 35% of reconstructed events. Best candidate is selected among those multiple B candidates with least χ^2, where

$$\chi^2 = \chi_V^2 + \left(\frac{\Delta E}{\sigma_{\Delta E}}\right)^2 + \left(\frac{M_{ll} - m_{J/\psi}}{\sigma_{J/\psi}}\right)^2 + \left(\frac{M_{\pi^+\pi^-\pi^0} - m_\omega}{\sigma_\omega}\right)^2$$
$$+ \left(\frac{M_{\gamma\gamma} - m_{\pi^0}}{\sigma_{\pi^0}}\right)^2 + \left(\frac{M_{\pi^+\pi^-} - m_{K_S}}{\sigma_{K_S}}\right)^2$$

Here, χ_V^2 is returned χ^2 from charge vertex fit and $\sigma_{\Delta E}$ is the width in ΔE. M, m, σ imply the reconstructed mass, PDG mass, mass width of the corresponding particle, respectively. Truthmatched signal reconstruction efficiency using this method is 68% for charged B meson and 57% for neutral B meson.

22.2.4 ΔE Optimization

We optimize the ΔE window for the candidates selected with best candidate selection to set the proper signal window for ΔE. We plot figure of merit (F_{OM}) as a function of ΔE, where $F_{OM} = N_{sig}/\sqrt{N_{sig} + N_{bkg}}$. Here, N_{sig} and N_{bkg} represent the number of signal and background events, respectively. Number of events from signal MC sample in a particular ΔE region are scaled by the branching fractions. We optimize the region $|\Delta E| < 20$ MeV as signal window.

22.2.5 ΔE and π^0 Mass Constrain Fit

ΔE should be zero for perfectly reconstructed events. We assume that our ΔE resolution is not good due to problem in π^0 reconstruction. Therefore, we force ΔE to be zero by keeping π^0 invariant mass fixed. So, new π^0 momentum is shifted by a factor of α, where $\alpha = \sqrt{(1 - (1 - s^2)E_{\pi^0}^2/\mathbf{P}_{\pi^0}^2)}$ with $s = [E_{beam} - (E_{\pi^+} + E_{\pi^-} + E_{K_S})]/E_{\pi^0}$. After performing this fit, we get ω candidate with better mass resolution.

22.3 Background Study

We use large $B \to J/\psi X$ inclusive MC sample (having 100 times statistics compared to data) to understand the sources of background. As we are interested in $M_{J/\psi\omega}$, we check the distribution for $M_{J/\psi\omega}$ and $M_{\omega K}$ (Fig. 22.1). One can clearly

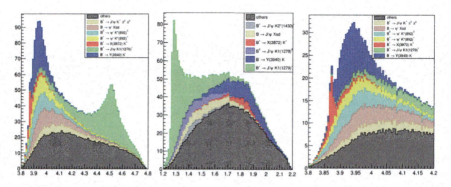

Fig. 22.1 Backgrounds in $M_{J/\psi\omega}$ and $M_{\omega K}$ for neutral B meson are plotted in left and middle plot respectively. Right side one is representing the $M_{J/\psi\omega}$ for whole MC sample after applying $\psi'K^*$ veto and $M_{\omega K}$ cut

see from the $M_{\omega K}$ distribution that, by applying a cut $M_{\omega K} > 1.4$ GeV most of background coming from $B \to J/\psi K_1(1270)$ decay can be removed. For extracting $X(3872)$ and $X(3915)$ signal, we look at $M_{J/\psi\omega}$ distribution from 3.81 to 4.2 GeV/c^2.

22.3.1 $\psi'K^*$ Veto

Background coming from the $B \to \psi'K^*$ decay is peaking around the signal peak. Here we expect $J/\psi\pi^+\pi^-$ coming from ψ' and $\pi^0 K$ from K^* to mimic our signal. Therefore, we apply $\psi'K^*$ veto, $3.67 \text{GeV}/c^2 < M_{J/\psi\pi^+\pi^-} < 3.72 \text{GeV}/c^2$ and $0.79 \text{GeV}/c^2 < M_{\pi^0 K} < 0.99 \text{GeV}/c^2$ to reduce such background.

22.4 Signal Extraction with Maximum Likelihood Fit

For extracting signal efficiency, we perform 1D unbinned maximum likelihood fit (UML) for $X(3872)$ and $X(3915)$ with signal MC sample. We model each of the signals with one Gaussian and two bifurcated Gaussians. For the peaking backgrounds, $B^\pm \to \psi'K^{*\pm}$ and $B^0 \to \psi'K^{*0}$, we use one Gaussian and two bifurcated Gaussian. The rest of the backgrounds have flat nature in the signal region. Therefore, we use a threshold function to model those backgrounds. Finally, we combine all the PDFs in a single PDF fixing all the parameters from signal MC including mean and sigma for $X(3872)$ and $X(3915)$, floating the yields of all three PDFs (Fig. 22.2).

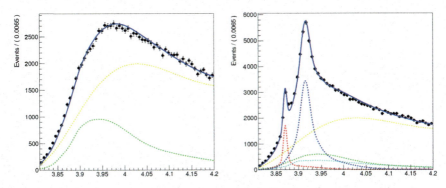

Fig. 22.2 1D UML fit to $M_{J/\psi\omega}$ distribution for the possible background estimated from $B \to J/\psi X$ sample (left) and total fit after signal inclusion for $B^\pm \to J/\psi\omega K^\pm$ decay modes (right)

22.5 Conclusion

A preliminary MC study for $B \to J/\psi\omega K$ is presented here. We tried different methods to reduce the cross feed and to improve the resolution of $M_{J/\psi\omega}$. Precise measurement of $\mathcal{B}[X(3872) \to J/\psi\omega]$ to $\mathcal{B}[X(3872) \to J/\psi\pi^+\pi^-]$ will help in understanding the nature of $X(3872)$.

References

1. S.-K. Choi et al., Belle collaboration. Phys. Rev. Lett. **91**, 262001 (2003)
2. T. Aushev et al., Belle collaboration. Phys. Rev. D **81**, 031103(R) (2010)
3. A. Vinokurova et al., Belle collaboration. J. High Energ. Phys. **2015**, 132 (2015)
4. P. del Amo Sanchez et al., (BABAR Collaboration). Phys. Rev. D **82**, 011101(R) (2010)
5. R. Brun et al., GEANT3.21, CERN Report No. DD/EE/84-1 (1984)

Chapter 23
$B_s^* \to l^+l^-$ Decays in Light of Recent B Anomalies

Suman Kumbhakar and Jyoti Saini

Abstract Some of the recent measurements in the neutral current sector $b \to sl^+l^-$ ($l = e$ or μ), as well as in the charged current sector $b \to c\tau\bar{\nu}$, show significant deviations from their Standard Model predictions. It has been shown that two different new physics solutions, in the form of vector and/or axial vector, can explain all the anomalies in $b \to sl^+l^-$ sector. We show that the muon longitudinal polarization asymmetry in $B_s^* \to \mu^+\mu^-$ decay is a good discriminant between the two solutions if it can be measured to a precision of $\sim 10\%$, provided the new physics Wilson coefficients are real. We also investigate the potential impact of $b \to c\tau\bar{\nu}$ anomalies on $B_s^* \to \tau^+\tau^-$ decay. We consider a model where the new physics contributions to these two transitions are strongly correlated. We find that two orders of magnitude enhancement in the branching ratio of $B_s^* \to \tau^+\tau^-$ is allowed by the present $b \to c\tau\bar{\nu}$ data.

23.1 Introduction

The recent anomalies in the charged current (CC) transition $b \to c\tau\bar{\nu}$ and in the flavor changing neutral current (FCNC) transitions $b \to sl^+l^-$ ($l = e$ or μ) provide tantalizing hints of physics beyond Standard Model (SM). In the SM, the above CC transition occurs at the tree level, whereas the FCNC transitions occur only at the loop level.

Some of the anomalies in $b \to sl^+l^-$ sector are: angular observables in $B \to K^*\mu^+\mu^-$ [1–3] particularly P_5' in 4.3-8.68 GeV^2 bin, the branching ratio of $B_s \to \phi\mu^+\mu^-$ and the corresponding angular observables [4, 5], the flavor ratio $R_K \equiv \Gamma(B^+ \to K^+\mu^+\mu^-)/\Gamma(B^+ \to K^+e^+e^-)$ in $1.0 \le q^2 \le 6.0$ GeV^2 [6], the ratio $R_{K^*} \equiv \Gamma(B^0 \to K^{*0}\mu^+\mu^-)/\Gamma(B^0 \to K^{*0}e^+e^-)$ in two different q^2 ranges,

S. Kumbhakar (✉)
Indian Institute of Technology, Bombay, India
e-mail: suman@phy.iitb.ac.in; kumbhakar.suman@gmail.com

J. Saini
Indian Institute of Technology, Jodhpur, India

© Springer Nature Singapore Pte Ltd. 2021
P. K. Behera et al. (eds.), *XXIII DAE High Energy Physics Symposium*,
Springer Proceedings in Physics 261,
https://doi.org/10.1007/978-981-33-4408-2_23

$(0.045 \leq q^2 \leq 1.1 \text{ GeV}^2)$ (low q^2) and $(1.1 \leq q^2 \leq 6.0 \text{ GeV}^2)$ (central q^2) [7]. In Moriond'19, the Belle collaboration has published their first measurements of R_{K^*} in both B^0 and B^+ decays. These measurements are reported in multiple q^2 bins and have comparatively large uncertainties [8]. Further, LHCb collaboration updated the value of R_K in Moriond'19 [9]. After Moriond'19, [10, 11] performed a global fit to identify the Lorentz structure of new physics (NP) which can account for all anomalies in $b \rightarrow s\mu^+\mu^-$ sector. In 1D scenario, there are two distinct solutions, one with the operator of the form $(\bar{s}\gamma^\alpha P_L b)(\bar{\mu}\gamma_\alpha\mu)$ and the other whose operator is a linear combination of $(\bar{s}\gamma^\alpha P_L b)(\bar{\mu}\gamma_\alpha\mu)$ and $(\bar{s}\gamma^\alpha P_L b)(\bar{\mu}\gamma_\alpha\gamma_5\mu)$.

It is interesting to look for new observables in the $b \rightarrow s\mu^+\mu^-$ sector in order to (a) find additional evidence for the existence of NP and (b) to discriminate between the two NP solutions. The branching ratio of $B_s^* \rightarrow \mu^+\mu^-$ is one such observable which is yet to be measured. In the SM, this decay mode is not subject to helicity suppression [12], unlike $B_s \rightarrow \mu^+\mu^-$. A model independent analysis of this decay was performed in [13] to identify the NP operators which can lead to a large enhancement of its branching ratio. It was found that such an enhancement is not possible due to the constraints from the present $b \rightarrow s\mu^+\mu^-$ data. In this work, we consider the longitudinal polarization asymmetry of muon in $B_s^* \rightarrow \mu^+\mu^-$ decay, $\mathcal{A}_{LP}(\mu)$. This asymmetry is theoretically clean because it has a very mild dependence on the decay constants unlike the branching ratio. We first calculate the SM prediction of $\mathcal{A}_{LP}(\mu)$ and then study its sensitivity to the NP solutions.

On the other hand, the discrepancies in the CC $b \rightarrow c\tau\bar{\nu}$ transition are: the ratios $R_{D^{(*)}}=\Gamma(B \rightarrow D^{(*)}\tau\bar{\nu})/\Gamma(B \rightarrow D^{(*)}\{e/\mu\}\bar{\nu})$ [14], $R_{J/\psi}=\mathcal{B}(B \rightarrow J/\psi\tau\bar{\nu})/\mathcal{B}(B \rightarrow J/\psi\mu\bar{\nu})$ [15]. References [16–18] identified the allowed NP solutions which can explain all anomalies in the $b \rightarrow c\tau\bar{\nu}$ sector and suggested methods to distinguish between various NP solutions. The NP WCs of these solutions are about 10% of the SM values. Since this transition occurs at the tree level in the SM, it is very likely that the NP operators also occur at the tree level. In [19], a model is constructed where the tree level FCNC terms due to NP are significant for $b \rightarrow s\tau^+\tau^-$ but are suppressed for $b \rightarrow sl^+l^-$ where $l = e$ or $l = \mu$. The branching ratios for the decay modes such as $B \rightarrow K^{(*)}\tau^+\tau^-$, $B_s \rightarrow \tau^+\tau^-$ and $B_s \rightarrow \phi\tau^+\tau^-$ will have a large enhancement in this model [19]. In this work, we study the effect of this NP on the branching ratio of $B_s^* \rightarrow \tau^+\tau^-$ and the τ polarization asymmetry $\mathcal{A}_{LP}(\tau)$.

23.2 Longitudinal Polarization Asymmetry for $B_s^* \rightarrow l^+l^-$ Decay

The decay $B_s^* \rightarrow l^+l^-$ is induced by the quark level transition $b \rightarrow sl^+l^-$. In the SM the corresponding effective Hamiltonian is

$$\mathcal{H}_{SM} = \frac{4G_F}{\sqrt{2\pi}} V_{ts}^* V_{tb} \left[\sum_{i=1}^{6} C_i(\mu) O_i(\mu) + C_7 \frac{e}{16\pi^2} [\bar{s}\sigma_{\mu\nu}(m_s P_L + m_b P_R)b] F^{\mu\nu} \right.$$

$$\left. + C_9 \frac{\alpha_{em}}{4\pi} (\bar{s}\gamma^\mu P_L b)(\bar{l}\gamma_\mu l) + C_{10} \frac{\alpha_{em}}{4\pi} (\bar{s}\gamma^\mu P_L b)(\bar{l}\gamma_\mu \gamma_5 l) \right],$$

where G_F is the Fermi constant, V_{ts} and V_{tb} are the Cabibbo-Kobayashi-Maskawa (CKM) matrix elements and $P_{L,R} = (1 \mp \gamma^5)/2$ are the projection operators. The effect of the operators O_i, $i = 1 - 6, 8$ can be embedded in the redefined effective Wilson coefficients as $C_7(\mu) \to C_7^{eff}(\mu, q^2)$ and $C_9(\mu) \to C_9^{eff}(\mu, q^2)$. The form factor parameterization of the $B_s^* \to l^+l^-$ decay amplitudes are given in [12]. These parameterization depend on the decay constants of B_s^* meson $f_{B_s^*}$ and $f_{B_s^*}^T$.

As the NP solutions to the $b \to sl^+l^-$ anomalies are in the form of vector and axial-vector operators, we consider the addition of these NP operators to the SM effective Hamiltonian of $b \to sl^+l^-$. Scalar and pseudo-scalar NP operators do not contribute to $B_s^* \to l^+l^-$ decay because $\langle 0|\bar{s}b|B_s^*(p_{B_s^*}, \epsilon)\rangle = \langle 0|\bar{s}\gamma_5 b|B_s^*(p_{B_s^*}, \epsilon)\rangle = 0$. The effective Hamiltonian now takes the form

$$\mathcal{H}_{eff}(b \to sl^+l^-) = \mathcal{H}_{SM} + \mathcal{H}_{VA}, \qquad (23.1)$$

where $\mathcal{H}_{VA}$ is

$$\mathcal{H}_{VA} = \frac{\alpha_{em} G_F}{\sqrt{2\pi}} V_{ts}^* V_{tb} \left[C_9^{NP} (\bar{s}\gamma^\mu P_L b)(\bar{l}\gamma_\mu l) + C_{10}^{NP} (\bar{s}\gamma^\mu P_L b)(\bar{l}\gamma_\mu \gamma_5 l) \right].$$

Here $C_{9(10)}^{NP}$ are the NP Wilson coefficients.

We define the longitudinal polarization asymmetry for the final state leptons in $B_s^* \to l^+l^-$ decay. The unit longitudinal polarization four-vector in the rest frame of the lepton (l^+ or l^-) is defined as

$$\bar{s}_{l\pm}^\alpha = \left(0, \pm \frac{\vec{p_l}}{|\vec{p_l}|} \right). \qquad (23.2)$$

In the dilepton rest frame (which is also the rest frame of B_s^* meson), these unit polarization vectors become

$$s_{l\pm}^\alpha = \left(\frac{|\vec{p_l}|}{m_l}, \pm \frac{E_l}{m_l} \frac{\vec{p_l}}{|\vec{p_l}|} \right), \qquad (23.3)$$

where E_l, $\vec{p_l}$, and m_l are the energy, momentum, and mass of the lepton (l^+ or l^-) respectively. We can define two longitudinal polarization asymmetries, $\mathcal{A}_{LP}^+$ for l^+ and $\mathcal{A}_{LP}^-$ for l^-, in the decay $B_s^* \to l^+l^-$ as [20]

$$\mathcal{A}_{LP}^{\pm} = \frac{[\Gamma(s_{l-}, s_{l+}) + \Gamma(\mp s_{l-}, \pm s_{l+})] - [\Gamma(\pm s_{l-}, \mp s_{l+}) + \Gamma(-s_{l-}, -s_{l+})]}{[\Gamma(s_{l-}, s_{l+}) + \Gamma(\mp s_{l-}, \pm s_{l+})] + [\Gamma(\pm s_{l-}, \mp s_{l+}) + \Gamma(-s_{l-}, -s_{l+})]}.$$

(23.4)

Within this NP framework, the branching ratio and $\mathcal{A}_{LP}$ are obtained to be [21]

$$\mathcal{B}(B_s^* \to l^+ l^-) = \frac{\alpha_{em}^2 G_F^2 f_{B_s^*}^2 m_{B_s^*}^3 \tau_{B_s^*}}{96\pi^3} |V_{ts} V_{tb}^*|^2 \sqrt{1 - 4m_l^2/m_{B_s^*}^2} \left[\left(1 + \frac{2m_l^2}{m_{B_s^*}^2}\right) \left| C_9^{eff} \right.\right.$$

$$\left. + \frac{2m_b f_{B_s^*}^T}{m_{B_s^*} f_{B_s^*}} C_7^{eff} + C_9^{NP} \right|^2 + \left(1 - \frac{4m_l^2}{m_{B_s^*}^2}\right) |C_{10} + C_{10}^{NP}|^2 \right], \quad (23.5)$$

$$\mathcal{A}_{LP}^{\pm}|_{NP} = \mp \frac{2\sqrt{1 - 4m_l^2/m_{B_s^*}^2} \, Re\left[\left(C_9^{eff} + \frac{2m_b f_{B_s^*}^T}{m_{B_s^*} f_{B_s^*}} C_7^{eff} + C_9^{NP} \right) \left(C_{10} + C_{10}^{NP} \right)^* \right]}{\left(1 + 2m_l^2/m_{B_s^*}^2\right) \left| C_9^{eff} + \frac{2m_b f_{B_s^*}^T}{m_{B_s^*} f_{B_s^*}} C_7^{eff} + C_9^{NP} \right|^2 + \left(1 - 4m_l^2/m_{B_s^*}^2\right) \left| C_{10} + C_{10}^{NP} \right|^2}.$$

(23.6)

23.3 Results and Discussion

23.3.1 $\mathcal{A}_{LP}(\mu)$ with NP Solutions

In this section, we first calculate $\mathcal{A}_{LP}(\mu)$ for the $B_s^* \to \mu^+\mu^-$ decay. The numerical inputs used for this calculation are $m_b = 4.18 \pm 0.03$ GeV, $m_{B_s^*} = 5415.4^{+1.8}_{-1.5}$ MeV [22], $f_{B_s^*}^T/f_{B_s} = 0.95$ [12] and $f_{B_s^*}/f_{B_s} = 0.953 \pm 0.023$ [23]. The SM prediction is given in Table 23.1. The uncertainty in this prediction (about 0.03%) is much smaller than the uncertainty in the decay constants (about 2%), making it theoretically clean.

From this Table 23.1, it is obvious that the prediction of $\mathcal{A}_{LP}(\mu)$ for the first solution deviates from the SM at the level of 3σ, whereas for the second solution, it is the same as that of the SM. Hence, any large deviation in this asymmetry can only

Table 23.1 Predictions of branching ratio and $\mathcal{A}_{LP}(\mu)$ for $B_s^* \to \mu^+\mu^-$ decay. The values of NP WCs are taken from [10]

NP type	NP WCs	$\mathcal{B}(B_s^* \to \mu^+\mu^-)$	$\mathcal{A}_{LP}^+(\mu) = -\mathcal{A}_{LP}^-(\mu)$
SM	0	$(1.10 \pm 0.60) \times 10^{-11}$	0.9955 ± 0.0003
(I) $C_9^{NP}(\mu\mu)$	-1.07 ± 0.18	$(0.82 \pm 0.50) \times 10^{-11}$	0.9145 ± 0.0246
(II) $C_9^{NP}(\mu\mu) = -C_{10}^{NP}(\mu\mu)$	-0.52 ± 0.09	$(0.80 \pm 0.49) \times 10^{-11}$	0.9940 ± 0.0038

be due to the first NP solution. We also provide the predictions for $\mathcal{B}(B_s^* \to \mu^+\mu^-)$ in Table 23.1. It is clear that neither of the two solutions can be distinguished from each other or from the SM via the branching ratio.

23.3.2   Effect of NP in $B_s^* \to \tau^+\tau^-$

As mentioned in the introduction, anomalies are also observed in the $b \to c\tau\bar{\nu}$ transitions. An NP model, which can account for these anomalies, is likely to contain NP amplitude for $b \to s\tau^+\tau^-$ transition also. Hence, the branching ratio of $B_s^* \to \tau^+\tau^-$ and τ longitudinal polarization asymmetry $\mathcal{A}_{LP}(\tau)$ will contain signatures of such NP. In the SM, the predictions for these quantities are: $\mathcal{B}(B_s^* \to \tau^+\tau^-) = (6.87 \pm 4.23) \times 10^{-12}$ and $\mathcal{A}_{LP}^+(\tau)|_{SM} = -\mathcal{A}_{LP}^-(\tau)|_{SM} = 0.8860 \pm 0.0006$.

The authors of [19] constructed a model of NP which accounts for the anomalies in $b \to c\tau\bar{\nu}$. This model contains tree level FCNC terms for $b \to s\tau^+\tau^-$ but not for $b \to sl^+l^-$ ($l = e, \mu$). The WCs for the $b \to s\tau^+\tau^-$ transition have the form $C_9(\tau\tau) = C_9^{SM} - C^{NP}(\tau\tau)$ and $C_{10}(\tau\tau) = C_{10}^{SM} + C^{NP}(\tau\tau)$, in this model, where

$$C^{NP}(\tau\tau) = \frac{2\pi}{\alpha} \frac{V_{cb}}{V_{tb}V_{ts}^*} \left(\sqrt{\frac{R_X}{R_X^{SM}}} - 1 \right). \tag{23.7}$$

The ratio R_X/R_X^{SM} is the weighted average of current experimental values of R_D, R_{D^*} and $R_{J/\psi}$. From the current world averages (after Moriond'19) of these quantities, we estimate this ratio to be $\simeq 1.14 \pm 0.05$. This, in turn, leads to $C^{NP}(\tau\tau) \sim \mathcal{O}(100)$. Thus, the NP contribution completely dominates the WCs and leads to greatly enhanced branching ratios for various B/B_s meson decays involving $b \to s\tau^+\tau^-$ transition [19].

We calculate $\mathcal{B}(B_s^* \to \tau^+\tau^-)$ and $\mathcal{A}_{LP}(\tau)$ as a function of R_X/R_X^{SM}. The plot of $\mathcal{B}(B_s^* \to \tau^+\tau^-)$ vs. R_X/R_X^{SM} is shown in left panel of Fig. 23.1. We note, from this plot, that $\mathcal{B}(B_s^* \to \tau^+\tau^-)$ can be enhanced up to 10^{-9} which is about two orders of magnitude larger than the SM prediction. The plot of $\mathcal{A}_{LP}(\tau)$ vs. R_X/R_X^{SM} is shown in the right panel of Fig. 23.1. It can be seen that $\mathcal{A}_{LP}(\tau)$ is suppressed by about 5% in comparison to its SM value.

After Moriond'19, the current world average of $R_{D^{(*)}}$ shows less tension with the SM which leads to smaller values of R_X/R_X^{SM}. As long as this ratio is greater than 1.03, the branching ratio of $B_s^* \to \tau^+\tau^-$ is enhanced by an order of magnitude at least. When $R_X/R_X^{SM} \sim 1.01$, $\mathcal{A}_{LP}(\tau)$ exhibits some very interesting behavior. In this case, the tree level FCNC NP contribution is similar in magnitude to the SM contribution (which occurs only at the loop level). Due to the interference between these two amplitudes, $\mathcal{A}_{LP}(\tau)$ changes sign and becomes almost (-1). Hence, a measurement of this asymmetry provides an effective tool for the discovery of tree level FCNC amplitudes of this model [19] when their magnitude becomes quite small.

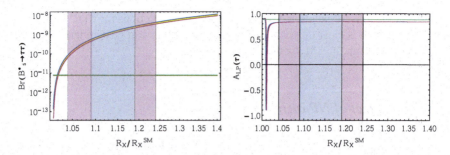

Fig. 23.1 Left and right panels correspond to $\mathcal{B}(B_s^* \to \tau^+\tau^-)$ and $\mathcal{A}_{LP}(\tau)$, respectively. In both the panels, the yellow band represents 1σ range of these observables. The 1σ and 2σ ranges of R_X/R_X^{SM} are indicated by blue and pink bands, respectively. The green horizontal line corresponds to the SM value

23.4 Conclusions

In this work, we consider the ability of the muon longitudinal polarization asymmetry in $B_s^* \to \mu^+\mu^-$ decay to distinguish between the two NP solutions, $C_9^{NP}(\mu\mu) < 0$ and $C_9^{NP}(\mu\mu) = -C_{10}^{NP}(\mu\mu) < 0$, which can account for all the measurements in $b \to sl^+l^-$ sector. This observable is theoretically clean because it has only a very mild dependence on the decay constants. For the case of real NP WCs, we show that this asymmetry has the same value as the SM case for the second solution but is smaller by $\sim 10\%$ for the first solution. Hence, a measurement of this asymmetry to $\sim 10\%$ accuracy can distinguish between these two solutions.

Further, we study the impact of the anomalies in $b \to c\tau\bar{\nu}$ transitions on the branching ratio of $B_s^* \to \tau^+\tau^-$ and $\mathcal{A}_{LP}(\tau)$. In [19], a model was constructed where tree level NP leads to both $b \to s\tau^+\tau^-$ and $b \to c\tau\bar{\nu}$ with moderately large NP couplings. Within this NP model, we find that the present data in $R_{D^{(*)}, J/\psi}$ sector imply about two orders of magnitude enhancement in the branching ratio of $B_s^* \to \tau^+\tau^-$ and a 5% suppression in $\mathcal{A}_{LP}(\tau)$ compared to their SM predictions. We also show that $\mathcal{A}_{LP}(\tau)$ undergoes drastic changes when the NP amplitude is similar in magnitude to the SM amplitude.

References

1. R. Aaij et al., LHCb collaboration. Phys. Rev. Lett. **111**, 191801 (2013)
2. R. Aaij et al., LHCb collaboration. JHEP **1602**, 104 (2016)
3. A. Abdesselam et al., [Belle Collaboration]. arXiv:1604.04042 [hep-ex]
4. R. Aaij et al., LHCb collaboration. JHEP **1307**, 084 (2013)
5. R. Aaij et al., LHCb collaboration. JHEP **1509**, 179 (2015)
6. R. Aaij et al., LHCb collaboration. Phys. Rev. Lett. **113**, 151601 (2014)
7. R. Aaij et al., LHCb collaboration. JHEP **1708**, 055 (2017)
8. A. Abdesselam et al., [Belle Collaboration]. arXiv:1904.02440 [hep-ex]

9. R. Aaij et al., [LHCb Collaboration]. Phys. Rev. Lett. **122**(19), 191801 (2019)
10. A.K. Alok, A. Dighe, S. Gangal, D. Kumar. arXiv: 1903.09617 [hep-ph]
11. M. Alguer, B. Capdevila, A. Crivellin, S. Descotes-Genon, P. Masjuan, J. Matias, J. Virto. arXiv:1903.09578 [hep-ph]
12. B. Grinstein, J. Martin Camalich, Phys. Rev. Lett. **116**(14), 141801 (2016)
13. D. Kumar, J. Saini, S. Gangal, S.B. Das, Phys. Rev. D **97**(3), 035007 (2018)
14. https://hflav-eos.web.cern.ch/hflav-eos/semi/spring19/html/RDsDsstar/RDRDs.html
15. R. Aaij et al., [LHCb Collaboration]. Phys. Rev. Lett. **120**(12), 121801 (2018)
16. A.K. Alok, D. Kumar, J. Kumar, S. Kumbhakar, S.U. Sankar, JHEP **1809**, 152 (2018)
17. A.K. Alok, D. Kumar, S. Kumbhakar, S. Uma Sankar, Phys. Lett. B **784**, 16 (2018)
18. A.K. Alok, D. Kumar, S. Kumbhakar, S. Uma Sankar. arXiv:1903.10486 [hep-ph]
19. B. Capdevila, A. Crivellin, S. Descotes-Genon, L. Hofer, J. Matias, Phys. Rev. Lett. **120**(18), 181802 (2018)
20. L.T. Handoko, C.S. Kim, T. Yoshikawa, Phys. Rev. D **65**, 077506 (2002)
21. S. Kumbhakar, J. Saini, Eur. Phys. J. C **79** (5), 394 (2019)
22. C. Patrignani et al., [Particle Data Group]. Chin. Phys. C **40**(10), 100001 (2016)
23. B. Colquhoun et al., [HPQCD Collaboration]. Phys. Rev. D **91**(11), 114509 (2015)

Chapter 24
In Search of New Physics with $B_s^0 \to l^+l^-$

S. Biswas and S. Sahoo

Abstract Rare leptonic $B_{s,d}^0$ decays are highly suppressed in the standard model (SM) and contribute to set different constraints for various models of new physics (NP). Among the $B_{s,d}^0 \to l^+l^-$ decays, so far only the branching ratio for the process $B_s^0 \to \mu^+\mu^-$ has been measured experimentally at the Large Hadron Collider. Other than branching ratio, sizeable decay width difference of B_s^0 mass eigen-states as well as the effective life time is being studied recently. On the other hand, $B_s^0 \to e^+e^-$ and $B_s^0 \to \tau^+\tau^-$ are the unnoticed part of $B_s^0 \to l^+l^-$ decays. The current experimental status of the above two decays are recently updated by LHCb. In this work, considering the effect of Z' boson we study of the branching ratio of $B_s^0 \to \mu^+\mu^-$ decay using χ^2 fitting. We hope this method can also be used to study $B_s^0 \to e^+e^-$ and $B_s^0 \to \tau^+\tau^-$ decays.

24.1 Introduction

In the SM, the decays. $B_{s,d}^0 \to l^+l^-$ ($l = e, \mu, \tau$) [1–3] are the result of quantum mechanical processes, i.e. the interchange of virtual particles at the loop level. These decays are extremely rare in the SM because they are loop and CKM suppressed, as well as helicity suppressed, since the two spin-1/2 leptons originate from a pseudoscalar B meson. These decays involve several observables which can be clearly elucidated from theoretical and experimental viewpoint. These rare B meson decays present vital base to analyse the flavour sector of the SM and also become potential source to dig out possible signatures of NP. The $B_s^0 \to l^+l^-$ decays involve $b \to sl^+l^-$ quark level transitions. In recent picture, several observables of rare B meson decays undergoing $b \to sl^+l^-$ transitions show intriguing pattern of deviations from the SM predictions which are [4–7]:

S. Biswas (✉) · S. Sahoo
Department of Physics, National Institute of Technology, Durgapur 713209, West Bengal, India
e-mail: getswagata92@gmail.com

© Springer Nature Singapore Pte Ltd. 2021   169
P. K. Behera et al. (eds.), *XXIII DAE High Energy Physics Symposium*,
Springer Proceedings in Physics 261,
https://doi.org/10.1007/978-981-33-4408-2_24

i.   Angular Observable P'_s of $B \to K^* \mu^+ \mu^-$ mode.
ii. Observation of 2.6σ deviation in $R_K = \frac{Br(B^+ \to K^+ \mu^+ \mu^-)}{Br(B^+ \to K^+ e^+ e^-)}$ in the $q^2 \in GeV^2$ bin.
iii. Observation of more than 3σ deviation in the decay rate of $B_s \to \varphi \mu^+ \mu^-$.
iv. Measurement of $R_{K^*} = \frac{Br(B \to K^* \mu^+ \mu^-)}{Br(B \to K^* e^+ e^-)}$.

In order to explain these anomalies, it is required to study B meson decays induced by flavour changing neutral current (FCNC) processes with the assistance of several new physics (NP) models. It is recently observed that most of the NP models involve the exchange of a Z' boson or a leptoquark [8]. NP models introduce additional couplings to new heavy mediators at both tree and loop level and these couplings could modify the values of branching ratios with respect to their SM values. Here we have investigated the various leptonic channels in Z' model considering its non-universality nature [9–11]. In this model, the NP is allowed to contribute at tree level by Z'-mediated flavour changing $b \to q (q = s, d)B \to \pi\pi$ decays where Z' boson couples to the flavour changing part, $\bar{q}b$ as well as to the leptonic part $\bar{l}l$.

So far, we paid our whole attention on the muonic decays only neglecting the other two channels. Not actually negligence, the electronic and tauonic channels are very hard to study experimentally. The helicity suppression is enlarged in e-channel by pseudoscalar contributions and very small mass. But in recent future, the search for $B_{s,d} \to e^+ e^-$ will be done with more dedication and interest exploiting the potential of physics of the LHC in its ATLAS, CMS and LHCb experiments and the BelleII experiments at KEK. On the other hand, the helicity suppression becomes less important for tauonic channels for its large lepton mass. Another most important complication in this experimental search is the presence of undetected neutrinos (at least two) originating while $B \to \tau\tau$ decays. Till now there are no experimental values for $B_{s,d} \to e^+ e^-$ and $B_{s,d} \to \tau^+ \tau^-$ decays though their branching fractions were constrained with 90 and 95% C.L., respectively. The recent updated average results of branching ratios measured by ATLAS, CMS and LHCb experiments are given by [12, 13],

$$Br(B_s \to \mu^+ \mu^-) = (3.0 \pm 0.6) \times 10^{-1}, \tag{24.1}$$

And the upper limits for the other two channels set as [13–15],

$$Br(B_s \to e^+ e^-) < 2.8 \times 10^{-7} (90\% \text{ C.L.})$$
$$Br(B_s \to \tau^+ \tau^-) < 6.8 \times 10^{-3} (95\% \text{ C.L.}) \tag{24.2}$$

In this paper, we have studied the muonic channel in Z' model by using the method of statistical analysis of x^2 fitting and have tested the goodness of fit. We can also apply this strategy for the other two channels to get acceptable predictions.

24.2 Theoretical Framework

The FCNC mediated $B_q \to l^+ l^- (q = s, d)$ decays are the most free-spoken B decays theoretically and are associated with $b \to s(d) l^+ l^-$ quark level transition. Neglecting the charm contributions, the effective Hamiltonian for these pure leptonic decays is given as [16, 17],

$$H_{\text{eff}}^{\text{SM}} = -\frac{G_F}{\sqrt{2}} \frac{\alpha}{2\pi \sin^2 \theta_W} V_{tb} V_{tb}^* Y(x_t) (\bar{q} b)_{v-A} (\bar{l} l)_{v-A} + h.c. \qquad (24.3)$$

Here $Y(x_t)$ is known as an appropriate loop function consists of Z-penguin and box-diagram contributions, QCD corrections and the leading electroweak corrections. Therefore, the branching ratio for the pure leptonic channel becomes

$$Br(B_q \to l^+ l^-) = \tau_{Bs} \frac{\sqrt{m_{B_q}^2 - 4m_l^2}}{16\pi m_{B_q}^2} |A_{SM}|^2. \qquad (24.4)$$

Before the formulation of NP in the pure leptonic decays we need to consider some simplifications: (i) we have neglected kinetic mixing, (ii) we have also neglected $Z - Z'$ mixing (as the mixing angle is constrained as less than 10^{-3} by Bandyopadhyay et al. [18, 19] and recently at LHC it was found as of the order 10^{-4} by Bobovnikov et al. [20]) for its very small mixing angle, (iv) we will consider the remarkable contribution of the flavour-off-diagonal left-handed couplings of quarks in the flavour changing $b - q - z'$ part. Considering all the simplifications for the non-universal couplings of Z', the effective Hamiltonian including the NP part becomes for $b \to q l^+ l^-$

$$H_{\text{eff}}^{z'} = -\frac{2G_F}{\sqrt{2}} V_{\tau b} V_{\tau q}^* \left[\frac{B_{qb}^L B_{ll}^R}{V_{tb} V_{tq}^*} (\bar{q} b)_{v-A} (\bar{l} l)_{v-A} - \frac{B_{qb}^L B_{ll}^R}{V_{tb} V_{tq}^*} (\bar{q} b)_{v-A} (\bar{l} l)_{v+A} \right] + h.c., \qquad (24.5)$$

where B_{qb}^L is the left-handed coupling of Z' boson with quarks, B_{ll}^L and B_{ll}^R are the left-handed and right-handed couplings with the leptons. The coupling parameter consists of a NP weak phase term also which is related as $B_{qb}^L = \left| B_{qb}^L \right| e^{i\varphi_q}$. Therefore, the branching ratio expression for the leptonic channel becomes as

$$Br(B_q \to l^+ l^-) = \tau_{B_q} \frac{G_F^2}{4\pi} f_{qB_q}^2 m_l^2 m_{B_q} \sqrt{1 - \frac{4m_l^2}{m_{B_q}^2}} |V_{tb} V_{tq}^*|^2$$

$$\left| \frac{\alpha}{2\pi \sin^2 \theta_W} Y(x_t) - 2 \frac{B_{qb}^L (B_{ll}^L - B_{ll}^R)}{V_{tb} V_{tq}^*} \right|^2 \qquad (24.6)$$

24.3 χ^2 Analysis and Goodness of Fit

To find the NP couplings, we have used the statistical method of χ^2 fitting with some specific constraints. The value of χ^2 is used to find the deviation of experimental values from the expected values. It can be defined as [21, 22]:

$$\chi^2 = \sum_i \frac{(f_i^{\text{th}} - f_i^{\text{exp}})}{(\Delta f_i)^2}. \tag{24.7}$$

By minimising this expression, we obtain several best fit values. We have used the tMinuit package (i.e. Ifit) of ROOT software for fitting. We have calculated the probability distribution for χ^2 which is defined as

$$P(\chi^2) = \int_{\chi^2}^{\infty} \frac{1}{2^{n/2}\Gamma(n/2)} (\chi^2) \left(\frac{n}{2} - 1\right)_{e^-} \chi^2/2 d(\chi^2) \tag{24.8}$$

This probability depends on the parameter 'n' 'n' 'n' 'n', i.e. degrees of freedom. For smaller values of degrees of freedom it is difficult to judge the fit quality by using $\chi^2_{min}/n \approx 1$ condition only. In that case the probability values (p-value) will help to determine the goodness of fit results where the p-value is preferred as 50%.

24.4 Numerical Analysis

In order to calculate the value of χ^2 we have used several observables as constraints involving the same $b \to sl^+l^-$ transition. The experimental results of the observables are recorded in Table 24.1 [13, 16, 23].

Using all these experimental data we have structured the χ^2 in terms of the couplings and minimised it. The best fit values are recorded in Table 24.2.

Taking couplings from Table 24.2, we calculate the branching ratio value for $B_s^0 \to \mu^+\mu^-$ decay with NP effects by using (24.6). From Table 24.3, we can see that the branching ratio value in Z' model enhances from the SM value and is in agreement with the experimental one. Here, the degree of freedom is one and the p-value is calculated from (24.8).

Table 24.1 Experimental data of the observables

Mode	Branching value
$Br(B_s \to \mu^+\mu^-)$	$(3.0 \pm 0.6) \times 10^{-9}$
$Br^{low}(B \to X_s\mu^+\mu^-)$	$(16 \pm 5.0) \times 10^{-7}$
$Br^{high}(B \to X_s\mu^+\mu^-)$	$(4.4 \pm 1.2) \times 10^{-7}$
$Br(B \to X_s\mu^+\mu^-)$	$(43 \pm 12.5) \times 10^{-7}$
$Br(B \to K\mu^+\mu^-)$	$(4.51 \pm 0.23) \times 10^{-7}$

Table 24.2 Best fit values for μ channel

$\chi^2_{min} = 0.233$, p-value $= 62.63\%$

Parameter	Best fit value
B_{sb}^L	$(2.8 \pm 0.38) \times 10^{-3}$
φ	(34.48 ± 2.13)
$B_{\mu\mu}^L$	$(-2.2 \pm 3.24) \times 10^{-2}$
$B_{\mu\mu}^R$	$(-0.80 \pm 0.54) \times 10^{-2}$

Table 24.3 Branching ratio values for μ channel in SM and in Z' model

$Br(B_s \to \mu^+\mu^-)$	
In SM	1.164×10^{-9}
In Z' model	3.071×10^{-9}

24.5 Conclusion

In this paper, considering the NP effect and using minimising technique, we have found the values of NP couplings and calculated the branching ratio of $B_s \to \mu^+\mu^-$ channel as (3.071×10^{-9}). Here, our obtained p-value (62.63%) is acceptable. So we can say that this fitting is very useful for $B_s \to \mu^+\mu^-$. We expect this technique can also be used to study $B_s^0 \to e^+ e^-$ and $B_s^0 \to \tau^+\tau^-$ decays. Several experiments established the fact that Z' boson does not behave in similar manner with all the generations of leptons and that infers the non-universality nature of Z' model [24]. By calculating the leptonic couplings of Z' boson for e and τ channels, it is possible to predict their branching ratios in future.

Acknowledgements S. Biswas acknowledges NIT Durgapur, for providing fellowship for her research. S. Sahoo would like to thank SERB, DST, Govt. of India, for financial support through grant no. EMR/2015/000817.

References

1. R. Fleischer, D.G. Espinosa, R. Jaarsma, G. Tetlalmatzi-Xolocotzi, Eur. Phys. J. C **78**, 1 (2018)
2. R. Mohanta, Phys. Rev. D **71**, 114013 (2005)
3. A.K. Giri, R. Mohanta, Eur. Phys. J. C **45**, 151 (2006)
4. R. Aaij et al., (LHCb Collaboration). Phys. Rev. Lett. **113**, 151601 (2014). [arXiv:1406.6482 [hep-ex]]
5. R. Aaij et al., (LHCb Collaboration). JHEP **1708**, 055 (2017). [arXiv:1705.05802 [hep-ex]]
6. R. Aaij et al., (LHCb Collaboration). JHEP **1509**, 179 (2015). [arXiv:1506.08777 [hep-ex]]
7. R. Aaij et al., (LHCb Collaboration). JHEP **1307**, 084 (2013). [arXiv:1305.2168 [hep-ex]]
8. R. Mohanta, *Phys. Rev. D* **89**, 014020 (2014) [arXiv:1310.0713[hep-ph]]
9. G. Hiller, (2018). [arXiv:1804.02011 [hep-ph]]
10. J. Alda, J. Guasch, S, Penaranda, (2018). [arXiv:1805.03636 [hep-ph]]
11. G. Hiller and M. Schmatz, *Phys. Rev. D* **90**, 054014 (2014) [arXiv:1408.1627 [hep-ph]]

12. R. Aaij et al. (LHCb Collaboration), *Phys. Rev. Lett.* **118**, 191801 (2017) [arXiv:1703.05747 [hep-ex]]

13. M. Tanabashi et al., Particle data group. Phys. Rev. D **98**, 030001 (2018)

14. R. Aaij et al. (LHCb Collaboration), *Phys. Rev. Lett.* **118**, 251802 (2017) [arXiv:1703.02508 [hep-ex]]

15. T. Aaltonen et al., (CDF Collaboration). Phys. Rev. Lett. **102**, 201801 (2009) [arXiv:0901.3803 [hep-ex]]

16. Q. Chang, X-Qiang Li and Y-Dong, JHEP **1002**, 082 (2010). [arXiv:0907.4408 [hep-ph]]

17. G. Buchalla, A. J. Buras and M. E. Lautenbacher, *Rev. Mod. Phys.* **68**, 1125 (1996) [arXiv:951 2380 [hep-ph]]

18. T. Bandyopadhyay, G. Bhattacharya, D. Das, A. Raychaudhuri, Phys. Rev. D **98**, 035027 (2018)

19. P. Nayek, P. Maji, S. Sahoo, Phys. Rev. D **99**, 013005 (2019)

20. I.D. Bobovnikov, P. Osland, A.A. Pankov, Phys. Rev. D **98**, 095029 (2018)

21. R. Andrae, T. Schulze-Hartung, P. Melchior, Dos and don'ts of reduced chi squared. [arXiv: 1012.3754 [astro-ph.IM]]

22. H.K. Dass, R. Verma, *Higher Mathematical Physics*, Chapter-52 (S. Chand & Company Pvt. Ltd., Ramnagar, New Delhi, India, 2014), pp. 1402–1488

23. J.P. Less et al., (BABAR Collaboration). Phys. Rev. Lett. **112,** 211802 (2014). [arXiv:1312. 5364 [hep-ph]]

24. R. Fleischer, R. Jaarsma and G. Tetlalmatzi-Xolocotzi, *JHEP* **05**, 156 (2017) [arXiv:1703. 10160 [hep-ph]]

Chapter 25
Notes on a Z'

Triparno Bandyopadhyay, Gautam Bhattacharyya, Dipankar Das, and Amitava Raychaudhuri

Abstract We reexamine anomaly free U(1) extensions of the standard model in the light of LHC Drell-Yan data, constraints from unitarity, and neutrino-electron scattering to put model-independent bounds in the parameter space populated by $M_{Z'}$, the Z-Z' mixing angle (α_z), and the extra U(1) effective gauge coupling (g'_x). We propose a formalism where any model dependence is absorbed into these three parameters.

25.1 Introduction

Given its central role in physics beyond the standard model, including, but not limited to, unified theories, left-right models [1], little higgs models, models of additional space dimensions, flavor physics models, and varied dark matter scenarios [2], there is enough motivation for the careful study of a heavier Z-like vector boson, Z'. In this work, we take up this matter for a massive Z' arising from a spontaneously broken $U(1)$, that we call $U(1)_X$. We show that the model specific features can be absorbed into the Z' mass, $M_{Z'}$, the Z-Z' mixing angle, α_z, and the coupling strength, g'_x,

T. Bandyopadhyay (✉)
Department of Theoretical Physics, Tata Institute of Fundamental Research, Mumbai 400005, India
e-mail: triparno@theory.tifr.res.in

G. Bhattacharyya
Saha Institute of Nuclear Physics, HBNI, 1/AF Bidhan Nagar, Kolkata 700064, India
e-mail: gautam.bhattacharyya@saha.ac.in

D. Das
Department of Astronomy and Theoretical Physics, Lund University, Solvegatan 14A, 223 62 Lund, Sweden
e-mail: dipankar.das@thep.lu.se

A. Raychaudhuri
Department of Physics, University of Calcutta, 92 APC Road, Kolkata 700009, India
e-mail: palitprof@gmail.com

© Springer Nature Singapore Pte Ltd. 2021
P. K. Behera et al. (eds.), *XXIII DAE High Energy Physics Symposium*,
Springer Proceedings in Physics 261,
https://doi.org/10.1007/978-981-33-4408-2_25

corresponding to the additional $U(1)$ for a class of anomaly free models. We proceed to give exclusions in this parameter space from LHC data, considerations of s-matrix unitarity, and e-ν_μ scattering data [3].

25.2 Formalism

We have two scalar multiplets, transforming under $SU(3)_C \times SU(2)_L \times U(1)_Y \times U(1)_X$ as: $\Phi \equiv (1, 2, 1/2, x_\Phi/2)$, $S \equiv (1, 1, 0, 1/2)$, where Φ is the usual $SU(2)_L$ doublet responsible for the standard model (SM) gauge symmetry breaking, as well as the Dirac masses of fermions. Note that, even if we start with $x_\Phi = 0$, Φ can develop a $U(1)_X$ charge due to gauge kinetic mixing among the two abelian field strength tensors. The covariant derivatives on the scalars are defined as

$$D_\mu \Phi = \left(\partial_\mu - ig\frac{\tau_a}{2} W_\mu^a - i\frac{g_Y}{2} B_\mu - i\frac{g_x}{2} x_\Phi X_\mu \right) \Phi ; \; D_\mu S = \left(\partial_\mu - i\frac{g_x}{2} X_\mu \right) S ,$$
(25.1)

where τ_a represents the Pauli matrices. The naming convention for the gauge fields follow standard conventions. In the original basis, the lagrangian contains the gauge kinetic mixing term $(\sin \chi /2) B_{\mu\nu} X^{\mu\nu}$ ($B_{\mu\nu}$ and $X_{\mu\nu}$ denote the $U(1)_Y$ and $U(1)_X$ field tensors respectively). We perform a general linear transformation to go to a basis where the lagrangian is canonically diagonal

$$\begin{pmatrix} B_\mu \\ X_\mu \end{pmatrix} \rightarrow \begin{pmatrix} B'_\mu \\ X'_\mu \end{pmatrix} = \begin{pmatrix} 1 & \sin \chi \\ 0 & \cos \chi \end{pmatrix} \begin{pmatrix} B_\mu \\ X_\mu \end{pmatrix} .$$
(25.2)

In this basis, the covariant derivatives take the following forms:

$$D_\mu \Phi = \partial_\mu \Phi - i\frac{g}{2} \left(\tau_a W_\mu^a + t_w B'_\mu + t_x x'_\Phi X'_\mu \right) \Phi ; \; D_\mu S = \left(\partial_\mu - i\frac{g'_x}{2} X'_\mu \right) S ,$$
(25.3)

where we have defined, $t_w \equiv \tan \theta_w = \frac{g_Y}{g}$, $t_x \equiv \tan \theta_x = \frac{g'_x}{g}$, with $g'_x = g_x \sec\chi$, and $x'_\Phi = x_\Phi - \frac{g_Y}{g_x} \sin \chi$. In the limit of zero kinetic mixing, t_x characterizes the strength of the $U(1)_X$ gauge coupling relative to the weak gauge coupling. After spontaneous symmetry breaking, we expand the scalar fields, in the unitary gauge, as

$$\Phi = \frac{1}{\sqrt{2}} \begin{pmatrix} 0 \\ v + \phi_0 \end{pmatrix} , \quad S = \frac{1}{\sqrt{2}} (v_s + s) ,$$
(25.4)

where v and v_s are the vevs for Φ and S, respectively. This leads to the neutral gauge boson mass matrix, in the basis where the gauge kinetic terms are diagonal, which can be written as follows:

$$\mathscr{M}_N^2 = \frac{g^2 v^2}{4} \begin{pmatrix} 1 & -\tan\theta_w & -x'_\Phi \tan\theta_x \\ -\tan\theta_w & \tan^2\theta_w & x'_\Phi \tan\theta_x \tan\theta_w \\ -x'_\Phi \tan\theta_x & x'_\Phi \tan\theta_x \tan\theta_w & \tan^2\theta_x \left(r^2 + x'^2_\Phi\right) \end{pmatrix}, \qquad (25.5)$$

with $r = v_s/v$. We go to the diagonal basis by performing the rotation

$$\begin{pmatrix} B'_\mu \\ W^3_\mu \\ X'_\mu \end{pmatrix} = \begin{pmatrix} \cos\theta_w & -\sin\theta_w \cos\alpha_z & \sin\theta_w \sin\alpha_z \\ \sin\theta_w & \cos\theta_w \cos\alpha_z & -\cos\theta_w \sin\alpha_z \\ 0 & \sin\alpha_z & \cos\alpha_z \end{pmatrix} \begin{pmatrix} A_\mu \\ Z_\mu \\ Z'_\mu \end{pmatrix}. \qquad (25.6)$$

This diagonalization gives the following relations:

$$M_{11}^2 \equiv M_Z^2 \cos^2\alpha_z + M_{Z'}^2 \sin^2\alpha_z = M_W^2/\cos^2\theta_w, \qquad (25.7a)$$
$$M_{Z'}^2 \cos^2\alpha_z + M_Z^2 \sin^2\alpha_z = M_W^2 \tan^2\theta_x \left(r^2 + x'^2_\Phi\right), \qquad (25.7b)$$
$$\left(M_{Z'}^2 - M_Z^2\right) \sin 2\alpha_z = 2x'_\Phi \tan\theta_x M_W^2/\cos\theta_w, \qquad (25.7c)$$

where $M_W = gv/2$ denotes the W-boson mass. We use (25.7) to replace θ_w, r, and x'_Φ in terms of $M_{Z'}$, α_z, and t_x. Note that we have not treated θ_w as the conventional weak angle under the implicit *a priori* assumption that α_z is small, instead, we have traded it in favor of $M_{Z'}$ and α_z. The gauge-scalar sector described here holds generally for different Z' models, but, the fermion charge assignments vary. Next, we develop a general formalism for the fermionic sector. We specifically look at models in which the SM is extended by a right-handed (RH) neutrino (N_R) per generation which get Majorana masses from their Yukawa interactions with S. Under the assumption of generation universality, the possible $U(1)_X$ charge options for the fermions are quite restricted.

With the requirements for Dirac mass for the SM fermions and Majorana mass for N_R, applying the constraints from chiral anomaly cancelation, all the $U(1)_X$ charges of the fermions can be determined in terms of one free parameter κ_x (also, [4]), as depicted in Table 25.1.

Different choices of κ_x give different models. In Table 25.2, we show some alternatives. For example, the $(B-L)$ extension of the SM corresponds to $\kappa_x = 1/4$.

Table 25.1 The $U(1)_X$-charge assignments of the multiplets, as a function of κ_x

Multiplet	Q_L	u_R	d_R	ℓ_L	e_R	N_R	Φ	S
Charge	$\frac{\kappa_x}{3}$	$\frac{4\kappa_x}{3} - \frac{1}{4}$	$-\frac{2\kappa_x}{3} + \frac{1}{4}$	$-\kappa_x$	$-2\kappa_x + \frac{1}{4}$	$-\frac{1}{4}$	$\kappa_x - \frac{1}{4}$	$\frac{1}{2}$

Table 25.2 κ_x for different $U(1)_X$ models

Model	$U(1)_{B-L}$	$U(1)_R$	$U(1)_X$	$U(1)_R \times U(1)_{B-L}$
Charge definition	$\frac{(B-L)}{4}$	$-\frac{T_{3R}}{2}$	$-Q_X/\sqrt{10}$	$\frac{1}{5}\left[(B-L) - \frac{1}{2}T_{3R}\right]$
κ_x	$\frac{1}{4}$	0	$\frac{3}{20}$	$\frac{1}{5}$

25.3 Exclusions in the Parameter Space

The cross section for resonant production of a Z' boson at the LHC and its subsequent decay into a pair of charged leptons can be conveniently expressed as (in the narrow width approximation) [5]

$$\sigma \left(pp \rightarrow Z'X \rightarrow \ell^+\ell^-X \right) = \frac{\pi}{6s} \sum_q C_q^\ell w_q \left(s, M_{Z'}^2 \right), \qquad (25.8)$$

where the sum is over all the partons. The co-efficients

$$C_q^\ell = \left[\left(g_L^q \right)^2 + \left(g_R^q \right)^2 \right] \mathrm{BR} \left(Z' \rightarrow \ell^+\ell^- \right) \qquad (25.9)$$

involve the chiral couplings, g_L^f and g_R^f of Z' and hence depend on the details of the fermionic sector of the model under consideration. The functions w_q contain all the information about the parton distribution functions (PDFs) and QCD corrections [3].

Using the most recent ATLAS limits [6] on the left-hand-side of (25.8) [6] for the $\ell \equiv e, \mu$ final state, we give limits in the C_u^ℓ-C_d^ℓ plane for different values of $M_{Z'}$.[1] The results have been displayed in the left panel of Fig. 25.1. For any chosen $M_{Z'}$, only the interior of the corresponding contour is allowed.

As the introduction of the new $U(1)$ redefines the SM Z boson, the Z' needs to be light enough to counter the fourth power of energy growth of the scattering amplitude for the process $W_L^+ W_L^- \rightarrow W_L^+ W_L^-$ (L denoting longitudinal) at leading order. This leads to an upper bound on $M_{Z'}$

$$\frac{M_{Z'}^4 \sin^2 \alpha_z}{\left(M_Z^2 \cos^2 \alpha_z + M_{Z'}^2 \sin^2 \alpha_z \right)} < 8\pi \times \frac{3}{32\sqrt{2}G_F}. \qquad (25.10)$$

Since this analysis does not depend on the details of the fermionic couplings, such a bound is quite general and can be applied to a wide class of Z' models.

The unitarity bound is lifted in the limit $\sin \alpha_z = 0$ as has been clearly depicted in Fig. 25.1 (right panel). It is possible to put lower bounds on $M_{Z'}$, in the limit of vanishing Z-Z' mixing, e.g., by using the data from low energy neutrino-electron scattering such as $\nu_\mu e \rightarrow \nu_\mu e$ which proceeds at the tree level purely via neutral current [7]. The dimension-six operator governing ν_μ-e scattering at low energies is written as

$$\mathcal{L}_{\nu e} = -\frac{G_F}{\sqrt{2}} \left[\bar{\nu}\gamma^\mu \left(1 - \gamma^5 \right) \nu \right] \left[\bar{e}\gamma_\mu \left(g_V^{\nu e} - g_A^{\nu e}\gamma^5 \right) e \right]. \qquad (25.11)$$

---

[1] See [3] for a similar analysis using the di-τ final state.

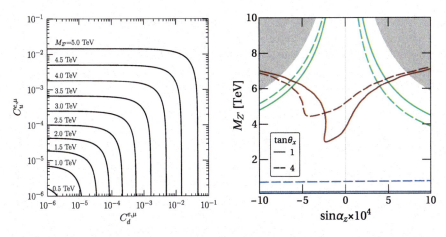

Fig. 25.1 *Left:* Exclusion contours at 95% C.L. in the $C_u - C_d$ plane for different values of $M_{Z'}$, derived using ATLAS DY data. *Right:* Consolidated bounds in the $(\sin\alpha_z\text{-}M_{Z'})$plane for anomaly free $U(1)_X$ models. Details in text

In the Z' models under consideration, the vector and axial couplings of the SM Z boson is modified, and using the global best fit values for g_V^{ve} and g_A^{ve}, $g_V^{ve} = -0.040 \pm 0.015$, $g_A^{ve} = -0.507 \pm 0.014$ [8] we draw the 2σ allowed regions in the $\sin\alpha_z\text{-}M_{Z'}$ plane.

25.4 Results

In Fig. 25.1, all the bounds have been displayed in the $\sin\alpha_z\text{-}M_{Z'}$ plane for any anomaly free $U(1)_X$ model for two typical choices of $\tan\theta_x$. The region excluded from unitarity has been shaded in gray and is independent of $\tan\theta_x$. The lower bounds on $M_{Z'}$, arising from the ATLAS (13 TeV, 36 fb^{-1}) exclusion of the DY production of Z', are depicted as red curves, whereas the region above the light blue curves denote the region consistent with v_μ-e scattering. Additionally, we also give contours that represent a constraint on the Z' decay width, as a guideline for the validity of a particle interpretation. The green lines in the figure arise from the consideration[2] $\Gamma_{Z'} \leq M_{Z'}/2$.

For all the colored contours, the solid (dashed) curves correspond to $\tan\theta_x = 1(4)$. Recall that $\tan\theta_x$ is proportional to the effective $U(1)_X$ coupling, g_x'. As it happens, the lower bounds on $M_{Z'}$ arising from low energy v_μ-e scattering are considerably weaker than those from direct searches. However, v_μ-e scattering can put important constraints for hydrophobic Z' models when the production of the Z' at the LHC is

---

[2] What constitutes an acceptable width of a heavy particle, or how far the narrow width approximation holds good can be a matter of discussion and hence we choose to veer on the conservative side, to illustrate what role the consideration of width might play in restricting the parameter space.

very suppressed. Combining the lower bound on $M_{Z'}$ from the direct searches with the corresponding upper bound coming from, e.g., unitarity, we are able to extract an upper limit on the magnitude of the Z-Z' mixing angle, α_z. Such bounds on $|\alpha_z|$ are at par with the corresponding limits from electroweak precision data [9].

25.5 Conclusions

We put constraints on the parameter space of the minimal extension of the SM with an additional gauged $U(1)$. We put forward a parametrization, in which, the constraints on different $U(1)_X$ models are expressed in a model-independent framework. We have not a priori assumed the Z-Z' mixing angle to be small or $M_{Z'} >> M_Z$. There are three important quantities which cover the extended parameter space, absorbing all model dependence for a non-anomalous $U(1)$ extension. These quantities are $M_{Z'}$, g'_x, and α_z. To constrain this space, we have employed three distinct pieces of information, the LHC (ATLAS) Drell-Yan data, results from low energy $\nu_\mu - e$ scattering, and consistency with s-wave unitarity in the $W_L^+ W_L^- \rightarrow W_L^+ W_L^-$ channel.

References

1. T. Bandyopadhyay, B. Brahmachari et al., JHEP **1602**, 023 (2016)
2. P. Langacker, Rev. Mod. Phys. **81**, 1199 (2009)
3. T. Bandyopadhyay, G. Bhattacharyya et al., Phys. Rev. D **98**(3), 035027 (2018)
4. A. Ekstedt, R. Enberg et al., JHEP **1611**, 071 (2016)
5. G. Paz, J. Roy, Phys. Rev. D **97**(7), 075025 (2018)
6. M. Aaboud et al., ATLAS collaboration. JHEP **1710**, 182 (2017)
7. G. Radel, R. Beyer, Mod. Phys. Lett. A **8**, 1067 (1993)
8. C. Patrignani et al., Particle Data Group. Chin. Phys. C **40**(10), 100001 (2016)
9. J. Erler, P. Langacker et al., JHEP **0908**, 017 (2009)

Chapter 26
Probing New Signature Using Jet Substructure at the LHC

Akanksha Bhardwaj

Abstract To explore interesting TeV scale BSM theories at the Large Hadron Collider (LHC), jet-substructure techniques can play a key role. This motivates us to look for challenging final states which can potentially be used for many hitherto unexplored interesting region of parameter space in different BSM scenarios. We exploit the characteristics of the jet-substructure techniques, as well as suitable kinematic variables to handle humongous Standard Model backgrounds. We also perform a comprehensive collider analysis to demonstrate the effectiveness of these new channels and techniques for different BSM models which significantly enhance the present reach at the 13 TeV LHC.

26.1 Introduction

The Standard Model (SM) of particle physics encapsulates our knowledge of fundamental interactions of the particle world with all its glory. Until now, apart from a few minor exceptions, the SM is in perfect agreement with all the high energy collider experiments like the Large Hadron Collider (LHC) experiments at CERN. The reputation of the SM being a complete theory gets tarnished when it cannot explain the presence of tiny yet nonzero masses of the neutrinos that are already established in neutrino oscillation experiments. Observations of cosmic microwave background radiation in various experiments unambiguously establish that 26% of the energy budget of our universe is made up of an inert, stable component, termed as the "dark matter" (DM). Inert doublet model (IDM) is proposed [1, 2] as a minimal extension of the SM that can provide an inert weakly interacting DM candidate, stabilized by the discrete symmetry of the model. The SM is extended with an extra $SU(2)_L$ scalar doublet which is odd under a discrete $\mathbb{Z}_2$ symmetry, and thus stabilizes the lightest neutral scalar of the model to be an ideal DM candidate. Simple extensions of the

A. Bhardwaj (✉)
Physical Research Laboratory (PRL), Ahmedabad 380009, Gujarat, India
e-mail: bhardwajakanksha22@gmail.com,akanksha@prl.res.in,akansha.bhardwaj@iitgn.ac.in

IIT Gandhinagar, Palaj, Gandhinagar 382 355, India

© Springer Nature Singapore Pte Ltd. 2021
P. K. Behera et al. (eds.), *XXIII DAE High Energy Physics Symposium*,
Springer Proceedings in Physics 261,
https://doi.org/10.1007/978-981-33-4408-2_26

181

Standard Model (SM) with additional Right Handed Neutrinos (RHNs) can elegantly explain the existence of small neutrino masses and their flavor mixings. Here we will consider both the extension of the SM and discuss new interesting phenomenology using jet-substructure.

26.2 Inverse Seesaw Scenario

In the inverse seesaw [3–5] framework, the SM particle content is extended by two SM singlet Majorana RHNs, N^β and S_L^β with same lepton number. β is the flavor index. The relevant part of the Lagrangian can again be written as

$$\mathcal{L} \supset -Y_D^{\alpha\beta} \overline{\ell_L^\alpha} H N_R^\beta - M^{\alpha\beta} \overline{S_L^\alpha} N_R^\beta - \frac{1}{2} \mu_{\alpha\beta} \overline{S_L^\alpha} S_L^{\beta^C} + \text{H.c.} . \qquad (26.1)$$

Here, ℓ^α and H are the SM lepton doublet and Higgs doublet, respectively. M_D is a Dirac mass matrix and μ is a small lepton number violating Majorana mass matrix. After EWSB we obtain the neutrino mass matrix

$$M_\nu = \begin{pmatrix} 0 & M_D & 0 \\ M_D^T & 0 & M^T \\ 0 & M & \mu \end{pmatrix} . \qquad (26.2)$$

Diagonalizing the mass matrix in (26.2), we obtain the inverse seesaw formula for the light neutrino masses as

$$M_\nu \simeq M_D M^{-1} \mu M^{-1^T} M_D^T . \qquad (26.3)$$

The Dirac mass $M_D = \frac{Y_D v}{\sqrt{2}}$ is generated after EWSB. In order to make our discussions simple we assume degenerate RHNs, with $M = M_N \times \mathbb{1}$. $\mathbb{1}$ is the unit matrix as before and M_N is the RHN mass eigenvalue. With these assumptions, the neutrino mass matrix may be simplified as

$$M_\nu = \frac{1}{M_N^2} M_D \mu M_D^T . \qquad (26.4)$$

Consider a typical flavor structure of the model where M_D and M_N are proportional to the unit matrix such as $M_D \to M_D \times \mathbb{1}$ and $M_N \to M_N \times \mathbb{1}$, respectively. Thus, the flavor structure is now fully encoded in the 3×3 matrix μ. We refer to this scenario as Flavor Diagonal (FD). It has been shown that the FD case in the inverse seesaw mechanism is also accommodated by neutrino oscillation data [6]. Another flavor structure possible in the inverse seesaw scenario is where M_D carries flavor structure while $\mu \to \mu \times \mathbb{1}$ and $M \to M_N \times \mathbb{1}$. This is called the Flavor Non-Diagonal (FND) scenario. This has been studied for different signals in [6, 7], under general parametrization [8]. Here, in the collider study, we are interested in a

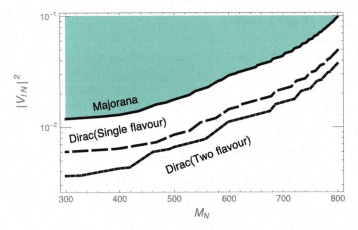

Fig. 26.1 The figure shows the 2-σ exclusion limits, in terms of heavy neutrino mass M_N and $|V_{\ell N}|^2$, at 3000 fb^{-1} of integrated luminosity at the 13 TeV LHC

very specific decay topology arising from the production and decay of heavy sterile neutrinos. We focus on opposite sign (OS) muon pair final states, in association with a reconstructed fatjet, at $\sqrt{s} = 13$ TeV LHC. For simplicity, we demonstrate explicitly our analysis assuming a simple, single flavor scenario where the light-heavy mixing is nonzero only for the muon flavor. This is also motivated by the fact that muons provide a clear detection at the LHC with high efficiency and hence is of primary interest. We will, however, also include the electron channel while discerning the final exclusion results. Here the major background arrises from $t\bar{t}$. We apply the following cut for the analysis (Fig. 26.1).

- Two opposite sign muons are selected with $p_T > 10$ GeV within the detector rapidity range $|\eta_\mu| < 2.4$, assuming a muon detection efficiency of 95%. We veto the event if any additional reconstructed lepton with $p_T > 10$ GeV is present.
- We demand at least one fatjet, reconstructed adopting the CA algorithm with radius parameter $R = 0.8$ and $|\eta^J| < 2.4$. We select events with the hardest reconstructed fatjet (J_0) having minimum transverse momentum $p_T^{J_0} > 150$ GeV and with N-subjettiness $\tau_{21}^{J_0} < 0.4$.
- The highest p_T muon is selected with $p_T > 100$ GeV and the next p_T ordered muon is selected with $p_T > 60$ GeV.
- To control the huge backgrounds coming from leptonic decays of Z bosons, we veto events if the opposite sign di-muon invariant mass ($M_{\mu^+\mu^-}$) is less than 200 GeV. The harder cut on $M_{\mu^+\mu^-}$ also reduces parts of the $t\bar{t}$ background further.
- We apply a b-veto to reduce the $t\bar{t}$ background without affecting signal acceptance.
- It is evident that our signal does not have any missing particle per se, hence should have relatively low missing transverse energy ($\not{E}_T$). The final $\not{E}_T$ would of course get contributions from measurements and uncertainties. Taking into account the unclustered towers, we consider only events with a maximum $\not{E}_T$ of 60 GeV. The figure [1] shows the 2-σ exclusion limits, in terms of heavy neutrino mass M_N and $|V_{\ell N}|^2$, at 3000 fb^{-1} of integrated luminosity at the 13 TeV LHC [9].

26.3   Inert Higgs Doublet

We first discuss the traditional IDM where one adds an additional $SU(2)_L$ complex scalar doublet Φ_2 apart from the SM Higgs doublet Φ_1, which are, respectively, odd and even under a discrete $\mathbb{Z}_2$ symmetry. The most general scalar potential that respects the electroweak symmetry $SU(2)_L \otimes U(1)_Y \otimes \mathbb{Z}_2$ of the IDM can be written as [10]

$$
V(\Phi_1, \Phi_2) = \mu_1^2 \Phi_1^\dagger \Phi_1 + \mu_2^2 \Phi_2^\dagger \Phi_2 + \lambda_1 (\Phi_1^\dagger \Phi_1)^2 + \lambda_2 (\Phi_2^\dagger \Phi_2)^2 + \lambda_3 \Phi_1^\dagger \Phi_1 \Phi_2^\dagger \Phi_2
$$
$$
+ \lambda_4 \Phi_1^\dagger \Phi_2 \Phi_2^\dagger \Phi_1 + \frac{\lambda_5}{2} \left[(\Phi_1^\dagger \Phi_2)^2 + h.c. \right], \tag{26.5}
$$

where Φ_1 and Φ_2 both are hypercharged, $Y = +1$, and can be written as

$$
\Phi_1 = \begin{pmatrix} G^+ \\ \dfrac{v + h + i G^0}{\sqrt{2}} \end{pmatrix}, \quad \Phi_2 = \begin{pmatrix} H^+ \\ \dfrac{H + i A}{\sqrt{2}} \end{pmatrix}. \tag{26.6}
$$

Here h is the SM Higgs with G^+, G^0 being the charged and neutral Goldstone bosons, respectively. The charged scalar H^+ is present in Φ_2, along with the neutral scalars, H, A, respectively, being CP-even and CP-odd. For the vacuum expectation values (VEVs) of the two doublets, we adopt the notation $\langle \Phi_1 \rangle = v/\sqrt{2}$, $\langle \Phi_2 \rangle = 0$, keeping in mind the exact nature of the $\mathbb{Z}_2$ symmetry. The zero VEV of Φ_2 is responsible for the inertness of this model. Here both H and A can be a viable candidate for the DM. We will consider H to be the DM candidate in the present analysis. The IDM parameter space is constrained from various theoretical, as well as experimental considerations. Here, we consider the light DM with hierarchical scalar mass spectrum, i.e., large mass differences (ΔM) with other heavy scalars, which satisfy all theoretical and experimental constraints (Table 26.1).

Table 26.1 Input parameters λ and the relevant scalar masses for some of the chosen benchmark points satisfying all the constraints coming from DM, Higgs, theoretical calculations, and low energy experimental data as discussed in the text. All the mass parameters are written in GeV unit. Standard choice of other two parameters are fixed at $M_H = 53.71$ GeV and $\lambda_L = 5.4 \times 10^{-3}$

Parameters	BP1	BP2	BP3	BP4	BP5	BP6	BP7
$M_{H^\pm}$(GeV)	255.3	304.8	350.3	395.8	446.9	503.3	551.8
M_A(GeV)	253.9	302.9	347.4	395.1	442.4	500.7	549.63
λ_2	1.27	1.07	0.135	0.106	3.10	0.693	0.285

26.3.1 Colider Analysis of IDM

We model the signal topology by designing a new final state of two-fatjet and large missing transverse energy. $2J_V + \not{E}_T$ channel: This final state can arise in the IDM for the aforementioned benchmarks in table [1] from the following three different channels:

$$pp \to AH^{\pm} \to (ZH)(W^{\pm}H) \equiv 2J_V + \not{E}_T$$
$$pp \to H^+H^- \to (W^+H)(W^-H) \equiv 2J_V + \not{E}_T \qquad (26.7)$$
$$pp \to AA \to (ZH)(ZH) \equiv 2J_V + \not{E}_T.$$

Here, A and $H^{\pm}$ decay to ZH and $W^{\pm}H$, respectively. As Z and W are originating from a heavy resonance, it is possible that they have sufficient boost to be reconstructed in a large radius jet. In this topology the following mono-fatjet process can also contribute significantly

$$pp \to H^{\pm}H \to (W^{\pm}H)H \equiv 1J_V + \not{E}_T$$
$$pp \to AH \to (ZH)H \equiv 1J_V + \not{E}_T. \qquad (26.8)$$

where extra jets can arise in the final state due to initial state radiation (ISR) and can form another fatjet. Here the major SM backgrounds comes from $Z + jets$, $W + jets$, $t\bar{t}, tW$ and Dibosons which is control by performing Multivariate analysis with the substructure variable. We use mass of leading fatjet M_{J1}, mass of sub leading fatjet M_{J0}, angular separation between two fatjet $\Delta R_{J1,J0}$, N-subjettiness for both the fatjet $\tau_{21}^{J0,J1}$, azimuthal angle $\Delta \Phi_{E,J0}$ and P_T^{J0} as input feature to train the boosted decision tree (BDT) network. The events are passes through the BDT after applying baseline cut as described in [11]. We show number of events after BDT analysis at $3000\ fb^{-1}$ for diffrent Benchmark points with the reach at the LHC (Table 26.2).

Table 26.2 Total number of signal events are, $\mathcal{N}_S^{bc}$ (including $1J_V$ and $2J_V$ topologies) and with number of background events N_{SM} before BDT_{opt} cut. The number of signal and background events after the BDT_{opt} cut are denoted by $\mathcal{N}_S$ and $\mathcal{N}_B$, respectively

BP	$\mathcal{N}_S^{bc}$	BDT_{opt}	$\mathcal{N}_S$	$\mathcal{N}_B$	$\mathcal{N}_S/\sqrt{\mathcal{N}_S + \mathcal{N}_B}$
1	1969	0.45	433	16439	3.3
2	2704	0.42	540	12329	4.7
3	3086	0.50	545	8799	5.6
4	2337	0.52	473	10274	4.5
5	1993	0.51	259	3698	4.1
6	1838	0.58	238	4109	3.6
7	1318	0.55	263	7397	3.0
N_{SM}	4109940	–	–	–	–

The best LHC sensitivity is obtained for the BP3 with $m_{H^\pm} \approx m_A \sim 350\,\text{GeV}$ and significance decreases both sides of the spectrum. With the increase of $m_{H^\pm}$, m_A, we get a higher boost for the decaying vector bosons, resulting in better discrimination power of jet-substructure variables.

26.4 Conclusions

We discuss two different model and their collider signatures which is modeled by utilizing the properties of jet-substructure. By looking into the jet-substructure characteristics, boosted jets reveal useful information on their origin and topology. We leveraged the same to achieve good discrimination between signal and background in the opposite sign di-lepton+fatjet channel for inverse seesaw case and Di-fatjet + missing transverse energy channel for inert doublet model.

References

1. Riccardo Barbieri, Lawrence J. Hall, Vyacheslav S. Rychkov, Improved naturalness with a heavy Higgs: an alternative road to LHC physics. Phys. Rev. D **74**, 015007 (2006)
2. Marco Cirelli, Nicolao Fornengo, Alessandro Strumia, Minimal dark matter. Nucl. Phys. **B753**, 178–194 (2006)
3. R.N. Mohapatra, Mechanism for understanding small neutrino mass in superstring theories. Phys. Rev. Lett. **56**, 561–563 (1986)
4. R.N. Mohapatra, J.W.F. Valle, Neutrino mass and baryon number nonconservation in superstring models. Phys. Rev. D **34**, 1642 (1986)
5. P. Arindam Das, S. Bhupal Dev, R.N. Mohapatra, Same Sign vs Opposite Sign Dileptons as a Probe of Low Scale Seesaw Mechanisms (2017)
6. Arindam Das, Nobuchika Okada, Inverse seesaw neutrino signatures at the LHC and ILC. Phys. Rev. D **88**, 113001 (2013)
7. Arindam Das, Nobuchika Okada, Bounds on heavy Majorana neutrinos in type-I seesaw and implications for collider searches. Phys. Lett. B **774**, 32–40 (2017)
8. J.A. Casas, A. Ibarra, Oscillating neutrinos and muon –> e, gamma. Nucl. Phys. B **618**, 171–204 (2001)
9. A. Bhardwaj, A. Das, P. Konar, A. Thalapillil, Looking for Minimal Inverse Seesaw scenarios at the LHC with Jet Substructure Techniques (2018)
10. Alexander Belyaev, Giacomo Cacciapaglia, Igor P. Ivanov, Felipe Rojas-Abatte, Marc Thomas, Anatomy of the inert two higgs doublet model in the light of the LHC and non-LHC dark matter searches. Phys. Rev. D **97**(3), 035011 (2018)
11. A. Bhardwaj, P. Konar, T. Mandal, S. Sadhukhan, Probing Inert Doublet Model using jet substructure with multivariate analysis (2019)

Chapter 27
Naturalness and Two Higgs Doublet Models

Ambalika Biswas and Amitabha Lahiri

Abstract In this talk, we have presented the study made on the implication of a criterion of naturalness for a broad class of two Higgs doublet models (2HDMs). We have considered the cancellation of quadratic divergences in what are called the type I, type II, lepton-specific and flipped 2HDMs. This results in a set of relations among masses of the physical scalars and coupling constants, a generalization of the Veltman conditions of the Standard Model. The model has been imposed with a softly broken $U(1)$ symmetry and we have studied the various limiting values of the scalar mixing angles α and β. These correspond to the Standard Model Higgs particle being the lighter CP-even scalar (alignment) or the heavier CP-even scalar (reverse alignment), and also the limit in which some of the Yukawa couplings of this particle are of the opposite sign from the vector boson couplings (wrong sign). Imposing further the constraints from the electroweak T-parameter (or ρ parameter), stability and perturbative unitarity conditions produce a range for the masses of each of the remaining physical scalars. We also calculate the $h \to \gamma\gamma$ decay rate in the wrong sign limit. The talk is based on the below two papers.

- 'Masses of physical scalars in two Higgs doublet models' by A.Biswas and A.Lahiri published in PHYSICAL REVIEW D 91, 115012 (2015).
- 'Alignment, reverse alignment, and wrong sign Yukawa couplings in two Higgs doublet models' by A. Biswas and A. Lahiri published in PHYSICAL REVIEW D 93, 115017 (2016).

A. Biswas (✉)
Department of Physics,Vivekananda College, Thakurpukur, India
e-mail: ani73biswas@gmail.com

A. Lahiri
SNBNCBS, Kolkata-98, India
e-mail: amitabha@bose.res.in

© Springer Nature Singapore Pte Ltd. 2021   187
P. K. Behera et al. (eds.), *XXIII DAE High Energy Physics Symposium*,
Springer Proceedings in Physics 261,
https://doi.org/10.1007/978-981-33-4408-2_27

27.1 Introduction

Issues regarding the origin of neutrino masses, dark matter and CP violation are among some of the unanswered questions that keep the door open for physics beyond the Standard Model. The simplest extensions of the Standard Model are two Higgs doublet models (2HDMs) [1]. Among the various motivations for 2HDMs, the one that is important to us is their use in models of dark matter.

We will work with the scalar potential [2]

$$V = \lambda_1 \left(|\Phi_1|^2 - \frac{v_1^2}{2} \right)^2 + \lambda_2 \left(|\Phi_2|^2 - \frac{v_2^2}{2} \right)^2 + \lambda_3 \left(|\Phi_1|^2 + |\Phi_2|^2 - \frac{v_1^2 + v_2^2}{2} \right)^2$$
$$+ \lambda_4 \left(|\Phi_1|^2 |\Phi_2|^2 - |\Phi_1^\dagger \Phi_2|^2 \right) + \lambda_5 \left| \Phi_1^\dagger \Phi_2 - \frac{v_1 v_2}{2} \right|^2 , \tag{27.1}$$

with real λ_i. This potential is invariant under the global U(1) symmetry $\Phi_1 \to e^{i\theta}\Phi_1$, $\Phi_2 \to \Phi_2$, except for a soft breaking term $\lambda_5 v_1 v_2 \Re(\Phi_1^\dagger \Phi_2)$. This in turn avoids FCNCs. Under U(1) symmetry the left-handed fermion doublets remain unchanged, $Q_L \to Q_L$, $l_L \to l_L$. The transformations of right-handed fermion singlets under U(1) determine the type of 2HDM: type I (none), type II ($d_R \to e^{-i\theta}d_R$, $e_R \to e^{-i\theta}e_R$), lepton-specific ($e_R \to e^{-i\theta}e_R$), flipped ($d_R \to e^{-i\theta}d_R$). The scalar doublets are parametrized as

$$\Phi_i = \begin{pmatrix} w_i^+(x) \\ \frac{v_i + h_i(x) + i z_i(x)}{\sqrt{2}} \end{pmatrix} , \quad i = 1, 2, \tag{27.2}$$

where the VEVs v_i may be taken to be real and positive without any loss of generality. Three of these fields get 'eaten' by the $W^\pm$ and Z^0 gauge bosons; the remaining five are physical scalar (Higgs) fields. The angle β diagonalizes both the CP-odd and charged scalar mass matrices, leading to the physical states A and $\xi^\pm$. The angle α diagonalizes the CP-even mass matrix leading to the physical states H and h, with

$$\tan \beta = \frac{v_2}{v_1} , \quad v = \sqrt{v_1^2 + v_2^2} = 246 \text{ GeV}. \tag{27.3}$$

27.2 Veltman Conditions

The Yukawa potential for the 2HDMs is of the form

$$\mathcal{L}_Y = \sum_{i=1,2} \left[-\bar{l}_L \Phi_i G_e^i e_R - \bar{Q}_L \tilde{\Phi}_i G_u^i u_R - \bar{Q}_L \Phi_i G_d^i d_R + h.c. \right] , \tag{27.4}$$

where l_L, Q_L are three vectors of isodoublets in the space of generations, e_R, u_R, d_R are three vectors of singlets, G_e^1, etc. are complex 3×3 matrices in generation space containing the Yukawa coupling constants, and $\tilde{\Phi}_i = i\tau_2\Phi_i^*$.

Cancellation of quadratic divergences in the scalar masses gives rise to four mass relations, which we may call the Veltman conditions [3] for the 2HDMs being considered. With g, g' being the $SU(2)$ and $U(1)_Y$ coupling constants, the Veltman conditions are

$$2\mathrm{Tr}G_e^1 G_e^{1\dagger} + 6\mathrm{Tr}G_u^{1\dagger}G_u^1 + 6\mathrm{Tr}G_d^1 G_d^{1\dagger} = \frac{9}{4}g^2 + \frac{3}{4}g'^2 + 6\lambda_1 + 10\lambda_3 + \lambda_4 + \lambda_5$$

$$2\mathrm{Tr}G_e^2 G_e^{2\dagger} + 6\mathrm{Tr}G_u^{2\dagger}G_u^2 + 6\mathrm{Tr}G_d^2 G_d^{2\dagger} = \frac{9}{4}g^2 + \frac{3}{4}g'^2 + 6\lambda_2 + 10\lambda_3 + \lambda_4 + \lambda_5.$$

27.3 Bounds on the Masses of Heavy and Charged Scalars

Further restrictions on the masses arise from stability [1, 4] and perturbative unitarity [5]. The conditions for the scalar potential to be bounded from below are

$$\lambda_1 + \lambda_3 > 0 \; , \;\; 2\lambda_3 + \lambda_4 + 2\sqrt{(\lambda_1 + \lambda_3)(\lambda_2 + \lambda_3)} > 0,$$
$$\lambda_2 + \lambda_3 > 0 \; , \;\; 2\lambda_3 + \lambda_5 + 2\sqrt{(\lambda_1 + \lambda_3)(\lambda_2 + \lambda_3)} > 0.$$

The perturbative unitarity condition puts some upper bounds on the absolute values of the combinations of the quartic coupling constants as shown below:

$$| 2\lambda_3 - \lambda_4 + 2\lambda_5| \le 16\pi \; , \;\; |2\lambda_3 + \lambda_4| \le 16\pi \; , \;\; |2\lambda_3 + \lambda_5| \le 16\pi \; ,$$
$$| 3(\lambda_1 + \lambda_2 + 2\lambda_3) \pm \sqrt{9(\lambda_1 - \lambda_2)^2 + (4\lambda_3 + \lambda_4 + \lambda_5)^2}| \le 16\pi,$$
$$| (\lambda_1 + \lambda_2 + 2\lambda_3) \pm \sqrt{(\lambda_1 - \lambda_2)^2 + (\lambda_4 - \lambda_5)^2}| \le 16\pi, \tag{27.5}$$
$$| 2\lambda_3 + 2\lambda_4 - \lambda_5| \le 16\pi \; , \;\; |(\lambda_1 + \lambda_2 + 2\lambda_3) \pm (\lambda_1 - \lambda_2)| \le 16\pi.$$

Further restrictions arise from the introduction of new physics. The electroweak rho parameter is defined as $\rho = \frac{m_W^2}{\cos^2\theta_w m_Z^2}$. New physics modifies this relation as $\rho = \frac{1}{1-\delta\rho}$. Recent bound on $\delta\rho$ is $\delta\rho = -0.0002 \pm 0.0007$ [6]. We impose this bound on the 2HDMs.

27.4 Alignment and Reverse Alignment Limits

The lighter CP-even Higgs in the alignment limit and the heavier CP-even Higgs in the reverse alignment limit are identified with the SM Higgs and hence $m_h = 125$ GeV and $m_H = 125$ GeV, respectively. The angles α and β are related as $\sin(\beta -$

Fig. 27.1 Type II 2HDM in alignment limit

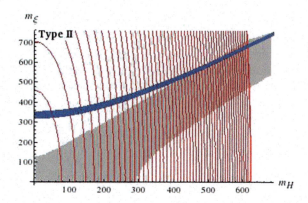

Fig. 27.2 Type II 2HDM in reverse alignment limit

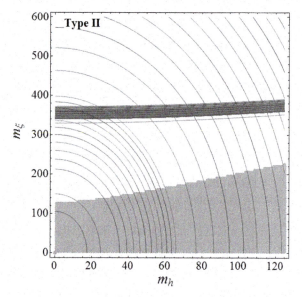

$\alpha) \approx 1 \implies \beta - \alpha = \frac{\pi}{2}$ in the alignment limit and $\cos(\beta - \alpha) \approx 1 \implies \beta \approx \alpha$ in the reverse alignment limit. Furthermore, the Yukawa coupling and the gauge coupling of h and H are same as that of the SM Higgs boson in the alignment and reverse alignment limits, respectively. The allowed mass range plots for the physical Higgs bosons have been plotted for type II 2HDM for $|\lambda_5| \leq 4\pi$ and $\tan \beta = 5$ by imposing the Veltman conditions, the restrictions coming from the stability conditions, the perturbative unitarity conditions and the corrections from new Physics in Fig. 27.1 for alignment limit and in Fig. 27.2 for reverse alignment limit.

In Fig. 27.1 for alignment limit, we see that the range of m_H lies between 450 and 620 GeV and that of m_ξ lies between 550 and 700 GeV. The above mass ranges vary between a few GeV for the various 2HDMs. Direct searches have shown that

$m_\xi > 100$ GeV and our results agree with this lower bound. The degeneracy in the masses of the physical Higgs bosons for large enough $\tan \beta$ is evident from our plots.

In case of reverse alignment limit, we find from the plot in Fig. 27.2 that there is no common region of intersection which obeys all the constraints. Thus, *Reverse alignment limit* is not a consistent limit with the *Naturalness condition* for 2HDMs.

27.5 Wrong-Sign Limit

The wrong-sign Yukawa coupling regime [7–9] is defined as the region of 2HDM parameter space in which at least one of the couplings of the SM-like Higgs to up-type and down-type quarks is opposite in sign to the corresponding coupling of SM-like Higgs to vectors bosons. The *wrong-sign limit* needs to be considered in conjunction with either the alignment limit or the reverse alignment limit. Since the reverse alignment limit is prohibited by *Naturalness*, we will now calculate the regions of parameter space when the wrong-sign limit is combined with the alignment limit.

Let us therefore consider a type II 2HDM where the Higgs–fermion Yukawa couplings normalized with respect to the Standard Model are

$$h\bar{D}D: \quad -\frac{\sin \alpha}{\cos \beta} = -\sin(\beta + \alpha) + \cos(\beta + \alpha)\tan \beta, \qquad (27.6)$$

$$h\bar{U}U: \quad \frac{\cos \alpha}{\sin \beta} = \sin(\beta + \alpha) + \cos(\beta + \alpha)\cot \beta. \qquad (27.7)$$

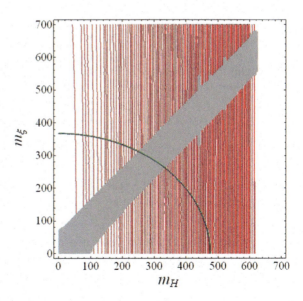

Fig. 27.3 Allowed mass range in wrong-sign and alignment limit

In the case when $\sin(\beta + \alpha) = 1$, the $h\bar{D}D$ coupling normalized to its SM value is -1, while the normalized $h\bar{U}U$ coupling is $+1$. Note that in this limiting case, $\sin(\beta - \alpha) = -\cos 2\beta$, which implies that the wrong-sign $h\bar{D}D$ Yukawa coupling can only be achieved for values of $\tan\beta > 1$. Likewise, in the case of $\sin(\beta + \alpha) = -1$, the normalized $h\bar{U}U$ coupling is -1, whereas the normalized $h\bar{D}D$ coupling is $+1$. Then $\sin(\beta - \alpha) = \cos 2\beta$, which implies that the wrong-sign $h\bar{U}U$ couplings can occur only if $\tan\beta < 1$. Thus, the realistic situation for wrong sign is $\sin(\beta + \alpha) = 1$. The wrong-sign limit approaches the alignment limit for $\tan\beta \approx 17$ as was displayed in [8, 9] for the allowed parameter space of the type II CP-conserving 2HDM based on the 8 TeV run of the LHC. We have plotted the Veltman conditions, stability, perturbative unitarity and new physics constraints on the $m_H - m_\xi$ plane for $\tan\beta = 17$ in Fig. 27.3. The range of m_H is approximately $(250, 330)$ GeV, and that of m_ξ is approximately $(260, 310)$ GeV. At higher values of $\tan\beta$, both ranges become narrower and move down on the mass scale.

27.6　Diphoton Decay Width

Diphoton decay width in wrong-sign and alignment limits is given by

$$\Gamma(h \to \gamma\gamma) = \frac{G_\mu \alpha^2 m_h^3}{128\sqrt{2}\pi^3} \Big| \sum_f N_c Q_f^2 g_{hff} A_{1/2}^h(\tau_f) + g_{hVV} A_1^h(\tau_W) \tag{27.8}$$

$$+ \frac{m_W^2 \lambda_{h\xi^+\xi^-}}{2c_W^2 M_{\xi^\pm}^2} A_0^h(\tau_{\xi^\pm}) \Big|^2 ,$$

where $g_{htt} = \frac{\cos\alpha}{\sin\beta}$, $g_{hbb} = -\frac{\sin\alpha}{\cos\beta}$, $g_{hWW} = \sin(\beta - \alpha)$ and $\lambda_{h\xi^+\xi^-} = \cos 2\beta \sin(\beta + \alpha) + 2c_W^2 \sin(\beta - \alpha) = \lambda_{hAA} + 2c_W^2 g_{hVV}$ ($c_W = \cos\theta_W$, θ_W being the Weinberg angle). The amplitudes A_i at lowest order for the spin 1, spin $\frac{1}{2}$ and spin 0 particle contributions are given by

$$A_{1/2}^h = 2\left[\tau + (\tau - 1)f(\tau)\right]\tau^{-2} , \quad A_1^h = -\left[2\tau^2 + 3\tau + 3(2\tau - 1)f(\tau)\right]\tau^{-2} ,$$

$$A_0^h = -\left[\tau - f(\tau)\right]\tau^{-2} \text{ where } \tau_x = m_h^2/4m_x^2 \text{ and}$$

$$f(\tau) = \begin{cases} \arcsin^2 \sqrt{\tau}, & \tau \le 1 \\ -\frac{1}{4}\left[\log \frac{1+\sqrt{1-\tau^{-1}}}{1-\sqrt{1-\tau^{-1}}} - i\pi\right]^2, & \tau > 1. \end{cases} \tag{27.9}$$

From the plot, Fig. 27.4 for diphoton decay width, we see that the relative diphoton decay width increases as m_A increases. Maximum value of about 6% is reached as compared to the SM value. Thus, this throws light on BSM Physics.

Fig. 27.4 Diphoton decay width in wrong-sign and alignment limit

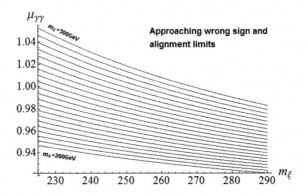

References

1. G.C. Branco, P.M. Ferreira, L. Lavoura, M.N. Rebelo, M. Sher, J.P. Silva, Theory and phenomenology of two-Higgs-doublet models. Phys. Rept. **516**, 1 (2012)
2. J.F. Gunion, H.E. Haber, G.L. Kane, S. Dawson, *Errata for the Higgs hunter's guide*, arXiv:hep-ph/9302272
3. M.J.G. Veltman, The infrared—ultraviolet connection. Acta Phys. Polon. B **12**, 437 (1981), C. Newton, T. T. Wu, Mass relations in the two Higgs doublet model from the absence of quadratic divergences. Z. Phys. C **62**, 253 (1994)
4. J.F. Gunion, H.E. Haber, The CP conserving two Higgs doublet model: the approach to the decoupling limit. Phys. Rev. D **67**, 075019 (2003)
5. A.G. Akeroyd, A. Arhrib, E.M. Naimi, Note on tree level unitarity in the general two Higgs doublet model. Phys. Lett. B **490**, 119 (2000)
6. K.A. Olive et al., Particle data group collaboration. Chin. Phys. C **38**, 090001 (2014)
7. P.M. Ferreira, J.F. Gunion, H.E. Haber, R. Santos, Phys. Rev. D **89**(11), 115003 (2014), arXiv:1403.4736 [hep-ph]
8. P.M. Ferreira, R. Guedes, M.O.P. Sampaio, R. Santos, JHEP **1412**, 067 (2014)
9. P.M. Ferreira, R. Guedes, J.F. Gunion, H.E. Haber, M.O.P. Sampaio, R. Santos, arXiv:1410.1926 [hep-ph]

Chapter 28
Extended Scalar Sectors, Effective Operators and Observed Data

Atri Dey, Jayita Lahiri, and Biswarup Mukhopadhyaya

Abstract The available data on the 125 GeV scalar h is analysed to explore new physics in the electroweak symmetry breaking sector. The first part of the study is model-independent, with h couplings to standard model particles scaled by some free parameters. At the same time, the additional loop contributions to $h \to \gamma\gamma$ and $h \to Z\gamma$, mediated by charged scalar contributions in the extended scalar sector, are treated in terms of gauge-invariant effective operators. We fit the existing data to obtain marginalized 1σ and 2σ regions in the space of the coefficients of such effective operators, by considering the correlation between, say, the gluon fusion and vector-boson fusion channels, as reflected in a non-diagonal covariance matrix. After thus obtaining model-independent fits, the allowed values of the coefficients are translated into permissible regions of the parameter spaces of several specific models.

28.1 Introduction

The ATLAS and CMS experiments at the Large Hadron Collider (LHC) have discovered a new boson mass 125 GeV. Though the properties of this particle are similar to those of the Higgs boson predicted in the standard model (SM) of electroweak interactions, everyone is on the lookout for any small difference that may reveal the participation of some new physics. It is thus imperative to closely examine all interactions (including supposedly 'effective' ones) of this particle with SM fermions and gauge bosons. Theoretical models extending the SM, including those augmenting the

A. Dey (✉) · J. Lahiri · B. Mukhopadhyaya
Regional Centre for Accelerator-based Particle Physics, Harish-Chandra Research Institute,
HBNI, Chhatnag Road, Jhunsi, Allahabad 211 019, India
e-mail: atridey1993@gmail.com; atridey@hri.res.in

J. Lahiri
e-mail: jayitalahiri@hri.res.in

B. Mukhopadhyaya
e-mail: biswarup@hri.res.in

© Springer Nature Singapore Pte Ltd. 2021
P. K. Behera et al. (eds.), *XXIII DAE High Energy Physics Symposium*,
Springer Proceedings in Physics 261,
https://doi.org/10.1007/978-981-33-4408-2_28

electroweak symmetry breaking sector, are thus explored. It is a natural endeavour to link the contributions of such new physics, in the form of modified Higgs interaction strengths as well as effective operators generated by them, to the departure from unity in the Higgs signal strengths in various final states f, defined as $\mu^f = \frac{\sigma^f}{\sigma^f_{SM}}$. Fitting the available data with various μ^f, therefore, enables one to analyse allowed strengths of effective operators generated in various models and ultimately constrain the model parameters themselves. This is particularly useful if the new particles belonging to any new physics scenario do not have a copious rate for direct production, but show up (i) by participating off-shell, and leading to higher dimensional effective operators, and (ii) by modifying the coupling strength(s) of the 125 GeV scalar to the other SM particles. The present study is devoted to such a situation.

Interestingly, significant constraints arise from the signal strengths for the loop-induced decay channel such as $h \to \gamma\gamma$. $h \to Z\gamma$, although yet unobserved, can also provide strong limits on new physics contributions, especially with accumulating luminosity. These limits can be coupled with those arising from tree-level decay modes such as $h \to WW, ZZ$, etc. where perceptible effects can come mostly via scaling of the SM coupling by a factor κ.

The effect of high-scale physics on low-energy processes can be formulated in terms of higher dimension operators in the Lagrangian, which will be suppressed by the new physics scale Λ. These higher dimensional operators can be derived from an $SU(2)_L \times U(1)_Y$-invariant basis, as they result from physics corresponding to a scale much higher than the electroweak scale.

Here we have adopted a slightly different formulation of new physics contributions to the aforementioned loop-induced decays. The consequences of new physics have been divided into two categories. The first of these is the scaling of the couplings $ht\bar{t}$, $hb\bar{b}$, $h\tau\bar{\tau}$, hVV, where modifications to the SM couplings are inevitable when additional fields mixing with the ones in SM are present. As we have already mentioned, such modifications usually override the effects of higher dimensional operators. Such scaling also affects loop effects such as $h \to \gamma\gamma, Z\gamma$, via modified vertices in the loop, where W's or top quarks are involved, and the corresponding amplitudes can be written down in terms of the scale factors κ. The second category consists in loop diagrams mediated by new particles such as charged scalars. Their contributions to loop amplitudes, we argue, can be expressed in isolation in terms of effective couplings. As shown in Sect. 28.4, these couplings, at least those ensuing from additional scalar fields, can be treated as ones derived by the aforesaid gauge-invariant dimension-6 operators, so long as the masses of the new particles are gauge-invariant and at least a little above the electroweak symmetry breaking (EWSB) scale.

Keeping this in mind, we perform a global fit of the currently available Higgs data to constrain the full parameter space including scale factors and the Wilson coefficients corresponding to various dimension-6 operators. While using the Higgs signal strength data in various channels we take into account the correlation between various production processes such as gluon fusion and vector-boson fusion, thus including non-diagonal covariant matrices in our analysis. Model-independent 2σ regions for the κ's well as the Wilson coefficients are thus obtained.

We select specific models in the next step of the analysis. The present study is restricted to additional scalars and includes various kinds of two Higgs doublet models (2HDM) as well as those with one and two scalar triplets of the kind introduced in Type-II see-saw mechanism. The marginalized 2σ regions in the space of dimension-6 operators are then recast, keeping track of the correlation between the scale factors κ and the Wilson coefficients for each model. These are finally translated into constraints in the space of masses and coupling strengths pertaining to all the models.

28.2 Modified Higgs Couplings

There can be an extended Higgs sector comprising of additional neutral and charged scalars. Their mixing may cause the coupling of the 125 GeV scalar to SM particles to be modified. Modification may occur in two ways. First, there can be scaling of the Higgs couplings, with unaltered Lorentz structure of the corresponding vertices, expressed as

$$\tilde{g}_{hVV} = \kappa_v \times g_{hVV}, \tag{28.1}$$

$$\tilde{g}_{ht\bar{t}} = \kappa_t \times g_{ht\bar{t}}, \tag{28.2}$$

$$\tilde{g}_{hb\bar{b}} = \kappa_b \times g_{hb\bar{b}} \tag{28.3}$$

$$\tilde{g}_{h\tau\bar{\tau}} = \kappa_\tau \times g_{h\tau\bar{\tau}}, \tag{28.4}$$

where g_{hVV}, $g_{ht\bar{t}}$, $g_{hb\bar{b}}$ and $g_{h\tau\bar{\tau}}$ are the couplings of the Higgs to the gauge bosons and the fermions in the SM. The couplings of Higgs to W boson and Z boson are scaled in the same way here.

Moreover, there can be heavy states running in the loop modifying Higgs couplings. A general approach to parametrize such modification is to express it in terms of gauge-invariant higher dimensional effective operators.

All Higgs interactions should, in principle, be modified via such operators. However, couplings which exist at the SM at tree level, namely, hWW, hZZ, $ht\bar{t}$, $hb\bar{b}$ and $h\tau\bar{\tau}$, are rather nominally affected by higher dimensional terms with (at least) TeV-scale suppression (Fig. 28.1).

Dimension-6 effective interactions involving a Higgs and two gauge bosons can be expressed in terms of the following gauge-invariant operators:
$\mathcal{O}_{\mathbf{BB}} = \frac{f_{BB}}{\Lambda^2} \Phi^\dagger \hat{B}_{\mu\nu} \hat{B}^{\mu\nu} \Phi$, $\mathcal{O}_{\mathbf{WW}} = \frac{f_{WW}}{\Lambda^2} \Phi^\dagger \hat{W}_{\mu\nu} \hat{W}^{\mu\nu} \Phi$, $\mathcal{O}_{\mathbf{B}} = \frac{f_B}{\Lambda^2} D_\mu \Phi^\dagger \hat{B}^{\mu\nu} D_\nu \Phi$, $\mathcal{O}_{\mathbf{W}} = \frac{f_W}{\Lambda^2} D_\mu \Phi^\dagger \hat{W}^{\mu\nu} D_\nu \Phi$.

The parts of the contributions from the effective operators $\mathcal{O}_{BB}$ and $\mathcal{O}_{WW}$ can be expressed in Table 28.1.

Table 28.1 The part of the amplitudes for $h \to \gamma\gamma$ and $h \to Z\gamma$ coming from new particle loops, and expressed in terms of the dimension-6 operators $\mathcal{O}_{BB}$ and $\mathcal{O}_{WW}$

	$h \to \gamma\gamma$	$h \to Z\gamma$
$\mathcal{O}_{BB}$	$-i\mathcal{M}_{BB} =$ $4\frac{f_{BB}}{\Lambda^2}\frac{g^2}{4}\sin^2\theta_w v \times$ $(k_1.k_2 g_{\mu\nu} - k_{1\mu}k_{2\nu})\epsilon^{*\mu}(k_2)\epsilon^{*\nu}(k_1)$	$-i\mathcal{M}_{BB} =$ $-4\frac{f_{BB}}{\Lambda^2}\frac{g^2}{2}\frac{\sin^3\theta_w}{\cos\theta_w} v \times$ $(k_1.k_2 g_{\mu\nu} - k_{1\mu}k_{2\nu})\epsilon^{*\mu}(k_2)\epsilon^{*\nu}(k_1)$
$\mathcal{O}_{WW}$	$-i\mathcal{M}_{WW} =$ $4\frac{f_{WW}}{\Lambda^2}\frac{g^2}{4}\sin^2\theta_w v \times$ $(k_1.k_2 g_{\mu\nu} - k_{1\mu}k_{2\nu})\epsilon^{*\mu}(k_2)\epsilon^{*\nu}(k_1)$	$-i\mathcal{M}_{WW} =$ $4\frac{f_{WW}}{\Lambda^2}\frac{g^2}{2}\sin\theta_w\cos\theta_w v \times$ $(k_1.k_2 g_{\mu\nu} - k_{1\mu}k_{2\nu})\epsilon^{*\mu}(k_2)\epsilon^{*\nu}(k_1)$

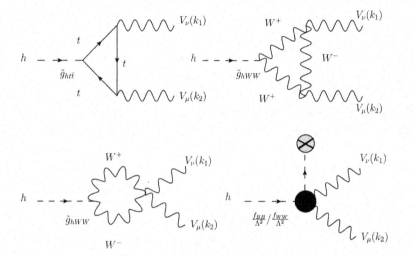

Fig. 28.1 Feynman diagrams for $h \to \gamma\gamma$ and $h \to Z\gamma$ in the most general situation. Contributions mediated by fields other than those in SM are lumped in the blob

28.3 The Global Fit

After parametrizing new physics effects, we investigate the region of eight-dimensional parameter space favoured by the 8 and 13 TeV results at the LHC, spanned by the four-scale factors κ_V, κ_t, κ_b and κ_τ (which parametrize the modification of SM tree-level hVV, $ht\bar{t}$, $hb\bar{b}$, $h\tau\bar{\tau}$ couplings) and f_{BB}, f_{WW}, f_B and f_W, which are the Wilson coefficients in the dimension-6 hVV operators. We also use the correlations between gluon fusion and vector-boson fusion production for each of the major Higgs decay channels. From the χ^2 minimization, one obtains the region allowed by the experimental data at the 1- and 2σ levels. Here we give some of them as an example.

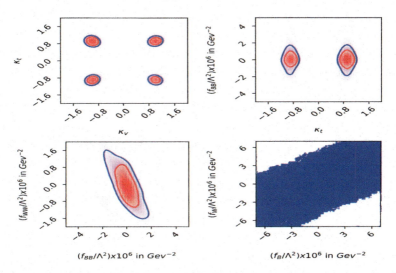

Fig. 28.2 Allowed regions at 1σ (red) and 2σ (blue) levels in the parameter space of scale factors

28.4   Extended Higgs Models and Dimension-6 Operators

Various new physics models predict extended electroweak symmetry breaking sectors. It is naturally of interest to link the model-independent analysis presented above to specific theoretical scenarios. Now we translate the results of the previous section to those pertaining to extended Higgs models with charge scalars, taking into account the additional constraints that connect model parameters. These charged scalars should contribute to loop-induced decays of the SM-like Higgs, namely, $h \rightarrow \gamma\gamma, h \rightarrow Z\gamma$. The corresponding decay widths deviate from the SM predictions due to (a) scaling of the hWW, $ht\bar{t}$ vertices and (b) the additional contributions from charge scalar loops. The contribution to the amplitude from the charged scalar loops for $h \rightarrow \gamma\gamma$ is of the form

$$-i\mathcal{M} = C_{vertex} \times (k_1.k_2 g_{\mu\nu} - k_{1\mu}k_{2\nu})\epsilon^{*\mu}(k_2)\epsilon^{*\nu}(k_1) \times A_H(\gamma\gamma)(m_{H^\pm}). \quad (28.5)$$

While that for $h \rightarrow Z\gamma$ is

$$-i\mathcal{M} = \tilde{C}_{vertex} \times (k_1.k_2 g_{\mu\nu} - k_{1\mu}k_{2\nu})\epsilon^{*\mu}(k_2)\epsilon^{*\nu}(k_1) \times A_H(Z\gamma)(m_{H^\pm}, m_Z), \quad (28.6)$$

where C_{vertex} and $\tilde{C}_{vertex}$ are the vertex factors for $h \rightarrow \gamma\gamma$ and $h \rightarrow Z\gamma$ in the two cases, while $A_H(\gamma\gamma)$ and $A_H(Z\gamma)$ are the loop integrals for $h \rightarrow \gamma\gamma$ and $h \rightarrow Z\gamma$, respectively.

We can compare this terms with those which comes from gauge-invariant operators when the terms coming from the charge loops are independent of EWSB. We can see that the loop amplitudes for $h \rightarrow \gamma\gamma$ does not involve m_Z, and the charged scalar masses arise from $SU(2)_L \times U(1)_Y$ invariant terms. Thus, there is no dependence

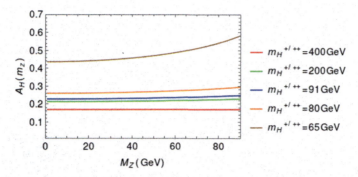

Fig. 28.3 Dependence of the additional scalar loop integral on the Z boson mass in $h \rightarrow Z\gamma$

on m_{EWSB} there. One would thus expect the same dependence (or lack of it) in the amplitude for $h \rightarrow Z\gamma$ unless there are highly fine-tuned boundary conditions in the RG running of parameters. A natural way of establishing consistency between the two amplitudes, therefore, is to have no m_Z-dependence in the loop amplitude for $h \rightarrow Z\gamma$ as well. As one can see from Fig. 28.3.

28.4.1 Two Higgs Doublet Models

We consider Type-I, Type-II, eptonSpecific and Flipped 2HDMs. In Type-I, from the Yukawa Lagrangian, the couplings of the fermions with the SM-like Higgs boson in this case are

$$C_{ht\bar{t}} = \frac{\cos\alpha}{\sin\beta} \times C_{ht\bar{t}}^{SM}$$

$$C_{hb\bar{b}} = \frac{\cos\alpha}{\sin\beta} \times C_{hb\bar{b}}^{SM}$$

$$C_{h\tau\bar{\tau}} = \frac{\cos\alpha}{\sin\beta} \times C_{h\tau\bar{\tau}}^{SM}, \tag{28.7}$$

while the gauge boson couplings are

$$C_{hVV} = \sin(\beta - \alpha) \times C_{hVV}^{SM}. \tag{28.8}$$

2σ allowed regions for them are found to be as in Fig. 28.4, which looks very similar to the corresponding contours in Fig. 28.2.

And we translate them to get the 2σ regions in model's parameter space directly in Fig. 28.5.

For other kind of 2HDMs, we do same kind of things.

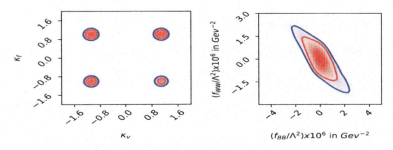

Fig. 28.4 Allowed regions at 1σ (red) and 2σ (blue) levels in the parameter space of scale factors and dimension-6 couplings in Type-I 2HDM

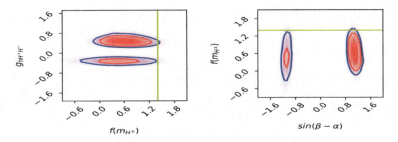

Fig. 28.5 Allowed regions at 1σ (red) and 2σ (blue) levels in the parameter space of $f(m_{H^\pm}) - g_{hH^+H^-}$ (left) and $\sin(\beta - \alpha) - f(m_{H^\pm})$ (right) in Type-I 2HDM

28.4.2 Higgs Triplet Models

Here we consider two kind of triplet models. For both of them there are single charged Higgs as well as doubly charged Higgs running in the loops. So in $h \to \gamma\gamma$ and $h \to Z\gamma$ there are extra four loop contribution which can be compared with the effective operators as well as modified couplings. SU(2) invariance of the Lagrangian plus the smallness demands of doublet–triplet mixing force masses of the triplet-dominated states to be nearly degenerate. In such a situation, SM-like Higgs couplings to fermions and gauge bosons are practically unaltered. Therefore, we can assume

$$\kappa_v \approx \kappa_t \approx \kappa_b \approx \kappa_\tau \approx 1. \qquad (28.9)$$

So the 1σ and 2σ regions in the f_{BB}-f_{WW} plane can now be obtained on the assumption that all κ's are unity. One can define an effective coupling as $g^{eff}_{hH^+H^-} = \tilde{g}_{hH^+H^-} + 4\tilde{g}_{hH^{++}H^{--}}$, which can be directly extracted (Figs. 28.6 and 28.7).

Similarly for two-triplet scenarios,

$$g^{eff}_{hH_1^+H_1^-} = \tilde{g}_{1hH_1^+H_1^-} + 4\tilde{g}_{1hH_1^{++}H_1^{--}}, \, g^{eff}_{hH_2^+H_2^-} = \tilde{g}_{2hH_2^+H_2^-} + 4\tilde{g}_{2hH_2^{++}H_2^{--}}$$

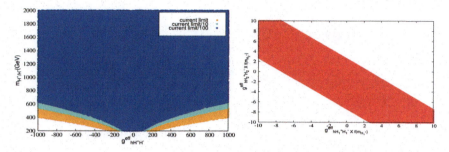

Fig. 28.6 Allowed regions at 1σ(red) and 2σ(blue) levels in the parameter space of scale factors and dimension-6 couplings in the single-triplet scenario (left) and two-triplet scenario (right)

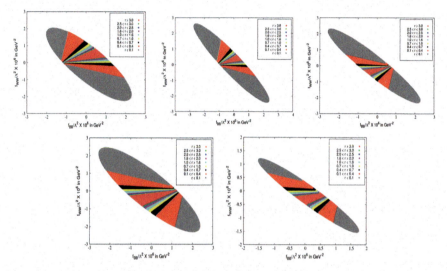

Fig. 28.7 The ratio r in f_{BB}-f_{WW} plane for Type-I (top left), Type-II (top middle), Lepton-specific (top right), Flipped 2HDM (bottom left) and triplet case (bottom right). The ellipses denote the regions allowed at 2σ level

28.4.2.1 Comparison Among Various 2HDMs

We next ask the question: while allowed regions for the various 2HDM and triplet scenarios are identified as above from available data, can some quantities be defined further, which may act as differentiators among them? We are essentially occupied with one such quantity as

$$r = \frac{\mu_{\gamma\gamma}}{\mu_{Z\gamma}} \tag{28.10}$$

These figures can help us to understand that if in near future the $fbb - fww$ plane can be more restricted what kind of model will be suitable from observed r constraints.

Chapter 29
Search for Dark Matter in the Mono−$W/Z(q\bar{q})$ Channel at the ATLAS Experiment

Bibhuti Parida

Abstract We present a search for Dark Matter (DM) particles production in the mono-$W/Z(q\bar{q})$ channel using pp collision data at a centre-of-mass energy of $\sqrt{s}$ = 13 TeV corresponding to an integrated luminosity of 36.1 fb^{-1}, recorded by the ATLAS detector at the Large Hadron Collider (LHC). No significant excess over the Standard Model (SM) prediction is observed. The search results are interpreted in terms of limits on invisible Higgs boson decays into DM particles, constraints on the parameter space of the simplified vector-mediator model and generic upper limits on the visible cross section for W/Z+DM production.

29.1 Introduction

Discovery of DM particles and the understanding of their interactions with SM particles is one of the greatest quests in particle physics and cosmology today. Different experimental approaches are being exploited. The indirect detection experiments search for signs of DM annihilation or decays in outer space whereas the direct detection experiments are related to low-energy recoils of nuclei induced by interactions with DM particles from the galactic halo. The interpretation of these searches is subject to astrophysical uncertainties in DM abundance and composition. Searches at particle colliders, for which these uncertainties are irrelevant, are complementary if DM candidates can be produced in particle collisions. DM signature which can be detected at the LHC experiments is a large overall missing transverse momentum (E_T^{miss}) from a pair of DM particles that recoil against one or more SM particles. Weakly Interacting Massive Particles (WIMPs), one of the leading DM candidates, could be produced in pp collisions at the LHC and detected by measuring the momentum imbalance associated with the recoiling SM particles. So far, several searches for

Bibhuti Parida, On behalf of the ATLAS Collaboration (Speaker), XXIII DAE-BRNS High Energy Physics Symposium, December 10–14, 2018, IIT Madras, India.

B. Parida (✉)
Tomsk State University, Tomsk, Russia
e-mail: bibhuti.parida@cern.ch

DM signatures were performed with LHC pp collision data at centre-of-mass energies of 7, 8 and 13 TeV with no significant deviations from SM predictions observed and set limits on various DM particle models. In this proceedings, we present the latest results of DM particles production in association with a hadronically decaying W or Z boson (mono-W/Z search).

29.2 Search Interpretation and Samples

Two signal models are used to describe DM production in the mono-W/Z final state. These are (i) simplified vector-mediator model, illustrated by the Feynman diagram in Fig. 29.1a, in which a pair of Dirac DM particles is produced via an s-channel exchange of a vector mediator (Z') [1, 2]. There are four free parameters in this model: the DM and the mediator masses (m_χ and $m_{Z'}$, respectively), and the mediator couplings to the SM and DM particles (g_{SM} and g_{DM}, respectively) and (ii) invisible Higgs boson decays in which a Higgs boson H produced in SM Higgs boson production processes decays into a pair of DM particles which escape detection as shown in Fig. 29.1b. The free parameter of this model is the branching ratio $B_{H \to \text{inv}}$. The cross sections for different Higgs boson production modes are taken to be given by the SM predictions.

Signal processes within the simplified Z' vector-mediator model are modelled at the Leading Order (LO) accuracy with the MadGraph5_aMCNLO v2.2.2 generator interfaced to the PYTHIA 8.186 and PYTHIA 8.210 parton shower models, respectively. The A14 set of tuned parameters are used together with the NNPDF23lo PDF set for these signal samples. The signal samples within the simplified vector-mediator model are generated in a grid of mediator and DM particle masses, with coupling values set to $g_{SM} = 0.25$ and $g_{DM} = 1$. The mediator mass $m_{Z'}$ and the DM particle mass m_χ range from 10 GeV to 10 TeV and from 1 GeV to 1 TeV, respectively. Processes in the mono-W/Z final state involving invisible Higgs boson decays originate from the VH, ggH and VBF SM Higgs boson production mechanisms and were all generated with the POWHEG-BOX v2 generator interfaced to PYTHIA 8.212 for the parton shower, hadronization and the underlying event modelling. The Higgs

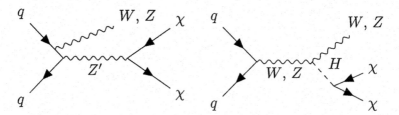

Fig. 29.1 Feynman diagrams for simplified vector-mediator model (left); invisible Higgs boson decay (right)

boson mass in these samples was set to $m_H = 125\,\text{GeV}$ and the Higgs boson was decayed through the $H \rightarrow ZZ^* \rightarrow \nu\nu\nu\nu$ process to emulate the decay of the Higgs boson into invisible particles with a branching ratio of $B_{H\rightarrow\text{inv}} = 100\%$.

The major sources of background are the production of top-quark pairs ($t\bar{t}$) and the production of W and Z bosons in association with jets (V +jets, where $V = W$ or Z). Other small background contributions include diboson (WW, WZ and ZZ) and single top-quark production. Their contribution is estimated from simulation. Events containing leptonically decaying W or Z bosons with associated jets were simulated using the Sherpa 2.2.1 generator. The NNPDF3.0 Next-to-Next-to-Leading Order (NNLO) PDF set was used in conjunction with dedicated parton shower tuning. For the generation of $t\bar{t}$ events, Powheg-Box v2 was used with the CT10 PDF set in the NLO matrix element calculations. Electroweak t-channel, s-channel and Wt-channel single-top-quark events were generated with Powheg-Box v1. Diboson events with one of the bosons decaying hadronically and the other leptonically were generated with the SHERPA 2.1.1 event generator. The CT10 PDF set was used in conjunction with dedicated parton shower tuning.

29.3 Object Reconstruction and Event Selection

The event selection of this analysis is based on dedicated 1-μ and 2ℓ ($\ell = \mu$ or e) control regions and that relies on the reconstruction and identification of jets, electrons and muons as well as on the reconstruction of the missing transverse momentum (E_T^{miss}). Three types of jets such as '*small-R*' with radius parameter $R = 0.4$, '*large-R*' with radius parameter $R = 1.0$ and '*track jets*' with radius parameter $R = 0.2$ using anti-k_t jet clustering algorithm are employed in this search. The small-R central jets containing b-hadrons are identified using b-tagging algorithm at an operating point with a 70% b-tagging efficiency measured in simulated $t\bar{t}$ events. As for the small-R jets, the track jets containing b-hadrons are identified using the MV2c10 algorithm at a working point with 70% efficiency. Electron candidates are reconstructed from energy clusters in the electromagnetic calorimeter that are associated to an inner detector track. The electron candidates are identified using a likelihood-based procedure in combination with additional track hit requirements. Muon candidates are primarily reconstructed from a combined fit to inner detector hits and muon spectrometer segments. In the middle part of the detector ($|\eta| < 0.1$), where the muon spectrometer coverage is suboptimal, muons are identified by matching a reconstructed inner detector track to calorimeter energy deposits consistent with a minimum ionizing particle. The vector missing transverse momentum, E_T^{miss}, is calculated as the negative vector sum of the transverse momenta of calibrated small-R jets and leptons, together with the tracks which are associated to the primary interaction vertex but not associated to any of these physics objects. A closely related quantity, $E_T^{\text{miss(no lepton)}}$, is calculated in the same way but excluding the reconstructed muons or electrons. The missing transverse momentum is given by the magnitude of these vectors, $E_T^{\text{miss}} = |E_T^{\text{miss}}|$ and $E_T^{\text{miss(no lepton)}} = |E_T^{\text{miss(no lepton)}}|$. In addition, the track-based missing transverse momentum vector, p_T^{miss}, and similarly $p_T^{miss(no\ lepton)}$, is calculated as the negative vector sum of the transverse momenta of tracks with $p_T > 0.5$ GeV and $|\eta| < 2.5$ originating from the primary vertex. Events studied in this search are accepted by a combination of E_T^{miss} triggers with thresholds between 70 and 110 GeV, depending on the data taking periods with the trigger efficiency is measured using events with large E_T^{miss} accepted by muon triggers.

Two distinct event topologies are considered depending on the Lorentz boost of the vector boson: a merged topology where the decay products of the vector boson are reconstructed as a single large-R jet, and a resolved topology where they are reconstructed as a pair of individual small-R jets. A priority merged selection procedure has been considered, i.e each event is first passed through the merged-topology selection and, if it fails, it is passed through the resolved-topology selection. Thus, there is no overlap of events between the two final state topologies. In the merged (resolved) event topology, at least one large-R jet (at least two small-R jets) and E_T^{miss} values above 250 GeV (above 150 GeV) are required in the final state. The events are again classified according to the number of b-tagged jets present in the events such as with exactly zero ($0b$), one ($1b$) and two ($2b$) b-tagged jets to improve the signal-to-background ratio and the sensitivity to $Z \to b\bar{b}$ decays. Small-R jets (track jets) are used for the b-tagging in the resolved (merged) category. The events in the $0b$ and $1b$ categories with merged topology are further classified into high-purity (HP) and low-purity (LP) regions; the former category consists of events satisfying the p_T-dependent requirements on the jet substructure variable $D_2^{\beta=1}$, allowing an improved discrimination for the jets counting $V \to q\bar{q}$ decays, while the latter one selects all the remaining signal events. Mass window requirements are imposed on the vector boson candidate in the $2b$ tag merged and $0b$, $1b$ and $2b$ tag resolved topologies whereas a W/Z tagger requirement has been applied for the $0b$ and $1b$ tag merged topology. More details of the object reconstruction and event selection used in the analysis can be found in the original published paper [3].

29.4 Background Estimation

The dominant background contribution originates from the $t\bar{t}$ and V+jets production and is estimated by the normalization factors extracted from the Control Regions (CRs). The sub-dominant backgrounds are constraints by theoretical values. The 1-μ CR is designed to distinguish W+jets and $t\bar{t}$ process whereas the 2ℓ CR is designed to estimate the contributions of the Z+jets process selecting $Z \to \mu^+\mu^-$ and e^+e^- decays. More details of the background estimation procedure can be found in [3].

29.5 Systematic Uncertainties

Several experimental and theoretical systematic uncertainties affect the final results of the analysis. Theoretical uncertainties on the signal yield are estimated to be about

10–15% for the simplified vector-mediator model and it is 5–10% for the invisible decays of the Higgs boson. The experimental systematics include large-R jet mass scale and resolution (10% on signal and 5% on background). This is the largest source of experimental systematic uncertainty in the merged topology. The uncertainty in the large-R jet energy resolution is 3% on signal and 1% on background. The uncertainty due to the $D_2^{\beta=1}$ substructure parameter is 5–10%. The uncertainties due to the small-R jet energy scale are 6% on signal and 10% on background. The small-R jet resolution uncertainties affect 2–5% and the b-tagging calibration uncertainty affects up to 10%. The uncertainties on the modelling of E_T^{miss} are 1–3% and 2–10% for the background and signal processes, respectively. The uncertainty in the combined 2015 + 2016 integrated luminosity is 2.1%.

29.6 Statistical Interpretation and Results

This search involves 40 analysis regions such as eight zero-lepton signal regions, six one-lepton and six two-lepton control regions, as well as the corresponding sideband regions for each of these twenty categories. A profile likelihood fit [4] is used in the interpretation of the data to search for DM production. The fit variables for different CR and SR are: E_T^{miss} (0ℓ SR), $E_{T,nomu}^{miss}$ (1ℓ CR) and p_T^V (2ℓ CR) respectively. There is a good agreement of data and background prediction in all signal regions. Out of the 40 analysis regions, Fig. 29.2 shows only the signal region distributions of missing transverse momentum, E_T^{miss} with the resolved (left) and merged (right) event topologies after the profile likelihood fit (with $\mu = 0$). The total background contribution before the fit to data is shown as a dotted blue line. The hatched area represents the total background uncertainty. The signal expectations for the simplified vector-mediator model with $m_\chi = 1$ GeV and $m_{Z'} = 600$ GeV (dashed red line) and for the invisible Higgs boson decays (dashed blue line) are shown for comparison. The inset at the bottom of each plot shows the ratio of the data to the total post-fit (dots) and pre-fit (dotted blue line) background expectation.

29.6.1 Limit Calculation

In the search for invisible Higgs boson decays, an observed (expected) upper limit of $0.83(0.58_{-0.16}^{+0.23})$ is obtained at 95% CL on the branching ratio $B_{H\to inv.}$, assuming the SM production cross sections and combining the contributions from VH, ggH and VBF production modes. The expected limit is a factor of about 1.5 better (while the observed is slightly worse) than the one reached by the previous analysis of Run 1 ATLAS data [5]. In the context of the mono-W/Z simplified vector-mediator signal model, an exclusion limit at 95% CL is calculated on the DM-mediator masses for Dirac DM particles and couplings, $g_{SM} = 0.25$ and $g_{DM} = 1$, which is shown in Fig. 29.3a. For the given coupling choices, vector-mediator masses, $m_{Z'}$, of up to

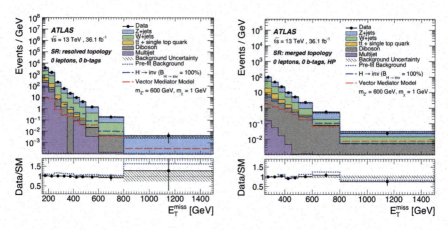

Fig. 29.2 0 *b*-tags resolved event topology (left) and 0 *b*-tags, HP merged event topology (right) [3]

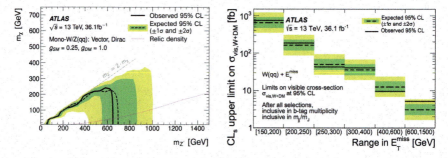

Fig. 29.3 Exclusion limits at 95% CL in the grid of $(m_\chi, m_{Z'})$ (left) and on the visible cross section $\sigma_{\mathrm{vis,W+DM}}$ (right) [3]

650 GeV are excluded at 95% CL for DM masses m∅ of up to 250 GeV, agreeing well with the expected exclusion of $m_{Z'}$ values of up to 700 GeV for m_χ of up to 230 GeV. The expected limits are improved by 15–30%, depending on the DM mass, compared to the analysis presented in [6]. In addition to these interpretations, the results are also expressed in terms of generic CL$_s$ upper limits at 95% CL on the allowed visible cross section σ_{vis} of potential W+DM or Z+DM production and is shown in Fig. 29.3b for the W+DM case. The limits on these two processes are evaluated separately to allow more flexibility in terms of possible reinterpretations, as the new models might prefer one of these two final states.

29.7 Summary

A search for DM has been performed in events having a large-R jet or a pair of small-R jets compatible with a hadronic W or Z boson decay, and large E_T^{miss}. It improves on previous searches by virtue of the larger dataset and further optimization of the selection criteria and signal region definitions. The results are in agreement with the SM predictions and are translated into exclusion limits on DM pair production. In the search for invisible Higgs boson decay, an upper limit of 0.83 is observed at 95% CL on the branching ratio $B_{H\rightarrow \text{inv.}}$, while the corresponding expected limit is 0.58. Limits are also placed on the visible cross section of the non-SM events with large E_T^{miss} and a W or a Z boson without extra model assumptions.

Acknowledgements The author wants to thank Prof. Dmitri Tsybychev for his valuable comments in the preparation of these proceedings.

References

1. O. Buchmueller, M.J. Dolan, S.A. Malik, C. McCabe, Characterising dark matter searches at colliders and direct detection experiments: vector mediators. JHEP **01**, 037 (2015). arXiv:1407.8257
2. D. Abercrombie et al., Dark Matter Benchmark Models for Early LHC Run-2 Searches: Report of the ATLAS/CMS Dark Matter Forum, arXiv:1507.00966
3. ATLAS Collaboration, Search for dark matter in events with a hadronically decaying vector boson and missing transverse momentum in pp collisions at $\sqrt{s}$= 13 TeV with the ATLAS detector. JHEP **10**(180) (2018). arXiv:1807.11471
4. G. Cowan, K. Cranmer, E. Gross, O. Vitells, Asymptotic formulae for likelihood-based tests of new physics. Eur. Phys. J. C **71** (2011) 1554 [Erratum ibid. C 73 (2013) 2501]. arXiv:1007.1727
5. ATLAS collaboration, Search for invisible decays of the Higgs boson produced in association with a hadronically decaying vector boson in pp collisions at $\sqrt{s}$ = 8 TeV with the ATLAS detector. Eur. Phys. J. C **75**(337) (2015) arXiv:1504.04324
6. ATLAS collaboration, Search for dark matter produced in association with a hadronically decaying vector boson in pp collisions at $\sqrt{s}$ = 13 TeV with the ATLAS detector. Phys. Lett. B **763**(251) (2016) arXiv:1608.02372
7. B. Parida, On behalf of the ATLAS Collaboration, Search for Dark Matter production in association with a hadronically decaying vector boson in pp collisions at $\sqrt{s}$ = 13 TeV with the ATLAS detector, ATL-PHYS-PROC-2018-158, https://cds.cern.ch/record/2647097, PoS(ICHEP2018)787

Chapter 30
Light Higgsinos at the LHC with Right-Sneutrino LSP

Arindam Chatterjee, Juhi Dutta, and Santosh Kumar Rai

Abstract We study an extension of the minimal supersymmetric standard model (MSSM) with additional right-handed singlet neutrino superfields in the context of a natural SUSY spectra, i.e., focusing on low values of the Higgsino mass parameter. While such an extension incorporates a mechanism for the neutrino mass, it also opens up the possibility of having right-sneutrinos ($\tilde{\nu}$) as the lightest supersymmetric particle (LSP). Considering prompt decays of the Higgsino-like states, we consider leptonic channels at the large Hadron collider (LHC) and conclude that mono-lepton and opposite-sign di-lepton channels with low hadronic activity would be extremely useful channels to look for sneutrino LSP while same-sign di-lepton would serve as a strong confirmatory channel for discovery.

30.1 Introduction

In the light of the increasing TeV scale limits on the strong sector supersymmetric particles (sparticles) from the LHC [1, 2], the electroweak sector of the minimal supersymmetric standard model (MSSM) has gained considerable attention in recent times. A light Higgsino sector is also favored to be an ingredient of a 'natural' supersymmetry (SUSY) [3, 4] and remains phenomenologically viable due to the compression in the Higgsino sector [1, 2]. Note that in R-parity conserving scenarios, the limits on the sparticle masses are crucially dependent on the nature of the lightest supersymmetric particle (LSP). This warrants a thorough search with different LSPs.

In this work, we consider simple extensions of the MSSM with right-handed neutrino superfields which address the issue of non-zero neutrino masses as established from oscillation experiments via the Type-I see-saw mechanism [5–7]. It also pro-

A. Chatterjee
Indian Statistical Institute, 203 B.T. Road, Kolkata 700108, India

J. Dutta (✉) · S. Kumar Rai
Regional Centre for Accelerator-based Particle Physics, Harish-Chandra Research Institute, HBNI, Chhatnag Road, Jhusi, Allahabad 211019, India
e-mail: dutta.juhi91@gmail.com

© Springer Nature Singapore Pte Ltd. 2021
P. K. Behera et al. (eds.), *XXIII DAE High Energy Physics Symposium*,
Springer Proceedings in Physics 261,
https://doi.org/10.1007/978-981-33-4408-2_30

vides a candidate for the LSP and dark matter (DM), the right-sneutrino $\tilde{\nu}$. Although the Yukawa interactions of a right-sneutrino are small ($y_\nu \sim 10^{-6} - 10^{-7}$), the presence of large soft-SUSY breaking trilinear coupling may lead to a large left-admixture in the LSP, therefore increasing the interaction strengths. In the presence of a right-sneutrino LSP candidate, new decay modes open up for the Higgsinos. This in turn helps one to probe the naturally compressed Higgsino sector using leptonic channels mainly due to the leptonic decay of the light chargino. Although leptonic channels provide a cleaner environment for the new physics searches at a hadron machine such as the LHC, one expects that the level of compression in the mass spectra of the electroweakinos would also play a major role in determining the efficacy of the leptonic channels.

30.2   The Model

We introduce right-handed neutrino superfields N in the MSSM. The new superpotential (generation indices suppressed) is

$$W = W_{MSSM} + y_\nu L H_u N^c + \frac{1}{2} M_R N^c N^c, \tag{30.1}$$

where L is the left-handed lepton doublet superfield, H_u is the up-type Higgs superfield, y_ν is the neutrino Yukawa coupling, M_R is the $\Delta L = 2$ Majorana mass term and N^c is the left-chiral right-handed neutrino superfield. After electroweak symmetry breaking occurs, H_u obtains a vev v_u thus giving a Dirac mass to the neutrinos, $m_D = y_\nu v_u$.

The relevant soft-supersymmetry breaking terms in the soft scalar potential are

$$V_{soft} = V_{MSSM}^{soft} + m_R^2 |\tilde{N}|^2 + \frac{1}{2} B_M \tilde{N}^c \tilde{N}^c + T_\nu \tilde{L} H_u \tilde{N}^c + h.c., \tag{30.2}$$

where m_R^2 is the soft-supersymmetry breaking mass parameter for the right-sneutrinos, B_M is the soft-breaking term corresponding to the Majorana mass term and T_ν refers to the soft-SUSY breaking trilinear term. Diagonalizing the sneutrino mass matrix, the mass eigenstates can be obtained with mass eigenvalues given by

$$m_{1,2}^{j2} = \frac{1}{2}\left(m_{LL}^2 + m_{RR}^{j\,2} \pm \sqrt{(m_{LL}^2 - m_{RR}^{j\,2})^2 + 4m_{LR}^{j\,4}}\right), \tag{30.3}$$

where

$$m_{LL}^2 = m_L^2 + \frac{1}{2}m_Z^2 \cos 2\beta + m_D^2$$

$$m_{RR}^{j2} = m_R^2 + m_D^2 + M_R^2 \pm B_M$$

$$m_{LR}^{j2} = (T_\nu \pm y_\nu M_R)v \sin \beta - \mu m_D \cot \beta. \tag{30.4}$$

Fig. 30.1 The $B_M - T_\nu$ plane constrained by mass of the heaviest neutrino as shown in the colored palette. The parameters of the scan are discussed in the text

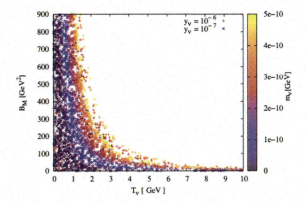

The left-right sneutrino mixing angle θ is given by

$$\sin 2\theta^j = \frac{(T_\nu \pm y_\nu M_R)v \sin\beta - \mu m_D \cot\beta}{m_2^{j2} - m_1^{j2}}, \tag{30.5}$$

where j denotes CP-even (e) or CP-odd (o) states. Note that the presence of a non-zero B_M term leads to radiative correction to the neutrino mass [8, 9] when the soft-trilinear mixing parameter T_ν, a soft-SUSY breaking parameter, is large leading to large left-right mixing in the sneutrino sector. We consider the effect of the neutrino mass by varying the relevant parameters $y_\nu \in 10^{-6} - 10^{-7}$, $\mu = 300$ GeV, $M_1 = 1.5$ TeV, $M_2 = 1.8$ TeV, and $m_{\tilde\nu}^{soft} = 100$ GeV. Figure 30.1 shows the regions of the parameter space with large B_M and large T_ν are severely constrained by the neutrino mass constraint [8, 10].

The other relevant sector for our study is a light Higgsino sector primarily motivated as an ingredient for natural SUSY spectrum [3, 4]. A low-lying Higgsino sector consists of a nearly degenerate pair of neutralinos $\tilde\chi_2^0$, $\tilde\chi_1^0$ and chargino $\tilde\chi_1^\pm$ with masses given by [11, 12]

$$m_{\tilde\chi_1^\pm} = |\mu| \left(1 - \frac{M_W^2 \sin 2\beta}{\mu M_2}\right) + \mathcal{O}(\frac{1}{M_2^2}) + \text{rad.corr.}$$

$$m_{\tilde\chi_{a,s}^0} = \pm\mu - \frac{M_Z^2}{2}(1 \pm \sin 2\beta)\left(\frac{\sin\theta_W^2}{M_1} + \frac{\cos\theta_W^2}{M_2}\right) + \text{rad.corr.} \tag{30.6}$$

Here the subscripts s (a) denote the symmetric (and anti-symmetric) states, respectively. We scan over the parameter space to realize distinct regions of the parameter space reflecting all possible mass hierarchy structures among the Higgsinos. The parameters for the scan are as discussed in Table 30.1.

From Fig. 30.2, we observe that the mass degeneracy among the Higgsinos, parametrized in terms of $\Delta m_1 = m_{\tilde\chi_1^\pm} - m_{\tilde\chi_1^0}$ and $\Delta m_2 = m_{\tilde\chi_2^0} - m_{\tilde\chi_1^\pm}$, increases with increasing M_1 and M_2. This signifies that with decreasing Gaugino fraction, the mass

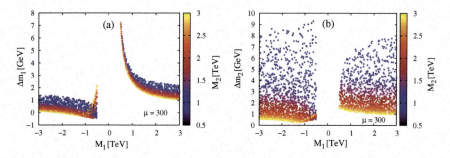

Fig. 30.2 Variation of Δm_1 and Δm_2 with respect to M_1 and M_2 in the colored palette, respectively

Table 30.1 Relevant input parameters for the parameter space scan have been presented. Other parameters kept at fixed values include $M_R = 100$ GeV, $B_M = 10^{-3}$ GeV^2, $M_3 = 2$ TeV, $M_{Q_3} = 1.3$ TeV, $M_{U_3} = 2$ TeV, $T_t = 2.9$ TeV, $M_{L_{1/2}} = 600$ GeV, $m_{\tilde{\nu}}^{soft} = 100$ GeV, $M_A = 2.5$ TeV and $y_\nu = 10^{-7}$

Parameters	$|M_1|$ (GeV)	$|M_2|$ (GeV)	$|\mu|$ (GeV)	$\tan\beta$	T_ν (GeV)
Values	(500–3000)	(500–3000)	300	5	0.5

$$
\begin{array}{ll}
\tilde{\chi}_2^0 & \rule{3cm}{0.4pt} \\
& \qquad \Delta m_2 \\
\tilde{\chi}_1^\pm & \rule{3cm}{0.4pt} \\
& \qquad \Delta m_1 \\
\tilde{\chi}_1^0 & \rule{3cm}{0.4pt} \\
\end{array}
$$

Fig. 30.3 Spectra of interest

degeneracy among the Higgsinos increases. Notably the Higgsino spectrum depends crucially on M_1, M_2 and $sign(\mu)$ as seen from (30.6). As we see in Fig. 30.2, there exists regions of parameter space where the $\tilde{\chi}_1^0$ is mostly the lightest Higgsino while in some regions of the parameter space where $M_1 < 0$ and $\mu > 0$, $\tilde{\chi}_1^\pm$ is the lightest. Thus, the relevant spectra of interest are summarized in Fig. 30.3.

We now discuss the possible branching ratios of the Higgsinos in the presence of a right-sneutrino LSP. Figure 30.4a and b, respectively, discusses the leptonic branching of $\tilde{\chi}_1^\pm \to l^\pm \tilde{\nu}$ and $\tilde{\chi}_2^0 \to \tilde{\chi}_1^\pm W^* \to l^\pm \tilde{\nu} W^*$. The features of the branching ratios are correlated with the mass splitting among the Higgsinos. Since the usual three-body MSSM decays of the Higgsinos, i.e., $\tilde{\chi}_1^\pm \to \tilde{\chi}_1^0 f \bar{f}'$, $\tilde{\chi}_2^0 \to \tilde{\chi}_1^\pm W^* \tilde{\chi}_2^0 \to$

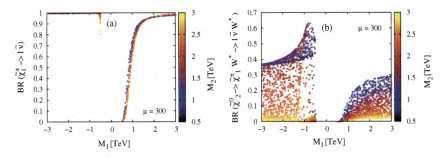

Fig. 30.4 Variation of the leptonic branching of $\widetilde{\chi}_1^\pm$ and $\widetilde{\chi}_2^0$, respectively

$\widetilde{\chi}_1^0 Z^*$ are controlled by the phase space (Δm^5), in the presence of the sneutrino LSP, the competing two-body decays may become relevant depending on the available phase space. In Fig. 30.4a, the branching ratio for the two-body decay modes $\widetilde{\chi}_1^\pm \to l^\pm \widetilde{\nu}$ is shown. Correlating from Fig. 30.2 we observe that with decreasing Gaugino fraction, the branching ratio to the sneutrino mode increases owing to the decrease in Δm_1. Similar features are observed for $\widetilde{\chi}_2^0 \to \widetilde{\chi}_1^\pm W^* \to l^\pm \widetilde{\nu} W^*$. Therefore, leptonic channels are of considerable interest in exploring this scenario.

30.3 Signal and Analysis

Our focus in this work has been on prompt decays of the Higgsinos. The leptonic modes of the charginos and neutralinos lead to the following signals of interest:

- Mono-lepton $+ \not{E}_T$,
- Di-lepton $+ \not{E}_T$,
  - Opposite-sign di-lepton $+ \not{E}_T$, and
  - Same-sign di-lepton $+ \not{E}_T$.

The mono-lepton signals arise from $\widetilde{\chi}_i^0 \widetilde{\chi}_1^\pm$ ($\widetilde{\chi}_i^0 \to \nu \widetilde{\nu}$, $\widetilde{\chi}_1^\pm \to l^\pm \widetilde{\nu}$) , $\widetilde{\chi}_1^+ \widetilde{\chi}_1^-$ ($\widetilde{\chi}_1^\pm \to l^\pm \widetilde{\nu}$, $\widetilde{\chi}_1^\pm \to \widetilde{\chi}_1^0 W^*$) as well as from the leptonic decay of either $\widetilde{\chi}_2^0 / \widetilde{\chi}_1^\pm$ from $\widetilde{\chi}_2^0 \widetilde{\chi}_1^0$. Di-lepton signals arise from leptonic decay of $\widetilde{\chi}_2^0$ and $\widetilde{\chi}_1^\pm$ from $\widetilde{\chi}_2^0 \widetilde{\chi}_1^\pm$ and $\widetilde{\chi}_1^+ \widetilde{\chi}_1^-$ production processes. This may give rise to opposite-sign and same-sign di-lepton signals due to the Majorana nature of $\widetilde{\chi}_2^0$. We choose some representative benchmarks of the parameter space and perform a collider analysis of the above signals at $\sqrt{s} = 13$ TeV LHC run. Our focus is primarily on the prompt decay of the chargino to hard leptons (small $\Delta m_{1/2}$ and large T_ν) which would be clean signals to observe at LHC. The following production channels are of interest to us:

$$p\,p \to \widetilde{\chi}_1^\pm\,\widetilde{\chi}_2^0,\ \widetilde{\chi}_1^\pm\,\widetilde{\chi}_1^0,\ \widetilde{\chi}_1^+\,\widetilde{\chi}_1^-,\ \widetilde{\chi}_1^0\,\widetilde{\chi}_2^0,\ \widetilde{\chi}_1^0\,\widetilde{\chi}_1^0\,\widetilde{\chi}_2^0\,\widetilde{\chi}_2^0,\ \widetilde{l}\,\widetilde{l},\ \widetilde{l}\,\widetilde{l}^*,\ \widetilde{l}\,\widetilde{\nu},\ \widetilde{\nu}\,\widetilde{\nu}.$$
$$(30.7)$$

Table 30.2 Low-energy input parameters and sparticle masses for the representative benchmarks used in the current study. All the parameters are in GeV except for $\tan\beta$ which is dimensionless

Benchmarks	Parameters									
	μ	$\tan\beta$	M_1	M_2	$m_{\tilde{\nu}}$	$m_{\tilde{\chi}_2^0}$	$m_{\tilde{\chi}_1^\pm}$	$m_{\tilde{\chi}_1^0}$	Δm_1	Δm_2
BP1	300	5	1500	1800	141.4	305.8	303.6	301.7	1.9	2.2
BP2	400	6.1	−1150	2500	331.7	407.5	407.2	407.3	−0.1	0.2

Table 30.3 Required luminosities for the 3σ excess observation in the signatures at 13 TeV LHC

Benchmark	Luminosity required for 3σ excess		
	$l + \not{E}_T$	$l^+l^- + \not{E}_T$	$l^\pm l^\pm + \not{E}_T$
BP1	254	568	–
BP2	448	160	1052

Our focus while studying the collider signals would be to suppress the usual SM backgrounds using key kinematic variables like *stransverse mass* M_{T_2}, $\not{E}_T$, *transverse mass* M_T, and low jet multiplicity $N_j \leq 1$. We observe that important backgrounds like $W + j$, $t\bar{t}$ and Drell Yan are effectively suppressed using appropriate cut values for the above kinematic variables [13]. For example, $M_T > 150$ GeV reduces contributions from $W + j$ background while $M_{T_2} > 90$ GeV substantially reduces backgrounds from top-quark and W bosons. A large missing energy $\not{E}_T > 100$ GeV reduces SM background over the SUSY signal. We present the results for two sample benchmarks with $\mu = 300 - 500$ GeV as discussed in Table 30.2 and present the results for them in all three signal regions.

We observe that conventional signals such as hadronically quiet mono-lepton and opposite-sign di-leptons, i.e., with at most one jet or no jet, would be extremely useful channels to look for cases of a sneutrino LSP. In addition, detecting same-sign di-lepton channels would serve as a strong confirmatory channel for a sneutrino LSP scenario and can exclude large portions of the regions of the parameter space with $M_1 < 0$. The required luminosities for observing a 3σ excess are shown in Table 30.3 with all the channels well within the discovery of the high-luminosity LHC run.

30.4   Summary and Conclusions

We have studied an extension of the MSSM with additional right-neutrino super-fields addressing the issue of mass generation of light neutrinos as well as yielding right-sneutrinos as the LSP candidate. The presence of right-sneutrinos as the LSP opens up new decay modes for the naturally compressed light Higgsino sector with the compression in the sector stringently controlled by the Gaugino fraction in the Higgsinos. We emphasize the crucial impact of the presence of a small Gaugino admixture ($\mathcal{O}(10^{-2})$) in the Higgsino-like states owing to the heavier bino and wino soft mass parameters, thereby affecting the decay of the Higgsino-like states significantly. Considering prompt decays of the Higgsino-like states, we consider leptonic channels at the large Hadron collider (LHC) and conclude that mono-lepton and opposite-sign di-lepton channels with low hadronic activity would be extremely useful channels to look for sneutrino LSP while same-sign di-lepton would serve as a strong confirmatory channel for discovery.

References

1. https://atlas.web.cern.ch/Atlas/GROUPS/PHYSICS/CombinedSummaryPlots/SUSY/
2. https://twiki.cern.ch/twiki/bin/view/CMSPublic/PhysicsResultsSUS
3. Howard Baer, Vernon Barger, Michael Savoy, Upper bounds on s-particle masses from naturalness or how to disprove weak scale supersymmetry. Phys. Rev. D **93**(3), 035016 (2016)
4. Howard Baer, Vernon Barger, James S. Gainer, Peisi Huang, Michael Savoy, Dibyashree Sengupta, Xerxes Tata, Gluino reach and mass extraction at the LHC in radiatively-driven natural SUSY. Eur. Phys. J. C **77**(7), 499 (2017). Jul
5. Peter Minkowski, $\mu \to e\gamma$ at a rate of one out of 109 μ on decays? Phys. Lett. B **67**(4), 421–428 (1977)
6. Tsutomu Yanagida, Horizontal Symmetry and Masses of Neutrinos. Conf. Proc. **C7902131**, 95–99 (1979)
7. Rabindra N. Mohapatra, Goran Senjanović, Neutrino mass and spontaneous parity nonconservation. Phys. Rev. Lett. **44**, 912–915 (1980). Apr
8. Yuval Grossman, Howard E. Haber, Sneutrino mixing phenomena. Phys. Rev. Lett. **78**, 3438–3441 (1997)
9. Athanasios Dedes, Howard E. Haber, Janusz Rosiek, Seesaw mechanism in the sneutrino sector and its consequences. JHEP **11**, 059 (2007)
10. P.A.R. Ade et al., Planck 2015 results. XIII. Cosmological parameters. Astron. Astrophys. **594**, A13 (2016)
11. M. Drees, M.M. Nojiri, D.P. Roy, Y. Yamada, Light Higgsino dark matter. Phys. Rev. **D56**, 276–290 (1997). [Erratum: Phys. Rev. D64, 039901 (2001)]
12. Gian F. Giudice, Alex Pomarol, Mass degeneracy of the Higgsinos. Phys. Lett. B **372**, 253–258 (1996)
13. Chatterjee, A., Dutta, J., Rai, S.K.: Natural SUSY at LHC with Right-Sneutrino LSP. J. High Energy Phys. **06**, 042 (2018). https://doi.org/10.1007/JHEP06(2018)042. arXiv: 1710.10617

Chapter 31
Search for Vector Boson Fusion Production of a Massive Resonance Decaying to a Pair of Higgs Bosons in the Four b-Quark Final State at the HL-LHC Using the CMS Phase-2 Detector

Alexandra Carvalho, Jyothsna Rani Komaragiri, Devdatta Majumder, and Lata Panwar

Abstract The search for a massive resonance produced by vector boson fusion and decaying into a pair of Higgs bosons, each decaying to a b quark-antiquark pair, at the High Luminosity Large Hadron Collider in proton-proton collisions at a centre-of-mass energy of 14 TeV is explored. The Higgs bosons are required to be sufficiently Lorentz-boosted for each to be reconstructed using a single large-area jet. We study the signal sensitivity for a narrow bulk graviton in extradimensional scenarios using a simulation of the upgraded CMS detector, assuming multiple proton-proton collisions in the same bunch crossing (up to 200), for data corresponding to an integrated luminosity of 3 ab^{-1}. The expected significance for different assumed masses of the bulk graviton is presented.

31.1 Introduction

The discovery of the Standard Model (SM) Higgs Boson in 2012, from data collected with CMS and ATLAS detector at CERN, Switzerland, has proven the Standard Model the most successful theory of the particle physics. The spontaneous symmetry breaking in the electroweak sector of the SM explains the Higgs boson's existence and how Standard Model particles acquire masses. However, the SM has also shortcom-

On behalf of the CMS collaboration.

A. Carvalho
Estonian Academy of Sciences, Tallinn, Estonia

J. R. Komaragiri · L. Panwar (✉)
Indian Institute of Science, Bengaluru, India
e-mail: panwarlsweet@gmail.com; lata@cern.ch

D. Majumder
University of Kansas, Lawrence, Kansas, USA

© Springer Nature Singapore Pte Ltd. 2021
P. K. Behera et al. (eds.), *XXIII DAE High Energy Physics Symposium*,
Springer Proceedings in Physics 261,
https://doi.org/10.1007/978-981-33-4408-2_31

Fig. 31.1 The vector boson
fusion production of a
resonance X decaying to a
pair of Higgs bosons, where
both Higgs bosons decay
into 4b final state

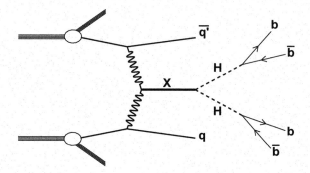

ings. It fails to provide a solution for e.g. the hierarchy problem, Gravity, Dark Matter, neutrino oscillation etc. Then we start exploring the physics beyond the SM (BSM).

Several BSM scenarios predict the existence of resonances decaying to a pair of Higgs bosons, such as Warped Extra Dimensional (WED) model [1], which has a spin-0 Radion and a spin-2 first Kaluza–Klein (KK) excitation of the graviton. These resonances may have a significant branching fraction to decay in a Higgs pair.

The search for KK graviton (X) in Vector Boson Fusion (VBF) production mode, as shown in Fig. 31.1, has not yet been explored. While the s-channel production cross section of a bulk graviton, assuming $\kappa/\bar{M}_{pl} = 0.5$, is in the range $0.05-5$ fb for masses between 1.5 and 3 TeV, the VBF production mode is expected to have a cross section an order of magnitude smaller than s-channel. The results published for s-channel [2] shows the negligible production rate of the signal which indicates highly suppressed couplings of X with the SM quarks and gluons. This makes VBF the dominant production process in pp collisions. Thus VBF production mode is interesting to study with large amount of data collected at the high luminosity LHC (HL-LHC).

In this report, we study the VBF production of WED spin-2 graviton resonance, which decays to pair of Higgs bosons, with 4b quarks in the final state, at the HL-LHC with the upgraded CMS detector.

A simulation [3] of the upgraded Phase-2 CMS detector was used for this study. The signal events for bulk graviton were simulated at leading order using MAD-GRAPH5_aMC@NLO [4], for masses in the range 1.5 to 3 TeV and for a fixed width of 1% of the mass. The main background is given by multijet events, and has been simulated using PYTHIA8 [5], for events containing two hard partons, with the invariant mass of the two partons is required to be greater than 1 TeV.

31.2 Analysis Strategy

For a very massive resonance, highly Lorentz-boosted Higgs bosons are more efficiently reconstructed as a single large-area jet (Higgs jet). In addition, a signal event will also have two energetic jets at large pseudorapidity η. Thus there are two sets of event selections, (a) Higgs Jets selection, (b) VBF Jets selection.

For (a), the two leading-p_T large-radius anti-k_T jets with a distance parameter of 0.8 (AK8 [6]) in the event, J1 and J2 , are required to have $p_T > 300$ GeV and $|\eta| <$ 3.0. To identify the two leading-p_T AK8 jets with the boosted H $\rightarrow$ ($b\bar{b}$) candidates from the X $\rightarrow$ HH decay, these jets are groomed to remove soft and wide-angle radiation effect using the soft-drop mass algorithm [7]. By undoing the last stage of the jet clustering, one gets two subjets each for J1 and J2 . The invariant mass of the two subjets is the soft-drop mass of each AK8 jet, which has a distribution with a peak near the Higgs boson mass $m_H = 125$ GeV. The soft-drop mass window selection was optimised using a figure of merit of $S/\sqrt{B}$ and required to be in the range $90-140$ GeV for both leading jets. The N-subjettiness [8] ratio $\tau_{21} = \tau_2/\tau_1$ has a value much smaller than unity for a jet with two subjets. For the signal selection, J1 and J2, both are required to have $\tau_{21} < 0.6$. The H-tagging of J1 and J2 further requires identifying their subjet pairs to be b-tagged with a probability of about 49% to contain at least one B-hadron, and the corresponding probability of about 1% of having no B/D-hadrons using DeepCSV [9] b-tagger. Events are classified into two categories: those having exactly three out of the four b-tagged subjets (3b category), and those which have all four subjets b-tagged (4b category).

For (b), events are required to have at least two AK4 jets j1 and j2, which are separated from the H jets by $\Delta R > 1.2$, with $p_T > 50$ GeV and $|\eta| < 5$. To pass the VBF selections, these jets must lie in opposite η regions of the detector, and a pseudo-rapidity difference $|\Delta\eta$ (j1 , j2) $| > 5$. The invariant mass m_{jj}, reconstructed using these AK4 jets, is required to pass $m_{jj} > 300$ GeV.

The bulk graviton invariant mass m_{JJ} is reconstructed from the 4-momenta of the two Higgs jets, in events passing the above mentioned full selection criteria. The main multijet background is smoothly falling above which the signal is searched as a localised excess of events for a narrow resonance X. It is expected that the multijet background component in a true search at the HL-LHC will rely on the data for a precise result. From the analysis of current LHC data at $\sqrt{s} = 13$ TeV [2], it was

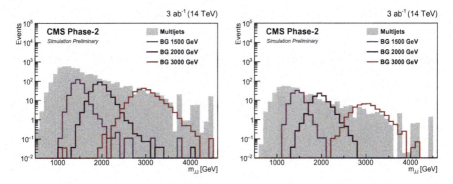

Fig. 31.2 The m_{JJ} distributions for bulk gravitons (BG) of masses 1.5, 2, and 3 TeV, assuming a signal cross section of 1 fb. The distributions on the left are for the 3b and those on the right are for the 4b subjet b-tagged categories and for an average pileup of 200 [10]

Table 31.1 Event yields and efficiencies for the signal and multijet background for an average pileup of 200. The product of the cross sections and branching fractions of the signals σ (pp $\rightarrow$ Xjj $\rightarrow$ HHjj) is assumed to be 1 fb [10]

Process	3b Category	3b Category	4b category	4b Category
	Events	Efficiency (%)	Events	Efficiency (%)
MultiJets	4755	1.6×10^{-3}	438	1.5×10^{-4}
BG(m_X=1.5 TeV)	326	11	95.2	3.2
BG(m_X=2.0 TeV)	316	11	81.2	2.7
BG(m_X=3.0 TeV)	231	7.7	41.4	1.4

found that the multijet backgrounds, measured in data, are a factor of 0.7 smaller than the estimated in simulation, hence the multijet background yield from simulation has been corrected by this factor, assuming this also holds for the simulations of the multijet processes at $\sqrt{s} = 14$ TeV . The m_{JJ} of the backgrounds thus obtained and are shown in Fig. 31.2, while the event yields are given in Table 31.1 after full selection. The requirement of additional VBF jets does not result in any appreciable gain in the signal sensitivity because of strong dependency of VBF Jets on pileup. It is anticipated that the developments in the rejection of pileup jets in the high η region will eventually help to suppress the multijets background and improve the signal sensitivity further.

31.3   Results

The expected significance of the signal, assuming a production cross section of 1 fb is estimated considering all the systematic uncertainties. These uncertainties are based on the projected values for the full data set at the HL-LHC [11]. In addition, several

Fig. 31.3 The expected signal significance for bulk graviton with explicit masses 1500, 2000, and 3000 GeV, assuming a production cross section of 1 fb. The data set corresponds to an integrated luminosity of 3 ab^{-1} and with a pileup of 200 [10]

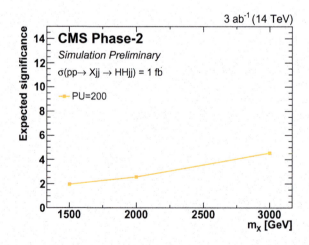

measurement uncertainties are considered based on the 2016 search for a resonance decaying to a pair of boosted Higgs bosons [2], scaled by 0.5. The expected signal significance of a bulk graviton with the mass 2000 GeV, produced through vector boson fusion, with an assumed production cross section of 1 fb, is found to be 2.6σ for an integrated luminosity of 3 ab^{-1}. With higher mass of bulk graviton, there might be possible evidence of it with this analysis strategy at HL-LHC (Fig. 31.3).

References

1. A. Carvalho, Gravity particles from Warped Extra Dimensions, predictions for LHC. (2014). arXiv: 1404.0102v4 [hep-ph]
2. CMS Collaboration, Search for a massive resonance decaying to a pair of Higgs bosons in the four b quark final state in proton-proton collisions at p s = 13 TeV. Phys. Lett. B **781**, 244 (2017). arXiv: 1710.04960v2 [hep-ex]
3. J. Allison et al., Geant4 developments and applications. IEEE Trans. Nucl. Sci. **53**, 270 (2006)
4. J. Alwall et al., The automated computation of tree-level and next-to-leading order differential cross sections, and their matching to parton shower simulations. JHEP **7**, 079 (2014). arXiv: 1405.0301 [hep-ph]
5. T. Sjöstrand et al., An introduction to PYTHIA 8.2. Comput. Phys. Commun. **191**, 159 (2015). arXiv: 1410.3012 [hep-ph]
6. M. Cacciari, G.P. Salam, G. Soyez., The anti-kt jet clustering algorithm. JHEP **4**, 63 (2008). arXiv: 0802.1189 [hep-ph]
7. A.J. Larkoski et al., Soft drop. JHEP **5**, 146 (2014). arXiv: 1402.2657 [hep-ph]
8. J. Thaler, K. Van Tilburg., Identifying Boosted Objects with N-subjettiness. JHEP **3**, 15 (2011). arXiv: 1011.2268 [hep-ph]
9. A.M. Sirunyan et al., Identification of heavy-flavour jets with the CMS detector in pp collisions at 13 TeV. J. Instrum. **13**(5) (2018). arXiv: 1712.07158 [physics.ins-det]
10. CMS Collaboration. Search for vector boson fusion production of a massive resonance decaying to a pair of Higgs bosons in the four b-quark final state at the HL-LHC using the CMS Phase-2 detector. Tech. rep. CMS-PAS-FTR-2018-003. CERN (2018)
11. CMS Collaboration. Expected performance of the physics objects with the upgraded CMS detector at the HL-LHC. Tech. rep. CMS-NOTE-2018-006. CERN (2018)

Chapter 32
Exotic Leptonic Solutions to Observed Anomalies in Lepton Universality Observables and More

Lobsang Dhargyal

Abstract In this chapter, I will present the work that we did in [1–5] related to observed lepton universality violation by Babar, Belle, and LHCb in R($D^{(*)}$) and $R_{K^{(*)}}$ as well as the reported deviation in muon (g-2) by BNL. We had shown that all these anomalies as well as Baryon-genesis, Dark-matter, and small neutrino masses could be explained by introducing new exotic scalars, leptons, and scalar-leptoquarks only. It turns out that some of these models have very peculiar signatures such as prediction of the existence of heavy stable charged particle [1, 2], vector-like fourth-generation leptons [3] or even scalar Baryonic DM candidates, etc. Some of these models turn out to have very unique collider signatures such as $ee/pp \rightarrow \mu\mu(\tau\tau) + missing\ energy\ (ME)$, see [1, 2, 4]. This is interesting in the sense that such peculiar signatures of these new particles can be searched in the upcoming HL-LHC or with an even better chance of observing these signatures are in the upcoming precision machines such as ILC, CEPC, etc.

In this chapter, I will present a short summary of interesting consequences of introducing new exotic leptons and scalars (LQ) to resolve the reported anomalies by Babar, Belle, LHCb, BNL in R($D^{(*)}$), $R_{K^{(*)}}$, muon (g-2). In this brief article, the emphasis is laid on the peculiar features (and interesting side observations) of the models that we proposed to resolve the mentioned anomalies, for more details we refer the readers to the sources [1–5]. In the following, we present the particle content, achievements, and peculiar features and observations about the respective models.

(1): In [1] we extended the inert-doublet 2HDM (IDM) by introducing three exotic leptons ($F_{iR,L}$ for i = 1, 2, 3) which is singlet under the SM $SU(2)_L$ and vector like under the SM $U(1)_Y$ beside a new $U(1)_F$ to which only the right-handed new exotic leptons are charged (all exotic leptons are odd under the Z_2 to avoid very stringent tree-level constrains). This model can explain the reported anomaly in muon (g-2) as

Based on a Talk given at the 23rd DAE-BRNS High Energy Physics Symposium 2018, IIT Chennai, 10–14 December 2018.

L. Dhargyal (✉)
Harish-Chandra Research Institute, HBNI, Chhatnag Road, Jhusi, Allahabad 211019, India
e-mail: dhargyal@hri.res.in; dhargyal2011@gmail.com

© Springer Nature Singapore Pte Ltd. 2021
P. K. Behera et al. (eds.), *XXIII DAE High Energy Physics Symposium*,
Springer Proceedings in Physics 261,
https://doi.org/10.1007/978-981-33-4408-2_32

well as why the anomaly is observed only in the muon sector and not in the electron sector (due to a peculiar solution choice of the γ_5 anomaly cancelation in [1]) besides small neutrino masses and Baryogenesis via leptogenesis (same as in the Scoto-genic model). But in [2], we realized that if let the left handed of the exotic fermions charged under the new $U(1)_F$ instead of the right-handed one as in [1], then besides the model explaining the anomalies explained by the model in [1], it is also able to incorporate the observed deviations in $R_{K^{(*)}}$ as well (via box loop diagrams). Due to a peculiar choice of γ_5 anomaly free conditions in [1, 2], which besides explaining why the anomaly is in the muon sector and not in the electron sector as mentioned before, it also predicts one stable (long lived[1]) charged exotic lepton which can be very heavy (in fact heavier than the unstable exotic lepton) which could turn up in LHC or HL-LHC data or even more prominently in ee colliders such as future ILC, CEPC, etc. If (somehow) this stable lepton has the same mass as the electron then only weak interaction charged current will be able to differentiate between the SM electron and the exotic electron properly.

(2): Taking inspirations from [1, 2], in [3] we built a NP model by introducing vector like exotic leptons and scalars (LQ) to explain the observed deviations in $R(D^{(*)})$ data at loop level. And in [4], we have introduced a pair of charged exotic lepton doublets (L_{1L} and L_{2R}) and a pair of charged exotic lepton singlets (E_{1R} and E_{2L}),[2] which is free of γ_5 anomaly, besides new LQ and scalars and shown that both the R($D^{(*)}$) and $R_{K^{(*)}}$ can be explained within the limits of present error estimates. One peculiar feature of the models in [3, 4] is that is to satisfy the very stringent constraints from the $K^0 - \bar{K}^0$ and $B^0 - \bar{B}^0$ mixings, we are forced to restrict the CKM angles in $\pi \leq \theta_{12} \leq \frac{3\pi}{2}$ and $\frac{3\pi}{2} \leq \theta_{13}, \theta_{23} \leq 2\pi$ and impose a constraint similar like GIM on the combinations of Yukawa couplings and CKM elements which lead to requirements of at least one of the Yukawa couplings must be complex which in fact predicts a small CP violation (in a particular choice of the Yukawa couplings here, but it could be made large too in other choices!) in $B_s^0 - \bar{B}_s^0$ mixing due to new exotic leptons (box loop level), for details see [3, 4].

(3): In [5], we have proposed an NP model (two pair of $SU(2)_L$ singlet right-handed leptons carrying opposite $U(1)_Y$ charges and their two left-handed counterparts which are neutral singlet leptons) where in the regime where the exotic leptons masses are in the electroweak (EW) scale, the model will be able to explain the $R_{K^{(*)}}$ and muon (g-2) as well as small neutrino masses via minimum-inverse seesaw scenario (MISS), for more details, see [5] and references therein. In this model, when the exotic fermion masses are well above the EW scale then we can have stable scalar baryon (singlet under strong interaction) with charge -3 which could explain the primordial Li problem by forming a hydrogen-like atom with the Li^{+3} nucleus with a peculiar absorption or emission line of the first excited state in the X-ray region at $E_2(Li) - E_1(Li) \approx 10.62$ MeV. Another peculiar but very interesting side observation about this model is that when the new fermion masses are well above the EW scale, then if we assign $U(1)_Y$ charges (unlike the way we did above) such

---

[1]Unless there also exists doubly charged scalar, in which case it need not be stable.

[2]Where doublets and singlets refer to SM gauge group $SU(2)_L$.

that there are vector like under $U(1)_Y$, then it will not be able to explain the $R_{K^{(*)}}$ and muon (g-2) as well as MISS is not possible now, but in this case, we can have the exotic leptons decay into stable scalar quarks. These stable scalar quarks (if we assign them a flavor $SU(3)_F$ similar/same like the SM quark u, d, and s flavors) can have electromagnetically neutral (as well as $Q = +2$ depending on the charges of the exotic fermions) stable scalar baryon and is expected to be suppressed under the strong interaction at the level of OZI rule or smaller (due to scalar baryon being singlet under both color and flavor and also due to heavy mass scale of scalar quarks and small size (due to the lack of exclusion force)) which can be a DM candidate and also in these kinds of models the origins of ordinary baryons and DM (scalar baryons here) could be linked.

References

1. L. Dhargyal, Eur. Phys. J. C78 (2018) no.2, 150 and references there in
2. L. Dhargyal, arXiv:1711.09772 and references there in
3. L. Dhargyal, S.K. Rai, arXiv:1806.01178 and references there in
4. L. Dhargyal, arXiv:1808.06499 and references there in
5. L. Dhargyal, arXiv:1810.10611 and references there in

Chapter 33
Constraints on Minimal Type-III Seesaw Model from Naturalness, Lepton Flavor Violation, and Electroweak Vacuum Stability

Srubabati Goswami, K. N. Vishnudath, and Najimuddin Khan

Abstract We study the minimal Type-III seesaw model to explain the origin of the non-zero neutrino masses and mixing. We show that the naturalness arguments and the bounds from lepton flavor violating decay ($\mu \rightarrow e\gamma$) provide very stringent bounds on the model along with the constraints on the stability of the electroweak vacuum up to high energy scale. We perform a detailed analysis of the model parameter space including all the constraints for both normal and inverted hierarchies of the light neutrino masses. We find that most of the regions that are allowed by naturalness and lepton flavor violating decay fall into the metastable region.

PACS numbers: 14.60.St · 12.60.-i · 12.15.Lk

33.1 Introduction

The LHC so far has failed to identify any new physics signature beyond the standard model. The Higgs signal strength data and measured value of the Higgs boson [1, 2] mass at 125.7 ± 0.3 GeV still can comfortably accommodate some new physics beyond the standard model. Various new physics scenarios have already been considered to address different issues like the Higgs hierarchy problem and generation of small neutrino masses, relic (dark matter) density of the Universe, etc. The quantum radiative corrections to the Higgs mass coming from its self-interaction and cou-

S. Goswami · K. N. Vishnudath
Theoretical Physics Division, Physical Research Laboratory, Ahmedabad 380009, India
e-mail: sruba@prl.res.in

K. N. Vishnudath
e-mail: vishnudath@prl.res.in

K. N. Vishnudath
Discipline of Physics, Indian Institute of Technology, Gandhinagar 382355, India

N. Khan (✉)
Centre for High Energy Physics, Indian Institute of Science, Bangalore 560012, India
e-mail: najimuddink@iisc.ac.in

© Springer Nature Singapore Pte Ltd. 2021
P. K. Behera et al. (eds.), *XXIII DAE High Energy Physics Symposium*,
Springer Proceedings in Physics 261,
https://doi.org/10.1007/978-981-33-4408-2_33

plings with fermions and gauge boson are being the primary cause for the hierarchy problem. Although the dimensional regularization [3–5] can be able to throw away the dangerous quadratic divergences, the presence of other finite and logarithmic contributions causes similar naturalness problem. This naturalness sets a stringent limit on the new physics mass scale as well as the coupling strength to the SM Higgs fields. The found values of the SM parameters, especially the Higgs mass at 125.7 ± 0.3 GeV, top mass M_t, and strong coupling constant α_s, have suggested that an extra deeper minimum resides near the Planck scale, threatening the stability of the present electroweak (EW) vacuum [6], i.e., the EW vacuum might tunnel into that true (deeper) vacuum. Using the state-of-the-art NNLO, the decay probability has been found to be less than *one*, which implies that the EW vacuum is metastable at 3σ (one-sided).

In this work, the minimal Type-III seesaw model is proposed in which the SM is extended by adding two hyperchargeless $SU(2)_L$ triplet fermions to explain the origin of the non-zero neutrino masses and U_{PMNS} mixing [7]. In this scenario, the lightest active neutrino will be massless. We here use the Casas-Ibarra parametrization for the neutrino Yukawa coupling matrix to explain the very tiny neutrino masses and mixing angles. We study this model in detail and put the bounds on these model parameters focusing on the impact of the naturalness and the EW vacuum metastability as well as lepton flavor violating (LFV) decays.

33.2 Type-III Seesaw Model and Present Bounds

In this section, we discuss the extended fermionic sectors of the models and present bounds on the model parameters. The Lagrangians that are relevant to neutrino mass generation are [7]

$$- \mathcal{L}_\Sigma = \tilde{\phi}^\dagger \overline{\Sigma}_R \sqrt{2} Y_\Sigma L + \frac{1}{2} \, \mathrm{Tr} \, [\overline{\Sigma}_R M \Sigma_R^c] \; + \; \mathrm{h.c.,} \tag{33.1}$$

where $L = (\nu_l \; l^-)^T$ is the lepton doublet and $\tilde{\phi} = i\sigma_2\phi^*$, σ's are the Pauli matrices, $\Sigma_{R_j} = (\Sigma_R^1, \Sigma_R^2, \Sigma_R^2)_j$, $j = 1, 2$ are the two hyperchargeless fermionic triplets, and

$$\Sigma_R = \frac{\Sigma_R^i \sigma^i}{\sqrt{2}} = \begin{bmatrix} \Sigma_R^0/\sqrt{2} & \Sigma_R^+ \\ \Sigma_R^- & -\Sigma_R^0/\sqrt{2} \end{bmatrix}, \tag{33.2}$$

where $\Sigma_R^\pm = (\Sigma_R^1 \mp i\Sigma_R^2)/\sqrt{2}$. M is proportional to the identity matrix in this case.

Table 33.1 The oscillation parameters in 3σ range, for both NH and IH, are taken from the global analysis of neutrino oscillation measurements with three light active neutrinos [10]

Parameter	NH	IH
$\Delta m^2_{21}/10^{-5}eV^2$	$7.03 \to 8.09$	$7.03 \to 8.09$
$\Delta m^2_{3l}/10^{-3}eV^2$	$+2.407 \to +2.643$	$-2.635 \to -2.399$
$\sin^2\theta_{12}$	$0.271 \to 0.345$	$0.271 \to 0.345$
$\sin^2\theta_{23}$	$0.385 \to 0.685$	$0.393 \to 0.640$
$\sin^2\theta_{13}$	$0.01934 \to 0.02392$	$0.01953 \to 0.02408$

33.2.1 Neutrino Mass and Mixing

Once the Higgs field ϕ acquires a vacuum expectation value (VEV), the 5×5 neutral fermion mass matrix M_ν [7] could be written as

$$M_\nu = \begin{pmatrix} 0 & M_D^T \\ M_D & M \end{pmatrix}. \tag{33.3}$$

Here, $m_D = Y_\Sigma v/\sqrt{2}$, where $v = 246$ GeV is the VEV of the SM Higgs. The given mass matrix M_ν could be diagonalized using a unitary matrix U_0 as $U_0^T M_\nu U_0 = M_\nu^{\mathrm{diag}} = \mathrm{diag}(m_1, m_2, m_3, M_\Sigma, M_\Sigma)$, where M_Σ is the degenerate mass for the two heavy neutral fermions. We have used the Casas-Ibarra parametrization for the new Yukawa coupling matrix Y_Σ, such that the stringent bounds on the light neutrino mixing angles as well as the mass squared differences as predicted from the oscillation data are automatically satisfied [7]. Casas-Ibarra parametrization and structure of the unitary matrix are discussed in detail in [8, 9]. The light neutrino masses for the normal and inverted hierarchies are $m_1 = 0$, $m_2 = \sqrt{\Delta m^2_{sol}}$, $m_3 = \sqrt{\Delta m^2_{atm}}$ and $m_1 = \sqrt{\Delta m^2_{atm}}$, $m_2 = \sqrt{\Delta m^2_{sol} + \Delta m^2_{atm}}$, $m_3 = 0$, respectively. We use the same parametrization of the PMNS matrix. In the numerical analysis, we have considered the values of mass squared differences and mixing angles in the ranges at 3σ as given in Table 33.1 and vary δ and α phases between $-\pi$ to $+\pi$. In this present work, we have taken the triplet common mass parameter $M_\Sigma \sim \mathcal{O}(10^4)$ GeV or higher, and hence these heavy fermions are out of reach of the present collider detector at LHC or ILC.

33.2.2 Naturalness

The heavy right-handed neutrino loop corrections to the running mass parameter μ are required to be smaller than $O(\mathrm{TeV}^2)$ for the Higgs naturalness. In the $\overline{\mathrm{MS}}$ scheme and taking the quantity $(\ln[\frac{M_\Sigma}{\mu_R}] - \frac{1}{2})$ to be unity (where μ_R is the renormalization scale), the correction using the Casas-Ibarra parametrization is $\delta\mu^2 \approx \frac{3}{4\pi^2} \text{Tr}[Y_\Sigma^\dagger D_\Sigma^2 Y_\Sigma] =$ $\frac{3M_\Sigma^3}{2\pi^2 v^2}\cosh(2\text{Im}[z])(m_2 + m_3)$ for (NH) and $\frac{3M_\Sigma^3}{2\pi^2 v^2}\cosh(2\text{Im}[z])(m_1 + m_1)$ for (IH), where z is a complex parameter [7]. Hence, one can see that the $\delta\mu^2$ values for NH an IH differ only by a factor $(m_1 + m_2)$ and $(m_2 + m_3)$.

33.2.3 The Lepton Flavor Violation

The decay width as well as the branching ratio (BR) for the lepton flavor violating decay $\mu \to e\gamma$ in this model has been worked out in [11]. This BR gives the strongest LFV bound on this minimal Type-III seesaw model parameter. In $m_\Sigma >> M_W$ at $O(\frac{Y_\Sigma v}{M_\Sigma})^2$ limit, the BR($\mu \to e\gamma$) can be written as

$$\text{BR}(\mu \to e\gamma) = \frac{3}{32}\frac{\alpha}{\pi}\,|(\frac{13}{3} + C)\epsilon_{e\mu} - \sum_i x_{vi}(U_{PMNS})_{ei}(U_{PMNS}^\dagger)_{i\mu}|^2, \quad (33.4)$$

where $x_{vi} = \frac{m_i^2}{m_W^2}$ and $C = -6.56$. The second term is the contribution from neutrino mixing, while the first one is the explicit contribution of the fermion triplets. The current experimental bound is Br($\mu \to e\gamma$) $< 4.2 \times 10^{-13} \Rightarrow \epsilon_{e\mu} < 1.7 \times 10^{-7}$ [12].

33.2.4 Vacuum Stability

In this analysis, we use two-loop [6] contributions to the effective Higgs potential from the standard model particles whereas extra fermion triplet is considered up to one loop only [7]. The contributions to the effective Higgs quartic coupling due to the extra fermionic triplet for $\mu(t) = h(t) >> v$ are

$$\lambda_{eff}^\Sigma(h) = -\sum_{i=1}^{N} \frac{3\,e^{4\Gamma(h)}}{64\pi^2}(Y_\Sigma^\dagger Y_\Sigma)_{ii}^2\left(\ln\frac{(Y_\Sigma'^\dagger Y_\Sigma')_{ii}}{2} - \frac{3}{2} + 2\Gamma(h)\right), \quad (33.5)$$

where $\Gamma(h) = \int_{M_t}^{h} \gamma(\mu)\,d\ln\mu$ indicates the wave function renormalization and $\gamma(\mu)$ is the anomalous dimension of the Higgs field [13]. The running energy scale μ is expressed in terms of a dimensionless parameter t as $\mu(t) = M_Z \exp(t)$. We have reproduced the SM couplings at M_t as in references [6] by using the threshold corrections [14–16]. We have written our own computer codes to compute these threshold corrections. The extra femionic Yukawa contributions are added after the threshold heavy fermionic mass scale. It is to be noted that we take care of the important effect on the g_2 gauge coupling due to the additional $SU(2)_L$ triplet [7].

Then we evolve all the couplings up to the Planck scale to find the scale at which the effective action $S = \int d^4x \, \mathcal{L}_{\text{Higgs}}$ of the Higgs potential becomes minimum. It is also to be noted that if triplet mass M_Σ is smaller than M_t, then the extra Yukawa starts to contribute after the energy scale M_Σ. However, this contribution is negligibly small for the running from M_t to M_Σ. The contribution is effective for $M_\Sigma > M_t$ up to the Planck scale.

A quantum tunneling to the new deeper vacuum may occur. It is because of the RGs running which make the quartic coupling λ negative at a high energy scale and at the same time the value of β_λ, i.e., the slope of the potential changes from negative to positive. If the decay probability $\mathcal{P}_0$ of the EW vacuum [17] is less than *one*, i.e., decay time is greater than the lifetime of the Universe $\tau_U \sim 10^{17}$ secs and in such a case, we say that the EW vacuum is metastable. In the other words, in the region with $\lambda_{eff}(\Lambda_B) < \lambda_{eff, \min}(\Lambda_B) = \frac{-0.06488}{1-0.00986 \ln(v/\Lambda_B)}$ [18] the EW vacuum becomes unstable. The EW vacuum is absolutely stable at $\lambda_{eff}(\Lambda_B) > 0$ where the probability of the EW vacuum decay is zero. The theory violates the perturbative unitarity at $\lambda_{eff}(\Lambda_B) > \frac{4\pi}{3}$ [18].

33.3 Results

The purple, green, and gray solid lines in Fig. 33.1(left) correspond to $M_t = 171.3$, 173.1, and 174.9 GeV, respectively, with fixed value of the new physics Yukawa coupling $\text{Tr}[Y_\Sigma^\dagger Y_\Sigma]^{\frac{1}{2}} = 0.283$ along with degenerate heavy fermion mass at $M_{\Sigma 1} = M_{\Sigma 2} = M_\Sigma = 10^7$ GeV. The Higgs quartic coupling λ remains positive up to the Planck scale for the first set of the parameter with top mass at $M_t = 171.3$ GeV, and hence the EW vacuum remains absolutely stable. The second (gray line) and third (red line) sets have been chosen such that the Higgs quartic coupling $\lambda \sim$

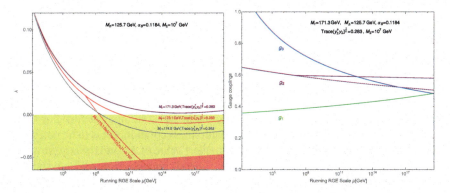

Fig. 33.1 (left) RG evolution of the Higgs quartic coupling for different values of $\text{Tr}[Y_\Sigma^\dagger Y_\Sigma]$ and (M_t) and other couplings. (right) The evaluation of the gauge coupling. The dashed purple line indicates the SM g_2 evolution

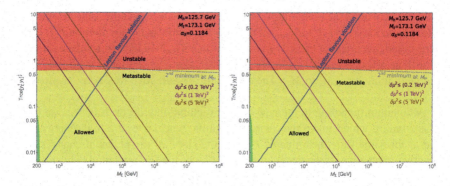

Fig. 33.2 Left plot stands for normal hierarchy whereas right one indicates inverted hierarchy

λ_{eff} becomes negative at the energy scale after $\sim 10^{10}$ GeV, which is known as instability scale Λ_I and remains negative up to M_{Pl}. However, we have checked that an extra deeper minimum is formed and the effective action $S = \int d^4x\, \mathcal{L}_{Higgs}$ becomes minimum around $\sim 10^{17}$ GeV. We find the EW vacuum corresponding to these BMPs is metastable, i.e., $\mathcal{P}_0 < 1$. The EW vacuum, on the other hand, remains unstable for the large values of the Yukawa coupling such as (red-dashed line) $\text{Tr}[Y_\Sigma^\dagger Y_\Sigma]^{\frac{1}{2}} = 0.707$. The purple line(s) in Fig. 33.1(right) shows the effect of the presence of extra fermion to the $SU(2)_L$ gauge coupling g_2.

In Fig. 33.2, we have given the phase diagram in the $\text{Tr}[Y_\Sigma^\dagger Y_\Sigma]^{\frac{1}{2}} - M_\Sigma$ plane to provide a quantitative measurement of (meta)stability in the new physics parameter space. We generate this plot for the fixed value of the SM parameters $M_t = 173.1$, $M_h = 125.7$, and $\alpha_s = 0.1184$. Here the line separating the unstable region (red) and the metastable (yellow) region is obtained when $\beta_\lambda(\mu) = 0$ along with $\lambda(\mu) = \lambda_{min}(\Lambda_B)$. The main result that we deduce from this plot is the parameter space with $\text{Tr}[Y_\nu^\dagger Y_\nu]^{\frac{1}{2}} \gtrsim 0.65$ with the heavy fermion mass scale $200 - 10^{10}$ GeV were excluded by instability of the EW vacuum. The red-dashed line separates the unstable and metastable regions of the EW vacuum. The gray-dashed line corresponds to the points for which the beta function of the quartic coupling λ is zero at the Planck scale, i.e., the second minima is situated at that scale. One can see a very small green region in bottom left corner of the plot for lower values of masses and couplings for which the EW vacuum due to the effect of g_2 is absolutely stable. However, this region is disfavored from the LFV constraints (left side of the blue dotted line) whereas the regions to the right of the purple, magenta, and brown solid lines are disallowed by the naturalness bounds depending on the naturalness condition. It is to be noted that the stability region will increase with the smallest value of M_t and the largest values of M_h and α_s [7]. Hence, the most stringent (liberal) bound is from vacuum stability with maximum (minimum) value of M_t and minimum (maximum) values of M_h and α_s from their allowed 3σ ranges [7].

33.4 Summary

The important and main goal of this work is to examine the stability of EW vacuum in the Type-III seesaw model [7]. If the Type-III seesaw model happens to be the only heavy particle, which explains the tiny slandered model neutrino masses, then can the presence of the new physics, i.e., extra fermionic triplet added at the TeV and/or high scale, alter the stability of the EW vacuum? It is also well known that if we have extra fermion, it destabilizes the EW vacuum. But if we are to solve both the neutrino mass, mixing angles, flavor, and EW vacuum stability problems in the context of this Type-III seesaw model, it is important to study the parameter space in detail which allows us to do so. In this work, we have analyzed the implications of naturalness of the Higgs mass, stability of the electroweak vacuum along with LFV decays in the context of Type-III seesaw model. Important result we have found is that in the parameter space which is allowed by both the LFV and naturalness constraints, the EW vacuum is metastable and the major part of the allowed parameter space lies in a region that could be tested in the future collider experiments.

Acknowledgements The work of Najimuddin Khan is supported by the Department of Science and Technology, Government of INDIA under the SERB-Grant PDF/2017/00372.

References

1. C.M.S. Collaboration, S. Chatrchyan et al., Observation of a new boson at a mass of 125 GeV with the CMS experiment at the LHC. Phys. Lett. B **716**, 30–61 (2012). https://doi.org/10.1016/j.physletb.2012.08.021, arXiv:1207.7235 [hep-ex]
2. ATLAS Collaboration, G. Aad et al., Observation of a new particle in the search for the Standard Model Higgs boson with the ATLAS detector at the LHC. Phys. Lett. **B716**, 1–29 (2012). https://doi.org/10.1016/j.physletb.2012.08.020, arXiv:1207.7214 [hep-ex]
3. J.A. Casas, J.R. Espinosa, I. Hidalgo, Implications for new physics from fine-tuning arguments. 1. Application to SUSY and seesaw cases. JHEP **11**, 057 (2004). https://doi.org/10.1088/1126-6708/2004/11/057, arXiv:hep-ph/0410298 [hep-ph]
4. F. Vissani, Do experiments suggest a hierarchy problem? Phys. Rev. D **57**, 7027–7030 (1998). https://doi.org/10.1103/PhysRevD.57.7027, arXiv:hep-ph/9709409 [hep-ph]
5. J.A. Casas, V. Di Clemente, A. Ibarra, M. Quiros, Massive neutrinos and the Higgs mass window. Phys. Rev. D **62**, 053005 (2000). https://doi.org/10.1103/PhysRevD.62.053005, arXiv:hep-ph/9904295 [hep-ph]
6. D. Buttazzo, G. Degrassi, P.P. Giardino, G.F. Giudice, F. Sala, A. Salvio, A. Strumia, Investigating the near-criticality of the Higgs boson. JHEP **12**, 089 (2013). https://doi.org/10.1007/JHEP12(2013)089, arxiv.org/abs/1307.3536 [hep-ph]
7. S. Goswami, K.N. Vishnudath, N. Khan, Constraining the minimal type-III seesaw model with naturalness, lepton flavor violation, and electroweak vacuum stability. Phys. Rev. D **99**(7), 075012 (2019). https://doi.org/10.1103/PhysRevD.99.075012, arXiv:1810.11687 [hep-ph]
8. J. Casas, A. Ibarra, Oscillating neutrinos and muon —> e, gamma. Nucl. Phys. **B618** (2001) 171–204, https://doi.org/10.1016/S0550-3213(01)00475-8, arXiv:hep-ph/0103065 [hep-ph]
9. A. Ibarra, G.G. Ross, Neutrino phenomenology: The Case of two right-handed neutrinos. Phys. Lett. **B591** (2004) 285–296, https://doi.org/10.1016/j.physletb.2004.04.037, arXiv:hep-ph/0312138 [hep-ph]

10. I. Esteban, M.C. Gonzalez-Garcia, M. Maltoni, I. Martinez-Soler, T. Schwetz, Updated fit to three neutrino mixing: exploring the accelerator-reactor complementarity. JHEP **01**, 087 (2017). https://doi.org/10.1007/JHEP01(2017)087, arXiv:1611.01514 [hep-ph]

11. A. Abada, C. Biggio, F. Bonnet, M.B. Gavela, T. Hambye, mu –> e gamma and tau –> l gamma decays in the fermion triplet seesaw model. Phys. Rev. D **78**, 033007 (2008). https://doi.org/10.1103/PhysRevD.78.033007, arXiv:0803.0481 [hep-ph]

12. MEG Collaboration, A.M. Baldini et al., Search for the Lepton Flavour Violating Decay $\mu^+ \rightarrow e^+\gamma$ with the Full Dataset of the MEG Experiment. http://arxiv.org/abs/1605.05081, arXiv:1605.05081 [hep-ex]

13. J. Casas, J. Espinosa, M. Quiros, Improved Higgs mass stability bound in the standard model and implications for supersymmetry. Phys. Lett. **B342** (1995) 171–179. https://doi.org/10.1016/0370-2693(94)01404-Z, arXiv:hep-ph/9409458 [hep-ph]

14. A. Sirlin, R. Zucchini, Dependence of the quartic coupling H(m) on M(H) and the possible onset of new physics in the higgs sector of the standard model. Nucl. Phys. B **266**, 389 (1986). https://doi.org/10.1016/0550-3213(86)90096-9

15. G. Degrassi, S. Di Vita, J. Elias-Miro, J.R. Espinosa, G.F. Giudice et al., Higgs mass and vacuum stability in the Standard Model at NNLO. JHEP **1208**, 098 (2012). https://doi.org/10.1007/JHEP08(2012)098, arXiv:1205.6497 [hep-ph]

16. F. Bezrukov, M.Y. Kalmykov, B.A. Kniehl, M. Shaposhnikov, Higgs Boson Mass and New Physics. JHEP **1210**, 140 (2012). https://doi.org/10.1007/JHEP10(2012)140, arXiv:1205.2893 [hep-ph]

17. S.R. Coleman, The Fate of the False Vacuum. 1. Semiclassical Theory. Phys. Rev. D **15**, 2929–2936 (1977). https://doi.org/10.1103/PhysRevD.15.2929, https://doi.org/10.1103/PhysRevD.16.1248

18. N. Khan, S. Rakshit, Study of electroweak vacuum metastability with a singlet scalar dark matter. Phys. Rev. D **90**(11), 113008 (2014). https://doi.org/10.1103/PhysRevD.90.113008, arXiv:1407.6015 [hep-ph]

Chapter 34
Flavor Violation at LHC in Events with Two Opposite Sign Leptons and a B-Jet

Nilanjana Kumar

Abstract Hints of flavor violation at both charged current and neutral current decays have been observed in experiments such as LHCb, Belle, and *Babar*. The anomalies in the result can be addressed in the Effective Field Theory (EFT) framework. The effective operators predict different Beyond Standard Model (BSM) signatures and the four-point interaction vertices can be probed at Large Hadron Collider (LHC). In this context, the discovery projection of two opposite sign leptons and a b-jet signature is studied in this paper at 13 TeV LHC.

34.1 Motivation

Recent experimental measurements at LHCb, Belle, and *Babar* present deviations in the SM prediction of B-meson decays and hints toward Lepton Flavor Violation (LFV). LFV has been observed in charged current decay at tree level, $b \to c\ell\nu$. Taking into account $R(D)$ and $R(D^*)$ measurements by *Babar* [1], Belle [2] and LHCb [3] and their correlations, the difference between SM and the data is nearly 3.8σ [4, 5]. The anomaly in $B_c \to J/\psi \ell\nu$ measurement is at the 2σ level. Whereas the neutral current transitions, namely, $b \to s\ell^+\ell^-$ show an opposite effect in the measurements of R_K and R_{K^*}. The recent results by LHCb Collaboration [6] and Belle [7], reflect that the data is more consistent with the SM. A deviation is also seen in $B_s \to \phi\mu\mu$ [8] that suggests that the discrepancies in R_K and R_{K^*} have been caused by a diminution of the $b \to s\mu^+\mu^-$ channel, rather than an enhancement in $b \to se^+e^-$.

To address these anomalies, one can choose the models with leptoquarks [9] or Z' [10]. Another way of addressing these anomalies would be to consider an Effective Field Theory (EFT) description with a set of Wilson coefficients. It is possible to construct such a theory with a few unknown parameters, if symmetry relations exist among the Wilson coefficients. As shown recently in [11–13], with a minimal set

N. Kumar (✉)
Department of Physics and Astrophysics, University of Delhi, Delhi 110007, India
e-mail: nilanjana.kumar@gmail.com

© Springer Nature Singapore Pte Ltd. 2021
P. K. Behera et al. (eds.), *XXIII DAE High Energy Physics Symposium*,
Springer Proceedings in Physics 261,
https://doi.org/10.1007/978-981-33-4408-2_34

of New Physics (NP) operators, accompanied by a single lepton mixing angle, it is possible to explain almost the flavor observables. The parameters of these models can be determined phenomenologically and if the scale of the new physics is a few TeVs, this leads to interesting collider signatures at LHC.

34.2 Theoretical Framework

The Hamiltonian for the new physics can be expressed in terms of two operators involving left-handed second- and third-generation quark doublets Q_{2L}, Q_{3L}, third-generation lepton doublet L_{3L}, and right-handed singlet τ_R as defined in [11, 12], where the terms that we are interested in are $(\bar{Q}_{2L}\gamma^\mu Q_{3L})_3 (\bar{L}_{3L}\gamma^\mu L_{3L})_3$ and $(\bar{Q}_{2L}\gamma^\mu Q_{3L})_1 (\bar{\tau}_R\gamma^\mu \tau_R)$, with coefficients $3A_1/4$ and A_5, following the literature [11, 12]. A_i are real unknown coefficients with dimension TeV^{-2}. The subscripts "3" and "1" represent the $SU(2)_L$ triplet and singlet currents, respectively. The flavor eigenstates can be expressed in terms of mass eigenstates by a field rotation

$$\tau = cos\theta(\tau') + sin\theta(\mu')$$
$$v_\tau = cos\theta(v'_\tau) + sin\theta(v'_\mu).$$

As a result of the mixing, the coupling with the second generation of leptons are induced. The magnitude of this mixing is found to be small (~ 0.02) [11, 12]. Also, for all class of models, the best fit values obtained can be approximated as $A_1 \sim 3.8$, $A_5 \sim 2.3$.

The flavor violating processes, generated by these operators, are listed in Table 34.1. For a process (a,b) $\rightarrow$ (c,d) with coefficient X in the operator, we can write the four-point coupling (λ^2) in the mass basis as

$$\frac{\lambda^*_{a,b}\lambda_{c,d}}{2M^2}(a, b)(c, d) = \epsilon^{abcd}\frac{4G_F}{\sqrt{2}}(a, b)(c, d),$$
$$\lambda^2 \sim \lambda^*_{a,b}\lambda_{c,d} \sim 2M^2 X,$$

where M is the mass of the integrated-out field and λ's are the dimensionless couplings. From perturbativity, the bound on λ is $\lambda^2/(4\pi)^2 \sim 1$.

Table 34.1 Operators and their effective coupling. For the notation of the operators, see [11]

Flavor basis	Mass basis	λ^2
$(3A_1/4)(s, b)(\tau\tau)$	$(3A_1/4)cos^2\theta(s, b)(\tau'\tau')_L$	$2M^2(3A_1/4)cos^2\theta$
–	$(3A_1/4)sin^2\theta(s, b)(\mu'\mu')_L$	$2M^2(3A_1/4)sin^2\theta$
–	$(3A_1/4)sin2\theta(s, b)(\mu'\tau')_L$	$2M^2(3A_1/4)sin2\theta$
$A_5(s, b)(\tau\tau)$	$A_5cos^2\theta(s, b)(\tau'\tau')_R$	$2M^2 A_5cos^2\theta$
–	$A_5sin^2\theta(s, b)(\mu'\mu')_R$	$2M^2 A_5sin^2\theta$
–	$A_5sin2\theta(s, b)(\mu'\tau')_R$	$2M^2 A_5sin2\theta$

34.3 Results

As can be seen from Table 34.1, there are three possible signatures:

- $(\mu^{\pm}\mu^{\mp})$ and a b-jet,
- $(\mu^{\pm}\tau^{\mp})$ and a b-jet, and
- $(\tau^{\pm}\tau^{\mp})$ and a b-jet.

These processes can be generated at 13 TeV LHC via g-g and g-s fusion in p-p collision, with the major contribution coming from g-s fusion. Now as the λ^2 is a function of A_1 and A_5, the cross section also varies with these parameters. We kept the value of A_1 fixed at the best fit 3.8 and varied at A_5. The range of values of the parameters is chosen such that the 95% C.L. upper bound of $Br(B_s \rightarrow \tau^{\pm}\mu^{\mp}) < 4.2 \times 10^{-5}$ [14] is satisfied.

The cross section of $(\mu^{\pm}\mu^{\mp})$ and a b-jet is very small, because their coupling is suppressed by $(\sin^2\theta)$ and hence we neglect it in this study. The cross section of $(\mu^{\pm}\tau^{\mp})$ and a b-jet will be suppressed by $\sin 2\theta$ and hence will be comparatively larger than the previous one, as shown in Fig. 34.1 (left). The cross section of $(\tau^{\pm}\tau^{\mp})$ and a b-jet is relatively very high as shown in Fig. 34.1 (right) with red line. The tau can also decay leptonically to a muon with branching ratio 0.174, enabling the final states with $(\mu^{\pm}\tau^{\mp})$ and a b-jet and $(\mu^{\pm}\mu^{\mp})$ and a b-jet, as shown in Fig. 34.1 by green and blue lines, respectively.

The major SM backgrounds for these channels are $t\bar{t}$, single top (Wt), diboson (W^+W^-, WZ, and ZZ), W+jets, WW+jets, and Z/γ+jets. We found that the background for the signal with opposite sign same flavor states ($\mu^{\pm}\mu^{\mp}$) or ($\tau^{\pm}\tau^{\mp}$) and a b-jet is higher than the opposite sign opposite flavor states. Also, the signal ($\tau^{\pm}\tau^{\mp}$)

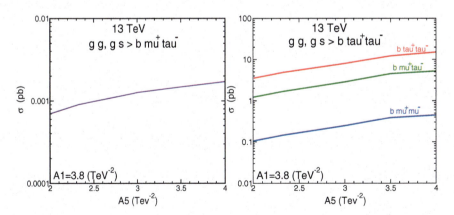

Fig. 34.1 (L) The total production of one μ and one τ in association with one b-jet in g-g fusion and g-s fusion. (R) The total production of two τ,s in association with one b-jet in g-g fusion and g-s fusion (red), also one μ and one τ + b-jet (green) and two μ's+b-jet (blue) when tau (one or both, respectively) decays leptonically

Fig. 34.2 The 5σ discovery projection at 13 TeV LHC in $(\mu^{\pm}\mu^{\mp}) + b$-jet and $(\mu^{\pm}\tau^{\mp})$ $+ b$-jet channel as a function of the integrated luminosity and model parameter A_5 with the assumption of 25% uncertainty in the background events

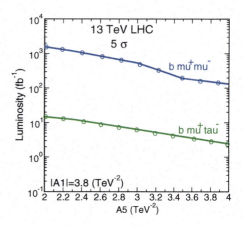

and a b-jet will suffer from tau tagging efficiency at LHC as both the tau decay hadronically. Hence, we study two channels, $(\mu^{\pm}\mu^{\mp})+b$-jet and $(\mu^{\pm}\tau^{\mp})+b$-jet In Fig. 34.2, we have shown the discovery projection of the these two channels as a function of the integrated luminosity at LHC. We have followed the search strategy as mentioned in [15, 16]. Figure 34.2 shows that the $(\mu^{\pm}\mu^{\mp})+b$-jet channel requires much larger luminosity than $(\mu^{\pm}\tau^{\mp})+b$-jet channel for 5σ discovery significance.

34.4   Conclusion

Recently observed anomalies in the decays of B-mesons hint toward new physics interaction which involves a b-quark, a s-quark, and a pair of opposite sign leptons. The four-point interactions can be probed at LHC p-p collision via the direct production of b-quark and two opposite sign leptons. The opposite sign lepton pair has either same or opposite flavor. In this study, 5σ discovery potential of $(\mu^{\pm}\mu^{\mp}) + b$-jet and $(\mu^{\pm}\tau^{\mp}) + b$-jet channels are discussed as a function of the model parameters. Overall, these channels have a very good detection prospect even with the currently collected data at LHC and a limit on the model parameter space can be set with the 13 TeV LHC data.

References

1. BaBar Collaboration, J.P. Lees et al., Phys. Rev. **D88**, 072012 (2013)
2. Belle Collaboration, M. Huschle et al., Phys. Rev. **D92**, 072014 (2015)
3. LHCb Collaboration, R. Aaij et al., Phys. Rev. Lett. **115**, 111803 (2015)
4. Heavy Flavor Averaging Group (HFAG) collaboration, Y. Amhis et al., arXiv: 1412.7515 [hep-ph]
5. https://hflav-eos.web.cern.ch/hflav-eos/semi/summer18/RDRDs.html

6. LHCb Collaboration collaboration, Tech. Rep. CERN-EP-2019-043. LHCB-PAPER-2019-009, CERN, Geneva, Mar, 2019
7. Belle Collaboration, A. Abdesselam et al
8. LHCb Collaboration, R. Aaij et al., JHEP **09**, 179 (2015)
9. D. Bečirevié, S. Fajfer, N. Košnik, O. Sumensari, Phys. Rev. D **94**, 115021 (2016)
10. P. Langacker, Rev. Mod. Phys. **81**, 1199–1228 (2009)
11. D. Choudhury, A. Kundu, R. Mandal, R. Sinha, Phys. Rev. Lett. **119**, 151801 (2017)
12. D. Choudhury, A. Kundu, R. Mandal, R. Sinha, Nucl. Phys. B **933**, 433–453 (2018)
13. S. Bhattacharya, A. Biswas, Z. Calcuttawala, S.K. Patra. http://arxiv.org/abs/1902.02796arXiv:1902.02796 [hep-ph]
14. LHCb collaboration, R. Aaij et al., *Search for the lepton-flavour-violating decays* $B_s^0 \to \tau^\pm \mu^\mp$ *and* $B^0 \to \tau^\pm \mu^\mp$. arXiv:1905.0661 [hep-ph]
15. D. Choudhury, N. Kumar, A. Kundu. http://arxiv.org/abs/1905.07982arXiv:1905.07982 [hep-ph]
16. Y. Afik, J. Cohen, E. Gozani, E. Kajomovitz, Y. Rozen, Establishing a Search for $b \to s\ell^+\ell^-$ Anomalies at the LHC. JHEP **08**, 056 (2018)

Chapter 35
Dark Matter in Leptoquark Portal

Rusa Mandal

Abstract A beyond the standard model portal scenario for Majorana fermion dark matter (DM) particle with leptoquark being the mediator field is of the main focus of this study. We explore the parameter space of the only unknown coupling in the model which is sensitive to all three main features of a DM model, namely, relic density, direct detection as well as indirect detection, while being consistent with the collider searches. The AMS-02 data for antiproton flux imposes stringent bound till date which excludes the DM mass up to 400 GeV at 95% C.L. The LUX 2016 data for DM-neutron scattering cross section allows the region compatible with relic density; however, the future sensitivity of LZ experiment can probe the model up to its perturbative limit.

35.1 Introduction

The only known interaction of the dark matter (DM) particle is gravitational in nature. In this work, we speculate the DM candidate interacts to the standard model (SM) sector through leptoquark; a colored particle having both baryon and lepton number. By suppressing the color and generation indices, we quote such possible interaction terms with the spin and SM quantum numbers ($SU(3)_C$, $SU(2)_L$, $U(1)_Y$) of the mediator leptoquark in Table 35.1. Here Q_L and u_R, d_R are the SM $SU(2)_L$ quark doublets and right-handed singlets, respectively. The boson $X_{(\mu)}$ denotes the scalar (vector) leptoquark and ψ is a SM singlet fermion that would be DM candidate.

For all three scalar and the first vector cases, the interaction of the leptoquark with other SM fields induces baryon number violating processes [1] and thus generally give rise to proton decay. Hence, we avoid such types from our considerations. It should be noted that in the case of Dirac DM particle, the spin-independent (SI) DM-nuclei cross-section measurements by XENON1T [2] and LUX [3] experiments almost completely exclude the parameter space at its minimal content. For instance,

R. Mandal (✉)
IFIC, Universitat de València-CSIC, Apt. Correus 22085, 46071 València, Spain
e-mail: Rusa.Mandal@ific.uv.es

© Springer Nature Singapore Pte Ltd. 2021
P. K. Behera et al. (eds.), *XXIII DAE High Energy Physics Symposium*,
Springer Proceedings in Physics 261,
https://doi.org/10.1007/978-981-33-4408-2_35

Table 35.1 Details of interaction terms with the SM quantum numbers and spin of the mediator leptoquark

Spin	Interaction	Quantum No.	Spin	Interaction	Quantum No.
0	$\bar{d}_R^C X \psi$	$(\bar{3}, 1, 1/3)$	1	$\bar{Q}_L^C \gamma^\mu X_\mu \psi$	$(\bar{3}, 2, -1/6)$
0	$\bar{u}_R^C X \psi$	$(\bar{3}, 1, -2/3)$	1	$\bar{u}_R \gamma^\mu X_\mu \psi$	$(3, 1, 2/3)$
0	$\bar{Q}_L X \psi$	$(3, 2, 1/6)$	1	$\bar{d}_R \gamma^\mu X_\mu \psi$	$(3, 1, -1/3)$

for the last two cases, if ψ is an $\mathcal{O}(100 \text{ GeV})$ Dirac DM particle, for $\mathcal{O}(\text{TeV})$ mediator mass, the coupling of the interaction term > 0.05 is excluded at 90% C.L. by XENON1T 2018 data [2], whereas the required coupling to satisfy observed relic density is one order higher. Hence, we prefer to consider Majorana DM candidate for the analysis of this work. Now we see in the next section that the thermal average annihilation cross section of Majorana DM pair is proportional to the square of the mass of final states quarks and thus in the case of DM pair annihilating to down-type quark–anti-quark pair, the annihilation rate is insufficient to produce the observed relic density within the perturbative limit of the interaction strength. Hence, the only reasonable choice among the six portals shown above reduces to the fifth portal, namely, a vector leptoquark with $(3, 1, 2/3)$ quantum numbers under the SM gauge group.

The Lagrangian can be written as

$$\mathcal{L} \subset -\frac{1}{2} U_{\mu\nu}^\dagger U^{\mu\nu} + m_U^2 \, U_\mu^\dagger U^\mu - \frac{1}{2} m_\chi \chi \chi$$
$$- y_L \bar{Q}_L \gamma_\mu U^\mu L_L - y_R \bar{d}_R \gamma_\mu U^\mu e_R - y_\chi \bar{u}_R \gamma_\mu U^\mu \chi + \text{h.c.}, \qquad (35.1)$$

where $U_{\mu\nu} = D_\nu U_\mu - D_\mu U_\nu$ with $D_\mu = \partial_\mu - i g_s \frac{\lambda^a}{2} G_\mu^a - i g' \frac{2}{3} B_\mu$. Here L_L and e_R are the SM $SU(2)_L$ lepton doublets and right-handed singlets, respectively. The new field U is a vector leptoquark with charges under the SM as $(3, 1, 2/3)$ whereas χ is a Majorana fermion being singlet under the SM gauge group. If $m_\chi < m_U$, the two-body decay of χ is forbidden at tree level. However, the interaction of U with the SM fields written in (35.1) will induce tree-level three-body decay and one-loop-induced decay of χ. To avoid this situation, we introduce an extra symmetry, namely, a Z_2 symmetry and assume only the leptoquark U and χ are odd under it. In such case, the fermion χ can serve as a cosmological stable DM candidate and only last term of (35.1) is the relevant interaction which connects the visible sector to the dark sector (χ).

Note that in (35.1), we have assumed minimal coupling scenario, i.e., the interaction term $i g_s U_\mu^\dagger \frac{\lambda^a}{2} U_\nu G_{\mu\nu}^a$ is absent. It should also be noted that the coupling of the DM particle χ to the three generations of up-type right-handed quarks can, in general, be different; however, for simplicity, we assume them to be identical for all three generations.

35.1.1 Relic Density

It can be seen from (35.1) that the DM candidate χ can annihilate into the SM up-type quark–anti-quark pair via a t-channel exchange of U. The thermal average annihilation cross section is given by

$$\langle \sigma v \rangle = \frac{3\, y_\chi^4 m_q^2}{8\pi \left(m_\chi^2 + m_U^2 - m_q^2\right)^2} \left(1 - \frac{m_q^2}{m_\chi^2}\right)^{1/2}, \tag{35.2}$$

where m_q is the mass of up-type quark. It is apparent that for $m_\chi > m_t$, $\chi\chi \to t\bar{t}$ is the most dominant annihilation mode.

For $m_\chi < m_t$, with $\mathcal{O}(1)$ coupling, the annihilation channels to $\chi\chi \to u\bar{u}$, $c\bar{c}$ are insufficient to explain the observed relic density at present Universe. In such parameter space, the processes like $\chi\chi \to gg$, Wtb, can also be important.

The model under consideration also poses co-annihilation channels such as $\chi U \to tg$ through a t-channel U exchange. This process is significant only when the DM and the mediator are very close in masses. Another interesting co-annihilation mode, through a s-channel top quark, is $\chi U \to Wb$ and will only be efficient near the top quark resonance.

The observed relic abundance by Planck data $\Omega h^2 = 0.1199 \pm 0.0027$ [4] can be achieved by the thermal freeze-out condition, $\langle \sigma v \rangle \approx 2 \times 10^{-9}\,\text{GeV}^{-2}$ and we explore the region in the next section.

35.1.2 Direct Detection

The direct detection signal for this type of model can arise as well. After integrating out the vector leptoquark and keeping terms only at leading order in ∂^2/m_U^2, the effective Lagrangian can be written as

$$\mathcal{L}_{\text{eff}} \simeq -\frac{y_\chi^2}{4\left(m_U^2 - m_\chi^2\right)} \bar{\chi}\,(1 - \gamma_5)\,\gamma_\mu u\;\bar{u}\,(1 - \gamma_5)\,\gamma^\mu \chi\,. \tag{35.3}$$

Due to the fact that for a Majorana fermion $\bar{\chi}\gamma_\mu\chi = 0$, after performing the Fierz transformation we are left only with the spin-dependent (SD) interaction given by

$$\mathcal{L}_{\text{eff}} \simeq d_u\,\bar{u}\gamma_\mu\gamma_5 u\;\bar{\chi}\gamma_\mu\gamma_5\chi\,;\quad d_u \equiv -\frac{y_\chi^2}{4\left(m_U^2 - m_\chi^2\right)}. \tag{35.4}$$

The DM-nucleon scattering cross section is expressed as [5]

$$\sigma_{SD} = \frac{16 m_\chi^2 m_N^2}{\pi \left(m_\chi + m_N\right)^2} \, d_u^2 \Delta_u^{N\,2} \, J_N \left(J_N + 1\right),$$ (35.5)

where $m_N \simeq 1\,\text{GeV}$ and $J_N = 1/2$ are the mass and spin of a nucleon, respectively. The factor Δ_u^N denotes the spin fraction carried by a u-quark inside a nucleon and the estimates are $\Delta_u^p = 0.78 \pm 0.02$ for proton and $\Delta_u^n = -0.48 \pm 0.02$ for neutron [6, 7].

35.2 Results

The discussion in the preceding sections led us to explore the parameter space of the model. The thermal averaged annihilation cross section (in (35.2)) and the DM-nucleon scattering cross sections (in (35.5)) are the two key predictions of the portal under consideration. It can be seen from (35.2) and (35.5), both the expressions depend on three parameters, namely, m_χ, m_U, and y_χ. As mentioned earlier for $m_\chi < m_t$, the annihilation cross section of the DM pair to lighter up-type quark anti-quark is inadequate to produce the observed relic density. Thus, we consider $m_\chi > m_t$ and vary up to $\mathcal{O}(\text{TeV})$ range.

Collider bounds: The leptoquarks can directly be searched for via pair and/or single production at the colliders. The signatures for the considered leptoquark portal model are $U\bar{U} \to t\bar{t}\chi\chi$, $jj\chi\chi$ topologies. A latest result from CMS with $35.9\,\text{fb}^{-1}$ data [8] imposes most stringent bound $m_U > 1.5\,\text{TeV}$ to date by considering 100% branching fraction to $t\nu$. On the other hand, the DM mass is constrained from the monojet $+\not{E}_T$ searches from ATLAS collaboration at 13 TeV center-of-mass energy data with an integrated luminosity of $36.1\,\text{fb}^{-1}$ [9]. To generate the parton-level cross section for the process $pp \to \chi\chi j$, we use MadGraph5 [10] where the model files are created by FeynRules [11]. We use NNPDF23LO [12] parton distribution function (PDF) with five flavor quarks in initial states. The basic cuts used are the following: $\not{E}_T > 250\,\text{GeV}$, a leading jet with transverse momentum $p_T > 250\,\text{GeV}$ and pseudorapidity $|\eta| < 2.4$. Due to large parton distribution probability of the gluon as compared to the quark or anti-quark in the proton, the $qg \to \chi\chi q$ process dominates. We find the region satisfying relic density with $y_\chi = 1$ for DM mass $m_\chi \lesssim 200\,\text{GeV}$ is excluded at 95% confidence level. A similar observation was made in [13] in context of a scalar leptoquark mediator with Majorana DM candidate using 8 TeV data from CMS collaboration [14].

Combining the above discussions, we highlight the allowed region in Fig. 35.1 for $m_U > 1\,\text{TeV}$ and $m_\chi > 200\,\text{GeV}$ for different choices of the coupling y_χ. The red curves satisfy the observed relic abundance by Planck data $\Omega h^2 = 0.1199 \pm 0.0027$ [4] where the textures solid, dot, and dash denote $y_\chi = 1$, 2, and $\sqrt{4\pi}$, respectively. The red shaded region is forbidden by perturbativity limit. The blue and green curves present current limit from SD DM-neutron scattering cross-section measurement by LUX [15] and future sensitivity of LUX-ZEPLIN (LZ) experiment [16], respectively, for the corresponding values of y_χ. The region below these

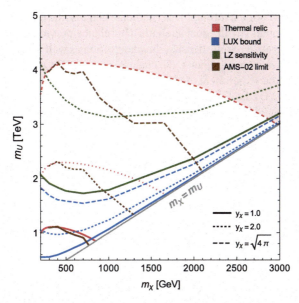

Fig. 35.1 The allowed region of the $m_\chi - m_U$ plane for different choices of coupling $\lambda_\chi = 1$ (solid), 2 (dotted) and perturbative limit $\sqrt{4\pi}$ (dashed) are shown. The red curves satisfy the observed relic density [4]. The blue and green curves represent current limit from spin-dependent DM-neutron scattering cross-section measurement by LUX [15] and future sensitivity of LZ experiment [16], respectively, for the corresponding values of y_χ. The region below these direct detection bounds is excluded at 90% confidence level. The brown curves depict the indirect detection bound from AMS-02 measurements [17, 18] which excludes $m_\chi \le 400\,\text{GeV}$ at 95% significance level

direct detection bounds is excluded at 90% confidence level. It can be seen that for the current limit from LUX data, starting from $y_\chi = 1$, all higher values are allowed and we have the entire parameter space compatible with relic abundance and direct detection limit. However, the future sensitivity of LZ will rule out all the parameter space up to the perturbative limit of y_χ. We also impose the bounds from [17] where the limits on DM pair annihilation cross section into different SM fields have been obtained by using AMS-02 measurements of antiproton flux [18]. For the case in our consideration, the dominant annihilation mode for the DM pair is to $t\bar{t}$ and the limits are depicted by brown curves for the respective y_χ values. This is currently the most stringent constraint on the model under consideration and for $m_\chi \le 400\,\text{GeV}$, the region satisfying thermal freeze-out condition is excluded at 95% confidence level.

35.3 Summary

- DM model mediated by colored particle—leptoquark—is discussed. By briefly reviewing the current status of the different models in leptoquark portal, we concentrate on a vector leptoquark portal with the leptoquark having charges under the SM group as (**3**, 1, 2/3).

- We consider a Majorana fermion DM candidate and assume the DM couples to all generations of quarks with equal strength. This choice provides one Yukawa-type coupling y_χ sensitive to relic density, direct detection as well as indirect detection.
- Collider searches from monojet + $\not{E}_T$ channel exclude DM mass < 200 GeV.
- Indirect detection experiment AMS-02 excludes DM mass < 400 GeV, which satisfies the observed relic density, at 95% confidence level.
- LUX 2016 data on DM-neutron SD cross-section measurements allow region compatible with relic density. The bound from DM-proton SD cross-section data from LUX is less stringent.
- The proposed sensitivity of LZ experiment can probe entire region up to the perturbativity limit of y_χ.

Acknowledgements This work has been supported in part by Grants No. FPA2014- 53631-C2-1-P, FPA2017-84445-P and SEV-2014-0398 (AEI/ERDF, EU) and by PROMETEO/2017/053.

References

1. I. Doršner, S. Fajfer, A. Greljo, J.F. Kamenik, N. Košnik, Phys. Rept. **641**, 1 (2016). arXiv:1603.04993 [hep-ph]
2. E. Aprile et al. [XENON Collaboration]. arXiv:1805.12562 [astro-ph.CO]
3. D.S. Akerib et al. [LUX Collaboration]. Phys. Rev. Lett. **118**, no. 2, 021303 (2017). arXiv:1608.07648 [astro-ph.CO]
4. P.A.R. Ade et al., Planck collaboration. Astron. Astrophys. **594**, A13 (2016). arXiv.org/abs/1502.01589 [astro-ph.CO]
5. P. Agrawal, Z. Chacko, C. Kilic, R.K. Mishra. arxiv.org/abs/1003.1912 [hep-ph]
6. G.K. Mallot, Int. J. Mod. Phys. A **15S1**, 521 (2000) [eConf C **990809**, 521 (2000)] [hep-ex/9912040]
7. J.R. Ellis, A. Ferstl, K.A. Olive, Phys. Lett. B **481**, 304 (2000). [hep-ph/0001005]
8. A.M. Sirunyan et al., CMS collaboration. arXiv:1805.10228 [hep-ex]
9. M. Aaboud et al., ATLAS collaboration. JHEP **1801**, 126 (2018). arXiv.org/abs/1711.03301 [hep-ex]
10. J. Alwall et al., JHEP **1407**, 079 (2014). arXiv.org/abs/1405.0301 [hep-ph]
11. A. Alloul, N.D. Christensen, C. Degrande, C. Duhr, B. Fuks, Comput. Phys. Commun. **185**, 2250 (2014). arXiv:1310.1921 [hep-ph]
12. R.D. Ball et al., Nucl. Phys. B **867**, 244 (2013). arXiv:1207.1303 [hep-ph]
13. H. An, L.T. Wang, H. Zhang, Phys. Rev. D **89**(11), 115014 (2014). arXiv:1308.0592 [hep-ph]
14. V. Khachatryan et al., CMS collaboration. Eur. Phys. J. C **75**(5), 235 (2015). arXiv:1408.3583 [hep-ex]
15. D.S. Akerib et al., LUX collaboration. Phys. Rev. Lett. **116**(16), 161302 (2016). arXiv:1602.03489 [hep-ex]
16. D.S. Akerib et al., LZ collaboration. arXiv:1509.02910 [physics.ins-det]
17. A. Cuoco, J. Heisig, M. Korsmeier, M. Krämer. arXiv:1711.05274 [hep-ph]
18. M. Aguilar et al., AMS collaboration. Phys. Rev. Lett. **117**(9), 091103 (2016)

Chapter 36
NSI in Electrophilic *ν*2HDM

Ujjal Dey, Newton Nath, and Soumya Sadhukhan

Abstract In the traditional neutrinophilic two Higgs doublet model (ν2HDM), there is no non-standard neutrino interaction (NSI), due to the tiny mixing of the two scalar doublets. Here, we generate significant NSI along with tiny Dirac neutrino mass, modifying ν2HDM with a negative charge to e_R under a global U(1) symmetry. We discuss constraints from the LEP experiments, tree-level lepton flavor violating processes, big bang nucleosynthesis, etc., on the electrophilic ν2HDM. Those constraints force this model to significantly restrict the range of permissible NSI parameters, putting a strict upper bound on different NSIs.

36.1 Introduction

The Standard Model (SM) epitomizes our knowledge of fundamental interactions of the particle world. Except for a few minor disagreements, SM is unblemished by the direct observations from LHC up to now. The neutrinos are the most elusive particles of SM which propels us toward physics beyond the SM to explain their tiny mass and their flavor oscillation, confirmed by various experiments.

Non-standard interactions (NSI) of neutrinos can be induced by the models beyond SM (BSM) when they provide corrections to the effective standard neutrino interactions, through the effects in the six- and eight-dimensional higher dimensional operators. These higher dimensional operators can be present in BSM scenarios like gauge extended models with new Z' bosons, models with single or multiple charged

U. Dey
Asia Pacific Center for Theoretical Physics, Pohang 37673, Korea

N. Nath
Institute of High Energy Physics, Chinese Academy of Sciences, Beijing, China

School of Physical Sciences, University of Chinese Academy of Sciences, Beijing, China

S. Sadhukhan (✉)
Department of Physics and Astrophysics, University of Delhi, Delhi, India
e-mail: physicsoumya@gmail.com

© Springer Nature Singapore Pte Ltd. 2021
P. K. Behera et al. (eds.), *XXIII DAE High Energy Physics Symposium*,
Springer Proceedings in Physics 261,
https://doi.org/10.1007/978-981-33-4408-2_36

heavy scalars, leptoquarks, R-parity violating supersymmetry, etc. For a detailed review and phenomenological consequences of these operators, check [2, 3] and the references therein. Neutral-current NSI in the presence of matter is described by the 6D four-fermion operators which are defined as [1]

$$\mathcal{L}_{\text{NSI}}^{NC} = -2\sqrt{2} G_F (\bar{\nu}_\alpha \gamma^\rho P_L \nu_\beta)(\bar{f} \gamma_\rho P_C f) \epsilon_{\alpha\beta}^{fC} + \text{h.c.} \qquad (36.1)$$

with NSI parameters $\epsilon_{\alpha\beta}^{fC}$, $\alpha, \beta = e, \mu, \tau$, $C = L, R$, and $f = e, u, d$.

In the standard ν2HDM, no NSI interaction is there due to the negligible left-handed neutrino interaction with the other leptons and quarks via the extra scalars. We construct an electrophilic ν2HDM, making the second scalar doublet Φ_2 couple only to the electron and neutrinos. We assign a negative charge to e_R under a global $U(1)$ symmetry. The ν_L couples to the charged leptons through a charged scalar messenger to contribute to NSI [4]. Along with the presence of NSI, here the fermion hierarchy is moderate compared to SM. The second Higgs doublet takes care of neutrinos and electron mass generation, with the first doublet providing mass to the rest. We apply the constraints from lepton flavor violation (LFV), oblique parameters, μ_{g-2}, big bang neucleosynthesis (BBN), etc., on this electrophilic ν2HDM.

36.2   Model Framework

We first discuss the standard neutrinophilic two Higgs doublet model (ν2HDM) [5, 6]. Then, we construct an electrophilic ν2HDM to get the NSI effects.

36.2.1   Standard ν2HDM

We add an extra SM-like scalar doublet Φ_2, three right-handed neutrinos (RHN) ν_{Ri} to generate the Dirac masses with the SM left-handed neutrinos. In this setup [5], a global $U(1)$ symmetry is invoked with the fields Φ_2 and ν_{Ri} having +1 charge under that, while all the SM fields are neutral. The Yukawa interaction of the neutrinos are $(-y_\nu^{ij} \bar{L}_{Li} \tilde{\Phi}_2 \nu_{Rj})$ where $L_L = (\nu_L, \ell_L)^T$, $\tilde{\Phi} = i\sigma_2 \Phi^*$. With an unbroken $U(1)$, Φ_2 has no vev and the neutrinos remain massless.

The most general scalar potential for the exact $U(1)$ symmetric case is given by

$$V(\Phi_1, \Phi_2) = m_{11}^2 \Phi_1^\dagger \Phi_1 + m_{22}^2 \Phi_2^\dagger \Phi_2$$
$$+ \frac{\lambda_1}{2}(\Phi_1^\dagger \Phi_1)^2 + \frac{\lambda_2}{2}(\Phi_2^\dagger \Phi_2)^2 + \lambda_3 \Phi_1^\dagger \Phi_1 \Phi_2^\dagger \Phi_2 + \lambda_4 \Phi_1^\dagger \Phi_2 \Phi_2^\dagger \Phi_1 .$$
$$(36.2)$$

We break the $U(1)$ to provide masses to the neutrinos, by introducing a soft-breaking term of the form $(-m_{12}^2 \Phi_1^\dagger \Phi_2)$. The two scalar doublets can be presented as

$$\Phi_a = \begin{pmatrix} \phi_a^+ \\ (v_a + h_a + i\eta_a)/\sqrt{2} \end{pmatrix}, \qquad a = 1, 2, \tag{36.3}$$

with VEVs $v_{\Phi_1} = v_1$, $v_{\Phi_2} = v_2$, and v_2 responsible for neutrino masses. The mass matrix is diagonalized when the CP-even and CP-odd charged scalars mix through angles α and β, respectively. For tiny neutrino mass, we have $v_2 \ll v_1$, which leads to $\alpha, \beta \ll 1$. The SM-like 125 GeV Higgs comes from Φ_1 whereas the BSM scalars are dominantly from Φ_2 and therefore neutrinophilic. The Yukawa couplings of the new scalars in the limit $v_2 \ll v_1$ are described as

$$\mathcal{L}_Y \supset \frac{m_{\nu_i}}{v_2} H \bar\nu_i \nu_i - i \frac{m_{\nu_i}}{v_2} A \bar\nu_i \gamma_5 \nu_i - \frac{\sqrt{2} m_{\nu_i}}{v_2} [U_{\ell i}^* H^+ \bar\nu_i P_L \ell + \text{h.c.}], \tag{36.4}$$

where m_{ν_i} are neutrino masses and $U_{\ell i}$ is the PMNS matrix. Due to $U(1)$ charge assignments of the different fields, ν_Ls cannot couple to l_R s through Φ_1. Similar couplings via Φ_2 are tiny, being proportional to $\sin \beta \approx v_2/v_1$, resulting in negligible NSI effects.

36.2.2 Electrophilic ν2HDM

Here we modify the model, giving mass to the electron along with the neutrinos through the second scalar doublet Φ_2. The right-handed electron e_R is given charge (-1) under the global $U(1)$ to arrange mass term for the electron. With this charge assignment, the Lagrangian for the $\nu - e$ sector becomes

$$\mathcal{L}_{\nu 2\text{HDM}}^m \supset y_e \bar{L}_e \Phi_2 e_R + y_\nu \bar{L}_e \tilde\Phi_2 \nu_R + \text{h.c.}, \tag{36.5}$$

where L_e is the SM electron doublet. The first term of the Lagrangian provides charged Higgs ($H^\pm$) related terms as

$$\mathcal{L}_{H^\pm}^{\text{Yuk}} \supset y_e \bar\nu_{eL} H^+ e_R + \text{h.c.}. \tag{36.6}$$

With an order one Yukawa coupling to get $m_e = 0.51$ MeV, we need $v_2 \sim$ MeV which is large enough to reduce the neutrino Yukawas to be $\sim 10^{-6}$. Thus the hierarchy is re-introduced in the Yukawa couplings to accommodate mass of ν, e together. If the neutrinos are given mass through the SM Higgs mechanism, we need a hierarchy of $\mathcal{O}(10^{12})$ in the Yukawas. In the standard ν2HDM, the neutrinos get mass from a second doublet, demanding only a hierarchy of $\mathcal{O}(10^6)$. In the electrophilic ν2HDM, a hierarchy of $\mathcal{O}(10^6)$ in Yukawa couplings explains the mass of the electron and neutrinos with $v_2 \approx$ MeV. Another hierarchy $\mathcal{O}(10^3)$ arranges for the masses of the rest of the fermions with $v_1 \approx 246$ GeV. Overall hierarchy reduces being distributed into two sectors. In modified ν2HDM, electrophilic BSM scalar couplings arise where H (A) couples as

$$\mathcal{L}_{LP} \supset \frac{y_e}{\sqrt{2}} H e_L e_R + i \frac{y_e}{\sqrt{2}} A e_L e_R + \text{h.c.}. \qquad (36.7)$$

36.3 Rise of NSI in Electrophilic ν2HDM

The t-channel process of Fig. 36.1 through the charged Higgs propagator generates non-standard interaction (NSI) terms between SM neutrinos and electrons. After integrating out the heavy charged Higgs, the effective Lagrangian looks like

$$\mathcal{L}_{\text{eff}} \supset \frac{y_e^2}{4 \, m_{H^\pm}^2} (\bar{\nu}_{eL} \gamma^\rho \nu_{eL}) \, (\bar{e}_R \gamma_\rho e_R) + \text{h.c.} \qquad (36.8)$$

Comparing this effective Lagrangian with the defined form of the NSI Lagrangian, (36.1), the NSI parameter ϵ_{ee} is written as

$$\epsilon_{ee} = \frac{1}{2\sqrt{2} G_F} \frac{y_e^2}{4 \, m_{H^\pm}^2}. \qquad (36.9)$$

In this model, only the right-handed electron contributes in (36.8) to provide $\epsilon_{ee} = \epsilon_{ee}^{eR}$. With the same $U(1)$ quantum number assignment, the extra relevant terms for the lepton sector that can be added are (Fig. 36.2):

$$\mathcal{L}_{\nu 2\text{HDM}}^m \supset y_1 \bar{L}_\mu \Phi_2 e_R + y_2 \bar{L}_\tau \Phi_2 e_R + \text{h.c.}, \qquad (36.10)$$

where $L_{\mu/\tau}$ are the SM lepton doublets $(\nu_{\mu/\tau} \mu/\tau)_L^T$. These terms will provide the NSI parameters $\epsilon_{e\mu}$, $\epsilon_{e\tau}$, $\epsilon_{\mu\tau}$, and $\epsilon_{\mu\mu}$. The complex nature of the Yukawa couplings y_1, y_2 results in phases in the NSI parameters.

Fig. 36.1 The Feynman diagram contributing to the NSI in the electrophilic ν2HDM

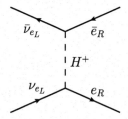

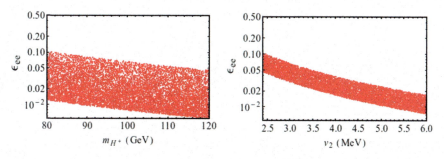

Fig. 36.2 Allowed range of ϵ_{ee}, with its distribution with charged Higgs mass, $m_{H\pm}$ and Φ_2 vev v_2

36.4 Constraints on Electrophilic ν2HDM

Along with the LEP and LFV constraints which are of more relevance due to Φ_2 being coupled only to the leptons, constraints from $(g - 2)$ as well as BBN are also discussed.

36.4.1 Constraints from LEP

LEP Γ_Z measurement allows no Z decays to BSM scalars, pushing their masses to $\geq m_Z/2$. The LEP charged Higgs search in $e^+e^- \to Z \to H^+H^-$ with $H^\pm \to \tau\nu$ limits the charged Higgs mass as $m_{H\pm} > 80$ GeV [7].

We recast LEP measurement of $e^+e^- \to e^+e^-$ cross section as a limit on an effective four-electron interaction, with its scale $\lambda > 9.1$ TeV [8]. This process $e^+e^- \to e^+e^-$ happens through both the CP-even (H) and CP-odd (A) scalars of this model to give rise to effective four-lepton operator that, for a global $U(1)$-symmetric ν2HDM with the degenerate CP-even and CP-odd scalars, translates to

$$\mathcal{L}_{\text{eff}} \supset \frac{y_e^2}{4\,m_H^2}(\bar{e}_L\gamma^\rho e_L)(\bar{e}_R\gamma_\rho e_R),$$

which is compared to the earlier effective operator to provide the bound on y_e as $y_e^2 \leq 8\pi m_H^2/\Lambda^2$. In global $U(1)$ symmetric electrophilic ν2HDM, the oblique parameters (S, T) allow the largest charged scalar-neutral scalar mass splitting for $m_{H\pm} \sim 100$ GeV along with neutral scalars of several hundred GeVs. We take $m_H = m_A = 500$ GeV. A large $m_H^\pm$, which decreases the mass splitting, constrains the Yukawa coupling tightly to reduce the NSI parameters to tiny values. This constraint translates to $y_e \leq 0.28$ which further puts the bound on Φ_2 vev as $v_2 \geq 2.5$ MeV. We fix the v_2 at 2.5 MeV and use LEP $e^+e^- \to \mu^+(\tau^+)\mu^-(\tau^-)$ measurement to get tight limits on other Yukawas y_1, y_2.

Another constraint from the LEP dark matter search in the mono-photon signal $e^+e^- \to$ DM DM γ. Mono-photon processes like $e^+e^- \to \nu_{e/\tau}\nu_{e/\tau}\gamma$ can occur in electrophilic ν2HDM through the charged Higgs exchange. The limit from LEP DM search is recast as

$$y_e^4 + 2y_e^2 y_2^2 + y_2^4 \leq \frac{16m_{H^{\pm}}^4}{\Lambda_{DM}^4}, \tag{36.11}$$

with DM scale $\Lambda_{DM} \approx 320$ GeV [9] for light DM particles.

36.4.2 Lepton Flavor Violation Constraints

The strongest bound on lepton flavor violating decay comes from the MEG experiment which gives the upper limit as $\text{BR}(\mu \to e\gamma) < 5.7 \times 10^{-13}$ [10]. Similar bounds on the other LFV decay channels ($\tau \to e\gamma$, $\tau \to \mu\gamma$) are weaker. Though the LFV processes like $\mu \to e\gamma$ are loop-driven ones, strong experimental constraints are there. The branching ratio for this charged Higgs mediated process is

$$\text{BR}(\mu \to e\gamma) = \text{BR}(\mu \to e\bar{\nu}\nu)\frac{\alpha_{\text{EM}}}{192\pi}|\langle m_{\mu e}^2 \rangle|^2 \rho^2, \tag{36.12}$$

where $\rho = (G_F m_{H^{\pm}}^2 v_2^2)^{-1} \lesssim 1.2\,\text{eV}^{-2}$, which translates to a limit such that $v_2 \lesssim 1$ eV for $m_H^{\pm} \gtrsim 250$ GeV. This is the tightest limit up to now on the v_2 and $m_{H^{\pm}}$ parameter space.

Due to the non-zero vev of Φ_2 in ν2HDM, lepton mass eigenstates e, μ, and τ mix at the tree level, resulting in LFV decays like $\tau \to 3e$ and $\mu \to 3e$ through the neutral scalar (H, A) mediation. The Yukawa couplings y_1 and y_2 get tight bounds from these. From the Belle measurement [11], we get

$$\frac{\Gamma(\tau \to 3e)}{\Gamma(\tau \to \mu\nu_\mu\nu_\tau)} \leq 1.58 \times 10^{-7},$$

which implies $y_e y_2 \leq \frac{(0.16\,m_H)^2}{(1\text{TeV})^2}$. This constrains $y_e - y_2$ plane tighter that the LEP constraints. $\text{BR}(\mu \to 3e) \leq 1 \times 10^{-12}$ put stronger bounds on y_e-y_1 plane as $y_e y_1 \leq \frac{(8.12 \times 10^{-3}\,m_H)^2}{(1\text{TeV})^2}$. For $y_e \sim 0.5$ with allowed m_H values, this bound reduces $y_1 \sim 10^{-6}$ which renders any NSI parameter involving y_1 insignificant (Fig. 36.3).

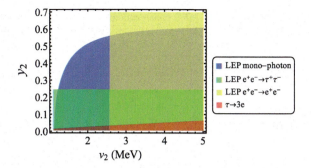

Fig. 36.3 Allowed region on the $y_2 - v_2$ plane for different LEP constraints. Here $m_H, m_A = 500$ GeV, $m_{H^\pm} \sim 100$ GeV. This is the maximal mass difference allowed from S, T, U and that maximal mass difference is crucial to generate large NSI values

36.4.3 g − 2 and BBN Constraints

The one-loop contribution to $(g-2)_{\mu/e}$ is negligible due to a suppression factor $m_l^4/m_{H^\pm}^2$ in the amplitude [6], along with the dependence on $y_1 \sim 10^{-6}$. Instead of significant two-loop contributions in 2HDM, $g-2$ constraints are negligible in ν2HDM due to tiny charged lepton couplings to H, A.

The right-handed neutrinos are very light, with mass $\sim eV$. These neutrinos can populate the early Universe through the charge scalar mediated process $\bar{l}l \to \nu_R \nu_R$. Big bang neucleosynthesis (BBN) limits new relativistic degrees of freedom as $\Delta N_{\text{eff}} \equiv N_{\text{eff}} - 3.046 = 0.10^{+0.44}_{-0.43}$ [12] that translates to $(T_{d,\nu_R}/T_{d,\nu_L})^3 \approx (\sigma_L/\sigma_R) = 4(v_2 m_{H^+}/(v_1 m_{\nu_i}|U_{li}|))^4$ [5] with decoupling temperatures $T_{d,\nu_R} \approx 200$ MeV and $T_{d,\nu_L} \approx 3$ MeV. This puts a bound on the neutrino Yukawa coupling y_{ν_i} as

$$y_{\nu_i} \leq 0.05 \times \left[\frac{m_{H^\pm}}{100 \text{ GeV}}\right]\left[\frac{1/\sqrt{2}}{|U_{ei}|}\right]. \tag{36.13}$$

This constraint is relaxed for electrophilic ν2HDM as $v_2 \sim 0.1$ MeV reduces y_{ν_i} values to be much smaller.

References

1. L. Wolfenstein, Phys. Rev. D **17**, 2369–2374 (1978)
2. T. Ohlsson, Rept. Prog. Phys. **76**, 044201 (2013). arXiv:1209.2710 [hep-ph]
3. O.G. Miranda, H. Nunokawa, New J. Phys. **17**(9), 095002 (2015). arXiv:1505.06254 [hep-ph]
4. U.K. Dey, N. Nath, S. Soumya, Phys. Rev. D98 **5**, 055004 (2018). arXiv:1804.05808 [hep-ph]
5. S.M. Davidson, H.E. Logan, Phys. Rev. D80, 095008 (2009). arXiv:0906.3335 [hep-ph]
6. P.A.N. Machado, Y.F. Perez, O. Sumensari, Z. Tabrizi, R. Funchal, Zukanovich. JHEP 12", 160 (2015). arXiv:1507.07550 [hep-ph]
7. G. Abbiendi, others, LEP collaboration. Eur. Phys. J. C73 (2013). arXiv:1301.6065 [hep-ex]
8. SLD Electroweak collaboration (2003). arXiv:0312023 [hep-ex]
9. P.J. Fox, R. Harnik, J. Kopp, Y. Tsai, Phys. Rev. D84, 014028 (2011). arXiv:1103.0240 [hep-ph]

10. J. Adam, others, MEG collaboration Phys. Rev. Lett. 110, 201801 (2013). arXiv:1303.0754 [hep-ex]
11. K. Hayasaka, others, Phys. Lett., B687, 139-143 (2010). arXiv:1001.3221 [hep-ex]
12. P.A.R. Ade, others, Planck collaboration, Astron. Astrophys., 594, A13 (2016). arXiv:1502.01589 [hep-ph]

Chapter 37
Search for Supersymmetry with a Compressed Mass Spectrum in the Vector-Boson Fusion Topology with Single Hadronic Tau Channel in Pp Collisions at P$\sqrt{s}$ = 13 TeV

Priyanka Kumari

Abstract A search for supersymmetric particles produced in the vector-boson fusion topology is presented. The search targets final states with one lepton, large missing transverse momentum (p_T^{miss}), and two jets with large separation in rapidity. The data sample corresponds to an integrated luminosity of 35.9 fb^{-1} of proton-proton collisions at $\sqrt{s} = 13$ TeV collected with the CMS detector during 2016 at the CERN LHC. The observed dijet invariant mass and transverse mass of lepton and missing transverse energy ($m_T(l, E_T^{\text{miss}})$) spectra are found to be consistent with the expected standard model predictions. Upper limits are set on the cross sections for chargino and neutralino production with two associated jets, assuming the supersymmetric partner of the lepton to be the lightest slepton and the lightest slepton to be lighter than the charginos. For a so-called compressed mass spectrum scenario, in which the mass difference between the lightest supersymmetric particle, $\tilde{\chi}_1^0$, and the next lightest, mass-degenerate, gaugino particle $\tilde{\chi}_2^0$ and $\tilde{\chi}_1^\pm$ is 30 GeV, a mass limit is set for these latter two particles, which are the most stringent limits to date.

37.1 Introduction

The Large Hadron Collider (LHC) machine has proved its remarkable performance with the discovery of Standard Model (SM) long-sought last particle "Higgs Boson" in July 2012, which is considered as the biggest success for the SM as well as for the CMS [1] and ATLAS [2] experiments. But there are few unsolved problems which SM cannot explain like the unification of forces, neutrino oscillations, matter-antimatter asymmetry, dark matter, *etc*. To solve such problems, various extensions of the SM have been developed by the physicists. One of them is known as Supersymmetry

P. Kumari (✉)
Panjab University, Chandigarh, India
e-mail: pri.nia293@gmail.com

© Springer Nature Singapore Pte Ltd. 2021
P. K. Behera et al. (eds.), *XXIII DAE High Energy Physics Symposium*,
Springer Proceedings in Physics 261,
https://doi.org/10.1007/978-981-33-4408-2_37

(SUSY) [3–5] that associates every SM fermion with its "super-partner" boson and vice versa. SUSY remains one of the best motivated of possible theories to explain simultaneously the particle nature of dark matter (DM) and solves the gauge hierarchy problem of the standard model (SM). However, for all of its attractive features, there is as yet no direct evidence of its existence. The strongly produced gluinos ($\tilde{g}$), as well as the squarks ($\tilde{q}$) of the first and second generations, have been ruled out below ~ 2 TeV in certain scenarios [6]. On the other hand, the limit is weaker on the masses of weakly produced charginos ($\tilde{\chi}_i^{\pm}$) and neutralinos ($\tilde{\chi}_i^0$), as may be expected at the CERN LHC where these particles suffer from much smaller production cross sections. The chargino-neutralino sector of SUSY plays a significant role in DM connection of SUSY models. The lightest neutralino $\tilde{\chi}_1^0$ is, as the Lightest Supersymmetric Particle (LSP), the canonical DM candidate in R-parity conserving SUSY extensions of the SM [7].

Various analyses on SUSY searches have been performed by CMS and ATLAS covering most of the parametric space. However, in this paper, we will present a search for the electroweak production of SUSY particles in the Vector-Boson Fusion (VBF) topology using data collected with the CMS detector in 2016 and corresponding to an integrated luminosity of 35.9 fb^{-1} of proton-proton collisions at a center-of-mass energy of $\sqrt{s} = 13$ TeV.

37.2 Analysis Strategy Through Vector-Boson Fusion Processes

A classic strategy to search for charginos and neutralinos is through cascade decays of heavier colored particles such as gluinos and squarks. Although a procedure must be developed to directly probe non-colored sectors where colored objects are heavy and production cross section is small. However, these searches are experimentally difficult in cases where the mass of the LSP is only slightly less than the masses of other charginos and neutralinos, making these so-called compressed spectrum scenarios important search targets using new techniques. So, to address these fascinating compressed SUSY scenarios, chargino and neutralino production via VBF processes of order α_{EW}^4 is very useful [8].

The charginos and neutralinos are produced in association with two forward jets, in the opposite hemisphere of a detector having large dijet invariant mass and large pseudorapidity. These pair-produced charginos and neutralinos further decay to sleptons ($\tilde{l}$). The sleptons subsequently decay to the leptons ($\tau_h/e/\mu$) and lightest neutralinos ($\tilde{\chi}_1^0$) which are LSP. The VBF jet and Missing Transverse Energy (MET) requirement reduces the background rate by a factor of $10^{-2} - 10^{-4}$. Figure 37.1 shows the Feynman diagrams for two of the possible VBF production processes: chargino-neutralino and chargino-chargino production. The SUSY search with 13 TeV data taken in 2016, presented in this paper, was performed on four final states, namely, $0ljj$, ejj, μjj, and $\tau_h jj$, where τ_h denotes a hadronically decaying τ lepton. The main focus of this paper will be on the τ_h channel.

Fig. 37.1 Feynman diagrams of (left) chargino-neutralino and (right) chargino-chargino pair production through vector-boson fusion, followed by their decays to leptons and the LSP $\tilde{\chi}_1^0$ via a light slepton

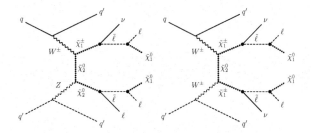

The event selection criteria for selecting signal events are divided into two parts: Central selections and VBF jet selections. Events are selected using a MET trigger with a threshold of 120 GeV on both $p_{T,trig}^{\text{miss}}$ and $H_{T,trig}^{\text{miss}}$. While the compressed mass spectrum SUSY models considered in this analysis results in final states with multiple leptons, the compressed mass spectra scenarios of interest also result in low-pt visible decay products, making it difficult to reconstruct and identify multiple leptons. For this reason, events are required to have zero or exactly one well-identified soft lepton. In the $\tau_h jj$ channel, an additional lepton veto is applied by rejecting events containing a second tau ($pt > 20\,\text{GeV}$), an electron ($pt > 10\,\text{GeV}$), or a muon candidate ($pt > 8$ GeV). Similarly, ejj and μjj channel events are required not to contain another electron, muon, or τ_h candidate. For $0ljj$ channel, a well-identified electron, muon, or τ_h candidate is rejected. The veto on additional leptons maintains high efficiency for compressed mass spectra scenarios and simultaneously reduces the SM backgrounds. Muon, electron, and τ_h candidates must have $8 < p_T < 40$ GeV, $10 < p_T < 40\,\text{GeV}$, and $20 < p_T < 40\,\text{GeV}$, respectively. The lower limit on τ_h p_T is larger because of the known fact that it is difficult to reconstruct low-p_T τ_h candidates, namely, that they do not produce a narrow jet in the detector, which makes them difficult to distinguish from quark or gluon jets. All leptons are required to have $|\eta| < 2.1$ in order to select high-quality and well-isolated leptons within the tracker acceptance. This requirement is 99% efficient for signal events. Lepton candidates are also required to pass the reconstruction and identification criteria, and they are well identified and isolated. In addition to the $0l$ or $1l$ selection, we impose the following requirements. The event is required to have $p_T^{\text{miss}} > 250$ GeV to suppress the DY$\rightarrow ll$ and QCD multijet backgrounds. Jets originating from the hadronization of bottom quarks (b-quark jets) are identified using the Combined Secondary Vertex (CSV) algorithm and are considered as b-jets with $p_T > 30$ GeV and $|\eta| < 2.4$. To reduce the $t\bar{t}$ background, we require the events not to have any jet identified as a b-quark jet. A cut on $m_T(l, E_T^{\text{miss}}) > 110$ GeV, i.e., beyond the Jacobian m_W peak to reduce the backgrounds coming from W boson. Up to here, we call it *Central Selections*.

The outgoing partons in the case of VBF SUSY signal processes have high p$_T$ as the VBF system is required to have enough momentum to produce the two SM vector-boson as well as pair of heavy SUSY particles (as shown in Fig. 37.1). Therefore, the *VBF selections* require the presence of two VBF jets with $p_T > 60$ GeV in opposite hemispheres ($\eta_1 \cdot \eta_2 > 0$), large pseudorapidity gap $|\Delta\eta| > 3.8$ and $|\eta| <$

5.0. Only jets separated from the leptons by $\Delta R > 0.4$ are considered. The VBF dijet candidates with the largest dijet mass are chosen and are required to have $m_{jj} > 1$ TeV.

The Signal Region (SR) is defined as the events that satisfy the central and VBF selection criteria.

37.3 Background Estimation

All the final states use the same general methodology for the background estimation in the SR and are based on both simulation and data. Background-Enriched Control Regions (CR) are constructed with some modification in nominal selections orthogonal to those of SR. These CRs are defined to measure the efficiencies of the VBF and central selections, to determine the correction factors to account for these efficiencies and to derive the shapes of the m_T and m_{jj} from data when possible or from Monte Carlo samples in such a way that distributions remain unbiased. The correction factors are determined by assessing the level of agreement in the yields between data and simulation. For both CR and SR, the same trigger is used for all final states.

One b-quark jet in addition to SR selections is used to construct $t\bar{t}$ CR, making sure that those control samples contain the same kinematics and composition of fakes as the SR so that the correction factors are not biased and can be used to correct the prediction from the simulation in the signal region. The $t\bar{t}$ background yields in the $1ljj$ channels are evaluated using the following equation:

$$N_{t\bar{t}}^{\text{pred}} = N_{t\bar{t}}^{\text{MC}} SF_{t\bar{t}}^{\text{CR}}, \tag{37.1}$$

where $N_{t\bar{t}}^{\text{pred}}$ is the predicted $t\bar{t}$ background (BG) yield in the signal region, $N_{t\bar{t}}^{\text{MC}}$ is the $t\bar{t}$ rate predicted by simulation for the SR selection, and $SF_{t\bar{t}}^{\text{CR}}$ is the data-to-simulation correction factor, given by the ratio of observed data events to the $t\bar{t}$ yield in simulation, measured in a $t\bar{t}$-enriched CR. The measured data-to-simulation correction factors $SF_{t\bar{t}}^{\text{CR}}$ are 0.8 ± 0.3, 0.8 ± 0.2, and 1.3 ± 0.5 for the $e$jj, μjj, and $\tau_h$jj channels, within statistical uncertainty, respectively. Since the m_T shape between data and MC is in well agreement within statistical uncertainty, the shapes for $t\bar{t}$ in the signal region are directly taken from the simulation.

The W+jets plays a significant role in the SR for $1ljj$ channels. The efficiency for the central selections is expected to be relatively well modeled by simulation (i.e., clean well-identified real lepton plus real E_T^{miss} from the W decay to a neutrino). Mismodeling of the W + jets background rate and shapes in the signal region is expected to come from the VBF selections and therefore evaluated by using two control regions CR1 and CR2 given in the 37.2. The CR1 is defined to measure a data-to-simulation scale factor to correct the modeling of all central selections and CR2 is defined to measure the VBF efficiency and shapes directly from data.

$$N_{\text{BG}}^{\text{pred}} = N_{\text{BG}}^{\text{MC}} \, SF_{\text{BG}}^{\text{CR1}} \, (\text{central}) \, SF_{\text{BG}}^{\text{CR2}} (\text{VBF}) \tag{37.2}$$

where $N_{\text{BG}}^{\text{pred}}$ is the predicted BG yield in the SR, $N_{\text{BG}}^{\text{MC}}$ is the rate predicted by simulation for the SR selection, $SF_{\text{BG}}^{\text{CR1}}$ (central) is the data-to-simulation correction factor for the central selection, given by the ratio of data to the BG simulation in control region CR1, and $SF_{\text{BG}}^{\text{CR2}}$ (VBF) the data-to-simulation correction factor for the efficiency of the VBF selections from CR2. CR1 is obtained with similar selections to the SR, except that the *VBF requirement is inverted* (0 dijet candidates passing all VBF cuts). Inverting the VBF requirement enhances the W+jets background yield by two orders of magnitude while suppressing the VBF signal contamination to negligible levels. To measure the VBF efficiency $\epsilon_{\text{W + jets}}$ (VBF cuts), we defined Z + jets-enriched events region (CR2) by taking the advantage of a relatively small difference in mass between W and Z bosons (compared to the energy scale of the search region). In CR2, VBF selections are also applied. The correction factors $SF_{\text{W+jets}}^{\text{CR1}}$ (central) obtained for $e$jj and μjj channels are 0.97 ± 0.10 and 1.10 ± 0.10 within statistical uncertainty, respectively.

For the $\tau_h$jj channel, it is difficult to obtain a control sample enriched in W+jets events because there is a significant contribution from QCD multijet events. Therefore, the average of the correction factors obtained for the $e$jj and μjj channels, 1.04 ± 0.13, are used to scale the W + jets prediction from the simulation in the $\tau_h$jj channel. This approach is justified since the $W(\to \tau\nu)$+jets prediction from simulation is corrected to account for slight differences in the τ_h identification efficiency observed in data. The correction factor $SF_{\text{W+jets}}^{\text{CR2}}$ (VBF) determined from the CR2 control sample is measured to be 1.18 ± 0.09.

The QCD multijet background plays a major role in the VBF $jj + \tau_h + E_T^{\text{miss}}$ and 0ℓjj channel and has a significant contribution in SR. To differentiate between QCD BG and signal, the discriminating variables are VBF selection criteria, the minimum separation between $\vec{E}_T^{\text{miss}}$ and any jet $|\Delta\phi_{\text{min}}(\vec{E}_T^{\text{miss}}, j)|$, and τ_h isolation. Therefore, to estimate the QCD multijet background, CRs are constructed by inverting these requirements. In the $\tau_h$jj channel, we estimate the QCD multijet background using a completely data-driven approach that relies on the matrix ("ABCD") method. The regions are defined as follows:

- CRA: inverted VBF selection, pass the nominal (tight) τ_h isolation;
- CRB: inverted VBF selection, fail the nominal τ_h isolation but pass loose τ_h isolation;
- CRC: pass the VBF selection, fail the nominal τ_h isolation but pass loose τ_h isolation; and
- CRD: pass the VBF selection, pass the nominal τ_h isolation.

We estimate the QCD multijet component N_{QCD}^i in regions $i = \text{CR}A, \text{CR}B, \text{CR}C$ by subtracting non-QCD backgrounds (predicted using simulation) from data ($N_{\text{QCD}}^i = N_{\text{Data}}^i - N_{\neq \text{QCD}}^i$). We then estimate the QCD multijet component in CRD (i.e., the SR) to be $N_{\text{QCD}}^{SR} = N_{\text{QCD}}^{\text{CR}A} \frac{N_{\text{QCD}}^{\text{CR}C}}{N_{\text{QCD}}^{\text{CR}B}}$, where $N_{\text{QCD}}^{\text{CR}C}/N_{\text{QCD}}^{\text{CR}B}$ is referred to as the "Pass-to-Fail VBF"

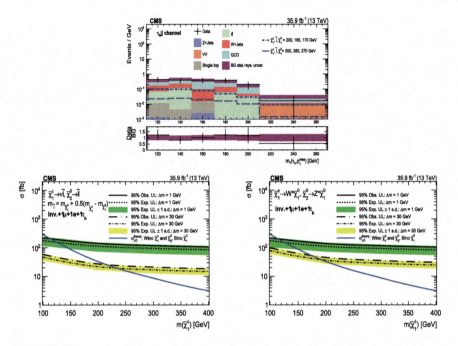

Fig. 37.2 The top plot shows the observed m_T distribution in the signal region for $\tau_h jj$, bottom left and right plots show the expected and observed 95% confidence level upper limit (UL) on the cross section as a function of $m_{\tilde{\chi}_1^\pm} = m_{\tilde{\chi}_2^0}$ for $\Delta m = 1$ and 30 GeV mass gaps between the chargino and the neutralino, assuming the light slepton model and W* and Z* model, respectively [9]

transfer factor (TF_{VBF}). The shape of the $m_T(\tau_h, p_T^{\mathrm{miss}})$ distribution is obtained from $\mathrm{CR}B$ (from the non-isolated τ_h plus inverted VBF control sample). This "ABCD" method relies on TF_{VBF} being unbiased by the τ_h isolation requirement.

37.4 Results and Conclusion

Figure 37.2 (top plot) shows the predicted SM background, expected signal, and observed data rates in bins of m_T for τ_h final state. The bin sizes in the distributions are chosen to maximize the signal significance of the analysis. No excess of events above the SM prediction in any of the search regions is observed. Therefore, the search does not reveal any evidence for new physics. For the R-parity conserving MSSM models, results are simplified in two scenarios: (*i*) the "light slepton" model where $\tilde{\ell}$ is the next-to-lightest SUSY particle and (*ii*) the "WZ" model where sleptons are too heavy and thus $\tilde{\chi}_1^\pm$ and $\tilde{\chi}_2^0$ decays proceed via W* and Z*.

The main difference between the two models is the branching ratio of $\tilde{\chi}_1^\pm$ and $\tilde{\chi}_2^0$ to leptonic final states. The bottom left distribution of Fig. 37.2 shows that for a compressed mass spectrum scenario, in which $\Delta m = m(\tilde{\chi}_1^\pm) - m(\tilde{\chi}_1^0) = 1\ (30)$ GeV and in which $\tilde{\chi}_1^\pm$ and $\tilde{\chi}_2^0$ branching fractions to light sleptons are 100%, $\tilde{\chi}_1^\pm$ and $\tilde{\chi}_2^0$ masses up to 112 (215) GeV are excluded at 95% CL. The bottom right plot shows that for the scenario where the sleptons are too heavy and decays of the charginos and neutralinos proceed via W^* and Z^* bosons, $\tilde{\chi}_1^\pm$ and $\tilde{\chi}_2^0$ masses up to 112 (175) GeV are excluded at 95% CL for $\Delta m = m(\tilde{\chi}_1^\pm) - m(\tilde{\chi}_1^0) = 1\ (30)$ GeV. This analysis obtains the most stringent limits to date on the production of charginos and neutralinos decaying to leptons in compressed mass spectrum scenarios defined by the mass separation $1 < \Delta m < 5$ GeV and $25 < \Delta m < 50$ GeV.

References

1. The CMS Collaboration, The CMS experiment at the CERN LHC. JINST **3**, S08004 (2008)
2. The ATLAS Collaboration, G. Aad et al., The ATLAS Experiment at the CERN Large Hadron Collider. JINST **3**, S08003 (2008)
3. P. Ramond, Dual theory for free fermions. Phys. Rev. D **3** (1971). https://doi.org/10.1103/PhysRevD.3.2415
4. J. Wess, B. Zumino, Supergauge transformations in four dimensions. Nucl. Phys. B **70** (1974). https://doi.org/10.1016/0550-3213(74)90355-1
5. L. Hall, J. Lykken, S. Weinberg, Supergravity as the messenger of supersymmetry breaking. Phys. Rev. D **27** (1983). https://doi.org/10.1103/PhysRevD.27.2359
6. The CMS Collaboration, Search for new phenomena with the M_{T2} variable in the all-hadronic final state produced in proton-proton collisions at $\sqrt{s} = 13$ TeV. Eur. Phys. J. C **77** (2017). https://doi.org/10.1140/epjc/s10052-017-5267-x
7. G.R Farrar, F. Pierre, Phenomenology of the production, decay, and detection of new hadronic states associated with supersymmetry. Phys. Lett. B **76** (1978). https://doi.org/10.1016/0370-2693(78)90858-4
8. A.G. Delannoy, B. Dutta, A. Gurrola, W. Johns, T. Kamon, E. Luiggi, A. Melo, P. Sheldon, K. Sinha, K. Wang, Probing dark matter at the LHC using vector boson fusion processes. Phys. Rev. Lett. **111** (2013). https://doi.org/10.1103/PhysRevLett.111.061801
9. The CMS Collaboration, Search for supersymmetry with a compressed mass spectrum in the vector boson fusion topology with 1-lepton and 0-lepton final states in proton-proton collisions at $\sqrt{s} = 13$ TeV. JHEP **08** (2019). https://doi.org/10.1007/JHEP08(2019)150

Chapter 38
Alignment in $A4$ Symmetric Three-Higgs-Doublet Model

Soumita Pramanick and Amitava Raychaudhuri

Abstract A model with three $SU(2)_L$ Higgs doublets transforming as a triplet under $A4$ was considered. It was shown that the alignment follows as a natural consequence of the discrete symmetry $A4$ without any fine-tuning for all the four global minima configurations. It was verified that the results were well in agreement with unitarity and positivity criteria.

38.1 Introduction

This talk is based on [1]. An $A4$ triplet comprising of three $SU(2)_L$ doublet scalars has been considered. We ensure $A4$ symmetry is preserved in the scalar potential. The terms conserving $A4$ symmetry were allowed only in the scalar potential and no soft breaking terms were entertained. It was found, owing to this $A4$ symmetry,[1] all the global minima of this scalar potential correspond to alignment of the vacuum [1]. No fine-tuning was needed for this. Needless to mention, such a scalar potential will lead to 3×3 charged scalar, neutral real scalar and neutral pseudoscalar mass-square matrices. After diagonalization, one of the modes of the neutral pseudoscalar and charged scalar matrices will be massless corresponding to the neutral Goldstone and charged Goldstone. Alignment implies that the above multiplet constitutes a mass eigenstate and the corresponding direction of the neutral real scalar mass-square matrix, in the mass-basis is the one that can be considered to be the analogue of the standard model (SM) Higgs boson. For each of the global minima configurations,

---

[1] Brief account of $A4$ group can be found in [1, 2].

S. Pramanick · A. Raychaudhuri
Department of Physics, University of Calcutta, 92 Acharya Prafulla Chandra Road,
Kolkata 700009, India
e-mail: palitprof@gmail.com

S. Pramanick (✉)
Harish-Chandra Research Institute, Chhatnag Road, Jhunsi, Allahabad 211019, India
e-mail: soumita509@gmail.com

© Springer Nature Singapore Pte Ltd. 2021
P. K. Behera et al. (eds.), *XXIII DAE High Energy Physics Symposium*,
Springer Proceedings in Physics 261,
https://doi.org/10.1007/978-981-33-4408-2_38

the mass-eigenstate basis of the physical scalars were same as the 'Higgs-basis' in which the vacuum expectation value (*vev*) was rotated to only one of three members of the scalar multiplet in the direction corresponding to that of the SM Higgs. This feature is studied case by case for all the four global minima configurations, and the viability of the results are tested in the light of the constraints put on the quartic couplings by demanding positivity and s-wave unitarity of the scalar potential in [1]. It is worth mentioning that this is a toy model. Models leading to realistic masses and mixing of quarks and leptons using these scalar fields are not studied in this work. Neutrino mass models based on $A4$ can be found in [2].

38.2 Objective

We consider, three colour singlet $SU(2)_L$ doublet scalar fields with hypercharge $Y = 1$. Together these three scalars form a triplet of $A4$. Let us represent these scalars collectively as

$$\Phi \equiv \begin{pmatrix} \Phi_1 \\ \Phi_2 \\ \Phi_3 \end{pmatrix} \equiv \begin{pmatrix} \phi_1^+ & \phi_1^0 \\ \phi_2^+ & \phi_2^0 \\ \phi_3^+ & \phi_3^0 \end{pmatrix} . \tag{38.1}$$

Here $A4$ is acting vertically, $SU(2)_L$ is acting horizontally. The neutral fields can be written in terms of scalar and pseudoscalar components as $\phi_i^0 = \phi_i + i\chi_i$.

After spontaneous symmetry breaking (SSB) the neutral components develop *vev*. The following four *vev* configurations correspond to global minima [3, 4] of the scalar potential to be studied in the following section:

$$\langle\Phi\rangle_1 = \frac{v}{\sqrt{2}} \begin{pmatrix} 0 & 1 \\ 0 & 0 \\ 0 & 0 \end{pmatrix} , \quad \langle\Phi\rangle_2 = \frac{v}{2} \begin{pmatrix} 0 & 1 \\ 0 & e^{i\alpha} \\ 0 & 0 \end{pmatrix} , \quad \langle\Phi\rangle_3 = \frac{v}{\sqrt{6}} \begin{pmatrix} 0 & 1 \\ 0 & 1 \\ 0 & 1 \end{pmatrix} , \quad \langle\Phi\rangle_4 = \frac{v}{\sqrt{6}} \begin{pmatrix} 0 & 1 \\ 0 & \omega \\ 0 & \omega^2 \end{pmatrix} , \tag{38.2}$$

where $v = v_{SM} \sim 246$ GeV.

To show that each of these four *vev* configurations automatically satisfy exact alignment is the goal of this enterprise [1]. From [5], we know there exists a unitary transformation U which leads to

$$U \begin{pmatrix} \Phi_1 \\ \Phi_2 \\ \Phi_3 \end{pmatrix} = \Psi \equiv \begin{pmatrix} \Psi_1 \\ \Psi_2 \\ \Psi_3 \end{pmatrix} \equiv \begin{pmatrix} \psi_1^+ & \psi_1^0 \\ \psi_2^+ & \psi_2^0 \\ \psi_3^+ & \psi_3^0 \end{pmatrix} , \tag{38.3}$$

where in the Ψ basis any one of the component acquires *vev*, i.e., $\langle\psi_i^0\rangle \neq 0$ and $\langle\psi_j^0\rangle = 0$ for $j \neq i$. In other words the unitary transformation U transports us to the 'Higgs basis'. Simultaneously, the components of Ψ_i, viz., ψ_i^+ and $\psi_i^0 \equiv \eta_i^0 + i\xi_i^0$ serve as the mass eigenstate with massless neutral states and charged states and a massive neutral state. Combinations of the other states Ψ_j ($j \neq i$) produce the other mass eigenstates. Thus Ψ_i can be treated analogous to the SM Higgs doublet. The notation we use for the physical scalar mass-square matrices in the Φ_i basis is given by

$$\mathcal{L}_{mass} = \frac{1}{2}(\chi_1\ \chi_2\ \chi_3)\ M^2_{\chi_i\chi_j}\begin{pmatrix}\chi_1\\\chi_2\\\chi_3\end{pmatrix} + \frac{1}{2}(\phi_1\ \phi_2\ \phi_3)\ M^2_{\phi_i\phi_j}\begin{pmatrix}\phi_1\\\phi_2\\\phi_3\end{pmatrix} + (\phi^-_1\ \phi^-_2\ \phi^-_3)\ M^2_{\phi^\mp_i\phi^\pm_j}\begin{pmatrix}\phi^+_1\\\phi^+_2\\\phi^+_3\end{pmatrix}$$

$$(38.4)$$

Here $M^2_{\chi_i\chi_j}$, $M^2_{\phi_i\phi_j}$ and $M^2_{\phi^\mp_i\phi^\pm_j}$ represents the pseudoscalar, neutral scalar and the charged scalar mass-square matrices. These matrices are diagonalized by the same U mentioned in (38.3). This was demonstrated case by case for each of the four global minima configurations in (38.2). To show that this alignment occurs as a natural consequence of the $A4$ symmetry and no fine-tuning is required, is the prime intent of this analysis [1].

38.3 The Scalar Potential, Positivity, Unitarity

The terms that conserve $A4$ symmetry and gauge symmetry are only allowed in the scalar potential. No terms breaking $A4$ symmetry even softly were allowed. This leads to the following potential for the scalar spectrum we have considered:

$$V(\Phi_i) = m^2\left(\sum_{i=1}^{3}\Phi^\dagger_i\Phi_i\right) + \frac{\lambda_1}{2}\left(\sum_{i=1}^{3}\Phi^\dagger_i\Phi_i\right)^2$$
$$+ \frac{\lambda_2}{2}\left(\Phi^\dagger_1\Phi_1 + \omega^2\Phi^\dagger_2\Phi_2 + \omega\Phi^\dagger_3\Phi_3\right)\left(\Phi^\dagger_1\Phi_1 + \omega\Phi^\dagger_2\Phi_2 + \omega^2\Phi^\dagger_3\Phi_3\right)$$
$$+ \frac{\lambda_3}{2}\left[\left(\Phi^\dagger_1\Phi_2\right)\left(\Phi^\dagger_2\Phi_1\right) + \left(\Phi^\dagger_2\Phi_3\right)\left(\Phi^\dagger_3\Phi_2\right) + \left(\Phi^\dagger_3\Phi_1\right)\left(\Phi^\dagger_1\Phi_3\right)\right]$$
$$+ \lambda_4\left[\left(\Phi^\dagger_1\Phi_2\right)^2 + \left(\Phi^\dagger_2\Phi_1\right)^2 + \left(\Phi^\dagger_2\Phi_3\right)^2 + \left(\Phi^\dagger_3\Phi_2\right)^2 + \left(\Phi^\dagger_3\Phi_1\right)^2 + \left(\Phi^\dagger_1\Phi_3\right)^2\right].$$

$$(38.5)$$

All the quartic couplings λ_i $(i = 1, \ldots 4)$ are taken to be real. Note in general, only λ_4 can be complex.

Using $1 + \omega + \omega^2 = 0$ we get

$$V(\Phi_i) = m^2\left(\sum_{i=1}^{3}\Phi^\dagger_i\Phi_i\right) + \frac{\lambda_1 + \lambda_2}{2}\left(\sum_{i=1}^{3}\Phi^\dagger_i\Phi_i\right)^2$$
$$- \frac{3\lambda_2}{2}\left[\left(\Phi^\dagger_1\Phi_1\right)\left(\Phi^\dagger_2\Phi_2\right) + \left(\Phi^\dagger_2\Phi_2\right)\left(\Phi^\dagger_3\Phi_3\right) + \left(\Phi^\dagger_3\Phi_3\right)\left(\Phi^\dagger_1\Phi_1\right)\right]$$
$$+ \frac{\lambda_3}{2}\left[\left(\Phi^\dagger_1\Phi_2\right)\left(\Phi^\dagger_2\Phi_1\right) + \left(\Phi^\dagger_2\Phi_3\right)\left(\Phi^\dagger_3\Phi_2\right) + \left(\Phi^\dagger_3\Phi_1\right)\left(\Phi^\dagger_1\Phi_3\right)\right]$$
$$+ \lambda_4\left[\left(\Phi^\dagger_1\Phi_2\right)^2 + \left(\Phi^\dagger_2\Phi_1\right)^2 + \left(\Phi^\dagger_2\Phi_3\right)^2 + \left(\Phi^\dagger_3\Phi_2\right)^2 + \left(\Phi^\dagger_3\Phi_1\right)^2 + \left(\Phi^\dagger_1\Phi_3\right)^2\right].$$

$$(38.6)$$

The form of the potential in (38.6) was used in the following analysis.

The potential in (38.6) has to be bounded from below. This puts constraints on the quartic couplings in (38.6). The 'copositivity' conditions for a general model with three $SU(2)_L$ doublet scalars are well studied in [6]. Modifying the results in [6] for $A4$ symmetric three-Higgs $SU(2)_L$ doublets case for real $vevs$, we have to look for the copositivity of the following matrix:

$$M_{cop} = \begin{pmatrix} \lambda_P & \lambda_Q & \lambda_Q \\ \lambda_Q & \lambda_P & \lambda_Q \\ \lambda_Q & \lambda_Q & \lambda_P \end{pmatrix} , \tag{38.7}$$

with

$$\lambda_P = (\lambda_1 + \lambda_2)/2 , \quad \lambda_Q = (2\lambda_1 - \lambda_2 + \lambda_3 + 4\lambda_4)/4 . \tag{38.8}$$

Thus one has to obey

$$\lambda_P \geq 0 , \quad \lambda_P + \lambda_Q \geq 0 , \text{ and } \sqrt{\lambda_P^3 + (3\lambda_Q)}\sqrt{\lambda_P} + \sqrt{2(\lambda_P + \lambda_Q)^3} \geq 0 . \tag{38.9}$$

Demanding $\lambda_P \geq 0$ and $\lambda_Q \geq -\frac{1}{2}\lambda_P$ is sufficient for the constraints in (38.9). This leads to

$$\lambda_1 + \lambda_2 \geq 0 , \quad 3\lambda_1 + \lambda_3 + 4\lambda_4 \geq 0 . \tag{38.10}$$

For complex vev, one has to look at the *positivity* matrix. For example, in case of $\langle\Phi\rangle_4$ in (38.2), where the phases associated with $\langle\Phi_1\rangle$, $\langle\Phi_2\rangle$, $\langle\Phi_3\rangle$ are $(0, 2\pi/3, 4\pi/3)$, the form of M_{cop} in (38.7) continues to be valid with the following replacement:

$$\lambda_Q \rightarrow \lambda_R = (2\lambda_1 - \lambda_2 + \lambda_3 - 2\lambda_4)/4 , \tag{38.11}$$

leading to

$$\lambda_1 + \lambda_2 \geq 0 , \quad 3\lambda_1 + \lambda_3 - 2\lambda_4 \geq 0 . \tag{38.12}$$

Unitarity for two $SU(2)_L$ doublet scalars was vividly studied in [7]. The results can be adopted for our case of $A4$ symmetric three-Higgs doublet model. In Table 38.1, the different scattering channels, the dimension of the scattering matrix and the eigenvalues are listed. Unitarity requires each of the eigenvalues to be bounded by $1/8\pi$. For example, the first row of Table 38.1 represents the matrix below:

$$8\pi S(1, 2)_{diag} = \begin{pmatrix} \lambda_1 & 2\lambda_4 & 2\lambda_4 \\ 2\lambda_4 & \lambda_1 & 2\lambda_4 \\ 2\lambda_4 & 2\lambda_4 & \lambda_1 \end{pmatrix} . \tag{38.13}$$

For a detailed discussion on the unitarity conditions of this model see [1]. With the unitarity and positivity constraints in hand, let us study alignment for each of the four global minima configurations in (38.2).

Table 38.1 Eigenvalues and dimension of the tree-level scattering matrices for different choices of $SU(2)_L$ and Y are shown. Apart from $SU(2)_L$ and Y properties the initial and final states have indices i, j corresponding to the two scalars participating in the scattering process belonging to Φ_i and Φ_j with $(i, j = 1, 2, 3)$. The processes with $i = j$ $(i \neq j)$ are represented by 'Diagonal' ('Off-diagonal') types. To satisfy unitarity constraints the eigenvalues shown in the last column has to be bounded by $1/8\pi$

Quantum numbers		Type	Matrix	Eigenvalues
$SU(2)_L$	Y		size	
1	2	Diagonal	3×3	$\|(\lambda_1 - 2\lambda_4)\|, \|(\lambda_1 + 4\lambda_4)\|$
1	2	Off-diagonal	3×3	$\|(\lambda_3 - 3\lambda_2)/2\|$
0	2	Off-diagonal	3×3	$\|(\lambda_3 + 3\lambda_2)/2\|$
1	0	Diagonal	3×3	$\|(\lambda_1 - \lambda_3/2)\|, \|(\lambda_1 + \lambda_3)\|$
1	0	Off-diagonal	6×6	$\|(3\lambda_2 + 4\lambda_4)/2\|, \|(3\lambda_2 - 4\lambda_4)/2\|$
0	0	Diagonal	3×3	$\|(6\lambda_1 + 6\lambda_2 - \lambda_3)/2\|, \|(3\lambda_1 - 6\lambda_2 + \lambda_3)\|$
0	0	Off-diagonal	6×6	$\|(-3\lambda_2 + 2\lambda_3 - 12\lambda_4)/2\|, \|(-3\lambda_2 + 2\lambda_3 + 12\lambda_4)/2\|$

38.4 Case 1: $\langle \phi_i^0 \rangle = \frac{v}{\sqrt{2}}(1, 0, 0)$

In this case $\langle \phi_1^0 \rangle = \frac{v}{\sqrt{2}}$ and $\langle \phi_2^0 \rangle = \langle \phi_3^0 \rangle = 0$. Note this is already in the 'Higgs basis' and the unitary transformation matrix U in (38.3) is simply identity matrix. The task we are left to do is to simply construct the physical scalar mass-square matrices which are expected to be diagonal as the unitary transformation matrix U is identity and identify the mode corresponding to the SM Higgs. The off-diagonal ij-th entry of the physical scalar mass-square matrices is proportional to the product $v_i v_j$ with $(i \neq j)$. The combination $v_i v_j$ with $(i \neq j)$ is always zero for this choice $\langle \phi_i^0 \rangle = \frac{v}{\sqrt{2}}(1, 0, 0)$, we get diagonal physical scalar mass-square matrices. The minimization equation given by

$$m^2 + \frac{v^2}{2}[\lambda_1 + \lambda_2] = 0 . \tag{38.14}$$

The physical scalar mass-square matrices are given by

$$M^2_{\phi_i^\mp \phi_j^\pm} = diag(0 , r_+ , r_+) \text{ where } r_+ = \frac{v^2}{4}(-3\lambda_2) . \tag{38.15}$$

$$M^2_{\chi_i \chi_j} = diag(0 , p , p) \text{ where } p = \frac{v^2}{4}(-3\lambda_2 + \lambda_3 - 4\lambda_4) . \tag{38.16}$$

$$M^2_{\phi_i \phi_j} = diag(q , r_0 , r_0) \text{ where } q = v^2(\lambda_1 + \lambda_2) , r_0 = \frac{v^2}{4}(-3\lambda_2 + \lambda_3 + 4\lambda_4) . \tag{38.17}$$

The physical scalar mass eigenvalues are positive consistent with the constraints put on the quartic couplings from the positivity and the unitarity constraints. Thus the first of the three directions corresponds to that of the SM Higgs boson in the neutral scalar mass-square matrix and the first direction for the charged scalar and the pseudoscalar mass-square matrix are the massless charged Goldstone and neutral Goldstone, respectively. Thus alignment has been achieved for this case.

Similar analysis was performed for vev configurations $\langle \phi_i^0 \rangle = \frac{v}{2}(1, e^{i\alpha}, 0)$ and $\langle \phi_i^0 \rangle = \frac{v}{\sqrt{6}}(1, 1, 1)$ and alignment could be achieved. Details can be found in [1].

38.5   Case 4: $\langle \phi_i^0 \rangle = \frac{v}{\sqrt{6}}(1, \omega, \omega^2)$

The minimization condition is

$$m^2 + \frac{v^2}{12}[3\lambda_1 + \lambda_3 - 2\lambda_4] = 0 . \tag{38.18}$$

The charged scalar mass-square matrix is

$$M^2_{\phi_i^\mp \phi_j^\pm} = \frac{v^2}{6}\begin{pmatrix} a & b & b^* \\ b^* & a & b \\ b & b^* & a \end{pmatrix} , \tag{38.19}$$

with $a = (2\lambda_4 - \lambda_3)/2$ and $b = (\omega^2\lambda_3 + 4\omega\lambda_4)/4$. The matrix in (38.19) is diagonalized by

$$U_3 = \frac{1}{\sqrt{3}}\begin{pmatrix} 1 & 1 & 1 \\ 1 & \omega^2 & \omega \\ 1 & \omega & \omega^2 \end{pmatrix} . \tag{38.20}$$

The eigenvectors are given by

$$\psi_1^\pm = (\phi_1^\pm + \phi_2^\pm + \phi_3^\pm)/\sqrt{3}, \quad \psi_2^\pm = (\phi_1^\pm = \omega\phi_2^\pm + \omega^2\phi_3^\pm)/\sqrt{3}, \quad \psi_3^\pm = (\phi_1^\pm + \omega^2\phi_2^\pm + \omega\phi_3^\pm)/\sqrt{3}. \tag{38.21}$$

The corresponding eigenvalues are $(a + 2Re(b)) = -\frac{v^2}{6}(3\lambda_3/4)$, $(a - Re(b) - \sqrt{3}Im(b)) = 0$, and $(a - Re(b) + \sqrt{3}Im(b)) = -\frac{v^2}{6}(3\lambda_3/4 - 3\lambda_4)$.

The vevs are complex. Hence there will be mixing between the neutral scalars and pseudoscalars. The mass matrix will now be a (6×6) matrix in the ($\chi_1, \chi_2, \chi_3, \phi_1, \phi_2, \phi_3$) basis given by

$$M^2_{\Phi_i^0 \Phi_j^0} = \frac{v^2}{6}\begin{pmatrix} 2\lambda_4 & -\lambda_4 & -\lambda_4 & 0 & \sqrt{3}\lambda_4 & -\sqrt{3}\lambda_4 \\ -\lambda_4 & f_2 & f_1 & g_1 & g_2 & g_3 \\ -\lambda_4 & f_1 & f_2 & -g_1 & -g_3 & -g_2 \\ 0 & g_1 & -g_1 & (\lambda_1 + \lambda_2) & h_1 & h_1 \\ \sqrt{3}\lambda_4 & g_2 & -g_3 & h_1 & h_3 & h_2 \\ -\sqrt{3}\lambda_4 & g_3 & -g_2 & h_1 & h_2 & h_3 \end{pmatrix} . \tag{38.22}$$

Here,

$$f_1 = -\frac{3}{4}[\lambda_1 - \lambda_2/2 + \lambda_3/2] - \lambda_4, \quad f_2 = \frac{3}{4}(\lambda_1 + \lambda_2) + \frac{1}{2}\lambda_4,$$

$$g_1 = \frac{\sqrt{3}}{4}\{2\lambda_1 - \lambda_2 + \lambda_3 - 4\lambda_4\}, \quad g_2 = -\frac{\sqrt{3}}{4}\{\lambda_1 + \lambda_2 - 2\lambda_4\}, \quad g_3 = -\frac{\sqrt{3}}{4}\{\lambda_1 - \lambda_2/2 + \lambda_3/2 - 4\lambda_4\},$$

$$h_1 = -\frac{1}{4}\{2\lambda_1 - \lambda_2 + \lambda_3 + 4\lambda_4\}, \quad h_2 = \frac{1}{8}\{2\lambda_1 - \lambda_2 + \lambda_3 - 8\lambda_4\}, \quad h_3 = \frac{1}{4}\{\lambda_1 + \lambda_2 + 6\lambda_4\}. \quad (38.23)$$

One of the eigenvalues of the matrix in (38.22) is zero corresponding to the neutral Goldstone. A 6×6 unitary transformation matrix

$$U_{6r} = \frac{1}{\sqrt{3}} \begin{pmatrix} 1 & 1 & 1 & 0 & 0 & 0 \\ 1 & -1/2 & -1/2 & 0 & -\sqrt{3}/2 & \sqrt{3}/2 \\ 1 & -1/2 & -1/2 & 0 & \sqrt{3}/2 & -\sqrt{3}/2 \\ 0 & 0 & 0 & 1 & 1 & 1 \\ 0 & \sqrt{3}/2 & -\sqrt{3}/2 & 1 & -1/2 & -1/2 \\ 0 & -\sqrt{3}/2 & \sqrt{3}/2 & 1 & -1/2 & -1/2 \end{pmatrix}. \quad (38.24)$$

block diagonalizes the 6×6 mass matrix in (38.22) into 2×2 blocks and transports $(\chi_1, \chi_2, \chi_3, \phi_1, \phi_2, \phi_3)$ basis to $(\xi_1, \xi_2, \xi_3, \eta_1, \eta_2, \eta_3)$ basis out of which ξ_2 corresponds to the neutral Goldstone. Needless to mention that the unitary transformation matrix in (38.24) is a 6×6 version of the matrix in (38.20) with the real and imaginary parts separated. Hence,

$$\begin{pmatrix} \xi_1 \\ \xi_2 \\ \xi_3 \\ \eta_1 \\ \eta_2 \\ \eta_3 \end{pmatrix} = \frac{1}{\sqrt{3}} \begin{pmatrix} \chi_1 + \chi_2 + \chi_3 \\ \chi_1 - (\chi_2 + \chi_3)/2 - \sqrt{3}(\phi_2 - \phi_3)/2 \\ \chi_1 - (\chi_2 + \chi_3)/2 + \sqrt{3}(\phi_2 - \phi_3)/2 \\ \phi_1 + \phi_2 + \phi_3 \\ (\sqrt{3}(\chi_2 - \chi_3)/2 + \phi_1 - (\phi_2 + \phi_3)/2) \\ -\sqrt{3}(\chi_2 - \chi_3)/2 + \phi_1 - (\phi_2 + \phi_3)/2 \end{pmatrix} \quad (38.25)$$

The 2×2 blocks to which the 6×6 mass matrix in (38.22) decomposes in the (ξ_1, ξ_3) and (η_1, η_3) basis are

$$M^2_{\xi_1,\xi_3} = \frac{v^2}{6} \begin{pmatrix} \lambda_A & -\lambda_A \\ -\lambda_A & \lambda_A + 18\lambda_4 \end{pmatrix}, \quad M^2_{\eta_1,\eta_3} = \frac{v^2}{6} \begin{pmatrix} \lambda_A & \lambda_A \\ \lambda_A & \lambda_A + 18\lambda_4 \end{pmatrix}, \quad (38.26)$$

with $\lambda_A = \frac{9}{4}\lambda_2 - \frac{3}{4}\lambda_3 - 3\lambda_4$. $m^2_{\xi_2} = 0$ and $m^2_{\eta_2} = v^2 (3\lambda_1/2 + \lambda_3/2 - \lambda_4)$. Defining $\tan 2\alpha = \lambda_A/9\lambda_4$, one can read off the eigenvalues and eigenvectors of the matrices given in (38.26) as

$$m_1^2 = \frac{v^2}{6}\left[\lambda_A + 9\lambda_4 + \sqrt{\lambda_A^2 + 81\lambda_4^2}\right] , \; \xi_1 = \chi_1 \cos\alpha - \chi_3 \sin\alpha , \; \eta_1 = \phi_1 \cos\alpha + \phi_3 \sin\alpha$$

$$m_3^2 = \frac{v^2}{6}\left[\lambda_A + 9\lambda_4 - \sqrt{\lambda_A^2 + 81\lambda_4^2}\right] , \; \xi_3 = \chi_1 \sin\alpha + \chi_3 \cos\alpha , \; \eta_3 = -\phi_1 \sin\alpha + \phi_3 \cos\alpha .$$

$$(38.27)$$

Also,

$$\frac{1}{\sqrt{3}}\begin{pmatrix} 1 & 1 & 1 \\ 1 & \omega^2 & \omega \\ 1 & \omega & \omega^2 \end{pmatrix} \frac{v}{\sqrt{6}}\begin{pmatrix} 0 & 1 \\ 0 & \omega \\ 0 & \omega^2 \end{pmatrix} = \frac{v}{\sqrt{2}}\begin{pmatrix} 0 & 0 \\ 0 & 1 \\ 0 & 0 \end{pmatrix} . \qquad (38.28)$$

Thus the second direction corresponds to the SM Higgs. The eigenvalues of the physical scalar mass-square matrices are positive consistent with the positivity and unitarity conditions. The detailed analysis can be found in [1]. Alignment is therefore achieved for this case also.

38.6 Conclusions

A triplet of $A4$ consisting of three $SU(2)_L$ doublet scalars is considered [1]. The $A4$ conserving scalar potential arising out of such a scalar spectrum has four possible vev configurations corresponding to the global minima. For each of these global minima configurations, it was verified that alignment follows as a consequence of the $A4$ symmetry.

Acknowledgements SP thanks organizers of XXIII DAE-BRNS High Energy Physics Symposium 2018. SP acknowledges CSIR (NET) Senior Research Fellowship. AR received partial funding from SERB Grant No. SR/S2/JCB-14/2009 and SERB grant No. EMR/2015/001989.

References

1. S. Pramanick, A. Raychaudhuri, JHEP **1801**, 011 (2018). http://arxiv.org/abs/1710.04433arXiv:1710.04433 [hep-ph]
2. S. Pramanick, A. Raychaudhuri, Phys. Rev. D **93**(3), 033007 (2016). http://arxiv.org/abs/1508.02330arXiv:1508.02330 [hep-ph], S. Pramanick, Phys. Rev. D **98**(7), 075016 (2018). http://arxiv.org/abs/1711.03510arXiv:1711.03510 [hep-ph]
3. A. Degee, I.P. Ivanov, V. Keus, JHEP **1302**, 125 (2013). http://arxiv.org/abs/1211.4989arXiv:1211.4989 [hep-ph]
4. R. Gonzalez Felipe, H. Serodio, J.P. Silva, Phys. Rev. D **88**, 015015 (2013). http://arxiv.org/abs/1304.3468arXiv:1304.3468 [hep-ph]
5. H.E. Haber, PoS CHARGED **2016**, 029 (2017). http://arxiv.org/abs/1701.01922arXiv:1701.01922 [hep-ph]
6. K. Kannike, Eur. Phys. J. C **76**, 324 (2016). http://arxiv.org/abs/1603.02680arXiv:1603.02680 [hep-ph]; J. Chakrabortty, P. Konar and T. Mondal, Phys. Rev. D 89, 095008 (2014). http://arxiv.org/abs/1311.5666arXiv:1311.5666 [hep-ph]
7. I.F. Ginzburg, I.P. Ivanov, Phys. Rev. D **72**, 115010 (2005). [hep-ph/0508020]

Chapter 39
Vector-Like Dark Matter and Flavor Anomalies with Leptoquarks

Suchismita Sahoo, Shivaramakrishna Singirala, and Rukmani Mohanta

Abstract We study vector-like fermionic dark matter and flavor anomalies in a standard model extension with vector-like multiplets of quark and lepton type. An admixture of lepton type doublet and singlet is examined in relic density and direct detection perspective. Furthermore, two leptoquarks, one scalar-type $(\bar{3}, 1, 1/3)$ and a vector-type $(\bar{3}, 1, 2/3)$ are introduced to study muon anomalous magnetic moment and the $B(\tau)$ anomalies.

39.1 Introduction

The failure of standard model (SM) in explaining the matter-antimatter asymmetry, neutrino mass, dark matter (DM) and dark energy indicates the presence of new physics (NP) beyond it. Additionally, the LHCb as well as Belle and BaBar have provided a collection of interesting observables associated with the flavor changing neutral currect (FCNC) and charge current (FCCC) transitions, whose measured data disagrees with their SM predictions. The measurements on decay rate of $B_{(s)} \rightarrow K^*(\phi)\mu^+\mu^-$ and P_5' observable of $B \rightarrow K^*\mu^+\mu^-$ disagree with their SM predictions at the level of 3σ [1]. Further, discrepancy at the level of 2.5σ [2] is observed in the lepton nonunisaity (LNU) ratios $R_K \equiv \Gamma(B^+ \rightarrow K^+\mu^+\mu^-)/\Gamma(B^+ \rightarrow K^+e^+e^-)$ along with $R_{K^*} \equiv \Gamma(B^0 \rightarrow K^{*0}\mu^+\mu^-)/\Gamma(B^0 \rightarrow K^{*0}e^+e^-)$. The measurements on $R_{D^{(*)}} \equiv \Gamma(B \rightarrow D^{(*)}\tau\bar{\nu})/\Gamma(B \rightarrow D^{(*)}l\bar{\nu})$ $(l = e, \mu)$ ratios also disagree with the SM at the level of $\sim 3.08\sigma$ [3]. Discrepancy of 1.7σ is also observed in

S. Sahoo (✉)
Physical Research Laboratory, Navrangpura, Ahmedabad 380009, India
e-mail: suchismita8792@gmail.com

S. Singirala
Indian Institute of Technology Indore, Khandwa Road, Simrol 453552, Madhya Pradesh, India
e-mail: krishnas542@gmail.com

R. Mohanta
University of Hyderbad, Hyderabad 500046, Telangana, India
e-mail: rukmani98@gmail.com

© Springer Nature Singapore Pte Ltd. 2021
P. K. Behera et al. (eds.), *XXIII DAE High Energy Physics Symposium*,
Springer Proceedings in Physics 261,
https://doi.org/10.1007/978-981-33-4408-2_39

the $R_{J/\psi}$ parameter [4]. In this concern, we would like to investigate these anomalies by including an extra vector-like (VL) fermions and leptoquarks (LQ) to the SM.

The paper is organized as follows. In Sect. 39.2, we discuss the particle content and Lagrangian of our model. The constraints on new parameters from the dark matter and flavor observables are presented in Sect. 39.3. Section 39.4 contains the implication of constrained couplings on $R_{D^{(*)}}$, $R_{J/\psi}$ ratios and our results are summarized in Sect. 39.5.

39.2 New Model with Leptoquarks

We extend SM with vector-like fermion multiplets, i.e., two doublets of quark (ζ) and lepton type (ψ), and also a lepton singlet (χ). We also introduce the model with a $(\bar{3}, 1, 1/3)$ scalar leptoquark (SLQ) and a $(\bar{3}, 1, 2/3)$ vector leptoquark (VLQ). All the new field content is assigned with odd charge under a discrete Z_2 symmetry. The particle content with their corresponding charges is shown in Table 39.1. The interaction terms are given as

$$
\begin{aligned}
\mathcal{L} = & \mathcal{L}_{\text{SM}} - (y_{Q\psi}^{S_1} \overline{Q_L}{}^C S_1 i\sigma^2 \psi + \text{h.c.}) - (y_{\ell\zeta}^{S_1} \overline{\zeta}{}^C S_1 \ell + \text{h.c.}) - (y_{d\chi}^{S_1} \overline{d_R}{}^C S_1 \chi + \text{h.c.}) \\
& - (y_{Q\psi}^{V_1} \overline{Q_L} \gamma_\mu V_1^\mu \psi + \text{h.c.}) - (y_{\ell\zeta}^{V_1} \overline{\zeta} \gamma_\mu V_1^\mu \ell + \text{h.c.}) - (y_{u\chi}^{V_1} \overline{u_R} \gamma_\mu V_1^\mu \chi + \text{h.c.}) \\
& - (y_D \overline{\psi} \tilde{H} \chi + \text{H.c.}) - M_\chi \overline{\chi} \chi - M_\psi \overline{\psi} \psi - M_\zeta \overline{\zeta} \zeta + \overline{\zeta} \gamma^\mu \left(i\partial_\mu - \frac{g}{2} \tau^a \cdot \mathbf{W}_\mu^a - \frac{g'}{6} B_\mu \right) \zeta \\
& + \overline{\chi} \gamma^\mu \left(i\partial_\mu \right) \chi + \overline{\psi} \gamma^\mu \left(i\partial_\mu - \frac{g}{2} \tau^a \cdot \mathbf{W}_\mu^a + \frac{g'}{2} B_\mu \right) \psi + \left| \left(i\partial_\mu - \frac{g'}{3} B_\mu \right) S_1 \right|^2 - V(H, S_1),
\end{aligned}
\tag{39.1}
$$

where the scalar potential is given by

$$
\begin{aligned}
V(H, S_1) = & \mu_H^2 H^\dagger H + \lambda_H (H^\dagger H)^2 + \mu_S^2 (S_1{}^\dagger S_1) \\
& + \lambda_S (S_1{}^\dagger S_1)^2 + \lambda_{HS} (H_2^\dagger H)(S_1{}^\dagger S_1).
\end{aligned}
\tag{39.2}
$$

Here, after electroweak symmetry breaking, the Higgs doublet can be written as $H = (0, (v+h)/\sqrt{2})^T$. The discrete symmetry forbids the mixing of the new fermions with the SM fermion content. Moreover, we consider that the LQs couple only to second and third generation fermions in the present model. Now, the new neutral fermions mixing takes the form

$$
M_N = \begin{pmatrix} M_\chi & \frac{y_D}{\sqrt{2}} \\ \frac{y_D}{\sqrt{2}} & M_\psi \end{pmatrix}.
\tag{39.3}
$$

Table 39.1 Fields and their charges under $SU(3)_c \times SU(2)_L \times U(1)_Y$ SM gauge group

	Field	$SU(3)_C \times SU(2)_L \times U(1)_Y$	Z_2
Fermions	$Q_L \equiv (u, d)_L^T$	$(\mathbf{3}, \mathbf{2},\ 1/6)$	$+$
	u_R	$(\mathbf{3}, \mathbf{1},\ 2/3)$	$+$
	d_R	$(\mathbf{3}, \mathbf{1},\ -1/3)$	$+$
	$\ell_L \equiv (v, e)_L^T$	$(\mathbf{1}, \mathbf{2},\ -1/2)$	$+$
	e_R	$(\mathbf{1}, \mathbf{1},\ -1)$	$+$
Vector-like fermions	$\zeta \equiv (\zeta_u, \zeta_d)^T$	$(\mathbf{3}, \mathbf{2},\ 1/6)$	$-$
	$\psi \equiv (\psi_v, \psi_l)^T$	$(\mathbf{1}, \mathbf{2},\ -1/2)$	$-$
	χ	$(\mathbf{1}, \mathbf{1},\ 0)$	$-$
Scalars	H	$(\mathbf{1}, \mathbf{2},\ 1/2)$	$+$
	S_1	$(\bar{\mathbf{3}}, \mathbf{1},\ 1/3)$	$-$
Vector	V_1	$(\bar{\mathbf{3}}, \mathbf{1},\ 2/3)$	$-$

One can diagonalize the above mass matrices by $U_\alpha^T M_N U_\alpha = \text{diag}\,[M_{N_-}, M_{N_+}]$, where $U_\theta = \begin{pmatrix} \cos\alpha & \sin\alpha \\ -\sin\alpha & \cos\alpha \end{pmatrix}$, with $\alpha = \frac{1}{2}\tan^{-1}\left(\frac{y_D v}{\sqrt{2}(M_\psi - M_\chi)}\right)$. The lightest mass eigenstate N_- is a probable dark matter in the present model.

39.3 Constraints on New Parameters from both Dark Matter and Flavor Phenomenology

The neutral fermion N_- communicates with the SM sector via Higgs, Z and LQs. Higgs portal s-channels include $f\bar{f}$, W^+W^-, ZZ, hh, and Z-portal gives $f\bar{f}$, Zh in the final state. In SLQ (VLQ) portal, the t-channel processes with $s\bar{s}$, $b\bar{b}$ ($c\bar{c}$, $t\bar{t}$) as output particles contribute to relic density. The formula for computing the abundance is given as

$$\Omega h^2 = \frac{2.14 \times 10^9 \text{ GeV}^{-1}}{M_{\text{pl}}\, g_*^{1/2}} \frac{1}{J(x_f)}, \quad J(x_f) = \int_{x_f}^\infty \frac{\langle \sigma v \rangle(x)}{x^2} dx. \quad (39.4)$$

Here the Planck mass, $M_{\text{pl}} = 1.22 \times 10^{19}$ GeV and $g_* = 106.75$ and $x_f \simeq 20$.

Moving to direct detection, singlet-doublet mixing gives the Z-portal spin-independent (SI) WIMP-nucleon cross section, re-scaled by $\sin^4\alpha$ factor [5], taking small mixing ($\sin\alpha \lesssim 0.035$) ensures the spin-independent contribution within stringent experimental bound [6]. In LQ portal, the contribution gets suppressed by quark mixing. For analysis, we fix $(y_D, \sin\alpha) = (1, 0.035)$, vary the new couplings within the perturbative limit and the LQs masses within $1.2 - 3$ TeV. We impose the Planck limit on relic density [7] in 3σ range, to constrain the new parameters. The obtained allowed parameter space from DM studies are shown in Fig. 39.1. We focus on the couplings that are relevant both in DM and flavor studies.

We also constrain the new parameters from both the quark and lepton sectors, which proceed through one loop box/penguin diagrams in the presence of an additional leptoquarks and vector-like fermions. The effective Hamiltonian mediating the $b \rightarrow sl^+l^-$ transitions is given by [8]

$$\mathcal{H}_{\text{eff}} = -\frac{4G_F}{\sqrt{2}} V_{tb} V_{ts}^* \left[\sum_{i=1}^{6} C_i(\mu)\mathcal{O}_i + \sum_{i=7,9,10} \left(C_i(\mu)\mathcal{O}_i + C_i'(\mu)\mathcal{O}_i' \right) \right], \quad (39.5)$$

where G_F is the Fermi constant, $V_{qq'}$ are the product of CKM matrix elements, $\mathcal{O}_i^{(\prime)}$'s are the effective operators and $C_i^{(\prime)}$'s are their corresponding Wilson coefficients. We obtain an additional coefficients as well as new contributions to the SM Wilson coefficients due to the presence of vector-like fermions and leptoquarks. We use the exist measured data on the branching ratios of $B_s \rightarrow ll$, $B^+ \rightarrow K^+\tau\tau$, $\bar{B} \rightarrow \bar{K}ll$, $\bar{B} \rightarrow \bar{K}^{(*)}\nu_l\bar{\nu}_l$, $\bar{B} \rightarrow X_s\gamma$, $\tau \rightarrow \mu\gamma$ and $\tau \rightarrow 3\mu(\mu\bar{\nu}_\mu\nu_\tau)$ processes to compute the allowed parameter space. We also further constrain from $R_{K^{(*)}}$ and the muon anomalous magnetic moment. For SM predictions, the CKM matrix elements, all the particle masses, life time of $B_{(s)}$ mesons are taken from [9] and the form factors from [10]. The constraints on $y_q^{S_1} - M_{S_1}$ (top-left), $y_q^{V_1} - M_{V_1}$ (top-right), $y_\mu^{S_1} - y_\tau^{S_1}$ (bottom-left) and $y_\mu^{V_1} - y_\tau^{V_1}$ (bottom-right) planes obtained by using the observables of both DM and flavor, are presented in Fig. 39.1. Here orange (magenta) and blue (cyan) bands represent the constraints from only flavor and from both DM and flavor observables in the presence of scalar (vector) LQ.

39.4   Impact on $R_{D^{(*)}}$ and $R_{J/\psi}$ Ratios

In this section, we would like to see whether the new parameters influencing the dark matter observables and the anomalies associated with quark and lepton sectors, also have an effect on $R_{D^{(*)}}$ and $R_{J/\psi}$ ratios, which are mediated by $b \rightarrow c\tau\bar{\nu}_l$ transitions. The effective Hamiltonian of $b \rightarrow c\tau\bar{\nu}_l$ decay modes with only the left handed neutrinos is given as [11]

$$\mathcal{H}_{\text{eff}} = \frac{4G_F}{\sqrt{2}} V_{cb} \left[(\delta_{l\tau} + V_L)\mathcal{O}_{V_L}^l + V_R\mathcal{O}_{V_R}^l + S_L\mathcal{O}_{S_L}^l + S_R\mathcal{O}_{S_R}^l + T\mathcal{O}_T^l \right], \quad (39.6)$$

where V_{cb} is the CKM matrix element. Though the Wilson coefficients, $X(= V_{L,R}, S_{L,R}, T)$ are absent in the SM, we find new V_L coefficients in the present model. All the required input parameters like particle masses, lifetime and CKM matrix elements are taken from [9] and the form factors from [12]. Using the allowed parameter space from Fig. 39.1, the variation of R_D (top-left), R_{D^*} (top-right) and $R_{J/\psi}$ (bottom) ratios with respect to q^2 is presented in Fig. 39.2. Here the blue

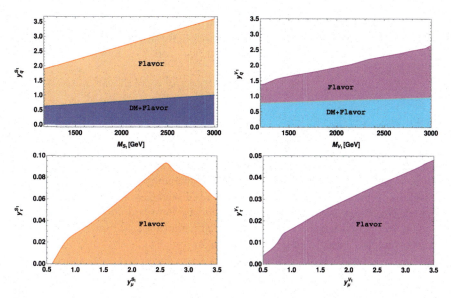

Fig. 39.1 Constraints on $y_q^{S_1} - M_{S_1}$ (top-left), $y_q^{V_1} - M_{V_1}$ (top-right), $y_\mu^{S_1} - y_\tau^{S_1}$ (bottom-left) and $y_\mu^{V_1} - y_\tau^{V_1}$ (bottom-right) planes obtained from DM and flavor observables

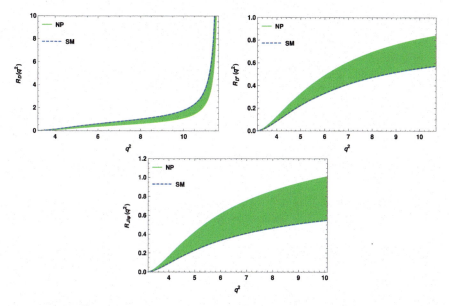

Fig. 39.2 The q^2 variation of R_D (top-left), R_{D^*} (top-right) and $R_{J/\psi}$ (bottom) parameters in the presence of new particles beyond the SM

dashed lines represent the SM predictions and the green bands are obtained by using the constrained from both DM and flavor studies. We observe that the presence of new particles provide significant deviation in $R_{D^{(*)}}$ and $R_{J/\psi}$ ratios from their corresponding SM results.

39.5 Conclusion

We build a model by extending the standard model with an additional vector-like fermions and leptoquark singlets. We constrain the new parameters consistent with Planck limit on relic density, PICO-60 and LUX bounds on direct detection cross section. We further constrain the parameter space from both the quark and lepton sectors. We then check the impact of allowed parameter space on $R_{D^{(*)}}$ and $R_{J/\psi}$ ratios, which are found to be significantly larger.

References

1. LHCb, R. Aaij et al., JHEP **06**, 133 (2014); LHCb, R. Aaij et al., Phys. Rev. Lett. **111**, 191801 (2013); LHCb, R. Aaij et al., JHEP **09**, 179 (2015)
2. LHCb, R. Aaij et al., Phys. Rev. Lett. **113**, 151601 (2014); LHCb, R. Aaij et al., arXiv:1903.09252; LHCb, R. Aaij et al., JHEP **08**, 055 (2017); Belle, M. Prim, talk given at Moriond, March 22 2019
3. Heavy Flavor Averaging Group (2019), https://hflav-eos.web.cern.ch/hflav-eos/semi/spring19/html/RDsDsstar/RDRDs.html
4. R. Aaij et al., LHCb. Phys. Rev. Lett. **120**, 121801 (2018). https://doi.org/10.1103/PhysRevLett.120.121801
5. S. Bhattacharya, N. Sahoo, N. Sahu, Phys. Rev. D **93**, 115040 (2016)
6. X. Cui et al., PandaX-II Collaboration. Phys. Rev. Lett. **119**, 181302 (2017)
7. N. Aghanim et al. (Planck Collaboration), arXiv:1807.06209
8. C. Bobeth, M. Misiak, J. Urban, Nucl. Phys. B **574**, 291 (2000)
9. M. Tanabashi et al., Particle data group. Phys. Rev. D **98**, 030001 (2018)
10. M. Beneke, T. Feldmann, D. Seidel, Eur. Phys. J. C **41**, 173 (2005); P. Colangelo, F. De Fazio, P. Santorelli, E. Scrimieri, Phys. Lett. B **395**, 339 (1997)
11. V. Cirigliano, J. Jenkins, M. Gonzalez-Alonso, Nucl. Phys. B **830**, 95 (2010). https://doi.org/10.1016/j.nuclphysb.2009.12.020
12. J. A. Bailey et al. (MILC), Phys. Rev. D **92**, 034506 (2015); I. Caprini, L. Lellouch, M. Neubert, Nucl. Phys. B **530**, 153 (1998); J. A. Bailey et al. (Fermilab Lattice, MILC), Phys. Rev. D **89**, 114504 (2014); Y. Amhis et al. (HFAG) (2014), 1412.7515; Z.-R. Huang, Y. Li, C.-D. Lu, M. A. Paracha, C. Wang, Phys. Rev. D **98**, 095018 (2018); R. Watanabe, Phys. Lett. B **776**, 5 (2018)

Chapter 40
Common Framework for Inflation, Dark Matter, Baryogenesis and Neutrino Masses in the Scotogenic Model

Debasish Borah, P. S. Bhupal Dev, and Abhass Kumar

Abstract We consider the scotogenic model involving an extra inert scalar doublet and three SM-singlet fermions, all under an additional $\mathbb{Z}_2$ symmetry, to explain inflation, dark matter, baryogenesis and neutrino masses simultaneously. The inert scalar doublet is coupled to gravity non-minimally and after a conformal transformation forms the inflation in the early Universe. Later its lightest scalar component freezes-out to give the dark matter candidate. Baryogenesis happens via leptogenesis by the decay of N_1 to SM leptons and the inert doublet particles. The N_i's and the inert doublet combine with the SM neutrinos and the Higgs vacuum expectation value (vev) to generate neutrino masses at the loop level. We show that with the scotogenic model, we get a very economical model for combining inflation with dark matter, baryogenesis and neutrino masses.

40.1 Introduction

The minimal scotogenic model with non-minimal gravity coupling contains an extra scalar doublet coupled to gravity non-minimally and three SM-singlet heavy fermions, both odd under an extra $\mathbb{Z}_2$ symmetry in which the SM particles are even. Due to the $\mathbb{Z}_2$ symmetry, the scalar doublet doesn't interact with SM fermions and is called inert. The model details are given below [1]

D. Borah
Department of Physics, Indian Institute of Technology Guwahati, Guwahati 781039, Assam, India
e-mail: dborah@iit.ac.in

P. S. B. Dev
Department of Physics, McDonnell Center for the Space Sciences, Washington University, St. Louis, MO 63130, USA
e-mail: bdev@wustl.edu

A. Kumar (✉)
Physical Research Laboratory, Ahmedabad 380009, India
e-mail: abhass@prl.res.in; abhasskumar@hri.res.in

Harish-Chandra Research Institute, Chhatnag Road, Jhunsi, Allahabad 211019, India

© Springer Nature Singapore Pte Ltd. 2021
P. K. Behera et al. (eds.), *XXIII DAE High Energy Physics Symposium*,
Springer Proceedings in Physics 261,
https://doi.org/10.1007/978-981-33-4408-2_40

$$S = \int d^4x \sqrt{-g} \left[-\frac{1}{2} M_{\mathrm{Pl}}^2 R - D_\mu \Phi_1 D^\mu \Phi_1^\dagger - D_\mu \Phi_2 D^\mu \Phi_2^\dagger - V(\Phi_1, \Phi_2) - \xi_1 \Phi_1^2 R - \xi_2 \Phi_2^2 R \right],$$
$$(40.1)$$

$$V(\Phi_1, \Phi_2) = \mu_1^2 |\Phi_1|^2 + \mu_2^2 |\Phi_2|^2 + \frac{\lambda_1}{2} |\Phi_1|^4 + \frac{\lambda_2}{2} |\Phi_2|^4 + \lambda_3 |\Phi_1|^2 |\Phi_2|^2$$
$$+ \lambda_4 |\Phi_1^\dagger \Phi_2|^2 + \left[\frac{\lambda_5}{2} (\Phi_1^\dagger \Phi_2)^2 + \mathrm{H.c.} \right]. \tag{40.2}$$

$$\Phi_1 = \frac{1}{\sqrt{2}} \begin{pmatrix} \chi \\ h \end{pmatrix}, \qquad \Phi_2 = \begin{pmatrix} H^\pm \\ \frac{H^0 + iA^0}{\sqrt{2}} \end{pmatrix}. \tag{40.3}$$

$$\mathcal{L} \supset \frac{1}{2} (M_N)_{ij} N_i N_j + \left(Y_{ij} \bar{L}_i \tilde{\Phi}_2 N_j + \mathrm{H.c.} \right), \tag{40.4}$$

where S is the scalar part of the full action, D_μ is the combined covariant derivative for the $SU(2)$ gauge sector of the SM and the space-time metric, Φ_1 is the SM Higgs, Φ_2 is the inert doublet and also the inflation and its neutral lightest particle is the dark matter, N_i, $(i = 1, 2, 3)$ are the SM-singlet fermions, L_i, $(i = 1, 2, 3)$ are the SM lepton doublets and Y_{ij} is the Yukawa matrix for the interaction between SM leptons, the inert doublet and the SM-singlet fermions. During inflation, all fields other than Φ_2 can be taken to be zero so that only Φ_2 has a non-minimal gravity coupling.

40.2 Inflation and Reheating

Due to the term $\xi_2 \Phi_2^2 R$, a canonical treatment of the action in (40.1) is not possible. We need to perform a conformal transformation to the so-called Einstein frame to remove such terms by redefining the metric $\tilde{g}_{\mu\nu} = \Omega^2 g_{\mu\nu}$ with $\Omega^2 \approx 1 + \frac{\xi}{M_{\mathrm{Pl}}^2} \left(2(H^\pm)^2 + H_0^2 \right)$ [2]. The inflationary potential obtained from the Φ_2 quartic self-coupling term after the transformation is

$$V_e \simeq \frac{\lambda_2 M_{\mathrm{Pl}}^4}{4 \xi_2^2} \left[1 - \exp \left(-\sqrt{\frac{2}{3}} \frac{X}{M_{\mathrm{Pl}}} \right) \right]^2, \tag{40.5}$$

where $X = \sqrt{\frac{3}{2}} M_{\mathrm{Pl}} \log (\Omega^2)$. The potential in (40.5) belongs to the Starobinsky class [3] and is sufficiently slow rolling when $X \gtrsim M_{Pl}$. The slow roll parameters $\epsilon = \frac{1}{2} M_{\mathrm{Pl}}^2 \left(\frac{V_e'}{V_e} \right)^2$ and $\eta = M_{\mathrm{Pl}}^2 \left(\frac{V_e''}{V_e} \right)$ can be calculated from the potential where $V_e' \equiv dV_e/dX$ and $V_e'' \equiv d^2 V_e/dX^2$. We can now find the tensor-to-scalar ratio $r = 16\epsilon = 0.0029$ and the scalar spectral index $n_s = 1 - 6\epsilon + 2\eta = 0.9678$ [2]. These are well within the Planck 2018 [4] results $r < 0.11$ at 95% C.L. and $n_s = 0.9649 \pm 0.0042$

at 68% C.L. The scalar power spectrum P_s is used to obtain a relationship between λ_2 and ξ_2 with $\xi_2 \simeq 5.33 \times 10^4 \lambda_2^{1/2}$.

The reheating era [5] starts when $X \ll M_{Pl}$ where potential is approximated by $V_e \simeq \frac{\lambda_2 M_{Pl}^2}{6\xi_2^2} X^2 \equiv \frac{1}{2}\omega^2 X^2$ where $\omega^2 = \frac{\lambda_2 M_{Pl}^2}{3\xi_2^2}$. The inflation can produce relativistic SM fermions either by decaying to the Higgs and the gauge bosons which further produce relativistic SM fermions or directly by decayin to the Sm-singlet fermions and the SM leptons. It was shown in [6] (see also [7, 8]) that significant relativistic energy density is formed only when the production of the mediator bosons enters a non-perturbative parametric resonance regime. It was also shown that the Higgs channel contribution is much slower than the gauge boson channel making it the dominant one. The requirement of parametric resonance imposes a lower bound on λ_2: $\lambda_2 \gtrsim \frac{1}{60}$ [2]. The relativistic energy density produced by the gauge boson and the direct decay channel are, respectively, [1]:

$$\rho_{r,\text{gauge}} \simeq \frac{1.06 \times 10^{57} \text{ GeV}^4}{\lambda_2}. \tag{40.6}$$

$$\rho_{r,\text{Yukawa}} \simeq \frac{6.16 \times 10^{49} \text{ GeV}^4}{\sqrt{\lambda_2}} \quad (\text{assuming } Y \sim 10^{-4}). \tag{40.7}$$

40.3   Dark Matter

Reheating ends when the quadratic approximation for the potential breaks down after which the remaining inert doublet particles become a part of the thermal plasma. These particles later freeze-out around the electroweak scales. The lightest neutral component which we take to be the CP even scalar becomes a relic dark matter candidate. Boltzmann equation is used to obtain the evolution of DM as it falls out of equilibrium and Planck 2018 [9] bounds for the relic abundance are used to find the relation between the mass μ_2 of the DM and the coupling $\lambda_s = \lambda_3 + \lambda_4 + \lambda_5$ [1]. The relationship is shown in Fig. 40.1. The DM spin-independent scattering cross section is of $\simeq 10^{-45} - 10^{-46}$ cm^2 which is within reach of near future DM direct detection experiments like LZ [10], XENONnT [11], DARWIN [12] and PandaX-20T [13].

40.4   Neutrino Masses

Because of the $\mathbb{Z}_2$ symmetry, Φ_2 does not get a vev and the usual see-saw mechanism does not work. However, at one loop the SM-singlet fermions along with Φ_2 and the Higgs vev can radiatively generate neutrino masses [14] as seen in Fig. 40.2. Since these neutrino masses and the Yukawa matrix elements are also used for leptogenesis which leads to baryogenesis, we work in the Casas-Ibarra parametrization for the

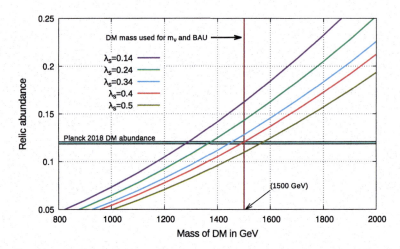

Fig. 40.1 The DM relic abundance. The horizontal band is the 68% C.L. observed DM relic abundance from Planck 2018 data [9]. The vertical line shows a benchmark value of the DM mass chosen for our subsequent analysis

Fig. 40.2 Radiative generation of neutrino masses

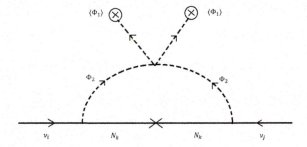

scotogenic model [15] which allows us to write the neutrino mass matrix in a form similar to the type I see-saw formula $M_\nu = Y\tilde{M}^{-1}Y^T$, where the diagonal matrix $\tilde{M}$ has elements:

$$\tilde{M}_i = \frac{2\pi^2}{\lambda_5}\zeta_i\frac{2M_i}{v^2}\,, \tag{40.8}$$

$$\text{and}\quad \zeta_i = \left(\frac{M_i^2}{8(m_{H^0}^2 - m_{A^0}^2)}\left[L_i(m_{H^0}^2) - L_i(m_{A^0}^2)\right]\right)^{-1}, \tag{40.9}$$

with M_i being the masses of the SM-singlet fermions and $L_k(m^2) = \frac{m^2}{m^2 - M_k^2}\ln\frac{m^2}{M_k^2}$.

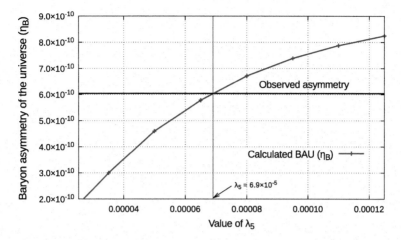

Fig. 40.3 The baryon-to-photon ratio as a function of $|\lambda_5|$ for benchmark DM mass of 1.5 TeV. The horizontal line gives the observed value

40.5 Baryogenesis

In vanilla leptogenesis, there exists an absolute lower bound on the lightest SM-singlet fermion mass $\gtrsim 10^9$ GeV [16, 17]. This can be lowered in the hierarchical SM-singlet fermion mass set-up to 10^4 GeV if the lightest SM neutrino has a mass of around $10^{-11} - 10^{-12}$ eV [18]. We use this to lower the leptogenesis scale in our model where we use $M_1 = 10$ TeV, $M_2 = 50$ TeV and $M_3 = 100$ TeV. Simultaneous Boltzmann equations are solved for the decay of N_1 and formation of $B - L$ number density n_{B-L} which is converted to the baryon asymmetry of the Universe by the sphaleron processes [1]. The results are shown in Fig. 40.3

40.6 Conclusion

We have used the scotogenic model with the extra scalar doublet coupled non-minimally to gravity to simultaneously explain inflation, dark matter, baryogenesis and neutrino masses. The inflationary parameters r and n_s obtained in the model are highly consistent with Planck 2018 results. After reheating, the remaining inflation particles become part of the equilibrium to later freeze-out and become the dark matter. The heavy SM-singlet fermions are used to generate masses for the SM neutrinos in conjunction with the inert doublet, thus combining inflation and dark matter with neutrino masses. Further, the decays of the SM-singlet fermions into SM lepton and the inert doublet particles become the source for a lepton asymmetry which is converted to a baryon asymmetry by the sphaleron processes.

References

1. P.S. Debasish Borah, B. Dev, A. Kumar, TeV scale leptogenesis, inflaton dark matter and neutrino mass in a scotogenic model. Phys. Rev. D **99**(5), 055012 (2019)
2. Sandhya Choubey, Abhass Kumar, Inflation and dark matter in the inert doublet model. JHEP **11**, 080 (2017)
3. A.A. Starobinsky, Spectrum of relict gravitational radiation and the early state of the universe. JETP Lett. **30**, 682–685 (1979). [Pisma Zh. Eksp. Teor. Fiz.30,719(1979)]
4. Y. Akrami et al., Planck 2018 results. X. Constraints on inflation (2018)
5. Rouzbeh Allahverdi, Robert Brandenberger, Francis-Yan Cyr-Racine, Anupam Mazumdar, Reheating in inflationary cosmology: theory and applications. Ann. Rev. Nucl. Part. Sci. **60**, 27–51 (2010)
6. F. Bezrukov, D. Gorbunov, M. Shaposhnikov, On initial conditions for the Hot Big Bang. JCAP **0906**, 029 (2009)
7. Juan Garcia-Bellido, Daniel G. Figueroa, Javier Rubio, Preheating in the Standard Model with the Higgs-Inflaton coupled to gravity. Phys. Rev. D **79**, 063531 (2009)
8. Jo Repond, Javier Rubio, Combined Preheating on the lattice with applications to Higgs inflation. JCAP **1607**(07), 043 (2016)
9. N. Aghanim et al., *Planck 2018 results*. VI, Cosmological parameters (2018)
10. D. S. Akerib et al. LUX-ZEPLIN (LZ) Conceptual Design Report. 2015
11. E. Aprile et al., Physics reach of the XENON1T dark matter experiment. JCAP **1604**(04), 027 (2016)
12. J. Aalbers et al., DARWIN: towards the ultimate dark matter detector. JCAP **1611**, 017 (2016)
13. Jianglai Liu, Xun Chen, Xiangdong Ji, Current status of direct dark matter detection experiments. Nat. Phys. **13**(3), 212–216 (2017)
14. Ernest Ma, Verifiable radiative seesaw mechanism of neutrino mass and dark matter. Phys. Rev. D **73**, 077301 (2006)
15. J.A. Casas, A. Ibarra, Oscillating neutrinos and muon $-\dot{\iota}$ e, gamma. Nucl. Phys. B **618**, 171–204 (2001)
16. Sacha Davidson, Alejandro Ibarra, A Lower bound on the right-handed neutrino mass from leptogenesis. Phys. Lett. B **535**, 25–32 (2002)
17. W. Buchmuller, P. Di Bari, M. Plumacher, Cosmic microwave background, matter - anti-matter asymmetry and neutrino masses. Nucl. Phys. B**643**, 367–390 (2002). [Erratum: Nucl. Phys.B793,362(2008)]
18. Thomas Hugle, Moritz Platscher, Kai Schmitz, Low-Scale Leptogenesis in the Scotogenic Neutrino Mass Model. Phys. Rev. D **98**(2), 023020 (2018)

Chapter 41
Fermion Singlet Dark Matter in Scotogenic B-L Model

Debasish Borah, Dibyendu Nanda, Nimmala Narendra, and Narendra Sahu

Abstract We study the possibility of right-handed neutrino Dark Matter (DM) in a gauged $U(1)_{B-L} \times Z_2$ extension of the Standard Model (SM) which is augmented by an additional scalar doublet, being odd under Z_2 symmetry to explain the scotogenic radiative neutrino masses. Due to lepton interactions, the right-handed neutrino DM can have additional channels apart from the usual annihilations through scalar and gauge portals, which give rise to much more allowed mass of DM from relic abundance criteria. This enlarged parameter space is consistent with the neutrino oscillation and relic density of DM. Due to the possibility of the Z_2 odd scalar doublet being the next to lightest stable particle (NLSP), one can have interesting signatures like displaced vertex or disappearing charged tracks provided that the mass splitting between DM and NLSP is small.

41.1 Introduction

Neutrino oscillation experiments confirmed that neutrinos have non-zero sub-eV masses. Although these two problems appear to have a different origin, it is highly appealing and economical to find a common origin. We note that this model was proposed by the authors of [1] with limited discussions on right-handed neutrino dark matter relic [2, 3]. In this model, we perform a more detailed study of right-handed neutrino dark matter, pointing out all possible effects that can affect its

D. Borah · D. Nanda
Department of Physics, Indian Institute of Technology Guwahati, Assam 781039, India
e-mail: dborah@iitg.ac.in

D. Nanda
e-mail: dibyendu.nanda@iitg.ac.in

N. Narendra (✉) · N. Sahu
Indian Institute of Technology Hyderabad, Kandi, Sangareddy 502285, Telangana, India
e-mail: ph14resch01002@iith.ac.in

N. Sahu
e-mail: nsahu@iith.ac.in

© Springer Nature Singapore Pte Ltd. 2021
P. K. Behera et al. (eds.), *XXIII DAE High Energy Physics Symposium*,
Springer Proceedings in Physics 261,
https://doi.org/10.1007/978-981-33-4408-2_41

relic abundance. Due to the existence of new Yukawa interactions, we find that the parameter space giving rise to correct relic abundance [4] is much larger than the resonance region $M_{\mathrm{DM}} \approx M_{Z_{B-L}}/2$ for usual right-handed neutrino DM in $U(1)_{B-L}$ model. This is possible due to additional annihilation and co-annihilation channels that arise due to Yukawa interactions. We also check the consistency of this enlarged DM parameter space with constraints from lepton flavour violation (LFV), as well as neutrino mass. Since the Z_2 odd scalar doublet can be the next to lightest stable particle (NLSP) in this case, it's charged component can be sufficiently produced at the Large Hadron Collider (LHC). Due to the possibility of small mass splitting between NLSP and DM, as well as within the components of the Z_2 odd scalar doublet, we can have interesting signatures like displaced vertex or disappearing charged track (DCT) which the LHC is searching for. We show that if the mass splitting between the DM and NLSP is less than τ lepton mass, then we can get displaced vertex upto 10 cm. In addition to that, the parameter space also remains sensitive to ongoing and near future experiments looking for lepton flavour violating charged lepton decay like $\mu \rightarrow e\gamma$.

41.2 The Model

We extend the minimal gauged $U(1)_{B-L}$ model by introducing an additional Z_2 symmetry and a scalar doublet η. The right-handed neutrino and scalar doublet η are odd under Z_2 symmetry. We assume lightest one among Z_2 odd sector particles to be N_1, hence a DM candidate. A singlet scalar χ is introduced in order to break the $U(1)_{B-L}$ gauge symmetry spontaneously after aquiring a non-zero vacuum expectation value (vev). The corresponding Lagrangian can be written as

$$\mathcal{L}_Y = \sum_{j,k=1}^{3} -y_{jk}\overline{\ell}_{jL} N_{kR}\, \tilde{\eta} - \lambda_{jk}\overline{(N_{jR})^c}\, N_{kR}\, \chi + h.c - V(H, \chi, \eta) \qquad (41.1)$$

where

$$\begin{aligned} V(H, \chi, \eta) &= -\mu_H^2\, H^\dagger H + \lambda_H (H^\dagger H)^2 - \mu_\chi^2 \chi^\dagger \chi + \lambda_\chi (\chi^\dagger \chi)^2 + \mu_\eta^2 \eta^\dagger \eta + \lambda_\eta (\eta^\dagger \eta)^2 \\ &\quad + \lambda_{H\chi}(H^\dagger H)(\chi^\dagger \chi) + \lambda_{H\eta}(H^\dagger H)(\eta^\dagger \eta) + \lambda_{\chi\eta}(\chi^\dagger \chi)(\eta^\dagger \eta) \\ &\quad + \lambda_1 (\eta^\dagger H)(H^\dagger \eta) + \frac{\lambda_2}{2}\Big[(H^\dagger \eta)^2 + h.c.\Big] \end{aligned} \qquad (41.2)$$

where $\mu_\eta^2 > 0$. The neutral component of only H, χ acquire non-zero VEV's v and u, respectively. Expanding around the VEV, we can write

$$H = \begin{bmatrix} 0 \\ \frac{v+h}{\sqrt{2}} \end{bmatrix}, \quad \chi = \frac{u+s}{\sqrt{2}} \quad \text{and} \quad \eta = \begin{bmatrix} \eta^+ \\ \frac{\eta R + i\eta I}{\sqrt{2}} \end{bmatrix}. \qquad (41.3)$$

After diagonalising the SM Higgs and singlet scalar mixing matrix, we get the mass eigenstates h_1 and h_2 as a linear combinations of h and s are $h_1 = h \cos \gamma - s \sin \gamma$, $h_2 = h \sin \gamma + s \cos \gamma$.

41.3 Neutrino Mass

The Z_2 symmetry forbids $\overline{N_R} \tilde{H}^\dagger L$. However, the term: $\frac{\lambda_2}{2}(H^\dagger \eta)^2$ allows us to get radiative neutrino mass at one-loop level [5]. The one-loop expression for neutrino mass is

$$(m_\nu)_{ij} = \sum_k \frac{y_{ik} y_{kj} M_k}{32\pi^2} \left[\frac{M_{\eta R}^2}{M_{\eta R}^2 - M_k^2} \log\left(\frac{M_{\eta R}^2}{M_k^2}\right) - \frac{M_{\eta I}^2}{M_{\eta I}^2 - M_k^2} \log\left(\frac{M_{\eta I}^2}{M_k^2}\right) \right] \tag{41.4}$$

where M_k is the right-handed neutrino mass. The 41.4 equivalently can be written as $(m_\nu)_{ij} \equiv (y^T \Lambda y)_{ij}$. Since the inputs from neutrino data are only in terms of the mass squared differences and mixing angles, it is often useful to express the Yukawa couplings in terms of light neutrino parameters. This is possible through the Casas-Ibarra parametrisation [6] extended to radiative seesaw model [7] which allows us to write the Yukawa couplings as

$$y = \sqrt{\Lambda}^{-1} R \sqrt{m_\nu} U_{\text{PMNS}}^\dagger. \tag{41.5}$$

where R is a real orthogonal matrix. U_{PMNS} is the leptonic mixing matrix.

41.4 Relic Density of N_1 DM in Scotogenic $B - L$ Model

Apart from usual annihilation contributions in minimal $U(1)_{B-L}$ model, there arises few more annihilation and co-annihilation contributions after extending the model in scotogenic fashion. We show the effects of co-annihilations on DM relic abundance by considering different mass splittings $\delta M_1 = M_{\text{NLSP}} - M_{N_1} = 50, 100, 300, 500$ GeV where NLSP is the scalar doublet η (and its components) and with the singlet scalar-SM Higgs mixing $\sin \gamma = 0.1$. The Yukawa couplings are generated through the Casas-Ibarra parametrisation so that they satisfy the current experimental bounds from solar and atmospheric mass squared differences and mixing angles. As can be seen from Fig. 41.1, the co-annihilation effects can change the relic abundance depending upon the mass splitting δM_1 and λ_{11}. In Fig. 41.1, allowing DM mass away from the resonance regions.

To generate this plot, we fixed the h_2 scalar mass and the $M_{Z_{B-L}}$ mass to be $M_{h_2} = 400$ GeV and $M_{Z_{B-L}} = 2000$ GeV, respectively. The gauge coupling is fixed at $g_{B-L} = 0.035$. Since the same Yukawa couplings also contribute to the charged

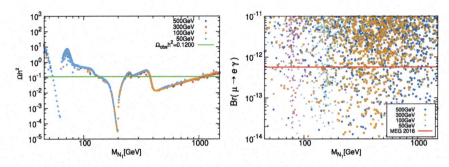

Fig. 41.1 For different mass splittings, the corresponding relic density and branching ratio Br($\mu \rightarrow e\gamma$) as function of DM mass are shown in the left and right panel, respectively

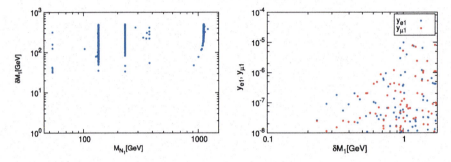

Fig. 41.2 $\delta M_1 = M_{\eta^\pm, \eta_I} - M_{N_1}$ (left panel) and non-zero values of y_{e1} and $y_{\mu1}$ (right panel) versus DM mass, which satisfy relic abundance, neutrino mass and LFV constraints. $\delta M_2 = M_{N_2} - M_{N_1} = 2000$ GeV, $\delta M_3 = M_{N_3} - M_{N_1} = 3000$ GeV

lepton flavour violation. The corresponding scattered plot is shown in Fig. 41.1 and has been compared with the MEG 2016 bound Br($\mu \rightarrow e\gamma$)=4.2 × 10^{-13} in the plane of Br($\mu \rightarrow e\gamma$) versus M_{N_1}. We then show the allowed parameter space in the plane of δM_1 versus M_{N_1} in Fig. 41.2 (left panel) that satisfies the constraints from observed DM abundance, neutrino mass, as well as LFV constraints from $\mu \rightarrow e\gamma$. We see that for a given M_{N_1}, relic density and LFV constraints can be satisfied in a large range of δM_1. We then consider the mass splitting in the range 0.5 MeV (electron mass) to 1.777 GeV (tau mass). We consider η-DM mass splitting to be below the tau lepton mass threshold so that $\eta^\pm$ can decay to first two generation leptons giving displaced vertex signatures if the Yukawa couplings are small. We allow y_{e1} and $y_{\mu1}$ to vary within the range 10^{-8}–10^{-5}, while other Yukawa couplings are generated through Casas-Ibarra parameterisation to obtain correct relic abundance while satisfying LFV constraints. The results are shown in Fig. 41.2 (right panel) in terms of y_{e1}, y_{μ_1} versus δM_1. We see that as δM_1 decreases, we need smaller and smaller y_{e1} and $y_{\mu1}$ values to satisfy relic density and LFV constraints. For further details [8].

41.5 Collider Signatures

The charged component of Z_2 odd scalar doubler η can be the NLSP, while right-handed neutrino is the LSP and it is a DM candidate. The decay width of $\eta^\pm$ can be written as

$$\Gamma_{\eta^\pm \to N_1 \mu} = \frac{y_{\mu 1}^2 \left(m_{\eta^\pm}^2 - (m_{N_1} + m_\mu)^2\right)}{8 m_{\eta^\pm} \pi} \sqrt{1 - \left(\frac{m_{N_1} - m_\mu}{m_\eta^\pm}\right)^2} \sqrt{1 - \left(\frac{m_{N_1} + m_\mu}{m_\eta^\pm}\right)^2}$$

(41.6)

where $y_{\mu 1}$ is the Yukawa coupling of the vertex $\eta^\pm N_1 \mu$. The decay length as a function of $\eta^\pm$ mass for different values of $y_{\mu 1}$ are given in Fig. 41.3 (left panel). Another possibility arises when the mass splitting between $\eta^\pm$ and η^0 is very small, of the order of 100 MeV. For such mass splitting, the dominant decay mode of $\eta^\pm$ can be $\eta^\pm \to \eta^0 \pi^\pm$, if the corresponding Yukawa coupling of $\eta^\pm N_1 l$ vertex is kept sufficiently small for the leptonic decay mode to be subdominant. The corresponding decay width is given by

$$\Gamma_{\eta^\pm \to \eta^0 \pi} = \frac{f_\pi^2 \, g^4}{m_W^4} \frac{\left(m_{\eta^\pm}^2 - m_{\eta^0}^2\right)^2}{512 m_{\eta^\pm} \pi} \sqrt{1 - \left(\frac{m_{\eta^0} - m_\pi}{m_\eta^\pm}\right)^2} \sqrt{1 - \left(\frac{m_{\eta^0} + m_\pi}{m_\eta^\pm}\right)^2}$$

(41.7)

where f_π, g, m_W are the form factor, gauge coupling and W boson mass, respectively. Such tiny decay width keeps the lifetime of $\eta_1^\pm$ considerably long enough that it can reach the detector before decaying. In the decay $\eta^\pm \to \eta^0 \pi^\pm$, the final state pion

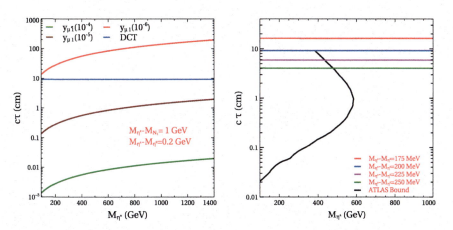

Fig. 41.3 Decay length corresponding to the pionic decay $\eta^\pm \to \eta^0 \pi^\pm$ leading to DCT and its comparison with the decay $\eta^\pm \to N_1 \mu$ responsible for displaced vertex signature (left panel). Decay length corresponding to the pionic decay $\eta^\pm \to \eta^0 \pi^\pm$ for fixed mass splitting of 200 MeV (left panel) and its comparison with the ATLAS bound for different benchmark values of mass splitting (right panel)

typically has very low momentum and it eventually decays into DM and a light neutrino, hence remains invisible throughout. Therefore, it gives a signature where a charged particle leaves a track in the inner part of the detector. The corresponding decay length as a function of $\eta^{\pm}$ mass is shown in Fig.41.3 with a blue line.

41.6   Conclusions

We have studied a simple extension of the minimal gauged $U(1)_{B-L}$ with three right-handed neutrinos in order to realise fermion singlet dark matter. The minimal model is extended by a scalar doublet η and an additional Z_2 symmetry so that the right-handed neutrinos and η are odd under this Z_2 symmetry. We studied the co-annihilation effects and we constrain the parameter space from the requirements of DM relic density, light neutrino masses and mixing, MEG 2016 bound on $\mu \rightarrow e\gamma$ and finally from the requirement of producing displaced vertex and disappearing charge track signatures at the LHC.

References

1. S. Kanemura, O. Seto, T. Shimomura, Phys. Rev. D **84**, 016004 (2011). arXiv:1101.5713 [hep-ph]
2. A. Vicente, C.E. Yaguna, JHEP **1502**, 144 (2015). arXiv:1412.2545 [hep-ph]
3. A.G. Hessler, A. Ibarra, E. Molinaro, S. Vogl, JHEP **1701**, 100 (2017). arXiv:1611.09540 [hep-ph]
4. N. Aghanim et al., [Planck Collaboration]. arXiv:1807.06209 [astro-ph.CO]
5. E. Ma, Phys. Rev. D **73**, 077301 (2006). [hep-ph/0601225]
6. J.A. Casas, A. Ibarra, Nucl. Phys. B **618**, 171 (2001). [hep-ph/0103065]
7. T. Toma, A. Vicente, JHEP **1401**, 160 (2014). arXiv:1312.2840 [hep-ph]
8. D. Borah, D. Nanda, N. Narendra, N. Sahu, arXiv:1810.12920 [hep-ph]d

Chapter 42
Mixing Dynamics of Dimension-Five Interactions (Scalar/Pseudoscalar-Photon) in Magnetized Medium

Ankur Chaubey, Manoj K. Jaiswal, and Avijit K. Ganguly

Abstract In many extentions of standard model, dimension-5 scalar di-photon ($g_{\gamma\gamma\phi}\phi\ F^{\mu\nu}F_{\mu\nu}$) or pseudoscalar di-photon ($g_{\gamma\gamma a}a\tilde{F}^{\mu\nu}F_{\mu\nu}$,) interaction materializes due to scale symmetry breaking or $U_A(1)$ symmetry breaking. In a magnetized vacuum (i.e., in an external background field $\bar{F}_{\mu\nu}$), the transverse degrees of freedom of the photons—for such systems—can be described in terms of the form factors constructed out of the background field strength tensor ($\bar{F}_{\mu\nu}$) and the same for dynamical photon ($f^{\mu\nu}$); they happen to be $\bar{F}_{\mu\nu}f^{\mu\nu}$ and $\bar{\tilde{F}}_{\mu\nu}f^{\mu\nu}$. These form factors transform differently under CP transformation. While $\bar{F}_{\mu\nu}f^{\mu\nu}$ (describing polarization orthogonal to B ($|\gamma_\parallel>$)) is CP even, the other one, $\bar{\tilde{F}}_{\mu\nu}f^{\mu\nu}$ (describing polarization along B ($|\gamma_\perp>$), is CP odd. In the interaction Lagrangian, if the scalar is interchanged with the pseudoscalar, the role of the two form factors just gets interchanged. Thus, for nearly degenerate strengths of the coupling constants ($g_{\gamma\gamma\phi}$ and $g_{\gamma\gamma a}$) and masses (m_ϕ and m_a) of the respective candidates, proper identification of one from the other may become very difficult in laboratory or astrophysics based experiments. The basic motivation of this investigation is to reduce this uncertainty through the incorporation of parity violating (*originating through magnetized medium effects*) part of the photon self-energy in the effective Lagrangian. This step, in turn, affects the (Pseudo) Scalar-Photon mixing dynamics drastically and brings out a significant change in the spectrum of the electromagnetic beam undergoing such interaction.

All queries about the content should be directed to the corresponding author Avijit K. Ganguly

A. Chaubey
Department of Physics, Institute of Science, Banaras Hindu University, Varanasi, India
e-mail: ankur.chaubey@bhu.ac.in

M. K. Jaiswal
Department of Physics, University of Allahabad, Prayagraj 211002, India
e-mail: mkjaiswalbhu@gmail.com

A. K. Ganguly (✉)
Department of Physics (MMV), Banaras Hindu University, Varanasi, India
e-mail: avijitkganguly@gmail.com

© Springer Nature Singapore Pte Ltd. 2021
P. K. Behera et al. (eds.), *XXIII DAE High Energy Physics Symposium*,
Springer Proceedings in Physics 261,
https://doi.org/10.1007/978-981-33-4408-2_42

293

42.1 Introduction

Pseudoscalar particles like axions a(x) are common to be associated with the breaking of chiral symmetries in many theories of unification (in physics beyond the standard model) [1–9] through quantum effects; and so is also the case with the Goldstone bosons of a spontaneously broken scale symmetry (dilaton) $\phi(x)$ [5, 6]. They both have remained possible candidates of Dark matter for some times now. The interaction dynamics of these exotic particles, i.e., scalars $(\phi(x))$ or pseudoscalars $(a(x))$ with photon (γ) is governed by Dim-5 operators $g_{\gamma\gamma\phi}\phi F^{\mu\nu}F_{\mu\nu}$ or $g_{\gamma\gamma a}a\tilde{F}^{\mu\nu}F_{\mu\nu}$.

The associated form factors $\bar{F}_{\mu\nu}f^{\mu\nu}$ and $\tilde{\bar{F}}_{\mu\nu}f^{\mu\nu}$, for the transverse degrees of freedom of the photons—in external background field $\bar{F}_{\mu\nu}$—for such systems have different CP transformation properties. As a result, in the equation of motion, the CP even form factor $\bar{F}_{\mu\nu}f^{\mu\nu}$ couples only to the CP even scalar field $\phi(x)$, while the other CP odd one, $\tilde{\bar{F}}_{\mu\nu}f^{\mu\nu}$, propagates freely. In other words, out of the three available degrees of freedom, the mixing is between only two degrees of freedom—having identical CP properties [10]. And most importantly: the mixing matrix is 2×2. On the other hand, for magnetized pseudoscalar photon system, the reverse happens, i.e., the roles of the form factors $\bar{F}_{\mu\nu}f^{\mu\nu}$ and $\tilde{\bar{F}}_{\mu\nu}f^{\mu\nu}$ get interchanged.

Furthermore, the presence of the external field compromises the Lorentz symmetry for both the systems identically. For an external magnetic field in the z direction (B_z), except for rotational and boost symmetry around and along B_z, all other space-time symmetries get compromised. This manifests itself by turning the vacuum into an optically active and dichroic medium, for the photons [10–12] passing through such region. Utilizing this, standard polarimetric observables like polarization or ellipticity angle can be measured and used to determine the magnitude of the coupling constant and mass $g_{\gamma\gamma\phi}$ and m_ϕ for $\phi - \gamma\gamma$ system or $g_{\gamma\gamma a}$ and m_a for $a - \gamma\gamma$ system. This process of determination is, however, subject to cross-correlated verification from other experiments, for example, [13].

However, the 2×2 nature of the mixing matrix for both $\phi - (x)\gamma\gamma$ and $a - (x)\gamma\gamma$ system poses a problem, when the magnitude of the masses m_a and m_ϕ, as well as the coupling constants $(g_{\gamma\gamma a})$ and $(g_{\gamma\gamma\phi})$ are close to each other. In such a scenario, the identification of one from the other is difficult using the polarimetric techniques. The reason being, as one moves from $\phi - (x)\gamma\gamma$ to $a - (x)\gamma\gamma$ system, the role of the two polarization form factors gets interchanged with each other. As a result, the absolute magnitude of the ellipticity and polarization angle remains the same. And the degree of polarization also remains insensitive to the underlying theory.

The main motivation of this study is to explore other physical corrections, such that the incorporation of them would eventually break the degeneracy in the 2×2 mixing pattern undergone by both $\phi(x) - \gamma\gamma$ and $a(x) - \gamma\gamma$ system in a magnetized vacuum.

It so happens that, as one incorporates the parity violating part of photon-self-energy-tensor (PSET), that appears once the effect of magnetized medium is incorporated in the evaluation of PSET, in the effective Lagrangian of the system, the apparent degeneracy in mixing gets lifted. This happens due to the discrete symmetries enjoyed by the respective form factors of the photon, as well as the scalar or pseudoscalar field. With the incorporation of such effect, the mixing matrix for $\phi F_{\mu\nu} F^{\mu\nu}$, type of interactions, turns out to be 3×3 and for $a \tilde{F}_{\mu\nu} F^{\mu\nu}$ interaction the mixing matrix is 4×4. That is there is the mixing of all four degrees of freedom—three degrees of freedom of the in-medium photon and one degree of freedom of the pseudoscalar, for $a(x) - \gamma\gamma$ system.

42.2 Mixing Dynamics of Scalars and Pseudosclars in Magnetized Plasma

The action for scalar photon system, as the quantum corrections due to ambient medium and an external magnetic field **eB** are taken into account [14], turns out to be

$$S = \int d^4 k \left[\frac{1}{2} A^\nu(-k) \left(-k^2 \tilde{g}_{\mu\nu} + \Pi_{\mu\nu}(k) + \Pi^p_{\mu\nu}(k) \right) A^\mu(k) \right.$$
$$\left. + i g_{\phi\gamma\gamma} \phi(-k) \bar{F}_{\mu\nu} k^\mu A^\nu(k) + \frac{1}{2} \phi(-k)[k^2 - m^2]\phi(k) \right]. \qquad (42.1)$$

Here $\Pi_{\mu\nu}(k)$ is the in-medium polarization tensor and $\Pi^p_{\mu\nu}(k)$ the parity violating part of the same evaluated in a magnetized medium. One can get the same for pseudoscalar/scalar-photon system from (42.1), by replacing $\phi(\pm k)$ by $a(\pm k)$ and $\bar{F}_{\mu\nu}$ by $\tilde{\bar{F}}_{\mu\nu}$. Derivation of the equations of motions follows next.

42.2.1 Mixing Matrix of Scalar Photon Interaction

The equations of motion, of scalar-photon system follows from (42.1), given by

$$\left[k^2 \mathbf{I} - \begin{pmatrix} \omega_p^2 & i \frac{\omega_p^2 e B_\parallel}{(\omega m_e)} & -i g_{\phi\gamma\gamma} B_\perp \omega \\ -i \frac{\omega_p^2 e B_\parallel}{(\omega m_e)} & \omega_p^2 & 0 \\ i g_{\phi\gamma\gamma} B_\perp \omega & 0 & m_\phi^2 \end{pmatrix} \right] \begin{bmatrix} A_\parallel(k) \\ A_\perp(k) \\ \phi(k) \end{bmatrix} = 0. \qquad (42.2)$$

The longitudinal degree of freedom doesn't couple to anything, it propagates freely. The same can be explained with the help of the discrete symmetries enjoyed by the form factor associated with the longitudinal degree of freedom of the photon. Hence, the mixing is between $A_\parallel(k)$, $A_\perp(k)$, and ϕ only. Where $A_\parallel(k)$, $A_\perp(k)$ are the form factors associated with the degrees of freedom of photon those are—parallel and perpendicular to the direction of magnetic field. We had obtained the solutions of (42.2) by diagonalizing the mixing matrix.

42.2.2 Mixing Matrix of Axion Photon Interaction

As before, the equations of motion for axion-photon system can be expressed in matrix notation as

$$[(\omega^2 + \partial_z^2)\mathbf{I} - \mathbf{M}'] \begin{pmatrix} A_\parallel(k) \\ A_\perp(k) \\ A_L(k) \\ a(k) \end{pmatrix} = 0, \tag{42.3}$$

where $\mathbf{I}$ is an identity matrix and matrix $\mathbf{M}'$ is the 4×4 mixing matrix. The same, in terms of its elements is given by

$$\mathbf{M}' = \begin{pmatrix} \Pi_T & -\Pi_p N_1 N_2 P_{\mu\nu} b^{(1)\mu} I^\nu & 0 & 0 \\ \Pi_p N_1 N_2 P_{\mu\nu} b^{(1)\mu} I^\nu & \Pi_T & 0 & -i g_{a\gamma\gamma} N_2 b_\mu^{(2)} I^\mu \\ 0 & 0 & \Pi_L & -i g_{a\gamma\gamma} N_L b_\mu^{(2)} \bar{u}^\mu \\ 0 & i g_{a\gamma\gamma} N_2 b_\mu^{(2)} I^\mu & i g_{a\gamma\gamma} N_L b_\mu^{(2)} \bar{u}^\mu & m_a^2 \end{pmatrix}. \tag{42.4}$$

We note that, projection operator $P_{\mu\nu}$ appearing in the M'_{12} and M'_{21} elements of the mixing matrix M' is a complex one, that makes the matrix M' a hermitian matrix, that is expected even otherwise on general grounds.

It is also important to note that for pseudoscalar-photon interaction, because of discrete symmetry considerations (PT symmetry to be specific), the form factor associated with longitudinal degree of freedom remains coupled with pseudoscalar field. Hence, the mixing matrix becomes 4×4. Therefore, the mixing dynamics for these two systems with incorporation of parity violating medium effect turns out to be completely different. Due to this, the identification of one from the other using polarimetric observables may become lot easier.

42.3 Optical Observables

Properties of polarized light waves can be described in terms of the Stokes parameters evaluated from the coherency matrix. The same is constructed from the solutions of the field equations; and is given by

$$\mathbf{D}'(z) = \begin{pmatrix} < A_\parallel(\omega, z) A_\parallel^*(\omega, z) > & < A_\parallel(\omega, z) A_\perp^*(\omega, z) > \\ < A_\perp(\omega, z) A_\parallel^*(\omega, z) > & < A_\perp(\omega, z) A_\perp^*(\omega, z) > \end{pmatrix}. \tag{42.5}$$

In (42.5) above, $<>$ represent the ensemble averages. The Stokes parameters are obtained from the elements of the coherency matrix by the following identifications: $\mathbf{I} = D'_{11}(z) + D'_{22}(z)$, $\mathbf{Q} = D'_{11}(z) - D'_{22}(z)$, $\mathbf{U} = 2Re\ D'_{12}(z)$, and $\mathbf{V} = 2Im\ D'_{12}(z)$.

The estimates of the other optical parameters, i.e., ellipticity angle, polarization angle, degree of linear polarization, degree of total polarization, follows from the expressions of $\mathbf{I}$, $\mathbf{U}$, $\mathbf{Q}$, and $\mathbf{V}$. The expressions for the polarization angle and ellipticity angle associated with an electromagnetic wave are provided below.

42.3.1   Polarization Angle & Ellipticity Angle

Polarization angle (represented by Ψ) is the angle between major and minor axis of ellipse, defined in terms of stokes parameters $\mathbf{U}$ and $\mathbf{Q}$, is given by

$$tan(2\Psi) = \frac{\mathbf{U}(\omega, z)}{\mathbf{Q}(\omega, z)}. \tag{42.6}$$

The ellipticity angle (denoted by χ) is defined in terms of the same parameters as

$$tan(2\chi) = \frac{\mathbf{V}(\omega, z)}{\sqrt{\mathbf{Q}^2(\omega, z) + \mathbf{U}^2(\omega, z)}}. \tag{42.7}$$

42.4   Results and Conclusions

Unlike polarization angle, the ellipticity angle remains invariant under rotation of the axes. So we have compared the magnitude of the ellipticity angle produced through axion-photon, as well as scalar-photon interaction, in the vicinity of a strongly mag-netized compact astrophysical source. The parameters, that we have considered for

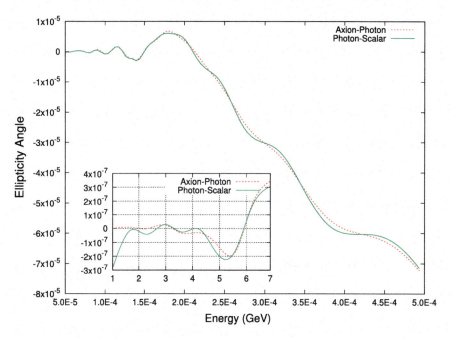

Fig. 42.1 Plot for ellipcity angle versus energy in case of coupled photon-axion system. The abscissa of the plot, in the inset is, in units of 10^{-5} GeV

the system are as follows: plasma frequency $\omega_p = 1.6 \times 10^{-10}$ GeV, coupling constants $g_{\gamma\gamma a} = g_{\gamma\gamma\phi} = 10^{-11} GeV^{-1}$ and mass m_a and m_ϕ both close to zero. The magnetic field is taken to be $B = 10^{12}$ Gauss and the path length considered here is 2.5 Km. The numerical estimates of the ellipticity angle for the two systems are plotted in, Fig. 42.1.

As can be seen in the plot that—for the values of the parameters chosen here—the numerical magnitudes of the angle for the $a\gamma$ and $\phi\gamma$ system, are extremely close to each other. However, there is some departure, that can be seen in the inset of Fig. 42.1. In the energy range of 1×10^{-5} GeV to 1.5×10^{-5} GeV, there is some visible difference in the ellipticity angle between axion photon and scalar photon systems. For energies close 1×10^{-5} GeV the difference is around 3×10^{-7} rad. Though this is little less for current sensitivity available for the detectors, however, we hope that the future detectors would have similar sensitivity to resolve this difference and shed light on the values of the parameters like $g_{\gamma\gamma\phi}$ or $g_{\gamma\gamma a}$ and m_ϕ or m_a. Studies along this direction are currently under progress and would be communicated elsewhere shortly.

References

1. R.D. Peccei, H.R. Quinn, Phys. Rev. Lett. **38**, 1440 (1977). Phys. Rev. D **16**, 1791 (1977). R.D. Peccei, Report No.hep-ph/9606475, for a more extensive review of the strong CP problem, see in CP Violation, edited by C. Jarlskog (World Scientific, Singapore, 1989)
2. S. Weinberg, Phys. Rev. Lett. **40**, 223 (1978)
3. F. Wilczek, Phys. Rev. Lett. **40**, 279 (1978)
4. Y. Chikashige, R. Mohapatra, R.D. Peccei, Phys. Rev. Lett. **45**, 1926 (1980). Phys. Lett. **98**B, 265 1981. G. Gelmini, M. Roncadelli, ibid. **99**B, 411 (1981)
5. Y.M. Cho, J.H. Kim, Phys. Rev. D 79 023504 (2009)
6. T. Damour, J.F. Donoghue, Phys. Rev. D **79**, 023504 (2009)
7. P. Svrcek, E. Witten, JHEP **06**, 061 (2006)
8. A. Das, J. Maharana, Phys. Lett. B **699**, 264 (2011). https://doi.org/10.1016/
9. D.J.E. Marsh, Phys. Rept. **643**, 1 (2016)
10. G. Raffelt, L. Stodolsky, Phys. Rev. D **37**, 1237 (1988)
11. L. Maiani, R. Petronzio, E. Zavattini, Phys. Lett. B **175**, 359 (1986)
12. A.K. Ganguly, M.K. Jaiswal, Phys. Rev. D **90**, 026004 (2014)
13. M. Lawson, A.J. Millar, E. Vitagliano, F. Wilczek, Tunable Axion Plasma haloscopes. arXiv:1904. (11872[hep-ph])
14. A.K. Ganguly, P. Jain, S. Mandal, Phys. Rev. D **79**, 115014 (2009)

Chapter 43
Late Time Cosmology with Viscous Self-interacting Dark Matter

Arvind Kumar Mishra, Jitesh R. Bhatt, and Abhishek Atreya

Abstract Self-interacting dark matter (SIDM) is an intriguing possibility to address the small scale problems faced by the collisionless cold dark matter. The self interaction between the dark matter (DM) particles can lead to viscosity. At late times, this viscosity can become strong enough to account for the present observed cosmic acceleration, and hence mimic the dark energy. In this work, using a power law form of the average peculiar velocity gradient of the DM, we calculate the Hubble expansion rate and the deceleration parameter for small redshifts ($0 \leq z \leq 2.5$). We then estimate the model parameters from χ^2 analysis and argue that the dissipational effect of viscous SIDM is small at the early times and become prominent at the late time. Later, we match our viscous SIDM model with the small redshift data and find that this model explains the data very well. Our analysis is independent of any SIDM particle Physics model.

43.1 Introduction

The observations suggest that the collisionless cold dark matter (CDM) paradigm works well on the large cosmological scales but fail to explain the small scale data. It has been argued that since the problems are related to the small length scale, hence including the baryonic physics at the small scale could possibily resolve the tensions. But whether the baryonic mechanism can ameliorate all the problems or not is still

A. K. Mishra (✉)
Theoretical Physics Division, Physical Research Laboratory, Navrangpura,
Ahmedabad 380009, India
e-mail: arvind@prl.res.in

Indian Institute of Technology Gandhinagar, Palaj, Gandhinagar 382355, India

J. R. Bhatt
Theoretical Physics Division, Physical Research Laboratory, Navrangpura, Ahmedabad 380009, India

A. Atreya
Center For Astroparticle Physics and Space Sciences, Bose Institute, Kolkata 700009, India

© Springer Nature Singapore Pte Ltd. 2021   299
P. K. Behera et al. (eds.), *XXIII DAE High Energy Physics Symposium*,
Springer Proceedings in Physics 261,
https://doi.org/10.1007/978-981-33-4408-2_43

under debate. For a recent review on small scale issues and their possible solutions, see [1].

Further, it has been suggested that instead of the collisionless dark matter, if the dark matter is self-interacting, then it can potentially solve the small scale issues [2]. For this, the dark matter scattering cross section should be mildly dependent on the DM velocity [3]. The beauty of SIDM lies in the fact that at the small scales, it solves the crisis by contributing sufficiently large self interaction, but on the large scales, it behaves like collisionless dark matter and respects the success of the standard cosmology.

43.2 SIDM Viscosity and Late Time Accelerated Expansion

The collisional nature of SIDM particles on the small scale suggests that the SIDM fluid can contribute to the cosmic viscosity. The origin of viscous coefficients is attributed to the self interaction between the DM particles. In order to calculate the SIDM viscosity, we use the kinetic theory in the relaxation time approximation. Using the non-relativistic Maxwell–Boltzmann distribution function and rest frame of fluid velocity, the shear, η and bulk viscous coefficient, ζ can be obtained as [4, 5]

$$\eta = \frac{1.18 \, m \langle v \rangle^2}{3 \quad \langle \sigma v \rangle} \tag{43.1}$$

$$\text{and } \zeta = \frac{5.9 \, m \langle v \rangle^2}{9 \quad \langle \sigma v \rangle}, \tag{43.2}$$

where $m, n, \langle v \rangle$ and $\langle \sigma v \rangle$ represents the DM mass, average number density, average velocity, and velocity weighted cross section average, respectively. We refer [6] for a review on the cosmic viscosity for late and early time cosmology and [7] for the recent constraints on the DM viscosity.

In order to study the effect of the SIDM viscosity, we need to set up the equation for cosmic evolution. In the Landau frame and first-order gradient expansion, the energy–momentum tensor for the viscous SIDM, $T^{\mu\nu}$ can be obtained as

$$T^{\mu\nu} = \epsilon u^\mu u^\nu + (P + \Pi_B)\Delta^{\mu\nu} + \Pi^{\mu\nu}, \tag{43.3}$$

where ϵ, P and u^μ corresponds for the energy density, kinetic pressure and four velocity of the SIDM particle. Also, $\Delta^{\mu\nu} = u^\mu u^\nu + g^{\mu\nu}$ is the projection operator that satisfy, $u_\mu \Delta^{\mu\nu} = u_\nu \Delta^{\mu\nu} = 0$. Further, the Π_B and $\Pi^{\mu\nu}$ represent the bulk and shear stress tensor, which are defined as

$$\Pi_B = -\zeta \nabla_\mu u^\mu \tag{43.4}$$

$$\text{and} \quad \Pi^{\mu\nu} = -\eta \left[\Delta^{\mu\alpha} \Delta^{\nu\beta} + \Delta^{\mu\beta} \Delta^{\nu\alpha} - \frac{2}{3} \Delta^{\mu\nu} \Delta^{\alpha\beta} \right] \nabla_\alpha u_\beta . \tag{43.5}$$

Further, equipped with the knowledge of $T^{\mu\nu}$, one can get the evolution equation of the energy density using the energy–momentum conservation equation, $\nabla_\mu T^{\mu\nu} = 0$ as [8]

$$\frac{1}{a} \langle \dot{\epsilon} \rangle_s + 3H \left[\langle \epsilon \rangle_s + \langle P \rangle_s - 3 \langle \zeta \rangle_s H \right] = D , \tag{43.6a}$$

where the D contains the information about SIDM viscous effect and given by

$$D = \frac{1}{a^2} \langle \eta \left[\partial_i v_j \partial_i v_j + \partial_i v_j \partial_j v_i - \frac{2}{3} \partial_i v_i \partial_j v_j \right] \rangle_s + \frac{1}{a^2} \langle \zeta [\nabla \cdot \mathbf{v}]^2 \rangle_s + \frac{1}{a} \langle \mathbf{v} \cdot \nabla (P - 6\zeta H) \rangle_s . \tag{43.6b}$$

In the above expression $\langle A \rangle_s$ represent the spatial average of A. Here, we consider the convention for denoting differentiation $\dot{A} \equiv \frac{dA}{d\tau}$ and $A' \equiv \frac{dA}{dz}$. Furthermore, the evolution equation for the Hubble expansion rate is estimated using the spatial average of the trace Einstein's equation, $\langle G^\mu_\mu \rangle_s = -8\pi G \langle T^\mu_\mu \rangle$. This gives [8]

$$\frac{\dot{H}}{a} + 2H^2 = \frac{4\pi G \langle \epsilon \rangle_s}{3} \left(1 - 3\hat{w}_{\text{eff}} \right) , \tag{43.7}$$

where the effective equation of state, $\hat{w}_{\text{eff}}$ is defined as $\langle P \rangle_s + \langle \Pi_B \rangle_s = \hat{w}_{\text{eff}} \langle \epsilon \rangle_s$. To study the late time cosmic evolution, we derive the equation for the deceleration parameter, q, which can be defined in terms of the dimensionless parameter $\bar{H} = H/H_0$, where H_0 is the present Hubble parameter, i.e., $H(z = 0) = H_0$, as

$$q(z) = -1 + (1+z) \frac{\bar{H}'}{\bar{H}} . \tag{43.8}$$

Thus, applying the (43.6a) and (43.7), we get the equation for q as [8]

$$-\frac{dq}{d \ln a} + 2(q - 1) \left[q - \frac{(1 + 3\hat{w}_{\text{eff}})}{2} \right] = \frac{4\pi G D}{3H^3} \left(1 - 3\hat{w}_{\text{eff}} \right) . \tag{43.9}$$

From the above equation, it becomes clear that the dynamics of q will depend on D. For sufficiently large values of the velocity gradient and viscosity, D becomes large and hence modifies the cosmic evolution. In the assumption $| - \frac{dq}{d \ln a} | \ll 1$, the condition for the accelerated expansion from (43.9) is given as

$$\frac{4\pi G D}{3 H^3} > \frac{1 + 3\hat{w}_{\text{eff}}}{1 - \hat{w}_{\text{eff}}} , \tag{43.10}$$

where $\hat{w}_{\text{eff}} \neq 1$. Thus, in order to study the dynamics of the Universe, we need to estimate the dissipation term D.

43.3   Calculation of D and Cosmic Evolution

To calculate the D term from the (43.6b), we use the following assumptions as discussed in [5]: (1) SIDM is cold, i.e., $\hat{w}_{\text{eff}} \approx 0$. (2) The viscous coefficients η and ζ depends on the thermalization scale, which will be decided by the mean free path calculation. The mean free path of the SIDM particle in the dilute gas approximation is $\lambda_{\text{SIDM}} \sim 10^9 (1/\rho)(m/\sigma)$ kpc, where σ/m is in cm^2/g and ρ is in $M_\odot$kpc^{-3} [4]. For cluster scale, $\lambda_{\text{SIDM}} \sim 1$ Mpc, which is order of the cluster size. But for the galaxy scale, the $\lambda_{\text{SIDM}} \sim 1$ MPc, but the size of the galaxy ~ 10 Kpc scale. Thus, we find that for SIDM, the fluid hydrodynamics will be valid only on the cluster to a larger scale. Hence, all the averaging and viscosity calculation should be done at least on the cluster scale. (3) The cluster scale has virialized and hence viscosities are fixed into the redshift range $0 \leq z \leq 2.5$. (4) The velocity gradients are defined on the scale larger than the hydrodynamic scales. We replace the velocity gradient from the average and to study their evolution into the range of our interest $0 \leq z \leq 2.5$, we consider its form as power law [5]

$$\langle \partial v \rangle_s \sim \frac{v_0}{L} (1+z)^{-n} ,$$ (43.11)

where $n \geq 0$ is free parameter and v_0 corresponds to fluid velocity.

Thus, using the above assumptions and from (43.6b), the D term simplifies as

$$D = (1+z)^2 \left(\frac{v_0}{L(1+z)^n} \right)^2 \left(\frac{4}{3}\eta + 2\zeta \right).$$ (43.12)

Hence, using (43.12) in (43.9), we get the simplified equation for the q as

$$\frac{dq}{dz} + \frac{(q-1)(2q-1)}{(1+z)} = \beta \left(\frac{1+z}{\bar{H}^3} \right) ,$$ (43.13)

where β is the new dissipation parameter defined as

$$\beta = 3.88 \times 10^{13} \left[\frac{100h\text{km/(sec-Mpc)}}{H_0} \right]^3 \left[\frac{(\text{cm}^2/\text{gm}) \, \text{km/sec}}{\langle \sigma v \rangle /m} \right] \left(\frac{\langle v \rangle}{c} \right)^2 \left(\frac{\text{Mpc}}{L} \right)^2 \left(\frac{v_0/c}{(1+z)^n} \right)^2 .$$ (43.14)

where $H_0 = 100h$ km/(sec-Mpc) and $h = 0.715$ and c is speed of light. Also the cluster scale velocity, $\frac{\langle v \rangle}{c} = \frac{10^{-2}}{3}$, $\frac{\langle \sigma v \rangle}{m}$ from [3] and fluid velocity is taken as the supercluster scale velocity, i.e., $v_0/c = 2 \times 10^{-2}$.

Fig. 43.1 The joint confidence region of model parameters n and L have been plotted. The region correspond to 68.3, 95.4 and 99.73% confidence limits. The best fit value is shown as a point. The plot is taken from the Ref. [5]

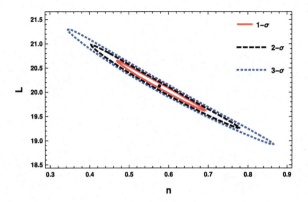

43.4   Estimation of Model Parameters

In order to study the cosmic evolution, we solve the coupled differential equations (43.8) and (43.13), with the initial conditions at present, $\bar{H}(z = 0) = 1$ and $q_0 = -0.60$ [9]. We thus find that the solution depends on the two parameters, n and L. To estimate theses parameters, we use χ^2 analysis with the cosmic chronometer data set [10]. The χ^2 is defined by

$$\chi^2(z, n, L) = \sum_{i=1}^{N} \left[\frac{H_{obs}(z_i) - H_{th}(z_i, n, L)}{\sigma_i} \right]^2. \qquad (43.15)$$

Here $H_{obs}(z_i), H_{th}(z_i, n, L)$ corresponds to the ith observational Hubble data and the theoretically predicted value for Hubble expansion rate and variance, respectively. Here N is total number of chronometer data points. The model parameters can be obtained from the χ^2 minimization. The quantity χ^2 per degree of freedom $\chi^2_{d.o.f} = \frac{\chi^2_{min}}{N-M}$, where M is the number of parameters.

In Fig. 43.2, we plot the regions corresponding to 68.3% (red solid), 95.4% (black dashed) and 99.73% (blue dotted) confidence limits and the best fit values as a point. Also, the χ^2_{min}, $\chi^2_{d.o.f}$, $1 - \sigma$ values and the best fit values of parameters (n, L) are listed in Table 43.1.

43.5   Results and Discussions

In this section, we will use the best fit values of the model parameters to study the late time cosmology. In Fig. 43.1a, we have fitted the supernovae (SN) data [11, 12] from the best fit model parameters of viscous SIDM model (red solid line) and also compared with the case when the velocity gradients are constant with the redshift,

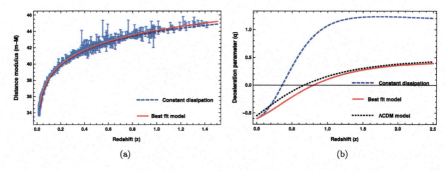

Fig. 43.2 In Fig. 43.2 (**a**), the SN data is plotted with the best fit model parameters and constant velocity gradient. In Fig. 43.2 (**b**), the q is plotted for best fit model parameters, constant velocity gradient and ΛCDM model. The plot suggests that the best fit model parameters explain the SN data and attain the correct value of q on large redshift. The plots are taken from the Ref. [5]

Table 43.1 The best fit model parameter

Data set	1-σ	χ^2_{min}	$\chi^2_{d.o.f}$	Best fit values
Cosmic chronometer	$0.5083 \leq 0.5770 \leq 0.6513$	22.0207	0.6116	$n = 0.5770$
	$19.8002 \leq 20.1265 \leq 20.4416$			$L = 20.1265$ Mpc

i.e., $n = 0$ (blue dashed line). We find that on the large redshift the best fit model differs from the constant velocity gradient model and explain the SN data comparably better.

Further, in Fig. 43.1b, we have also plotted the deceleration parameter for both the best fit model parameters (red solid line) and constant velocity gradient with the redshift (blue dashed line) and compare with the standard ΛCDM model (black dotted line). We see that that the best fit model obtained the correct value of q at large redshift and matches with the ΛCDM model very well but the constant velocity gradient with the redshift doesn't. Here we also note that at present time, i.e., $z = 0$, the $\beta = 4\pi GD/3H_0^3 \approx 4.1$. Hence, from (43.10), it is clear that the viscosity may cause to present observed cosmic acceleration.

43.6 Conclusion

Self-interacting dark matter can solve the small scale problems through dark matter scattering and also leads to the bulk and shear viscosity. The effect of the viscosity becomes prominent at late time and modify the cosmic evolution. At present, the contribution of these viscous effect becomes large and hence create the accelerated expansion of the Universe. In this work, we have studied the late time cosmic evolution within the framework of the viscous SIDM model.

Assuming the cluster scale as a virialized structure at redshift range of our interest and the SIDM viscosity to be constant, we study the cosmic evolution for the power law form of the average velocity gradient. To find the values of model parameters, we apply the χ^2 analysis. The best fit values of the model parameters suggest that the average velocity gradient and hence the viscous dissipation decreases at earlier times (at large redshift). This explains the low redshift observation such as supernova data and also produces the correct value of the deceleration parameter at large redshift.

Thus, we conclude that the viscous SIDM model can produce low redshift observations and produce the correct value of the cosmic acceleration without any necessary of the extra dark energy component.

References

1. S. Tulin, H.B. Yu, Phys. Rept. **730**, 1 (2018)
2. D.N. Spergel, P.J. Steinhardt, Phys. Rev. Lett. **84**, 3760 (2000)
3. M. Kaplinghat, S. Tulin, H.B. Yu, Phys. Rev. Lett. **116**(4), 041302 (2016)
4. A. Atreya, J.R. Bhatt, A. Mishra, JCAP **1802**(02), 024 (2018)
5. A. Atreya, J.R. Bhatt, A.K. Mishra, JCAP **1902**, 045 (2019)
6. I. Brevik, Ø. Grøn, J. de Haro, S.D. Odintsov, E.N. Saridakis, Int. J. Mod. Phys. D **26**(14), 1730024 (2017)
7. J.R. Bhatt, A.K. Mishra, A.C. Nayak, Phys. Rev. D **100**, 063539 (2019) http://arxiv.org/abs/1901.08451. arXiv: 1901.08451
8. S. Floerchinger, N. Tetradis, U.A. Wiedemann, Phys. Rev. Lett. **114**(9), 091301 (2015)
9. P.A.R. Ade et al., Planck collaboration. Astron. Astrophys. **571**, A16 (2014)
10. O. Farooq, F.R. Madiyar, S. Crandall, B. Ratra, Astrophys. J. **835**(1), 26 (2017)
11. R. Amanullah et al., Astrophys. J. **716**, 712 (2010)
12. N. Suzuki et al., Supernova cosmology project collaboration. Astrophys. J. **746**, 85 (2012)

Chapter 44
Constraining Axion Mass from Cooling of Neutron Star

Avik Paul and Debasish Majumdar

Abstract Neutron star cooling takes place generally via the neutrino and gamma emissions. Here additionally we studied the cooling of neutron star through axion emission via the nucleon-nucleon axion bremsstrahlung processes. We consider the nature of cooling of neutron star including the nucleon-nucleon axion bremsstrahlung process for both the degenerate and non-degenerate cases. The effect of axion cooling is demonstrated through the variation of luminosity with time and variation of surface temperature with time. The thermal evolution of a neutron star is discussed with two neutron star masses (1.4, 1.8 $M_\odot$) and for each of the cases we consider three axion masses namely $m_a = 10^{-5}$ eV, 10^{-3} eV, 10^{-2} eV. The results are compared with the case when the neutron star is cooled only by neutrino and gamma emissions and no axion induced cooling is considered. We use three data points from the observations of pulsars namely PSR B0656+14, Geminga, and PSR B1055-52 and compared these with the cooling curve obtained from our calculations. From these analyses, we derive an upper bound on axion mass limit to be $m_a \leq 10^{-3}$ eV, which indicates that the axion decay constant $f_a \geq 0.6 \times 10^{10}$ GeV.

44.1 Introduction

Cooling of a neutron star (NS) generally takes place via the emission of photons and neutrinos [1]. Several studies have indicated that axions can also be emitted from a NS in addition to photon and neutrino and affects the cooling processes of the NS [2, 3]. Axions are pseudo Nambu-Goldstone bosons arise out of the Peccei-Quinn (PQ) solution of strong CP problem [4] of QCD Lagrangian where a chiral $U(1)_A$ symmetry is introduced which is spontaneously broken at the PQ energy scale.

A. Paul (✉) · D. Majumdar
Astroparticle Physics and Cosmology Division, Saha Institute of Nuclear Physics, HBNI,
1/AF Bidhannagar, Kolkata 700064, India
e-mail: avik.paul@saha.ac.in

D. Majumdar
e-mail: debasish.majumdar@saha.ac.in

© Springer Nature Singapore Pte Ltd. 2021
P. K. Behera et al. (eds.), *XXIII DAE High Energy Physics Symposium*,
Springer Proceedings in Physics 261,
https://doi.org/10.1007/978-981-33-4408-2_44

Axions are related to the various processes in astrophysics and cosmology. They can also be a particle candidate of dark matter in the Universe. Due to small couplings of axion with photons, nucleons, and electrons, they can be produced inside a NS and also can escape freely from the star. In this work, we consider the nucleon-nucleon axion bremsstrahlung process for the axion production in the interior of the NS. We calculate the axion energy loss rate for two cases namely degenerate and non-degenerate limits. We obtain the energy loss rate of NS as a function of time, and also the variation of temperature with time for the cases without axion and with axion in addition to neutrinos and photons. From these analyses, we find that the axion emission effects significantly on the cooling of neutron star when axion mass is $\sim 10^{-5}$ eV and higher.

44.2 Formalism

The Hamiltonian for the interaction of nucleons with the axions can be expressed as [5]

$$\mathcal{H}_{int} = -\frac{C_N}{2f_a}\bar{\psi}_N\gamma_\mu\gamma_5\psi_N\partial^\mu a, \qquad (44.1)$$

where C_N denotes dimensionless model dependent coupling constant, f_a represents the PQ energy scale for axions, the ψ_N is the nucleon Dirac fields, where N can be a proton p or a neutron n, and "a" denotes the axion field. Due to axion-nucleon coupling, axion can be produced inside the neutron star by nucleon-nucleon axion bremsstrahlung process $N + N \rightarrow N + N + a$ and can escape freely from the star. In this axion emitting process, nucleons are interacting through one-pion exchange (OPE) potential. The expressions of the spin-summed squared matrix element for the pure processes $nn \rightarrow nn + a$ and $pp \rightarrow pp + a$ are given as [6]

$$\sum_{spins}\|\mathcal{M}\|_{NN}^2 = \frac{16(4\pi)^3\alpha_\pi^2\alpha_a}{3m_N^2}\left[\left(\frac{\mathbf{K}^2}{\mathbf{K}^2+m_\pi^2}\right)^2 + \left(\frac{\mathbf{l}^2}{\mathbf{l}^2+m_\pi^2}\right)^2 + \frac{\mathbf{K}^2\mathbf{l}^2 - 3\left(\mathbf{K}.\mathbf{l}\right)^2}{(\mathbf{K}^2+m_\pi^2)(\mathbf{l}^2+m_\pi^2)}\right],$$

(44.2)

where m_N represents the nucleon mass, m_π denotes the pion mass, axion-nucleon "fine-structure constant" $\alpha_a \equiv (C_N m_N/f_a)^2/4\pi = g_{aN}^2/4\pi$, where $g_{aN} = (C_N m_N/f_a)$ is the axion-nucleon coupling constant, the pion-nucleon "fine-structure constant" $\alpha_\pi \equiv (f2m_N/m_\pi)^2/4\pi \approx 17$ with the pion-nucleon coupling $f \approx 1.05$. In 44.2 $\mathbf{K} = \mathbf{p}_2 - \mathbf{p}_4$ and $\mathbf{l} = \mathbf{p}_2 - \mathbf{p}_3$ where p_i's are the momenta of the nucleons N_i. The expressions of the spin-summed squared matrix element for "mixed" process $np \rightarrow np + a$ is given as [6]

$$\sum_{spins} |\mathcal{M}|^2_{np} = \frac{256\pi^2\alpha_\pi^2}{3m_N{}^2} \frac{(g_{an} + g_{ap})^2}{4} \left[2\left(\frac{l^2}{l^2 + m_\pi^2}\right)^2 - \frac{4\left(\mathbf{K}.\mathbf{l}\right)^2}{(\mathbf{K}^2 + m_\pi^2)(l^2 + m_\pi^2)} \right]$$

$$+ \frac{256\pi^2\alpha_\pi^2}{3m_N{}^2} \frac{(g_{an}^2 + g_{ap}^2)}{2} \left[\left(\frac{\mathbf{K}^2}{\mathbf{K}^2 + m_\pi^2}\right)^2 + 2\left(\frac{l^2}{l^2 + m_\pi^2}\right)^2 + 2\frac{\mathbf{K}^2 l^2 - \left(\mathbf{K}.\mathbf{l}\right)^2}{(\mathbf{K}^2 + m_\pi^2)(l^2 + m_\pi^2)} \right]. \tag{44.3}$$

In order to obtain the axion energy loss rate per unit volume, one needs to perform the phase-space integration which has the following form [5]:

$$Q_a = \int \frac{d^3\mathbf{K}_a}{2\omega_a(2\pi)^3}\omega_a \int \prod_{i=1}^{4} \frac{d^3\mathbf{P}_i}{2E_i(2\pi)^3} f_1 f_2(1 - f_3)(1 - f_4)$$

$$\times (2\pi)^4\delta^4(P_1 + P_2 - P_3 - P_4 - K_a)S \sum_{spins} |\mathcal{M}|^2_{NN}, \tag{44.4}$$

where K_a is the four-momentum of the axion, f_i's are the occupation numbers for the nucleons N_i's. In 44.4, for identical particles in the initial and final states $S = 1/4$ for pure processes and $S = 1$ for mixed processes. In this work, we consider degenerate and non-degenerate limits for the above mentioned axion emission processes and the simplified expressions for the axion energy loss rate are furnished below.

44.2.1 Non-degenerate Limit

In the Non-degenerate (ND) limit [6] ($\beta \equiv 3\langle(\hat{\mathbf{K}}.\hat{\mathbf{l}})^2\rangle = 1.3078$) total energy loss rate per unit volume Q_a using 44.2–44.4 is simplified as [7]

$$Q_a^{ND} = \frac{32}{105}\xi(T)\frac{\alpha_\pi^2 n_B^2 T^{7/2}}{m_N^{9/2}\pi^{3/2}}g_{ND}^2, \tag{44.5}$$

where $g_{ND} = \left(Y_n^2\tilde{g}_{nn}^2 + Y_p^2\tilde{g}_{pp}^2 + 4Y_nY_p\tilde{g}_{np}^2\right)$ is the total effective axion-nucleon coupling constant for the non-degenerate limit where Y_p and Y_n are the proton and neutron number fractions, respectively. Using the numerical values of $Y_p \approx 0.1$, $Y_n \approx 0.9$, $\xi(T) \approx 0.5$ and with the relation $m_a = 6\text{ eV}\left(\frac{10^{12}\text{GeV}}{f_a}\right)$ the values of g_{ND} can be calculated numerically as

$$g_{ND} = 4.71 \times 10^{-8}\left(\frac{m_a}{\text{eV}}\right). \tag{44.6}$$

Using the above relation for g_{ND}, the axion energy loss rate per unit volume (44.5) has the following form:

$$Q_a^{ND} = 2.90166 \times 10^{31} \ \text{erg cm}^{-3} \ \text{yr}^{-1} \ T_9^{3.5} \ \rho_{12}^2 \ m_{eV}^2 \tag{44.7}$$

where $m_{eV} \equiv m_a/eV$, $T_9 \equiv T/10^9 K$, and $\rho_{12} \equiv \dfrac{\rho}{10^{12}g/cm^3} = n_B m_N$ where n_B is the nucleon (baryon) density.

44.2.2 Degenerate Limit

In the degenerate (D) limit [6] ($\beta = 0$) total energy loss rate per unit volume Q_a using 44.2–44.4 is simplified as [7]

$$Q_a^D = \frac{31\pi^{5/3}(3n_B)^{1/3}\alpha_\pi^2 T^6}{3780m_N^2} \tilde{g}_D^2, \tag{44.8}$$

where $g_D = \left(Y_n^{1/3} F(Y_n)\tilde{g}_{an}^2 + Y_p^{1/3} F(Y_p)\tilde{g}_{ap}^2 + Y_{np}^{1/3} F(Y_{np})\tilde{g}_{np}^2\right)$ is the total effective axion-nucleon coupling constant for the degenerate limit where $F(u)$ ($u = Y_{np}, Y_p, Y_n$) has the following form [5]:

$$F(u)=1 - \frac{5u}{6}\arctan\left(\frac{2}{u}\right) + \frac{u^2}{3(u^2+4)} + \frac{u^2}{6\sqrt{2u^2+4}} \times \arctan\left(\frac{2\sqrt{2u^2+4}}{u^2}\right). \tag{44.9}$$

In the above $u = m_\pi/p_{F,N}$ and in this degenerate limit ($\rho_B \approx 2\rho_{\text{nuc}}$) it's value is approximated as $u \approx 0.32Y_N^{-1/3}$. We calculate the values of $F(u)$ for the pure and mixed processes by considering the numerical values of nucleon number fractions $Y_p = 0.01$, $Y_n = 0.99$ and $Y_{np} = 0.06$ and found $F(Y_n) \approx 0.64$, $F(Y_p) \approx 0.12$ and $F(Y_{np}) \approx 0.31$. With the above values, the effective coupling constant g_D can be calculated as

$$g_D = 2.04 \times 10^{-8}\left(\frac{m_a}{eV}\right). \tag{44.10}$$

Using 44.8 and 44.10, the axion energy loss rate per unit volume for the degenerate limit reduces to

$$Q_a^D = 4.84244 \times 10^{30} \ \text{erg cm}^{-3} \ \text{yr}^{-1} \ T_9^6 \ m_{eV}^2 \left(\frac{\rho_{NS}}{\rho_{nuc}}\right)^{1/3} \tag{44.11}$$

where ρ_{NS} denotes the neutron star density, ρ_{nuc} refers the nuclear density, $m_{eV} \equiv m_a/eV$ and $T_9 \equiv T/10^9K$.

44.2.3 Neutron Star Cooling

One can calculate the thermal evolution of the neutron star by solving the energy balance equation for the neutron star which can be written as [2]

$$\frac{dE_{\text{th}}}{dt} = C_v \frac{dT}{dt} = -L_\nu(T) - L_a(T) - L_\gamma(T_e), \tag{44.12}$$

where E_{th} denotes the thermal energy content of the star, T is the internal temperature, T_e is the effective temperature of the NS and C_v represents the specific heat capacity of the core. The quantities L_ν, L_a and L_γ are the neutrino, axion and photon luminosities, respectively, from the bulk of the NS. Several neutrino emitting processes contribute in the cooling of neutron stars. In the present work, we used the NSCool code [8] for calculating the neutrino and axion luminosities.

44.3 Calculations and Results

The thermal evolution of neutron star is computed by using NSCool numerical code for studying the effect of axion emission on the cooling of NSs. For the equation of state (EOS), we adopt Akmal-Pandharipande-Ravenhall (APR) EOS [9]. We have chosen two different masses for the neutron stars namely $1.4M_\odot$ and $1.8M_\odot$ and consider three axion masses namely $m_a = 10^{-5}$eV, 10^{-3}eV, and 10^{-2}eV in our calculations. In Fig. 44.1, we plot the variation of luminosity with time while in Fig. 44.2, variation of temperature with time are shown for both the degenerate and non-degenerate cases. We compare our theoretical cooling curves with the obserrvational data of three pulsars namely PSR B0656+14, Geminga, and PSR B1055-52. These observational data are shown by dots with error bars in the Figs. 44.1–44.2. For the NS mass $1.4M_\odot$ the observational data points (PSR B0656+14, Geminga) barely agree with these variations while for the higher NS mass of $1.8M_\odot$ these observational data points agree better with the theoretical calculations. In Fig. 44.3, we compare the two cases namely degenerate and non-degenerate limits for $1.4M_\odot$ NS with the axion mass of 10^{-3}eV. From Fig. 44.3 it can be seen that the cooling patterns for the two cases are different.

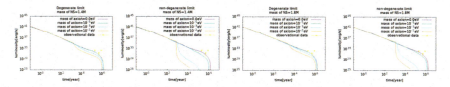

Fig. 44.1 Variation of luminosity with time for M $= 1.4M_\odot$ and M $= 1.4M_\odot$ with axion masses $m_a = 0$eV, 10^{-5}eV, 10^{-3}eV, 10^{-2}eV (from top to bottom). The theoretical cooling curves are compared with the observational data of three pulsars namely PSR B0656+14, Geminga, and PSR B1055-52 shown by dots with error bars from left to right in each plot. (Reproduced from Avik Paul et al., Pramana 92, 44 (2019))

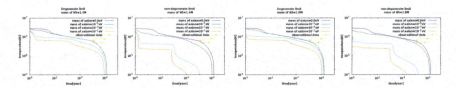

Fig. 44.2 Variation of temperature with time using the same conditions as in Fig. 44.1. (Reproduced from Avik Paul et al., Pramana 92, 44 (2019))

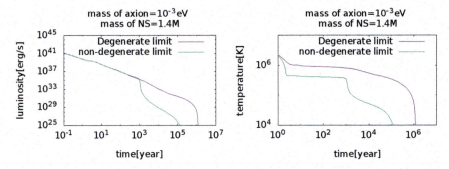

Fig. 44.3 Comparision between degenerate and non-degenerate cases with M = 1.4M$_\odot$ and m_a=10^{-3}eV. (Reproduced from Avik Paul et al., Pramana 92, 44 (2019))

44.4 Summary and Discussions

In this work, we explore the nature of cooling of neutron star due to axion emission by nucleon-nucleon axion bremsstrahlung process from the NS. We calculate axion loss rate for both the degenerate and non-degenerate cases. We demonstrate the thermal evolution of NS by calculating the variations of luminosity with time, as well as the variation of temperature of the NS with time. We have performed the calculations with considering three axion masses, namely 10^{-5}eV, 10^{-3}eV, and 10^{-2}eV and two neutron star masses of $1.4M_\odot$ and $1.8M_\odot$. We also adopt the APR equation of state for our calculations. We have compared our results with observational data of three pulsars namely PSR B0656+14, Geminga, and PSR B1055-52. We find that the emission of axions from the NS affects the cooling of NS through other emission processes gamma rays and neutrinos. We also find that the form of cooling curves are different for the degenerate and non-degenerate cases. From these analyses, we derive an upper bound on axion masses $m_a \leq 10^{-3}$ eV which indicates that the decay constant $f_a \geq 0.6 \times 10^{10}$GeV.

References

1. D. Page, J.M. Lattimer, M. Prakash, A.W. Steiner, Astrophys. J. **707**, 1131 (2009)
2. A. Sedrakian, Phys. Rev. D **93**(6), 065044 (2016)
3. H. Umeda, N. Iwamoto, S. Tsuruta, L. Qin, K. Nomoto, astro-ph/9806337
4. R.D. Peccei, Lect. Notes Phys. **741**, 3 (2008)
5. G.G. Raffelt, Stars as laboratories for fundamental physics. *The Astrophysics of Neutrinos, Axions, and Other Weakly Interacting Particles* (Chicago University Press, USA, 1996), 664 pages
6. G. Raffelt, D. Seckel, Phys. Rev. D **52**, 1780 (1995)
7. K. Bocker, hep-ph/0006337
8. http://www.astroscu.unam.mx/neutrones/NSCool/. We use the files Crust-EOS-Cat-HZD-NV.dat and APR-EOS-Cat.dat for the equation of state input
9. A. Akmal, V.R. Pandharipande, D.G. Ravenhall, Phys. Rev. C **58**, 1804 (1998)

Chapter 45
Constraining Kähler Moduli Inflation from CMB Observations

Mayukh Raj Gangopadhyay and Sukannya Bhattacharya

Abstract In models of inflation motivated by string theory, moduli vacuum misalignment leads to an epoch in the post-inflationary history of the universe when the energy density is dominated by cold moduli particles. This modifies the number of e-foldings (N_{pivot}) between horizon exit of the CMB modes and the end of inflation. Taking Kähler moduli inflation as a case, the shift in e-foldings is determined as a function of the model parameters which also determines the inflationary observables. The scenario is studied in detail and confronted with the latest *Planck+BICEP2/Keck array* data. We advocated a careful consideration of any post-inflationary non-standard epoch and the effects of reheating in the era of precision cosmology.

45.1 Introduction

In cosmology Inflation is described as the rapid exponential expansion of the universe after the big bang. Inflation is a necessary add-on to the statndard Big Bang model to describe the observed universe. Intense efforts are to probe the Cosmic Microwave Background (CMB) for tensor modes and non-Gaussianity (f_{nl}). Observations of these will strengthen the support for the inflationary paradigm. There are programs to measure the spectral tilt of scalar modes (n_s) with greater accuracy are being planned. The ground-based CMB-S4 experiment [1], the LiteBIRD satellite [2], and the CORE

M. R. Gangopadhyay (✉) · S. Bhattacharya
Theory Divison, Saha Institute of Nuclear Physics, Bidhannagar, Kolkata 64, India
e-mail: mayukh@ctp-jamia.res.in; mayukhraj@gmail.com

S. Bhattacharya
e-mail: sukannya@prl.res.in

M. R. Gangopadhyay
Centre for Theoretical Physics, Jamia Millia Islamia, New Delhi 25, India

S. Bhattacharya
Theoretical Physics Division, Physics Research Laboratory, Ahmedabad 09, India

© Springer Nature Singapore Pte Ltd. 2021
P. K. Behera et al. (eds.), *XXIII DAE High Energy Physics Symposium*,
Springer Proceedings in Physics 261,
https://doi.org/10.1007/978-981-33-4408-2_45

315

satellite [3] can significantly reduce the uncertainty in the measurement of n_s. Th standard procedure to constrain model of inflation is to express the parameters such as n_s, r, and f_{nl} and to compare with the observed values from CMB fluctuations. This is the general procedure followed by the Planck collaboration to obtain constraints on several inflation models [4]. In this methodology, it is important to remember that the observables need to be calculated at certain e-folds back (N_{pivot}) from the end of inflation when CMB modes go outside the horizon. Thus, in this era of precision cosmology, it is very important to precisely predict N_{pivot} for the model of inflation in question.

To constrain a particular model of inflation with data, a robust approach was developed in [5] where one takes the coefficients of the inflation potential and the parametrisation of the reheating epoch as the 'model inputs'. Observational predictions are examined directly in terms of the coefficients of the potential; estimates and errors for the coefficients of the potential are directly obtained. One of the ways, this can be achieved is by making use of MODECHORD which provides a numerical evaluation of the inflationary perturbation spectrum taking the potential coefficients as input; which is then used as a plug-in for CAMB [7] and COSMOMC [8]. The parameters are then estimated using a nested sampling method [9].

The inflationary models are sensitive to the ultraviolet degrees of freedom. Thus to embed this paradigm in a ultraviolet complete theory is more realistic, and string theory being the best of hope for such makes it very important to study string inflationary models in great detail. The compactifications of large dimensions in string theory is related to the massless moduli fields (see, e.g., [10]). The volume misalignment of such fields leads to non-standard cosmological history. The most interesting effect of such an era is the change of effective number of N_{pivot} which has dependence on the model parameters. Given this, a complete numerical methodology using MODECHORD+COSMOMC is carried out in [11].

45.2   Theoretical Framework

We begin by briefly reviewing Kähler moduli inflation, the reader should consult [12] for further details. Kähler moduli inflation is set in the Large Volume Scenario (LVS) for moduli stabilisation of IIB flux compactification [13].

The simplest models of LVS are the ones in which the volume of the Calabi-Yau takes the Swiss-cheese form: $\mathcal{V} = \alpha \left(\tau_1^{3/2} - \sum_{i=2}^{n} \lambda_i \tau_i^{3/2} \right)$. τ_1; the moduli τ_2, ..., τ_n are blow-up modes and correspond to the size of the holes in the compactification. The potential for the scalars in the regime $\mathcal{V} \gg 1$ and $\tau_1 \gg \tau_i$ (for $i > 1$) is

$$V_{\text{LVS}} = \sum_{i=2}^{n} \frac{8(a_i A_i)^2 \sqrt{\tau_i}}{3\mathcal{V}\lambda_i} e^{-2a_i \tau_i} - \sum_{i=2}^{n} \frac{4a_i A_i W_0}{\mathcal{V}^2} \tau_i e^{-a_i \tau_i}$$

$$+ \frac{3\hat{\xi} W_0^2}{4\mathcal{V}^3} + \frac{D}{\mathcal{V}^\gamma}. \tag{45.1}$$

Here A_i, a_i are the pre-factors and coefficients in the exponents of the non-perturbative terms in the superpotential and W_0 is the vacuum expectation value of flux superpotential. The uplift term is $V_{\text{up}} = \frac{D}{\mathcal{V}^\gamma}$ with $D > 0$, $1 \leq \gamma \leq 3$.

Integrating out the heavy directions and canonically normalising the inflaton (σ), the potential (in Planck units) is

$$V = \frac{g_s}{8\pi} \left(V_0 - \frac{4W_0 a_n A_n}{\mathcal{V}_{\text{in}}^2} \left(\frac{3\mathcal{V}_{\text{in}}}{4\lambda_n} \right)^{2/3} \sigma^{4/3} \right.$$

$$\left. \times \exp\left[-a_n \left(\frac{3\mathcal{V}_{\text{in}}}{4\lambda_n} \right)^{2/3} \sigma^{4/3} \right] \right), \tag{45.2}$$

where

$$\frac{\sigma}{M_{\text{pl}}} = \sqrt{\frac{4\lambda_n}{3\mathcal{V}_{\text{in}}}} \, \tau_n^{\frac{3}{4}} \text{ with } V_0 = \frac{W_0^2}{\mathcal{V}_{\text{in}}^3}. \tag{45.3}$$

$\mathcal{V}_{\text{in}}$ is the value of the volume during inflation and $\beta = \frac{3}{2}\lambda_n a_n^{-3/2} (\ln \mathcal{V})^{3/2}$. Phenomenological considerations put the volume at $\mathcal{V}_{\text{in}} \approx 10^5 - 10^7$. Vacuum misalignment and the resulting post-inflationary moduli dynamics in this model was studied in detail in [14], During inflation, the volume modulus gets displaced from its global minimum. Thus an epoch in the post-inflationary history in which the energy density is dominated by cold moduli particles is unavoidable. The number of e-foldings that the universe undergoes in this epoch is [14]

$$N_{\text{mod}} = \frac{2}{3} \ln \left(\frac{16\pi a_n^{2/3} \mathcal{V}^{5/2} Y^4}{10\lambda_n (\ln \mathcal{V})^{1/2}} \right). \tag{45.4}$$

The presence of this epoch reduces the number of e-foldings between horizon exit of the pivot mode and the end of inflation by an amount $\frac{1}{4} N_{\text{mod}}$.

45.3   Methodology Of Analysis

For an inflationary scenario with instantaneous reheating (IRH) where the universe goes to radiation domination instantly after inflation, the number of e-foldings at the pivot scale is given by [6]

$$N_{\text{pivot}}^{\text{IRH}} = 55.75 - \log\left[\frac{10^{16}\text{Gev}}{V_{\text{pivot}}^{1/4}}\right] + \log\left[\frac{V_{\text{pivot}}^{1/4}}{V_{\text{end}}^{1/4}}\right]. \tag{45.5}$$

Here, V_{pivot} is the value of the inflation potential at which the pivot scale leaves the horizon and V_{end} is the potential at the end of inflation. From the observational upper limit of the strength of the gravitational wave ($r < 0.11$ [4]), the second term in the above equation is negative, whereas the third term is positive definite, but it can be very small for observationally favoured flat inflaton potential. The uncertainties associated with reheating are accounted for by varying N_{pivot} between $20 < N_{\text{pivot}} < N_{\text{pivot}}^{\text{IRH}}$, coined as the general reheating (GRH) scenario. The upper limit comes from the fact that after inflation there is a direct transition to radition domination. The lower limit comes from the requirement that at the end of inflation, all the cosmologically relevant scales are well outside of the horizon. Here, the assumption is the absence of any $w_{re} > 1/3$.

If there is an epoch of moduli domination in the post-inflationary history, then (45.5) gets modified. For Kähler moduli inflation N_{mod} is given by (45.4), and in this case $N_{\text{pivot}}^{\text{IRH}}$ is

$$N_{\text{pivot}}^{\text{IRH}} = 55.75 - \log\left[\frac{10^{16}\text{Gev}}{V_{\text{pivot}}^{1/4}}\right] + \log\left[\frac{V_{\text{pivot}}^{1/4}}{V_{\text{end}}^{1/4}}\right]$$
$$- \frac{1}{6}\ln\left(\frac{16\pi a_n^{2/3} V^{5/2} Y^4}{10\lambda_n(\ln V)^{1/2}}\right). \tag{45.6}$$

Note the additional dependence on the model parameters that arises from the last term in (45.6). In our analysis for $-1/3 < w_{re} \le 1/3$, we will vary N_{pivot} between 20 and $N_{\text{pivot}}^{\text{IRH}}$.

N_{pivot} is determined by $N_{\text{pivot}}^{\text{IRH}}$, w_{re} and N_{re}:

$$N_{\text{pivot}} = N_{\text{pivot}}^{\text{IRH}} - \frac{1}{4}(1 - 3w_{re})N_{re}. \tag{45.7}$$

The most general reheating case for the modulus can be treated with considering $-1/3 < w_{re} < 1$, where the upper bound comes from the positivity conditions in general relativity. In this work, we have analysed two different scenarios: *Case- i* for GRH scenario and *Case- ii* for three different w_{re} including exotic scenarios.

45.4 Results

CASE- I
First for the GRH scenario, the model parameters are varied in the following ranges: $W_0 : 0.001$ to 130, $\log_{10} V : 5$ to 8 and $A_n : 1.80$ to 1.95. $g_s = 0.06$, $\lambda_n = 1$ and

Fig. 45.1 Favoured regions in the W_0- $\log_{10} V$ plane. The 1-σ region is shaded as dark blue, the 2-σ region is shaded as light blue

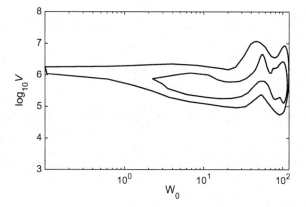

Fig. 45.2 Favoured region in the n_s-r plane. The 1-σ region is shaded as dark blue, the 2-σ region is shaded as light blue

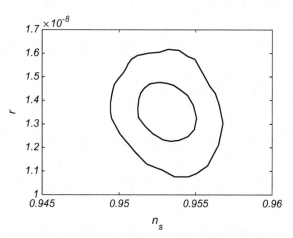

$a_n = 2\pi$ are kept fixed as the observables depend mildly on these parameters. The likelihoods used are *Planck TT+TE+EE, Planck lowP*, estimated using commander, *Planck lensing* and *Planck+BICEP2/Keck array* joint analysis likelihood [15–17]. Figures 45.1 and 45.2 shows the 1-σ and 2-σ bounds on the model parameters and the inflationary observables, repectively. The detailed result is summarised in Table 45.1.

CASE- II

For w_{re} ($w_{re} = 0, 2/3, 1$), the range for variation of N_{pivot} is obtained using the expression for N_{pivot}^{IRH} and the requirement of successful BBN.

The N_{re} can be expressed in terms of the model parameters, the effective equation of state during reheating and the reheating temperature:

$$N_{re} = -\frac{2}{3}\left(\frac{1}{1+w_{re}}\right)\ln\left[\frac{16\pi^2 g_*^{1/2} V^{9/2}(\ln V)^{3/2} T_{re}^2}{\sqrt{90} M_{pl}^2 W_0^3}\right]. \qquad (45.8)$$

Table 45.1 Constraints on the model parameters and the cosmological parameters. Data combination used: $Planck\, TT + TE + EE + low\, P + lensing + BK\, Planck\, 14$

Parameters	Central Value	1σ
W_0	57	46
$log_{10}\mathcal{V}$	5.9	0.3
A_n	1.87	0.04
n_s	0.953	0.002
$r/10^{-8}$	1.34	0.1
N_{pivot}	43	2

Fig. 45.3 1-σ and 2-σ confidence levels in the n_s-r plane for $w_{re} = 0$ (blue contours), $w_{re} = 2/3$ (green contours) and $w_{re} = 1$ (red contours)

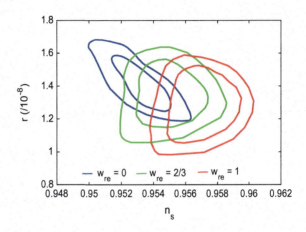

Table 45.2 Constraints on the model parameters and cosmological parameters for $w_{re} = 2/3, 1, 0$. Data combination used: $Planck\, TT + TE + EE + low\, P + lensing + BK\, Planck\, 14$

	$w_{re} = 0$	$w_{re} = 2/3$	$w_{re} = 1$
Parameters	Best-fit$\pm 1\sigma$	Best-fit$\pm 1\sigma$	Best-fit$\pm 1\sigma$
W_0	56.9$\pm$46.5	58$\pm$45	59$\pm$48
$log_{10}\mathcal{V}$	5.9$\pm$0.3	5.9$\pm$0.3	5.9$\pm$0.3
A_n	1.87$\pm$0.04	1.867$\pm$0.03	1.865$\pm$0.05
n_s	0.9535$\pm$0.002	0.9555$\pm$0.003	0.9575$\pm$0.003
$r/10^{-8}$	1.34$\pm$0.1	1.33$\pm$0.1	1.31$\pm$0.1
N_{pivot}	43$\pm$2.5	45.2$\pm$2.25	47.7$\pm$2

Successful nucleosynthesis requires $T_{re} > T_{BBN} = 5.1$ MeV [18]. Using all these conditions the MCMC analysis is carried out which is shown in the Fig. 45.3 exhibiting the constraints on the observables n_s and r for three different $w_r e$. The results are summarised in Table 45.2. The variation of T_{re} with $w_r e$ is shown in Fig. 45.4

Fig. 45.4 Reheating
temperature(T_{re}) as a
function of equation of
states(w_{re})

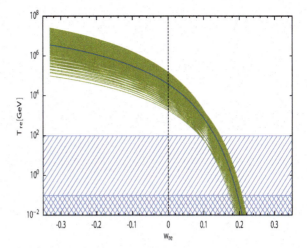

45.5 Conclusion

In this work, we have constrained a string motivated model of inflation directly from
data and concluded the importance of precise determination of N_{pivot} while calculating the observable parameters from those models. There are several interesting
directions to pursue from this. One such case would be to cross correlate with particle physics observables and dark radiation [19, 20] in LVS. An elaborate analysis in
that direction is left for future analysis.

References

1. K.N. Abazajian et al. [CMB-S4 Collaboration]. arXiv: 1610.02743 [astro-ph.CO]
2. T. Matsumura et al. arXiv: 1311.2847 [astro-ph.IM]]
3. F. Finelli et al. [CORE Collaboration]. arXiv: 1612.08270 [astro-ph.CO]]
4. P.A.R. Ade et al. Planck collaboration. Astron. Astrophys. **594**, A20 (2016)
5. J. Martin et al. arXiv: astro-ph/0605367
6. M.J. Mortonson et al. arXiv: 1007.4205
7. A. Lewis et al., Astrophys. J. **538**, 473 (2000)
8. A. Lewis, S. Bridle. arXiv: astro-ph/0205436
9. F. Feroz et al. arXiv: 0809.3437
10. G. Kane et al. arXiv: 1502.07746
11. S. Bhattacharya et al. arXiv: 1711.04807
12. J.P. Conlon, F. Quevedo. arXiv: hep-th/0509012
13. S.B. Giddings et al. arXiv: hep-th/0105097
14. M. Cicoli et al. arXiv: 1604.08512
15. Keck A, BICEP2 Collaborations: P.A.R. Ade et al. arXiv: 1510.09217

16. Planck Collaboration, Astron. Astrophys. **594**, A10 (2016)
17. P.A.R. Ade et al. arXiv: 1502.00612
18. P.F. de Salas et al. arXiv: 1511.00672
19. N. Sasankan et al. arXiv: 1607.06858
20. N. Sasankan et al. arXiv: 1706.03630

Chapter 46
Estimation of the Parameters of Warm Inflationary Models

Richa Arya and Raghavan Rangarajan

Abstract Observations of the temperature anisotropies in the Cosmic Microwave Background (CMB) radiation reveal key information about the early Universe phenomenon that possibly generated them, known as inflation. Here we study some single field monomial potential models of warm inflation and estimate their parameters that are consistent with the observations. We also obtain the n_s and r values for these models and find them to be within the *Planck* allowed range, implying that these are observationally viable models in the context of warm inflation. Also, the values of the tensor-to-scalar ratio, r, for some of our models is within the sensitivity of the future CMB polarisation experiments, that can serve for the testability of these models.

46.1 Introduction

Inflation is a phase of accelerated expansion of the early Universe that lasted for a very brief period of time. This phase in the cosmic evolution is assumed to resolve many shortcomings of the standard cosmology, like the horizon, flatness, monopole problem, etc. [1, 2].

In the standard description of inflation, also known as cold inflation, the number densities of all species dilute away during inflation and the Universe becomes super-cooled. However, there is a separate phase after inflation, known as reheating, in which the inflation oscillates and decays into other particles, leading to a generation of thermal bath in the Universe. In this description, it is assumed that the inflaton's coupling to the other particles is negligible during inflation and particle production takes place only during the reheating phase.

R. Arya (✉)
Physical Research Laboratory, Ahmedabad 380009, India
e-mail: richaarya@prl.res.in

Indian Institute of Technology Gandhinagar, Palaj, Gandhinagar 382355, India

R. Rangarajan
School of Arts and Sciences, Ahmedabad University, Ahmedabad 380009, India

© Springer Nature Singapore Pte Ltd. 2021
P. K. Behera et al. (eds.), *XXIII DAE High Energy Physics Symposium*,
Springer Proceedings in Physics 261,
https://doi.org/10.1007/978-981-33-4408-2_46

As an alternative to this, there is another description of inflation in which one presumes that the inflation is sufficiently coupled with the other fields during inflation, such that it dissipates its energy into them. These produced particles constitute the thermal bath in the Universe and maintain temperature even during inflation. This description of inflation is known as warm inflation [3, 4].

46.2 Basics of Warm Inflation

In warm inflation, the homogeneous inflaton field ϕ evolves as

$$\ddot{\phi} + (3H + \Upsilon)\dot{\phi} + V'(\phi) = 0, \tag{46.1}$$

where H is the Hubble expansion rate, $\Upsilon\dot{\phi}$ is the dissipative term due to inflaton dissipation into other fields and $\Upsilon(\phi, T)$ is known as the dissipation coefficient. In our notation, overdot and $'$ represent the derivative w.r.t time and ϕ, respectively. We define a dissipation parameter $Q \equiv \Upsilon/3H$ and rewrite Eq. (46.1) as

$$\ddot{\phi} + 3H(1+Q)\dot{\phi} + V'(\phi) = 0. \tag{46.2}$$

The regime with $Q \gg 1$ is termed as the strong dissipative regime and $Q \ll 1$ is termed as the weak dissipative regime of warm inflation.

The energy dissipated by the inflaton is transferred to the radiation, and hence the continuity equation can be written as

$$\dot{\rho}_r + 4H\rho_r = \Upsilon\dot{\phi}^2. \tag{46.3}$$

In the slow roll approximation, we can assume $\ddot{\phi} \approx 0$ in Eq. (46.2) and $\dot{\rho}_r \approx 0$ in Eq. (46.3) and obtain

$$\dot{\phi} \approx \frac{-V'(\phi)}{3H(1+Q)}, \rho_r \approx \frac{\Upsilon}{4H}\dot{\phi}^2 = \frac{3}{4}Q\dot{\phi}^2. \tag{46.4}$$

46.2.1 Forms of the Dissipation Coefficient

We consider the following two forms of dissipation coefficient [5, 6] in this study.

1. Dissipation coefficient with a cubic dependence on the temperature

$$\Upsilon(\phi, T) = C_\phi \frac{T^3}{\phi^2}. \tag{46.5}$$

This form of dissipation coefficient contributes in the low temperature limit, when the inflaton is coupled with the heavy intemediate X fields ($m_X \gg T$), which are then coupled to the radiation Y. The constant $C_\phi = 0.02\, h^2 N_Y N_X$ where h is the coupling between X and Y, and $N_{X,Y}$ are the multiplicities of the X and Y fields.

2. Dissipation coefficient with a linear dependence on the temperature

$$\Upsilon(\phi, T) = C_T T. \tag{46.6}$$

This form of dissipation coefficient arises in the high temperature limit of warm inflation when the inflaton is coupled to light intermediate X fields ($T \gg m_X$). Here the constant $C_T = 0.97 g^2 / h^2$, where g is the coupling between ϕ and X.

46.2.2 Primordial Power Spectrum

The primordial power spectrum for the warm inflation is given in Refs. [7, 8] as

$$P_\mathcal{R}(k) = \left(\frac{H_k^2}{2\pi\dot{\phi}_k}\right)^2 \left[1 + 2n_k + \left(\frac{T_k}{H_k}\right)\frac{2\sqrt{3}\pi Q_k}{\sqrt{3 + 4\pi Q_k}}\right] G(Q_k), \tag{46.7}$$

where the subscript k refers to the epoch when the kth mode of cosmological perturbations leaves the horizon during inflation. Here n_k is the Bose-Einstein distribution of the inflaton particles. The growth factor $G(Q_k)$ arises because of the inhomogeneous perturbations in the radiation contributing to the inflaton perturbations and is given in Ref. [9] as

For $\Upsilon \propto T$ $\qquad G(Q_k)_{\text{linear}} = 1 + 0.0185\, Q_k^{2.315} + 0.335\, Q_k^{1.364}.$

For $\Upsilon \propto T^3$ $\qquad G(Q_k)_{\text{cubic}} = 1 + 4.981\, Q_k^{1.946} + 0.127\, Q_k^{4.330}.$

If $Q \ll 1$, $G(Q_k)$ is almost 1. But for $Q \gg 1$, $G(Q_k)$ contributes significantly to the power spectrum.

46.3 Results and Discussion

In our study [10], we parameterize the primordial power spectrum for our models of warm inflation in terms of two model parameters, the inflaton self-coupling λ, and the dissipation parameter, Q_P (subscript P stands for the pivot scale, $k_P = 0.05$ Mpc^{-1}) and run the numerical code CosmoMC (Cosmological Monte Carlo) to estimate them. We also calculate the spectral index, n_s and the tensor-to-scalar ratio, r for our models and compare it with the allowed values from *Planck* 2015 (TT, TE, EE+ lowP) dataset [11].

46.3.1 $V(\phi) = \lambda\phi^4$ with Dissipation Coefficient $\Upsilon = C_T T$

We list the priors and the obtained values of the parameters in Tables 46.1, 46.2 and the joint probability distribution for the Q_P, λ in Fig. 46.1. We find that the values of n_s and r for the mean values of λ and Q_P are within the *Planck* 95% C.L., which implies that this model of warm inflation is consistent with the observations. We also find that the value of r for the weak dissipative regime is within the sensitivity of the future CMB polarisation experiments.

For the mean values of λ and Q_P, the constant C_T corresponds to a value of 1.75×10^{-4} in the weak dissipative regime, and a value equal to 3.66×10^{-2} in the strong dissipative regime for the upper limit of Q_P. These values signify that the ratio of couplings g/h is $\mathcal{O}(10^{-2})$ in the weak dissipative regime and $\mathcal{O}(10^{-1})$ in the strong dissipative regime. Previously, in Ref. [12], we have also studied $\lambda\phi^4$ potential of warm inflation with a cubic dissipation coefficient.

46.3.2 $V(\phi) = \lambda\dfrac{\phi^6}{M_{Pl}^2}$ with Dissipation Coefficient $\Upsilon = C_T T$

We list the priors and the obtained values of the parameters in Tables 46.3 and 46.4 and the joint probability distribution for the Q_P, λ in Fig. 46.2. For the mean values of λ and Q_P, the values of n_s and r are obtained to be within the 95% C.L. allowed from *Planck* 2015 data, thus making this model a viable model. Here also,

Table 46.1 $V(\phi) = \lambda\phi^4$ with $\Upsilon = C_T T$ in the weak dissipative regime

Parameter	Priors	68% C.L.
$\Omega_b h^2$	[0.005,0.1]	0.02168 ± 0.00014
$\Omega_c h^2$	[0.001,0.99]	0.1217 ± 0.0010
100θ	[0.50,10.0]	$1.04027^{+0.00029}_{-0.00033}$
τ	[0.01,0.8]	$0.048^{+0.016}_{-0.031}$
$-\log_{10}\lambda$	[13.7,15.5]	$14.39^{+0.34}_{-0.24}$
$-\log_{10} Q_P$	[0.0,5.4]	$3.64^{+0.76}_{-1.1}$

Mean value of $\lambda = 4.07 \times 10^{-15}$
Mean value of $Q_P = 2.29 \times 10^{-4}$
For these values, we obtain
$n_s = 0.967$
$r = 0.0330$

Table 46.2 $V(\phi) = \lambda\phi^4$ with $\Upsilon = C_T T$ in the strong dissipative regime

Parameter	Priors	68% C.L.
$\Omega_b h^2$	[0.005,0.1]	0.02174 ± 0.00013
$\Omega_c h^2$	[0.001,0.99]	0.1200 ± 0.0011
100θ	[0.50,10.0]	1.04044 ± 0.00029
τ	[0.01,0.8]	0.061 ± 0.024
$-\log_{10}\lambda$	[15.0,15.6]	$15.166^{+0.036}_{-0.056}$
$\log_{10} Q_P$	[0.0,0.6]	< 0.156

Mean value of $\lambda = 6.82 \times 10^{-16}$
Upper limit of $Q_P = 1.43$
For the upper limit, we obtain
$n_s = 0.973$
$r = 0.000214$

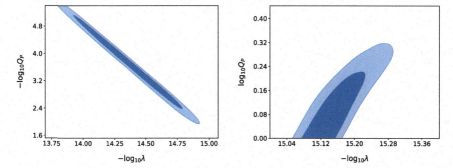

Fig. 46.1 Joint probability distribution for the model $V(\phi) = \lambda\phi^4$ and $\Upsilon = C_T T$ in the (*Left*) weak dissipative regime and (*Right*) strong dissipative regime. Here we count the number of efolds from the end of inflation such that at the pivot scale, $N_P = 60$

the tensor-to-scalar ratio for the weak dissipative regime can be testified with future experiments. The constant C_T corresponds to a value of 2.04×10^{-4} in the weak dissipative regime, and a value equal to 4.81×10^{-2} in the strong dissipative regime, which signifies that the ratio of couplings g/h is $\mathcal{O}(10^{-2})$ in the weak dissipative regime and $\mathcal{O}(10^{-1})$ in the strong dissipative regime of warm inflation.

46.3.3 $V(\phi) = \lambda\dfrac{\phi^6}{M_{Pl}^2}$ with a Cubic Dissipation Coefficient $\Upsilon = C_\phi\dfrac{T^3}{\phi^2}$

We list the priors and the obtained values of the parameters for the Q_P and λ in the weak dissipative regime in Table 46.5. and their joint probability distribution in Fig. 46.3. For the mean values of λ and Q_P, the n_s and r values are obtained to be consistent with the *Planck* bounds, signifying that this is a viable model of

Table 46.3 $V(\phi) = \lambda \frac{\phi^6}{M_{Pl}^2}$ with $\Upsilon = C_T T$ in the weak dissipative regime

Parameter	Priors	68% C.L.
$\Omega_b h^2$	[0.005,0.1]	0.02157 ± 0.00013
$\Omega_c h^2$	[0.001,0.99]	0.12484 ± 0.00099
100θ	[0.50,10.0]	1.03989 ± 0.00029
τ	[0.01,0.8]	0.056 ± 0.020
$-\log_{10}\lambda$	[15.4,16.6]	$16.07^{+0.27}_{-0.19}$
$-\log_{10} Q_P$	[1.8,5.4]	$3.54^{+0.68}_{-0.82}$

Mean value of $\lambda = 8.51 \times 10^{-17}$
Mean value of $Q_P = 2.88 \times 10^{-4}$
For these values, we obtain
$n_s = 0.956$
$r = 0.0451$

Table 46.4 $V(\phi) = \lambda \frac{\phi^6}{M_{Pl}^2}$ with $\Upsilon = C_T T$ in the strong dissipative regime

Parameter	Priors	68% C.L.
$\Omega_b h^2$	[0.005,0.1]	0.02170 ± 0.00014
$\Omega_c h^2$	[0.001,0.99]	0.1206 ± 0.0015
100θ	[0.50,10.0]	1.04037 ± 0.00030
τ	[0.01,0.8]	0.066 ± 0.022
$-\log_{10}\lambda$	[14.8,15.9]	15.253 ± 0.029
$\log_{10} Q_P$	[0,1.5]	0.596 ± 0.048

Mean value of $\lambda = 5.59 \times 10^{-16}$
Mean value of $Q_P = 3.94$
For these values, we obtain
$n_s = 0.970$
$r = 0.0000426$

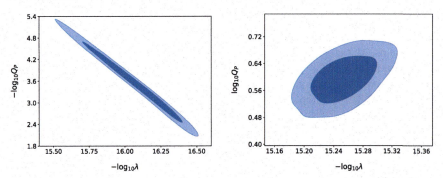

Fig. 46.2 Joint probability distribution for the model $V(\phi) = \lambda \frac{\phi^6}{M_{Pl}^2}$ and $\Upsilon = C_T T$ in the *Left:* weak dissipative regime and *Right:* strong dissipative regime. Here $N_P = 60$

Table 46.5 $V(\phi) = \lambda \frac{\phi^6}{M_{Pl}^2}$ with $\Upsilon = C_\phi \frac{T^3}{\phi^2}$ in the weak dissipative regime

Parameter	Priors	68% C.L.
$\Omega_b h^2$	[0.005,0.1]	0.02170 ± 0.00013
$\Omega_c h^2$	[0.001,0.99]	0.1207 ± 0.0014
100θ	[0.50,10.0]	1.04036 ± 0.00030
τ	[0.1,0.8]	0.061 ± 0.023
$-\log_{10} \lambda$	[15.8,17.0]	16.064 ± 0.38
$-\log_{10} Q_P$	[0,1.5]	$0.799^{+0.068}_{-0.10}$

Mean value of $\lambda = 8.63 \times 10^{-17}$
Mean value of $Q_P = 0.1588$
For these values, we obtain
$n_s = 0.969$
$r = 0.00480$

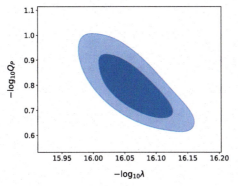

Fig. 46.3 Joint probability distribution for the model $V(\phi) = \lambda \frac{\phi^6}{M_{Pl}^2}$ with $\Upsilon = C_\phi \frac{T^3}{\phi^2}$ in the weak dissipative regime. Here $N_P = 60$

inflation. The constant C_ϕ corresponds to a value of 4.87×10^7, which signifies a huge multiplicity of X of order 10^9. We did not consider the strong dissipative regime for this model, as for Q_P greater than 1, the calculated value of the spectral index does not lie in the *Planck* allowed range.

46.4 Summary

Warm inflation is a description of inflation, in which the inflation dissipates into radiation during inflation, and therefore, the Universe has a temperature during the inflationary phase. We study some models of warm inflation and estimate their model parameters, the inflation self-coupling λ and the dissipation parameter, Q_P. These parameters signify the couplings and multiplicities of the fields coupled to the inflaton, and hence their estimation is necessary for the inflation model building. In our study, we argue that the $\lambda\phi^4$ and $\lambda\phi^6$ potentials of warm inflation are consistent with the present observations and can be tested with the next generation CMB polarisation experiments.

References

1. A.H. Guth, Phys. Rev. D **23**, 347 (1981)
2. A.D. Linde, Phys. Lett. **108B**, 389 (1982)
3. A. Berera, L.Z. Fang, Phys. Rev. Lett. **74**, 1912 (1995)
4. A. Berera, Phys. Rev. Lett. **75**, 3218 (1995)
5. I.G. Moss, C. Xiong, hep-ph/0603266
6. M. Bastero-Gil, A. Berera, R.O. Ramos, JCAP **1109**, 033 (2011)
7. R.O. Ramos, L.A. da Silva, JCAP **1303**, 032 (2013)
8. S. Bartrum, M. Bastero-Gil, A. Berera, R. Cerezo, R.O. Ramos, J.G. Rosa, Phys. Lett. B **732**, 116 (2014)
9. M. Benetti, R.O. Ramos, Phys. Rev. D **95**, 023517 (2017)
10. R. Arya, R. Rangarajan. Int. J. Mod. Phys. D **29**(08), 2050055 (2020)
11. P.A.R. Ade et al., Planck collaboration. Astron. Astrophys. **594**, A13 (2016)
12. R. Arya, A. Dasgupta, G. Goswami, J. Prasad, R. Rangarajan, JCAP **1802**(02), 043 (2018)

Chapter 47
One-Loop Effective Action for Nonminimal Natural Inflation Model

Sandeep Aashish and Sukanta Panda

Abstract Recent and upcoming experimental data, as well as the possibility of rich phenomenology, has spiked interest in studying the quantum gravitational effects in cosmology at low (inflation-era) energy scales. While Planck scale physics is under development, it is still possible to incorporate quantum gravity effects at relatively low energies using quantum field theory in curved spacetime, which serves as a low-energy limit of Planck scale physics. We use the Vilkovisky-DeWitt's covariant effective action formalism to study quantum gravitational corrections to a recently proposed Natural Inflation model with periodic nonminimal coupling. We present the 1-loop effective action for this theory valid in the flat-potential region, considering perturbative corrections up to quadratic order in background scalar fields.

47.1 Introduction

Natural inflation was first introduced by Freese et al. [1] as an approach where inflation arises dynamically (or *naturally*) from particle physics models. In natural inflation models, a flat potential is effected using pseudo Nambu-Goldstone bosons arising from breaking the continuous shift symmetry of Nambu-Goldstone modes into a discrete shift symmetry. As a result, the inflation potential in a Natural inflation model takes the form

$$V(\phi) = \Lambda^4 \left(1 + \cos(\phi/f)\right) \tag{47.1}$$

where the magnitude of parameter Λ^4 and periodicity scale f are model dependent. However, majority of natural inflation models are in tension with recent Planck 2018 results [2].

S. Aashish (✉) · S. Panda
Department of Physics, Indian Institute of Science Education and Research,
Bhopal 462066, India
e-mail: sandeepaashish@gmail.com; sandeepa16@iiserb.ac.in

S. Panda
e-mail: sukanta@iiserb.ac.in

© Springer Nature Singapore Pte Ltd. 2021
P. K. Behera et al. (eds.), *XXIII DAE High Energy Physics Symposium*,
Springer Proceedings in Physics 261,
https://doi.org/10.1007/978-981-33-4408-2_47

Recently, an extension of the original natural inflation model was proposed that introduces a new periodic nonminimal coupling to gravity [3]. This modification leads to a better fit with the observed data, with n_s and r values well within 95% C.L. region in the combined Planck 2018+BAO+BK14 data. Moreover, f becomes sub-Planckian, contrary to a super-Planckian f in the original natural inflation model, and addresses issues related to gravitational instanton corrections.

Our objective here is to study one-loop quantum gravitational corrections to the natural inflation model with nonminimal coupling, using Vilkovisky-DeWitt's covariant effective action approach [4]. A caveat is that effective action cannot be calculated exactly, so only perturbative results are feasible. Hence, we apply a couple of approximations. First, we work in the regime where potential is flat, i.e., $\phi \ll f$. Second, the background metric is Minkowski.

The action for the nonminimal natural inflation in the Einstein frame is given by

$$S = \int d^4x \sqrt{-g} \left(-\frac{2R}{\kappa^2} + \frac{1}{2} K(\phi)\phi_{;a}\phi^{;a} + \frac{V(\phi)}{(\gamma(\phi))^4} \right) \tag{47.2}$$

where

$$\gamma(\phi)^2 = 1 + \alpha \left(1 + \cos\left(\frac{\phi}{f}\right) \right), \tag{47.3}$$

and

$$K(\phi) = \frac{1 + 24\gamma'(\phi)^2/\kappa^2}{\gamma(\phi)^2}. \tag{47.4}$$

$V(\phi)$ is as in Eq. (47.1). In the region where potential is flat, $\phi/f \ll 1$, and we expand all periodic functions in Eq. (47.2) upto quartic order in ϕ

$$S = \int d^4x \sqrt{-g} \left(-\frac{2R}{\kappa^2} + \frac{1}{2}m^2\phi^2 + \frac{1}{24}\lambda\phi^4 + \frac{1}{2}(k_0 + k_1\phi^2)\phi_{;a}\phi^{;a} \right) + \mathcal{O}(\phi^5) \tag{47.5}$$

where parameters m, λ, k_0, and k_1 have been defined out of α, f, and Λ^4 from Eq. (47.2)

$$m^2 = \frac{\Lambda^4(2\alpha - 1)}{(1 + 2\alpha)^3 f^2};$$

$$\lambda = \frac{\Lambda^4(8\alpha^2 - 12\alpha + 1)}{(1 + 2\alpha)^4 f^4};$$

$$k_0 = \frac{1}{1 + 2\alpha};$$

$$k_1 = \frac{\alpha(\kappa^2 f^2 + 96\alpha^2 + 48\alpha)}{2\kappa^2 f^4(1 + 2\alpha)^2}. \tag{47.6}$$

We have also omitted a constant term appearing in (47.5) because such terms are negligibly small in early universe.

47.2 Effective Action Formalism

The quantization of a theory $S[\varphi]$ with fields φ^i is performed about a classical background $\bar{\varphi}^i$: $\varphi^i = \bar{\varphi}^i + \zeta^i$, where ζ^i is the quantum part. In our case, $\varphi^i = \{g_{\mu\nu}, \phi\}$; $\bar{\varphi}^i = \{\eta_{\mu\nu}, \bar{\phi}\}$, where $\eta_{\mu\nu}$ is the Minkowski metric; and $\zeta^i = \{\kappa h_{\mu\nu}, \delta\phi\}$.

The 1-loop effective action is given by

$$\Gamma = -\ln \int [d\zeta] \exp \left[\zeta^i \zeta^j \left(S_{,ij}[\bar{\varphi}] - \bar{\Gamma}^k_{ij} S_{,k}[\bar{\varphi}] \right) + \frac{1}{2\alpha} \chi^2_\beta \right] - \ln \det Q_{\alpha\beta}[\bar{\varphi}] \quad (47.7)$$

as $\alpha \longrightarrow 0$ (Landau-DeWitt gauge). Here, $S_{,i}$ and $S_{,ij}$ are first and second functional derivative w.r.t the fields $(g_{\mu\nu}, \phi)$ at background $(\eta_{\mu\nu}, \bar{\phi})$, respectively. $\bar{\Gamma}^k_{ij}$ are Vilkovisky-DeWitt connections that ensure covariance. χ_β is the gauge condition for the GCT symmetry and $Q_{\alpha\beta}$ is the ghost term that appears during quantization. Fortunately, upto quadratic order in background fields, and upto κ^2 order in quartic order terms, ghost determinant does not contribute. Hence, we will omit writing $Q_{\alpha\beta}$ in our calculations.

For convenience, we write the exponential in the first term of Γ as

$$\exp[\cdots] = \exp \left(\tilde{S}[\bar{\varphi}^0] + \tilde{S}[\bar{\varphi}^1] + \tilde{S}[\bar{\varphi}^2] + \tilde{S}[\bar{\varphi}^3] + \tilde{S}[\bar{\varphi}^4] \right)$$
$$\equiv \exp(\tilde{S}_0 + \tilde{S}_1 + \tilde{S}_2 + \tilde{S}_3 + \tilde{S}_4) \quad (47.8)$$

Treating $\tilde{S}_1, ..., \tilde{S}_4$ terms as perturbative, the final contribution to Γ at each order of $\bar{\varphi}$ is

$$\mathcal{O}(\bar{\varphi}) = 0$$
$$\mathcal{O}(\bar{\varphi}^2) = \langle \tilde{S}_2 \rangle - \frac{1}{2} \langle \tilde{S}_1^2 \rangle$$
$$\mathcal{O}(\bar{\varphi}^3) = 0$$
$$\mathcal{O}(\bar{\varphi}^4) = \langle \tilde{S}_4 \rangle - \langle \tilde{S}_1 \tilde{S}_3 \rangle + \mathcal{O}(\kappa^4) - \ln \det Q_{\alpha\beta} \quad (47.9)$$

The correlators are calculated using Wick's theorem and basic propagator relations

$$\langle h_{\mu\nu}(x) h_{\rho\sigma}(x') \rangle = G_{\mu\nu\rho\sigma}(x.x');$$
$$\langle \delta\phi(x) \delta\phi(x') \rangle = G(x, x');$$
$$\langle h_{\mu\nu}(x) \delta\phi(x') \rangle = 0, \quad (47.10)$$

where, $G(x, x')$ and $G_{\mu\nu\rho\sigma}(x, x')$ are respective green's functions for gravity and scalar field.

334

S. Aashish and S. Panda

47.3 Results and Conclusions

In what follows, an integration over coordinates ($\int d^4x$) and coordinate dependence is assumed unless stated otherwise. Our results for $S_{0,1,2}$ are as follows:

$$S_0 = \tfrac{1}{2}m^2(\delta\phi)^2 + \tfrac{1}{2}k_0\delta\phi_{,a}\delta\phi^{,a} + h^{ab}h_a{}^c{}_{,b,c} - \frac{h^{ab}h_a{}^c{}_{,b,c}}{\alpha} - h^a{}_a h^{bc}{}_{,b,c}$$

$$+ \frac{h^a{}_a h^{bc}{}_{,b,c}}{\alpha} - \tfrac{1}{2}h^{ab}h_{ab}{}^{,c}{}_{,c} + \tfrac{1}{2}h^a{}_a h^b{}_{b}{}^{,c}{}_{,c} - \frac{h^a{}_a h^b{}_{b}{}^{,c}{}_{,c}}{4\alpha} \tag{47.11}$$

$$S_1 = \tfrac{1}{2}m^2\kappa\delta\phi h^a{}_a\phi - \frac{1}{4}m^2\kappa\nu\delta\phi h^a{}_a\phi - \frac{1}{2}k_0\kappa\delta\phi h^b{}_b\phi^{,a}{}_{,a} + \tfrac{1}{4}k_0\kappa\nu\delta\phi h^b{}_b\phi^{,a}{}_{,a}$$

$$- \tfrac{1}{2}k_0\kappa\delta\phi h^b{}_{b,a}{}^{,a} + \frac{\kappa\omega\delta\phi h^b{}_{b,a}\phi^{,a}}{2\alpha} + k_0\kappa\delta\phi\phi^{,a}\,h_a{}^b{}_{,b} - \frac{\kappa\omega\delta\phi\phi^{,a}h_a{}^b{}_{,b}}{\alpha}$$

$$+ k_0\kappa\delta\phi h_{ab}\phi^{,a,b} \tag{47.12}$$

$$S_2 = -\tfrac{1}{8}m^2\kappa^2 h_{ab}h^{ab}\phi^2 + \tfrac{1}{16}m^2\kappa^2 h^a{}_a h^b{}_b\phi^2 + \tfrac{1}{4}\lambda\phi^2(\delta\phi)^2 - \tfrac{1}{8}m^2\kappa^2\nu\phi^2(\delta\phi)^2$$

$$+ \tfrac{1}{2}k_1\phi^2\delta\phi_{,a}\delta\phi^{,a} + 2k_1\delta\phi\phi_{,a}\delta\phi^{,a} - \tfrac{1}{8}k_0\kappa^2 h_{bc}h^{bc}\phi_{,a}\phi^{,a} + \tfrac{1}{16}k_0\kappa^2\nu h_{bc}h^{bc}\phi_{,a}\phi^{,a}$$

$$+ \tfrac{1}{16}k_0\kappa^2 h^b{}_b h^c{}_c\phi_{,a}\phi^{,a} - \tfrac{1}{32}k_0\kappa^2\nu h^b{}_b h^c{}_c\phi_{,a}\phi^{,a} + \tfrac{1}{2}k_1(\delta\phi)^2\phi_{,a}\phi^{,a} - \tfrac{1}{16}k_0\kappa^2\nu(\delta\phi)^2\phi_{,a}\phi^{,a}$$

$$+ \frac{\kappa^2\omega^2(\delta\phi)^2\phi_{,a}\phi^{,a}}{4\alpha} + \tfrac{1}{2}k_0\kappa^2 h_a{}^c h_{bc}\phi^{,a}\phi^{,b} - \tfrac{1}{4}k_0\kappa^2\nu h_a{}^c h_{bc}\phi^{,a}\phi^{,b} - \tfrac{1}{4}k_0\kappa^2 h_{ab}h^c{}_c\phi^{,a}\phi^{,b}$$

$$+ \tfrac{1}{8}k_0\kappa^2\nu h_{ab}h^c{}_c\phi^{,a}\phi^{,b} \tag{47.13}$$

In our case, S_0 leads to the free theory propagators for gravity and massive scalar field, which are well known. S_1 contains cross terms of $h_{\mu\nu}$ and $\delta\phi$, and does not contribute at $\mathcal{O}(\bar\phi)$, since $\langle S_1(x)\rangle = 0$. Contribution from S_1 comes at $\mathcal{O}(\bar\phi^2)$ from terms consisting of $\langle S_1(x)S_1(x')\rangle$ (Eq. 47.9). Expectation value of $S_2(x)$ describes contributions from tadpole diagrams. $\langle S_1(x)S_1(x')\rangle$ encompasses interaction terms between gravity and scalar field. For bookkeeping, parameters like ω and ν have been introduced to track gauge dependent terms and Vilkovisky-DeWitt terms, respectively. Playing with these parameters is an interesting exercise to check the gauge invariant nature of Vilkovisky-DeWitt approach [5]. We present here the divergent part of *Gamma* at $\mathcal{O}(\bar\phi^4)$, obtained after regularizing these path integrals using dimensional regularization (a factor of $1/(n-4)$ is assumed with the limit $n \to 4$)

$$divp(S_2) = \frac{ik_1m^4\phi^2}{16k_0^2\pi^2} - \frac{im^2\lambda\phi^2}{32k_0\pi^2} + \frac{im^4\kappa^2\nu\phi^2}{64k_0\pi^2} + \frac{ik_1m^2\phi\phi^{,a}{}_{,a}}{16k_0\pi^2} - \frac{im^2\kappa^2\nu\phi\phi^{,a}{}_{,a}}{128\pi^2} + \frac{im^2\kappa^2\omega^2\phi\phi^{,a}{}_{,a}}{32k_0\pi^2\alpha} \tag{47.14}$$

$$divp(S_1S_1') = -\frac{3im^4\kappa^2\phi^2}{16\pi^2} + \frac{im^4\alpha\kappa^2\phi^2}{16\pi^2} + \frac{3im^4\kappa^2\nu\phi^2}{16\pi^2} - \frac{im^4\alpha\kappa^2\nu\phi^2}{16\pi^2}$$

$$- \frac{3im^4\kappa^2\nu^2\phi^2}{64\pi^2} + \frac{im^4\alpha\kappa^2\nu^2\phi^2}{64\pi^2} + \frac{3ik_0m^2\kappa^2\phi\phi^{,a}{}_{,a}}{16\pi^2} - \frac{9ik_0m^2\kappa^2\nu\phi\phi^{,a}{}_{,a}}{32\pi^2}$$

$$+ \frac{3ik_0m^2\alpha\kappa^2\nu\phi\phi^{,a}{}_{,a}}{32\pi^2} + \frac{3ik_0m^2\kappa^2\nu^2\phi\phi^{,a}{}_{,a}}{32\pi^2} - \frac{ik_0m^2\alpha\kappa^2\nu^2\phi\phi^{,a}{}_{,a}}{32\pi^2} - \frac{im^2\kappa^2\omega\phi\phi^{,a}{}_{,a}}{8\pi^2}$$

$$+ \frac{im^2\kappa^2\omega^2\phi\phi^{,a}{}_{,a}}{16k_0\pi^2\alpha} - \frac{ik_0^2\kappa^2\nu\phi\phi^{,b}{}_{,b}{}^{,a}{}_{,a}}{32\pi^2} + \frac{ik_0^2\alpha\kappa^2\nu\phi\phi^{,b}{}_{,b}{}^{,a}{}_{,a}}{32\pi^2} - \frac{ik_0m^2\alpha\kappa^2\phi\phi_{,a}{}^{,a}}{16\pi^2}$$

$$+\frac{ik_0 m^2 \alpha \kappa^2 \nu \phi \phi_{,a}{}^{,a}}{32\pi^2} + \frac{im^2 \kappa^2 \omega \phi \phi_{,a}{}^{,a}}{16\pi^2} - \frac{im^2 \kappa^2 \nu \omega \phi \phi_{,a}{}^{,a}}{32\pi^2} - \frac{ik_0^2 \alpha \kappa^2 \phi \phi_{,a}{}^{,b}{}_{,b}{}^{,a}}{16\pi^2}$$

$$-\frac{ik_0^2 \alpha \kappa^2 \nu \phi \phi_{,a}{}^{,b}{}_{,b}{}^{,a}}{32\pi^2} + \frac{ik_0 \kappa^2 \omega \phi \phi_{,a}{}^{,b}{}_{,b}{}^{,a}}{16\pi^2} + \frac{ik_0 \kappa^2 \nu \omega \phi \phi_{,a}{}^{,b}{}_{,b}{}^{,a}}{32\pi^2} + \frac{ik_0^2 \alpha \kappa^2 \phi \phi^{,a}{}_{,a}{}^{,b}{}_{,b}}{16\pi^2}$$

$$+\frac{ik_0^2 \kappa^2 \nu \phi \phi^{,a}{}_{,a}{}^{,b}{}_{,b}}{8\pi^2} - \frac{ik_0^2 \alpha \kappa^2 \nu \phi \phi^{,a}{}_{,a}{}^{,b}{}_{,b}}{16\pi^2} - \frac{3ik_0^2 \kappa^2 \nu^2 \phi \phi^{,a}{}_{,a}{}^{,b}{}_{,b}}{64\pi^2} + \frac{ik_0^2 \alpha \kappa^2 \nu^2 \phi \phi^{,a}{}_{,a}{}^{,b}{}_{,b}}{64\pi^2}$$

$$\tag{47.15}$$

Note that there are gauge dependent terms in both Eqs. (47.14) and (47.15), two of which diverge as $\alpha \to 0$ (Landau gauge). However, when all contributions are added to evaluate Γ, these terms vanish so that the final result is gauge invariant. Final result for divergent part of Γ obtained from Eq. (47.9) after removing bookkeeping parameters ($\omega \to 1$, $\nu \to 1$) in the Landau gauge ($\alpha \to 0$)

$$div p(\Gamma) = L \int d^4 x \left[\frac{ik_1 m^4 \phi^2}{2k_0^2} + \frac{3}{16} im^4 \kappa^2 \phi^2 + \frac{im^4 \kappa^2 \phi^2}{8k_0} - \frac{im^2 \lambda \phi^2}{4k_0} + \frac{ik_1 m^2 \phi \phi^{,a}{}_{,a}}{2k_0} + \frac{5}{16} im^2 \kappa^2 \phi \phi^{,a}{}_{,a} \right.$$

$$\left. - \frac{3}{8} ik_0 \kappa^2 \phi \phi^{,a}{}_{,a}{}^{,b}{}_{,b} - \frac{3}{16} ik_0^2 \kappa^2 \phi \phi^{,a}{}_{,a}{}^{,b}{}_{,b} \right] \tag{47.16}$$

where $L = \dfrac{1}{8\pi^2(n-4)}$.

One can, in principle, construct counterterms from the classical action functional to absorb the divergent part in Eq. (47.16), which will inturn induce 1-loop corrections to parameters of the theory (47.2). Unfortunately, no counterterms can be found for the last two terms in Eq. (47.16) which are indeed fourth order derivative terms, and highlight the issue of renormalizability of gravity theories. A subsequent work will contain more detailed calculations upto quartic order in background fields, and the effect of quantum corrections in the context of inflation [6].

Acknowledgements This work was partially funded by DST (Govt. of India), Grant No. SERB/PHY/2017041. Calculations were performed using xAct packages for Mathematica.

References

1. K. Freese, J.A. Frieman, A.V. Olinto, Phys. Rev. Lett. **65**, 3233 (1990). https://doi.org/10.1103/PhysRevLett.65.3233
2. Y. Akrami et al. (Planck) (2018). arXiv:1807.06211 [astro-ph.CO]
3. R.Z. Ferreira, A. Notari, G. Simeon, J. Cosmol. Astropart. Phys. **2018**, 021 (2018). https://doi.org/10.1088/1475-7516/2018/11/021
4. L.E. Parker, D. Toms,*Quantum Field Theory in Curved Spacetime*. Cambridge Monographs on Mathematical Physics (Cambridge University Press, 2009). https://doi.org/10.1017/CBO9780511813924
5. P.T. Mackay, D.J. Toms, Phys. Lett. B **684**, 251 (2010). https://doi.org/10.1016/j.physletb.2009.12.032
6. S. Aashish, S. Panda, **JCAP06**, 009 (2020). https://iopscience.iop.org/article/10.1088/1475-7516/2020/06/009

Chapter 48
Majorana Dark Matter, Neutrino Mass, and Flavor Anomalies in $L_\mu - L_\tau$ Model

Shivaramakrishna Singirala, Suchismita Sahoo, and Rukmani Mohanta

Abstract We investigate Majorana dark matter in a new variant of $U(1)_{L_\mu - L_\tau}$ gauge extension of Standard Model, where the scalar sector is enriched with an inert doublet and a $(\bar{3}, 1, 1/3)$ scalar leptoquark. We constrain the parameter space consistent with Planck limit on relic density, PICO-60 bound on spin-dependent direct detection cross section. Further, we constrain the new couplings from the present experimental data on $\mathrm{Br}(\tau \to \mu \nu_\tau \bar{\nu}_\mu)$, $\mathrm{Br}(B \to X_s \gamma)$, $\mathrm{Br}(\bar{B}^0 \to \bar{K}^0 \mu^+ \mu^-)$, $\mathrm{Br}(B^+ \to K^+ \tau^+ \tau^-)$, and $B_s - \bar{B}_s$ mixing. Using the allowed parameter space, we estimate the $P'_{4,5}$ observables and the lepton nonuniversality parameters, $R_{K^{(*)}}$ and R_ϕ. We also briefly discuss about the neutrino mass generation at one-loop level.

48.1 Introduction

Though Standard Model (SM) is considered as a beautiful theory of particle physics, it fails to answer some of the open questions like matter-antimatter asymmetry, neutrino mass, nature dark matter (DM), and dark energy, thus indicating the presence of new physics (NP) beyond it. Recently, discrepancy of $\sim 3\sigma$ is observed in the decay rate and P'_5 of $\bar{B} \to \bar{K}^* \mu\mu$ [1]. The $B_s \to \phi\mu\mu$ also has tension [2]. The lepton nonuniversality (LNU) ratio, R_K [3]

$$R_K^{\mathrm{Exp}} = 0.846^{+0.060\ +0.0016}_{-0.054\ -0.014}, \tag{48.1}$$

S. Singirala (✉)
Indian Institute of Technology Indore, Khandwa road, Simrol 453552, Madhya Pradesh, India
e-mail: krishnas542@gmail.com

S. Sahoo
Physical Research Laboratory, Navrangpura, Ahmedabad 380009, India
e-mail: suchismita8792@gmail.com

S. Singirala · R. Mohanta
University of Hyderbad, Hyderabad 500046, India
e-mail: rukmani98@gmail.com

© Springer Nature Singapore Pte Ltd. 2021
P. K. Behera et al. (eds.), *XXIII DAE High Energy Physics Symposium*,
Springer Proceedings in Physics 261,
https://doi.org/10.1007/978-981-33-4408-2_48

show a deviation of 2.5σ from its SM result [4] and the measurements on R_{K^*} ratios [5]

$$R_{K^*}^{\text{LHCb}} = 0.660_{-0.070}^{+0.110} \pm 0.024, \quad (0.045 \le q^2 \le 1.1)\ \text{GeV}^2,$$
$$= 0.685_{-0.069}^{+0.113} \pm 0.047, \quad (1.1 \le q^2 \le 6.0)\ \text{GeV}^2, \qquad (48.2)$$

has $2.2\sigma(2.4\sigma)$ discrepancy from the respective SM predictions [6] in $q^2 \in [0.045, 1.1]([1.1, 6])$ bins, respectively. Hence, we would like to resolve these issues in an $U(1)_{L_\mu - L_\tau}$ gauge extension of SM with a scalar leptoquark (SLQ).

The paper is organized as follows. In Sect. 48.2, we present the particle masses and Lagrangian of our model. Sections 48.3 and 48.4 discuss the neutrino mass generation and the parameter space consistent with DM (flavor) study. Section 48.5 describes the impact of NP on LNU ratios and our results are summarized in Sect. 5.

48.2 New $L_\mu - L_\tau$ Model with a Scalar Leptoquark

We study $U(1)_{L_\mu - L_\tau}$ [7] extension of SM with three additional neutral fermions N_e, N_μ, N_τ, with $L_\mu - L_\tau$ charges 0, 1 and -1, respectively. A scalar singlet ϕ_2, an inert doublet η and a scalar leptoquark S_1 $(\bar{3}, 1, 1/3)$ are added to the scalar content of the model. We impose an additional Z_2 symmetry under which all the new fermions, η and the leptoquark are odd and rest are even. The particle content and their corresponding charges are displayed in Table 48.1. The Lagrangian of the present model can be written as

$$\begin{aligned}
\mathcal{L} = \mathcal{L}_{\text{SM}} &- \frac{1}{4} Z'_{\mu\nu} Z'^{\mu\nu} - g_{\mu\tau} \overline{\mu}_L \gamma^\mu \mu_L Z'_\mu - g_{\mu\tau} \overline{\mu}_R \gamma^\mu \mu_R Z'_\mu + g_{\mu\tau} \overline{\tau}_L \gamma^\mu \tau_L Z'_\mu + g_{\mu\tau} \overline{\tau}_R \gamma^\mu \tau_R Z'_\mu \\
&+ \overline{N_e} i\partial\!\!\!/ \, N_e + \overline{N_\mu} \left(i\partial\!\!\!/ - g_{\mu\tau} Z'_\mu \gamma^\mu \right) N_\mu + \overline{N_\tau} \left(i\partial\!\!\!/ + g_{\mu\tau} Z'_\mu \gamma^\mu \right) N_\tau - \frac{f_\mu}{2} \left(\overline{N_\mu^c} N_\mu \phi_2^\dagger + \text{h.c.} \right) \\
&- \frac{f_\tau}{2} \left(\overline{N_\tau^c} N_\tau \phi_2 + \text{h.c.} \right) - \frac{1}{2} M_{ee} \overline{N_e^c} N_e - \frac{1}{2} M_{\mu\tau} (\overline{N_\mu^c} N_\tau + \overline{N_\tau^c} N_\mu) - \sum_{q=d,s,b} (y_{qR} \overline{d_{qR}^c} S_1 N_\mu + \text{h.c.}) \\
&- \sum_{l=e,\mu,\tau} (Y_{\beta l} (\overline{\ell_L})_\beta \tilde{\eta} N_{lR} + \text{h.c.}) + \left| \left(i\partial_\mu - \frac{g}{2} \tau^a \cdot \mathbf{W}_\mu^a - \frac{g'}{2} B_\mu \right) \eta \right|^2 + \left| \left(i\partial_\mu - \frac{g'}{3} B_\mu + g_{\mu\tau} Z'_\mu \right) S_1 \right|^2 \\
&+ \left| \left(i\partial_\mu - 2g_{\mu\tau} Z'_\mu \right) \phi_2 \right|^2 - V(H, \eta, \phi_2, S_1). \qquad (48.3)
\end{aligned}$$

$$\begin{aligned}
V(H, \eta, \phi_2, S_1) &= \mu_H^2 H^\dagger H + \lambda_H (H^\dagger H)^2 + \mu_\eta (\eta^\dagger \eta) + \lambda_\eta (\eta^\dagger \eta)^2 + \lambda'_{H\eta} (H^\dagger \eta)(\eta^\dagger H) + \frac{\lambda''_{H\eta}}{2} \left[(H^\dagger \eta)^2 + \text{h.c.} \right] \\
&+ \mu_2^2 (\phi_2^\dagger \phi_2) + \lambda_2 (\phi_2^\dagger \phi_2)^2 + \mu_S^2 (S_1^\dagger S_1) + \lambda_S (S_1^\dagger S_1)^2 + \lambda_{S2} (\phi_2^\dagger \phi_2)(S_1^\dagger S_1) + \lambda_{\eta 2} (\phi_2^\dagger \phi_2)(\eta^\dagger \eta) \\
&+ \left[\lambda_{H2} (\phi_2^\dagger \phi_2) + \lambda_{HS} (S_1^\dagger S_1) + \lambda_{H\eta} (\eta^\dagger \eta) \right] (H^\dagger H) + \lambda_{S\eta} (S_1^\dagger S_1)(\eta^\dagger \eta).
\end{aligned}$$

The $U(1)_{L_\mu - L_\tau}$ symmetry gets spontaneously broken by assigning VEV v_2 to ϕ_2. Then the SM Higgs doublet breaks the SM gauge group to low energy theory. The fermion and scalar mass matrices take the form

Table 48.1 Fields and their charges of the proposed $U(1)_{L_\mu - L_\tau}$ model

	Field	$SU(3)_C \times$ $SU(2)_L \times U(1)_Y$	$U(1)_{L_\mu - L_\tau}$	Z_2
Fermions	$Q_L \equiv (u, d)_L^T$	$(3, 2, \ 1/6)$	0	+
	u_R	$(3, 1, \ 2/3)$	0	+
	d_R	$(3, 1, \ -1/3)$	0	+
	$e_L \equiv (\nu_e, \ e)_L^T$	$(1, 2, \ -1/2)$	0	+
	e_R	$(1, 1, \ -1)$	0	+
	$\mu_L \equiv (\nu_\mu, \ \mu)_L^T$	$(1, 2, \ -1/2)$	1	+
	μ_R	$(1, 1, \ -1)$	1	+
	$\tau_L \equiv (\nu_\tau, \ \tau)_L^T$	$(1, 2, \ -1/2)$	-1	+
	τ_R	$(1, 1, \ -1)$	-1	+
	N_e	$(1, 1, \ 0)$	0	−
	N_μ	$(1, 1, \ 0)$	1	−
	N_τ	$(1, 1, \ 0)$	-1	−
Scalars	H	$(1, 2, \ 1/2)$	0	+
	η	$(1, 2, \ 1/2)$	0	−
	ϕ_2	$(1, 1, \ 0)$	2	+
	S_1	$(\bar{3}, 1, \ 1/3)$	-1	−

$$M_N = \begin{pmatrix} \frac{1}{\sqrt{2}} f_\mu v_2 & M_{\mu\tau} \\ M_{\mu\tau} & \frac{1}{\sqrt{2}} f_\tau v_2 \end{pmatrix}, \quad M_S = \begin{pmatrix} 2\lambda_H v^2 & \lambda_{H2} v v_2 \\ \lambda_{H2} v v_2 & 2\lambda_2 v_2^2 \end{pmatrix}. \quad (48.4)$$

One can diagonalize them using a 2×2 rotation matrix as $U_{\alpha(\zeta)}^T M_{N(S)} U_{\alpha(\zeta)} =$ diag $[M_{N_-(H_1)}, M_{N_+(H_2)}]$. The lightest mass eigenstate N_- qualifies as DM and H_1 is considered to be the observed Higgs with $M_{H_1} = 125$ GeV. The neutrino mass generated at one-loop level is given by

$$(\mathcal{M}_\nu)_{\beta\gamma} = \frac{\lambda_{H\eta}'' v^2}{16\pi^2} \sum_{l = e, \mu, \tau} \frac{Y_{\beta l} Y_{\gamma l} M_{Dl}}{m_0^2 - M_{Dl}^2} \left[1 + \frac{M_{Dl}^2}{m_0^2 - M_{Dl}^2} \ln \frac{M_{Dl}^2}{m_0^2} \right]. \quad (48.5)$$

Here $M_{Dl} = (U^T M_N U)_l = \text{diag}(M_{ee}, M_-, M_+)$.

48.3 Dark Matter Phenomenology

Relic Density

The annihilations include s-channel processes mediated by H_1, H_2 with $f\bar{f}$, W^+W^-, ZZ, $Z'Z'$, $H_i H_j$ in the final state, Z' portal giving $Z'H_i$, pair of leptons of μ and τ

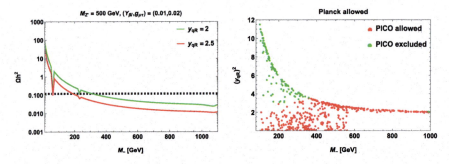

Fig. 48.1 Left panel depicts the relic density as a fucntion of DM mass for $(M_{S_1}, M_{\eta^+}, M_{\eta_{e,o}}, M_{H_2}) = (1.2, 2, 1.5, 2.2)$ TeV. Horizontal dashed lines correspond to Planck limit [8]. Right panel shows the paramter space consistent with 3σ range of Planck central value

type in output. Finally, in S_1-portal $d\bar{d}$, $s\bar{s}$, $b\bar{b}$ arise as output and in η-portal, pair of charged and neutral leptons arise in the final state. The model parameters to be analyzed are $(M_-, g_{\mu\tau}, M_{Z'}, Y_{\beta l}, y_{qR})$. Left panel of Fig. 48.1 shows relic density with DM mass for various values of Yukawa coupling.

Direct Searches

In the scalar portal, one can obtain contribution from spin-dependent (SD) interaction mediated by SLQ, of the form $\overline{N}_- \gamma^\mu \gamma^5 N_- \overline{q} \gamma_\mu \gamma^5$. The corresponding cross section is given by [9]

$$\sigma_{\text{SD}} = \frac{\mu_r^2}{\pi} \frac{\cos^4\alpha}{(M_{S_1}^2 - M_-^2)^2} \left[y_{dR}^2 \Delta_d + y_{sR}^2 \Delta_s \right]^2 J_n (J_n + 1), \qquad (48.6)$$

where, $J_n = \frac{1}{2}$, $\mu_r = \frac{M_- M_n}{M_- + M_n}$ with $M_n \simeq 1$ GeV for nucleon. The values of quark spin functions Δ_q are provided in [9]. Right panel of Fig. 48.1, depicts the parameter space consistent with 3σ range of Planck limit, with green data points violate the PICO-60 bound [10].

48.4 Flavor Phenomenology

After obtaining the allowed parameter space from DM study, we further constrain the new parameters from the present experimental limits on $\text{Br}(\tau \to \mu\nu_\tau\bar{\nu}_\mu)$, $\text{Br}(B \to X_s\gamma)$, $\text{Br}(\bar{B}^0 \to \bar{K}^0\mu^+\mu^-)$, $\text{Br}(B^+ \to K^+\tau^+\tau^-)$ and $B_s - \bar{B}_s$ mixing. The effective Hamiltonian responsible for $b \to sll$ quark level transition is given by [11]

$$\mathcal{H}_{\text{eff}} = -\frac{4G_F}{\sqrt{2}} \lambda_t \left[\sum_{i=1}^{6} C_i(\mu)\mathcal{O}_i + \sum_{i=7,9,10} \left(C_i(\mu)\mathcal{O}_i + C_i'(\mu)\mathcal{O}_i' \right) \right], \quad (48.7)$$

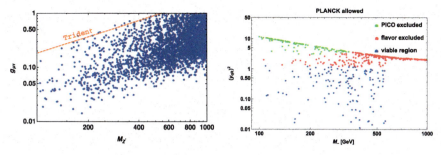

Fig. 48.2 Constraints on $M_{Z'} - g_{\mu\tau}$ (left panel) and $M_- - y_{qR}^2$ (right panel) planes

where G_F is the Fermi constant, $\lambda_t = V_{tb}V_{ts}^*$ is the product of CKM matrix elements, $\mathcal{O}_i^{(\prime)}$'s are the effective operators and $C_i^{(\prime)}$'s are their corresponding Wilson coefficients. We use the CKM matrix elements, life time, and masses of particles from [12] and the required form factors from [13], in order to compute the SM values of the observables. Now correlating the theoretical predictions with their respective 3σ experimental data, we compute the allowed parameter space of $M_{Z'} - g_{\mu\tau}$ (left) and $M_- - y_{qR}^2$ (right), presented in Fig. 48.2. In the right panel, the blue points represent the allowed space obtained from both DM and flavor observables (DM+Flavor). Both the red and blue points represent the constraints from only DM study, which are denoted as DM-I and DM-II region in the next section.

48.5   Effect on $R_{K^{(*)}}$ and R_ϕ Lepton Nonuniversality Ratios

In this section, we discuss the effect of constrained new parameters obtained from DM and flavor phenomenology on the lepton nonuniversality ratios associated with $b \to sll$ decay modes, $R_{K^{(*)}}$ and R_ϕ. For numerical analysis, all the required input parameters are taken from [12] and the form factors from [13]. Using the allowed parameter space from Fig. 48.2, we showed the q^2 variation of R_K (top-left), R_{K^*} (top-right), and R_ϕ (bottom) ratios in Fig. 48.3 in the full q^2 region excluding the intermediate dominant $c\bar{c}$ resonance region. Here blue dashed lines stand for SM predictions, cyan (magenta) bands represent the DM-I (DM-II) contributions, and the orange bands are obtained by using the DM+Flavor allowed parameter space. The experimental data are presented in black color [3, 5]. We observe that though the LNU ratios deviate significantly from their SM predictions for all the regions of the allowed parameter space, the constraint obtained from DM observables comparatively has a larger impact on these ratios.

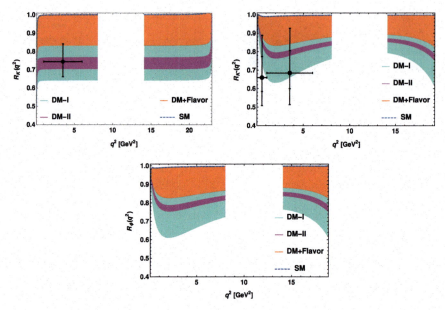

Fig. 48.3 The q^2 variation of R_D (top-left), R_{D^*} (top-right) and R_ϕ (bottom)

48.6 Conclusion

We scrutinized the dark matter, neutrino mass, and flavor anomalies in the $U(1)_{L_\mu - L_\tau}$ extended SM. We constrained the new parameters from both the dark matter and flavor phenomenology. We then checked the effects of allowed parameter space on $R_{K^{(*)}}$, R_ϕ. We found that the constraint obtained from the only DM study showed a comparatively good impact than the flavor observables.

References

1. LHCb, R. Aaij et al., Phys. Rev. Lett. **111**, 191801 (2013)
2. LHCb, R. Aaij et al., JHEP **09**, 179 (2015)
3. LHCb, R. Aaij et al., Phys. Rev. Lett. **113**, 151601 (2014). arXiv:1903.09252
4. C. Bobeth, G. Hiller, G. Piranishvili, JHEP **0712**, 040 (2007)
5. LHCb, R. Aaij et al., JHEP **08**, 055 (2017)
6. B. Capdevila, A. Crivellin, S. Descotes-Genon, J. Matias, J. Virto, JHEP **01**, 093 (2018)
7. X.G. He, G.C. Joshi, H. Lew, R.R. Volkas, Phys. Rev. D **43**, R22 (1991)
8. N. Aghanim et al. (Planck Collaboration). arXiv:1807.06209
9. P. Agrawal, Z. Chacko, C. Kilic, R.K. Mishra, arXiv:1003.1912
10. C. Amole et al. (PICO Collaboration), Phys. Rev. Lett. **118**, 251301 (2017)
11. C. Bobeth, M. Misiak, J. Urban, Nucl. Phys. B **574**, 291 (2000)
12. M. Tanabashi et al. (Particle Data Group), Phys. Rev. D **98**, 030001 (2018)
13. P. Colangelo, F. De Fazio, P. Santorelli, E. Scrimieri, Phys. Lett. B **395**, 339 (1997), M. Beneke, T. Feldmann, D. Seidel, Eur. Phys. J. C **41**, 173 (2005)

Chapter 49
Goldstone Inflation in Non-canonical Settings

Sukannya Bhattacharya and Mayukh Raj Gangopadhyay

Abstract Latest cosmic microwave background (CMB) observations favor a flat inflationary potential, which can be realized by the symmetry breaking of a pseudo Nambu-Goldstone boson (pNGB). A general extension of pNGB natural inflation is Goldstone inflation, which requires fine-tuning of the model parameters to be observationally allowed. A non-canonical kinetic term in the Goldstone inflation Lagrangian effectively slows down the inflation field. As a result, the prediction of inflationary observables also changes, so that it may even be possible to obtain observationally viable predictions for subPlanckian symmetry breaking scales. This work analyzed non-canonical Goldstone inflation phenomenologically using the latest CMB data.

49.1 Introduction

CMB observations constrain the early universe with the primordial spectrum of scalar and tensor perturbations, generated from the quantum fluctuations during inflation. Although models of inflation are plenty in number and construction in the current literature [1], the growing precision of current and forthcoming CMB experiments put tight constraints on the inflationary observables and on model parameters. In this light, the phenomenological analysis of theoretically well-motivated models of inflation is gaining much focus.

S. Bhattacharya (✉) · M. R. Gangopadhyay
Theory Divison, Saha Institute of Nuclear Physics, Bidhannagar, Kolkata 700064, India
e-mail: sukannya.bh@gmail.com

M. R. Gangopadhyay
e-mail: mayukh@ctp-jamia.res.in

S. Bhattacharya
Theoretical Physics Division, Physical Research Laboratory,
Navrangpura, Ahmedabad 380009, India

M. R. Gangopadhyay
Centre for Theoretical Physics, Jamia Millia Islamia, New Delhi 100025, India

© Springer Nature Singapore Pte Ltd. 2021   343
P. K. Behera et al. (eds.), *XXIII DAE High Energy Physics Symposium*,
Springer Proceedings in Physics 261,
https://doi.org/10.1007/978-981-33-4408-2_49

Natural inflation (NI) [2] arises from the broken Peccei-Quinn symmetry of a pNGB. Despite being an excellent bridge between particle physics theory and inflationary cosmology, vanilla NI model is refuted by recent CMB observations by Planck 2018 [3]. A more general form of the Goldstone potential

$$V(\phi) = \Lambda^4(C_\Lambda + \alpha \cos(\phi/f) + \beta \sin^2(\phi/f)), \tag{49.1}$$

where $\alpha = 1, \beta = 0$ returns the natural inflation potential, has been introduced in the context of inflationary model building in [4]. But the canonical Goldstone inflation model can conform with observational constraints only by fine tuning of the parameters of the model.

The introduction of a non-canonical kinetic energy term in the Lagrangian, in general, slows down the roll of the inflaton field and modifies the dynamics of inflation. This affects the inflationary observables in CMB, most important of which are the scalar spectral index n_s and the tensor-to-scalar ratio r.

Current and upcoming precision observations of CMB motivate the analysis of Goldstone inflation in the non-canonical domain. A full numerical study of the phenomenological deviation of non-canonical Goldstone inflation from canonical natural inflation is done in [5].

49.2 Non-canonical Inflation

Depending on the inflaton potential $V(\phi)$ and the excursion $\Delta\phi$ of the inflaton field, the field velocity $\dot\phi$ may have a nontrivial evolution when the action contains a non-canonical form of the kinetic energy, still providing enough number of inflationary e-folds. The generic action here can be written as [6]

$$S = \int \sqrt{-g}\, p(\phi, X) d^4 x \,, \tag{49.2}$$

where $p(\phi, X) = K(\phi, X) - V(\phi)$ and $X \equiv \frac{1}{2}\partial_\mu\phi\partial^\mu\phi$. For a homogeneous background inflaton field, $X \equiv \frac{\dot\phi^2}{2}$. In the most general case, $K(\phi, X)$ is an arbitrary function of ϕ and $\dot\phi$. To simplify, let

$$K(X, \phi) = K_1(\phi) K_2(X) \,, \tag{49.3}$$

where $K_1(\phi) = 1$ and $K_2(X) = X$ returns the canonical action. However, nontrivial arbitrary forms of these functions have been studied in the context of several inflationary models in literature.

49.2.1 Case 1: Arbitrary $K_1(\phi)$ and $K_2(X) = X$

The Lagrangian in this case is $\mathcal{L} = K_1(\phi)X - V(\phi)$. The inflaton Equation of Motion (EoM) is then

$$\ddot{\phi} + 3H\dot{\phi} + \frac{K_{1,\phi}}{2K_1}\dot{\phi}^2 + \frac{V_{,\phi}}{K_1} = 0, \qquad (49.4)$$

where $V_{,\phi} = dV/d\phi$ and $K_{1,\phi} = dK_1/d\phi$. The canonical field ψ is such that $\psi = \int d\phi\sqrt{K_1(\phi)}$. The modified velocity profile of ϕ also affects the slow roll parameters such that

$$\epsilon_V = \frac{M_{Pl}^2}{2}\left(\frac{V_{,\psi}}{V}\right)^2 = \frac{M_{Pl}^2}{2K_1}\left(\frac{V_{,\phi}}{V}\right)^2 = \frac{\epsilon_V^{(c)}}{K_1}, \qquad (49.5)$$

$$\eta_V = M_{Pl}^2\left(\frac{V_{,\psi\psi}}{V}\right) = \frac{M_{Pl}^2}{V}\left(\frac{V_{,\phi\phi}}{K_1} - \frac{V_{,\phi}K_{1,\phi}}{2K_1^2}\right) = \frac{\eta_V^{(c)}}{K_1} - \sqrt{\frac{\epsilon_V^{(c)}}{2}}\frac{K_{1,\phi}}{K_1^2}, \qquad (49.6)$$

where $\epsilon_V^{(c)}$ and $\eta_V^{(c)}$ are the slow roll parameters for the canonical kinetic form of ϕ. The number of e-folds of inflation are

$$N = \frac{1}{M_{Pl}}\int_{\phi_i}^{\phi_e}\frac{V}{V_{,\phi}\sqrt{K_1}}d\phi. \qquad (49.7)$$

Here, ϕ_i and ϕ_e represents the values of the inflaton field at the horizon exit of the CMB print scale and end of inflation, respectively. The inflationary observables in this case are

$$n_s - 1 = 2\eta_V - 6\epsilon_V \text{ and } r = 16\epsilon_V. \qquad (49.8)$$

The above relations are true for any inflaton potential $V(\phi)$ and arbitrary $K_1(\phi)$. To study Goldstone potential[1] (Eq. 49.1) in this non-canonical setting, we chose the form $K_1(\phi) = V(\phi)/\Lambda^4$ to attain an effective flatness of the potential.[2]

49.2.2 Case 2: $K_1(\phi) = 1$ and $K_2(X) = k_{n+1}X^n$

The most generic function for the non-canonical kinetic energy as function of the derivatives of ϕ is a power law form $K_2(X) = k_{n+1}X^n$ (k-inflation), where $n = 1$ is the canonical limit and k_{n+1} is a dimensionful constant. Here, the Lagrangian is $\mathcal{L} = k_{n+1}X^n - V(\phi)$ and the energy density of the system is [7]

---

[1] This work assumes $C_\Lambda = 1$ since it can be absorbed into the scale of inflation.
[2] Other functional forms of $K_1(\phi)$, arising from non-minimal gravitational couplings of the inflaton, would result in a different set of points in the n_s-r plane but this is not included in this work.

$$\rho = 2X p_{,X} - p(\phi, X). \tag{49.9}$$

The speed of sound is

$$c_s^2 = \frac{p_{,X}}{\rho_{,X}} = \frac{1}{(2n-1)}. \tag{49.10}$$

The EoM for the inflaton in this case is modified as

$$\ddot{\phi} + \frac{3H}{2n-1}\dot{\phi} + \frac{V_{,\phi}}{(2n^2-n)k_{n+1}X^{n-1}} = 0. \tag{49.11}$$

The potential slow roll parameters in this case are

$$\epsilon_V = \frac{1}{2}\gamma(n)\left(\frac{V_{,\phi}^{2n}}{V^{(3n-1)}}\right)^{\frac{1}{2n-1}} \text{ and } \eta_V = \gamma(n)\left(\frac{V_{,\phi\phi}^{(2n-1)}}{V^n V_{,\phi}^{(2n-2)}}\right)^{\frac{1}{2n-1}}, \tag{49.12}$$

where $\gamma(n) = \left(\frac{6^{n-1}}{nk_{n+1}}M_{Pl}^{2n}\right)^{\frac{1}{2n-1}}$. The inflationary observables are calculated in terms of ϵ_V and η_V as

$$n_s - 1 = \frac{1}{2n-1}[2n\eta_V - 2(5n-2)\epsilon_V] \text{ and } r = 16c_s\epsilon_V. \tag{49.13}$$

The inflationary number of e-folds

$$N = \int_{\phi_i}^{\phi_e} \frac{1}{\gamma(n)}\left(\frac{V^n}{V_{,\phi}}\right)^{\frac{1}{2n-1}} d\phi. \tag{49.14}$$

Interestingly, here ϵ_V, η_V, and N depend on the constants Λ and k_{n+1}, which is not the case for a canonical picture. So, one has to take care of Λ and k_{n+1} to obtain enough slow roll ($\epsilon_V c_s \ll 1$). The scalar power spectra $\mathcal{P}_s \propto \frac{V}{\epsilon_V c_s}|c_s k=aH$ also gets affected by Λ and k_{n+1} so that

$$\frac{\mathcal{P}_s^{(n)}}{\mathcal{P}_s^{(n=1)}} = \frac{1}{c_s\gamma(n)}\left(\frac{V_{\phi}^{2n-2}}{V^{n-1}}\right)^{\frac{1}{2n-1}}. \tag{49.15}$$

For example, in case of natural inflation for $n = 2$, the ratio of the scalar power for general n to the power for a canonical case is $\frac{\mathcal{P}_s^{(n=2)}}{\mathcal{P}_s^{(n=1)}} \propto \frac{(k_3\Lambda^4)^{1/3}}{f^{2/3}}$. Since current bounds on the pivot scale amplitude of the scalar power are very precise, the allowed values for Λ and k_{n+1} are to be chosen very carefully to obtain a viable natural inflation.

Similarly, for Goldstone inflation potential (Eq. 49.1) in such a kinetic setting, we obtain the inflationary observables as functions of f and β/α. Here, we analyze the model only for $n = 2$, primarily because our motivation is to study whether Goldstone

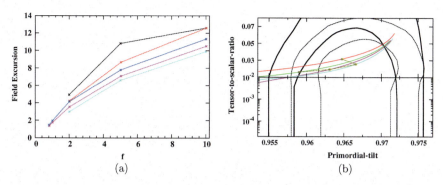

Fig. 49.1   In **a**, the field excursion $\Delta\phi$ (in M_{Pl} units) of the inflaton is shown as a function of f (in M_{Pl} units) for non-canonical Goldstone inflation case 1. Black, red, blue, magenta, and cyan curves signify $\beta/\alpha = 0$ (canonical), 0 (non-canonical), 0.25, 0.5, and 0.75, respectively. The predictions for n_s and r in this case are shown in **b** with $\beta/\alpha = 0$ (non-canonical natural inflation in red), 0.25 (cyan), 0.5 (magenta), and 0.75 (green). The green dot-dashed line connects the points with $f = 5M_{Pl}$ in all the curves. The solid black lines signify 68% and 96% confidence limits (CL) for Planck TT,TE,EE+lowE+lensing data (2018) [3], whereas the dashed black lines signify 68% and 96% CL for Planck TT,TE,EE+lowE (2018)+lensing+BK14 [8]+BAO data [9]

inflation is allowed by CMB data, even for the first order deviations from the canonical case. Moreover, $n \geq 2$ already introduces problems in the renormalisability of the system and higher n is constrained from the bound on c_s from recent CMB data [3].

49.3   Results

Thorough numerical analysis was done to evaluate the dynamics for Goldstone inflation for the two cases (Sects. 49.2.1 and 49.2.2) of the non-canonical kinetic term. For case 1, Fig. 49.1a shows the modification in the field excursion from the natural inflation case in terms of the breaking scale for different values of β/α. As discussed earlier, the field excursion $\Delta\phi$ monotonously decreases with increasing β/α due to increased friction for the inflaton.[3]

Figure 49.1b shows the prediction of Goldstone inflation in the non-canonical regime for case 1. Each curve represents a particular β/α for maximum symmetry breaking scale at $f = 16M_{Pl}$. The minimum f up to which the effective potential remains flat enough to provide 50–60 e-folds of inflation varies with the value of β/α. This is expected since the shape of the potential gets modified by changing β/α. All of the curves in Fig. 49.1b go out of the observationally allowed region for

---

[3]The cases with $\beta/\alpha = 0, 0.75$ are shown for only three points here because further decrease in f makes the potential steep enough to prevent obtaining enough number of inflationary e-folds. The points corresponding to $f \leq M_{Pl}$ predict n_s and r in the observationally excluded region, but they are plotted in Fig. 49.1a for demonstration purpose.

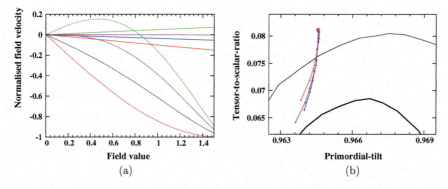

Fig. 49.2 Plot **a** shows the slow roll phase space of kinetic Goldstone inflation (case 2, $n = 2$) for $\beta/\alpha = 0$ (red), 0.25 (blue), 0.5 (magenta), and 0.75 (green). The solid curves are plotted for $f = 10M_{Pl}$ and the dashed ones are for $f = M_{Pl}$. Oscillations after the end of inflation are not shown here. The quantity in the y-axis, derived for slow roll from Eq. 49.4, is dimensionless. In **b**, the n_s-r plot for kinetic inflation. Kinetic natural inflation curve is plotted in red, whereas kinetic Goldstone inflation curves for $\beta/\alpha = 0.2$ (magenta), 0.5 (blue). The Planck 2018 CL for two different data combinations are shown in black solid and dashed curves similar to Fig. 49.1

$f > M_{Pl}$. Therefore, for case 1 of non-canonical kinetic term, it is not possible to obtain a subPlanckian symmetry breaking scale for the Goldstone boson that is also consistent with CMB observations.

The non-canonical Goldstone inflation for case 2 has been numerically analyzed for $n = 2$. The slow roll inflationary attractor in phase space of ϕ is shown in Fig. 49.2a where the y-axis is $\phi(\frac{k_3}{\Lambda^2})^{1/3}$. As discussed earlier, the attractor dynamics depend on the constants of the system k_3 and Λ, which are kept fixed for each β/α while evaluating n_s and r to obtain enough number of inflationary e-folds. The predictions for the inflationary observables in this case is given in Fig. 49.2b, where the lowest value of r of each curve is for $f = 0.5M_{Pl}$.

49.4 Discussions

It is evident from Fig. 49.1 that the non-canonical Goldstone inflation for case 1 predicts viable values of n_s and r, but only for the superPlanckian symmetry breaking scales. Case 1 of non-canonical Goldstone inflation predicted lower values of tensor power than that of canonical and non-canonical natural inflation. Case 2 of non-canonical Goldstone inflation ($n = 2$) can be viable even for subPlanckian f, although with an r value at the edge of the tightest 1σ confidence limit. k-inflation may produce high primordial non-Gaussianity for large n values [10], but for $n = 2$ studied here, the non-Gaussianity parameter $f_{NL} \sim 0.5$, i.e., quite small, for all β/α.

An interesting future problem is to check the effects of arbitrary $K_1(\phi)$ and $K_2(X)$ in a phenomenologically constructive way to obtain Goldstone inflation potential with subPlanckian f. The viability of non-canonical Goldstone potential can also be checked with upcoming CMB experiments (e.g., CMB-S4 [11], CORE, etc., [12]), which propose to constrain r with very high precision ($\sigma(r) \sim 10^{-3}$).

References

1. J. Martin et al., arXiv:1303.3787 [astro-ph.CO]
2. K. Freese et al., Phys. Rev. Lett. **65**, 3233 (1990)
3. Y. Akrami et al. [Planck Collaboration], arXiv:1807.06211 [astro-ph.CO]
4. D. Croon et al., arXiv:1503.08097 [hep-ph]
5. S. Bhattacharya et al., arXiv:1812.08141 [astro-ph.CO]
6. C. Armendariz-Picon et al., [hep-th/9904075]
7. S. Li et al., arXiv:1204.6214 [astro-ph.CO]
8. P.A.R. Ade et al. [BICEP2 and Keck Array Collaborations], arXiv:1510.09217 [astro-ph.CO]
9. L. Anderson et al. [BAO], arXiv:1203.6594 [astro-ph.CO]
10. G. Panotopoulos, arXiv:0712.1713 [astro-ph]
11. K. N. Abazajian et al. [CMB-S4 Collaboration], arXiv:1610.02743 [astro-ph.CO]
12. F. Finelli et al. [CORE Collaboration], arXiv:1612.08270 [astro-ph.CO]

Chapter 50
Infrared Finiteness of Theories with Bino-Like Dark Matter at T = 0

Pritam Sen, D. Indumathi, and Debajyoti Choudhury

Abstract We use the technique of Grammer and Yennie to show that the field theories of dark matter particles interacting with charged scalars and fermions are infrared (IR) finite to all orders in perturbation theory. This has important consequences for the consistency of such models.

50.1 Introduction

Dark matter (DM) has become a central element in modern precision cosmology on account of a multitude of astrophysical observations and theoretical arguments trying to explain these. Models trying to incorporate DM, hence should satisfy the stringent relic abundance constraints available in this era of precision cosmology. Among quite a handful of candidates eligible for this dark matter paradigm, we consider a simplified model of bino-like dark matter, which can be described by the Lagrangian density

$$
\begin{aligned}
\mathcal{L} = &-\frac{1}{4} F_{\mu\nu} F^{\mu\nu} + \overline{f} \left(i \slashed{D} - m_f \right) f + \frac{1}{2} \overline{\chi} \left(i \slashed{\partial} - m_\chi \right) \chi \\
&+ (D^\mu \phi)^\dagger \left(D_\mu \phi \right) - m_\phi^2 \phi^\dagger \phi + \left(\lambda \overline{\chi} P_L f^- \phi^+ + \text{h.c.} \right) ,
\end{aligned}
\tag{50.1}
$$

which is an extension of the Standard Model (SM) containing left-handed lepton doublets, $f = (f^0, f^-)^T$, together with an additional scalar doublet, $\phi = (\phi^+, \phi^0)^T$, and the $SU(2) \times U(1)$ singlet Majorana fermion χ being the dark matter candidate.

P. Sen (✉) · D. Indumathi
The Institute of Mathematical Sciences, Tharamani, Chennai 600113, Tamil Nadu, India
e-mail: pritamsen@imsc.res.in

Homi Bhabha National Institute, Anushaktinagar, Mumbai 400094, Maharashtra, India
e-mail: indu@imsc.res.in

D. Choudhury
Department of Physics and Astrophysics, University of Delhi, Delhi 110 007, India
e-mail: debajyoti.choudhury@gmail.com

© Springer Nature Singapore Pte Ltd. 2021
P. K. Behera et al. (eds.), *XXIII DAE High Energy Physics Symposium*,
Springer Proceedings in Physics 261,
https://doi.org/10.1007/978-981-33-4408-2_50

Although this may seem to be a specific choice, it actually captures the essence of IR behaviour of a wide class of models in MSSM without much loss of generality [1].

We assume the bino to be a TeV scale particle, such that the freeze-out occurs after the electroweak phase transition. Thus, only electromagnetic interactions are relevant at these scales and we keep only the electromagnetic term in the covariant derivatives. Hence, only the charged (s)fermion interactions with χ has been considered in the above Lagrangian; as those with the neutral component will not receive any higher order electromagnetic corrections and are, hence not germane to the issue at hand.

In such a model, the DM candidate (presumably coming into existence during the post-inflation reheating phase) χ stays in equilibrium with the SM sector via interactions of the form

$$\chi + \overline{\chi} \leftrightarrow \mathcal{F}_{SM} + \overline{\mathcal{F}}_{SM} , \qquad \chi + \mathcal{F}_{SM} \leftrightarrow \chi + \mathcal{F}_{SM} , \qquad (50.2)$$

etc., where $\mathcal{F}_{SM}$ is a generic SM particle. When χ couples to the SM sector with strength comparable to the weak gauge coupling, and has mass in the range of the weak scale, the relic abundance turns out to be of the correct order.

A typical diagram contributing to DM annihilation cross section according to this model is shown in Fig. 50.1.

Usually, most of the DM relic density computations involve estimating DM annihilation cross section at zero temperature. But, recently, these calculations were extended [2] to incorporate thermal corrections at next-to-leading (NLO) order. Infrared (IR) divergences were found to cancel out in such collision processes at NLO involving both charged fermions and changed scalars.

Here, in this paper, our attention is focussed on the proof of IR finiteness of such models, which include both charged scalars and charged fermions, along with neutral DM fields, to all orders at zero temperature. We discuss the case of finite temperature in a companion paper.

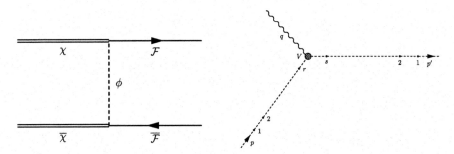

Fig. 50.1 L: A typical dark matter annihilation process; here the scalars and SM fermions are assumed to be charged. R: Schematic diagram of a nth order graph of the scalar QED subprocess, $\gamma^*\phi \rightarrow \phi$, with r vertices on the p leg and s on the p' leg, such that $r + s = n$. Here V labels the special but arbitrary photon–scalar vertex

To analyze the IR finiteness of such processes, both the higher order virtual corrections involving photons and the real (soft) photon emissions have to be considered. To prove the IR finiteness to all orders, we will use the approach of Grammer and Yennie [4] (henceforth, referred to as GY) by rearranging the photon polarization sum in terms of K and G photons. The photon propagator will then be expressed as

$$-i\frac{g_{\mu\nu}}{k^2 + i\epsilon} = \frac{-i}{k^2 + i\epsilon} \left[\left(g_{\mu\nu} - b_k(p_f, p_i)k_\mu k_\nu\right) + \left(b_k(p_f, p_i)k_\mu k_\nu\right) \right]$$
$$\equiv \frac{-i}{k^2 + i\epsilon} \left[G_{\mu\nu} + K_{\mu\nu} \right], \tag{50.3}$$

where b_k depends on the momenta p_f, p_i, where the final and initial vertices are inserted

$$b_k(p_f, p_i) = \frac{1}{2} \left[\frac{(2p_f - k) \cdot (2p_i - k)}{((p_f - k)^2 - m^2)((p_i - k)^2 - m^2)} + (k \leftrightarrow -k) \right]. \tag{50.4}$$

With this definition [3], the G photon insertions lead to finite corrections whereas the K-photon terms contain all the IR divergent pieces. With the understanding that the effect of inserting $(n + 1)$th K or G photon can be considered independently in the fermion and scalar sector, the IR behaviour of both the fermionic QED and the scalar QED turns out to be crucial while determining the IR behaviour of the theory involving bino-like dark matter. Fermionic QED is well known to be IR finite [4–6]. Thus, we proceed to prove the IR finiteness of scalar QED in the next section.

50.2 Infrared Behaviour of Scalar QED

To analyze the IR behaviour of scalar QED, we will start with the higher order corrections (both virtual and real) to a generic n-photon diagram (related to the hard scattering process $\gamma^{(*)} + \phi^{(*)} \to \phi^{(*)}$), which is schematically represented in Fig. 50.1, having r vertices on the p leg and s on the p' leg such that $r + s = n$.

Apart from the usual trilinear (scalar-scalar-photon) vertices, we have also the possibility of quadrilinear (scalar-scalar-photon-photon) vertices in scalar QED. Therefore, if m number of photons participated in quadrilinear insertions in the n-photon diagram then the number of vertices would have reduced to $(n - m/2)$ from n [7]. This is the major difference between the fermionic and scalar case. The results provided below are inclusive of this factor.

50.2.1 Insertion of Virtual K Photon

Virtual K photon insertions can be grouped into three broad classes,
Insertion between p and p' legs: Upon considering all the possible insertions of the K photon, the matrix element factorises and can be written in terms of the lower order matrix element, namely

$$\mathcal{M}_{n+1}^{K\gamma,p'p} = -ie^2 \int \frac{d^4k}{(2\pi)^4} \frac{1}{k^2 + i\epsilon} b_k(p', p) \mathcal{M}_n . \qquad (50.5)$$

Both insertions are on p' leg: This is the most complicated calculation, as tadpole insertions are allowed too. Apart from this, both the insertions being on the same leg proliferate the number of higher order diagrams. After including all the higher order diagrams, the matrix element becomes

$$\mathcal{M}_{n+1} \propto \{\mathcal{O}(k)_{k\ deno} + \text{Seagull}\} + \mathcal{O}(k)_{no\ k\ deno} + \{\mathcal{O}(k^2) + \text{tadpole}\} + \mathcal{M}_n ,$$
$$\propto \mathcal{M}_n , \qquad (50.6)$$

where the $\mathcal{O}(k)$ and $\mathcal{O}(k^2)$ denote, respectively, the finite linear and quadratic remainders other than $\mathcal{M}_n$, and the suffixes '$k\ deno$' and '$no\ k\ deno$' suggest, respectively, the presence and absence of photon momentum k dependence in the denominator of remainder terms. While the $\mathcal{O}(k)_{no\ k\ deno}$ term vanishes, the integrand being odd, the rest of the finite remainders cancel *exactly* against the seagull and tadpole contributions, leading to the vanishing of the sums enclosed by the curly braces. This leaves us again with a term proportional to the lower order matrix element as before.
Both insertions are on p leg: A similar exercise for p leg can be done, keeping in mind that the outermost self energy insertion has to be neglected to compensate for wave function renormalisation. After symmetrising the result between the p and p' legs, we have

$$\mathcal{M}_{n+1}^{K\gamma,\text{tot}} = ie^2 \int \frac{d^4k}{(2\pi)^4} \frac{1}{2(k^2 + i\epsilon)} [b(p', p') - 2b(p', p) + b(p, p)] \mathcal{M}_n ,$$
$$= ie^2 \int \frac{d^4k}{(2\pi)^4} \frac{1}{2(k^2 + i\epsilon)} [J^2(k)] \mathcal{M}_n \equiv [B] \mathcal{M}_n . \qquad (50.7)$$

50.2.2 Insertion of Virtual G Photons

The leading divergence structure after including all the possible G photons is

$$\mathcal{M}_{n+1}^{G\gamma;\text{fermion}} \sim \{g_{\mu\nu} - b(p_f, p_i)k_\mu k_\nu\} \times p_f^\mu p_i^\nu = 0 + \mathcal{O}(k) . \qquad (50.8)$$

As the leading divergence was already logarithmic, the $\mathcal{O}(k)$ terms are IR finite. The terms arising from 4-point insertions give a factor proportional to $-2g_{\mu\mu_q}(2p' \cdot k + k^2)$ and, following the above argument, are also IR finite.

50.2.3 Emission of Real $\widetilde{K}$ and $\widetilde{G}$ Photons

The case of emission of real photons is a little bit different as a physical momentum is carried away. After realizing that the photon polarization sum can be written as $\sum_{\text{pol}} \epsilon_\mu^*(k)\,\epsilon_\nu(k) = -g_{\mu\nu}$; and using the technique of GY it can be rearranged as contribution of $\widetilde{K}$ and $\widetilde{G}$ as

$$-g_{\mu\nu} = -\left[\left(g_{\mu\nu} - \tilde{b}_k(p_f, p_i)k_\mu k_\nu\right) + \tilde{b}_k(p_f, p_i)k_\mu k_\nu\right] \equiv -\left\{\widetilde{G}_{\mu\nu} - \widetilde{K}_{\mu\nu}\right\},$$
$$(50.9)$$

where $\tilde{b}_k(p_f, p_i) = b_k(p_f, p_i)|_{k^2=0}$. Following a similar analysis as above, the $\widetilde{K}$ photon contributions turn out to be

$$\left|\mathcal{M}_{n+1}^{\widetilde{K}\gamma,\text{tot}}\right|^2 \propto -e^2\left[\tilde{b}_k(p, p) - 2\tilde{b}_k(p', p) + \tilde{b}_k(p', p')\right] \equiv -e^2\widetilde{J}^2(k). \quad (50.10)$$

The diverging factor here has a form exactly similar and consistent with 'Weinberg's soft photon theorem'. An investigation of the universality of this factor may turn out to be interesting. All the contributions for the $\widetilde{G}$ photon emissions also turn out to be finite.

The total cross section, after fully symmetrising the matrix elements and including correct phase space factor $d\phi_{p'}$ for real photon emissions turn out to be [7]

$$d\sigma^{\text{tot}} = \int d^4x\, e^{-i(p+q-p')\cdot x}\, d\phi_{p'}\, \exp\left[B + B^* + \widetilde{B}\right]\sigma^{\text{finite}}(x), \quad (50.11)$$

where $\sigma^{\text{finite}}(x)$ contains all the finite parts, and all the divergences are inside $\exp\left[B + B^* + \widetilde{B}\right]$ term. At the IR limit, these terms in the exponential combine to give an IR finite sum

$$(B + B^*) + \widetilde{B} = e^2 \int d\phi_k\left[J(k)^2 - \widetilde{J}(k)^2 e^{ik\cdot x}\right] \xrightarrow{k\to 0} 0 + \mathcal{O}(k). \quad (50.12)$$

Hence, the theory of scalar QED is IR finite.

50.3 The IR Finiteness of the Dark Matter Interactions

We can now generalize the result obtained in the above section with the under-standing that both charged scalars and charged fermions are involved in the DM scattering/annihilation. The key to the factorisation in the case of the generic scattering process with *two* vertices, $\chi(q + q')\mathcal{F}(p) \rightarrow \mathcal{F}(p')\chi(q')$, where $p + q = p'$, is the identification of the special vertex V that separates the p and p' legs. We identify the p' leg with the final state fermion and the p leg with both the initial fermion and intermediate scalar line. With this definition, it can be shown [1] that the matrix element factorises exactly in the similar fashion as above, having a non-trivial double cancellation to achieve this factorisation.

50.4 Conclusion

In this age of precision cosmology, the extremely accurate measured values of relic abundance are imposing strong constraints on the parameter space of an eligible model of dark matter (DM). Treating the MSSM as a prototype theory, we here consider the DM to be a (bino-like) Majorana fermion, interacting with SM fields mediated by sfermions (charged scalars). Armed with the IR finiteness of fermionic QED and after an explicit proof of IR finiteness of scalar QED, we find that the higher order corrections due to virtual and real (soft) photon emissions, to a generic hard scattering process involving DM particles emerge to be IR finite at all orders. Seagull diagrams and tadpole contributions played a crucial role in the scalar sector to achieve this clear factorisation and resummation of soft contributions for a generic amplitude in the full theory of DM.

References

1. P. Sen, D. Indumathi, D. Choudhury, Eur. Phys. J. C **80**, 972 (2020)
2. M. Beneke, F. Dighera, A. Hryczuk, JHEP 1410 (2014) 45. Erratum: JHEP 1607 (2016) 106. arXiv:1409.3049 [hep-ph]
3. D. Indumathi, Ann. Phys. **263**, 310 (1998)
4. G. Grammer Jr., D.R. Yennie, Phys. Rev. D **8**, 4332 (1973)
5. D.R. Yennie, S.C. Frautschi, H. Suura, Ann. Phys. (NY) **13**, 379 (1961)
6. L. Faddeev, P. Kulish, Theor. Math. Phys. **4**, 745 (1970)
7. P. Sen, D. Indumathi, D. Choudhury, Eur. Phys. J. C **79**, 532 (2019)

Chapter 51
Evaluating Performance of Usual ILC and Global ILC Approach in Pixel Space Over Large Angular Scales of the Sky

Vipin Sudevan and Rajib Saha

Abstract During the implementation of usual Internal Linear Combination (ILC) method in pixel space, the foreground minimization is performed without taking into consideration the covariance structure of the CMB anisotropies. A new foreground model independent method, the Global ILC method, is developed where we incorporate the CMB theoretical covariance matrix in the foreground minimization algorithm. In this paper, we evaluate the performance of the usual ILC and the new Global ILC methods implemented in pixel space over large angular scales of the sky. We use the observed CMB maps provided by WMAP and Planck satellite missions. We also perform detailed Monte Carlo simulations to validate our results. We find that the cleaned CMB map and the CMB angular power spectrum obtained using the Global ILC method has significantly lower reconstruction error as compared to those obtained by usual ILC method.

51.1 Introduction

Among various foreground minimization techniques available, the Internal Linear Combination (ILC) method [1–3] is a model independent method where a foreground minimized CMB map is obtained by making minimal physical assumptions on the spectral distribution of foregrounds. The cleaned CMB map is obtained by linearly combining the observed CMB maps with some amplitude terms known as weight factors. The method is based on the assumption that the foreground spectra is different from the CMB spectrum which follows a black-body spectrum. The ILC method in pixel space has been implemented by the WMAP team using high resolution data to obtain a cleaned CMB map by employing local information of the spectral distribution of foregrounds [1]. The usual ILC method does not take into account the covariance structure of the CMB maps while estimating the weights. In [4, 5], the authors have developed a Global ILC method where the prior information about

V. Sudevan (✉) · R. Saha
Indian Institute of Science Education and Research, Bhopal 462066, MP, India
e-mail: vipins@iiserb.ac.in

© Springer Nature Singapore Pte Ltd. 2021
P. K. Behera et al. (eds.), *XXIII DAE High Energy Physics Symposium*,
Springer Proceedings in Physics 261,
https://doi.org/10.1007/978-981-33-4408-2_51

the CMB theoretical covariance matrix is used during the weight estimation. In this work, we evaluate the performance of the usual ILC method and the Global ILC method in the pixel space over large angular scales of the sky.

51.2 Formalism

In the pixel space usual ILC method, a cleaned CMB, $\hat{s}$, is defined as

$$\hat{s} = \sum_{i=1}^{n} w_i \mathbf{X}_i. \tag{51.1}$$

Here $\hat{s}$ and $\mathbf{X}_i$ are $N \times 1$ vectors describing full sky HEALPix [6] maps with N pixels for a pixel resolution parameter N_{side} ($N = 12 N_{side}^2$). $\mathbf{X}_i$ denotes the observed CMB map at ith frequency channel with $i = 1, 2, \ldots, n$. The weights, w_i, follow a constrain equation $\sum_{i=1}^{n} w_i = 1$. In the usual ILC method, the weights are obtained by minimizing the variance in the cleaned maps, $\sigma^2 = \hat{s}^T \hat{s}$, while in the new Global ILC method, we minimize

$$\sigma^2_{red} = \hat{s}^T \mathbf{C}^\dagger \hat{s}. \tag{51.2}$$

Here, σ^2_{red} is the CMB covariance weighted variance or reduced variance, $\mathbf{C}^\dagger$ is the Moore-Penrose generalized inverse of the CMB theoretical covariance matrix. The structure of the CMB theoretical covariance matrix is given in [4]. The weights which minimize the variance are obtained after following a Lagrange's method of undetermined multipliers, i.e.

$$\mathbf{W} = \frac{\mathbf{eA}^T}{\mathbf{eA}^T \mathbf{e}^T}, \tag{51.3}$$

where, $\mathbf{W} = (w_1, w_2, \ldots, w_n)$ is a $1 \times n$ row vector of weights corresponding to different frequency maps, $\mathbf{e}$ is a $1 \times n$ unit row vector representing the shape vector of CMB in thermodynamic temperature unit. The elements of $n \times n$ matrix $\mathbf{A}$ is given as

$$A_{ij} = \begin{cases} \mathbf{X}_i \mathbf{X}_j & \text{for usual ILC method,} \\ \mathbf{X}_i \mathbf{C}^\dagger \mathbf{X}_j & \text{for Global ILC method.} \end{cases} \tag{51.4}$$

51.3 Methodology

We use Planck 2015 and WMAP 9 year difference assembly (DA) maps in our analysis. Since we are working at large angular scales of the sky, we downgrade all the high resolution input maps to a pixel resolution $N_{side} = 16$ and smoothed by a Gaussian beam of $9°$ after properly taking care of the beam and pixel window functions of the individual maps [4]. We convert all the input maps to μK (thermodynamic) temperature units and then subtract from each frequency map its corresponding mean temperature. Using these low pixel resolution input maps, we perform the foreground minimization employing the following methods. In Method 1, we use Global ILC method in a single iteration where we estimate the weights (51.3) using the full sky CMB theoretical covariance matrix. The iterative version of Global ILC method is employed in Method 2 to obtain a cleaned CMB map. The details about the iterative implementation of the Global ILC method are given in [4]. Finally, we perform a foreground minimization over large angular scales of the sky using usual ILC approach in pixel space in Method 3. We discuss the results in the next section.

51.4 Results

We show the cleaned CMB map (hereafter CMap1) obtained after following Method 1 in the top panel of the Fig. 51.1, followed by the difference plots obtained after taking the difference between COMMANDER CMB map (at $N_{side} = 16$ and smoothed by a Gaussian beam of FWHM = 9°) and CMap1, and COMMANDER CMB map and cleaned map from usual ILC method (hereafter ILC CMap) in second and third panels, respectively. A visual inspection of the difference plots shows that the cleaned map obtained after using Method 1 and the COMMANDER CMB map agrees well with each other as compared to the ILC CMap which contains residual foreground contaminations present not only in the Galactic region, but also in higher Galactic regions. In the last panel of Fig. 51.1, we show the difference between the cleaned maps obtained after following Methods 1 and 2. The CMap2 (from Method 2) matches very closely with CMap1 and it has slightly less foreground residuals along both sides of the galactic plane at the expense of some additional detector noise residuals along the ecliptic plane. Similarly, we show, in the top panel of Fig. 51.2, the CMB angular power spectrum obtained after correcting the beam and pixel effects from the CMap2 and the ILC CMap from Method 3. We compare both the power spectra against the Planck 2015 [7] theoretical CMB angular power spectrum and the angular power spectrum estimated from COMMANDER CMB map. In the second and third panel of the Fig. 51.2, we show the difference between the COMMANDER and CMap2 angular power spectrum and COMMANDER and ILC CMap angular power spectrum, respectively. We see a close match between the COMMANDER and CMap2 angular power spectrum. The difference between CMap1 and CMap2 CMB angular power spectrum is plotted in the bottom panel of the Fig. 51.2. An interesting fact to note here is the close match between the cleaned maps CMap1 and CMap2 and their corresponding angular power spectra shows that using the Global ILC method, where prior information from the CMB theoretical covariance matrix is incorporated into the cleaning algorithm, we do not require more number of iterations to minimize the foregrounds present in the observed CMB maps. In order to understand the error

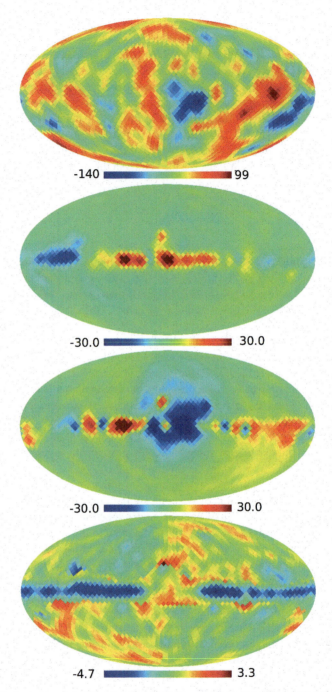

Fig. 51.1 In the top panel we show the cleaned CMB map (CMap1) from Method 1. In second and third panels, we plot COMMNDER—Method 1 and COMMANDER—Method 3 difference maps, respectively, and Methods 1–3 difference map is plotted in the bottom panel. All the plots are in μK units

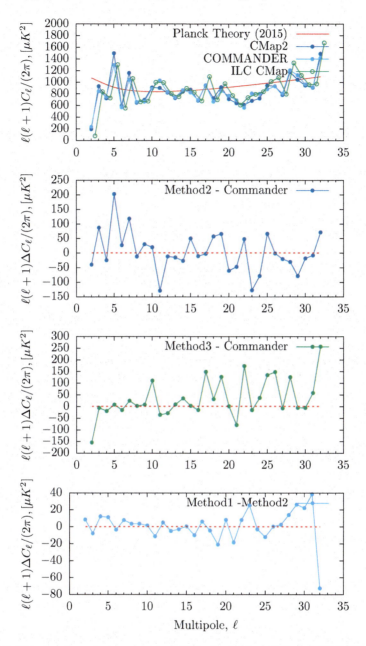

Fig. 51.2 Top panel shows the power spectrum (PS) estimated from CMap1 and ILC CMap along with PS from COMMANDER CMB map. (For visual purpose, ILC CMap PS is slightly shifted along the horizontal axis from the actual positions of the integer multipoles). We plot Method 2—COMMANDER and Method 3—COMMANDER PS in second and third panels, respectively, and finally Methods 1–2 PS in the last panel

while reconstructing the cleaned CMB map and its angular power spectrum using the different methods, we perform detailed Monte Carlo simulations after simulating a realistic model of foreground contaminated CMB maps at different Planck and WMAP frequency channels.

51.5 Monte Carlo Simulations

We simulate 200 sets of all 12 foreground contaminated maps corresponding to different WMAP and Planck frequency bands following the procedure as outlined in [4, 8]. Our foreground model consists of synchrotron emission with constant spectral index of -3.0, free-free and thermal dust emission. Using these sets of simulated foreground contaminated maps, we perform foreground minimization after implementing methods 1, 2, and 3. Using all the three sets of 200 such difference maps we obtain the standard deviation maps which are shown in Fig. 51.3. The pixel value corresponding to each pixel in the standard deviation map gives the reconstruction error in the final cleaned map. Upon inspecting the standard deviation plots for Methods 1 and 2 shown in the first and second panel of Fig. 51.3 respectively, and the plot corresponding to the usual ILC method in the bottom panel, we see that cleaned maps obtained after following Methods 1 and 2 have lower reconstruction error as compared to the cleaned map from usual ILC method. We show the mean cleaned CMB angular power spectrum obtained from Methods 2 and 3 in the top panel of Fig. 51.4 and the mean difference between the cleaned CMB angular power spectrum and corresponding input angular power spectrum for Methods 2 and 3 in the second and third panels, respectively. We see from the first panel of Fig. 51.4, the presence of negative bias [2, 9] at lower multipole ($\ell < 5$) in the mean angular power spectrum obtained using Method 3 (the usual ILC approach), while the same is absent in the power spectra estimated using the cleaned maps from Method 2. At higher multipoles $\ell > 5$, we observe excess power in the spectrum obtained from usual ILC method. This excess power in mean power spectrum is due to the presence of residual foreground contamination in the final cleaned maps from Method 3. In the last panel of Fig. 51.4, we show the close match between the CMap1 and CMap2 mean angular power spectra.

51.6 Conclusions

Using the new foreground minimization method, the Global ILC approach where the prior information from the CMB theoretical covariance matrix is used, provides better cleaned CMB maps (both Methods 1 and 2) as compared to the cleaned map provided by the usual ILC method (Method 3) implemented in pixel space over large angular scales of the sky. The difference plot between CMap1 and CMap2 shows that the results from Methods 1 and 2 agree very closely which is further evident

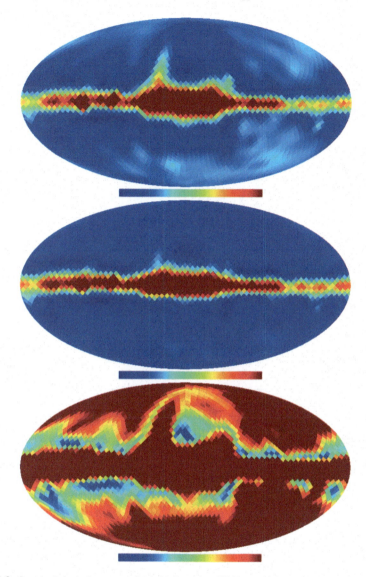

Fig. 51.3 Standard deviation maps obtained from the difference of foreground minimized CMB map and corresponding randomly generated input CMB map using 200 Monte Carlo simulations of foreground minimization following Methods 1, 2, and 3 are plotted in the top, middle, and bottom panels, respectively. All the plots are in $(\mu K)^2$ units

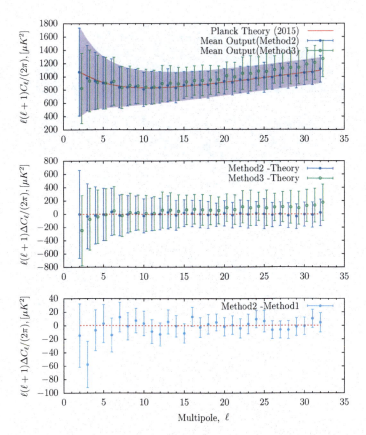

Fig. 51.4 Mean PS from Methods 1 and 3 along with their corresponding error bars is plotted in top panel (For visual purpose, the PS from Method 3 is slightly shifted along the horizontal axis from the actual positions of the integer multipoles). We have shown the Planck 2015 theoretical PS as the red line and the filled color band is the cosmic variance. In the middle panel, we show Method 2—Planck PS and Method 3—Planck PS. Methods 1–2 PS is shown in the bottom panel

after comparing their respective angular angular power spectra. This shows that the Global ILC method is truly global since by incorporating the prior information from the CMB theoretical covariance matrix in the minimization algorithm we do not require to perform more number of iterations to minimize the foregrounds effectively. The results obtained after performing detailed Monte Carlo simulations validate the results obtained using the observed CMB maps. We see that, from the simulation results, both the cleaned maps provided by Methods 1 and 2 have lower reconstruction error as compared to cleaned map from Method 3. We also see the presence of negative bias in the lower multipoles of the mean angular power spectrum obtained using Method 3, while the same is absent in the case of mean angular power spectra obtained from both the Methods 1 and 2.

References

1. C. Bennett et al. [WMAP Collaboration], Astrophys. J. Suppl. **148**, 97 (2003). https://doi.org/10.1086/377252 [astro-ph/0302208]
2. R. Saha, P. Jain, T. Souradeep, Astrophys. J. **645**, L89 (2006). https://doi.org/10.1086/506321 [astro-ph/0508383]
3. M. Tegmark and G. Efstathiou, Mon. Not. Roy. Astron. Soc. **281**, 1297 (1996). https://doi.org/10.1093/mnras/281.4.1297 [astro-ph/9507009]
4. V. Sudevan, R Saha, Astrophys. J. **867**(1), 74 (2018). https://doi.org/10.3847/1538-4357/aae439, arXiv:1712.09804 [astro-ph.CO]
5. V. Sudevan, R. Saha, arXiv:1810.08872 [astro-ph.CO]
6. K.M. Gorski et al., Astrophys. J. **622**, 759 (2005). https://doi.org/10.1086/427976 [astro-ph/0409513]
7. P.A.R. Ade et al., [Planck Collaboration], Astron. Astrophys. **594**, A13 (2016). https://doi.org/10.1051/0004-6361/201525830, arXiv:1502.01589 [astro-ph.CO]
8. V. Sudevan, P.K. Aluri, S.K. Yadav, R. Saha, T. Souradeep, Astrophys. J. **842**(1), 62 (2017). https://doi.org/10.3847/1538-4357/aa7334, arXiv:1612.03401 [astro-ph.CO]
9. R. Saha, S. Prunet, P. Jain, T. Souradeep, Phys. Rev. D **78**, 023003 (2008). https://doi.org/10.1103/PhysRevD.78.023003, arXiv:0706.3567 [astro-ph]

Chapter 52
Inflation with an Antisymmetric Tensor Field

Sandeep Aashish, Abhilash Padhy, Sukanta Panda, and Arun Rana

Abstract We investigate the possibility of inflation with models of antisymmetric tensor field having minimal and nonminimal couplings to gravity. Although the minimal model does not support inflation, the nonminimal models, through the introduction of a nonminimal coupling to gravity, can give rise to stable de-Sitter solutions with a bound on the coupling parameters. The values of field and coupling parameters are sub-planckian. Slow roll analysis is performed and slow roll parameters are defined which can give the required number of e-folds for sufficient inflation. Stability analysis has been performed for perturbations to antisymmetric tensor field while keeping the metric unperturbed, and it is found that only the sub-horizon modes are free of ghost instability for de-Sitter space.

52.1 Introduction

The theory of inflation, have played a crucial role in explaining the evolution of the universe which is consistent with the experimental observations [1, 2]. But the source of inflation still remains debatable. As ordinary matter or radiation cannot source inflation, several models have been built to describe inflation where a hypothetical field may it be scalar, vector or tensor drives the inflation [3]. Most of the scalar field models having the simple form of potential are ruled out as they are not compatible with the Planck's observational data for the cosmic microwave background [3]. Simi-

S. Aashish · A. Padhy (✉) · S. Panda · A. Rana
Department of Physics, Indian Institute of Science Education and Research, Bhopal 462066, India
e-mail: abhilash92@iiserb.ac.in

S. Aashish
e-mail: sandeepa16@iiserb.ac.in

S. Panda
e-mail: sukanta@iiserb.ac.in

A. Rana
e-mail: arunrana@iiserb.ac.in

© Springer Nature Singapore Pte Ltd. 2021
P. K. Behera et al. (eds.), *XXIII DAE High Energy Physics Symposium*,
Springer Proceedings in Physics 261,
https://doi.org/10.1007/978-981-33-4408-2_52

larly [4–6], almost all of the vector field inflation models suffer from instabilities like ghost instability [7] and gradient instability [8] which leads to an unstable vacuum.

In such a case, string theory provides an interesting alternative to the traditional model building. A particular theory of interest is that of rank 2 antisymmetric tensor field.[9, 10].

Our research interest is to study the possibility of inflation with antisymmetric tensor field by considering nonminimal models originally considered in Altschul et al. [11]. The nonminimal coupling terms we incorporate here are part of a general action constructed in [11] and are inspired by spontaneous Lorentz violation theories. We also set up a perfect slow roll scenario for this inflationary model, prior to developing a full perturbation theory for antisymmetric tensor in future works. However, an instability analysis only for the perturbations to the antisymmetric tensor field is performed, where as the metric is kept unperturbed.

52.2 Background Dynamics

An obvious choice for the background metric is the FriedmannLematreRobertson-Walker (FLRW) metric, motivated by the cosmological principle that imposes homogeneity and isotropy symmetries on the background universe. With the choice of metric signature $(-+++)$, the (background) metric components $g_{\mu\nu}$ read

$$g_{00} = -1, \quad g_{ij} = a(t)^2 \delta_{ij}, \tag{52.1}$$

where $a(t)$ is the scale factor for expansion of universe.

We are interested in a theory where the inflation-driving field is an antisymmetric tensor $B_{\mu\nu}$

$$B_{\mu\nu} = -B_{\nu\mu}. \tag{52.2}$$

In general, $B_{\mu\nu}$ has six independent components and a structure similar to that of the electromagnetic field strength tensor. A convenient representation of $B_{\mu\nu}$, analogous to the electrodynamic decomposition of field strength into electric and magnetic fields, is given by [11]

$$B_{0j} = -\Sigma^j, \quad B_{jk} = \epsilon_{jkl} \Xi^l. \tag{52.3}$$

For setting up the background dynamics, we can exploit the freedom to choose a structure for $B_{\mu\nu}$ that simplifies the calculations of the present work without losing generality. This choice of $B_{\mu\nu}$ structure manifests in the constraint equations for off-diagonal components of spatial part of energy momentum tensor, ensuring homogeneity and isotropy of background metric $g_{\mu\nu}$. For our convenience, we choose $\Sigma^j = 0$, and $\Xi^l = B(t), l = 1, 2, 3$, so that

$$B_{\mu\nu} = \begin{pmatrix} 0 & 0 & 0 & 0 \\ 0 & 0 & B(t) & -B(t) \\ 0 & -B(t) & 0 & B(t) \\ 0 & B(t) & -B(t) & 0 \end{pmatrix}. \tag{52.4}$$

If the minimal coupling with the curvature term is considered only, then that will lead to a negative acceleration of the scale factor here by contrasting the idea of inflation. The requirement of positive acceleration of the scale factor is met by a simple extension of theory consisting of a nonminimal coupling to gravity [11]. The action can be expressed as

$$S = \int d^4x \sqrt{-g} \left[\frac{R}{2\kappa} - \frac{1}{12} H_{\lambda\mu\nu} H^{\lambda\mu\nu} - \frac{m^2}{4} B_{\mu\nu} B^{\mu\nu} + \mathcal{L}_{NM} \right], \tag{52.5}$$

where $\mathcal{L}_{NM}$ is the nonminimal coupling term. In the most general case, the $\mathcal{L}_{NM}$ can be expressed as

$$\mathcal{L}_{NM} = \frac{1}{2\kappa} \xi B^{\mu\nu} B_{\mu\nu} R + \frac{1}{2\kappa} \zeta B^{\lambda\nu} B^{\mu}_{\nu} R_{\lambda\mu} + \frac{1}{2\kappa} \gamma B^{\kappa\lambda} B^{\mu\nu} R_{\kappa\lambda\mu\nu} \tag{52.6}$$

We will consider only the case, with $\mathcal{L}_{NM} = \frac{1}{2\kappa} \xi B^{\mu\nu} B_{\mu\nu} R$ for the sake of convenience in order to make our calculation easy. The nonminimal coupling term $\mathcal{L}_{NM}$, is parametrized by ξ for couplings with R. The parameters ξ has dimension of M_{pl}^{-2}.

With $\mathcal{L}_{NM} = \frac{1}{2\kappa} \xi B^{\mu\nu} B_{\mu\nu} R$ in (52.5), the corresponding energy momentum tensor is given by

$$T_{\mu\nu} = T^M_{\mu\nu} + T^{\xi}_{\mu\nu}, \tag{52.7}$$

where,

$$T^{\xi}_{\mu\nu} = \frac{\xi}{\kappa} \left[\nabla_{\mu}\nabla_{\nu}(B_{\alpha\beta}B^{\alpha\beta}) - g_{\mu\nu}\nabla^{\lambda}\nabla_{\lambda}(B_{\alpha\beta}B^{\alpha\beta}) - G_{\mu\nu}(B_{\alpha\beta}B^{\alpha\beta}) - 2RB^{\alpha}_{\mu}B_{\alpha\nu} \right]. \tag{52.8}$$

Following the steps of previous section, we write the Einstein equations

$$G_{00} = \kappa T_{00} \implies H^2 + 6\xi(2H\phi\dot{\phi} + H^2\phi^2) = \frac{\kappa}{2}[(\dot{\phi} + 2H\phi)^2 + m^2\phi^2], \tag{52.9}$$

$$G_{ij} = \kappa T_{ij} \implies 2\dot{H} + 3H^2 + 6\xi(2\phi\ddot{\phi} + 2\dot{\phi}^2 - 2\dot{H}\phi^2 - 5H^2\phi^2 + 4H\phi\dot{\phi})$$
$$= \frac{\kappa}{2}[(\dot{\phi} + 2H\phi)^2 - m^2\phi^2], \quad i = j, \tag{52.10}$$

Similarly, the constraint equation for off-diagonal components of T_{ij} becomes

$$\frac{\kappa}{2}[(\dot{\phi} + 2H\phi)^2 - m^2\phi^2] = -6\xi(\dot{H} + 2H^2)\phi^2. \tag{52.11}$$

The question of whether an exact de-Sitter space exists boils down to finding non-zero solutions (ϕ_0, H_0) to the Einstein equations (52.9)–(52.11) in the de-Sitter limit, $\dot{H} = \dot{\phi} = 0$. First, using the constraint (52.11) in (52.10), we get for the coupling with R

$$2\dot{H} + 3H^2 + 12\xi(\phi\ddot{\phi} + \dot{\phi}^2 + 2H\phi\dot{\phi} - \frac{1}{2}\dot{H}\phi^2 - \frac{3}{2}H^2\phi^2) = 0. \tag{52.12}$$

Then, applying the de-Sitter limit to (52.9) and (52.12), de-Sitter solutions ϕ_0 and H_0 are obtained as

$$\phi_0^2 = \frac{1}{6\xi} \tag{52.13}$$

$$H_0^2 = \frac{\kappa m^2}{4(6\xi - \kappa)} \tag{52.14}$$

from the above results, it can be observed that the de-Sitter inflation is possible in this model under a condition on the coupling parameter ξ, that is $\xi > \frac{\kappa}{6}$.

We now consider a nearly de-Sitter spacetime for building an inflationary model. For a successful inflation, the duration of inflation should be more than 70 e-folds. One of the slow roll conditions relevant for the acceleration is ϵ, given in terms of Hubble parameter

$$\epsilon = -\frac{\dot{H}}{H^2}. \tag{52.15}$$

It can be seen that ϵ has to be small in order for acceleration to be positive. A second slow roll parameter in terms of ϕ must be introduced to control the duration of inflation. $\delta \equiv \frac{\dot{\phi}}{H\phi}$ can act as a possible slow roll parameter which can be related to ϵ as follows:

$$\epsilon = \delta\left[\frac{(6\xi - 2\kappa)\phi^2}{1 + (6\xi - 2\kappa)\phi^2 + \delta(12\xi - 2\kappa)\phi^2} - \frac{\phi V_\phi}{2V}\right]. \tag{52.16}$$

where $V_\phi = dV/d\phi$. An explicit relation between ϵ and δ is obtained by using the flat potential condition, $V_\phi << V$. In the above expression, the condition on the constituent parameters can lead to the approximation $\delta \approx \epsilon$. The small δ indicates that the background field should be nearly constant which eventually leads to flat potential satisfying the requirement of slow roll. The duration of inflation can be expressed by the number of e-folds. The number of e-folds can be calculated to be

$$N = \int_{t_i}^{t} H \, dt = \int_{\phi_i}^{\phi} d\phi \frac{H}{\dot{\phi}} = \frac{1}{\delta} \int_{\phi_i}^{\phi} \frac{d\phi}{\phi} = \frac{1}{\delta} \ln\left(\frac{\phi}{\phi_i}\right). \qquad (52.17)$$

Clearly, it is feasible now to get 70 or more e-folds since δ is the only controlling parameter, and its smallness ensures sufficient duration of slow-rolling inflation.

52.3   Stability of Perturbations to $B_{\mu\nu}$

Although this model is able to provide a stable de-sitter type inflation with a lightly tuned nonminimal coupling with curvature terms, it should be free from the instabilities in order to give a sustainable inflationary model. A complete stability analysis would include perturbations to $B_{\mu\nu}$ and the metric. However, as an initial check, we consider here only the perturbations to the background antisymmetric tensor field $B_{\mu\nu}$, leaving the metric unperturbed. The choice of background structure of $B_{\mu\nu}$ remains the same as in (52.4). The perturbed field is given by $B_{\mu\nu} + \delta B_{\mu\nu}$, where

$$\delta B_{0i} = -E_i, \qquad \delta B_{ij} = \epsilon_{ijk} M_k. \qquad (52.18)$$

Substituting this perturbation in the action (52.5) results in the perturbed action containing terms upto quadratic order in perturbation. The second order part of the perturbed action reads

$$\begin{aligned}
S_2 = \int d^4x \Bigg[&\frac{1}{2a} \left(\vec{\dot{M}} \cdot \vec{\dot{M}} + 2\vec{\dot{M}} \cdot (\vec{\nabla} \times \vec{E}) + (\vec{\nabla} \times \vec{E}) \cdot (\vec{\nabla} \times \vec{E}) \right) - \frac{1}{2a^3}(\vec{\nabla} \cdot \vec{M})^2 \\
&+ \left(\frac{m^2}{2} - \frac{6\xi}{\kappa}\dot{H} - \frac{12\xi}{\kappa}H^2 \right) a(\vec{E} \cdot \vec{E}) \\
&- \left(\frac{m^2}{2} - \frac{6\xi}{\kappa}\dot{H} - \frac{12\xi}{\kappa}H^2 \right) \frac{(\vec{M} \cdot \vec{M})}{a} \Bigg].
\end{aligned} \qquad (52.19)$$

From (52.19), it can be observed that E_i's are merely auxiliary fields, whose equations of motion give unique solutions to E_i in terms of the dynamical modes M_i. To proceed, it is convenient to transform to 3-momentum space in order to get rid of the spatial derivatives. A further simplification is introduced by choosing the z-axis along the direction of 3-momentum $\mathbf{k}$. Consequently, the nondynamic modes E_is are replaced with the dynamical modes M_is. Finally, the kinetic part of the effective action is obtained to be

$$\left(S_{eff}\right)_{Kin.} = \int dt\, d^3k \left[\frac{N}{2a(N - \kappa k^2)} \dot{M}_x^{\dagger} \dot{M}_x + \frac{N}{2a(N - \kappa k^2)} \dot{M}_y^{\dagger} \dot{M}_y + \frac{1}{2a} \dot{M}_z^{\dagger} \dot{M}_z \right], \qquad (52.20)$$

where, $N = \kappa(2k^2 + m^2 a^2) - 12\xi a^2 \dot{H} - 24\xi a^2 H^2$.

S. Aashish et al.

S. Aashish et al.

Clearly, (52.20) implies that there is no ghost instability in the longitudinal mode $\tilde{M}_z$, whereas the coefficients of the remaining two transverse modes may come with a negative sign and hence give rise to instability. If we demand that S_{eff} be free of ghosts, then the following condition needs to be satisfied:

$$\frac{k^2}{a^2} + m^2 > \frac{H^2}{\kappa}(24\xi - 12\xi\epsilon). \implies k^2 > 4a^2 H_0^2. \tag{52.21}$$

Equation (52.21) indicates that there will be no ghost in the action for sub-horizon modes only. While for super-horizon modes the action will encounter a ghost.

52.4 Conclusion with a Possible Resolution

The possibility of inflation is analyzed in a model with nonminimal coupling with ricci scalar, where tensor field $B_{\mu\nu}$ acted as the source for inflation. As a result, it is observed that an exact de-Sitter solution is possible in this model with some constraints on the coupling parameter. In addition to that from the slow roll analysis of the model, it is verified that sufficient e-folds for inflation can be achieved. A notable feature of the present analysis is that the values of ξ and ϕ are sub-planckian in these models.

The ghost instability analysis has been performed for perturbations to $B_{\mu\nu}$ (keeping the metric unperturbed). We find that while the longitudinal modes are ghost free, the transverse modes may admit ghosts. For a special case of exact de-Sitter space, only the sub-horizon modes are ghost free. It is noteworthy that the conditions encountered here are common in vector field models as well [7, 12].

An interesting possibility arises by adding a $U(1)$ symmetry breaking kinetic term to (52.5). The possible addendum to the kinetic term is

$$S_T = \int d^4x \sqrt{-g}(\tau(\nabla_\lambda B^{\lambda\nu})(\nabla_\mu B^\mu_\nu))$$
$$\equiv \int d^4x \left[\frac{1}{2a}\left(\vec{M}.\vec{M} + \tau a^2 \vec{E}.\vec{E}\right) + \mathcal{L}_{\text{nonkinetic}}\right]$$

The contribution of this term to the energy momentum tensor is

$$T^T_{\mu\nu} = \tau\left[g_{\mu\nu}\left((\nabla_\lambda B^{\sigma\lambda})(\nabla_\rho B^\rho_\sigma) + 2B^{\sigma\lambda}\nabla_\lambda\nabla_\rho B^\rho_\sigma\right) + 2(\nabla_\lambda B^\lambda_\mu)(\nabla_\rho B^\rho_\nu)\right.$$
$$\left. + 2\left(B_\mu^\lambda\nabla_\lambda\nabla_\rho B_\nu^\rho + B_\nu^\lambda\nabla_\lambda\nabla_\rho B_\mu^\rho\right)\right]$$

the important points to be noted here are that the additional term does not play role in the background dynamics, where as in the perturbed scenario it adds dynamics to the previously nondynamical E_i modes making coefficients of both the kinetic term positive, here by putting forward a probable resolution to remove the ghost from this model. The complete perturbation scenario has to be investigated to construct a stable inflationary model.

Acknowledgments This work is partially supported by DST (Government of India) Grant No. SERB/PHY/2017041. The authors thank Tomi Koivisto for pointing out the results of [13].

References

1. A.H. Guth, Phys. Rev. D **23**, 347 (1981)
2. A. Starobinsky, Phys. Lett. B **91**, 99 (1980)
3. J. Martin, C. Ringeval, V. Vennin, Phys. Dark Universe **5–6**, 75 (2014), hunt for Dark Matter
4. L.H. Ford, Phys. Rev. D **40**, 967 (1989)
5. A. Golovnev, V. Mukhanov, V. Vanchurin, JCAP **0806**, 009 (2008), arXiv:0802.2068 [astro-ph]
6. O. Bertolami, V. Bessa, J. Pramos, Phys. Rev. **D93**, 064002 (2016), arXiv:1511.03520 [gr-qc]
7. B. Himmetoglu, C. R. Contaldi, M. Peloso, Phys. Rev. **D80**, 123530 (2009), arXiv:0909.3524 [astro-ph.CO]
8. R. Emami, S. Mukohyama, R. Namba, Y. li Zhang, J. Cosmol. Astroparticle Phys. **2017**, 058 (2017)
9. R. Rohm, E. Witten, Ann. Phys. **170**, 454 (1986)
10. A.M. Ghezelbash, JHEP **8**, 45 (2009), arXiv:0901.1670 [hep-th]
11. B. Altschul, Q.G. Bailey, V.A. Kostelecky, Phys. Rev. **D81**, 065028 (2010), arXiv:0912.4852 [gr-qc]
12. A. Padhy, S. Panda, Establishing a stable vector ination model. Under preparation
13. T.S. Koivisto, D.F. Mota, C. Pitrou, J. High Energy Phys. **2009**, 092 (2009)

Chapter 53
IR Finiteness of Theories with Bino-Like Dark Matter at Finite Temperature

Pritam Sen, D. Indumathi, and Debajyoti Choudhury

Abstract We consider a theory of bino-like dark matter particles interacting with charged scalar and fermion fields at a finite temperature T. We show that such a theory is infrared finite to all orders in thermal field theory. Such a result is important for calculations of DM relic densities, including thermal corrections.

53.1 Introduction

With the advent of precision cosmology, the contribution to the energy budget of the universe in the form of the Dark Matter (DM) has been measured with high precision. Consequently, theoretical predictions of the relic abundance of DM need to be very accurate and the inclusion of higher order calculations is a must. Although some effort has been made to this end, these seldom include corrections due to finite temperature. Initial efforts in this direction stopped at the next-to-leading order (NLO) [1] alone, demonstrating the cancellation of the Infrared divergences to this order. In this paper, we present an all order proof of the IR finiteness of such models. Although we focus here on a bino-like DM candidate, the analysis captures the essence of a broad class of models [2].

P. Sen (✉) · D. Indumathi
The Institute of Mathematical Sciences, Tharamani, Chennai 600113, Tamil Nadu, India
e-mail: pritamsen@imsc.res.in

Homi Bhabha National Institute, Anushaktinagar, Mumbai 400094, Maharashtra, India
e-mail: indu@imsc.res.in

D. Choudhury
Department of Physics and Astrophysics, University of Delhi, Delhi 110007, India
e-mail: debajyoti.choudhury@gmail.com

© Springer Nature Singapore Pte Ltd. 2021   375
P. K. Behera et al. (eds.), *XXIII DAE High Energy Physics Symposium*,
Springer Proceedings in Physics 261,
https://doi.org/10.1007/978-981-33-4408-2_53

The Lagrangian density relevant to this model is described by

$$
\begin{aligned}
\mathcal{L} = {} & -\frac{1}{4} F_{\mu\nu} F^{\mu\nu} + \overline{f} \left(i \slashed{D} - m_f \right) f + \frac{1}{2} \overline{\chi} \left(i \slashed{\partial} - m_\chi \right) \chi \\
& + \left(D^\mu \phi \right)^\dagger \left(D_\mu \phi \right) - m_\phi^2 \phi^\dagger \phi + \left(\lambda \overline{\chi} P_L f^- \phi^+ + \text{h.c.} \right),
\end{aligned}
\tag{53.1}
$$

where χ, the dark matter candidate, is a $SU(2) \times U(1)$ singlet Majorana fermion, along with left-handed fermion doublet, $f = (f^0, f^-)^T$, with an additional scalar doublet, $\phi = (\phi^+, \phi^0)^T$ which is the supersymmetric partner of f. We assume the bino to be a TeV-scale particle such that freeze-out happens after electroweak phase transition. Consequently, only electromagnetic interactions are of relevance as far as IR finiteness is concerned. For analyzing the IR finiteness of this model, we will start with the simplest of processes like $\chi + \overline{\chi} \to \mathcal{F} + \overline{\mathcal{F}}$ or $\chi + \mathcal{F} \to \chi + \mathcal{F}$ (where $\mathcal{F}$ is a generic Standard Model fermion), and consider all higher order contributions to these. As both the charged scalars and charged fermions participate in these processes, the IR finiteness of both of these play a crucial role in determining the IR finiteness of the whole model. The thermal fermionic QED is already known to be IR finite [3, 4]. Hence, we will proceed via proving the IR finiteness of thermal scalar QED at all order. But at first, we will address some complications arising due to the presence of temperature in the next section.

53.2 Real-Time Formulation of the Thermal Field Theory and Additional IR Divergences

In the real-time formalism [5–7] of thermal field theory, fields (scalar, fermion and photon) in equilibrium with a heat bath of temperature T obeys periodic and anti-periodic boundary conditions given by $\phi(t_0) = \pm \phi(t_0 - i\beta)$, respectively, for boson and fermion fields. These conditions result in the well-known field doubling, with the propagators acquiring a 2×2 matrix form. The original (and physical) fields (hereafter called Type-1) can appear as both external (asymptotic) and internal lines, whereas their mimics, the thermal ghosts (Type-2), can occur only as internal lines. The thermal part of the propagators now carry a number operator corresponding to the field statistic. In particular, the photon propagator (in Feynman gauge) corresponding to a momentum k can be expressed as $i\mathcal{D}_{\mu\nu}^{ab}(k) = -g_{\mu\nu} i D^{ab}(k)$, with

$$
i\mathcal{D}^{ab}(k) \sim \left[\frac{i}{k^2 + i\epsilon} \delta^{ab} \pm 2\pi \, \delta(k^2) \, N(|k^0|) \, D_T^{ab} \right],
\tag{53.2}
$$

where the first and second term, respectively, correspond to the zero and finite temperature contribution. In the soft limit, the bosonic number operator worsens the IR divergence (the fermionic number operator is well behaved in the soft limit and converges to $1/2$).

$$N(|k^0|) \equiv \frac{1}{\exp\{|k^0|/T\} - 1} \xrightarrow{k \to 0} \frac{T}{|k^0|}, \qquad (53.3)$$

leading to a linear leading IR divergence at finite temperature rather than a logarithmic one as in $T = 0$ case. Hence, subleading logarithmic divergences (which were not present at $T = 0$) also arise at finite temperature. Exactly similar factors also arise from thermal scalar fields. All these extra diverging factors make the proof of IR finiteness of the scalar QED highly non-trivial.

53.3 Infrared Behaviour of Thermal Scalar QED

To analyze the IR behaviour of thermal scalar QED, we will include higher order contributions due to photon insertion (both virtual and real) to a generic n-photon diagram related to the hard scattering process $\gamma^{(*)}(q) + \phi^{(*)}(p) \to \phi^{(*)}(p')$, schematically represented in Fig. 53.1, having r and s number of vertices, respectively, on the p and p' leg such that $r + s = n$.

While adding higher order photon contributions, we will follow the technique due to Grammer and Yennie [8] (henceforth mentioned as GY) by separating the virtual (real) photon contributions into K ($\widetilde{K}$) and G ($\widetilde{G}$) photon contributions. For virtual photons, we will rearrange the $g_{\mu\nu}$ factor in the propagator as

$$g_{\mu\nu} = \left[\left(g_{\mu\nu} - b_k(p_f, p_i)k_\mu k_\nu\right) + \left(b_k(p_f, p_i)k_\mu k_\nu\right)\right] \equiv \left[G_{\mu\nu} + K_{\mu\nu}\right], \quad (53.4)$$

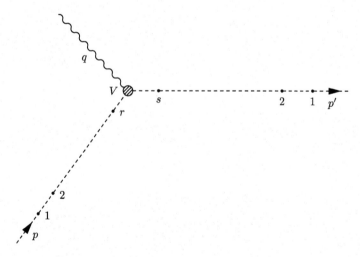

Fig. 53.1 Schematic diagram of a nth order graph of $\gamma^*\phi \to \phi$, with $r + s = n$. Here V labels the special but arbitrary scalar–photon vertex

whereas for real photons, we reexpress the polarization sum $\sum_{\text{pol}} \epsilon_\mu^*(k)\, \epsilon_\nu(k) = -g_{\mu\nu}$ as

$$-g_{\mu\nu} = -\left[\left(g_{\mu\nu} - \tilde{b}_k(p_f,\, p_i)k_\mu k_\nu\right) + \tilde{b}_k(p_f,\, p_i)k_\mu k_\nu\right] \equiv -\tilde{G}_{\mu\nu} - \tilde{K}_{\mu\nu}.$$
(53.5)

Here, b_k ($\tilde{b}_k = b_k|_{k^2=0}$) [3] is defined in such a way that the G ($\tilde{G}$)-photon terms are IR finite and the K ($\tilde{K}$)-photon terms contain all the IR divergent pieces

$$b_k(p_f,\, p_i) = \frac{1}{2}\left[\frac{(2p_f - k)\cdot(2p_i - k)}{((p_f - k)^2 - m^2)((p_i - k)^2 - m^2)} + (k \leftrightarrow -k)\right].$$
(53.6)

In contrast to thermal fermionic QED, we, now, also have 4-point scalar-scalar-photon-photon vertices and, consequently, seagull and tadpole diagrams. Only with the inclusion of these diagrams, we will be able to obtain a neat factorization leading to resummation. In fact, we find that the IR finite tadpole contribution is crucial in obtaining the factorization and subsequent resummation of IR divergent pieces to all orders.

53.3.1 Insertion of Virtual K Photon

After allowing for all possible insertions of K photons and realizing that the external scalar legs and the special scalar-photon hard vertex V is of type-1, the whole contribution due to K photon can be written as

$$\mathcal{M}_{n+1}^{K\gamma,\text{tot}} = \frac{ie^2}{2}\int\frac{d^4k}{(2\pi)^4}\, D^{11}(k)\left[b_k(p',\, p') - 2b_k(p',\, p) + b_k(p,\, p)\right]\mathcal{M}_n,$$

$$\equiv \frac{ie^2}{2}\int\frac{d^4k}{(2\pi)^4}\, D^{11}(k)\left[J^2(k)\right]\mathcal{M}_n \equiv [B]\,\mathcal{M}_n.$$
(53.7)

53.3.2 Insertion of Virtual G Photon

G photon contributions, after including all the possible insertions, are IR finite as

$$\mathcal{M}_{n+1}^{G\gamma} \sim \int d^4k\left[\frac{i}{k^2 + i\epsilon}\delta_{t_\mu,t_\nu} \pm 2\pi\delta(k^2)N(|k|)D_{t_\mu,t_\nu}\right]$$

$$\times \left[0(p_f \cdot p_i) + 2(p_f + p_i)\cdot k\right]\left[\mathcal{O}(1) + \mathcal{O}(k) + \mathcal{O}(k^2) + \cdots\right].$$
(53.8)

From the above expression, it is easily seen that the leading linear divergence arising from the finite temperature part vanishes [9]. The leading logarithmic divergence arising from $T = 0$ part cancels trivially from the construction of b_k. While a part of the subleading logarithmic divergence is seen to vanish as well, the detailed proof of IR finiteness of the G photon contribution is cumbersome; see [2]. We find that the G photon contribution is IR finite, just as in the zero temperature case.

53.3.3 Emission and Absorption of Real $\widetilde{K}$ and $\widetilde{G}$ Photon

The separation into $\widetilde{K}$ and $\widetilde{G}$ photon contributions when including additional real photons is achieved by the definitions in (53.5). Unlike for zero temperature field theory, in thermal field theory the real $\widetilde{K}$ and $\widetilde{G}$ photons can be both emitted and absorbed with respect to the heat bath. And as the real photons carry physical momenta, the phase space gets modified to $d\phi_i = (d^4 k_i/(2\pi)^4)\, 2\pi \delta(k_i^2)\, [\theta(k_i^0) + N(|k_i|)]$. The resultant $\widetilde{K}$ photon contribution is proportional to

$$\left| \mathcal{M}_{n+1}^{\widetilde{K}\gamma,\text{tot}} \right|^2 \propto -e^2 \left[\tilde{b}_k(p, p) - 2\tilde{b}_k(p', p) + \tilde{b}_k(p', p') \right] \equiv -e^2 \widetilde{J}^2(k). \quad (53.9)$$

The $\widetilde{G}$ photon calculations after following a similar approach as G, photon calculation turns out to be IR finite.

After including the correct phase space factors $d\phi_{p'}$ for real photon emission and absorption and after fully symmetrizing the matrix elements, the total cross section turns out to be [9]

$$d\sigma^{\text{tot}} = \int d^4 x\, e^{-i(p+q-p')\cdot x}\, d\phi_{p'} \exp\left[B + B^* + \widetilde{B} \right] \sigma^{\text{finite}}(x), \quad (53.10)$$

where all the divergences (from virtual and real contributions) are contained inside the exponential term and combine to give an IR finite sum in the soft limit

$$(B + B^*) + \widetilde{B} = e^2 \int d\phi_k \left[J(k)^2 \left\{ 1 + 2N(|k^0|) \right\} \right.$$
$$\left. - \widetilde{J}(k)^2 \left\{ \left(1 + N(|k^0|)\right) e^{ik\cdot x} + N(|k^0|) e^{-ik\cdot x} \right\} \right],$$
$$\xrightarrow{k\to 0} 0 + \mathcal{O}(k^2).$$

Therefore, the theory of thermal scalar QED is IR finite to all orders.

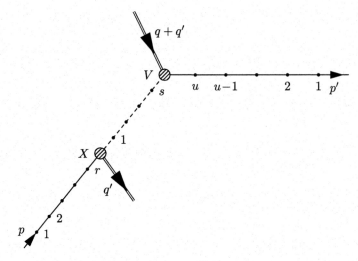

Fig. 53.2 Schematic diagram to define the p and p' legs for the process, $\chi\mathcal{F} \to \mathcal{F}\chi$. Here, $r + s + u = n$

53.4   The IR Finiteness of the Dark Matter Interactions

With both the results of thermal scalar and fermionic QED at hand now, we will be interested in the higher order corrections to a generic scattering process $\chi(q + q')\mathcal{F}(p) \to \mathcal{F}(p')\chi(q')$ as schematically represented in Fig. 53.2 where, $p + q = p'$. The final state fermion has been identified with the p' leg, and both the initial fermion and intermediate scalar line has been identified with p leg and an analysis similar to that above effected. On inclusion of each of the higher order K, G, $\widetilde{K}$ and $\widetilde{G}$ contributions, the matrix element can be shown to factorize exactly [2] in a fashion similar to that above, benefitting from a non-trivial double cancellation around the vertices V and X as seen in Fig. 53.2.

The G and $\widetilde{G}$ photon contributions turn to be IR finite here too. Together with the cancellation of IR divergent terms between K and $\widetilde{K}$ terms in the soft limit, this ensures that the thermal theory of bino-like dark matter is infrared finite to all orders.

53.5   Conclusion

Finite temperature corrections can turn out to be of prime importance when delineating the parameter space allowed to a particular model of dark matter (DM). Armed with the IR finiteness of thermal fermionic QED, and after providing an explicit proof of IR finiteness of scalar QED we obtain the theory of bino-like DM to be IR safe to all orders. The inclusion of seagull and tadpole contributions along with both emission and absorption of real photons with respect to the heat bath are essential to obtain this IR safety. This result is important in the context of obtaining relic densities for dark matter including thermal effects.

References

1. M. Beneke, F. Dighera, A. Hryczuk, JHEP 1410 (2014) 45, Erratum: JHEP**1607**, 106 (2016), arXiv:1409.3049 [hep-ph]
2. P. Sen, D. Indumathi, D. Choudhury, Eur. Phys. J. C **80**, 972 (2020), arXiv:1812.06468
3. D. Indumathi, Ann. Phys. **263**, 310 (1998)
4. H.A. Weldon, Phys. Rev. D **49**, 1579 (1994)
5. R.L. Kobes, G.W. Semenoff, Nucl. Phys. **B260**, 714 (1985)
6. A.J. Niemi, G.W. Semenoff, Nucl. Phys. **B230**, 181 (1984)
7. R.J. Rivers, *Path Integral Methods in Quantum Field Theory* (Cambridge University Press, Cambridge, 1987), Chapter 15
8. G. Grammer Jr., D.R. Yennie, Phys. Rev. D **8**, 4332 (1973)
9. P. Sen, D. Indumathi, D. Choudhury, Eur. Phys. J. C **79**, 532 (2019)

Chapter 54
Modified Higgs Couplings in the Minimal Composite Higgs Model and Beyond

Avik Banerjee

Abstract Composite Higgs scenario, where the Higgs boson emerges as a pseudo-Nambu-Goldstone boson (pNGB), is well motivated as an approach to solve the Hierarchy problem of the Standard Model. One of the main phenomenological consequences of this setup is sizable deviations of the Higgs couplings from their Standard Model predictions. We discuss how the modification of the Higgs couplings with weak gauge bosons and quarks arise in the minimal composite Higgs model and its impact on the scale of compositeness. We take an effective field theoretic approach to illustrate our results. The coupling modifications in models beyond the minimal scenario where the scalar sector is extended with Standard Model singlets and triplets are also discussed.

54.1 Introduction

The discovery of the Higgs boson at the Large Hadron Collider (LHC) paved the path towards the precision study of the properties and origin of the Higgs. Several motivated beyond the Standard Model (BSM) scenarios predict deviations of the Higgs couplings compared to their Standard Model (SM) values. These deviations can in principle be tested at the LHC and proposed future colliders and yield significant information about the nature of the Higgs boson. Composite Higgs models [1, 2] provide an alternative to supersymmetry to solve the Hierarchy problem of SM. It consists of a framework where the Higgs boson originates as a pseudo-Nambu-Goldstone boson (pNGB) of a strong sector with spontaneously broken global symmetry (analogous to the pions in QCD) [3–5]. The scale at which the strong sector condenses sets the geometric dimension of the composite Higgs and is known as the scale of compositeness (f). The SM gauge bosons and fermions, on the other hand,

---

Avik Banerjee—Speaker. Work done in collaboration with Gautam Bhattacharyya, Nilanjana Kumar and Tirtha Sankar Ray.

A. Banerjee (✉)
Saha Institute of Nuclear Physics, HBNI, 1/AF Bidhan Nagar, Kolkata 700064, India
e-mail: avik92kol@gmail.com

P. K. Behera et al. (eds.), *XXIII DAE High Energy Physics Symposium*,
Springer Proceedings in Physics 261,
https://doi.org/10.1007/978-981-33-4408-2_54

are elementary and their interaction with the strong sector *resonance* states via linear mixing explicitly breaks the global symmetry. This endows the Higgs boson with a potential and is responsible for the electroweak symmetry breaking. The couplings of the Higgs in this scenario are modified due to the strong dynamics compared to the corresponding SM values, and the measurements of the Higgs couplings at LHC, in turn, put stringent limits on the scale of compositeness. However, in the absence of any signal of new physics at the LHC, effective field theoretic frameworks become more popular in the study of Higgs boson [6]. The idea behind using an effective theory is that the effect of new dynamics appearing at a high scale can be captured by constructing higher dimensional operators, which respect symmetries at low energies. The higher dimensional operators contribute to the modifications of the couplings of the Higgs boson with other SM particles.

The other yet unknown aspect of the electroweak symmetry breaking is the full constituents of the scalar sector. Apart from the usual $SU(2)_L$ doublet Higgs, additional singlet, doublet or triplet scalars can exist. The presence of the additional scalars also modifies the couplings of the 125 GeV Higgs boson. Moreover, these scenarios contain other neutral and charged scalars with different CP eigenstates providing interesting phenomenological consequences.

In what follows, we discuss the modifications of Higgs coupling in composite Higgs setup (Sect. 54.2). We also show the use of the strongly interacting light Higgs framework to construct the higher dimensional operators and calculate the coupling modifiers. Then we discuss some models with extended scalar sectors and show how the couplings of the Higgs boson changes in those cases (Sect. 54.3). Finally, we conclude (Sect. 54.4).

54.2 Minimal Composite Higgs Model (MCHM)

The minimal realization of the composite Higgs setup, compatible with electroweak precision constraints, consists of a coset $SO(5)/SO(4)$ producing four pNGBs [7]. In the unitary gauge, the pNGB Higgs can be parametrized as

$$\Sigma = (0, \ 0, \ 0, \ s_h, \ c_h)^T \,, \tag{54.1}$$

where $s_h \equiv \sin(h/f)$, $c_h \equiv \cos(h/f)$. The gauge interactions of the pNGB Higgs with SM gauge bosons are given by

$$\mathcal{L}_{\text{gauge}} \simeq \frac{g^2 f^2}{4} \sin^2\left(\frac{h+V}{f}\right) \left[W_\mu^+ W^{\mu-} + \frac{1}{2\cos^2\theta_w} Z_\mu Z^\mu \right]. \tag{54.2}$$

While the masses of gauge bosons set the definition of the electroweak vacuum expectation value (vev) as

$$v_{\text{EW}} = f \sin\frac{V}{f} \,, \tag{54.3}$$

the couplings of the physical (125 GeV) Higgs boson with the weak gauge bosons $(V = W^{\pm}, Z)$ get modified as follows:

$$k_V = \frac{g_{hVV}}{g_{hVV}^{SM}} = \sqrt{1 - \xi},$$

(54.4)

where $\xi = v^2/f^2$. It is worth noting that the modification factor depends only on the scale of the compositeness and this feature is in general true for any $SO(N)/SO(N - 1)$ coset.

The modifications of the Yukawa couplings, on the other hand, depends on the representations of $SO(5)$ in which the SM fermions are embedded. The *partial compositeness paradigm* implies that the SM fermions couples to the Higgs boson through a linear mixing with some strong sector operators. This means, after the condensation of the strong sector, the mass eigenstates are the linear superposition of elementary and composite resonance states. To fulfill the requirement of assigning the correct hypercharge of the SM fermions, an additional unbroken $U(1)_X$ is customarily introduced. For the purpose of illustration, we consider only the embeddings of the SM fermions (specifically top quark) in the fundamental **5** and symmetric **14** of $SO(5)$. Decomposition of $\mathbf{5}_{2/3}$ of $SO(5) \times U(1)_X$ under the SM gauge group is given by

$$\mathbf{5}_{2/3} \rightarrow \mathbf{2}_{7/6} \oplus \mathbf{2}_{1/6} \oplus \mathbf{1}_{2/3}.$$

(54.5)

We present the relevant incomplete multiplets for left- and right-handed top quarks in the so-called $MCHM_{5_L - 5_R}$ model as [4]

$$Q_L^5 = \frac{1}{\sqrt{2}}(-ib_L, -b_L, -it_L, t_L, 0)^T, \quad T_R^5 = (0, 0, 0, 0, t_R)^T.$$

(54.6)

The only $SO(5)$ invariant term (involving Q_L^5, T_R^5 and Σ), constituting the effective low energy Yukawa Lagrangian can be written as

$$\mathcal{L}_{Yuk} = \Pi_{LR}(q^2)(\overline{Q}_L^5.\Sigma)(\Sigma^T.T_R^5) + h.c. \quad \Rightarrow \quad \mathcal{L}_{Yuk} = \Pi_{LR}(q^2)s_h c_h \bar{t}_L t_R + h.c.,$$

(54.7)

where we assume, for simplicity, the momentum dependence of the form factor $\Pi_{LR}(q^2)$ can be approximated with a constant value $\Pi_{LR}(q^2 = 0)$. Clearly, $\Pi_{LR}(q^2 = 0)$ can be absorbed in the definition of the top quark mass, while the modification of the Yukawa coupling is given as [8]

$$k_t = \frac{g_{ht\bar{t}}}{g_{ht\bar{t}}^{SM}} = \frac{1 - 2\xi}{\sqrt{1 - \xi}} \simeq 1 - \frac{3}{2}\xi.$$

(54.8)

In the last equality, we make an expansion around small ξ. In the case of $MCHM_{14_L - 14_R}$ model (both left- and right-handed top quarks in **14** of $SO(5)$), however, more than one Yukawa invariants can be constructed. As a result of that, the form factors cannot be completely absorbed in the definition of top mass. The low energy Lagrangian in MCHM$_{14_L-14_R}$ with two invariants is given by

$$\mathcal{L}_{\text{Yuk}} = \Pi_{LR}^{(1)}(\Sigma^T.\overline{Q}_L^{14}.T_R^{14}.\Sigma) + \Pi_{LR}^{(2)}(\Sigma^T.\overline{Q}_L^{14}.\Sigma)(\Sigma^T.T_R^{14}.\Sigma) + \text{h.c.},$$
$$= \left(\Pi_{LR}^{(1)} + \Pi_{LR}^{(2)}s_h^2\right)s_h c_h \bar{t}_L t_R + \text{h.c.} \tag{54.9}$$

Modification of Yukawa coupling in this case is [9]

$$k_t = \frac{g_{ht\bar{t}}}{g_{ht\bar{t}}^{\text{SM}}} \simeq 1 - \left(2\frac{\Pi_{LR}^{(2)}}{\Pi_{LR}^{(1)}} - \frac{3}{2}\right)\xi. \tag{54.10}$$

Evidently, in this case, the coupling modification depends on the dynamics of the strong sector resonance states through the form factors. We comment in passing that, if either of t_L or t_R is embedded in the **14**, two Yukawa invariants can be constructed. The limit on f in MCHM$_{5_L-5_R}$ model at 95% CL, as obtained using the LHC data on the Higgs coupling measurements [10], is rather strong ($f \geq 1$ TeV), because k_V and k_t depends solely on a single parameter ξ. However, a relaxation on the limit on f is observed in models where more than one Yukawa invariants are present ($f \geq 640$ GeV) [9].

The modification of the Higgs couplings can also be captured using an effective field theory approach. In the strongly interacting light Higgs framework [6], the SM Lagrangian is extended with a set of gauge invariant dimension-six operators. For the illustrative purpose, we present a few of such operators which directly contributes to modify the tree level Higgs couplings. The kinetic term of the Higgs boson in this scenario containing dimension-4 and dimension-6 terms is given by

$$\mathcal{L}_{\text{kin}} = \left(D_\mu H\right)^\dagger (D^\mu H) + \frac{c_H}{2f^2}\partial_\mu(H^\dagger H)\partial^\mu(H^\dagger H), \tag{54.11}$$

while the gauge and the Yukawa couplings of the Higgs boson is

$$\mathcal{L}_{\text{gauge}} = \frac{g^2}{2}(H^\dagger H)\left(W_\mu^+ W^{-\mu} + \frac{1}{2\cos^2\theta_W}Z_\mu Z^\mu\right), \tag{54.12}$$

$$\mathcal{L}_{\text{Yuk}} = -y_f \bar{Q}_L H t_R - \Delta\left(\frac{H^\dagger H}{f^2}\right)y_f \bar{Q}_L H t_R. \tag{54.13}$$

The modifications of the Higgs couplings upon electroweak symmetry breaking is then given by

$$k_V = \sqrt{1 - c_H \xi}, \quad k_t \simeq 1 + \left(\Delta - \frac{c_H}{2}\right)\xi. \tag{54.14}$$

Clearly, in MCHM$_{5_L-5_R}$ model $c_H = 1$ and $\Delta = -1$, while in MCHM$_{14_L-14_R}$ model Δ depends on the strong sector dynamics via the form factors.

54.3 Models with Extended Scalar Sector

In this section, we discuss the models where the SM particle content is extended with additional scalar fields. In general, singlet, doublet or triplet scalars can be accommodated with the existing Higgs doublet, with completely different phenomenological implications. In the context of composite Higgs scenario, non-minimal coset structures can lead to the presence of these additional scalars. For example, in next-to-minimal composite Higgs model ($SO(6)/SO(5)$ coset) a singlet CP-odd particle is present [11]. Needless to mention that neutral component of the standard Higgs can mix with this additional singlet, and therefore, modifies the couplings of the 125 GeV Higgs boson. For example, the deviation of the Higgs couplings with the weak gauge bosons are suppressed by an additional factor of mixing angle θ_{mix} as given by [12]

$$k_V = \cos\theta_{\text{mix}}\sqrt{1-\xi}. \tag{54.15}$$

For the Yukawa part we write an effective operator as

$$\Delta\mathcal{L}_\eta \sim -y_t(\Delta_t^\eta)'\frac{\eta^2}{f^2}\bar{q}_L H^c t_R, \tag{54.16}$$

which leads to the coupling modification

$$k_t = \cos\theta_{\text{mix}}\left[1+\left(\Delta-\frac{c_H}{2}\right)\xi\right]+\sin\theta_{\text{mix}}\Delta_t^\eta\sqrt{\xi}. \tag{54.17}$$

This implies that LHC data can provide constraints in the plane of $\theta_{\text{mix}} - \xi$ plane, as shown in [9]. For details of effective field theory analysis in the singlet extended models and two Higgs doublet models see [13, 14]. On the contrary, adding a single triplet scalar with either hypercharge $Y = 0$ or $Y = 1$ leads to a severely constrained scenario from the electroweak precision data. However, in a Georgi-Machacek like model [15], where the $Y = 0$ and $Y = 2$ triplets are embedded in a $(3, 3)$ under the $SU(2)_L \times SU(2)_R$, this constraint can be somewhat relaxed. We use such a setup and include dimension-5 operators in the Yukawa sector as [16]

$$-\mathcal{L}_{\text{Yuk}} = \frac{c_5^t}{\Lambda}y_t\,\bar{Q}_L\chi^\dagger\phi t_R + \frac{c_5^b}{\Lambda}y_b\bar{Q}_L\chi\phi^c b_R + \frac{d_5^t}{\Lambda}y_t\,\bar{Q}_L\xi\phi^c t_R$$

$$+ \frac{d_5^b}{\Lambda}y_b\bar{Q}_L\xi\phi b_R + \text{h.c.} \tag{54.18}$$

Note that, the inclusion of these additional terms modifies the 125 GeV Higgs couplings, as well as the couplings of the charged Higgs boson with the third generation quarks. This implies that the limits on the charged Higgs masses can be changed if we admit the existence of higher dimensional operators in the Gerogi-Machacek model, as shown in [16].

54.4   Conclusions

Composite Higgs is an interesting non-supersymmetric alternative to address the hierarchy problem of SM. One of the main features of these models is modifications in Higgs couplings which can be tested at LHC and proposed future colliders. Higher dimensional operators can capture the nonlinearity of pNGBs (e.g. strongly interacting light Higgs framework), which in turn, is responsible for the coupling modifications. The hVV modifications are generically universal, while the Yukawa coupling modifiers depend on the representation in which the fermions are embedded. Going beyond the minimal scenario one finds more pNGB scalars, for example, in the next-to-minimal model extra singlet scalar gives additional modifications due to neutral scalar mixing. Finally, dimension-five operators can have a significant impact on the constraints from flavour physics observables on the charged Higgs sector in triplet extended models (e.g. Georgi-Machacek model).

Acknowledgements This talk is based on the works done with Gautam Bhattacharyya, Nilanjana Kumar and Tirtha Sankar Ray, and I thank them for their valuable comments and discussions. I acknowledge the support provided by the Department of Atomic Energy, Government of India. Finally, I thank the organizers of XXIII DAE-BRNS High Energy Physics Symposium 2018, for the invitation.

References

1. D.B. Kaplan, H. Georgi, SU(2) x U(1) breaking by vacuum misalignment. Phys. Lett. B **136**, 183–186 (1984)
2. M.J. Dugan, H. Georgi, D.B. Kaplan, Anatomy of a composite Higgs model. Nucl. Phys. B **254**, 299–326 (1985)
3. R. Contino, The Higgs as a composite Nambu-Goldstone Boson, in *Physics of the large and the small, TASI 09, proceedings of the Theoretical Advanced Study Institute in Elementary Particle Physics*, Boulder, Colorado, USA, 1–26 June 2009 (2011), pp. 235–306
4. G. Panico, A. Wulzer, The composite Nambu-Goldstone Higgs. Lect. Notes Phys. **913**, 1–316 (2016) [1506.01961]
5. C. Csaki, P. Tanedo, Beyond the standard model, in *Proceedings, 2013 European School of High-Energy Physics (ESHEP 2013)*, Paradfurdo, Hungary, 5–18 June 2013
6. G.F. Giudice, C. Grojean, A. Pomarol, R. Rattazzi, The strongly-interacting light Higgs. JHEP **6**, 45 (2007) [hep-ph/0703164]
7. K. Agashe, R. Contino, A. Pomarol, The minimal composite Higgs model. Nucl. Phys. B **719**, 165–187 (2005) [hep-ph/0412089]
8. M. Montull, F. Riva, E. Salvioni, R. Torre, Higgs couplings in composite models. Phys. Rev. D **88**, 095006 (2013) [1308.0559]
9. A. Banerjee, G. Bhattacharyya, N. Kumar, T.S. Ray, *Constraining composite Higgs models using LHC data*. JHEP **1803**, 62 (2018) [1712.07494]
10. ATLAS, CMS collaboration, G. Aad et al., Measurements of the Higgs boson production and decay rates and constraints on its couplings from a combined ATLAS and CMS analysis of the LHC pp collision data at $\sqrt{s} = 7$ and 8 TeV. JHEP **8** 45 (2016) [1606.02266]
11. B. Gripaios, A. Pomarol, F. Riva, J. Serra, Beyond the minimal composite Higgs model. JHEP **04**, 070 (2009) [0902.1483]

12. A. Banerjee, G. Bhattacharyya, T.S. Ray, Improving fine-tuning in composite Higgs models. Phys. Rev. **D96**, 035040 (2017) [1703.08011]
13. M. Chala, G. Durieux, C. Grojean, L. de Lima, O. Matsedonskyi, Minimally extended SILH. JHEP **6**, 88 (2017) [1703.10624]
14. S. Karmakar, S. Rakshit, Higher dimensional operators in 2HDM. JHEP **10**, 48 (2017) [1707.00716]
15. H. Georgi, M. Machacek, Doubly charged Higgs Bosons. Nucl. Phys. B **262**, 463–477 (1985)
16. A. Banerjee, G. Bhattacharyya, N. Kumar, Impact of Yukawa-like dimension-five operators on the Georgi-Machacek model. Phys. Rev. **D99** 035028 (2019) [1901.01725]

Chapter 55
Connecting Loop Quantum Gravity and String Theory via Quantum Geometry

Deepak Vaid

Abstract We argue that String Theory and Loop Quantum Gravity can be thought of as describing different regimes of a single unified theory of quantum gravity. LQG can be thought of as providing the pre-geometric exoskeleton out of which macroscopic geometry emerges and String Theory then becomes the *effective* theory which describes the dynamics of that exoskeleton. The core of the argument rests on the claim that the Nambu-Goto action of String Theory can be viewed as the expectation value of the LQG area operator evaluated on the string worldsheet. A concrete result is that the string tension of String Theory and the Barbero-Immirzi parameter of LQG turn out to be proportional to each other.

ArXiv ePrint: arXiv:1711.05693

55.1 Strings Versus Loops

There are several competing candidates for a theory of quantum gravity. Two of the strongest contenders are Loop Quantum Gravity (LQG) [2, 3, 14] and String Theory [12, 21, 22]. Of these, string theory has been around for much longer, is more mature, and has a far greater number of practitioners. LQG is younger, with fewer adherents, but still with the potential to present a serious challenge to the supremacy of String Theory.

In this note, we would like to suggest that rather than an "either/or" situation, one can instead have the best of both worlds. The commonalities between the two approaches are far greater than their apparent differences and that both frameworks provide essential conceptual constructs that will go into any final theory of quantum gravity. The following observations hold true for both fields:

D. Vaid (✉)
Department of Physics National Institute of Technology Karnataka (NITK),
Mangalore, India
e-mail: dvaid79@gmail.com

© Springer Nature Singapore Pte Ltd. 2021
P. K. Behera et al. (eds.), *XXIII DAE High Energy Physics Symposium*,
Springer Proceedings in Physics 261,
https://doi.org/10.1007/978-981-33-4408-2_55

391

1. **fundamental degrees of freedom are the same**—extended one-dimensional objects referred to as "strings" by string theorists and as "holonomies" by LQG-ists.

2. **identical predictions** [9, 18] **for the Bekenstein-Hawking entropy of black holes** are obtained,[1] albeit both fields follow different routes to get there.

3. **geometry becomes discrete, or more appropriately, "quantized", as one approaches the Planck scale** and the continuum approximation of a smooth background spacetime breaks down. In string theory, this happens because closed strings cannot shrink to zero size due to quantum fluctuations. In LQG, there is a natural and explicit construction [4, 13, 15] of quantum operators for area and volume which have a discrete spectrum and whose minimum eigenvalues are greater than zero.

4. **coherent, consistent description of matter degrees of freedom is missing** in both LQG and string theory. Though one can always add matter *by hand* to either theory, it would be much more satisfying if the particles of the Standard Model were to arise naturally as geometrical objects which are part and parcel of either theory.

5. **state space of LQG and string theory can be mapped onto each other**. In LQG the kinematical Hilbert space consists of graphs with edges labeled by representations of $SU(2)$ and edges labeled by invariant $SU(2)$ tensors. As it so happens, string theory also contains very similar structures which go by the name of "string networks". This similarity was alluded to in an older paper by Sen [17]:

> in future a manifestly SL(2,Z) invariant non-perturbative formulation of string theory may be made possible by regarding the string network, instead of string loops, as fundamental objects. This would be similar in spirit to recent developments in canonical quantum gravity, in which loops have been replaced by spin networks.

Apart from the five points outlined above, there is a very crucial aspect which connects both string theory and LQG. This has to do with the definition of the action associated with a string in string theory and the quantum operator for the area which arises in LQG and will form the core of our argument for the existence of a clear connection between LQG and string theory.

The plan of this work is as follows. In Sect. 55.2, we provide a lightning introduction to LQG and introduce the area operator. In Sect. 55.3, we discuss how the low-energy effective field theory emerges from the string action and what this entails for the relation between quantum geometry and string theory. In Sect. 55.4, we argue that the Nambu-Goto string action can be seen as arising from quantum geometric constructs which are present in LQG, and finally in Sect. 55.5, we conclude with some thoughts on the present situation and future developments.

---

[1]Though note [18, 19] where disagreements between the two sets of calculations are pointed out. Though certain factors are not the same, the overall form of the entropy area relation including logarithmic corrections is the same. The differences could possibly be traced to the use of Euclidean geometry to determine black hole entropy in string theory. This introduces ingredients which are missing in the LQG calculation.

55.2 LQG Area Operator

In LQG the basic dynamical variables are [3] a $\mathfrak{su}(2)$ connection A_a^i (where a, b, c three dimensional spatial indices and i, j, k are Lie algebra indices) and the triad e_i^a (which determines the three dimensional metric h_{ab}, of the 3-manifold Σ, via the relation: $h_{ab} = e_a^i e_b^j \delta_{ij}$). These satisfy the Poisson bracket

$$\{A_a^i, e_j^b\} = \kappa \delta_a^b \delta_j^i \tag{55.2.1}$$

where $\kappa = 8\pi G_N$. Associated with each of these variables, one can construct operators which correspond to gauge invariant observables. The connection can be smeared along one-dimensional curves to obtain holonomies, the trace of which is gauge invariant. Holonomies are nothing more than the Wilson loops of field theory. Similarly, the triad fields are smeared over two-dimensional surfaces to obtain the generalized momentum variables. It turns out that using these observables it is possible to construct an operator acting on the kinematical Hilbert space which measures the areas of two-dimensional surfaces. The area of a two-dimensional surface S with intrinsic metric h_{AB} is given by

$$A_S = \int d^2x \sqrt{\det(h_{AB})} \tag{55.2.2}$$

Divide S into N cells S_I, with $I = 1, 2, \ldots, N$. Then A_S can be approximated by

$$A_S = \sum_{I=1}^{N} \sqrt{\det(h_{AB}^I)} \tag{55.2.3}$$

where h_{AB}^I is the two-dimensional metric in the Ith cell. Now as N increases in (55.2.3), A_S will become a better and better approximation to the actual area of S. Of course, it should be kept in mind that when working in the regime where a classical geometry is yet to emerge, the "actual" or "exact" area of a surface is not a well-defined quantity.

Using the well-understood procedure of quantization of the Einstein-Hilbert action in terms of spin-network states, we obtain the following expression for the quantized area operator:

$$\hat{A}_S \Psi_\Gamma = 8\pi l_{PL}^2 \sum_k \sqrt{j_k(j_k + 1)} \Psi_\Gamma \tag{55.2.4}$$

where the Planck length $l_{PL} = \sqrt{\hbar G_N}$ (if $c = 1$), the sum is over all the points where edges of Γ intersect S and j_k is the eigenvalue of the Casimir operator τ^2 along that edge.

55.2.1 Minimum Area and Conformal Symmetry

The quantum operator corresponding to A_S turns out to possess a minimum eigenvalue

$$A_{S_I} = 8\pi l_{PL}^2 \sqrt{j_I(j_I + 1)} \tag{55.2.5}$$

where j_I denotes the representation of $\mathfrak{su}(2)$ assigned to the Ith cell of the surface S. Since the smallest eigenvalue of the angular momentum operator is $j = 1/2$, the minimum quantum of area permitted by LQG is

$$A_{\min} = 2\sqrt{3}\pi l_{PL}^2 \tag{55.2.6}$$

This has an immediate implication for any field theories living on two-dimensional surfaces—the presence of a length scale implies that in such theories conformal symmetry will always be broken. We will see a bit later what this might imply when viewed from the perspective of string theory.

55.3 String Theory and Quantum Geometry

Now, recall that the Nambu-Goto action for a $1 + 1$ dimensional string worldsheet Σ embedded in some $D + 1$ dimensional *worldvolume* $\mathcal{M}$ is given by [20–22]

$$S_{NG} = -T \int d\tau d\sigma \sqrt{-\det(h_{AB})} \tag{55.3.1}$$

where T is the string tension, τ, σ are the timelike and spacelike co-ordinates respective on the string worldsheet and h_{AB} is the two-dimensional metric induced on Σ due to its embedding in $\mathcal{M}$. There is the only term in this action $\sqrt{-h_{AB}}$ and that is precisely the area of the string worldsheet (the $1 + 1$ dimensional surface the string sweeps out as it evolves in spacetime). The system that (55.3.1) represents is a single string evolving in a flat background spacetime.

55.3.1 Gravity from String Theory

Of course, just as the action for a single free particle cannot capture the complexity of a many body system, the action for a single free string is unlikely to capture the complexity of gravitational and particle physics. Now, we know that the spectrum of a free string embedded in D spacetime dimensions (the "worldvolume") contains, in addition to an infinite tower of left and right moving massive modes, three massless fields described, respectively, by a traceless symmetric tensor $h_{\mu\nu}(X)$, an antisymmetric tensor $B_{\mu\nu}(X)$ and a scalar $\Phi(X)$. These objects are identified with, respectively, a graviton, a Kalb-Ramond field, and a dilaton [21, Sect. 2.3.2]. Here μ, ν are worldvolume indices and $X^\mu(\tau, \sigma)$ are worldvolume co-ordinates which describe the embedding of the string in the worldvolume. The $\{X^\mu\}$ can also be thought of as scalar fields living on the string worldsheet. This can be seen by writing the string action in the form associated with Polyakov's name

$$S_{\text{Polya}} = -\frac{T}{2} \int d\tau d\sigma \sqrt{-g} g^{ab} \partial_a X^\mu \partial_b X^\nu \eta_{\mu\nu} \qquad (55.3.2)$$

where $a, b \in (1, 2)$ are worldsheet indices, g_{ab} is the intrinsic metric of the worldsheet and $\eta_{\mu\nu}$ is the flat worldvolume metric. In this form, it is clear that the string action can be thought of describing $D + 1$ massless, non-interacting scalar fields living on a two-dimensional manifold.

The traceless symmetric tensor $h_{\mu\nu}(X)$ has spin-2 and is, therefore, the source of the claim that gravity is already present in string theory. The argument [6] is that the only consistent interacting theory that can be constructed from a spin-2 field has to be General Relativity.[2] An alternate route to obtain the low-energy effective action is to replace the flat metric $\eta_{\mu\nu}$ with a general curved metric $G_{\mu\nu}$ in (55.3.2)

$$S_{\text{Polya}} = -\frac{T}{2} \int d\tau d\sigma \sqrt{-g} g^{ab} \partial_a X^\mu \partial_b X^\nu G_{\mu\nu}(X) \qquad (55.3.3)$$

The requirement that (55.3.3) satisfy Weyl invariance implies [12, Sect. 3.7], [21, Sect. 7.2] that the beta functions for the graviton, the antisymmetric tensor and the dilaton, must be zero. This, in turn, implies that the background in which the string is propagating must satisfy Einstein's field equations with source terms coming from the antisymmetric tensor and the dilaton. For example, the low-energy effective action for the bosonic string (in $D = 26$ spacetime dimensions) takes the form [21]

$$S_{\text{eff}} = \frac{1}{2\kappa_0^2} \int d^{26} X \sqrt{-G} e^{-2\Phi} \left(R - \frac{1}{12} H_{\mu\nu\lambda} H^{\mu\nu\lambda} + 4\partial_\mu \Phi \partial^\mu \Phi \right) \qquad (55.3.4)$$

where $H_{\mu\nu\lambda}$ is the curvature of the antisymmetric tensor field $B_{\mu\nu}$.

Let us pause for a moment to appreciate what a remarkable result this is. Starting from nothing but a free string propagating in a flat spacetime, the requirement of worldsheet Weyl invariance implies that the background geometry must necessarily satisfy Einstein's equations. As David Tong puts it so eloquently [21, p. 175]:

> That tiny string is seriously high-maintenance: its requirements are so stringent that they govern the way the whole universe moves.

---

[2]This does not, however, appear to be an entirely settled point [7, 11].

55.3.2   Geometry Versus Pre-geometry

While the elegance and power of the string theoretical arguments cannot be disputed, seen from the perspective of LQG, string theory has a fundamental flaw. LQG leads us to a concrete notion of *quantum geometry*. It allows us to construct a framework in which we can do away with the notion of classical geometries—whether flat or curved—entirely, and instead of starting with the *atoms* of spacetime [10, 16] using which we can build up almost any kind of geometry we can think of.[3] Using the terminology of John Wheeler, starting with *pre-geometry*, we can construct geometry.

In much the same way that the Pauli principle and the theory of linear combination of atomic orbitals (LCAO) shows us that atoms can be brought together to form molecules only in certain combinations, the kinematical constraints of quantum geometry determine how the atoms of quantum geometry can be "glued" together to form more complicated structures. Thus, in LQG, we make no presumption of a classical background spacetime. LQG is not only a theory of quantum gravity which does not depend on the background geometry, but also a theory *in which there is no background* to begin with! This is in sharp contrast to string theory, where *in the very first step itself* (55.3.2) we assume that there exists a smooth, flat background spacetime on which the string can propagate. While it *is* true, that starting from a string in a flat background, theoretical consistency ultimately requires the background to ultimately satisfy Einstein's equations, this does not obviate the fact that the existence of smooth, continuum background spacetime is taken for granted in conventional formulations of string theory.

It seems clear that any theory which claims to be a theory of "quantum gravity" must explain how spacetime arises in the first place rather than putting it in by hand at the very beginning. String theory fails this test. However, and this is crucial, this fact does not invalidate the results obtained in string theory. It only motivates us to try and understand whether the tools of LQG can be harnessed to provide the missing link between pre-geometry and string theory.

55.4   Geometry from Pre-geometry

Let us make the assumption that, *a priori*, the spacetime manifold has no structure, no metric, no way to measure distances and areas. In that case, the question arises as to how are we to define the integrand of the Nambu-Goto action in (55.3.1). The clue lies in LQG. Recall that in Sect. 55.2, we explained how the edges of a graph state in LQG carry angular momenta and how these angular momenta endow surfaces pierced by those edges with quanta of area. This leads to the expression (55.2.5) for the area of a surface in terms of the angular momenta j_i carried by each edge which pierces that surface

---

[3]Though not all such geometries will be stable against perturbations. The formalism of Causal Dynamical Triangulations (CDT), closely related in spirit to LQG allows one to study this question in detail [1].

$$A_{S_l} = 8\pi l_{PL}^2 \sqrt{j_l(j_l + 1)} \tag{55.4.1}$$

Now, imagine that there are many strings moving around in this structureless manifold, with each string carrying some angular momentum in some representation of $\mathfrak{su}(2)$. Each string traces out a two-dimensional surface—or "worldsheet", but as yet without any notion of area—as it moves through the manifold. We would expect that the worldsheets generated by the different strings would inevitably intersect. If seen from the perspective of just a single string, its worldsheet will be pierced by the other strings at various points. According to the LQG prescription (55.2.5), at each such point, the worldsheet of the original string will be endowed with one quantum of area determined by the angular momentum carried by the corresponding string. The total area of the worldsheet will then be given by

$$\langle \hat{A}_S \rangle = 8\pi l_{PL}^2 \sum_k \sqrt{j_k(j_k + 1)} \tag{55.4.2}$$

How can we understand string evolution from this perspective? One would expect that strings would tend to avoid running into each other. This expectation can be converted into a mathematical statement by requiring that (55.4.2) be minimized. But this is nothing more than the statement of the extremization of the Nambu-Goto action (55.3.1) of the string worldsheet! Finally, we are left with the following conjecture. The Nambu-Goto action for the bosonic string can be expressed in terms of the expectation value of the LQG area operator acting on a pre-geometric graph state

$$S_{NG} \propto \langle \Psi | \hat{A} | \Psi \rangle \tag{55.4.3}$$

Notice that we have used proportionality instead of equality in the above expression. This is because, a priori, the two quantities on either side have different units. S_{NG} has units of action or energy per unit time and the area has units of L^2 or E^{-2}. In order to equate both sides, suitable proportionality constants must be introduced. This will lead us to the relationship between the Barbero-Immirzi parameter β of LQG [5, 8] and the string tension T (or alternatively the Regge slope $\alpha = 1/4\pi T$). This is the topic of the next section.

In passing, let us note that one consequence of the quantized nature of the string worldsheet is that, as mentioned in Subsect. 55.2.1, the conformal invariance of the worldsheet would no longer be an exact symmetry. This would have immediate implications for the number of spacetime dimensions in which one could define a consistent string theory.

55.4.1 Barbero-Immirzi Parameter as String Tension

In the definition of the classical phase space, and therefore, of the kinematical Hilbert space of LQG, there is a one-parameter ambiguity [5, 8] known as the Barbero-

Immirzi parameter. At the classical level, theories with different values of β are physically equivalent, however, in the *quantum* theory, different values of β correspond to physically distinct Hilbert spaces with inequivalent spectra of the fundamental geometric operators. In particular, the spectrum of the area operator (55.4.1) is modified to become

$$A_{S_I} = 8\pi\beta\, l_{PL}^2 \sqrt{j_I(j_I+1)} \tag{55.4.4}$$

In light of this modification, let us reconsider the two sides of (55.4.3). The left side has the form

$$S_{NG} \simeq \text{tension} \times \text{area}$$

and the right side

$$\langle\Psi|\hat{A}|\Psi\rangle \simeq \beta \times \text{area}$$

Demanding equality of the two sides would therefore imply that the string tension T_{string} and the Barbero-Immirzi parameter β are related to each other by some, as yet undetermined, constant T_{loop}

$$T_{\text{string}} = \beta\, T_{\text{loop}} \tag{55.4.5}$$

Since β is dimensionless, the proportionality constant will also have units of tension. Hence, the nomenclature T_{loop} is used to indicate that this quantity is a tension associated with the 1D graph edges of LQG states.

55.5   Discussion and Future Work

We have presented a rough outline of a method in which one can resolve the central challenges of both LQG and string theory. The problem with LQG is the lack of a clear method to obtain a semiclassical spacetime geometry in the limit of large graph states. The flaw with string theory, in the author's humble opinion, is the presumption of the existence of a background geometry on which the string can propagate. Both of these problems can be cured by viewing the Nambu-Goto action, and the geometry of the string worldsheet, as arising from a pre-geometry which can be described in the language of LQG. From this perspective, String Theory is the glue which connects a quantum theory of geometry (LQG) to a classical theory of gravity with matter such as the one given in (55.3.4).

Acknowledgements The author would like to dedicate this article to his wife on the occasion of her birthday. The author wishes to acknowledge the support of a visiting associate fellowship from the Inter-University Centre For Astronomy And Astrophysics (IUCAA), Pune, India, where a portion of this work was completed.

References

1. J. Ambjorn, A. Goerlich, J. Jurkiewicz, R. Loll, Quantum gravity via causal dynamical triangulations, February 2013
2. A. Ashtekar, *Lectures on non-perturbative canonical gravity* (1991)
3. A. Ashtekar, J. Lewandowski, Background independent quantum gravity: a status report, **21**(15), R53–R152 (2004)
4. A. Ashtekar, C. Rovelli, L. Smolin, Weaving a classical metric with quantum threads, **69**(2), 237–240, July 1992
5. F. Barbero, From euclidean to lorentzian general relativity: the real way, June 1996
6. D. Boulware, S. Deser, Classical general relativity derived from quantum gravity, **89**(1), 193–240, January 1975
7. S. Deser, Gravity from self-interaction redux, November 2009
8. G. Immirzi, Real and complex connections for canonical gravity, December 1996
9. R.K. Kaul, Entropy of quantum black holes, February 2012
10. K. Krasnov, C. Rovelli, Black holes in full quantum gravity, May 2009
11. T. Padmanabhan, From gravitons to gravity: Myths and reality, September 2004
12. J. Polchinski, *String Theory (Cambridge Monographs on Mathematical Physics)*, vol. 1, 1st edn. (Cambridge University Press, October 1998)
13. C. Rovelli, Area is the length of ashtekar's triad field. **47**, 1703–1705 (1993)
14. C. Rovelli, Zakopane lectures on loop gravity, February 2011
15. C. Rovelli, L. Smolin, Discreteness of area and volume in quantum gravity, November 1994
16. C. Rovelli, S. Speziale, A semiclassical tetrahedron, August 2006
17. A. Sen, String network, **1998**(3), 005 (1997)
18. A. Sen, Logarithmic corrections to schwarzschild and other non-extremal black hole entropy in different dimensions, May 2012
19. A. Sen, Microscopic and macroscopic entropy of extremal black holes in string theory, February 2014
20. G. 't Hooft, Introduction to string theory, May 2004
21. D. Tong, *Lectures on String Theory*, June 2010
22. B. Zwiebach, *A First Course in String Theory*, 2nd edn. (Cambridge University Press, 2009)

Chapter 56
Perturbativity Constraints on $U(1)_{B-L}$ and Left-Right Models

Garv Chauhan, P. S. Bhupal Dev, R. N. Mohapatra, and Yongchao Zhang

Abstract We derive theoretical perturbative constraints on models with extended gauge sectors, whose generators contribute to the electric charge. We also constrain gauge couplings by imposing the condition for them to be perturbative up to the grand unification scale. This leads to strong constraints on the masses of the corresponding gauge bosons. In this work, we specifically focus on the $SU(2)_L \times U(1)_{I_{3R}} \times U(1)_{B-L}$ and the left-right symmetric models based on $SU(2)_L \times SU(2)_R \times U(1)_{B-L}$, and discuss the implications of the perturbativity constraints for new gauge boson searches at current and future hadron colliders.

56.1 Introduction

Standard Model (SM) has been highly successful in explaining the particle interactions in the universe. However, we must extend the SM to explain the empirical evidence for new physics, such as observed neutrino oscillations, dark matter, and baryon asymmetry. Beyond the Standard Model (BSM), physics can be hidden at any energy scale, but if they are present in the multi-TeV scale, one can probe their predictions in current and planned future experiments. Many BSM scenarios include extended gauge sectors with extra $U(1)$ or $SU(2) \times U(1)$. We focus on a subclass of these gauge extensions of the SM, namely, the $U(1)_{B-L}$ and minimal left-right symmetric models (LRSM), where the generators of the extra gauge groups contribute to the electric charge [1, 2]. In these models, we study the renormalization group (RG) evolution of the gauge couplings combining both theoretical and perturbativity constraints. we show that demanding perturbativity of the gauge couplings,

G. Chauhan (✉) · P. S. Bhupal Dev · Y. Zhang
Department of Physics and McDonnell Center for the Space Sciences, Washington University, St. Louis, MO 63130, USA
e-mail: garv.chauhan@wustl.edu; garv.chauhan@gmail.com

R. N. Mohapatra
Department of Physics, Maryland Center for Fundamental Physics, University of Maryland, College Park, MD 20742, USA

© Springer Nature Singapore Pte Ltd. 2021   401
P. K. Behera et al. (eds.), *XXIII DAE High Energy Physics Symposium*,
Springer Proceedings in Physics 261,
https://doi.org/10.1007/978-981-33-4408-2_56

i.e., $g_i < \sqrt{4\pi}$ up to the Grand Unification Theory (GUT) or Planck scale imposes both lower and upper bounds on the new gauge couplings. This has important consequences for the masses of the associated gauge bosons that are being searched for in high-energy collider experiments.

56.2 Theoretical Constraints on Gauge Couplings

Let us consider an electroweak gauge extension of the SM with the gauge group $SU(2)_L \times U(1)_X \times U(1)_Z$. We also assume that this group undergoes spontaneous symmetry breaking to $U(1)_{em}$ with SM electroweak gauge group as an intermediate stage. Given this, the generators of the group will obey the relation $Q = I_{3L} + I_X + \frac{Q_Z}{2}$. Given that $U(1)_X \times U(1)_Z \to U(1)_Y$, we have the following relation among the corresponding gauge couplings [3]: $g_Y^{-2} = g_X^{-2} + g_Z^{-2}$. This relation should also hold even if the coupling g_X belongs to an $SU(2)$ group. Then requiring that the coupling g_Z should remain perturbative at the symmetry breaking scale yields a theoretical lower bound on the coupling g_X [4].

$$r_g \equiv \frac{g_X}{g_L} > \tan\theta_w \left(1 - \frac{4\pi}{g_Z^2} \frac{\alpha_{em}}{\cos^2\theta_w} \right)^{-1/2}, \qquad (56.1)$$

where θ_w is the weak mixing angle. Note that the above relation for the gauge couplings only holds at the symmetry breaking scale.

56.3 $U(1)_{B-L}$

We first discuss the $SU(2)_L \times U(1)_{I_{3R}} \times U(1)_{B-L}$ model [5, 6] that has two BSM $U(1)$ groups which break down to $U(1)_{em}$. For anomaly cancelation, the model requires three generations of right-handed neutrinos (RHNs) which help generate neutrino masses via the seesaw mechanism [7–10]. In the scalar sector, a Higgs singlet Δ is introduced to give mass to the RHNs. The particle content of the $U(1)_{B-L}$ model is given in Table 56.1.

The β-functions for the one-loop RG evolution of the new $U(1)$ gauge couplings are given by

$$16\pi^2 \beta(g_R) = \frac{9}{2} g_R^3, \quad 16\pi^2 \beta(g_{BL}) = 3 g_{BL}^3. \qquad (56.2)$$

This model can be regarded as an effective low-energy theory of LRSM with the $SU(2)_R$ breaking scale and the mass of the heavy W_R bosons much higher than the $U(1)_{B-L}$ scale.

We start by setting the $U(1)$-breaking scale $v_R = 5$ TeV, and run the SM coupling g_Y up to this scale where it is related to $g_{I_{3R}}$ and g_{B-L}. Then we run the couplings

Table 56.1 Particle content and their representations in the $SU(2)_L \times U(1)_{I_{3R}} \times U(1)_{B-L}$ model

	$SU(2)_L$	$U(1)_{I_{3R}}$	$U(1)_{B-L}$
Q	**2**	0	$\frac{1}{3}$
u_R	**1**	$+\frac{1}{2}$	$\frac{1}{3}$
d_R	**1**	$-\frac{1}{2}$	$\frac{1}{3}$
L	**2**	0	-1
N	**1**	$+\frac{1}{2}$	-1
e_R	**1**	$-\frac{1}{2}$	-1
H	**2**	$-\frac{1}{2}$	0
Δ_R	**1**	-1	2

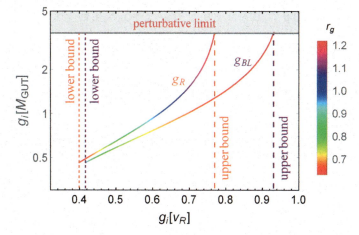

Fig. 56.1 Correlation of $g_{R,BL}(v_R)$ and $g_{R,BL}(M_{GUT})$ in the $U(1)_{B-L}$ model as a function of $r_g \equiv g_R/g_L$ at the breaking scale v_R

up to the GUT scale based on the RGE's given in (56.2). The correlation of couplings $g_{R,BL}$ at v_R and M_{GUT} as function of g_R/g_L is shown in Fig. 56.1. We find that the gauge couplings are constrained to lie within a narrow window at the v_R scale

$$0.398 < g_R < 0.768 \quad \text{and} \quad 0.416 < g_{BL} < 0.931, \quad \text{with} \quad 0.631 < r_g < 1.218.$$

Using these perturbativity constraints, we show in Fig. 56.2 the constraints on the new Z_R gauge boson mass for different v_R scales. Also shown are the current constraints from LHC13 on the Z_R mass and future prospects at the HL-LHC and the 100 TeV collider FCC-hh [11]. The lower bounds on the Z_R boson mass derived from this analysis are also listed in Table 56.2.

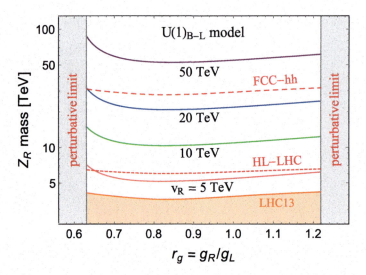

Fig. 56.2 Current constraints from LHC13 on Z_R mass in $U(1)_{B-L}$ model as a function of r_g, as well as future prospects at the HL-LHC, with an integrated luminosity of 3000 fb^{-1} (short-dashed red) and the 100 TeV collider FCC-hh with a luminosity of 30 ab^{-1} (long-dashed red)

Table 56.2 The lower bounds on the Z_R boson mass M_{Z_R} and v_R scale in the $U(1)_{B-L}$ model from current LHC13 data and future prospects from HL-LHC and FCC-hh. The range in each case corresponds to the allowed range of r_g from perturbativity constraints as shown in Fig. 56.1

Collider	M_{Z_R}[TeV]	v_R[TeV]
LHC13	[3.6,4.2]	[3.02,3.57]
HL-LHC	[6.0,6,6]	[4.60,5.82]
FCC-hh	[27.9,31.8]	[19.9,26.8]

56.4 Minimal Left-Right Symmetric Model

The minimal LRSM is based on the gauge group $SU(2)_L \times SU(2)_R \times U(1)_{B-L}$ [12–14] and aims to explain the asymmetric chiral structure of electroweak interactions in the SM. It can also account for the observed smallness of neutrino masses through the seesaw mechanism [7–10]. The minimal scalar content of the LRSM consists of two triplets and one bidoublet Higgs. The fermion sector is same as in the SM, except for the RHNs, which naturally appear as the parity partner of the left-handed neutrinos. The charge assignments for the particles under the gauge group for this model is given in Table 56.3.

The one-loop RG evolution for the gauge couplings in minimal LRSM are captured by the following β-functions:

$$16\pi^2 \beta(g_L) = -3\,g_L^3, \quad 16\pi^2 \beta(g_R) = -\frac{7}{3}\,g_R^3, \quad 16\pi^2 \beta(g_{BL}) = \frac{11}{3}\,g_{BL}^3.$$

Table 56.3 Particle content and their representations under the gauge group of LRSM

	$SU(3)_c$	$SU(2)_L$	$SU(2)_R$	$U(1)_{B-L}$
Ψ_L^Q	3	2	1	1/3
Ψ_R^Q	3	1	2	1/3
Ψ_L^l	1	2	1	-1
Ψ_R^l	1	1	2	-1
ϕ	1	2	2	0
Δ_L	1	3	1	2
Δ_R	1	1	3	2

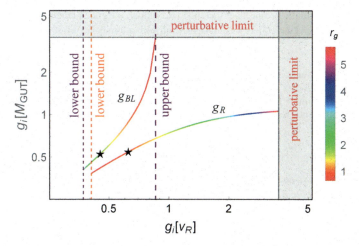

Fig. 56.3 Correlation of $g_{R,BL}(v_R)$ and $g_{R,BL}(M_{\mathrm{GUT}})$ in the minimal LRSM as function of r_g at the breaking scale v_R

For our numerical analysis, we set the breaking scale $v_R = 10$ TeV (Note that v_R cannot be smaller, due to stringent flavor-changing neutral current constraints from the bidoublet sector [15]) and run the gauge couplings up to the GUT scale. The correlation of couplings $g_{R,BL}$ at v_R and M_{GUT} as a function of r_g is shown in Fig. 56.3. We find that the allowed ranges of the gauge couplings in the minimal LRSM at the scale of v_R are

$$0.406 < g_R < \sqrt{4\pi} \text{ and } 0.369 < g_{BL} < 0.857 \text{ with } 0.648 < r_g < 5.65.$$

Including the scalar perturbativity constraints, the range of r_g is further restricted to $0.65 < r_g < 1.6$.

Similarly, Fig. 56.4 shows current constraints from LHC13, as well as future prospects at the HL-LHC and the 100 TeV collider FCC-hh, on W_R and Z_R masses in the minimal LRSM as a function of g_R/g_L. The corresponding lower bounds on the Z_R and W_R boson masses and the breaking scale from perturbativity is shown in

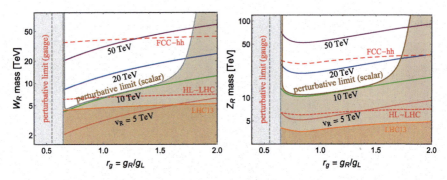

Fig. 56.4 Current constraints from LHC13 on W_R and Z_R masses in the minimal LRSM as a function of r_g and future prospects at the HL-LHC with an integrated luminosity of 3000 fb^{-1} (short-dashed red) and the 100 TeV collider FCC-hh with a luminosity of 30 ab^{-1} (long-dashed red)

Table 56.4 The lower bounds on the Z_R and W_R boson masses and v_R scale in the minimal LRSM from current LHC13 data and future prospects from HL-LHC and FCC-hh, with both gauge and scalar perturbativity limits taken into consideration. The missing entries mean that the corresponding maximum experimental reach has been already excluded by the scalar perturbativity constraints

	W_R searches		Z_R searches	
	M_{W_R} [TeV]	v_R [TeV]	M_{Z_R} [TeV]	v_R [TeV]
LHC13	–	–	–	–
HL-LHC	[6.09, 6.47]	[10.3, 14.8]	–	–
FCC-hh	[35.6, 42.2]	[38.3, 87.5]	[27.9, 35.4]	[21.8, 26.8]

Table 56.4. It is remarkable that the perturbativity constraints from the scalar sector supersede the current LHC constraints on the W_R and Z_R bosons in the minimal LRSM, and even the projected Z_R sensitivity at the HL-LHC.

56.5 Conclusion

We have derived stringent limits on the gauge couplings from the theoretical requirement of perturbativity up to the GUT scale. This has important consequences for the associated heavy gauge boson searches at colliders. For the $U(1)_{B-L}$ model, we found that the allowed parameter space can be almost completely probed at HL-LHC for the breaking scale $v_R = 5$ TeV. For the minimal LRSM, we found that the W_R and Z_R bosons could not have been seen at LHC13, and there exists a very narrow window for them to be accessible at the HL-LHC.

Acknowledgments This work was supported by the US Department of Energy under Grant No. DE-SC0017987 (GC, BD, YZ) and National Science Foundation under Grant No. PHY1620074 (RNM).

References

1. R.E. Marshak, R.N. Mohapatra, Phys. Lett. **91B**, 222 (1980)
2. R.N. Mohapatra, R.E. Marshak, Phys. Rev. Lett. **44**, 1316 (1980) [Erratum: Phys. Rev. Lett. **44**, 1643 (1980)]
3. H. Georgi, S. Weinberg, Phys. Rev. D **17**, 275 (1978)
4. P.S.B. Dev, R.N. Mohapatra, Y. Zhang, JHEP **1605**, 174 (2016)
5. N.G. Deshpande, D. Iskandar, Nucl. Phys. B **167**, 223 (1980)
6. P. Galison, A. Manohar, Phys. Lett. **136B**, 279 (1984)
7. P. Minkowski, Phys. Lett. B **67**, 421 (1977)
8. R.N. Mohapatra, G. Senjanović, Phys. Rev. Lett. **44**, 912 (1980)
9. T. Yanagida, Conf. Proc. C **7902131**, 95 (1979)
10. M. Gell-Mann, P. Ramond, R. Slansky, Conf. Proc. C **790927**, 315 (1979)
11. G. Chauhan, P.S.B. Dev, R.N. Mohapatra, Y. Zhang, JHEP **1901**, 208 (2019)
12. J.C. Pati, A. Salam, Phys. Rev. D **10**, 275 (1974)
13. R.N. Mohapatra, J.C. Pati, Phys. Rev. D **11**, 2558 (1975)
14. G. Senjanovic, R.N. Mohapatra, Phys. Rev. D **12**, 1502 (1975)
15. Y. Zhang, H. An, X. Ji, R.N. Mohapatra, Nucl. Phys. B **802**, 247 (2008)

Chapter 57
Fragmentation of Pseudo-Scalar Mesons

H. Saveetha and D. Indumathi

Abstract A complete analysis of an entire nonet of pseudo-scalar mesons ($\pi(\pi^+$, π^-, π^0), $K(K^+, K^-, K^0, \overline{K}^0)$, η and η') fragmentation has been performed for the first time using a model with broken SU(3) symmetry at next-to-leading order (NLO) for $e^+ e^-$ annihilation. The model parameterises three input fragmentation functions: valence $V(x, Q^2)$, sea $\gamma(x, Q^2)$, and gluon $D_g(x, Q^2)$, with a SU(3) breaking parameter λ, at an initial scale of $Q_0^2 = 1.5$ GeV^2, with which the model is able to fit the pure octet (π, K) data from $e^+ e^-$ scattering at the Z-pole. The model has been further extended to octet-singlet mixing, with the inclusion of few parameters in order to describe the mixing in pseudo-scalar mesons. Despite contamination by prompt decay of other heavy mesons, π meson production has been successfully studied.

57.1 Introduction

Fragmentation process [1] plays a vital role in hadron production in various collisions like $e^+ e^-$, ep, pp and the fragmentation functions corresponding to such processes are determined through phenomenological studies. Several works have been carried out to investigate the individual global fragmentation functions at both leading order (LO) and NLO for baryons (p/p^-, $\Lambda/\overline{\Lambda}$) and some mesons like π and K [2]. Also few works are available for η meson and η' mesons [3, 4].

Investigation on fragmentation of complete pseudo-scalar meson nonet at NLO approximation has not been done so far. A model with broken SU(3) symmetry was used to describe the leading order hadron production of baryonic and mesonic systems [4, 5]. A successful explanation of the entire vector meson nonet were carried out [6] for two different processes ($e^+ e^-$ and pp) at LO and NLO and their relevant fragmentation functions were published. In this paper, we would like to investigate the fragmentation of the entire pseudo-scalar meson nonet for $e^+ e^-$ as it is a pure channel with no quarks in the initial state and to get the corresponding fragmentation

H. Saveetha (✉) · D. Indumathi
Institute of Mathematical Sciences, Chennai, India
e-mail: saveehari@gmail.com

© Springer Nature Singapore Pte Ltd. 2021
P. K. Behera et al. (eds.), *XXIII DAE High Energy Physics Symposium*,
Springer Proceedings in Physics 261,
https://doi.org/10.1007/978-981-33-4408-2_57

functions at NLO. One of the motivations to do this analysis is to study the π meson, lightest of all hadrons, an interesting and challenging candidate to understand, since it has both direct production and decay contribution from heavy mesons as well.

57.2 Model with Broken SU(3) Symmetry

To determine the fragmentation function of the meson octet, we use the SU(3) flavour symmetry since it can describe the meson octet with three light quarks as the fundamental representation. The SU(3) model [6] introduces three input fragmentation functions $(\alpha(x, Q^2), \beta(x, Q^2), \gamma(x, Q^2))$ for quarks and three other functions for the anti-quarks and one for the gluon fragmentation $D_g(x, Q^2)$. Hence in total we have 56 unknown fragmentation functions for the meson which makes it very difficult. Employing the SU(3) symmetry, we can reduce the number of unknown fragmentation function from 56 to *three*-independent quark fragmentation functions and a gluon fragmentation function.

    Through the use of the isospin and charge conjugation symmetry we can further reduce the fragmentations into just two functions: valence, $V(x, Q^2)$ and sea, $\gamma(x, Q^2)$ for quarks (Table 57.1) apart from a gluon fragmentation function $D_g(x, Q^2)$ to describe the pure octet part (π's with non-strange valence). However, for K meson case, with valence having massive strange quark, the symmetry is broken through strangeness suppression parameter. Several features like production of η and η' mesons, mixture of unphysical octet (η_8) and singlet (η_1) state can be explained through few additional parameters like mixing angle, sea suppression, and singlet constants.

Table 57.1 Quark fragmentation functions in terms of the functions, V and γ for pseudo-scalar mesons with their valence content

fragmenting quark	$K^+(u\bar{s})$	fragmenting quark	$K^0(d\bar{s})$
u	: $V + 2\gamma$	u	: 2γ
d	: 2γ	d	: $V + 2\gamma$
s	: 2γ	s	: 2γ
fragmenting quark	η'	fragmenting quark	π^0
u	: $\frac{1}{6}V + 2\gamma$	u	: $\frac{1}{2}V + 2\gamma$
d	: $\frac{1}{6}V + 2\gamma$	d	: $\frac{1}{2}V + 2\gamma$
s	: $\frac{4}{6}V + 2\gamma$	s	: 2γ
fragmenting quark	$\pi^+(u\bar{d})$	fragmenting quark	$\pi^-(\bar{u}d)$
u	: $V + 2\gamma$	u	: 2γ
d	: 2γ	d	: $V + 2\gamma$
s	: 2γ	s	: 2γ
fragmenting quark	$\bar{K}^0(\bar{u}s)$	fragmenting quark	$K^-(\bar{d}s)$
u	: 2γ	u	: 2γ
d	: 2γ	d	: 2γ
s	: $V + 2\gamma$	s	: $V + 2\gamma$

57.3 Analysis and Results

The whole analysis is carried out for entire mesons by comparison with the data from $e^- e^+$ annihilation (LEP(ALEPH, DELPHI, OPAL, SLD) [7]) at Z-pole. The functional form of V, γ, D_g at Q_0^2 used to fit the data is

$$F_i(x) = a_i x^{b_i} (1 - x)^{c_i} (1 + d_i \sqrt{x} + e_i x) \, ;$$

where a, b, c, d and e are the parameters to be determined from the fit.

Hence the octet mesons are described in terms of valence V and sea S parts with light quarks (u, d, s) at a starting scale of $Q_0^2 = 1.5$ GeV^2 where the fragmentation of charm and bottom flavours are zero. With the help of DGLAP evolution equations, QCD evolve these fragmentation functions to Z-pole of $Q^2 = (91.2)^2$ GeV^2. The contribution of heavier flavours is added appropriately during evolution. Thus all the parameters including the strangeness suppression in the input fragmentation functions are completely determined from best fits to the π and K meson data.

The power of the method lies in the idea that we can relate the singlet fragmentation function to one of the octet fragmentation function through an ansatz and hence no new fragmentation function is needed. Including the η and η' mesons, the mixing angle θ, and other few parameters are determined at NLO.

For K meson the SLD uds data is being used to remove contamination from heavy flavours that contribute both directly through fragmentation as well as indirectly through production and decay of heavy flavour mesons.

Table 57.2 Best fit values with 1-σ error of various input parameters defining the input fragmentation functions at $Q_0^2 = 1.5$ GeV^2

		Central value	Error bars
V	a	3.18	0.12
	b	-0.31	0.03
	c	1.47	0.13
	d	-0.13	0.28
	e	-0.78	0.08
γ	a	9.26	0.25
	b	-0.47	0.01
	c	9.79	0.14
	d	-3.15	0.10
	e	3.69	0.22
D_g	a	4.47	0.56
	b	-0.32	0.07
	c	5.92	0.57
	d	-10.84	1.20
	e	-19.92	2.89

	Central value	Error bars
λ	0.10	0.01
θ	-17.19	1.27
f_{sea}^{η}	0.10	0.02
$f_{sea}^{\eta'}$	0.36	0.13
$f_1^u(\eta)$	2.30	0.42
$f_1^s(\eta')$	0.43	0.04
f_g^K	0.02	0.04
f_g^{η}	0.89	0.12
$f_g^{\eta'}$	0.0002	0.06

Analysis of π meson production is bit challenging as it is contaminated by prompt decay of other (strange, charm, etc.) heavy mesons and the model has been successfully tailored to account for this.

Good fits with best parameter values (Table 57.2) and reasonable χ^2 (Table 57.3) reflect the consistency and reliability of the model (Figs. 57.1 and 57.2).

Table 57.3 χ^2 for fits to inclusive pseudo-scalar meson production data on the Z-pole from LEP and SLD experiments

Data set	No. of data points	χ^2
π^0	49	190.0
π^{+-}	27	26.3
K^0	17	12.3
K^{+-}	21	26.9
η	24	12.3
η'	19	14.3

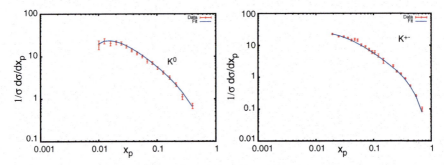

Fig. 57.1 Fits for $K^0 (= K^0 + \overline{K^0})$ (L) and $K^{+-} (= K^+ + K^-)$ (R) meson for SLD uds data at $\sqrt{s} = 91$ GeV at NLO

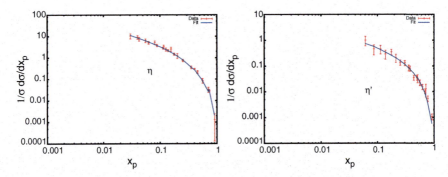

Fig. 57.2 $\eta \& \eta'$ fits for LEP data at $\sqrt{s} = 91$ GeV at NLO

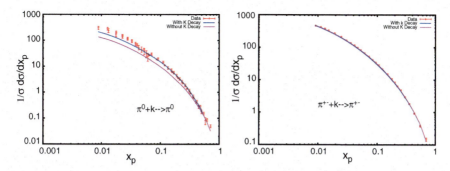

Fig. 57.3 Fits for π meson in terms of fragmentation functions with π^0 LEP data (L) and $\pi^{+,-}$ SLD data (R) at $\sqrt{s} = 91$ GeV at NLO

57.4 Conclusion

Using simple SU(3) model and NLO QCD evolution, the quark fragmentation functions are fitted for pseudo-scalar mesons π, K, η, and η' for e^+e^- process. The model with three light flavours uses universal functions, the valence $V(x, Q^2)$, sea $\gamma(x, Q^2)$ quark fragmentation functions and a gluon fragmentation function $D_g(x, Q^2)$. No new additional fragmentation function is introduced in order to explain the singlet sector which shows the efficiency of the model.

Extending this NLO study to pp collision in future will help to understand the η meson production in particular which forms the base-line to understand the nature of QGP (quark-gluon plasma) studies. Pseudoscalar meson fragmentation function (especially π fragmentation) will be a new and relevant result to the field of neutrino physics where there is a lot of interest in studies of low energy π production in neutrino-nucleus collisions (Fig. 57.3).

References

1. R.D. Field, R.P. Feymann, Phys. Rev. D **15**, 2590 (1977)
2. S. Kretzer, Phys. Rev. **D 62**, 054001 (2000). D. de Florian, R. Sassot, M. Stratmann, Phys. Rev. **D 76**, 074033 (2007). S. Albino, B.A. Kniehl, G. Kramer, Nucl. Phys. **B803**, 42 (2008). M. Hirai, S. Kumano, T.-H. Nagai, K. Sudoh, Phys. Rev. **D 75**, 094009 (2007)
3. M. Greco, S. Rolli, A.Z. Vicini, Phys. **C65**, 277 (1995). C.A. Aidala, F. Ellinghaus, R. Sassot, J.P. Seele, M. Stratmann, Phys. Rev. **D 83**, 034002 (2011)
4. D. Indumathi, B. Misra, (2009). arXiv: 0901.0228
5. D. Indumathi, H.S. Mani, A. Rastogi, Phys. Rev. D **58**, 094014 (1998)
6. D. Indumathi, H. Saveetha, Int. J. Mod. Phys. **A 27**, 19, 1250103 (2012). H. Saveetha, D. Indumathi, S. Mitra, Int. J. Mod. Phys. **A 29**, 7, 1450049 (2014). H. Saveetha, D. Indumathi, Int. J. Mod. Phys. **A 32**, 33, 1750199 (2017)
7. Abe et al., Phys. Rev. **D 59**, 052001 (1999). Barate et al., Eur. Phys. J. **C 16**, 613 (2000). Heister et al., Phys. Lett. **B 528**, 19 (2002). Abbiendi et al., Eur. Phys. J. **C 17**, 373 (2000). Ackerstaff et al., Eur. Phys. J. **C 5**, 411 (1998). Abreu et al., Eur. Phys. J. **C 5**, 585 (1998). Acciarri et al., Phys. Lett. **B 393**, 465 (1997)

Chapter 58
The Role of Global Monopole in Joule–Thomson Effect of AdS Black Hole

A. Naveena Kumara, C. L. Ahmed Rizwan, and K. M. Ajith

Abstract We study the throttling process of the AdS black hole with a global monopole in the extended phase space. In the approach followed, the cosmological constant and the black hole mass are identified with the thermodynamic pressure and enthalpy, respectively. We investigate the dependency of the inversion temperature and isenthalpic curves on the global monopole parameter η. Our study shows a close resemblance between the phase transition of the black hole in the extended phase space and Van der Waals fluid. The presence of global monopole plays an important role in the throttling process.

58.1 Introduction

In gravitational physics, black hole thermodynamics continues to be one of the important subject. Even though it is well-known for a long time that black holes have temperature and entropy which are proportional surface gravity and area, respectively, and obey the first law of black hole thermodynamics, the subject is still not completely explored. After the initial proposals of black hole critical behaviour, i.e. Hawking-Page phase transitions, asymptotically anti-de Sitter (AdS) geometries have been investigated in great detail. The interpretation that the ADM mass as the enthalpy of the spacetime opened new gateways in this subject. This interpretation has emerged from the geometrical derivation of Smarr formula which insisted us to treat the cosmological constant as a thermodynamic variable that plays the role of pressure in the first law. As a result, what we have is an extended phase space with additional thermodynamic variables. In the extended phase space, one can establish the similarities between the thermodynamic behaviour of AdS black holes and van der Waals fluid [1].

Joule Thomson expansion of RN-AdS black hole was studied by Ökcü and Aydiner [2]. Later it was extended to Kerr AdS black hole. In these studies, the inversion curve

A. Naveena Kumara (✉) · C. L. Ahmed Rizwan · K. M. Ajith
Department of Physics, National Institute of Technology Karnataka, Mangalore, India
e-mail: naviphysics@gmail.com

© Springer Nature Singapore Pte Ltd. 2021
P. K. Behera et al. (eds.), *XXIII DAE High Energy Physics Symposium*,
Springer Proceedings in Physics 261,
https://doi.org/10.1007/978-981-33-4408-2_58

and isenthalpic curves are analysed and the similarities with the van der Waals fluid are established. The origin of global monopole that we consider is due to the gauge symmetry breaking. Barriola and Vilenkin derived the black hole solution with global monopole having a distinct topological structure.

58.2   Joule Thomson Effect

In elementary thermodynamics expansion of a gas through a porous plug leading to heating and cooling effects known as *Joule–Thomson expansion*. For the better description of this process, a state function called *enthalpy* $H = U + PV$ is defined, which remains unchanged in the end states. The locus of all points with the same molar enthalpy representing initial and final equilibrium states, forms an *isenthalpic curve*, which characterises Joule–Thomson effect (Fig. 58.1).

58.3   The RN-AdS Black Hole with Global Monopole

AdS black hole with global monopole is defined by the metric [3, 4]

$$d\tilde{s}^2 = -\tilde{f}(\tilde{r})d\tilde{t}^2 + \tilde{f}(\tilde{r})^{-1}d\tilde{r}^2 + \tilde{r}^2 d\Omega^2, \qquad (58.1)$$

where $d\Omega^2 = d\theta^2 + \sin^2\theta d\phi^2$. And $\tilde{f}(\tilde{r})$ is given by

$$\tilde{f}(\tilde{r}) = 1 - 8\pi\eta_0^2 - \frac{2\tilde{m}}{\tilde{r}} + \frac{\tilde{q}^2}{\tilde{r}^2} + \frac{\tilde{r}^2}{l^2} \ , \quad \tilde{A} = \frac{\tilde{q}}{\tilde{r}}d\tilde{t}. \qquad (58.2)$$

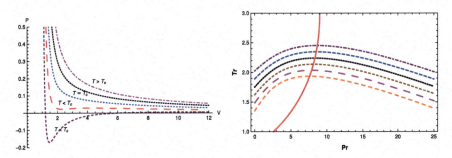

Fig. 58.1 The left figure shows $P - V$ isotherms showing critical behaviour below T_C. In the right figure isenthalpic and inversion curves for van der Waals gas are shown

Here $\tilde{m}$, $\tilde{q}$, and l are parameters related to the mass, electric charge, and AdS length of the black hole, respectively. The appropriate coordinate transformation and redefinition of parameters are done to obtain metric in the form similar to RN-AdS.

Then ADM mass is calculated from that metric,

$$M = \frac{(1 - \eta^2)}{2}r_+ + \frac{Q^2}{2r_+(1 - \eta^2)} + \frac{r_+^3(1 - \eta^2)}{2l^2}. \tag{58.3}$$

The first law of thermodynamics for the RN-AdS black hole is given by

$$dM = TdS + \Phi dQ + VdP. \tag{58.4}$$

The calculation of entropy S of the black hole is straight forward, which is related to the area of the event horizon (A). We work in extended phase space where the role of pressure is played by the cosmological constant (Λ). The thermodynamic volume is given by the conjugate quantity of Λ.

$$S = \frac{A}{4} = \pi(1 - \eta^2)r_+^2 \ , \quad P = -\frac{\Lambda}{8\pi} = \frac{3}{8\pi l^2} \ , \quad V = \frac{4}{3}\pi(1 - \eta^2)r_+^3. \tag{58.5}$$

The temperature of black hole is obtained as follows:

$$T = \left(\frac{\partial M}{\partial S}\right)_{P,Q} = \frac{1}{4\pi r_+}\left(1 + \frac{3r_+^2}{l^2} - \frac{Q^2}{(1 - \eta^2)^2 r_+^2}\right). \tag{58.6}$$

The equation of state reads as follows (Fig. 58.2),

$$P = \frac{T}{2r_+} - \frac{1}{8\pi r_+^2} + \frac{Q^2}{8\pi(1 - \eta^2)^2 r_+^4}. \tag{58.7}$$

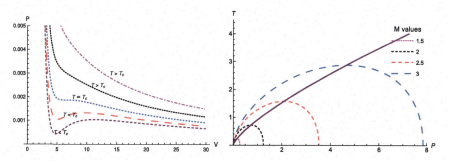

Fig. 58.2 The first figure shows $P - V$ isotherms for the black hole. In the second figure crossing diagrams between isenthalpic and and inversion curves are shown

58.4 Joule–Thomson Effect of RN-AdS Blackhole with Monopole Term

The expression for Joule–Thomson coefficient

$$\mu_J = \left(\frac{\partial T}{\partial P}\right)_M = \frac{1}{C_P}\left[T\left(\frac{\partial V}{\partial T}\right)_P - V\right]. \tag{58.8}$$

The vanishing condition of μ_J gives the invesion temperature

$$T_i = V\left(\frac{\partial T}{\partial V}\right)_P. \tag{58.9}$$

Simple calculation gives the inversion temperature for the black hole,

$$T_i = v\frac{Q^2}{4\pi r_+^3(1 - \eta^2)^2} + \frac{2}{3}Pr_+ - \frac{1}{12\pi r_+}. \tag{58.10}$$

From equation of state we get

$$T_i = -\frac{Q^2}{4\pi r_+^3(1 - \eta^2)^2} + 2Pr_+ + \frac{1}{4\pi r_+}. \tag{58.11}$$

From equation (58.10) and equation (58.11) we get

$$8\pi(1 - \eta^2)^2 Pr_+^4 + 2(1 - \eta^2)^2 r_+^2 - 3Q^2 = 0. \tag{58.12}$$

Solving the above equation and substituting the appropriate root for r_+ into equation (58.11)

$$T_i = \frac{\sqrt{P_i}\left(1 + \frac{16\pi P_i Q^2}{(1-\eta^2)^2} - \frac{\sqrt{24P_i\pi Q^2 + (1-\eta^2)^2}}{(1-\eta^2)}\right)}{\sqrt{2\pi}\left(-1 + \frac{\sqrt{24P_i\pi Q^2 + (1-\eta^2)^2}}{(1-\eta^2)}\right)^{3/2}}. \tag{58.13}$$

Using this expression the inversion curves are obtained and analysed [5].

58.5 Conclusion

Joule–Thomson effect similar to van der Waals gas is observed in the AdS black hole. Monopole term plays an important role in the thermodynamics and Joule Thomson expansion of the AdS black hole. It was noticed that the monopole parameter increases the inversion temperature and pressure monotonically.

Acknowledgements Author N.K.A. acknowledge University Grants Commission, Government of India for financial support under UGC-NET-JRF scheme.

References

1. D. Kubizňák, RB. Mann, P-V criticality of charged AdS black holes. J. High Energy Phys. (7) (2012)
2. Özgür Ökcü, E.A, Joule-Thomson expansion of the charged AdS black holes. Eur. Phys. J. C **77** (2017)
3. Manuel Barriola, Alexander Vilenkin, Gravitational field of a global monopole. Phys. Rev. Lett. **63**, 341–343 (1989). Jul
4. G.-M. Deng, J. Fan, X. Li, Y.-C. Huang, Thermodynamics and phase transition of charged AdS black holes with a global monopole. Int. J. Mod. Phys. A **33**(03), 1850022 (2018)
5. C.L. Ahmed Rizwan, A. Naveena Kumara, D. Vaid, K.M. Ajith, Joule-Thomson expansion in AdS black hole with a global monopole (2018)

Chapter 59
Scattering Amplitudes from Positive Geometries

Pinaki Banerjee

Abstract We describe how one can obtain certain planar scattering amplitudes from the geometry of some positive geometries (e.g., polytopes) living in the kinematic space. For massless scalar ϕ^4 theory these particular positive geometries are known as Stokes polytopes which were introduced by Baryshnikov (New developments in singularity theory, vol 21, pp 65–86 (2001), [1]). The canonical form on these Stokes polytopes contains all the information of the planar amplitudes.

59.1 Introduction

Scattering amplitudes are one of the most fundamental observables in physics. In experiments a bunch of incoming particles come from "infinity," then scatter to a bunch of outgoing particles and finally are detected by some detectors at "infinity." Scattering amplitudes measure the probabilities of such events. But it is not at all clear what happens in-between. People build up their own favorite "stories" that explain the phenomenon. The most popular "story" is that of Feynman diagrams—the standard algorithm of computing amplitudes perturbatively in couplings. Although this program is remarkably successful it has the following limitations:

1. It works for perturbative theories,
2. There are huge redundancies, e.g., actions related by mere field redefinition which gives different set of diagrams but at the end somehow "miraculously" give the same amplitude, and
3. It is not efficient practically, e.g., 6-gluon amplitude has 220 diagrams!

All these combined suggest that Feynman diagrams are not suitable for analyzing amplitudes. Various other avenues to obtain scattering amplitudes have been explored. Treating amplitude as "differential form" instead of functions on kinematical space and identifying it as the "canonical form" of certain "positive geometries"

P. Banerjee (✉)
ICTS-TIFR, Bengaluru 560 089, India
e-mail: pinaki.banerjee@icts.res.in

© Springer Nature Singapore Pte Ltd. 2021
P. K. Behera et al. (eds.), *XXIII DAE High Energy Physics Symposium*,
Springer Proceedings in Physics 261,
https://doi.org/10.1007/978-981-33-4408-2_59

is one such novel way. Locality, causality, unitarity, etc. are then evident from the geometric properties of the corresponding positive geometry. In [2], planar scattering amplitude for bi-adjoint ϕ^3 scalar theory was recovered from the canonical form of positive geometries called associahedra [4, 5]. The philosophy of the "positive geometry program" (which is, needless to say, rather ambitious) is as follows. Given an interacting theory there should exist a positive geometry (or a collection of positive geometries) whose canonical top form contains the complete information of the scattering amplitude of the theory.

59.2 Associahedron: The Positive Geometry for Cubic Interactions

In [2], it was shown that tree-level planar amplitude for massless ϕ^3 theory can be obtained from a positive geometry known as the associahedron [4, 5] sitting inside the kinematic space. Let us start by asking, what is a *positive geometry*? Positive geometry is a closed geometry with boundaries of *all* co-dimensions. Polytopes are the most famous examples. A positive geometry ($\mathcal{A}$) has a unique differential form $\Omega(\mathcal{A})$, known as the *canonical form*—a complex differential form defined by the following properties.

1. It has logarithmic singularities at the boundaries of $\mathcal{A}$.
2. Its singularities are recursive, i.e., at every boundary $\mathcal{B}$, $\text{Res}_{\mathcal{B}}\Omega(\mathcal{A}) = \Omega(\mathcal{B})$.

The above two criteria and *projectivity*[1] uniquely fix canonical form which has the complete information about full tree-level planar amplitude. For massless scalar ϕ^3, the positive geometry is associahedron. What is an *associahedron*?

The associahedron of dimension $(n-3)$ is a polytope whose co-dimension d boundaries are in one-to-one correspondence with the partial triangulation[2] by d diagonals. The vertices represent complete triangulations and k-faces represent k-partial triangulations of the n-gon. The total number of ways to triangulate a convex n-gon by non-intersecting diagonals is the $(n-2)$th Catalan number, $C_{n-2} = \frac{1}{n-1}\binom{2n-4}{n-2}$. The dimension of the associahedron corresponding to a n-gon is $n-3$ (Fig. 59.1). So far everything looks mathematical/combinatorial. What do all these things have to do with something physical and experimentally observable like the scattering amplitude?

We use planar kinematic variables, $X_{i,j} = (p_i + p_{i+1} + \ldots + p_{j-1})^2$; $1 \leq i < j \leq n$. They visualized as the diagonals between the ith and jth vertices of the dual n-gon. Each $X_{i,j}$ cuts the internal propagator of a Feynman diagram once (see Fig. 59.2). For $n = 4$, the number of complete triangulations is 2. For $n = 5$, it is 5 and $n = 6$, it is 14 and so on. This is the same counting as for complete triangulations of an n-gon and thus the associahedron appears from Feynman diagrams! For any

---

[1] Projectivity implies $\Omega(\mathcal{A})$ can only be a function of ratios of Mandelstam variables.
[2] A partial triangulation of regular n-gon is a set of non-crossing diagonals.

Fig. 59.1 Two-dimensional associahedron $\mathcal{A}_5$: 5 partial triangulations are represented by five diagonals. Five complete triangulations are represented by five vertices

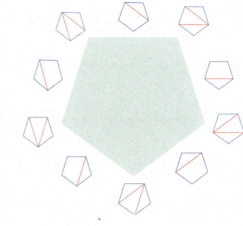

Fig. 59.2 Two channels of four-point scattering. The red diagonal always cuts the propagator once. This holds true for higher point planar diagrams as well. These are also the two possible triangulations of a 4-gon

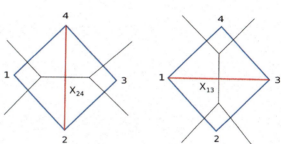

given n (i.e., number of particles), one can associate an associahedron to the n-gon which is made out of momenta $p_1, p_2, \ldots p_n$. The vertices of the associahedron represent complete triangulations and its k-faces represent k-partial triangulations of the n-gon. One can directly write down the canonical form $\Omega(\mathcal{A})$.

$$\Omega(\mathcal{A}_n) = \sum_{\text{vertex } Z} \text{sign}(Z) \bigwedge_{a=1}^{n-3} d \log X_{i_a, j_a} \qquad (59.1)$$

Note that summing over the vertices (Z) is equivalent to summing over all possible planar cubic graphs (or channels). Thus $\Omega(\mathcal{A}_n)$ should have the information of the amplitude for planar cubic theory. Finally all one needs to do is to pull Ω back to $\mathcal{A}_n$ (i.e., to restrict the Ω on the polytope $\mathcal{A}_n$). One can directly read off the amplitude from this. Let us look at a simple example—four particle scattering in massless cubic theory. For $n = 4$, the dimension of kinematic space, $dim(\mathcal{K}_4) = \frac{4(4-3)}{2} = 2$. From standard QFT, we know one can construct three Mandelstam variables s, t and u. But they are not independent and for massless theory they satisfy, $s + t + u = 0$. Thus $\mathcal{K}_4$ is really two-dimensional spanned by s and t, say. These are the two channels shown in the above diagram. The corresponding associahedron $\mathcal{A}_4$ in $\mathcal{K}_4$ is $4 - 3 = 1$ dimensional. We need to embed it inside $\mathcal{K}_4$ which is two-dimensional s-t plane. This

Fig. 59.3 This is how the
1-d associahedron $\mathcal{A}_4$ is
embedded in $\mathcal{K}_4$, i.e., s-t
plane

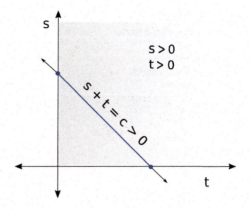

can be done by demanding, $s \geq 0$ and $t \geq 0$ and imposing the other constraint $s + t = -u = c$ where $c > 0$. Thus one gets $\mathcal{A}_4$ which is a line segment in $\mathcal{K}_4$.

Once one has $\mathcal{A}_4$ inside $\mathcal{K}_4$ one can write down the canonical form which has logarithmic singularities at the two end points of the segment (represent two channels for four-point scattering, namely, s and t channels) : $\Omega(\mathcal{A}_4) = \frac{ds}{s} - \frac{dt}{t}$. The minus sign might look annoying but it comes from the demand of projectivity and is very important. How to get the amplitude, then? One needs to perform the last step of the prescription, i.e., to pull back the form to the line segment, i.e., to impose $s + t = -u = c \implies ds = -dt$. Finally we can read off the planar 4-point amplitude from the pulled back canonical form as follows:

$$\Omega(\mathcal{A}_4) = \underbrace{\left(\frac{1}{s} + \frac{1}{t}\right)}_{\text{4-pt amplitude}} ds. \tag{59.2}$$

This can be done for all higher point amplitudes. The only "difficulty" is that the associahedron of n-particle scattering is $(n-3)$-dimensional, and therefore hard to visualize. But mathematically/operationally they are as simple as 4-point case (Figs. 59.3 and 59.4).

The immediate natural question would be, can this be generalized to higher point interactions (e.g., quartic case)?

59.3 Stokes Polytope: The Positive Geometry for Quartic Interactions

For ϕ^4 interactions [3] one can only have even-point amplitudes. The natural way forward would be tilling the corresponding n-gon by quadrilaterals. Then there will be a 1–1 correspondence between planar tree-level diagrams of ϕ^4 theory and complete

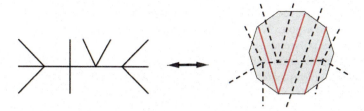

Fig. 59.4 1–1 correspondence between planar tree-level diagrams of ϕ^4 theory and complete quadrangulations of a polygon

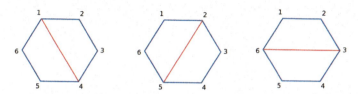

Fig. 59.5 Three planar scattering channels for the six-point diagram

quadrangulations of the n-gon. The total number of complete quadrangulations of an $n = (2N + 2)$-gon is given by the Fuss-Catalan number, $F_N = \frac{1}{2N+1}\binom{3N}{N}$. The immediate natural question is—Is there a polytope whose dimension is $\frac{n-4}{2}$ (i.e., no. of propagators) and number of vertices are same as F_N? Here we immediately run into an obstacle. For example, for six-point scattering (i.e., $N = 2$) we should get a one-dimensional polytope, which can only be a line segment with two boundaries. But there are three (see Fig. 59.5) planar scattering channels!

So, the only way to define a polytope is to exclude one of the channels using some systematic rule. This rule will be called Q-compatibility. What is Q-compatibility?

At an operational level, every diagonal is Q-compatible with every alternate diagonal when we move clockwise (i.e., 14 with 36 , 25 with 41 and 36 with 52). It is clear from the example that Q-compatability is not an equivalence relation and is very much dependent on the reference quadrangulation Q. We use this rule just as a filter which selects, among the set of all quadrangulations of a polygon, a subset which will be in 1-1 correspondence with vertices of the anticipated positive geometry which turns out to be the Stokes polytope [1] discovered by Yuliy Baryshnikov rather recently in the context of studying singularities of quadratic forms. It may look pretty much the same story as of the associahedra. But it is not. We have already noticed two key differences from the associahedron program.

1. Definition of Stokes polytope depends on the reference quadrangulation Q, and for each Q one has a Stokes polytope $\mathcal{S}_n^Q$.
2. Vertices of $\mathcal{S}_n^Q$ are in 1-1 correspondence only with a *specific sub-set* of quadrangulations, namely Q-compatible quadrangulations.

For ϕ^4 case we have the notion of Q-compatibility, using which we can define a new operation on the n-gon : *Flip*. This operation helps us uniquely fixing the differential form on Stokes polytopes.

Any n-point diagram with $n \geq 8$ will have two or more hexagons inside it. Flip is an operation of replacing a diagonal of any such hexagon inside the quadrangulation of the polygon with its Q-compatible diagonal, (see Fig. 59.6). Flip helps us assigning particular signs ($\sigma = \pm 1$) to each complete quadrangulation relative to its Q-compatible diagrams. Let Q be a quadrangulation of an n-gon which is associated to a planar Feynman diagram with propagators given by $X_{i_1}, \ldots, X_{i_{\frac{n-4}{2}}}$. Then we define the ($Q$-dependent) planar scattering form [3],

$$\Omega^Q n = \sum_{\text{graphs}} \text{flips}(-1)^{\sigma(\text{flip})} d \ln X_{i_1} \wedge \ldots d \ln X_{i_{\frac{n-4}{2}}}, \tag{59.3}$$

where $\sigma(\text{flip}) = \pm 1$ depending on whether the quadrangulation $X_{i_1}, \ldots, X_{i_{\frac{n-4}{2}}}$ can be obtained from Q by even or odd number of flips.

Consider the simplest case, i.e., $n = 2$. Let us start with $Q = 14$. Then the set of Q compatible quadrangulations are $\{(14, +), (36, -)\}$. We have attached a sign to each of the quadrangulation which measures the number of flips needed to reach it, starting from reference $Q = 14$. The form Ω_6^Q on the kinematic space is given by

$$\Omega_6^{Q=14} = (d \ln X_{14} - d \ln X_{36}). \tag{59.4}$$

It is evident that it does not capture the singularity associated to X_{25} channel. We can get around this problem by considering other possible Qs whose forms on Kinematic space are given by $\Omega_6^{Q=36} = (d \ln X_{36} - d \ln X_{25})$ and $\Omega_6^{Q=25} = (d \ln X_{25} - d \ln X_{14})$. We can embed Stokes polytopes $(d = \frac{n-4}{2})$ inside corresponding associahedra $(d = n - 3)$, and we pull Ω_n^Q back to the $\hat{S}_n^Q$ and then sum over all possible Qs with particular weights. For six particle case, after embedding inside the three-dimensional associahedron, $\mathcal{A}_6$, we can pull back all three "distinct" Stokes polytopes, i.e., $Q = 14, 36$ and 25, and sum them with weights,

$$\widetilde{\mathcal{M}}_6 := \alpha_{Q_1} \left(\frac{1}{X_{14}} + \frac{1}{X_{36}} \right) + \alpha_{Q_2} \left(\frac{1}{X_{25}} + \frac{1}{X_{14}} \right) + \alpha_{Q_3} \left(\frac{1}{X_{36}} + \frac{1}{X_{25}} \right). \tag{59.5}$$

It is evident that if and only if $\alpha_{Q_1} = \alpha_{Q_2} = \alpha_{Q_3} = \frac{1}{2}, \widetilde{\mathcal{M}}_6 = \mathcal{M}_6$ (planar amplitude). There are different avenues that one can follow from this work. Here we list a few.

1. One obvious question would be how to go beyond planar and/or tree level.
2. Another interesting avenue is to explore its relationship with CHY formalism [6]—the meaning of $\frac{n-4}{2}$ form in the worldsheet.
3. This program can be generalized to planar ϕ^p, $p > 4$. The notion of Q-compatible quadrangulations has an immediate extension to p-gulations of a polygon.

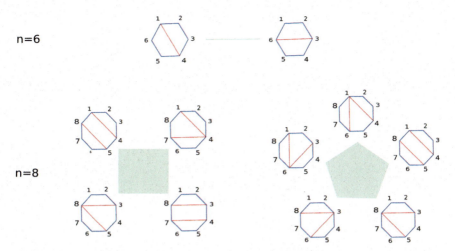

Fig. 59.6 Stokes polytopes for 6 and 8 point scattering. Notice that for n = 8 there are two types of positive geometries. To obtain the full planar tree-level amplitude one needs to sum over all such geometries with correct weightage

Higher point amplitudes can be obtained in similar way [3] but some interesting different classes polytopes will appear (see Fig. 59.6).

Acknowledgements The author would like to thank Alok Laddha and Prashanth Raman for the collaboration in the project [3] and numerous fruitful discussions.

References

1. Y. Baryshnikov, On stokes sets. New developments in singularity theory, vol. 21, pp. 65–86 (2001)
2. N. Arkani-Hamed, Y. Bai, S. He, G. Yan, Scattering forms and the positive geometry of Kinematics. Color and the worldsheet. JHEP **05**, 096 (2018). arXiv:1711.09102 [hep-th]
3. P. Banerjee, A. Laddha, P. Raman, Stokes polytope : the positive geometry for ϕ^4 interactions, arXiv:1811.05904 [hep-th]
4. J. D. Stasheff, Homotopy associativity of h-spaces. i. Trans. Am. Math. Soc. **108**(2), 275–292 (1963)
5. J. D. Stasheff, Homotopy associativity of h-spaces. ii. Trans. Am. Math. Soc. **108**(2), 293–312 (1963)
6. F. Cachazo, S. He, E.Y. Yuan, Scattering of massless particles in arbitrary dimensions. Phys. Rev. Lett. **113**(17), 171601 (2014). arXiv:1307.2199 [hep-th]

Chapter 60
Scale of Non-commutativity and the Hydrogen Spectrum

Pulkit S. Ghoderao, Rajiv V. Gavai, and P. Ramadevi

Abstract In view of the current null results for the popular models of physics beyond the standard model, it has become necessary to explore other plausible alternatives for such physics. Non-commutative spaces, which can arise in specific string theory models, is one such alternative. We explore the possibility that quarks and leptons exist in such spaces, and obtain a bound on the scale of non-commutativity.

60.1 Introduction

While the Standard Model has been very successful in explaining an impressive amount and variety of data, the need for physics beyond the standard model (BSM) has been argued for eloquently over the past many years. Many different paths have been suggested. Ultimately, the data will hopefully shed some light on which is the correct one to follow. Non-commutative field theories form one such path. The aim of this exercise is to examine how feasible this may be from the point of view of existing experimental results, and the sort of deviations one may expect to observe, or at least put some bound on. Non-commutative (NC) spaces are spaces in which there exists a non-vanishing commutator, $[x_\mu, x_\nu] = \iota\theta_{\mu\nu}$, between the coordinates themselves along with the ordinary commutation relation between coordinates and momenta. The antisymmetric matrix $\theta_{\mu\nu}$ contains real parameters which are of dimension $[L^2]$. Just as $\hbar$ captures the fuzziness of phase space, the θ captures the fuzziness of space itself. This fuzziness or "unknowability" of space is well motivated by certain String Theory configurations [1] and is expected to be a feature of a general quantum theory

P. S. Ghoderao · P. Ramadevi
Indian Institute of Technology Bombay, Powai 400 076, Mumbai, India
e-mail: pulkitsg@iitb.ac.in

P. Ramadevi
e-mail: ramadevi@phy.iitb.ac.in

R. V. Gavai (✉)
Tata Institute of Fundamental Research, Colaba 400 005, Mumbai, India
e-mail: gavai@tifr.res.in

© Springer Nature Singapore Pte Ltd. 2021
P. K. Behera et al. (eds.), *XXIII DAE High Energy Physics Symposium*,
Springer Proceedings in Physics 261,
https://doi.org/10.1007/978-981-33-4408-2_60

of gravity as one approaches the Planck length. It is thus an interesting question to ask what the scale of $\theta_{\mu\nu}$ is.

A simple prescription to convert quantities on NC space to a corresponding expression in ordinary space is by replacing ordinary multiplication with a star product. The Moyal (star) product is defined as

$$(f \star h)(x) = f(x) \exp\left(\frac{\iota\theta^{\mu\nu}}{2}\overleftarrow{\partial_\mu}\overrightarrow{\partial_\nu}\right)h(x). \tag{60.1}$$

The hydrogen atom spectrum has played a canonical role in serving as the test bed for NC physics, and there has been some work on the possible impact NC spaces have on the Lamb shift. However, corrections to the spectrum have remained a bone of contention. In this talk, we first shed light on some of the competing proposals to describe the NC hydrogen spectrum. Then we investigate the role of composite operators in NC spaces to resolve the issue surrounding treatment of proton as a composite particle. Along with the charge quantisation constraint in NC spaces, we are then led to a bound on the NC scale which allows us to have a final say on the NC hydrogen atom problem.

60.2 The NC Hydrogen Atom Problem

In this section, we will briefly review the development of NC hydrogen spectrum in the literature and the issues of contention arising from each approach.

60.2.1 Non-commutative Corrections to Lamb Shift

The corrections to Lamb Shift in NC space were first derived by Chaichian et al. [2] through a quantum electrodynamics perspective. However, this result can also be obtained using a non-relativistic treatment [3] which we shall summarise here.

Considering the proton as *fixed* at the origin of a Cartesian NC coordinate system and an electron with coordinates x_i', where $i = 1, 2, 3$ surrounding it, the hydrogen atom potential in NC space takes the form[1], $V = -1/\sqrt{x_i' x_i'}$. It turns out that under the following change of variables,[2]

$$x_i' \to x_i - \sum_{j=1}^{3}\frac{\theta_{ij}}{2}p_j; \qquad p_i' \to p_i \tag{60.2}$$

---

[1] Here and throughout, we work in natural units so that $e = \hbar = 4\pi\epsilon_0 = 1$.

[2] This coordinate change is the same as expected from Moyal product if exponential is considered as a translation operator.

the commutator between the unprimed coordinates vanishes, $[x_i, x_j] = 0$ while the coordinate-momenta commutator becomes the ordinary, $[x_i, p_j] = \iota \delta_{ij}$ provided $[p_i', p_j'] = 0$.

In terms of the new coordinates the potential becomes

$$V = -1 \bigg/ \sqrt{\left(x_i - \sum_{j=1}^{3} \frac{\theta_{ij}}{2} p_j\right)\left(x_i - \sum_{k=1}^{3} \frac{\theta_{ik}}{2} p_k\right)}. \tag{60.3}$$

Performing a binomial expansion and keeping terms upto order θ we obtain

$$V = -\frac{1}{r} - \frac{1}{4r^3}(\vec{L} \cdot \vec{\Theta}), \tag{60.4}$$

where $L_k = \sum_{i,j=1}^{3} \epsilon_{ijk} x_i p_j$ is the angular momentum operator and $\vec{\Theta} = (\theta_x, \theta_y, \theta_z)$. Thus, we find a perturbation term to the classical $1/r$ potential. Using first-order perturbation theory the shift in energy levels can be found to be [3],

$$\Delta E = \langle n, l, m | - \frac{\theta_z L_z}{4r^3} | n, l, m \rangle \tag{60.5}$$

$$\Delta E = -\frac{\theta_z m}{4}\left(\frac{1}{n^3 l(l+1/2)(l+1)}\right), \tag{60.6}$$

where we have assumed $\vec{\Theta} = (0, 0, \theta_z)$.

In a relativistic treatment one also needs to take into account the spin of the wavefunction. In that case the perturbation can be modified to be [2],

$$\Delta E = \langle n, j, j_z | - \frac{\theta_z \hat{L}_z}{4r^3} | n, j, j_z \rangle \tag{60.7}$$

$$\Delta E = \frac{\theta_z j_z}{4}\left(1 \mp \frac{1}{2l+1}\right)\left(\frac{1}{n^3 l(l+1/2)(l+1)}\right) \text{ for } j = l \pm \frac{1}{2}. \tag{60.8}$$

The above result is interesting because it splits the $^2P_{1/2}$ level into $^2P_{-1/2}$ and $^2P_{+1/2}$ due to the presence of a j_z term. Thus there are corrections to Lamb Shift $(^2S_{1/2} \longrightarrow {}^2P_{1/2})$ at tree level itself, which are linear in the non-commutative parameter θ_z. Hence precise measurements of Lamb Shift can potentially reveal the NC scale.

60.2.2   Cancellation of Non-commutative Effects

Treating the hydrogen atom as a two body problem with a non-stationary proton, the Hamiltonian takes the form,

$$H = \frac{p'^2}{2\mu} - \frac{1}{r'}, \tag{60.9}$$

where μ is the reduced mass and r' is the relative separation. As usual, primed variables reside in NC space. Consider the relative coordinates, $r'_i = x_i^{\prime \text{electron}} - x_i^{\prime \text{proton}}$ such that $r' = \sqrt{\sum_{i=1}^{3}(r'_i r'_i)}$. The commutator between relative coordinates is then,

$$[r'_i, r'_j] = [x_i^{\prime \text{electron}} - x_i^{\prime \text{proton}}, x_j^{\prime \text{electron}} - x_j^{\prime \text{proton}}] \tag{60.10}$$

$$= [x_i^{\prime \text{electron}}, x_j^{\prime \text{electron}}] + [x_i^{\prime \text{proton}}, x_j^{\prime \text{proton}}] \tag{60.11}$$

$$= \iota \theta_{ij} + \iota(-\theta_{ij}) = 0. \tag{60.12}$$

Here we have used the fact that particles carrying opposite charge have opposite sign of the non-commutativity parameter [4]. Thus the NC effects in hydrogen cancel at commutator level itself [5], leading one to expect no effect of non-commutativity on the hydrogen atom spectrum.

60.2.3 Non-commutative Corrections from Composite Nature of Proton

As we know, proton is a bound state of quarks, and is described by Quantum Chromodynamics (QCD). This argument has been used [6] to modify the potential to first order as

$$V = V_{u_1} + V_{u_2} + V_d = -\frac{2}{3}\frac{1}{r'_{u_1}} - \frac{2}{3}\frac{1}{r'_{u_2}} + \frac{1}{3}\frac{1}{r'_d}, \tag{60.13}$$

where prime indicates that the variables reside in NC space. Such a potential can then give rise to non-trivial effects in perturbation theory which will result in a correction to the Lamb Shift similar to Sect. 60.2.1.

60.3 Composite Operators in NC Space

Ideally, we should consider the proton as a composite operator of quarks, especially since quarks are permanently confined inside it. The above simple modification to the potential comprising three separate quarks appears too simplistic and ignores confinement. Accordingly, in this section, we will consider the general description of composite particles in NC space. As usual we shall assume that the quarks are confined into composites via the SU(3) colour gauge theory and write the composite proton operator as $\psi_p = \psi_u \star \tilde{\psi}_u \star \psi_d$, where the quark fields reside in NC space.

The finite NC electromagnetic transformation with gauge parameter λ is given by

$$U = \exp\left(\iota\lambda(x)\right)_\star = 1 + \iota\lambda + \frac{(\iota)^2}{2!}(\lambda \star \lambda) + \dots \tag{60.14}$$

Under this gauge transform we expect, $\psi'_p = U \star \psi_p$, that is, the proton composite transforms as an operator with aggregate charge $+1$. However, the transformation of quark fields gives us,

$$\psi'_p = (U^{2/3} \star \psi_u) \star (U^{2/3} \star \tilde{\psi}_u) \star (U^{-1/3} \star \psi_d). \tag{60.15}$$

The infinitesimal version of the above is

$$\psi'_p = \psi_u \star \tilde{\psi}_u \star \psi_d - \frac{\iota}{3}(\psi_u \star \tilde{\psi}_u \star \lambda \star \psi_d)$$
$$+ \frac{2\iota}{3}(\psi_u \star \lambda + \lambda \star \psi_u) \star (\tilde{\psi}_u \star \psi_d). \tag{60.16}$$

For a consistent gauge transformation, we expect this to be equal to an infinitesimal version of $\psi'_p = U \star \psi_p$ which is

$$\psi'_p = \psi_u \star \tilde{\psi}_u \star \psi_d + \iota(\lambda \star \psi_u \star \tilde{\psi}_u \star \psi_d). \tag{60.17}$$

Clearly (60.16) and (60.17) fail to become equal since there is no way to commute the gauge parameter λ and any of the quark fields. This argument can be extended to other hadrons as well. Thus composite particles appear to be forbidden in NC space.

60.3.1   Tolerating Compositeness

Note that if λ in (60.16) could be commuted, it will match (60.17). Since $\lambda \star \psi - \psi \star \lambda = \iota\theta^{\mu\nu}\partial_\mu\lambda\ \partial_\nu\psi + \mathcal{O}(\theta^3) \longrightarrow 0$ and by the uncertainty principle $\partial \sim 1/r$ where r is the radius of the composite particle, the above requirement is satisfied if

$$\theta << r^2. \tag{60.18}$$

Hence for a composite particle to exist in NC space, the NC scale must lie well below its radius.

60.4   NC Scale and Consequences for Hydrogen Atom

It is a well-known result that NC spaces severely restrict the allowed particle charges.[3] This so-called charge quantisation constraint can be summarised as [8, 9],

In NC quantum electrodynamics, if the basic photon–photon coupling is g, the only allowed charges which the fields can carry are $\pm g$, 0 and no other multiples of g are permissible.

Armed with our condition for existence of composite particles (60.18) and the above charge constraint, we can revisit the various proposals for the hydrogen atom spectrum outlined in Sect. 60.2.

1. In Sect. 60.2.2, the electron and proton are considered to be point particles in NCQED. By (60.18), this means $\sqrt{\theta} \gtrsim 10^{-15} m$ can be possible. But we know that particles like Δ^{++} have a charge magnitude 2 as opposed to charge magnitude of 1 for electron and proton. Hence this approach does not satisfy charge quantisation and we need to consider protons as composite particles.

2. In Sect. 60.2.3, the proton is considered to be a composite particle. By (60.18), we have $\sqrt{\theta} << 10^{-15} m$. For modifying the potential, the quarks and electron were assumed to be point particles in NCQED. Since quarks have charges $-1/3$, $2/3$ and the electron has charge -1, charge quantisation condition will be violated and we need to consider electrons and quarks as composite particles.

3. The bound on compositeness scale for quarks and leptons as suggested by LHC experiments is between $10 - 25\ TeV$ [10], or about $7.9 - 19.7 \times 10^{-21} m$. Accordingly by (60.18), if we take $\sqrt{\theta} << 2 \times 10^{-20} m$ then we can hope to construct a composite model of quarks and leptons which satisfies charge quantisation. As an indicative example, if we take two new "fundamental" particles with charge $\pm 1/6$ to be point particles in NCQED, then they satisfy charge quantisation. Quarks can be constructed via a combination of four while leptons via a combination of six of these particles. The hydrogen spectrum in this case can be expected to obtain corrections similar to the ones derived in Sect. 60.2.1, with a magnitude dependent on the model of preons.

60.5   Conclusion

It was shown that a composite particle is allowed in NC space only if the NC scale lies well below its radius. This condition along with the known charge quantisation constraint in NC spaces implies that quarks and leptons must be composite particles for NC spaces to be a physical reality. We also fix the NC scale to lie well below $2 \times 10^{-20} m$. The final word on the hydrogen atom problem in NC space depends on the construction of an explicit model of preons near the NC scale.

---

[3]For a detailed derivation see Pulkit et al. [7].

References

1. N. Seiberg, Edward witten, sring theory and noncommutative geometry. JHEP **09**, 032 (1999)
2. M. Chaichian, M.M. Sheikh-Jabbari, A. Tureanu, Hydrogen atom spectrum and the lamb shift in noncommutative qed. Phys. Rev. Lett. **86**, 2716 (2001)
3. P. S. Ghoderao, P. Ramadevi. Are we living in non-commutative space? Revisiting the classic hydrogen atom system. In preparation
4. M.M. Sheikh-Jabbari, C, P, and T invariance of noncommutative gauge theories. Phys. Rev. Lett. **84**, 5265 (2000)
5. Pei-Ming Ho, Hsien-Chung Kao, Noncommutative quantum mechanics from noncommutative quantum field theory. Phys. Rev. Lett. **88**, 151602 (2002)
6. M. Chaichian, M.M. Sheikh-Jabbari, A. Tureanu, Non-commutativity of space-time and the hydrogen atom spectrum. Eur. Phys. J. C Particles Fields **36**(2), 251 (2004)
7. P. S. Ghoderao, R. V. Gavai, P. Ramadevi, Probing the scale of non-commutativity of space. Modern Phys. Lett. A, **34**, 195019 (2019)
8. M. Hayakawa. Perturbative analysis on infrared and ultraviolet aspects of noncommutative QED on R**4. hep-th/9912167 (1999)
9. M. Chaichian, P. Presnajder, M.M. Sheikh-Jabbari, A. Tureanu, Non-commutative standard model: model building. Eur. Phys. J. Particles Fields **29**(3), 413 (2003)
10. V. Khachatryan et al., Search for narrow resonances decaying to dijets in proton-proton collisions at $\sqrt{s} = 13$TeV. Phys. Rev. Lett. **116**, 071801 (2016)

Chapter 61
Inverse Seesaw, Singlet Scalar Dark Matter and Vacuum Stability

Ila Garg, Srubabati Goswami, K. N. Vishnudath, and Najimuddin Khan

Abstract We study the stability of the electroweak vacuum in the context of an inverse seesaw model extended by a singlet scalar dark matter. We show that even though these two sectors seem disconnected at low energy, the coupling constants of both the sectors get correlated at a high energy scale by the constraints coming from the perturbativity and stability/metastability of the electroweak vacuum. The new Yukawa couplings try to destabilize the electroweak vacuum while the additional scalar quartic couplings aid the stability. In fact, the electroweak vacuum may attain absolute stability even up to the Planck scale for suitable values of the parameters. We analyze the parameter space for the singlet fermion and the scalar couplings for which the electroweak vacuum remains stable/metastable and at the same time giving the correct relic density and neutrino masses and mixing angles as observed.

I. Garg
Department of Physics, Indian Institute of Technology Bombay, Powai 400 076, Mumbai, India
e-mail: ila.garg@iitb.ac.in

S. Goswami · K. N. Vishnudath (✉)
Theoretical Physics Division, Physical Research Laboratory, Ahmedabad 380009, India
e-mail: vishnudath@prl.res.in; vishnudathkn@gmail.com

S. Goswami
e-mail: sruba@prl.res.in

K. N. Vishnudath
Discipline of Physics, Indian Institute of Technology, Gandhinagar 382355, India

N. Khan
Centre for High Energy Physics, Indian Institute of Science, C. V. Raman Avenue, Bangalore 560012, India
e-mail: khanphysics.123@gmail.com

© Springer Nature Singapore Pte Ltd. 2021
P. K. Behera et al. (eds.), *XXIII DAE High Energy Physics Symposium*,
Springer Proceedings in Physics 261,
https://doi.org/10.1007/978-981-33-4408-2_61

61.1   Introduction

The two major experimental motivations entailing scenarios beyond Standard Model (SM) are neutrino mass and dark matter. For neutrino mass, the most natural approach is the seesaw mechanism and from the point of view of testability at the colliders, the TeV seesaw mechanism has become an extensive topic of research of late. On the other hand, among the various models of dark matter that are proposed in the literature, the most minimal renormalizable extension of SM are the so-called Higgs portal models [1–3]. These models include a scalar singlet that couples only to the SM Higgs. An additional Z_2 symmetry is imposed to prevent the decay of the DM and safeguard its stability. The coupling of the singlet with the Higgs provides the only portal for its interaction with SM. Nevertheless, there can be testable consequences of this scenario which can put constraints on its coupling and mass. These include constraints from searches of invisible decay of the Higgs boson at the LHC, direct and indirect detections of dark matter as well as compliance with the observed relic density. Implications of such an extra scalar for the LHC have also been studied. Combined constraints from all these have been discussed most recently in [4]. See [5] for more references.

In addition, the singlet scalar can also affect the stability of the EW vacuum and it has been seen from various studies in this direction that the singlet scalar can help in stabilizing the EW vacuum by adding a positive contribution which prevents the Higgs quartic coupling from becoming negative. On the other hand, as also seen in the previous chapter, the extra fermions can affect the stability adversely, and for TeV seesaw models the effect can be appreciable because of low mass thresholds and large Yukawa couplings. See [5] to see details and for more references. In this work, we extend SM by adding extra fermions as well as scalar singlets and see to what extend the additional scalar singlet can ameliorate the stability problem introduced by fermionic singlets and at the same time explaining the origin of neutrino mass as well as the existence of dark matter [5]. Here, the real singlet scalar is the dark matter candidate where we have imposed an additional Z_2 symmetry which ensures its stability. For the generation of neutrino mass at the TeV scale, we consider the inverse seesaw model with three right-handed neutrinos and three additional singlets. These two sectors are disconnected at low energy. However, the consideration of the stability of the electroweak (EW) vacuum and perturbativity induces a correlation between the two sectors. We study the stability of the EW vacuum in this model and explore the effect of the two opposing trends—singlet fermions trying to destabilize the vacuum further and the singlet scalar trying to oppose this. We find the parameter space, which is consistent with the constraints from relic density and neutrino oscillation data and at the same time can cure the instability of the EW vacuum. In addition to absolute stability, we also explore the parameter region which gives metastability in the context of this model. We investigate the combined effect of these two sectors and obtain the allowed parameter space consistent with observations and vacuum stability/metastability and perturbativity.

61.2 Model

In this section, we discuss the fermionic and the scalar sectors of the models that we have studied including the scalar potential in the presence of a singlet scalar.

61.2.1 Inverse Seesaw Model

In the inverse seesaw model, three right-handed neutrinos N_R and three gauge-singlet sterile neutrinos ν_s are added to SM [6]. ν_s and N_R are assigned with lepton numbers -1 and $+1$, respectively. The corresponding Yukawa Lagrangian responsible for neutrino masses before SSB is

$$- L_\nu = \bar{l}_L Y_\nu \, H^c N_R \; + \overline{N_R^c} \, M_R \, \nu_s \; + \; \frac{1}{2} \overline{\nu_s^c} M_\mu \nu_s \; + \; \text{h.c.,} \tag{61.1}$$

where l_L and H are the SM lepton and Higgs doublets, respectively. After the spontaneous symmetry breaking where the Higgs field acquires a vacuum expectation value $v \sim 246$ GeV, the above Lagrangian becomes

$$- L_{\nu \text{ mass}} = \bar{\nu}_L M_D N_R \; + \overline{N_R^c} \, M_R \, \nu_s \; + \; \frac{1}{2} \overline{\nu_s^c} M_\mu \nu_s \; + \; \text{h.c.,} \tag{61.2}$$

where $M_D = Y_\nu v / \sqrt{2}$. Since the mass term M_R is not subject to the $SU(2)_L$ symmetry breaking and the mass term M_μ violates the lepton number, the scales corresponding to the three sub-matrices of the neutral fermion mass matrix may naturally have a hierarchy $M_R >> M_D >> M_\mu$. In this case, the effective light neutrino mass matrix in the seesaw approximation is given by

$$M_{\text{light}} = M_D (M_R^T)^{-1} M_\mu M_R^{-1} M_D^T. \tag{61.3}$$

In the heavy sector, there will be three pairs of degenerate pseudo-Dirac neutrinos of masses of the order $\sim M_R \pm M_\mu$. Note that the smallness of M_{light} is naturally attributed to the smallness of both M_μ and $\frac{M_D}{M_R}$. For instance, $M_{\text{light}} \sim \mathcal{O}(0.1)$ eV can easily be achieved for $\frac{M_D}{M_R} \sim 10^{-2}$ and $M_\mu \sim \mathcal{O}(1)$ keV. Thus, the seesaw scale can be lowered down considerably assuming $Y_\nu \sim \mathcal{O}(0.1)$, such that $M_D \sim 10$ GeV and $M_R \sim 1$ TeV.

61.2.2 Scalar Sector

As mentioned earlier, in addition to the extra fermions, we also add an extra real scalar singlet S to SM. The potential for the scalar sector with an extra Z_2 symmetry under $S \to -S$ is given by

$$V(S, H) = -m^2 H^\dagger H + \lambda(H^\dagger H)^2 + \frac{\kappa}{2} H^\dagger H S^2 + \frac{m_S^2}{2} S^2 + \frac{\lambda_S}{24} S^4. \quad (61.4)$$

In this model, we take the vacuum expectation value (vev) of S as 0, so that Z_2 symmetry is not broken. Thus, the scalar sector consists of two particles h and S, where h is the SM Higgs boson with a mass of ~ 126 GeV, and the mass of the extra scalar is given by

$$M_{DM}^2 = m_S^2 + \frac{\kappa}{2} v^2. \quad (61.5)$$

As the Z_2 symmetry is unbroken up to the Planck scale, $M_{Planck} = 1.22 \times 10^{19}$ GeV, the potential can have minima only along the Higgs field direction and also this symmetry prevents the extra scalar from acquiring a vacuum expectation value. This extra scalar field does not mix with the SM Higgs boson. Also, an *odd* number of this extra scalar does not couple to the SM particles and the new fermions. As a result, this scalar is stable and serves as a viable weakly interacting massive dark matter particle. The scalar field S can annihilate to the SM particles as well as to the new fermions only via the Higgs boson exchange. So it is called Higgs portal dark matter.

61.3 Numerical Analysis and Results

For the inverse seesaw model, the input parameters are the entries of the matrices Y_ν, M_R and M_μ. Here Y_ν is a complex 3×3 matrix. M_R is a real 3×3 matrix and M_μ is a 3×3 diagonal matrix with real entries. We vary the entries of various mass matrices in the range $10^{-2} < M_\mu < 1$ keV and $0 < M_R < 5 \times 10^4$ GeV. This implies a heavy neutrino mass of maximum up to a few TeV. With these input parameters, we search for parameter sets consistent with the low energy data using the downhill simplex method [7]. We have considered the bounds from the oscillation data [8, 9], cosmological constraints on the sum of light neutrino masses ($\Sigma m_i < 0.14$ eV) [10], constraints on the non-unitarity of the PMNS mixing matrix [11] and the collider bounds on the masses of heavy neutrinos [12].

In addition, we have also considered the constraints from dark matter relic density and the recent detection experiments, in particular, the LUX-2016 [13] data and the indirect Fermi-LAT data [14]. We use FeynRules along with micrOMEGAs to compute the relic density of the scalar dark matter. Also, scalar sector couplings are constrained by the requirement of perturbative unitarity [15]. At very high field values, one can obtain the scattering matrix a_0 for the $J = 0$ partial wave by considering the various scalar–scalar scattering amplitudes. Using the equivalence theorem, we have reproduced the perturbative unitarity bounds on the eigenvalues of the scattering matrix for this model. These are given by [15]

$$|\kappa(\Lambda)| \leq 8\pi, \quad \text{and} \quad \left|6\lambda + \lambda_S \pm \sqrt{4\kappa^2 + (6\lambda - \lambda_S)^2}\right| \leq 16\pi. \qquad (61.6)$$

We have evaluated the SM coupling constants at the top quark mass scale and then run them using the renormalization group equations (RGEs) from M_t to M_{Planck} where we have taken into account the various threshold corrections at M_t. To evaluate the couplings from M_t to M_{Planck}, we have used three-loop RGEs for the SM couplings, two-loop RGEs for the extra scalar couplings and one-loop RGEs for the extra neutrino Yukawa couplings. We have used the package SARAH for studying the RGEs. In addition, in our analysis, we have taken two-loop (one-loop) contributions to the effective potential from the SM particles (extra singlet scalar and fermions).

The present central values of the SM parameters, especially the top Yukawa coupling y_t and strong coupling constant α_s with the Higgs boson mass $M_h \approx 125.7$ GeV, suggest that the beta function of the Higgs quartic coupling $\beta_\lambda (\equiv dV(h)/dh)$ goes from negative to positive around 10^{15} GeV [16, 17]. This implies that there is an extra deeper minima situated at that scale. So there is a finite probability that the EW vacuum might tunnel into that true (deeper) vacuum. But this tunneling probability is not large enough and hence the lifetime of the EW vacuum remains larger than the age of the universe. This implies that the EW vacuum is metastable in the SM. The expression for the tunneling probability at zero temperature is given by [18]

$$\mathcal{P}_0 = V_U \Lambda_B^4 \exp\left(-\frac{8\pi^2}{3|\lambda(\Lambda_B)|}\right) \qquad (61.7)$$

where Λ_B is the energy scale at which the action of the Higgs potential is minimum. V_U is the volume of the past light cone taken as τ_U^4, where τ_U is the age of the universe ($\tau_U = 4.35 \times 10^{17}$ s) [19]. In this work, we have neglected the loop corrections and gravitational correction to the action of the Higgs potential [20]. For the vacuum to be metastable, we should have $\mathcal{P}_0 < 1$ which implies that

$$0 > \lambda(\mu) > \lambda_{\min}(\Lambda_B) = \frac{-0.06488}{1 - 0.00986\ln(v/\Lambda_B)}, \qquad (61.8)$$

whereas the situation $\lambda(\mu) < \lambda_{\min}(\Lambda_B)$ leads to the unstable EW vacuum. In these regions, κ and λ_S should always be positive to get the scalar potential bounded from below. In our model, the EW vacuum shifts toward stability/instability depending upon the new physics parameter space for the central values of $M_h = 125.7$ GeV, $M_t = 173.1$ GeV and $\alpha_s = 0.1184$, and there might be an extra minima around 10^{12-17} GeV.

In Fig. 61.1, we have given the phase diagram in the $\text{Tr}[Y_\nu^\dagger Y_\nu] - \kappa$ plane. The line separating the stable region and the metastable region is obtained when the two vacua are at the same depth, i.e., $\lambda(\mu) = \beta_\lambda(\mu) = 0$. The unstable and the metastable regions are separated by the boundary line where $\beta_\lambda(\mu) = 0$ along with $\lambda(\mu) = \lambda_{\min}(\Lambda_B)$, as defined in (61.8). For simplicity, we have plotted Fig. 61.1 by fixing all the eight entries of the 3×3 complex matrix Y_ν, but varying only the

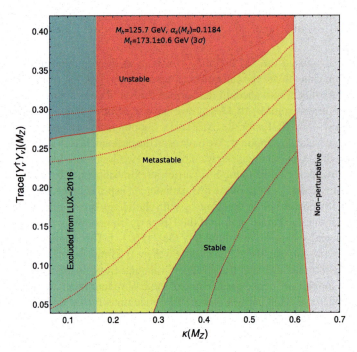

Fig. 61.1 Phase diagram in the $\text{Tr}[Y_\nu^\dagger Y_\nu]$ - κ plane. We have fixed all the entries of Y_ν except $(Y_\nu)_{33}$. The three boundary lines (two dotted and a solid) correspond to $M_t = 173.1 \pm 0.6$ GeV (3σ), and we have taken $\lambda_S(M_Z) = 0.1$. The dark matter mass is dictated by $\kappa(M_z)$ to give the correct relic density. See text for details

$(Y_\nu)_{33}$ element to get a smooth phase diagram. From Fig. 61.1, it can be seen that the values of κ beyond ~ 0.58 are disallowed by perturbativity bounds, and those below ~ 0.16 are disallowed by the direct detection bounds from LUX-2016 [13]. The value of the dark matter mass in this allowed range is thus ~ 530–2100 GeV. Note that the vacuum stability analysis of the inverse seesaw model done in reference [21] had found that the parameter space with $\text{Tr}[Y_\nu^\dagger Y_\nu] > 0.4$ were excluded by vacuum metastability constraints whereas, in our case, Fig. 61.1 shows that the parameter space with $\text{Tr}[Y_\nu^\dagger Y_\nu] \gtrsim 0.25$ are excluded for the case when there is no extra scalar. The possible reasons could be that we have kept the maximum value of the heavy neutrino mass to be around a few TeV, whereas the authors of [21] had considered heavy neutrinos as heavy as 100 TeV. Obviously, considering larger thresholds would allow us to consider the large value of $\text{Tr}[Y_\nu^\dagger Y_\nu]$ as the corresponding couplings will enter into RG running only at a higher scale. Another difference with the analysis of [21] is that we have fixed 8 of the 9 entries of the Yukawa coupling matrix Y_ν. Also, varying all the 9 Yukawa couplings will give us more freedom and the result is expected to change. The main result that we deduce from this plot is the effect of κ on the maximum allowed value of $\text{Tr}[Y_\nu^\dagger Y_\nu]$, which increases from 0.26 to 0.4 for a value of κ as large as 0.6. In addition, we see that the upper bound on $\kappa(M_Z)$

from perturbativity at M_{Planck} decreases from 0.64 to 0.58 as the value of $\text{Tr}[Y_\nu^\dagger Y_\nu]$ changes from 0 to 0.44. This is because $[Y_\nu^\dagger Y_\nu]$ affect the running of κ positively. Since $M_{DM} \sim 3300\,\kappa$ GeV for $M_{DM} >> M_t$, the mass of dark matter for which perturbativity is valid decreases with increase in the value of the Yukawa coupling.

61.4 Summary

We have studied the stability of the EW vacuum in the context of the TeV scale inverse seesaw model extended with a scalar singlet dark matter. We have studied the interplay between the contribution of the extra singlet scalar and the singlet fermions to the EW vacuum stability. We have shown that the coupling constants in these two seemingly disconnected sectors can be correlated at high energy by the vacuum stability/metastability and perturbativity constraints. Overall, we find that it is possible to find parameter spaces for which the EW vacuum remains absolutely stable for the inverse seesaw model in the presence of the extra scalar particle. We find an upper bound from metastability on $\text{Tr}[Y_\nu^\dagger Y_\nu]$ as 0.25 for $\kappa = 0$ which increases to 0.4 for $\kappa = 0.6$.

References

1. V. Silveira, A. Zee, Scalar phantoms. Phys. Lett. B **161**, 136–140 (1985)
2. J. McDonald, Gauge singlet scalars as cold dark matter. Phys. Rev. **D50**, 3637–3649 (1994). arXiv:hep-ph/0702143 [HEP-PH]
3. C.P. Burgess, M. Pospelov, T. ter Veldhuis, The Minimal model of nonbaryonic dark matter: A Singlet scalar. Nucl. Phys. B619, 709–728 (2001). arXiv:hep-ph/0011335 [hep-ph]
4. GAMBIT Collaboration, P. Athron et al., Status of the scalar singlet dark matter model. arXiv:1705.07931 [hep-ph]
5. I. Garg, S. Goswami, V.K. N., N. Khan, Electroweak vacuum stability in presence of singlet scalar dark matter in TeV scale seesaw models. Phys. Rev. D96(5), 055020 (2017).arXiv:1706.08851 [hep-ph]
6. R. Mohapatra, J. Valle, Neutrino Mass and Baryon Number Nonconservation in Superstring Models. Phys. Rev. D **34**, 1642 (1986)
7. W.T.V.W.H. Press, S.A. Teukolsky, B.P. Flannery, *Numerical Recipes in Fortran 90*, 2nd edn. (Cambridge University, Cambridge , 1996)
8. P.F. de Salas, D.V. Forero, C.A. Ternes, M. Tortola, J.W.F. Valle, Status of neutrino oscillations, 3σ hint for normal mass ordering and improved CP sensitivity. Phys. Lett. B **782**, 633–640 (2018). arXiv:1708.01186 [hep-ph]
9. I. Esteban, M.C. Gonzalez-Garcia, A. Hernandez-Cabezudo, M. Maltoni, T. Schwetz, Global analysis of three-flavour neutrino oscillations: synergies and tensions in the determination of θ_{23}, $\delta_C P$, and the mass ordering. JHEP **01**, 106 (2019). arXiv:1811.05487 [hep-ph]
10. Planck Collaboration, N. Aghanim et al., Planck 2018 results. VI. Cosmological parameters. arXiv:1807.06209 [astro-ph.CO]
11. S. Antusch, O. Fischer, Probing the nonunitarity of the leptonic mixing matrix at the CEPC. Int. J. Mod. Phys. A **31**(33), 1644006 (2016). arXiv:1604.00208 [hep-ph]

12. C.M.S. Collaboration, A.M. Sirunyan et al., Search for heavy neutral leptons in events with three charged leptons in proton-proton collisions at $\sqrt{s} = 13$ TeV. Phys. Rev. Lett. **120**(22), 221801 (2018). arXiv:1802.02965 [hep-ex]

13. L.U.X. Collaboration, D.S. Akerib et al., Results from a search for dark matter in the complete LUX exposure. Phys. Rev. Lett. **118**(2), 021303 (2017). arXiv:1608.07648 [astro-ph.CO]

14. Fermi-LAT Collaboration, M. Ajello et al., Fermi-LAT Observations of High-Energy γ-Ray Emission Toward the Galactic Center. Astrophys. J. 819(1), 44 (2016). arXiv:1511.02938 [astro-ph.HE]

15. G. Cynolter, E. Lendvai, G. Pocsik, Note on unitarity constraints in a model for a singlet scalar dark matter candidate. Acta Phys. Polon. B36, 827–832 (2005). arXiv:hep-ph/0410102 [hep-ph]

16. S. Alekhin, A. Djouadi, S. Moch, The top quark and Higgs boson masses and the stability of the electroweak vacuum. Phys. Lett. B **716**, 214–219 (2012). arXiv:1207.0980 [hep-ph]

17. D. Buttazzo, G. Degrassi, P.P. Giardino, G.F. Giudice, F. Sala, A. Salvio, A. Strumia, Investigating the near-criticality of the Higgs boson. JHEP **12**, 089 (2013). arXiv:1307.3536 [hep-ph]

18. G. Isidori, G. Ridolfi, A. Strumia, On the metastability of the standard model vacuum. Nucl.Phys. B609, 387–409 (2001). arXiv:hep-ph/0104016 [hep-ph]

19. Planck Collaboration, P.A.R. Ade et al., Planck 2015 results. XIII. Cosmological parameters. Astron. Astrophys. 594, A13 (2016). arXiv:1502.01589 [astro-ph.CO]

20. G. Isidori, V.S. Rychkov, A. Strumia, N. Tetradis, Gravitational corrections to standard model vacuum decay. Phys. Rev. D **77**, 025034 (2008). arXiv:0712.0242 [hep-ph]

21. L. Delle Rose, C. Marzo, A. Urbano, On the stability of the electroweak vacuum in the presence of low-scale seesaw models. arXiv:1506.03360 [hep-ph]

Chapter 62
Study of Atmospheric Neutrino Oscillation Parameters at the INO-ICAL Detector Using $\nu_e + N \to e + X$ Events

Aleena Chacko, D. Indumathi, James F. Libby, and Prafulla Kumar Behera

Abstract The India-based Neutrino Observatory will host a 50 kton magnetised tracking iron calorimeter with resistive plate chambers as its active detector element. We present the direction reconstruction of electron neutrino events with ICAL and the sensitivity of these events to neutrino oscillation parameters θ_{23} and δ_{CP}. We find that ICAL has adequate sensitivity to the CP violating phase δ_{CP}, with regions ranging $\delta_{CP} \sim 130\text{--}295°$ being excluded at 1σ for $\delta_{CP,\text{true}} = 0°$, from the sub-dominant electron neutrino oscillation channels. We also obtain a relative 1σ precision of 20% on the mixing parameter $\sin^2 \theta_{23}$. We neither discuss the possible backgrounds to ν_e interaction in ICAL nor investigate the effect of systematic uncertainties.

62.1 Introduction

Neutrino experiments over the past few decades [1–7] have been successful in measuring most of the neutrino oscillation parameters, viz., neutrino mixing angle (θ_{12}, θ_{23}, θ_{13}), their mass squared differences (Δm_{12}^2, Δm_{32}^2) and CP violating phase (δ_{CP}), although their mass hierarchy is yet to be determined. One of the experiments of this kind is the India-based Neutrino Observatory (INO) which aims to study the atmospheric neutrinos to probe the mass hierarchy, independent of δ_{CP}. The pro-

A. Chacko (✉) · J. F. Libby · P. K. Behera
Indian Institute of Technology Madras, Chennai 600 036, India
e-mail: aleenachacko@physics.iitm.ac.in

J. F. Libby
e-mail: libby@iitm.ac.in

P. K. Behera
e-mail: behera@iitm.ac.in

D. Indumathi
The Institute of Mathematical Sciences, Chennai 600 113, India
e-mail: indu@imsc.res.in

Homi Bhabha National Institute, Training School Complex, Anushakti Nagar, Mumbai 400085, India

© Springer Nature Singapore Pte Ltd. 2021
P. K. Behera et al. (eds.), *XXIII DAE High Energy Physics Symposium*,
Springer Proceedings in Physics 261,
https://doi.org/10.1007/978-981-33-4408-2_62

posed detector in INO is a magnetised iron calorimeter (ICAL) [8], built in three modules, with a resistive plate chamber (RPC) as its active detector element. The RPCs will be interleaved with iron layers (interaction medium) and pick-up strips are placed orthogonal to each other on either side of the RPC. ICAL is primarily optimised for muons.

The main signal of interest in ICAL will be charge current (CC) interactions of ν_μ (CCμ), but this paper focuses on the sub-dominant signal (nearly half of the ν_μ flux), namely the CC interactions of ν_e (CCe). These interactions are simulated for a 50 kton ICAL detector with 100-year exposure time by using the NUANCE [9] neutrino generator and incorporating the HONDA three-dimensional flux [10]. In Sects. 62.2 and 62.3, we study these NUANCE generated events. In Sects. 62.4 and 62.5, we describe the reconstruction of these events and their sensitivity to neutrino oscillation parameters θ_{23} and δ_{CP}.

62.2 Oscillation Probabilities

The neutrino oscillation probabilities of interest for CCe events are P_{ee} (electron survival probability) and $P_{\mu e}$ (muon disappearance probability) [11]. Figure 62.1 shows the effect of varying Δm^2_{32}, θ_{23} and δ_{CP}, for P_{ee} and $P_{\mu e}$. We see that the effect of varying Δm^2_{32} is opposite for P_{ee} and $P_{\mu e}$, which means the CCe events will provide very little sensitivity to Δm^2_{32}. Though not shown here, P_{ee} does not vary with different values of θ_{23} and δ_{CP}, but from Fig. 62.1 (bottom panel) $P_{\mu e}$ does. Therefore in this paper, we study only the ICAL sensitivity to $\sin^2 \theta_{23}$ and δ_{CP} from CCe events.

62.3 Ultimate Sensitivity Study

We first examine in the regions of true neutrino energy (E_ν) and direction ($\cos\theta_\nu$) that have significant oscillation probabilities. We find that (Fig. 62.2), for $P_{ee} < 0.8$ and $P_{\mu e} > 0.1$ (to see significant oscillation signature), both probabilities have sensitivity in regions where $E_\nu > 2$ and $\cos\theta_\nu > 0$ (up-going neutrinos). The values for oscillation parameters are taken from [12]. Throughout this paper normal hierarchy is assumed.

Next, we use an ideal ICAL detector (100% efficiency and perfect resolution) to study the maximum sensitivity CCe events can provide to the oscillation parameters θ_{23} and δ_{CP}. We take a sample corresponding to 5 years of NUANCE generated events using unoscillated ν_e and ν_μ flux and incorporate oscillations on these events with the "accept-reject" method. From Fig. 62.3, we see the oscillation signatures in the same regions as in Fig. 62.2.

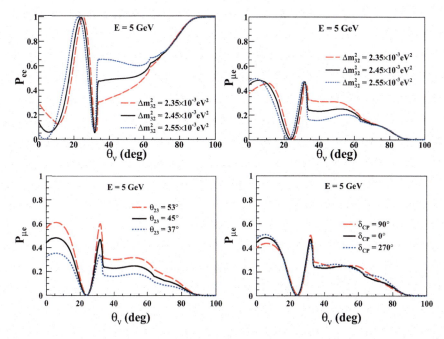

Fig. 62.1 P_{ee} and $P_{\mu e}$ (top panel) as a function of zenith angle, shown for three values of Δm_{32}^2 (2.55×10^{-3}eV^2 [dotted blue line], 2.45×10^{-3}eV^2 [solid black line] and 2.35×10^{-3}eV^2 [dashed red line]). $P_{\mu e}$ (bottom panel) as a function of zenith angle, shown for three values of θ_{23} [left] ($53°$ [dotted blue line], $45°$ [solid black line], $37°$ [dashed red line]) and three values of δ_{CP} [right] ($270°$ [dotted blue line], $0°$ [solid black line] and $90°$ [dashed red line]), with $\theta_{13} = 8.33°$ and assuming the normal hierarchy

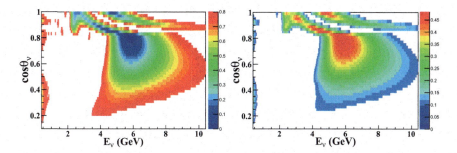

Fig. 62.2 $P_{ee} < 0.8$ (left) and $P_{\mu e} > 0.1$ (right) as a function of E_ν and $\cos \theta_\nu$

62.4   Reconstruction of CC*e* Events

To study the actual sensitivity that can be extracted from CC*e* events in ICAL, NUANCE generated unoscillated ν_e and ν_μ events are processed by a GEANT4 [13, 14] -based detector simulation of the ICAL detector. These simulated events have to be reconstructed to obtain E_ν and $\cos \theta_\nu$ from the final state particles (electrons and

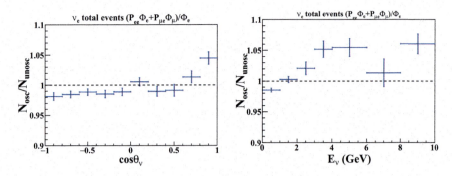

Fig. 62.3 Ratio of oscillated to unoscillated CCe events as a function of $\cos\theta_\nu$ (left) and E_ν (right), corresponding to 5 years of data

hadrons) in CCe interactions. Since electrons and hadrons only leave hits (shower) in the detector, unlike muons which leave a trail (track), an algorithm has to be developed to reconstruct the E_ν and $\cos\theta_\nu$ from the hit information.

62.4.1 Direction Reconstruction

The hit information in ICAL consist of the (x, y, z) positions and timing t of the hit. The x and y co-ordinates are the centres of the X- and Y-strips respectively, and the z co-ordinate is the centre of the RPC air-gap. We use the *raw-hit* method [15] which utilises this hit information to reconstruct the direction of the shower. In this method, the hit positions are plotted in two separate planes x-z and y-z, to avoid *ghost-hits* [15]. A straight line is fit to the hit positions in x-z and y-z planes, and from the slope of these fits $m_{x(y)}$, the average direction of the shower can be calculated as follows:

$$\theta = \tan^{-1}\left(\sqrt{m_x^2 + m_y^2}\right); \qquad \phi = \tan^{-1}\left(\frac{m_y}{m_x}\right). \tag{62.1}$$

The hits used for the reconstruction have to pass certain selection criteria. The timing window in which the hits are collected is restricted to 50 ns to ensure the hits are from the event under consideration. The hits have to be found in at least two layers and there must be a minimum of three hits in each event, to enable a straight line fit to hit positions. Around 54% of events are discarded due to this restriction. To pin the direction of the shower as up- or down-going, we make use of the slopes $m_{tx(ty)}$ of straight line fits to hit time in t_x-$z(t_y$-$z)$ graphs. If m_{tx} and m_{ty} have opposite signs, those events are discarded and about 10% of the events are removed due to this restriction. The reconstruction efficiency ϵ_{reco} is defined as the percentage of reconstructed events (N_{reco}) in total CCe events (N) and relative directional efficiency ϵ_{dir}, is defined as percentage of correctly reconstructed events

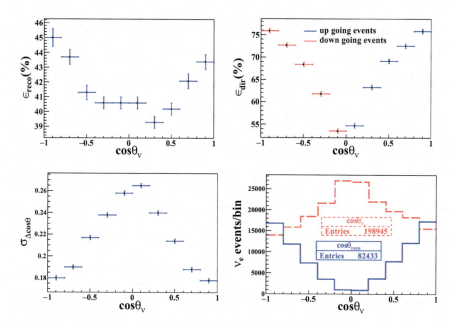

Fig. 62.4 Reconstruction efficiency, ϵ_{reco} (top left) and the relative directional efficiency ϵ_{dir} (top right) as a function of $\cos \theta_\nu$. $\cos \theta_\nu$ resolution (bottom left) and the distribution (bottom right) of the $\cos \theta_\nu$ (dashed red line) and reconstructed $\cos \theta_{\text{reco}}$ (solid blue line)

(N'_{reco}) as up- or down-going in total reconstructed events (N_{reco}) (62.2). The E_ν and $\cos \theta_\nu$ averaged values of ϵ_{reco} (Fig. 62.4, top left) and ϵ_{dir} (Fig. 62.4, top right) are $(41.7 \pm 0.2)\%$ and $(66.8 \pm 0.2)\%$, respectively, showing that we can distinguish an up-going event from a down-going event, which is crucial for the oscillation studies. The $\cos \theta_\nu$ resolution (Fig. 62.4, bottom left) improves for vertical events ($|\cos \theta_\nu > 0.5|$), as events traverse more layers in this direction. Figure 62.4 (bottom right) compares the $\cos \theta_\nu$ distribution before and after reconstruction.

$$\epsilon_{\text{reco}} = \frac{N_{\text{reco}}}{N} \, , \qquad \epsilon_{\text{dir}} = \frac{N'_{\text{reco}}}{N_{\text{reco}}} \, . \qquad (62.2)$$

62.4.2 Energy Reconstruction

Unlike direction, E_ν cannot be reconstructed by directly using the hit information, rather we calibrate the total number of hits (n_{hits}) in an event to its E_ν. The calibration is done by grouping n_{hits} in different E_ν bins. The mean value of n_{hits} ($\overline{n}(E_\nu)$) in each of these distributions is plotted against the mean value $\overline{E}_\nu$ of the corresponding E_ν bin. This data is then fitted with

$$\bar{n}(E) = n_0 - n_1 \exp(-\bar{E}/E_0) \tag{62.3}$$

to obtain the values of constants n_0, n_1 and E_0 (Fig. 62.5[left]). Once we have the values of these constants, (62.3) is inverted to estimate reconstructed energy E_{reco} (Fig. 62.5[right]). The E_ν resolution improves with E_ν.

62.5 Oscillation Parameter Sensitivity

We perform a χ^2 analysis to assess the sensitivity of CCe events to oscillation parameters. We bin the 100-year "data" set (scaled down to 10 years for the fit) simulated with true oscillation parameters in the reconstructed observables of $\cos\theta_{reco}$ (ten bins of equal width) and E_{reco} (seven bins of unequal width in 0–10 GeV range). We define the Poissonian χ^2 as

$$\chi^2 = 2 \sum_i \sum_j \left[(T_{ij} - D_{ij}) - D_{ij} \ln\left(\frac{T_{ij}}{D_{ij}}\right) \right], \tag{62.4}$$

where T_{ij} and D_{ij} are the "theoretically expected" and "observed number" of events respectively, in the ith $\cos\theta_{reco}$ bin and jth E_{reco} bin. Figure 62.6 shows $\Delta\chi^2$ as a function of $\sin^2\theta_{23}$ (left) and δ_{CP} (right), comparing binning in $\cos\theta_{reco}$, E_{reco} and in both. By binning in $\cos\theta_{reco}$ alone, we have a relative 1σ precision [8] of 20% on $\sin^2\theta_{23}$ and we are able to exclude $\delta_{CP} \sim 130$–$295°$ at 1σ for $\delta_{CP,true} = 0°$.

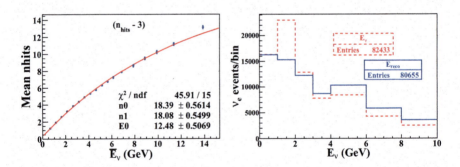

Fig. 62.5 Left: $\bar{n}(E)$ versus $\overline{E}_\nu$ and Right: the distribution of true E_ν (dashed red lines) and reconstructed E_{reco} (solid blue lines)

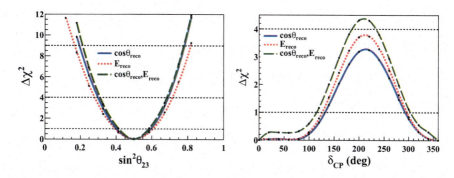

Fig. 62.6 $\Delta\chi^2$ as a function of $\sin^2\theta_{23}$ (left) and δ_{CP} (right) with bins in $\cos\theta_{\text{reco}}$ (solid blue lines) alone, E_{reco} (dotted red lines) alone and in both (dashed green lines) $\cos\theta_{\text{reco}}$ and E_{reco}. "Data" were generated with true $\sin^2\theta_{23} = 0.5$ and $\delta_{CP} = 0°$

62.6   Conclusion

In this paper, we have presented the reconstruction and oscillation parameter sensitivity of a pure sample of CCe events in ICAL. In reality, there are other types of events, like the neutral current events from both ν_μ and ν_e, which can be easily mis-identified as CCe events in ICAL. A significant fraction of CCμ events for which a track could not be reconstructed also mimics CCe hit patterns in ICAL. Hence, the next step would be finding selection criteria to separate CCe events from other types and analysing oscillation parameter sensitivity after including the mis-identified events. With CCe events alone, we find that ICAL has sufficient sensitivity to both oscillation parameters.

References

1. A. Gando et al., Phys. Rev. D **88**, 033001 (2013)
2. Daya Bay Collaboration, F. An et al., Phys. Rev. D **95** (2017)
3. Double Chooz Collaboration, Y. Abe et al., JHEP **1601** (2016)
4. RENO Collaboration, J.H. Choi et al., Phys. Rev. Lett. **116** (2016)
5. The KM3NeT collaboration, Adrián-Martínez et al., J. High Energ. Phys. **2017** (2017)
6. IceCube Collaboration, M.G. Aartsen et al., Phys. Rev. Lett. **120** (2018)
7. DUNE Collaboration, B. Abi et al., arXiv:1807.10334
8. ICAL Collaboration, S. Ahmed et al., Pramana **88**(5), 79 (2017)
9. D. Casper, Nucl. Phys. Proc. Suppl. **112**, 161 (2002)
10. M. Honda, T. Kajita, K. Kasahara, S. Midorikawa, Phys. Rev. D **83**, 123001 (2011)
11. D. Indumathi, M.V.N. Murthy, G. Rajasekaran, N. Sinha, Phys. Rev. D **74**, 053004 (2006)
12. C. Patrignani et al., Chin. Phys. C, **40**, 100001 (2016) and 2017 update
13. GEANT4 collaboration, S. Agostinelli et al., Nucl. Instrum. Meth. A **506**, 250 (2003)
14. J. Allison, K. Amako, J. Apostolakis, H. Araujo, P. Dubois et al., IEEE Trans. Nucl. Sci. **53**, 270 (2006)
15. M.M. Devi et al., JINST **13** (2018)

Chapter 63
Effect of Sterile Neutrinos on Degeneracy Resolution Capacities of NOvA and DUNE

Akshay Chatla, Sahithi Rudrabhatla, and Bindu A. Bambah

Abstract We investigate the implications of sterile neutrinos on the physics potential of the proposed experiment DUNE and future runs of NOvA using the latest NOvA results. Using a combined analysis of the disappearance and appearance data, NOvA reported three degenerate best-fit solutions where two are of normal hierarchy (NH) and one inverted hierarchy (IH). These degeneracies are expected to be resolved after an anti-neutrino run of NOvA. But in the presence of sterile neutrino, the degeneracy resolution capacity is reduced due to the new degrees of freedom. We study the chances of resolving parameter degeneracies with future runs of NOvA and DUNE in the light of this degraded degeneracy resolution power.

63.1 Introduction

The discovery of neutrino oscillations by Super-Kamiokande [1], SNO [2], and KAMLAND [3] was the first evidence for the physics beyond the standard model (SM). The 3-flavour neutrino model considered as the standard theory of neutrino flavour oscillations is able to explain the observed oscillation data with six parameters (three mixing angles, one CP phase, and two mass-squared differences). But, the oscillation probability equations derived allow different sets of oscillation parameters to have the same value of oscillation probability. Thus, wrong parameters can mimic the true solutions causing parameter degeneracy. The current unknowns in 3-flavour model are octant degeneracy of θ_{23} and mass hierarchy degeneracy (MH) and the CP violating phase. Resolving these degeneracies is an important objective of NOvA [4] and DUNE [5] experiments.

The 3-flavour model fits with experimental results from the solar, atmospheric, reactor and long-baseline experiments very well. But, there are some anomalous

A. Chatla (✉) · B. A. Bambah
School of Physics, University of Hyderabad, Hyderabad 500046, India
e-mail: chatlaakshay@gmail.com

S. Rudrabhatla
Department of Physics, University of Illinois at Chicago, Chicago, IL 60607, USA

© Springer Nature Singapore Pte Ltd. 2021   453
P. K. Behera et al. (eds.), *XXIII DAE High Energy Physics Symposium*,
Springer Proceedings in Physics 261,
https://doi.org/10.1007/978-981-33-4408-2_63

results at short baseline (SBL) by LSND [6] and MiniBooNE [7], which can be explained by introducing new m^2 $\sim$1 eV^2. Since the LEP [8] experiment limits the number of active neutrino flavours to three, the new neutrino must be a sterile (no weak interaction) neutrino (ν_s). The sterile neutrino introduces new oscillation parameters (three mixing angles and two CP phases) to the 3-flavour model, which is now the 3 + 1 model. These new parameters increase the degrees of freedom and will affect degeneracy resolution capabilities of NOvA and DUNE. We attempt to find the effect of one sterile neutrino on degeneracy resolution capabilities of NOvA and DUNE and the extent to which these degeneracies are resolved in the future runs of NOvA and DUNE.

63.2 Theoretical Framework

In this paper, we worked with one light sterile neutrino (3 + 1 model). In this model, the flavour and mass eigenstates are coupled with a 4 × 4 mixing matrix. A suitable parametrization of the mixing matrix is

$$U_{\text{PMNS}_{3+1}} = R_{34} \tilde{R}_{24} \tilde{R}_{14} \boldsymbol{R_{23}} \tilde{R}_{13} R_{12}. \tag{63.1}$$

Here, R_{ij} and $\tilde{R}_{ij}$ represent real and complex 4 × 4 rotation in the plane containing the 2 × 2 sub-block in (i, j) sub-block

$$R_{ij}^{2\times2} = \begin{pmatrix} c_{ij} & s_{ij} \\ -s_{ij} & c_{ij} \end{pmatrix} \qquad \tilde{R}_{ij}^{2\times2} = \begin{pmatrix} c_{ij} & \tilde{s}_{ij} \\ -\tilde{s}_{ij}{}^* & c_{ij} \end{pmatrix} \tag{63.2}$$

where $c_{ij} = \cos\theta_{ij}$, $s_{ij} = \sin\theta_{ij}$, $\tilde{s}_{ij} = s_{ij}e^{-i\delta_{ij}}$, and δ_{ij} are the CP phases.

The bold matrices in (63.1) represent the standard 3-flavour model. We see that the addition of one sterile neutrino introduces 3 new mixing angles and 2 new CP phases. The measurement of the new parameters is important for the study of sterile neutrinos. We know that the SBL experiments are sensitive to sterile mixing angles. But, they are not sensitive to new CP phases introduced by ν_s as they need longer distances to become measurable. We use long baseline (LBL) experiments to study CP phases. The oscillation probability, $P_{\mu e}$ for LBL experiments in the 3 + 1 model, after averaging Δm_{41}^2 oscillations and neglecting MSW effects [9], can be expressed as a sum of the four terms [10]

$$P_{\mu e}^{4\nu} \simeq P_1 + P_2(\delta_{13}) + P_3(\delta_{14} - \delta_{24}) + P_4(\delta_{13} - (\delta_{14} - \delta_{24})). \tag{63.3}$$

We see that CP phases introduced by sterile neutrinos persist in the $P_{\mu e}$ even after averaging out Δm_{41}^2 lead oscillations. The last two terms of (63.3) give the sterile CP phase dependence terms. $P_3(\delta_{14} - \delta_{24})$ depends on the sterile CP phases δ_{14} and

δ_{24}, while P_4 depends on a combination of δ_{13} and $\delta_{14} - \delta_{24}$. The amplitudes of P_2 and P_4 terms of (63.3) are of the same order. This new interference terms reduce the sensitivity of experiments to the standard CP phase (δ_{13}).

63.3 Results

We used General Long Baseline Experiment simulator (GLoBES) [11, 12] to simulate the data for NOvA and DUNE. The simulation and experimental details we used are in [10] and the references therein. The recent NOvA analysis [13], taking both appearance and disappearance channel data for 3 years of neutrino run, gave 2 best-fit points for normal hierarchy(NH), and inverted hierarchy (IH) is disfavored at the 95% confidence level. We take these best fit points and try to find the extent to which these degeneracies can be resolved in the future runs of NOvA and DUNE. In Fig. 63.1, we plot the oscillation probability $P_{\mu e}$ take in account of matter effects as a function of energy for the two best-fit values of NOvA with different values of δ_{14} (δ_{24}), from $-180°$ to $180°$ while keeping δ_{24} (δ_{14}) = 0. We observe in Fig. 63.1 that δ_{14} and δ_{24} bands are mirror reflections of each other. This implies that even after the matter effects are considered, δ_{14} and δ_{24} act as a single entity ($\delta_{14} - \delta_{24}$). Using this result, we can reduce our computation effort considerably.

In Fig. 63.2, the allowed regions of $\sin^2\theta_{23} - \delta_{cp}$ plane from NOvA and DUNE simulation data with different run-times, considering the latest NOvA results as true values, are plotted. The test values are taken from both NH and IH, for 3 and 3+1 neutrino models. The contour denotes the region where the test hypothesis is not excluded at a given confidence level (CL). For example, contour titled NH-IH denotes the region where IH (test hypothesis) could not be excluded at mentioned CL when

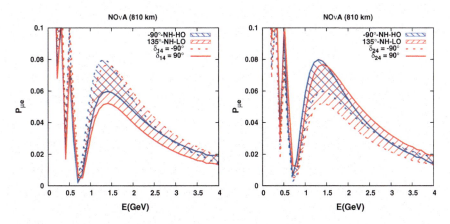

Fig. 63.1 The oscillation probability $P_{\mu e}$ as a function of energy. The bands correspond to different values of δ_{14} (δ_{24}), from $-180°$ to $180°$ while keeping δ_{24} (δ_{14}) = 0

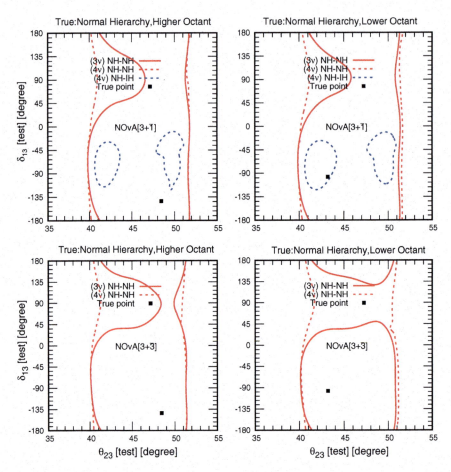

Fig. 63.2 Contour plots of allowed regions in the test plane, θ_{23} versus δ_{13}, at 2σ CL with top and bottom rows for NOvA runs of $3 + \bar{1}$ and $3 + \bar{3}$ years, respectively

NH is the true hypothesis. The different test cases we used are NH-NH(3ν), NH-IH(3ν), NH-NH(4ν) and NH-IH(4ν). The absence of a contour in the plot implies that its test hypothesis is excluded at the mentioned CL.

In the first row of Fig. 63.2, the allowed areas for NOvA[$3+\bar{1}$] are shown. In the first column, 2σ CL allowed regions are plotted for true values of $\delta_{13} = 262.8°$ and $\theta_{23} = 43.2°$ and normal hierarchy. It is seen that NOvA[3+1] has a wrong hierarchy (WH), i.e. (NH-IH) contour for 4ν case which was absent in the 3ν case. For NOvA[3+3], WH (NH-IH) is excluded even for 4ν. In the second column, 2σ CL allowed regions are plotted for true values of $\delta_{13} = 217.8°$ and $\theta_{23} = 48.4°$ and normal hierarchy. It is seen that NOvA[3+1] has WH contour for the 4ν case which was absent in the 3ν case. For NOvA[3+3], WH (NH-IH) is excluded even for the 4ν case. It is seen that the allowed contour sizes reduce for NOvA[3+3] due to increased statistics. Since

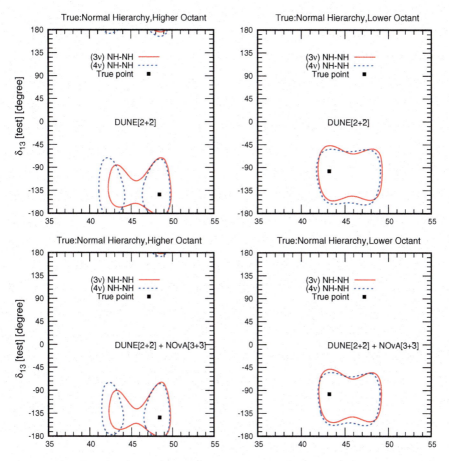

Fig. 63.3 Contour plots of allowed regions in the test plane, θ_{23} versus δ_{13}, at 2σ CL with top and bottom rows for DUNE[$2 + \bar{2}$] and DUNE[$2 + \bar{2}$] + NOvA[$3 + \bar{3}$], respectively

the true values of θ_{23} of both cases are near maximal mixing (MM) angle, we need more data to exclude MM angle as seen.

In Fig. 63.3, it is seen that DUNE has considerably better sensitivity compared to NOvA. WH (NH-IH) is excluded for both 3ν and 4ν cases at 2σ CL. A slight improvement in degeneracy resolution is observed for combined statistics of DUNE[$2 + \bar{2}$] + NOvA[$3 + \bar{3}$] over just DUNE[$2 + \bar{2}$].

In summary, we discussed how the presence of a sterile neutrino will affect the physics potential of the proposed experiment DUNE and future runs of NOvA, in the light of latest NOvA results [13]. It is seen that for the current best-fit values of NOvA, small IH degeneracy introduced by 3+1 model gets resolved at 2σ level with increased run-time of the experiment. Since the best-fit values are close to MM angle, more data is required to exclude MM angle.

References

1. Y. Fukuda et al., Super-Kamiokande Collaboration. Phys. Rev. Lett. **81**, 1562 (1998). https://doi.org/10.1103/PhysRevLett.81.1562 [hep-ex/9807003]
2. Q.R. Ahmad et al., SNO Collaboration. Phys. Rev. Lett. **87**, 071301 (2001). https://doi.org/10.1103/PhysRevLett.87.071301 [nucl-ex/0106015]
3. K. Eguchi et al., KamLAND Collaboration. Phys. Rev. Lett. **90**, 021802 (2003). https://doi.org/10.1103/PhysRevLett.90.021802 [hep-ex/0212021]
4. P. Adamson et al., [NOvA Collaboration], Phys. Rev. Lett. **116**(15), 151806 (2016). https://doi.org/10.1103/PhysRevLett.116.151806, arXiv:1601.05022 [hep-ex]
5. R. Acciarri *et al.* [DUNE Collaboration], arXiv:1601.05471 [physics.ins-det]
6. C. Athanassopoulos et al., LSND Collaboration. Phys. Rev. Lett. **75**, 2650 (1995). https://doi.org/10.1103/PhysRevLett.75.2650 [nucl-ex/9504002]
7. A.A. Aguilar-Arevalo et al., MiniBooNE Collaboration. Phys. Rev. Lett. **102**, 101802 (2009). https://doi.org/10.1103/PhysRevLett.102.101802, arXiv:0812.2243 [hep-ex]
8. S. Schael et al., ALEPH and DELPHI and L3 and OPAL and SLD Collaborations and LEP electroweak working group and SLD electroweak group and SLD heavy flavour group. Phys. Rep. **427**, 257 (2006). https://doi.org/10.1016/j.physrep.2005.12.006 [hep-ex/0509008]
9. A.Y. Smirnov, Phys. Scripta T **121**, 57 (2005). https://doi.org/10.1088/0031-8949/2005/T121/008 [hep-ph/0412391]
10. A. Chatla et al., Adv. High Energy Phys. **2018**, 2547358 (2018)
11. P. Huber et al., Comput. Phys. Commun. **167**, 195 (2005). [hep-ph/0407333]
12. P. Huber et al., Comput. Phys. Commun. **177**, 432 (2007). [hep-ph/0701187]
13. M.A. Acero et al., NOvA Collaboration. Phys. Rev. D **98**, 032012 (2018). https://doi.org/10.1103/PhysRevD.98.032012. arXiv:1806.00096 [hep-ex]

Chapter 64
Current Status for the Inclusive Neutral Current π^0 Production Cross-Section Measurement with the NOvA Near Detector

D. Kalra

Abstract The NuMI Off-axis ν_e Appearance (NOvA) experiment is a long-baseline neutrino oscillation experiment. It uses two functionally identical detectors, the NOvA near detector (ND) at Fermilab and the NOvA far detector (FD) at a distance 810 km in northern Minnesota to measure ν_e appearance in a narrow-band beam of ν_μ peaked at 2 GeV in energy. Neutrino induced Neutral Current (NC) interactions with a π^0 in the final state are a significant background in the ν_e appearance measurement. The π^0 decay into two photons can fake the ν_e appearance signal either due to the merging of two photon showers or one of the two photons escaping the detection. Therefore, a complete understanding of NC interactions with π^0 in the final state is very important. To constrain this background, NOvA will perform cross-section measurement of inclusive NC π^0 production using data from the NOvA ND. It will also help in reducing the background uncertainties for current and future long-baseline neutrino oscillation experiments.

64.1 Introduction

The neutrino-Nucleus (ν-N) interactions have been studied intensively for decades [1]. ν_μ-induced neutral current (NC) interactions with a π^0 in the final state are the dominant background for experiments looking for the ν_e appearance such as NOvA and DUNE [2]. The signal for the ν_e appearance channel is an electron in the final state that showers electromagnetically. Neutral pion decay into two photons can fake the ν_e appearance signal in two ways: either 2 γ's can merge together or one of them may escape detection and hence behave like an electron shower. Therefore, a complete understanding of NC π^0 production is very important.

On behalf of the NOVA Collaboration.

D. Kalra (✉)
Department of Physics, Panjab University, Chandigarh 160014, India
e-mail: daisykalra89@gmail.com

© Springer Nature Singapore Pte Ltd. 2021   459
P. K. Behera et al. (eds.), *XXIII DAE High Energy Physics Symposium*,
Springer Proceedings in Physics 261,
https://doi.org/10.1007/978-981-33-4408-2_64

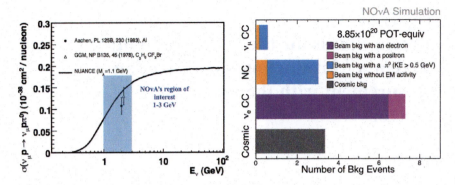

Fig. 64.1 (Left) Existing measurements of the cross section for the NC process as a function of neutrino energy. (Right) Plot shows the NOvA ν_e appearance backgrounds

There exist a very few measurements for this channel [3] as can be seen in Fig. 64.1 (Left). A~10% uncertainty on the NC background for the NOvA ν_e appearance is dominated by π^0 production as shown in Fig. 64.1 (Right). So, it is very important to constrain this background.

64.2 Simulation and Reconstruction Details

NOvA uses GEANT4 [4] to simulate the detector geometry and GENIE (v2-12-10b) [5] to simulate neutrino interactions. Neutrino interactions in the NOvA ND are reconstructed into slices (clusters of cell hits that are closely related in space and time) [6]. The slices are examined to find the particle paths using a Hough transformation [7]. The information from the intersection of the paths is used to find a neutrino interaction vertex (a point where the primary neutrino interaction takes place). The clusters that correspond to the same shower are reconstructed as prongs. Thus, a prong is defined as a collection of cell hits with a starting point and direction. The prongs are sorted by energy which means the leading prong has most of the energy and is referred to as prong1 whereas, the second-most energetic prong is called prong2. Figure 64.2 shows a ND MC event display with two reconstructed prongs.

The simulated sample used for this study has ~4x more statistics than data. So, all the distributions with the simulated sample are normalized to 8.09×10^{20} POT which reflects the NOvA ND data POT.

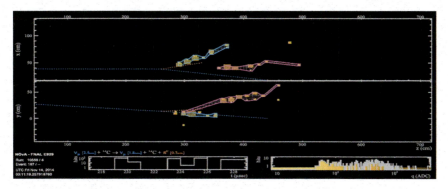

Fig. 64.2 An event display that shows two reconstructed prongs in both the detector views, XZ and YZ views. The NuMI beam is coming in from the left side

64.3 Signal and Background

Signal for this analysis is defined as neutrino-induced NC interactions with at least one π^0 in the final state with true π^0 K. E > 0.1 GeV.

The background comes from the neutrino-induced CC interactions (CC background) and NC interactions (NC background). The CC background consists of interactions in which the outgoing μ is not identified, and can contain a π^0 in the final state or not. The NC background consists of NC interactions without a π^0 in the final state and with π^0 below true K. E 0.1 GeV.

64.4 Pre-Selection

Pre-selection starts by applying some quality cuts to reject the noise hits [8]. Then, the reconstructed interaction vertex is required to be inside the NOvA ND fiducial volume and all showers must be contained [8]. Pre-selection cuts also include the events with exactly two 3D prongs. We also include the oscillation analysis muonID (also called as Reconstructed muon identification (ReMId)) [9] in the pre-selection. Figure 64.3 (left) shows a distribution of ReMID for 2-prong events which shows a good separation of CC events from NC. To reject these CC background events, we choose a cut value on ReMId based on its statistical figure of merit as shown in Fig. 64.3 (right). The FOM is maximized at 0.36, so we include MuonID/ReMId < 0.36 in the pre-selection.

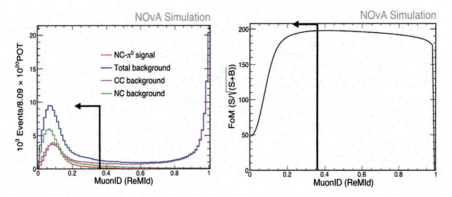

Fig. 64.3 (Left) MuonID/ReMId variable distribution for the signal and background events. (Right) Distribution of figure of merit (FOM) evaluated from MuonID/ReMId distribution

64.5   Event Identification

An eventID is developed based on the Boosted Decision Tree algorithm. The network is trained using the variables that characterize the electro-magnetic shower properties. Further, the variables associated with prong1 are selected as they showed comparatively better separation between the signal and background than the subleading prong variables [8]. The distribution of eventID aka NC π^0 ID for the signal and background events with pre-selection cuts is shown in Fig. 64.4.

Fig. 64.4 BDTG output, NC π^0 ID, distributions for the signal (Red) and background (Blue) with pre-selection cuts. The total background is broken down into CC background (Magenta) and NC background (Green)

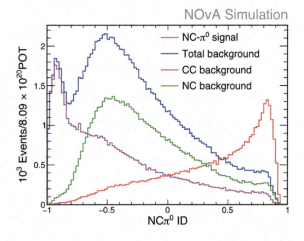

64.6 Prerequisites For Cross-Section Measurement

The differential cross section w.r.t. the π^0 kinematics is written as

$$\frac{d\sigma}{dx} = \frac{U(N^{sel}(x) - N^{bkgd}(x))}{N_{Target}\phi\epsilon(x)dx}, \qquad (64.1)$$

where N^{sel} and N^{bkgd} are the numbers of selected events and background events, respectively. The event ID is used to estimate the signal and background which is a very important step in making the cross-section measurement and is discussed in the next section. U is unfolding matrix that corrects the reconstructed quantities for detector resolution and smearing, N_{Target} is the number of target nucleons, ϕ is the flux, and ϵ is the signal selection efficiency.

x is the variable w.r.t. which the cross section is measured and in this study x is π^0 K. E, and angle and the distributions of both these variables in analysis bins are shown in Fig. 64.5.

64.7 Background Estimation

Background estimation is done in each of the π^0 K. E and angle bins, separately and independent of the neighboring bins, taking all the systematics (detector response, flux, and cross section) into account. The background is estimated by fitting the total signal, NC background, and CC background components of NC π^0 ID to the fake-data and to determine each of these parameter values, we minimize chi-square which is defined as

$$\chi^2 = (Data_i - MC_i)^T V_{ij}^{-1}(Data_j - MC_j), \qquad (64.2)$$

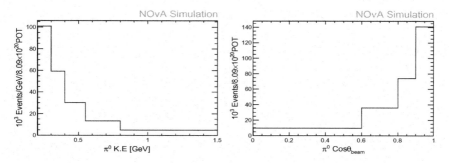

Fig. 64.5 The distribution of events for π^o K. E (left) and π^o $cos\theta$ (right) is shown in the analysis bins. The events in each bin are divided by the bin width

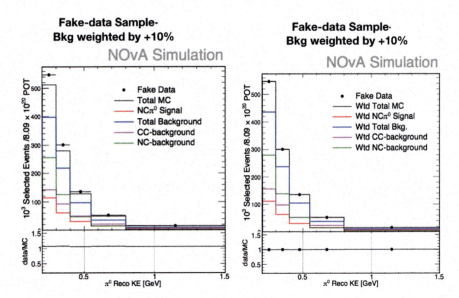

Fig. 64.6 Reconstructed kinetic energy distributions before and after the fit with background weighted fake-data sample

where i runs over the number of NC π^0 ID bins. V_{ij} is the covariance matrix, a simple linear addition of statistical and systematic covariance matrices [8]. The fit results with this procedure are used to check the π^0 kinematics before and after the fit as shown in Fig. 64.6 with background weighted fake-data sample. The fit results gave reasonable results and the details can be found in [8].

64.8   Summary

The current status of the analysis to measure NC π^0 production cross section is discussed. An eventId (NC π^0 Id) is developed to select NC π^0 events of interest. A data-driven technique including all the systematics is used to estimate the background which is a very important step in making a cross-section measurement. We aim to produce world-class differential cross-section measurement with uncertainties $\sim 10\%$.

Acknowledgements NOvA is supported by US Department of Energy; US National Science Foundation; Department of Science and Technology, India; European Research Council; MSMT CR, Czech Republic; RAS, RMES, and RFBR, Russia; CNPq and FAPEG, Brazil; and the State and University of Minnesota. We are grateful for the contributions of the staff of the University of Minnesota module assembly facility and NOvA FD Laboratory, Argonne National Laboratory, and Fermilab. Fermilab is operated by Fermi Research Alliance, LLC under Contract No. DeAC02-07CH11359 with the US DOE.

References

1. D.A. Harris, The State of the Art of Neutrino Cross Section Measurements. Fermilab-Conf-15-254-ND
2. K.S. McFarland, Neutrino Interactions Conf:C06-08-08
3. J.A. Formaggio, From eV to EeV: Neutrino Cross Section Across Energy Scales, Fermilab-Pub-12-785-E
4. S. Agostinelli et al., GEANT4 Collaboration. Nucl. Instrum. Meth. **A506**, 250 (2003)
5. C. Andreopoulos et al., The GENIE Neutrino Monte Carlo Generator (2015). ArXiv:1510.05494 [hep-ex]
6. M. Baird, Slicing module comparison technical note. NOVA Internal Document **9195** (2013)
7. M. Baird, Global vertex reconstruction beginning with a modified hough transform. NOVA Internal Document **8241** (2012)
8. D. Kalra, Technical note on NC π^0 cross-section measurement. NOVA Internal Document **36381** (2018)
9. N. Raddatz, ReMId technical note. NOVA Internal Document **11206** (2014)

Chapter 65
Exploring Partial μ–τ Reflection Symmetry in DUNE and Hyper-Kamiokande

K. N. Deepthi, Kaustav Chakraborty, Srubabati Goswami, Anjan S. Joshipura, and Newton Nath

Abstract In this work, we study the consequences of the 'partial μ–τ' reflection symmetry and the testability of the corresponding symmetry predictions at the upcoming experiments: Deep Underground Neutrino Experiment (DUNE) and Hyper-Kamiokande (HK) experiment. Each prediction $|U_{\mu i}| = |U_{\tau i}|$ ($i = 1, 2, 3$) when applied to a single column of the leptonic mixing matrix U gives rise to different correlations between θ_{23} and δ_{CP}. We tested the correlations from the two leading cases of partial μ–τ reflection symmetry, namely $|U_{\mu 1}| = |U_{\tau 1}|$ and $|U_{\mu 2}| = |U_{\tau 2}|$ using the experiments, and also examined the sensitivity of these experiments to distinguish between the two cases.

K. N. Deepthi (✉)
Mahindra Ecole Centrale, Hyderabad 500043, India
e-mail: nagadeepthi.kuchibhatla@mahindrauniversity.edu.in; kdeepthin@gmail.com

K. Chakraborty · S. Goswami · A. S. Joshipura
Theoretical Physics Division, Physical Research Laboratory, Ahmedabad 380009, India
e-mail: kaustav@prl.res.in

S. Goswami
e-mail: sruba@prl.res.in

A. S. Joshipura
e-mail: anjan@prl.res.in

K. Chakraborty
Discipline of Physics, Indian Institute of Technology, Gandhinagar 382355, India

N. Nath
Institute of High Energy Physics, Chinese Academy of Sciences, Beijing 100049, China
e-mail: newton@ihep.ac.cn

School of Physical Sciences, University of Chinese Academy of Sciences, Beijing 100049, China

© Springer Nature Singapore Pte Ltd. 2021   467
P. K. Behera et al. (eds.), *XXIII DAE High Energy Physics Symposium*,
Springer Proceedings in Physics 261,
https://doi.org/10.1007/978-981-33-4408-2_65

65.1   Introduction

The major goal of the current and upcoming neutrino oscillation experiments is to determine the neutrino mass hierarchy (i.e. whether $\Delta m_{31}^2 > 0$—normal hierarchy (NH) (or) $\Delta m_{31}^2 < 0$—inverted hierarchy (IH)), the octant of atmospheric mixing angle θ_{23} ($\theta_{23} < 45°$ called lower octant (LO) (or) $\theta_{23} > 45°$ called higher octant (HO)) and the CP violating phase δ_{CP}. Symmetry-based approaches can aid in constraining the parameter space of these unknown parameters by providing correlations among them. They also predict the structure of leptonic mixing matrix [1–5].

In the standard PDG parameterization, the leptonic mixing matrix U is given by

$$U = \begin{bmatrix} c_{12}c_{13} & s_{12}c_{13} & s_{13}e^{-i\delta_{CP}} \\ -s_{12}c_{23} - c_{12}s_{23}s_{13}e^{i\delta_{CP}} & c_{12}c_{23} - s_{12}s_{23}s_{13}e^{i\delta_{CP}} & s_{23}c_{13} \\ s_{12}s_{23} - c_{12}c_{23}s_{13}e^{i\delta_{CP}} & -c_{12}s_{23} - s_{12}c_{23}s_{13}e^{i\delta_{CP}} & c_{23}c_{13} \end{bmatrix}. \quad (65.1)$$

One of the well-motivated symmetries which is in good agreement with the current oscillation parameters is the μ–τ reflection symmetry [6] which states

$$|U_{\mu i}| = |U_{\tau i}|, \quad (65.2)$$

for all the columns $i = 1, 2, 3$ of the leptonic mixing matrix U. Using (65.1) and (65.2), one can obtain two predictions

$$\theta_{23} = \frac{\pi}{4}, \quad s_{13} \cos \delta_{CP} = 0. \quad (65.3)$$

Equation (65.3) implies $\theta_{23} = \frac{\pi}{4}$ and $\delta_{CP} = \pm\frac{\pi}{2}$ which is in accord with the current global fit of neutrino oscillation data. However, θ_{23} deviates from the maximal value depending on whether the neutrino mass hierarchy obeys NH or IH and a range of δ_{CP} values are allowed at 3σ. This deflection from the model predicted values can be attributed to a deviation from the μ–τ reflection symmetry. A well-known model that can achieve this is the 'partial μ–τ' reflection symmetry [7]. According to this model, (65.2) applies only to a single column of the matrix U. One should note that the unitarity of U requires that if the condition holds for two columns, then it is also valid for the third one. Applying this condition to the first ($|U_{\mu 1}| = |U_{\tau 1}|$) and second ($|U_{\mu 2}| = |U_{\tau 2}|$) columns gives correlations between two major unknowns—δ_{CP} and the octant of θ_{23}

$$\cos \delta_{CP} = \frac{(c_{23}^2 - s_{23}^2)(c_{12}^2 s_{13}^2 - s_{12}^2)}{4c_{12}s_{12}c_{23}s_{23}s_{13}}, \quad c_{12}^2 c_{13}^2 = \frac{2}{3} \quad : \quad C_1, \quad (65.4)$$

$$\cos \delta_{CP} = \frac{(c_{23}^2 - s_{23}^2)(c_{12}^2 - s_{12}^2 s_{13}^2)}{4c_{12}s_{12}c_{23}s_{23}s_{13}}, \quad s_{12}^2 c_{13}^2 = \frac{1}{3} \quad : \quad C_2. \quad (65.5)$$

In this work, we refer to the prediction in (65.4) as C_1 and that in (65.5) as C_2 as they can be obtained from the symmetries Z_2 and $\overline{Z_2}$ [8, 9] whereas, $|U_{\mu 3}| = |U_{\tau 3}|$ predicts maximal θ_{23} and the CP violating phase δ_{CP} is unrestricted. A thorough study of these model predictions at the long baseline experiments will provide some insight into the unknown oscillation parameters—neutrino mass hierarchy, octant of θ_{23} and δ_{CP}. In [10–12], a similar study has been performed in the context of T2K and NOνA experiments.

In this work, we study the testability of the predictions of the 'partial μ–τ' reflection symmetry C_1 and C_2 at the upcoming long baseline experiments DUNE and HK. The detailed experimental and the simulation details are given in [13].

65.2 Results

65.2.1 Testing the Model Predictions at DUNE and HK

In this section, we present the contour plots in the true $\sin^2 \theta_{23}$(true)–δ_{CP}(true) plane for DUNE and Hyper-Kamiokande (HK) experiments. HK has proposed two alternative options for the location of the far detector. The first option Tokai-to-Hyper-Kamiokande (T2HK) is to have two 187 kt water-cherenkov detectors placed at 295 km in Kamioka while the second—T2HKK—proposes to place one 187 kt detector at Kamioka and the other at 1100 km in Korea. The analysis has been performed using General Long Baseline Experiment Simulator (GLoBES) [14, 15] and the necessary auxiliary files are obtained from [16, 17]. To do the χ^2 analysis, we obtain the true events by considering the values of the oscillation parameters as given in Table 65.1. The test events are evaluated by considering the symmetry predictions on δ_{CP} as given in (65.4) (C_1) and (65.5) (C_2). Moreover, we have marginalized over $\sin^2 \theta_{13}$, $|\Delta m_{31}^2|$, $\sin^2 \theta_{23}$ and $\sin^2 \theta_{12}$ in the test plane (ranges given in Table 65.1). We then minimize the χ^2 and obtain 1, 2 and 3 σ contours in the $\sin^2 \theta_{23}(true) - \delta_{CP}(true)$ plane for three proposed experiments DUNE, T2HK and T2HKK in Figs. 65.1, 65.2 and 65.3, respectively.

The yellow, green and the orange bands in these figures represent 1σ, 2σ and 3σ regions in the $\sin^2 \theta_{23} - \delta_{CP}$ plane, respectively. Here, the red curve represents the 3σ allowed parameter space as obtained by Nu-fit collaboration [18, 19]. They show

Table 65.1 Oscillation parameters considered in this work unless otherwise mentioned

Osc. param.	True values	Test values
$\sin^2 \theta_{13}$	0.0219	0.0197–0.0244
$\sin^2 \theta_{12}$	0.306	0.272–0.346
θ_{23}	39–51°	39–51°
Δm_{21}^2 (eV^2)	7.50×10^{-5}	Fixed
Δm_{31}^2 (eV^2)	2.50×10^{-3}	$(2.35$–$2.65) \times 10^{-3}$
δ_{CP}	$(0$–$360)°$	Symmetry predictions

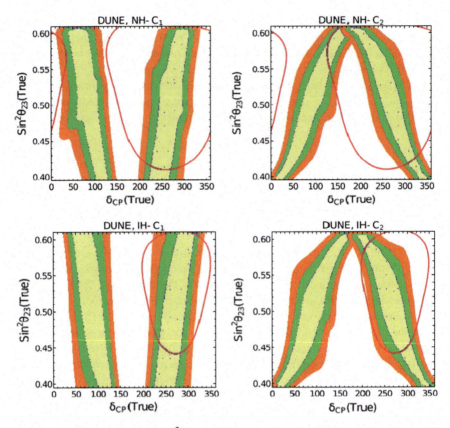

Fig. 65.1 Contour plots in the true $\sin^2\theta_{23}$(true)–δ_{CP}(true) plane for DUNE as predicted by $C_1(C_2)$ in the left and right columns. The hierarchy is fixed to NH(IH) in the upper(lower) panel. The yellow, green and orange shaded contours correspond to 1σ, 2σ and 3σ, respectively. And the red contour represents the 3σ allowed region from the global neutrino oscillation data [18, 19]

to what extent the experiments DUNE and HK can test the correlation between δ_{CP} and $\sin^2\theta_{23}$ when the symmetry predictions are taken into consideration. The left panel of each figure tests the prediction C_1 and the right panel is for testing C_2. For the plots in the upper panel of each figure, we have assumed hierarchy to be fixed and NH, whereas we have fixed IH for the plots in the lower panel.

By comparing Figs. 65.1, 65.2 and 65.3, one can infer that T2HK and T2HKK experiments constrain the δ_{CP} parameter space better than the DUNE experiment. This can be seen from the contours getting thinner as we go from Fig. 65.1 to Fig. 65.3 irrespective of whether the neutrino mass hierarchy is NH or IH. However, one should note that the allowed parameter region given by Nu-fit collaboration is more constrained in the case of IH and it is further restricted by the symmetry predictions. Also, certain regions of $\sin^2\theta_{23} - \delta_{CP}$ have been omitted by all the three experiments because of the symmetry predictions.

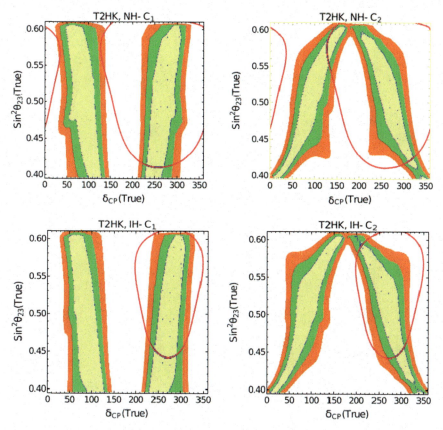

Fig. 65.2 Contour plots in the true $\sin^2\theta_{23}$(true)–δ_{CP}(true) plane for T2HK as predicted by $C_1(C_2)$ in the left and right columns. The hierarchy is fixed to NH(IH) in the upper(lower) panel. The yellow, green and orange shaded contours correspond to 1σ, 2σ and 3σ, respectively. And the red contour represents the 3σ allowed region from the global neutrino oscillation data [18, 19]

In Fig. (65.4), we plot $\Delta\chi^2$ versus true θ_{23} showing the capability of the three experiments to differentiate between the symmetries C_1 and C_2. True event spectra are obtained by varying $\sin^2\theta_{13}$ and $\sin^2\theta_{12}$ in the 3σ ranges as allowed by (65.4) and the corresponding δ_{CP} ranges are obtained using the same equation leading to two sets δ_{CP} and $(360° - \delta_{CP})$. The rest of the oscillation parameters are fixed as per Table 65.1. To obtain the test events, we consider δ_{CP} values as obtained from C_2 and marginalize over $|\Delta m_{31}^2|$, $\sin^2\theta_{13}$ and $\sin^2\theta_{23}$. The corresponding $\Delta\chi^2$ versus true θ_{23} for DUNE (left plot), T2HK (middle plot) and T2HKK (right plot) are obtained and plotted in Fig. 65.4. Here, we have assumed the hierarchy to be fixed and normal (we have verified that IH plots give similar results).

The solid (dashed) blue curves are obtained for the range $\delta_{CP} \in (0° < \delta_{CP} < 180°)$ $(360° - \delta_{CP} \in (180° < \delta_{CP} < 360°))$ as predicted by the correlations. The brown solid line represents the 3σ C.L. corresponding to $\Delta\chi^2 = 9$. It can be seen

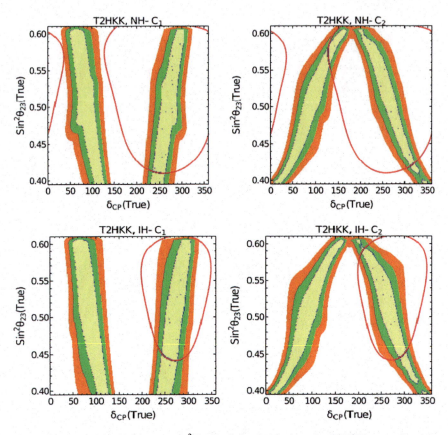

Fig. 65.3 Contour plots in the true $\sin^2\theta_{23}$(true)–δ_{CP}(true) plane for T2HKK as predicted by $C_1(C_2)$ in the left and right columns. The hierarchy is fixed to NH(IH) in the upper(lower) panel. The yellow, green and orange shaded contours correspond to 1σ, 2σ and 3σ, respectively. And the red contour represents the 3σ allowed region from the global neutrino oscillation data [18, 19]

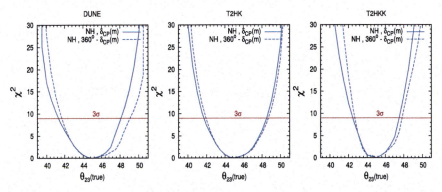

Fig. 65.4 The sensitivity of DUNE, T2HK and T2HKK experiments to differentiate between C_1 and C_2 correlations (for known hierarchy—NH)

from Fig. (65.4) that both the predictions are indistinguishable for maximal θ_{23} for all the three experiments.

In conclusion, we have studied the testability of the symmetry predictions arising from the 'partial μ–τ' reflection symmetry at the forthcoming neutrino oscillation experiments DUNE and HK. We have also evaluated the capability of these experiments to distinguish between different scenarios C_1 and C_2.

References

1. G. Altarelli, F. Feruglio, Rev. Mod. Phys. **82**, 2701 (2010). 1002.0211
2. G. Altarelli, F. Feruglio, L. Merlo, Fortsch. Phys. **61**, 507 (2013) 1205.5133
3. A. Yu. Smirnov, J. Phys. Conf. Ser. **335**, 012006 (2011). 1103.3461
4. H. Ishimori, T. Kobayashi, H. Ohki, Y. Shimizu, H. Okada, M. Tanimoto, Prog. Theor. Phys. Suppl. **183**, 1 (2010). 1003.3552
5. S.F. King, C. Luhn, Rept. Prog. Phys. **76**, 056201 (2013). 1301.1340
6. P.F. Harrison, W.G. Scott, Phys. Lett. B **547**, 219 (2002). hep-ph/0210197
7. Z.-Z. Xing, S. Zhou, Phys. Lett. **B737**, 196 (2014). 1404.7021
8. S.-F. Ge, D.A. Dicus, W.W. Repko, Phys. Lett. **B702**, 220 (2011). 1104.0602
9. S.-F. Ge, D.A. Dicus, W.W. Repko, Phys. Rev. Lett. **108**, 041801 (2012). 1108.0964
10. R. de Adelhart Toorop, F. Feruglio, C. Hagedorn, Phys. Lett. **B703**, 447 (2011). 1107.3486
11. A.D. Hanlon, S.-F. Ge, W.W. Repko, Phys. Lett. **B729**, 185 (2014a). 1308.6522
12. A.D. Hanlon, W.W. Repko, D.A. Dicus, Adv. High Energy Phys. **2014**, 469572 (2014b). 1403.7552
13. K. Chakraborty, K.N. Deepthi, S. Goswami, A.S. Joshipura, N. Nath, Phys. Rev. **D98**, 075031 (2018). 1804.02022
14. P. Huber, M. Lindner, W. Winter, Comput. Phys. Commun. **167**, 195 (2005). hep-ph/0407333
15. P. Huber, J. Kopp, M. Lindner, M. Rolinec, W. Winter, Comput. Phys. Commun. **177**, 432 (2007). hep-ph/0701187
16. M.D. Messier, Ph.D. Thesis (Advisor: James L. Stone) (1999)
17. E. Paschos, J. Yu, Phys. Rev. D **65**, 033002 (2002). hep-ph/0107261
18. I. Esteban, M.C. Gonzalez-Garcia, M. Maltoni, I. Martinez-Soler, T. Schwetz, JHEP **01**, 087 (2017). 1611.01514
19. NuFIT, NuFIT 3.2 (2018). http://www.nu-fit.org/

Chapter 66
Effect of Event-By-Event Reconstruction and Low Event Statistics on the Sensitivity of Oscillation Parameters in the INO-ICAL Detector

Karaparambil Rajan Rebin, James F. Libby, D. Indumathi, and Lakshmi S. Mohan

Abstract We study the sensitivity of the proposed INO-ICAL in determining the neutrino-oscillation parameters θ_{23} and Δm_{32}^2 using full event-by-event reconstruction for the first time. Low event statistics is a common feature among neutrino experiments. Hence, for the first time in INO, we study the fluctuations arising from low event statistics and their effect on the parameter sensitivities and mass-hierarchy determination. We obtain a mean resolution of $\Delta \chi^2 \approx 2.9$ from an ensemble of 60 experiments, which differentiates the correct mass hierarchy of the neutrinos with a significance of approximately $1.7\,\sigma$.

66.1 Introduction

The [1] Standard Model (SM) does not contain any right-handed neutrinos, and hence the neutrinos are massless by definition. Evidence for neutrino oscillations [2, 3] has proved that the neutrinos are massive, and it requires an extension of SM or theories beyond SM to explain the origin of neutrino mass. The sign of Δm_{23}^2 which determines the mass-hierarchy (MH) of neutrinos, *i.e.*, whether the MH is ($m_1 < m_2 < m_3$) normal (NH) or ($m_3 < m_1 < m_2$) inverted (IH), is yet to be determined. However, the recent results from NOνA [4] have disfavored the entire inverted mass hierarchy region at 95% CL.

The Iron Calorimeter detector (ICAL) to be built at India-based Neutrino Observatory (INO) [5] will principally measure atmospheric neutrinos. The main goals of ICAL are to determine the neutrino MH via earth matter effects and to precisely

K. R. Rebin (✉) · J. F. Libby · L. S. Mohan
Indian Institute of Technology Madras, Chennai 600 031, India
e-mail: rebinraj.k@gmail.com

D. Indumathi
The Institute of Mathematical Sciences, Chennai 600 113, India

© Springer Nature Singapore Pte Ltd. 2021
P. K. Behera et al. (eds.), *XXIII DAE High Energy Physics Symposium*,
Springer Proceedings in Physics 261,
https://doi.org/10.1007/978-981-33-4408-2_66

measure the atmospheric oscillation parameters $\sin^2 \theta_{23}$ and Δm_{32}^2. The detailed description of ICAL can be found in [5]. The most important property of the ICAL will be its ability to discriminate the charge of particles using the magnetic field, hence it can distinguish between ν_μ and $\bar{\nu}_\mu$ events by observing the charge of the final state muons produced in charged current (CC) ν_μ interactions. Thus, the ICAL could study the MH by observing earth-matter effects independently on ν_μ and $\bar{\nu}_\mu$ events.

In the following sections, we will briefly discuss the methodology and the analysis procedure followed to determine the oscillation parameters $\sin^2 \theta_{23}$ and Δm_{32}^2. The detailed description can be found in [1].

66.2 Determination of the Oscillation Parameter Sensitivity

NUANCE [6], a neutrino event generator, is used to generate neutrino events corresponding to an exposure of 50 kton $\times$ 1000 years, and they are simulated within a virtual ICAL detector using the GEANT4-based [7] simulation toolkit. The information on energy loss and momentum of the secondary particles is obtained from GEANT4, which are then digitized to form (x, z) or (y, z) and time t of the signal, referred to as hits. The $\mu^\pm$ forms a well-defined track within the detector, and is fit to obtain the direction and momentum of the muon as the observables.

Separate sub-samples corresponding to 5 and 995 years are created out of the 1000-year sample, where the 5 years of data are used as the experimentally simulated sample and the remaining 995 years of data are used to construct probability distribution functions (PDF) that are used in the χ^2 fit. Hence, the PDFs that are used to fit the pseudo-data are completely uncorrelated, and the 5-year sample is naturally fluctuated due to the low event statistics.

66.2.1 Event Selection

The reconstruction of muons is adversely affected by the non-uniform magnetic field and dead spaces within the detector. Hence, we apply event selection to obtain a better reconstructed sample of data. A selection based on the χ^2 estimate obtained from the fit to the muon tack is used to remove poorly reconstructed events. Further, the event selection is also applied on the basis of magnetic field strength, where the entire ICAL is classified into central, side, and peripheral regions. A detailed description on the magnetic field strength, dimension, and the selection applied in each region can be found in [1].

The muon energy and the zenith angle resolutions show an overall improvement of 23% and 19%, respectively, after the event selection. The muon charge identification efficiency also shows an improvement of $\sim 6 - 10\%$ at all muon energies. However, the reconstruction efficiency decreases, as the event selection removes $\approx 40\%$ of the reconstructed events [1]. Hence, we also study the effect of event selection on the parameter determination.

66.2.2 Binning and χ^2 Analysis

The oscillation probabilities calculated from numerically evolving the neutrino flavor eigenstates [8] are used to apply the oscillations via the accept or reject method [1]. The 5-year pseudo-data would have negligible contribution from ν_e flux (Φ_{ν_e}) compared to ν_μ flux (Φ_{ν_μ}). Hence, only Φ_{ν_μ} events are binned in our analysis. To observe the matter effects separately in ν and $\bar{\nu}$ events, the information on the reconstructed muons with negative and positive charges are binned separately in $Q_\mu E_\mu$ and $\cos\theta_z$ bins after applying oscillations, where $Q_\mu = \pm 1$ for $\mu^\pm$.

After binning, the 5-year simulated data set is fit by defining the following χ^2 [1]:

$$\chi^2 = \min_{\{\xi_k\}} \sum_{i=1}^{n_{\cos\theta_z}} \sum_{j=1}^{n_{E_\mu}} 2 \left[\left(N_{ij}^{\text{pdf}} - N_{ij}^{\text{data}} \right) - N_{ij}^{\text{data}} \ln \left(\frac{N_{ij}^{\text{pdf}}}{N_{ij}^{\text{data}}} \right) \right] + \sum_{k=1}^{2} \xi_k^2, \quad (66.1)$$

where

$$N_{ij}^{\text{pdf}} = R \left[f T_{ij}^{\bar{\nu}} + (1-f) T_{ij}^{\nu} \right] \left[1 + \sum_{k=1}^{2} \pi_{ij}^k \xi_k \right]. \quad (66.2)$$

Here, the observed and the expected number of muon events are given by N_{ij}^{data} and N_{ij}^{pdf}, respectively. The true values of oscillation parameters are used to calculate N_{ij}^{data}, whereas N_{ij}^{pdf} is obtained by combining ν_μ and $\bar{\nu}_\mu$ PDFs given by T_{ij}^{ν} and $T_{ij}^{\bar{\nu}}$ in (66.2). Here, f is the free parameter which describes the relative fraction of $\bar{\nu}_\mu$ and ν_μ in the sample, with R being a normalization factor. The theoretical and systematic uncertainties are parametrized in terms of variables $\{\xi_k\}$ called pulls. We have considered a 5% uncertainty on the zenith angle dependence of the flux and another 5% on the energy dependent tilt error [1].

66.2.3 Parameter Determination

The parameters $\sin^2\theta_{23}$ and Δm_{32}^2 are correlated; Fig. 66.1 compares the correlated precision reach obtained by fitting a 5-year pseudo-data set with (WS) and without (WOS) event selection.

We define the significance of the fit as significance $= \sqrt{\chi_{\text{input}}^2 - \chi_{\text{min}}^2}$, where χ_{input}^2 and χ_{min}^2 are the χ^2 values at the true and observed values of the parameter, respectively. The best-fit point of the fit without event selection is obtained within a

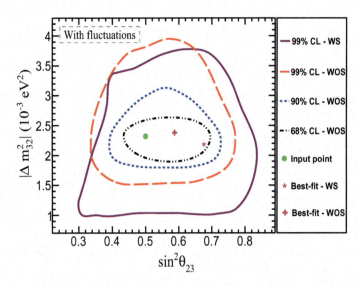

Fig. 66.1 Precision reach obtained from the fit to a 5-year fluctuated pseudo-data set in $\sin^2 \theta_{23}$ − Δm^2_{32} plane [1]

significance of 1σ from the input value, whereas the fit with event selection converges within a significance of 2σ. Note that the fit with event selection shows larger coverage at 99% CL due to larger statistical uncertainty, as the sample size was reduced by 40%.

66.2.3.1 Effect of Fluctuations

Earlier analyses [5] nullified the effect of fluctuations by scaling the 1000-year sample to a size corresponding to 5 years of data. Figure 66.2a compares the fit without (WOF) fluctuations to the fit to three independent fluctuated data sets (WF: 1, WF: 2, and WF: 3), in the $\sin^2 \theta_{23}$ − Δm^2_{32} plane. The fluctuations in the data induce fluctuations in the resultant best-fit point and the coverage area obtained from the fit. The significance of the convergence also changes along with the result of each fluctuated pseudo-data set.

The analysis was repeated for sixty different fluctuated data sets, and Fig. 66.2b shows the significance of convergence in terms of the standard deviation σ. Almost 68% of times, the fit converges within 1σ of the input value. Hence, it evidently shows the Gaussian nature of the fit, and confirms that there are no biases in the experiment or analysis procedure. Secondly, it also shows the range of best-fit values that is feasible for a 5-year run of ICAL.

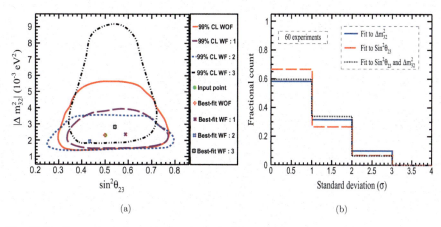

Fig. 66.2 **a** Comparison of precision reach obtained from the fit with and without fluctuations [1], and **b** significance of convergence obtained from sixty different data sets [1]

66.2.4 Mass Hierarchy Determination

Oscillations are applied on the 5-year pseudo-data set assuming NH (IH), and fit to true NH (IH) and false IH (NH) PDFs, where $\Delta\chi^2_{MH} = \chi^2_{false} - \chi^2_{true}$ is the observed resolution to identify and differentiate the correct hierarchy. The procedure is repeated for sixty different data sets to see the effect of fluctuations. Figure 66.3 shows the distribution of $\Delta\chi^2_{MH}$ obtained from an ensemble of sixty experiments. The mean resolution of $\Delta\chi^2_{MH} = 2.9$ rules out the wrong hierarchy with a significance of $\approx 1.7\sigma$. We also obtain a 15% probability of identifying the wrong MH [1].

Fig. 66.3 Distribution of $\Delta\chi^2_{MH}$ obtained from fit to sixty fluctuated data sets [1]

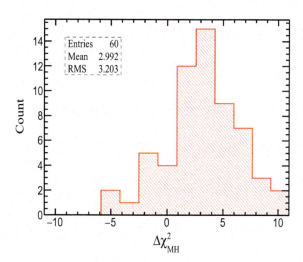

66.3 Conclusions

We have incorporated a realistic analysis procedure of ICAL data for the first time by applying event-by-event reconstruction. Also for the first time, we study the effect of low event statistics on the precision and MH measurements, by introducing fluctuations in the data. Also for the first time, we show the effect of event selection criterion on the parameter sensitivities, and show that we can include all reconstructed muons to get better sensitivity of parameters. A combined analysis including all the CC and NC events along with the hadron information will give us the maximum sensitivity the ICAL can attain, and it is an ongoing effort which is likely to improve our results.

References

1. K.R. Rebin, J. Libby, D. Indumathi, L.S. Mohan, Study of neutrino oscillation parameters at the INO-ICAL detector using event-by-event reconstruction. Eur. Phys. J. C **79**, 295 (2019)
2. B. Aharmim et al. [SNO Collaboration], Low-energy-threshold analysis of the phase I and phase II data sets of the Sudbury Neutrino Observatory, Phys. Rev. C **81**, 055504 (2010). [nucl-ex] arXiv:0910.2984
3. R. Wendell et al. [The Super-Kamiokande Collaboration], Atmospheric neutrino oscillation analysis with subleading effects in Super-Kamiokande I, II, and III, Phys. Rev. D **81**, 092004 (2010). arXiv:1002.3471 [hep-ex]
4. M.A. Acero et al., [NOvA Collaboration], New constraints on oscillation parameters from ν_e appearance and ν_μ disappearance in the NOvA experiment. Phys. Rev. D **98**, 032012 (2018)
5. A. Kumar et al. [ICAL Collaboration], Invited review: Physics potential of the ICAL detector at the India-based Neutrino Observatory (INO), Pramana **88**, 79 (2017). [physics.ins-det] arXiv:1505.07380
6. D. Casper, The nuance neutrino physics simulation, and the future. Nuclear Phys. B - Proc. Suppl. **112**, 161 (2002). arXiv:hep-ph/0208030
7. S. Agostinelli, Geant4—a simulation toolkit. Nucl. Instrum. Meth. A **506**, 250 (2003)
8. D. Indumathi, M.V.N. Murthy, G. Rajasekaran, N. Sinha, Neutrino oscillation probabilities: Sensitivity to parameters. Phys. Rev. D **74**, 053004 (2006). arXiv:hep-ph/0603264

Chapter 67
Constraints on Millicharged Particles and Bosonic Dark Matter Using Germanium Detectors With Sub-keV Sensitivity

Lakhwinder Singh

Abstract Germanium ionization detectors with their unique features and diverse applications in fundamental research are novel candidates for the search of exotic particles. The TEXONO Collaboration aims to progressively improve the sensitivities toward exotic particles like low energy neutrinos, light dark matter candidates, and relativistic millicharged particles at the Kuo-Sheng Neutrino Laboratory (KSNL) in Taiwan. Relativistic millicharged particles (χ_q) have been proposed in various extensions to the Standard Model of particle physics. We present the direct constraints on χ_q with low threshold point-contact germanium detectors under the scenarios of χ_q produced at (i) nuclear power reactors, (ii) as products of cosmic-ray interactions, and (iii) as dark matter particle accelerated by supernova shock. The atomic ionization cross sections of χ_q with matter are derived with the equivalent photon approximation. Smoking-gun signatures with significant enhancement in the differential cross section are identified. We also report results from searches of pseudoscalar and vector bosonic super-weakly interacting massive particles (super-WIMP) using 314.15 kg days of data from an n-type Point-Contact Germanium detector.

67.1 Introduction

Several astrophysical and cosmological independent observations on a wide range of length scales conclude that exotic dark matter is one of the basic ingredients of the universe. The fundamental nature of exotic dark matter has not been established beyond its gravitational effects. The identification of the nature of dark matter is one of the most important challenges in the post-Higgs era. Weakly interacting massive particles (WIMPs), axions or axionlike-particles (ALPs), sterile neutrinos, millicharged particles, and bosonic super-weakly interacting particles are exotic can-

---

On behalf of the TEXONO Collaboration.

L. Singh (✉)
Institute of Physics, Academia Sinica, Taipei, Taiwan
e-mail: lakhwinder@gate.sinica.edu.tw

© Springer Nature Singapore Pte Ltd. 2021
P. K. Behera et al. (eds.), *XXIII DAE High Energy Physics Symposium*,
Springer Proceedings in Physics 261,
https://doi.org/10.1007/978-981-33-4408-2_67

didates of dark matter, which naturally arise in many extensions of the standard model (SM) of particle physics. Searches for these leading candidates are in full swing, but an experimental verification via direct, indirect detection or production from LHC is still awaited. The millicharged particles denoted by χ_q with mass m_{χ_q} can be obtained via including an extra abelian gauge $U_{HS}(1)$ (the subscript denotes "Hidden Sector") into the Standard Model (SM) gauge groups [1]. The SM particles are not charged under this new gauge group, while the χ_q under $U_{HS}(1)$ acquire small electric charge (δe_0) due to the kinetic mixing of SM photon and HS dark photon, where δ is the charge fraction of χ_q and e_0 is the standard electron charge.

Point-Contact Germanium (PGe) detectors with their excellent energy resolution, sub-keV threshold, and low intrinsic radioactivity background are the best candidates to study exotic physics beyond SM [2]. The TEXONO Collaboration [3] is pursuing the research programs on neutrino electromagnetic properties [4, 5], νN coherent scattering [6], and beyond SM at the Kuo-Sheng Neutrino Laboratory (KSNL). A detailed description of the KSNL facilities can be found in [7, 8].

67.2 Millicharged Particles

A hidden sector with massless gauge boson allows possibilities of multicomponent dark matter (DM). Its ionic constituents can acquire small charges (millicharge, δe_0) under the hidden sector gauge group. We consider the three scenarios where they can be produced: (1) The light-χ_q can be produced through Compton-like processes, where γ-rays of $\mathcal{O}$(MeV) energy scatter off electrons in the nuclear reactor core. The differential χ_q-flux (ϕ_{χ_q}) is determined from the convolution of reactor γ-ray spectrum and differential production cross section normalized by the total cross section (σ_{tot}) [9],

$$\frac{\mathrm{d}\phi_{\chi_q}}{\mathrm{d}E_{\chi_q}} = \frac{2}{4\pi R^2} \int \frac{1}{\sigma_{tot}} \frac{\mathrm{d}\sigma}{\mathrm{d}E_{\chi_q}} \frac{\mathrm{d}N_\gamma}{\mathrm{d}E_\gamma} \mathrm{d}E_\gamma, \qquad (67.1)$$

where R is a distance of the detector from the center of the reactor core. The factor 2 in (67.1) comes from the fact that χ_q particles produce in pairs and both can interact in the detectors. (2) High energy cosmic-rays can produce relativistic-χ_q when they interact with a nucleus in the earth's atmosphere. The investigation of cosmic-rays can provide constraints on χ_q via interactions between χ_q and detectors. The energy loss of χ_q through excitation and ionization is proportional to δ^2, which is much lower than the minimum ionizing particles of unit charge under similar conditions. The mass range of cosmogenic produced χ_q is unknown due to unknown production conditions. Therefore, the experimental sensitivity is usually expressed in terms of the integral incoming flux (I_{χ_q}) in the units of cm^{-2} s^{-1} sr^{-1} as a function of δ. (3) Millicharged particles could also be candidates for DM, and become relativistic through acceleration by supernova explosion shock waves. The multicomponent dark sector is an interesting scenario which may have both neutral and ionized components.

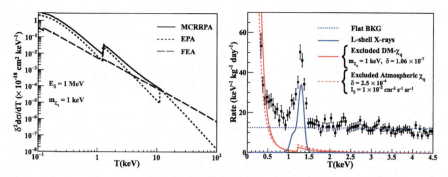

Fig. 67.1 a The differential scattering cross sections of Ge-ionization by χ_q with $m_{\chi_q} = 1$ keV, $\delta = 1$ and monochromatic $E_{\chi_q} = 1$ MeV are derived for FEA (solid line) and EPA (dashed line). **b** The total $AC^- \otimes CR^-$ spectrum showing a flat background due to ambient high-energy γ-rays and the L-shell X-rays from internal radioactivity. Excluded scenarios of atmospheric and DM-χ_q at specified ($\delta = 2.5 \times 10^{-4}$, $I_\delta = 1 \times 10^{-2}$ cm^{-2}s^{-1}sr^{-1}) and ($m_{\chi_q} = 1$ keV, $\delta = 1.06 \times 10^{-7}$), respectively, are superimposed

The latter component may be due to incomplete recombination of primordial DM gas and re-ionization by sources such as starlight and supernova explosions which can efficiently overcome the binding energy of dark atoms. The ionized components of dark matter (χ_q) can be accelerated in the shock wave of supernova. The maximum energy attained by χ_q of charge fraction δ is simply a product of the rate of energy gain and time spent in shock.

The millicharged χ_q's are relativistic and can interact electromagnetically with matter via atomic ionization

$$\chi_q + A \rightarrow \chi_q + A^+ + e^-, \qquad (67.2)$$

through the t-channel process. The main challenging part of the calculation of the differential cross section is the transition matrix elements which involve many-body initial and final states. The transition matrix elements can, in principle, be calculated by many-body wave functions. In practice, for most cases, the calculations are highly non-trivial and schemes such as free electron approximation (FEA) and equivalent photon approximation (EPA) provide good estimations at certain kinematic regions as shown in Fig. 67.1a. Although the EPA is a good approximation in the most interesting sub-keV region of T, it underestimates the scattering cross section above a few keV regions of T where FEA works well. The EPA and FEA schemes therefore serve as conservative approximations in the region near and away from ionization thresholds, respectively.

The expected differential count rates due to possible χ_q interaction with matter are obtained by integrating the χ_q-flux with the differential cross sections:

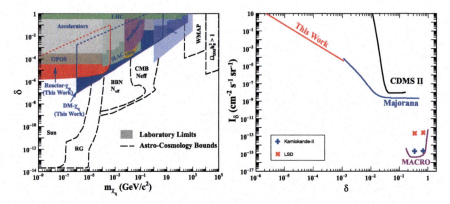

Fig. 67.2 a Exclusion regions at 90% C.L. in (m_{χ_q}, δ) parameter space for millicharged particles with a massless dark photon. The excluded regions of this work with χ_q from the reactor and dark primary cosmic-rays are shown as red and blue shaded areas, respectively. The dotted lines correspond to the upper bounds of the exclusion regions, due to complete attenuation of χ_q before reaching the detector. Cosmological and astrophysical bounds are denoted as dotted lines. The direct laboratory limits from other benchmark experiments are represented as shaded regions. **b** Excluded parameter space at 90% C.L. on incoming flux of χ_q from secondary dark cosmic-rays versus its charge fraction δ

$$\frac{dR}{dT} = \rho_A \int_{E_{\min}}^{E_{\max}} \left[\frac{d\sigma}{dT}\right] \left[\frac{d\phi_{\chi_q}}{dE_{\chi_q}}\right] dE_{\chi_q}, \tag{67.3}$$

where ρ_A is atomic number density per unit target mass and $(E_{\min}, E_{\max})$ are the (minimum, maximum) energy of χ_q. Constraints from each of the three discussed χ_q channels are derived from the measured $AC^- \otimes CR^-$ spectra after subtraction of (i) internal radioactivity due to K/L-shell X-rays from cosmogenically activated isotopes in the Ge-target and (ii) a flat background estimated from ambient high-energy γ-rays, following background understanding and analysis procedures from earlier work on similar detectors and configurations [10, 11]. Improved limits at 90% (confidence level) CL on δ as a function of particle mass are derived using the KSNL data for all three scenarios. The new excluded regions of this work with χ_q from the reactor and dark primary cosmic-rays are shown in Fig. 67.2a as red and blue shaded areas, respectively [9]. The upper limits on I_{χ_q} as a function of δ are depicted in Fig. 67.2b and constraints from previous experiments are also superimposed. The sub-keV sensitivity of the PGe detector leads to improved direct limits at a small mass of millicharged particles and extend the lower reach of δ to 10^{-6}.

67.3 Bosonic Dark Matter

Bosonic super-WIMP is a well-motivated class of DM with coupling smaller than the weak scale. These particles are experimentally very interesting due to their absorption via the ionization or excitation of an electron in the target-atom of the detector. Bosonic super-WIMP would deposit energy equivalent to their rest mass, which manifest as a photo-peak. Therefore, a good energy resolution device like PGe detectors has advantages to study such class of DM. Pseudoscalar, scalar, and vector are three generic possible candidates of nonrelativistic LDM that may have a superweak coupling with SM particles. The correct relic density of bosonic LDM in a wider mass range could be obtained via either thermal or non-thermal misalignment mechanism. The bosonic pseudoscalar (χ_s) are excellent candidates of LDM. The phenomenology behind χ_s is similar to nonrelativistic ALPs. The χ_s have coupling to atoms through the axioelectric effect which is analogous to the photoelectric effect with the absorption of χ_s instead of a photon. The absorption cross section σ_{abs} (axio-electric effect) for χ_s can be written as

$$\sigma_{abs} \simeq \frac{3m_{ps}^2}{4\pi \alpha f_a^2 \beta} \sigma_{pe}(w = m_{ps}), \tag{67.4}$$

where σ_{pe} is the photoelectric cross section with the photon energy ω replaced by the mass of χ_s (m_{ps}), $f_a = 2m_e/g_{aee}$ is the dimensionless coupling strength of χ_s to SM particles and $\beta \equiv v_\chi/c$. The absorption cross section of χ_s is directly proportional to m_{ps}^2.

The best motivated model for bosonic vector dark matter (χ_v) is the kinetic mixing model, in which an extra U(1)$_D$ gauge group is introduced into the SM gauge group. The kinetic mixing with the hypercharge field strength is responsible for the interaction between the ordinary matter and χ_v. The absorption cross section of χ_v (σ_{abs}) can be expressed in the photoelectric effect with the replacement of photon energy ω by the mass m$_v$ of χ_v, and the coupling constant is scaled appropriately as

$$\frac{\sigma_{abs}}{\sigma_{pe}(\omega = m_v)} \simeq \frac{\alpha'}{\alpha} \frac{1}{\beta}, \tag{67.5}$$

where $\alpha' = (e\kappa)^2/4\pi$ is similar to a vector-electric fine-structure constant. κ is the vector hypercharge.

The theoretically expected interaction rate of χ in a direct detection experiment can be expressed as

$$R_\chi = \rho_d \sigma_{abs} \Phi_\chi, \tag{67.6}$$

where $\rho_d = N_A/A$ is the atomic number density per unit target mass of detector and N_A is Avogadro's number. σ_{abs} is the absorption cross section of χ. $\Phi_\chi = \rho_\chi v_\chi/m_\chi$ is total average flux of χ with the assumption that the bosonic DM constitutes all of the galactic DM with local DM density $\rho_\chi = 0.3$ GeV/cm^3.

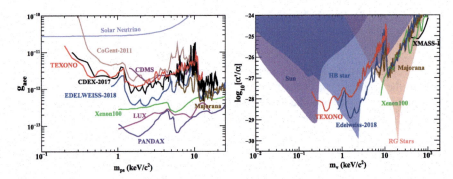

Fig. 67.3 **a** Limits on the coupling of pseudoscalar super-WIMP with electron as a function of mass from various benchmark experiments at 90% C.L. **b** The 90% C.L. bounds on vector bosonic DM coupling from different astrophysical sources as well as other benchmark experiments

Using the low-background and low-threshold data from KNSL with PGe detectors, improved limits at 90% CL on the coupling constant of pseudoscalar (g_{aee}) and vector super-WIMP (α'/α) as a function of particle mass are derived [12]. The upper bounds on coupling constant g_{aee} and α'/α at various mass are shown in Fig. 67.3a, b, respectively

67.4 Conclusions

Ge detectors with sub-keV threshold and excellent energy resolution have been used to study neutrino properties and interactions, neutrino nucleus coherent scattering, WIMP, millicharged particle, and bosonic DM searches. Data taking and analysis are continuing at KSNL. Intensive R&D programs are being pursued to understand and suppress the sub-keV background.

References

1. B. Holdom Phys. Lett. B 166, 196 (1986)
2. A.K. Soma et al., (TEXONO Collaboration) Nucl. Instrum. Meth. **A836**, 67–82 (2016)
3. H.T. Wong, Int J Modern Phys A **33**, 1830014 (2018)
4. J.-W. Chen et al., Phys. Rev. D **90**, 011301 (2014)
5. J.-W. Chen et al., Phys. Rev. D **91**, 013005 (2015)
6. S. Kerman et al., Phys. Rev. D **93**, 113006 (2016)
7. H.B. Li et al., (TEXONO Collaboration) Phys. Rev. Lett. **90**, 131802 (2003)
8. H.T. Wong et al., (TEXONO Collaboration) Phys. Rev. D **75**, 012001 (2007)
9. L. Singh et al., (TEXONO Collaboration). Phys. Rev. D **99**, 032009 (2019)
10. H.B. Li et al., (TEXONO Collaboration) Phys. Rev. Lett. **110**, 261301 (2013)
11. H. Jiang et al., (CDEX Collaboration) Phys. Rev. Lett. **120**, 241301 (2018)
12. M.K. Singh et al., (TEXONO Collaboration) Chin. J. Phys. **58**, 63–74 (2019)

Chapter 68
Probing Leptonic δ_{CP} Using Low Energy Atmospheric Neutrinos

D. Indumathi, S. M. Lakshmi, and M. V. N. Murthy

Abstract The possibility of probing leptonic δ_{CP} via oscillations of low energy atmospheric neutrinos is explored. The measurement of δ_{CP} is not very easy if the hierarchy of neutrino masses is unknown. Dedicated accelerator long baseline neutrino experiments like DUNE can determine δ_{CP} without ambiguity with hierarchy. But low energy atmospheric neutrino events also can be used to determine δ_{CP} independent of neutrino mass hierarchy in the energy range 0.1–1.5 GeV. This is also important since the atmospheric neutrino flux peaks in the low energy region and hence will provide a significant number of events for the study. A simple analytical derivation is used to show how the δ_{CP} sensitivity arises and what the relevant energies and baselines are. Though the ν_e and $\overline{\nu}_e$ events are mainly sensitive to δ_{CP}, we also show that in the low energy region, even the ν_μ and $\overline{\nu}_\mu$ events can contribute to δ_{CP}.

68.1 Introduction

There is a hint that the value of the leptonic Dirac CP violating phase δ_{CP} is in the range $\approx -145°$ $(-76°)$ for normal (inverted) hierarchy [1]. Many accelerator-based long baseline experiments (LBL) are taking data/are being installed to measure this parameter precisely [2–4]. In addition to these, low energy atmospheric neutrinos

D. Indumathi · M. V. N. Murthy
The Institute of Mathematical Sciences, Chennai 600 113, India
e-mail: indu@imsc.res.in

M. V. N. Murthy
e-mail: murthy@imsc.res.in

D. Indumathi
Homi Bhabha National Institute, Training School Complex, Anushakti Nagar, Mumbai 400085, India

S. M. Lakshmi (✉)
Indian Institute of Technology Madras, Chennai 600 036, India
e-mail: slakshmi@physics.iitm.ac.in

© Springer Nature Singapore Pte Ltd. 2021
P. K. Behera et al. (eds.), *XXIII DAE High Energy Physics Symposium*,
Springer Proceedings in Physics 261,
https://doi.org/10.1007/978-981-33-4408-2_68

can be used to measure δ_{CP}. Although the flux of atmospheric neutrinos cannot be controlled and are less than those of the accelerator neutrinos, they provide an independent way of determining δ_{CP}. The major advantage of atmospheric neutrinos is that they span a large range of L/E, where L is the distance travelled by the neutrino in km and E is its energy in GeV. Atmospheric neutrino flux peaks in the sub-GeV energy range [5–7], and there will be a good number of events in this energy range which can be used to study various oscillation parameters including δ_{CP}. In this proceeding, we discuss the possibility of probing δ_{CP} using low energy atmospheric neutrinos. An analytic calculation as well as the event spectra binned in the final state lepton energy and direction illustrate that δ_{CP} can be measured irrespective of neutrino mass hierarchy. A preliminary χ^2 analysis assuming a perfect detector with 100% efficiencies, perfect resolutions and ability to separate ν_e, $\bar{\nu}_e$, ν_μ and $\bar{\nu}_\mu$ events is done.

68.2   Hierarchy Independence at Low Energies

At low energies, there is no hierarchy ambiguity for atmospheric neutrinos and hence δ_{CP} can be measured irrespective of hierarchy. The 3-flavour vacuum oscillation probability of a flavour $\nu_\alpha \to \nu_\beta$ is given by

$$\overset{(-)vac}{P}_{\alpha\beta} = \delta_{\alpha\beta} - 4\sum_{i>j} Re\left[U_{\alpha i}U_{\beta i}^* U_{\alpha j}^* U_{\beta j}\right]\sin^2\left(\frac{1.27\Delta m_{ij}^2 L}{E}\right) \tag{68.1}$$

$$\pm 2\sum_{i>j} Im\left[U_{\alpha i}U_{\beta i}^* U_{\alpha j}^* U_{\beta j}\right]\sin\left(\frac{2.53\Delta m_{ij}^2 L}{E}\right),$$

where $\alpha, \beta = e, \mu, \tau$ are flavour indices, and the $\pm$ sign corresponds to neutrinos and anti-neutrinos, respectively; $i, j = 1, 2, 3$ represent the mass eigenstates, $\Delta m_{ij}^2 = m_i^2 - m_j^2$ ($j < i$), m_i the mass of ν_i.

The 3×3 mixing matrix in vacuum is

$$U_{\alpha i}^{vac} = \begin{pmatrix} c_{12}c_{13} & s_{12}c_{13} & s_{13}e^{-i\delta} \\ -c_{23}s_{12} - s_{23}c_{12}s_{13}e^{i\delta} & c_{23}c_{12} - s_{23}s_{12}s_{13}e^{i\delta} & s_{23}c_{13} \\ s_{23}s_{12} - c_{23}c_{12}s_{13}e^{i\delta} & -s_{23}c_{12} - c_{23}s_{12}s_{13}e^{i\delta} & c_{23}c_{13} \end{pmatrix},$$

where $c_{ij} = \cos\theta_{ij}, s_{ij} = \sin\theta_{ij}$; θ_{ij} are the mixing angles and δ_{CP} is the leptonic CP violation phase. Here, L (in km) is the distance travelled by a neutrino of energy E (in GeV).

While the survival probability $P_{\alpha\alpha}$ has no imaginary part, the transition probabilities $\alpha \neq \beta$ have different signs for the the imaginary part with $P_{\alpha\beta} = \overline{P}_{\beta\alpha}$, the corresponding anti-neutrino probability. For small E, of the order of a few hundred MeV, the corresponding oscillatory terms average out whenever L/E is large compared to Δm_{ij}^2. Since $|\Delta m_{3j}^2| \sim 2.4 \times 10^{-3}$ eV$^2 \gg \Delta m_{21}^2 \sim 7.6 \times 10^{-5}$ eV^2, $j = 1, 2$, this applies to the "atmospheric" terms:

$$1.27\Delta m_{3j}^2 \frac{L}{E} \approx \pi \frac{(L/100\ \text{km})}{(E/0.1\ \text{GeV})}, \tag{68.2}$$

rather than to "solar" terms:

$$1.27\Delta m_{21}^2 \frac{L}{E} \approx \pi \frac{(L/3000\ \text{km})}{(E/0.1\ \text{GeV})}. \tag{68.3}$$

Hence the atmospheric event rates at these low energies with $L \geq$ a few 100 km are independent of Δm_{32}^2 and Δm_{31}^2 and hence of mass ordering. Δm_{21}^2 remains, but its magnitude and sign are well known.

The transition probability in vacuum can be expressed as

$$P_{\alpha\beta}^{vac} = -4Re[U_{\alpha2}U_{\beta2}^*U_{\alpha1}^*U_{\beta1}]\sin^2(1.27\Delta m_{21}^2 L/E) \tag{68.4}$$
$$-2Re[U_{\alpha3}U_{\beta3}^*(\delta_{\alpha\beta} - U_{\alpha3}^*U_{\beta3})] \tag{68.5}$$
$$+2Im[U_{\alpha2}U_{\beta2}^*U_{\alpha1}^*U_{\beta1}]\sin(2.53\Delta m_{21}^2 L/E). \tag{68.6}$$

Since the probability is independent of Δm_{32}^2, there is no hierarchy ambiguity.

$$P_{e\mu} = A + B\cos\delta - C\sin\delta = \overline{P}_{\mu e}; \quad P_{\mu e} = A + B\cos\delta + C\sin\delta = \overline{P}_{e\mu}, \tag{68.7}$$

where,

$$A = c_{13}^2 \sin^2(2\theta_{12})(c_{23}^2 - (s_{23}s_{13})^2)\sin^2(\delta_{21}/2) + \frac{1}{2}s_{23}^2\sin^2(2\theta_{13}),$$
$$B = (1/4)c_{13}\sin(4\theta_{12})\sin(2\theta_{13})\sin(2\theta_{23})\sin^2(\delta_{21}/2),$$
$$C = (1/4)c_{13}\sin(2\theta_{12})\sin(2\theta_{13})\sin(2\theta_{23})\sin(\delta_{21}),$$
$$\delta_{21} = 2.534\Delta m_{21}^2 L/E.$$

A, B, C are limited only by precision measurements of oscillation parameters. The CP asymmetry can be expressed as

$$A_{CP} = \frac{P_{e\mu} - P_{\mu e}}{P_{e\mu} + P_{\mu e}} = -\frac{C}{A + B\cos\delta}\sin\delta; \quad \overline{A}_{CP} = \frac{\overline{P}_{e\mu} - \overline{P}_{\mu e}}{\overline{P}_{e\mu} + \overline{P}_{\mu e}} = \frac{C}{A + B\cos\delta}\sin\delta \tag{68.8}$$

for ν and $\overline{\nu}$, respectively. The oscillation parameters and hence the probabilities will get modified in presence of Earth matter.

68.2.1 Events Spectra at Low and Higher Energies

The oscillated events when binned in the energy of the incident neutrino E_ν (GeV) follow the oscillation probabilities as in Fig. 68.1.

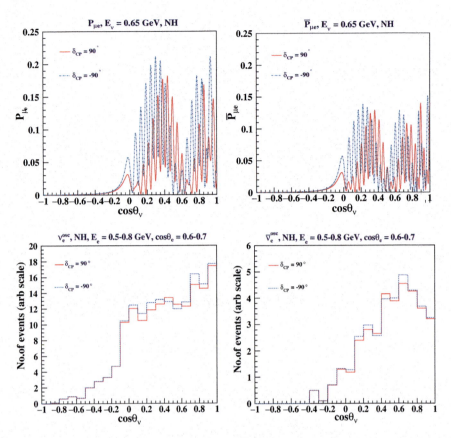

Fig. 68.1 Transition probabilities (left-set) $P_{\mu e}$ and $\overline{P}_{\mu e}$ as a function of $\cos\theta_\nu$ for $E_\nu = 0.65$ GeV and normal mass hierarchy and $\delta_{CP} = +90°, -90°$ and NH. (Right-set) Oscillated ν_e and $\overline{\nu}_e$ events as a function of neutrino direction, $\cos\theta_\nu$, for events with final lepton energy, $E_l = 0.5$–0.8 GeV and direction $\cos\theta_l = 0.6$–0.7, with $\delta_{CP} = \pm 90°$ and NH. Note that y-axes are different

When the event spectra are binned in final state lepton energy E_l as shown in Fig. 68.2, the hierarchy (δ_{CP}) independence at lower (higher) energies can be seen clearly. This property can be made use of in determining δ_{CP} irrespective of hierarchy at low energies and vice versa. The top (bottom) panels show variation w.r.t. δ_{CP} at lower (higher) energies. It can also be seen that for a given δ_{CP}, the NH and IH spectra overlap indicating hierarchy independence. In the higher energy range, due to matter effects, hierarchy can be determined. For a given hierarchy, spectra with all δ_{CP} values overlap showing that hierarchy can be determined irrespective of δ_{CP}.

The distribution of oscillated events with different δ_{CP} values as a function of $\cos\theta_l$, the final state lepton direction clearly indicates the effect of various δ_{CP} values as shown in Fig. 68.3. At very low E_l, i.e., 0.2–0.4 GeV, the distinction between various δ_{CP} values is present in the entire $\cos\theta_l$ range. As the energy increases to

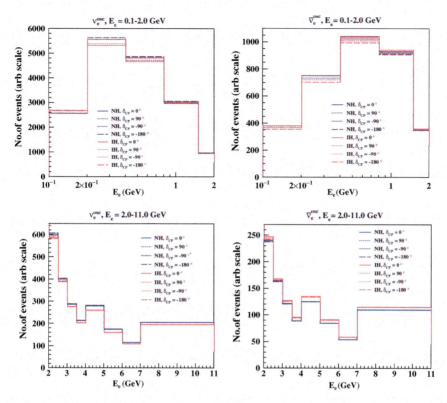

Fig. 68.2 Oscillated events with different δ_{CP} values in the E_e range (left-set) 0.1–2.0 GeV and (right-set) 2.0–11.0 GeV for normal and inverted hierarchies. The left panels are for ν events and the right ones are for $\overline{\nu}$ events. Note that the y-axes are kept different for visibility

0.5–0.8 GeV, the distinction is more prominent in the up direction $\cos\theta_l = [0, 1]$. Hence, low energy neutrinos travelling large L can help us probe δ_{CP}.

68.3 χ^2 Analysis With the Ideal Case

A sensitivity study to δ_{CP} with atmospheric neutrinos in the energy range 0.1–30 GeV is conducted assuming a detector with perfect resolutions and 100% efficiency. This detector is also assumed to be able to separate ν_e, $\overline{\nu}_e$, ν_μ and $\overline{\nu}_\mu$ events which are relevant for the analysis. No systematic uncertainties are considered here. 100 years for unoscillated events generated using the NUANCE [8] neutrino generator are scaled down to 10 years after applying oscillations event by event. The central values and 3σ of the oscillation parameters used are listed in Table 68.1.

The number of charged current (CC) ν_e events detected is given by

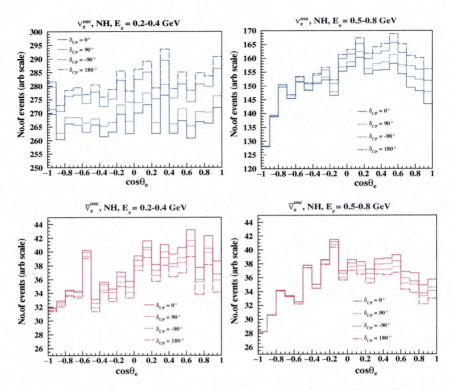

Fig. 68.3 ν_e and $\bar{\nu}_e$ events for different true δ_{CP} values as a function of $\cos\theta_l^{obs}$ for $E_l^{obs} = 0.2$–0.4 and 0.5–0.8 GeV; $l = e$. The Y-axes are not the same

Table 68.1 True values and 3σ ranges of parameters used to generate oscillated events. Values except that of δ_{CP} are taken as in [9]. For the oscillation analysis, $\Delta m_{31}^2 = \Delta m_{\text{eff}}^2 + \Delta m_{21}^2 \left(\cos^2\theta_{12} - \cos\delta_{CP}\sin\theta_{13}\sin 2\theta_{12}\tan\theta_{23}\right)$; $\Delta m_{32}^2 = \Delta m_{31}^2 - \Delta m_{21}^2$, for normal hierarchy when $\Delta m_{\text{eff}}^2 > 0$. When $\Delta m_{\text{eff}}^2 < 0$, $\Delta m_{31}^2 \leftrightarrow -\Delta m_{32}^2$ for inverted hierarchy

Parameter	True value	Marginalisation range
θ_{13}	$8.5°$	$[7.80°, 9.11°]$
$\sin^2\theta_{23}$	0.5	$[0.39, 0.64]$
Δm_{eff}^2	2.4×10^{-3} eV^2	$[2.3, 2.6]\times 10^{-3}$ eV^2
$\sin^2_{12}$	0.304	Not marginalised
Δm_{21}^2	7.6×10^{-5} eV^2	Not marginalised
δ_{CP}	$0°, -90°$	$[-180°, 180°]$

$$\frac{d^2 N^e}{dE_e \, d\cos\theta_{ev}} = t \times n_d \times \int dE_\nu \, d\cos\theta_\nu \, d\phi_\nu$$

$$\times \left[P_{ee} \frac{d^3 \Phi_{\nu_e}}{dE_\nu \, d\cos\theta_\nu \, d\phi_\nu} + P_{\mu e} \frac{d^3 \Phi_{\nu_\mu}}{dE_\nu \, d\cos\theta_\nu \, d\phi_\nu} \right] \times \frac{d\sigma_{\nu e}(E_\nu)}{dE_e \, d\cos\theta_{ev}},$$

$$\text{(68.9)}$$

where t is the exposure time, n_d is the number of targets in the detector, $d\sigma_e$ is the differential neutrino interaction cross section, and Φ_{ν_μ} and Φ_{ν_e} are the ν_μ and ν_e fluxes. The oscillated events are binned in (E_l^{obs}, $\cos\theta_l^{\text{obs}}$, $E_{\text{had}}^{\prime\text{obs}}$), and the energy and direction of the lepton in the final state, $l = e, \mu$; and $E_{\text{had}}^{\prime\text{obs}}$ is the observed final state hadron energy. A Poissonian χ^2

$$\chi_{l\pm}^2 = \sum_i \sum_j \sum_k 2 \left[\left(T_{ijk}^{l\pm} - D_{ijk}^{l\pm} \right) - D_{ijk}^{l\pm} \ln\left(\frac{T_{ijk}^{l\pm}}{D_{ijk}^{l\pm}} \right) \right], \qquad \text{(68.10)}$$

where i, j, k are the indices corresponding to E_l, $\cos\theta_l$, E^{had} bins, respectively. Here $l = e, \mu$ are the final state leptons; $T_{ijk}^{l\pm}$ and $D_{ijk}^{l\pm}$ are the theory and "data" events respectively; $+$ stands for anti-neutrino events and $-$ for neutrino events. For the perfect separation of ν from $\bar{\nu}$, the corresponding χ^2s can be found out separately. The total χ_l^2 for l type of events is

$$\chi_{l\delta_{CP}}^2 = \chi_{l+}^2 + \chi_{l-}^2. \qquad \text{(68.11)}$$

The δ_{CP} sensitivity χ^2 versus δ_{CP} for fixed and marginalised cases as well as when ν and $\bar{\nu}$ can and cannot be separated are is shown in Fig. 68.4. The figure also shows that the sensitivity is higher when ν and $\bar{\nu}$ events can be separated from each other. For $\delta_{CP}^{\text{true}} = -90°$, the parameter space except that from $-135°$ to $-60°$ can be excluded above 2σ with electron like events alone.

The use of a magnetic field can help the separation of ν_μ and $\bar{\nu}_\mu$, while as of now there are no existing techniques to readily separate ν_e and $\bar{\nu}_e$ events. Proposals have been made for the Gd doping of water Cherenkov detectors like Super-K [10–15]. This, if implemented can be used for identifying and studying atmospheric ν_e and $\bar{\nu}_e$ and hence enhance the sensitivity to δ_{CP}. Future detectors like DUNE can also be used to probe low energy atmospheric neutrino events since they has very good energy and direction resolutions [16].

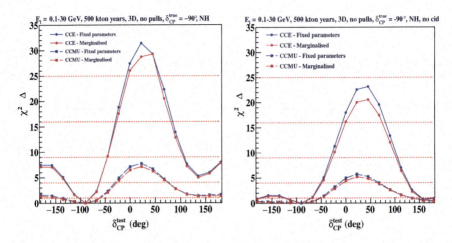

Fig. 68.4 $\Delta\chi^2$ versus δ_{CP}^{test} (deg) from CC $\nu_e + \overline{\nu}_e$ (solid) and CC $\nu_\mu + \overline{\nu}_\mu$ (dot dashed) events obtained with 500 kton year of an isoscalar detector and (left) with cid and (right) no cid. Here $\delta_{CP}^{true} = -90°$ (deg) and NH are assumed. Here cid implies the separation of ν and $\overline{\nu}$ events

68.4 Conclusions

Low energy (sub-GeV) atmospheric neutrinos can be used to probe the Dirac CP violating phase δ_{CP} in the leptonic sector *unambiguous* of neutrino mass hierarchy. This was shown analytically. Oscillated events binned in final state lepton direction illustrate the consistent distinction between various values of δ_{CP} at these energies. A sensitivity study assuming a perfect detector with 100% efficiencies and resolutions along with the ability to separate ν_e, $\overline{\nu}_e$, nu_μ, and $\overline{\nu}_\mu$ was performed. It was found that for 500 kton year exposure of the detector, a 2σ rejection of the parameter space from $[-180°, -135°]$ and $[-60°, -180°]$ can be obtained with ν_e and $\overline{\nu}_e$ alone, for true $\delta_{CP} = -90°$, when ν_e and $\overline{\nu}_e$ events can be separated from each other. Currently, there are no atmospheric neutrino detectors with $\nu_e - \overline{\nu}_e$ separation, but water Cherenkov detectors like Super-K can be doped with gadolinu (Gd). Proposals have already been made for supernova detection, but it should be studied if sub-GeV atmospheric $\nu_e - \overline{\nu}_e$ separation can be obtained with this technique. Though realistic detector resolutions and efficiencies may dilute the sensitivity to δ_{CP} from low energy atmospheric neutrinos, events accumulated over a large exposure can contribute to the overall sensitivity to δ_{CP}, and further studies can be conducted on how to improve the lepton reconstruction at low energies to obtain better sensitivity.

Acknowledgements Professor G. Rajasekaran and Prof. Rahul Sinha, IMSc, Chennai, for many discussions; IMSc journal club, Prof. T. Kajita. LSM thanks Prof. James Libby, IIT Madras, Chennai, and acknowledges Nandadevi cluster, a part of IMSc computing facility, with which the simulations were performed.

References

1. I. Esteban et al., Global analysis of three-flavour neutrino oscillations: synergies and tensions in the determination of θ_{23}, δ_{CP}, and the mass ordering. JHEP **01**, 106 (2019)
2. K. Abe et al., (T2K Collaboration), Search for CP violation in neutrino and anti-neutrino oscillations by the T2K experiment with 2.2×10^{21} protons on target. Phys. Rev. Lett. **121**, 171802 (2018)
3. M.A. Acero et al., (NOvA Collaboration), New constraints on oscillation parameters from ν_e appearance and ν_μ disappearance in the NOvA experiment. Phys. Rev. D **98**, 032012 (2018)
4. B. Abi et al., The DUNE far detector interim design report Volume 1: Physics, technology and strategies. arXiv:1807.10334 (2018)
5. M. Honda et al., Calculation of atmospheric neutrino flux using the interaction model calibrated with atmospheric muon data. Phys. Rev. D **75**, 043006 (2007)
6. M. Honda et al., Improvement of low energy atmospheric neutrino flux calculation using the JAM nuclear interaction model. Phys. Rev. D **83**, 123001 (2011)
7. M. Honda et al., Atmospheric neutrino flux calculation using the NRLMSISE-00 atmospheric model. Phys. Rev. D **92**, 023004 (2015)
8. D. Casper, The nuance neutrino physics simulation, and the future. Phys. Proc. Suppl. **112**, 161–170 (2002)
9. S. Choubey et al., Sensitivity to neutrino decay with atmospheric neutrinos at the INO-ICAL detector. Phys. Rev. D **97**, 033005 (2018)
10. Takaaki Mori, Status of the Super-Kamiokande gadolinium project. NIM A **732**, 316–319 (2013)
11. L. Labarga, The SuperK-gadolinium project. PoS (HQL 2016) 007
12. John F. Beacom, Mark R. Vagins, Antineutrino spectroscopy with large water Čerenkov detectors. Phys. Rev. Lett. **93**, 171101 (2004)
13. P. Fernandez, Status of GADZOOKS!: Neutron Tagging in Super-Kamiokande, in *Nuclear and Particle Physics Proceedings* (2016), pp. 273–275; pp. 353–360
14. T. Mori, *Development of a gadolinium-doped water cherenkov detector for the observation of supernova relic neutrinos*, Ph.D. Thesis, The University of Okayama (2015)
15. C. Xu, *Study of a 200 ton gadolinium-loaded water cherenkov detector for Super-KamiokaNDE gadolinium project*, Master Thesis, Okayama University (2016)
16. R. Acciarri et al., Long-Baseline Neutrino Facility (LBNF) and Deep Underground Neutrino Experiment (DUNE) (2016). arXiv:1601.05471
17. M.M. Devi et al., Enhancing sensitivity to neutrino parameters at INO combining muon and hadron information. JHEP **10**, 189 (2014)
18. L.S. Mohan, D. Indumathi, Pinning down neutrino oscillation parameters in the 2–3 sector with a magnetised atmospheric neutrino detector: a new study. Eur. Phys. J. C **77**, 54 (2017)

Chapter 69
Probing the Effects of Unparticle Decay of Neutrinos on the Possible Ultrahigh Energy Neutrino Signals at a Km² Detector for 4-Flavour Scenario

Madhurima Pandey

Abstract We address a possibility in which the ultrahigh energy (UHE) neutrinos undergo unparticle decays. In this framework, a neutrino decays to an unparticle and a lighter neutrino. The idea of unparticle has been proposed by Georgi almost a decade back by invoking the concept of the probable existence of a scale-invariant sector at high energies, and at low energies this scale invariant sector manifests itself by non-integral number $(d_{\mathcal{U}})$ of massless invisible particles called "unparticles" below a dimensional transmutation scale $\Lambda_{\mathcal{U}}$. In the 4-flavour framework, we explore here the neutrino signals at a km² detector like IceCube when the UHE neutrinos from cosmic astrophysical sources undergo possible unparticle decays.

69.1 Introduction

The probable existence of a scale-invariant sector has been introduced by Georgi [1] almost a decade back. The scale-invariant sector and the Standard Model (SM) sector may coexist at a very high energy scale, and the fields of these two sectors can interact among themselves via a mediator messenger field having mass scale $M_{\mathcal{U}}$. In the real world, the scale invariance feature of SM is manifestly broken by the masses of SM particles. Below the mass scale $M_{\mathcal{U}}$, the coupling is non-renormalizable and the interactions between this scale-invariant sector and SM are suppressed by inverse powers of $M_{\mathcal{U}}$. Georgi observed that at low energies, the scale-invariant sector manifests itself by non-integral number $(d_{\mathcal{U}})$ of massless invisible particles called "unparticles". A prototype model of such scale-invariant sector can be obtained from the Banks–Zaks theory [2] and the scale-invariant section in this theory can flow to a lower energy scale $\Lambda_{\mathcal{U}}$ having infrared fixed points, through dimensional transmutation [3]. Below $\Lambda_{\mathcal{U}}$, unparticle physics manifests itself as a field of various fractional scaling dimensions $(d_{\mathcal{U}})$. Unparticle operators can interact with SM fields

M. Pandey (✉)
Astroparticle Physics and Cosmology Division, Saha Institute of Nuclear Physics, HBNI, 1/AF Bidhannagar, Kolkata 700064, India
e-mail: madhurima.pandey@saha.ac.in

© Springer Nature Singapore Pte Ltd. 2021
P. K. Behera et al. (eds.), *XXIII DAE High Energy Physics Symposium*,
Springer Proceedings in Physics 261,
https://doi.org/10.1007/978-981-33-4408-2_69

497

via the exchange of some heavy fields, which when integrated out induce the effective operators by which the unparticle interacts with SM fields at low energy. We have taken a scalar unparticle operator and scalar interactions with neutrinos that enable a heavy neutrino to decay to a lighter neutrino and another invisible unparticle ($\mathcal{U}$). We consider unparticle decay of ultrahigh energy (UHE) neutrinos for a 4-flavour scenario, where an extra sterile neutrino is introduced to the three families of active neutrinos, from a distant extragalactic sources such as Gamma-Ray Bursts (GRBs) and estimate the detection yield of these neutrinos at a kilometre square detector like IceCube [4].

69.2 Formalism

69.2.1 Four-Flavour and Three-Flavour Neutrino Oscillations

In general, the oscillation probability $P_{\nu_\alpha \to \nu_\beta}$ for a neutrino $|\nu_\alpha\rangle$ of flavour α oscillates to a neutrino $|\nu_\beta\rangle$ of flavour β is given by (there is no CP violation in neutrino sector) [5]

$$P_{\nu_\alpha \to \nu_\beta} = \delta_{\alpha\beta} - 4 \sum_{j>i} U_{\alpha i} U_{\beta i} U_{\alpha j} U_{\beta j} \sin^2 \left(\frac{\pi L}{\lambda_{ij}} \right), \qquad (69.1)$$

where i, j indicate the mass index, L denotes the baseline distance and the oscillation length is symbolized as λ_{ij}. The oscillatory part in the probability equation is averaged to half for UHE neutrinos from distant GRBs (L is large, $\Delta m^2 L/E \gg 1$). Then the probability equation (69.1) reduces to the form

$$P_{\nu_\alpha \to \nu_\beta} = \sum_j | U_{\alpha j} |^2 | U_{\beta j} |^2 . \qquad (69.2)$$

The present 4-flavour scenario is considered to be the minimal extension of the 3-flavour case by a sterile neutrino. The form of the mass-flavour mixing matrix $\tilde{U}_{(4\times4)}$ is adopted from [6]. For the 3-flavour case, the usual Pontecorvo–Maki–Nakagawa–Sakata (PMNS) [7] matrix has been considered.

69.2.2 Unparticle Decay of Neutrinos and Its Consequences for UHE Neutrinos From a Single GRB

We consider an unparticle decay hypothesis, where a neutrino with mass eigenstate ν_j decays to an another light neutrino with mass eigenstate ν_i and the invisible unparticle $(\mathcal{U})$ [8],

$$\nu_j \rightarrow \mathcal{U} + \nu_i \ . \tag{69.3}$$

In the low energy regime, the effective Lagrangian for scalar interaction can be written as $L_s = \dfrac{\lambda_\nu^{\alpha\beta}}{\Lambda_\mathcal{U}^{d_\mathcal{U}-1}} \bar{\nu}_\alpha \nu_\beta \mathcal{O}_\mathcal{U}$, where $\alpha, \beta = e, \mu, \tau, s$ are defined as the flavour indices, $d_\mathcal{U}$ signifies the fractional scaling dimension of the unparticle operator $\mathcal{O}_\mathcal{U}$. The scale invariance sets in at the dimension transmutation scale $\Lambda_\mathcal{U}$. The relevant coupling constant is indicated by $\lambda_\nu^{\alpha\beta}$.

In the present context, it is convenient to work with neutrino mass eigenstates rather than the flavour eigenstates. The relation between the mass and the flavour eigenstate is $|\nu_i\rangle = \sum_\alpha U_{\alpha i}^* |\nu_\alpha\rangle$, and $U_{\alpha i}$'s are the elements of the mass-flavour mixing matrix. In the mass basis, the neutrino unparticle scalar interaction term takes the form $L = \lambda_\nu^{ij} \bar{\nu}_i \nu_j \mathcal{O}_\mathcal{U} / \Lambda_\mathcal{U}^{d_\mathcal{U}-1}$, where $\lambda_\nu^{ij} = \sum_{\alpha,\beta} U_{\alpha i}^* \lambda_\nu^{\alpha\beta} U_{\beta j}$ is the coupling constant expressed in the mass eigenstate basis.

The most relevant quantity of the unparticle decay process is the total decay rate (Γ_j) or equivalently the lifetime of neutrino $(\tau_\mathcal{U})$, which can be expressed as

$$\frac{\tau_\mathcal{U}}{m_j} = \frac{16\pi^2 d_\mathcal{U}(d_\mathcal{U}^2 - 1)}{A_d |\lambda_\nu^{ij}|^2} \left(\frac{\Lambda_\mathcal{U}^2}{m_j^2}\right)^{d_\mathcal{U}-1} \frac{1}{m_j^2} \ , \tag{69.4}$$

where m_j indicates the decaying neutrino mass. In the above, the normalization constant A_d can be written as [1]

$$A_d = \frac{16\pi^{5/2}}{(2\pi)^{2d_\mathcal{U}}} \frac{\Gamma(d_\mathcal{U} + 1/2)}{\Gamma(d_\mathcal{U} - 1)\Gamma(2d_\mathcal{U})} \ . \tag{69.5}$$

For the UHE neutrinos in the 4-flavour framework, which undergo unparticle decays, the lightest mass state $|\nu_1\rangle$ is considered to be stable as it does not decay and all other states $|\nu_2\rangle$, $|\nu_3\rangle$ and $|\nu_4\rangle$ are unstable. By considering (69.2), the flux of the neutrino at the detector for neutrino flavour α undergoing unparticle decay is given as

$$\phi_{\nu_\alpha}^{\text{detector}}(E_\nu) = \sum_i \sum_\beta [\phi_{\nu_\beta}^s |U_{\beta i}|^2 |U_{\alpha i}|^2 \exp(-4\pi L(z)/(\lambda_d)_i)]a_1 \ , \tag{69.6}$$

where $L(z) = \dfrac{c}{H_0} \displaystyle\int_0^z \dfrac{dz'}{(1+z')^2 \sqrt{\Omega_\Lambda + \Omega_m (1+z')^3}}$ is the source distance (at a red-

shift z) and $a_1 = \dfrac{1}{4\pi L^2(z)}(1+z)$. The 4×4 mass-flavour mixing matrix $\tilde{U}_{(4\times 4)}$ for

the 4-flavour case has been adopted from [6]. The decay length $((\lambda_d)_i)$ in (69.6)
is given by $(\lambda_d)_i = \dfrac{4\pi E_\nu}{\alpha_i} = 2.5\,\mathrm{Km}\dfrac{E_\nu}{\mathrm{GeV}}\dfrac{\mathrm{eV}^2}{\alpha_i}$, where $\alpha_i = m_i/\tau_\mathcal{U}$, $\tau_\mathcal{U}$ being, as mentioned, the neutrino decay lifetime in the rest frame.

By using (69.6) and taking into account that the lightest mass state $|\nu_1\rangle$ is stable,
the fluxes of of each flavour (for the 4-flavour case) neutrinos undergoing unparticle
decays on reaching the Earth can be expressed as

$$
\begin{aligned}
\phi_{\nu_e}^{\text{detector}} &= ([|\,\tilde{U}_{e1}\,|^2(1+|\,\tilde{U}_{\mu 1}\,|^2 - |\,\tilde{U}_{\tau 1}\,|^2 - |\,\tilde{U}_{s1}\,|^2) + \\
&\quad |\,\tilde{U}_{e2}\,|^2(1+|\,\tilde{U}_{\mu 2}\,|^2 - |\,\tilde{U}_{\tau 2}\,|^2 - |\,\tilde{U}_{s2}\,|^2)\exp(-4\pi L(z)/(\lambda_d)_2) + \\
&\quad |\,\tilde{U}_{e3}\,|^2(1+|\,\tilde{U}_{\mu 3}\,|^2 - |\,\tilde{U}_{\tau 3}\,|^2 - |\,\tilde{U}_{s3}\,|^2)\exp(-4\pi L(z)/(\lambda_d)_3) + \\
&\quad |\,\tilde{U}_{e4}\,|^2(1+|\,\tilde{U}_{\mu 4}\,|^2 - |\,\tilde{U}_{\tau 4}\,|^2 - |\,\tilde{U}_{s4}\,|^2)\exp(-4\pi L(z)/(\lambda_d)_4)]\phi_{\nu_e}^s)a_1 \ , \\[4pt]
\phi_{\nu_\mu}^{\text{detector}} &= ([|\,\tilde{U}_{\mu 1}\,|^2(1+|\,\tilde{U}_{\mu 1}\,|^2 - |\,\tilde{U}_{\tau 1}\,|^2 - |\,\tilde{U}_{s1}\,|^2) + \\
&\quad |\,\tilde{U}_{\mu 2}\,|^2(1+|\,\tilde{U}_{\mu 2}\,|^2 - |\,\tilde{U}_{\tau 2}\,|^2 - |\,\tilde{U}_{s2}\,|^2)\exp(-4\pi L(z)/(\lambda_d)_2) + \\
&\quad |\,\tilde{U}_{\mu 3}\,|^2(1+|\,\tilde{U}_{\mu 3}\,|^2 - |\,\tilde{U}_{\tau 3}\,|^2 - |\,\tilde{U}_{s3}\,|^2)\exp(-4\pi L(z)/(\lambda_d)_3) + \\
&\quad |\,\tilde{U}_{\mu 4}\,|^2(1+|\,\tilde{U}_{\mu 4}\,|^2 - |\,\tilde{U}_{\tau 4}\,|^2 - |\,\tilde{U}_{s4}\,|^2)\exp(-4\pi L(z)/(\lambda_d)_4)]\phi_{\nu_e}^s)a_1 \ , \\[4pt]
\phi_{\nu_\tau}^{\text{detector}} &= ([|\,\tilde{U}_{\tau 1}\,|^2(1+|\,\tilde{U}_{\mu 1}\,|^2 - |\,\tilde{U}_{\tau 1}\,|^2 - |\,\tilde{U}_{s1}\,|^2) + \\
&\quad |\,\tilde{U}_{\tau 2}\,|^2(1+|\,\tilde{U}_{\mu 2}\,|^2 - |\,\tilde{U}_{\tau 2}\,|^2 - |\,\tilde{U}_{s2}\,|^2)\exp(-4\pi L(z)/(\lambda_d)_2) + \\
&\quad |\,\tilde{U}_{\tau 3}\,|^2(1+|\,\tilde{U}_{\mu 3}\,|^2 - |\,\tilde{U}_{\tau 3}\,|^2 - |\,\tilde{U}_{s3}\,|^2)\exp(-4\pi L(z)/(\lambda_d)_3) + \\
&\quad |\,\tilde{U}_{\tau 4}\,|^2(1+|\,\tilde{U}_{\mu 4}\,|^2 - |\,\tilde{U}_{\tau 4}\,|^2 - |\,\tilde{U}_{s4}\,|^2)\exp(-4\pi L(z)/(\lambda_d)_4)]\phi_{\nu_e}^s)a_1 \ , \\[4pt]
\phi_{\nu_s}^{\text{detector}} &= ([|\,\tilde{U}_{s1}\,|^2(1+|\,\tilde{U}_{\mu 1}\,|^2 - |\,\tilde{U}_{\tau 1}\,|^2 - |\,\tilde{U}_{s1}\,|^2) + \\
&\quad |\,\tilde{U}_{s2}\,|^2(1+|\,\tilde{U}_{\mu 2}\,|^2 - |\,\tilde{U}_{\tau 2}\,|^2 - |\,\tilde{U}_{s2}\,|^2)\exp(-4\pi L(z)/(\lambda_d)_2) + \\
&\quad |\,\tilde{U}_{s3}\,|^2(1+|\,\tilde{U}_{\mu 3}\,|^2 - |\,\tilde{U}_{\tau 3}\,|^2 - |\,\tilde{U}_{s3}\,|^2)\exp(-4\pi L(z)/(\lambda_d)_3) + \\
&\quad |\,\tilde{U}_{s4}\,|^2(1+|\,\tilde{U}_{\mu 4}\,|^2 - |\,\tilde{U}_{\tau 4}\,|^2 - |\,\tilde{U}_{s4}\,|^2)\exp(-4\pi L(z)/(\lambda_d)_4)]\phi_{\nu_e}^s)a_1 \ .
\end{aligned}
$$
$$(69.7)$$

In a GRB, the UHE neutrinos are produced in the proportion $\nu_e : \nu_\mu : \nu_\tau : \nu_s = 1 :$
$2 : 0 : 0$. Therefore, $\phi_{\nu_e}^s = \dfrac{1}{6}\mathcal{F}(E_\nu^s)$, $\phi_{\nu_\mu}^s = \dfrac{2}{6}\mathcal{F}(E_\nu^s) = 2\phi_{\nu_e}^s$, $\phi_{\nu_\tau}^s = 0$, $\phi_{\nu_s}^s = 0$,
where $\phi_{\nu_e}^s$, $\phi_{\nu_\mu}^s$, $\phi_{\nu_\tau}^s$ and $\phi_{\nu_s}^s$ represent the fluxes of ν_e, ν_μ, ν_τ and ν_s, respectively, at
the source. In the absence of CP violation, $\mathcal{F}(E_\nu^s) = \dfrac{\mathrm{d}N}{\mathrm{d}E_\nu^s} = \dfrac{\mathrm{d}N_{\nu+\bar\nu}}{\mathrm{d}E_\nu^s}$. The spectra for
neutrinos (ν) will be therefore $0.5\mathcal{F}(E_\nu^s)$. At the source, which in this case is a single

GRB, the neutrino spectrum can be expressed as $\dfrac{dN}{dE_\nu^s} = N \times \min\left(1, \dfrac{E_\nu}{E_\nu^{\mathrm{brk}}}\right)\dfrac{1}{E_\nu^2}$,

where $N = \dfrac{E_{\mathrm{GRB}}}{1 + \ln(E_{\nu\,\mathrm{max}}^s/E_\nu^{\mathrm{brk}})}$ is the normalization constant and E_ν^s represents

the neutrino energy. The neutrino spectrum break energy, $E_\nu^{\mathrm{brk}} \approx 10^6 \dfrac{\Gamma_{2.5}^2}{E_{\gamma,\mathrm{MeV}}^{\mathrm{brk}}}$ GeV,

where $\Gamma_{2.5} = \Gamma/10^{2.5}$, Γ being the Lorentz boost factor and $E_{\gamma,\mathrm{MeV}}^{\mathrm{brk}}$ denotes the photon spectral break energy.

69.3 Detection of UHE Neutrinos at IceCube From a Single GRB

The detection of UHE neutrinos is mainly through upgoing neutrino induced muons inside the Earth's rock and in the detector material (ice in the present) following the interactions $\nu_\mu + N \rightarrow \mu + X$, $\nu_\tau + N \rightarrow \tau + X \rightarrow \mu + \nu_\mu + \nu_\tau + X$.

The event rate (S) of muons is expressed as

$$S = \int_{E_{\mathrm{th}}}^{E_{\nu\mathrm{max}}} dE_\nu \phi_{\nu_\alpha}^{\mathrm{detector}} P_{\mathrm{shadow}}(E_\nu) P_\mu(E_\nu, E_{\mathrm{th}}) . \tag{69.8}$$

For a particular GRB, the zenith angle θ_z is fixed and the shadow factor is given as

$P_{\mathrm{shadow}}(E_\nu) = \exp[-X(\theta_z)/L_{\mathrm{int}}(E_\nu)]$, where $L_{\mathrm{int}}(E_\nu) = \dfrac{1}{\sigma_{\mathrm{tot}}(E_\nu)N_A}$, N_A and σ_{tot}

denote the Avogadro number and the total cross-section. In (69.8), the effective path length for incident zenith angle $\theta(z)$ is defined as $X(\theta_z) = \int \rho(r(\theta_z, l))dl$, where $\rho(r(\theta_z, l))$ denotes the matter density profile inside the Earth. In the above equation, the flux $\phi_{\nu_\alpha}^{\mathrm{detector}}$ (for flavour α) is computed from (69.7). The probability P_μ in (69.8) (the probability that a neutrino induced muon, with energy above the detector threshold, reaches the detector) is computed following the formalism given in [9].

69.4 Calculations and Results

The main motivation of our work is to probe the effects of the unparticle decay of UHE neutrinos at the IceCube detector. In Fig. 69.1a, b, we have shown how the neutrino induced muon events vary with the unparticle parameters ($d_{\mathcal{U}}$ and λ_ν^{ij}) in comparison to the case where only mass-flavour oscillations (no decay) are considered. In Fig. 69.1a, we have observed that the muon yield is depleted by $\sim$70% from what is expected for only the mas flavour oscillation case and from this observation we can comment that the decay effect would be significant.

Table 69.1 The ratio of muon yields at IceCube for UHE neutrinos from a GRB with and without unparticle decay in a 4-flavour neutrino framework. See text for details

θ_{14}	θ_{24}	θ_{34}	$d_{\mathcal{U}}$	$|\lambda_\nu^{ij}|$	R
3°	5°	20°	1.2	10^{-4}	0.9
			1.3	10^{-2}	0.6

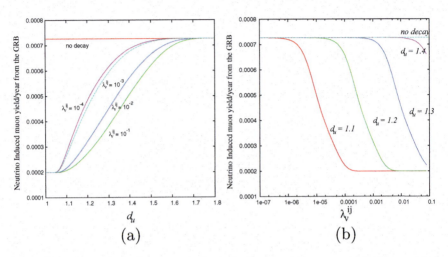

Fig. 69.1 The predicted upward going neutrino induced muon yields per year at the detector, when the UHE neutrinos suffer unparticle decay and its comparison with the no decay case. The variation of the yield with $d_{\mathcal{U}}$ for different fixed values of couplings ($|\lambda_\nu^{ij}|$). Similar variation with the coupling $|\lambda_\nu^{ij}|$ for different chosen values of $d_{\mathcal{U}}$ (Figure reproduced from M. Pandey, JHEP **1901**, 066 (2019))

In order to understand further the effect of decay, we have considered a ratio between the expected neutrino induced muon events per year with and without unparticle decay in addition to the usual mass-flavour oscillations in the 4-flavour framework. The ratio is defined as $R = \frac{\text{Rate}_{(\text{with decay})}}{\text{Rate}_{(\text{no decay})}}$. For a fixed set of values of the 4-flavour mixing angles and for two sets of values for each of the quantities $d_{\mathcal{U}}, |\lambda_\nu^{ij}|$, we demonstrate the deviation of R from the value 1 in Table 69.1 (reproduced from M. Pandey, JHEP **1901**, 066 (2019)). From Table 69.1, it can be seen that for the parameter set $d_{\mathcal{U}} = 1.3$ and $|\lambda_\nu^{ij}| = 10^{-2}$, (that may cause considerable decay effect (Fig. 69.1)), the ratio R can be depleted to as low a value as 0.6.

69.5 Summary and Discussions

In a 4-flavour (3 active + 1 sterile) neutrino framework, we consider unparticle decay of neutrinos and apply it to UHE neutrinos from a distant GRB. We calculate the neutrino induced muon yields in such a framework in case of a square kilometre detector such as IceCube. We then investigate the effects of fractional unparticle dimension $d_{\mathcal{U}}$ as also the coupling λ_ν^{ij} on the muon detection yields and compare them with the case where normal oscillation/suppression of GRB UHE neutrino flavours are considered without any unparticle decay. We found that the unparticle decay effects could be quite prominent for certain values of $d_{\mathcal{U}}$ and λ_ν^{ij}.

References

1. H. Georgi, Phys. Rev. Lett. **98**, 221601 (2007)
2. T. Banks, A. Zaks, Nucl. Phys. B **196**, 189 (1982)
3. S. Coleman, E. Weinberg, Phys. Rev. D **7**, 1888 (1973)
4. J. Ahrens et al., (IceCube Collaboration), in *Proceedings of the 27th International Cosmic Ray Conference* (Hamburg, Germany, 2001) (unpublished), p. 1237. http://icecube.wisc.edu
5. D. Majumdar, A. Ghosal, Phys. Rev. D **75**, 113004 (2007)
6. M. Pandey, D. Majumdar, A.D. Banik, Phys. Rev. D **97**, 103015 (2018)
7. Z. Maki, M. Nakagawa, S. Sakata, Prog. Theo. Phys. **28** (1962)
8. S. Zhou, Phys. Lett. B **659** (2008)
9. R. Gandhi, C. Quigg, M.H. Reno, I. Sarcevic, Astropart. Phys. **5**, 81 (1996). arXiv:hep-ph/9512364

Chapter 70
Phenomenological Study of Two-Zero Textures of Neutrino Mass Matrices in Minimal Extended Seesaw Mechanism

Priyanka Kumar and Mahadev Patgiri

Abstract In this chapter, we study the phenomenology of two-zero textures of 4×4 neutrino mass matrices $M_\nu^{4 \times 4}$ in the context of the Minimal Extended Seesaw (MES) mechanism. The MES mechanism is an extension to the canonical type-I seesaw mechanism which incorporates an extra gauge singlet field 'S' apart from the three right-handed neutrinos. The MES mechanism deals with 3×3 forms of Dirac neutrino mass matrix (M_D), right-handed Majorana mass matrix (M_R) and 1×3 row matrix (M_S) which couples the right-handed neutrinos and the sterile singlet 'S'. In our work, we realize the two-zero textures of $M_\nu^{4 \times 4}$ within the context of the MES mechanism by considering a $(5 + 4)$ scheme, where the digits in the pair represent the number of zeros of M_D and M_R, along with a one-zero texture of M_S. We find that out of 15 allowed two-zero textures, only 6 two-zero textures can be realized under the $(5 + 4)$ scheme. On enforcing zeros, the neutrino mass matrix $M_\nu^{4 \times 4}$ yields a number of correlations. We check the viability of each texture by scanning their correlations under recent neutrino oscillation data. We find that certain textures are allowed only for some selected ranges of values of $\sin \theta_{34}$. We present scatter plots as a viability check for each of the textures.

70.1 Introduction

From the proposition of massless neutrinos by Wolfgang Pauli to massive neutrinos confirmed by a number of solar, atmospheric and reactor experiments, neutrino physics have appreciably progressed with time. Untiring efforts from experimentalists have succeeded in providing solid and precise information about the mass-squared differences (Δm_{21}^2, Δm_{31}^2) and mixing angles ($\theta_{12}, \theta_{23}, \theta_{13}$) in case of the three active neutrinos. There are, however, a number of problems which are still

P. Kumar (✉) · M. Patgiri
Department of Physics, Cotton University, Guwahati 781001, Assam, India
e-mail: prianca.kumar@gmail.com

M. Patgiri
e-mail: mahadevpatgiri@mail.com

© Springer Nature Singapore Pte Ltd. 2021
P. K. Behera et al. (eds.), *XXIII DAE High Energy Physics Symposium*,
Springer Proceedings in Physics 261,
https://doi.org/10.1007/978-981-33-4408-2_70

striving for solutions, for instance, the origin of sub-eV scale neutrino mass, exact nature of neutrinos—whether Dirac or Majorana, CP violation, to name a few. Apart from these anomalies in the three active neutrino scenario, there are a number of experiments which hint towards the presence of a fourth flavor of neutrino. The LSND [1] experiment observed an anomalous oscillation mode $\bar{\nu}_\mu - \bar{\nu}_e$ which corresponds to squared mass difference of eV^2, which could not be explained within the three neutrino scheme. The LSND anomaly was later supported by the MiniBooNE experiment [2]. Similar kind of discrepancies were also observed in the Gallium solar neutrino experiment [3, 4], reactor neutrino experiment [5] and cosmological observations [6]. All these anomalies hint toward the existence of a light sterile neutrino. From the cosmological front, the recent Planck data suggests that accommodating one light sterile neutrino requires deviation from the standard ΛCDM model [7]. In August 2018, the MinibooNE collaboration [8] glorified once again the presence of a light sterile neutrino. They reported their data remains to be consistent with the excess of events reported by the LSND experiment. Again the ANITA experiment, in August 2018, observed discrepancies in their upgoing air shower events which are in contradiction with the standard neutrino-matter interaction models [9]. Accommodating a light sterile neutrino and explaining their mixing with the active neutrinos as well as having a consistent cosmological model will serve as a new challenge from the theoretical point of view.

A light sterile neutrino can be added to the Standard Model (SM) in a number of ways. However, the (3+1) scheme, that is, three active and one light sterile neutrino, serves to be the minimal extension [10, 11]. The admixtures of the three active neutrinos with one sterile neutrino have been studied in the Minimal Extended Seesaw Mechanism (MES) [12]. MES is an extension of the type-I seesaw mechanism whereby SM is extended by including one gauge singlet chiral field 'S'. In MES, eV-scale sterile neutrino results naturally without needing to insert any tiny Yukawa couplings or mass scale for ν_s.

In our work, we shall study the two-zero textures of 4×4 neutrino mass matrix in the context of the MES mechanism. Within the three neutrino scenario, zeros of neutrino mass matrix can be realized within the context of the type-I seesaw mechanism, whereby zeros of Dirac neutrino mass matrix (M_D) and right-handed Majorana mass matrix (M_R) propagate as zeros in $m_\nu^{3\times3}$ [13]. The 4×4 MES neutrino mass matrix consists of 3×3 form of M_D and M_R along with 1×3 form of M_S which couples the singlet 'S' with the three right-handed neutrinos. In our work, we shall consider the (5+4) scheme, where digits in the pair represent the number of zeros of M_D and M_R, respectively, along with zero textures of M_S to realize the two-zero textures of $M_\nu^{4\times4}$. We find that out of 15 allowed two-zero textures [14] of $M_\nu^{4\times4}$, only 6 textures can be realized within the (5+4) scheme. On realizing the textures, we arrive at certain correlations between the different parameters of the mass matrix. We show the viability of each texture by means of scatter plots whereby correlations are plotted against $\sin\theta_{34}$ which is bounded by < 0.4. In our work, we consider the lower limit to be 0. We find that certain correlations are not allowed for all values of $\sin\theta_{34} = (0\text{--}0.4)$, while some are allowed for all ranges of $(0\text{--}0.4)$. While doing so, CP phases are kept unconstrained, that is, $0\text{--}2\pi$.

The chapter is organized as follows: Section 70.2 includes a brief discussion on the MES mechanism. In Sect. 70.3, we present the six two-zero textures that can be realized in the (5+4) scheme. In Sect. 70.4, the two-zero textures are realized in the context of the MES mechanism, and correlations are plotted under recent neutrino oscillation data for each texture. Finally, we conclude in Sect. 70.5.

70.2 Minimal Extended Seesaw (MES) Mechanism

In the MES mechanism, the Lagrangian representing the neutrino mass term takes the form[12]

$$- \mathcal{L}_m = \bar{\nu}_L M_D \nu_R + \bar{S}^c M_S \nu_R + \frac{1}{2} \bar{\nu}_R^c M_R \nu_R + h.c..$$ (70.1)

Using the seesaw approximation $M_R \gg M_S > M_D$, one arrives at the 4×4 square matrix $M_\nu^{4\times4}$ as

$$M_\nu^{4\times4} = - \begin{pmatrix} M_D M_R^{-1} M_D^T & M_D M_R^{-1} M_S^T \\ M_S (M_R^{-1})^T M_D^T & M_S M_R^{-1} M_S^T \end{pmatrix}.$$ (70.2)

The mass matrix $M_\nu^{4\times4}$ can have at most rank 3 since $\det(M_\nu^{4\times4}) = 0$.

In our work, we shall consider the 4×4 form of neutrino mass matrix in (70.2) for the realization of the two-zero textures.

70.3 Two-Zero Textures of $M_\nu^{4\times4}$

In the flavor basis, the 4×4 Majorana neutrino mass matrix can be expressed as

$$M_\nu^{4\times4} = \begin{pmatrix} m_{ee} & m_{e\mu} & m_{e\tau} & m_{es} \\ m_{e\mu} & m_{\mu\mu} & m_{\mu\tau} & m_{\mu s} \\ m_{e\tau} & m_{\mu\tau} & m_{\tau\tau} & m_{\tau s} \\ m_{es} & m_{\mu s} & m_{\tau s} & m_{ss} \end{pmatrix}.$$ (70.3)

There are 15 viable two-zero textures of $M_\nu^{4\times4}$[14]. Out of 15, only six textures can be realized with 5 zeros in M_D, 4 zeros in M_R and one-zero of M_S. The six textures are $A_1 : ee, e\mu = 0$; $A_2 : ee, e\tau = 0$; $B_3 : e\mu, \mu\mu = 0$; $B_4 : e\tau, \tau\tau = 0$; $D_1 : \mu\mu, \mu\tau = 0$; $D_2 : \mu\tau, \tau\tau = 0$. Texture A_1, A_2 allows only normal hierarchical (NH) mass spectrum, while the other textures allow both normal and inverted (IH) mass spectrum.

70.4   Realization of Two-Zero Textures

The general form of 3×3 non-symmetric M_D having nine independent entries, 3×3 symmetric right-handed Majorana mass matrix M_R having six independent entries and 1×3 matrix M_S with three independent entries can be represented in the following form:

$$M_D = \begin{pmatrix} a & b & c \\ d & e & f \\ g & h & i \end{pmatrix}, \qquad M_R = \begin{pmatrix} A & B & C \\ B & D & E \\ C & E & F \end{pmatrix}, \qquad M_S = (s_1 \ \ s_2 \ \ s_3). \quad (70.4)$$

In this section, we shall present the realization and corresponding scatter plots for each of the 6 two-zero textures.

Texture A_1: The following combination

$$M_D = (b, d, g, i \neq 0), \qquad M_R = (A, E \neq 0), \qquad M_S = (s_1 \ 0 \ s_3) \quad (70.5)$$

in (70.2) leads to the following correlation:

$$\frac{m_{\mu\mu}}{m_{\mu\tau}} = \frac{m_{\mu\tau}}{m_{\tau\tau}} = \frac{m_{\mu s}}{m_{\tau s}} = \sqrt{\frac{m_{\mu\mu}}{m_{\tau\tau}}}. \quad (70.6)$$

On plotting the correlation in (70.6) against $\sin\theta_{34}$, we find that overlapping appears for values of $\sin\theta_{34} > 0.04$ (Fig. 70.1). This sets a lower limit on the value of θ_{34}.

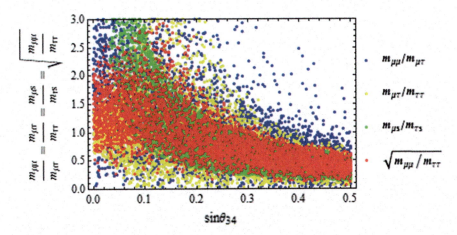

Fig. 70.1 Scatter plot for (70.6) against $\sin\theta_{34}$

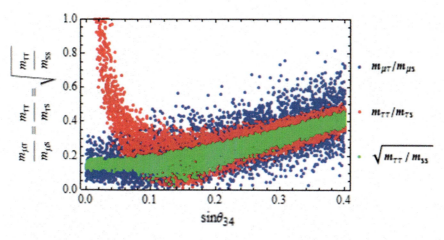

Fig. 70.2 Scatter plot for texture A_2

Texture A_2: Combinations of M_D, M_R and M_S and their corresponding correlation are as follows:

$$M_D = (b, d, f, g \neq 0), \quad M_R = (A, E \neq 0), \quad M_S = (s_1 \ \ 0 \ \ s_3), \quad \frac{m_{\mu\tau}}{m_{\mu s}} = \frac{m_{\tau\tau}}{m_{\tau s}} = \sqrt{\frac{m_{\tau\tau}}{m_{ss}}}. \tag{70.7}$$

Correlation plot shows that the texture is allowed for the range of $\sin \theta_{34} > 0.08$ (Fig. 70.2). For $\sin \theta_{34} < 0.08$, the overlapping vanishes and the texture is not allowed.

Texture B_3: The combination of M_D, M_R and M_S and their corresponding correlations are listed in (70.8)

$$M_D = (a, e, g, i \neq 0), \quad M_R = (A, E \neq 0), \quad M_S = (s_1 \ \ 0 \ \ s_3), \quad \frac{m_{ee}}{m_{e\tau}} = \frac{m_{e\tau}}{m_{\tau\tau}} = \sqrt{\frac{m_{ee}}{m_{\tau\tau}}}. \tag{70.8}$$

Figure 70.3 shows that for NH, the texture is allowed for values of $\sin \theta_{34} > 0.1$ while for IH the texture is allowed for all ranges of $\sin \theta_{34} = (0\text{--}0.4)$. Thus, the texture behaves differently in the case of NH and IH.

Texture B_4: The combination of M_D, M_R and M_S and their corresponding correlations are listed in (70.9)

$$M_D = (a, d, e, i \neq 0), \quad M_R = (A, E \neq 0), \quad M_S = (s_1 \ \ s_2 \ \ 0), \quad \frac{m_{ee}}{m_{e\mu}} = \frac{m_{e\mu}}{m_{\mu\mu}} = \sqrt{\frac{m_{ee}}{m_{\mu\mu}}}. \tag{70.9}$$

From Fig. 70.4, it is evident that the texture is allowed for all the ranges of $\sin \theta_{34} = (0 - 0.4)$ for both NH and IH spectra.

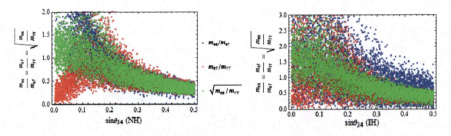

Fig. 70.3 Scatter plots for NH (left) and IH (right) for texture B_3

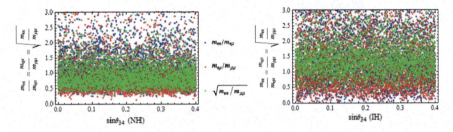

Fig. 70.4 Scatter plots for NH (left) and IH (right) for texture B_4

Texture D_1: The combination of M_D, M_R and M_S and their corresponding correlations are listed in (70.10).

$$M_D = (a, b, f, g \neq 0), \quad M_R = (A, E \neq 0), \quad M_S = (s_1 \quad s_2 \quad 0), \quad \frac{m_{ee}}{m_{e\tau}} = \frac{m_{e\tau}}{m_{\tau\tau}} = \sqrt{\frac{m_{ee}}{m_{\tau\tau}}}. \tag{70.10}$$

Texture D_2: The combination of M_D, M_R and M_S and their corresponding correlations are listed in (70.11).

$$M_D = (a, b, d, i \neq 0), \quad M_R = (A, E \neq 0), \quad M_S = (s_1 \quad s_2 \quad 0), \quad \frac{m_{ee}}{m_{e\mu}} = \frac{m_{e\mu}}{m_{\mu\mu}} = \sqrt{\frac{m_{ee}}{m_{\mu\mu}}}. \tag{70.11}$$

The correlation of textures D_1 and D_2 is similar to that of textures B_3 and B_4, respectively. The phenomenologies are therefore similar as shown in Figs. (70.3) and (70.4).

70.5 Conclusions

In this work, we have explored the two-zero textures of $M_\nu^{4\times4}$ with 5 zeros in M_D, 4 zeros in M_R (5+4 Scheme) and one-zero in M_S within the context of the MES mechanism. We have found that only 6 out of 15 two-zero textures can be realized within the (5+4) scheme. On enforcing zeros, we have found that each texture generates some constrained conditions called correlations where different elements of the mass matrix $M_\nu^{4\times4}$ are correlated. We have checked the viability of each texture by scanning their respective correlations under recent neutrino oscillation data and then within the common range, we have plotted the correlation against $\sin\theta_{34}$ which is bounded by an upper limit of < 0.4. In our analysis, we have taken its lower limit to be 0. We have found that texture A_1 is allowed for values of $\sin\theta_{34} > 0.04$. For values < 0.04, the overlapping ceases and the texture is not allowed. Texture A_2 is allowed for $\sin\theta_{34} > 0.1$. For texture B_3, we have observed that the correlation shows different phenomenology for NH and IH spectra, whereby the allowed ranges are $\sin\theta_{34} > 0.1$ and $\sin\theta_{34} = (0\text{–}0.4)$, respectively. Similar phenomenology for NH and IH for texture B_4 has been observed (allowed for all ranges of $\sin\theta_{34}$). Textures D_1 and D_2 generate similar correlations as B_3 and B_4, respectively, and therefore show similar phenomenology.

References

1. LSND: C. Athanassopoulos *et al., Phys. Rev. Lett.* **77**, 3082(1996); A. Aguilar et al., *Phys. Rev. D* **64**, 112007 (2001)
2. MiniBooNE Collaboration: A. A. Aguilar-Arevalo *et al.,Phys. Rev. Lett.* **110**, 161801 (2013)
3. M.A. Acero, C. Giunti, M. Laveder, Phys. Rev. D **78**, 073009 (2008)
4. C. Giunti, M. Laveder, Phys. Rev. C **83**, 065504 (2011)
5. G. Mention et al., Phys. Rev. D **83**, 073006 (2011)
6. K. N. Abazajian, M. A. Acero, S. K. Agarwalla, A. A. Aguilar-Arevalo, C. H. Albright, S. Antusch, C. A. Arguelles and A. B. Balantekin et al., [hep-ph] arXiv:1204.5379
7. P. A. R. Ade et al. [Planck Collaboration] [astro-ph.CO],arXiv:1502.01589
8. MiniBooNE Collaboration: A. A. Aguilar-Arevalo *et al.*; *Phys. Rev. Lett.* **121**, 221801 (2018) [hep-ex]. arXiv:1805.12028, 2018
9. Guo-yuan Huang; Phys, Rev. D **98**, 043019 (2018)
10. J.J.Gomez-Cadenas, M.C. Gonzalez-Garcia, Z. Phys, C **71**, 443–454 (1996). [hep-ph/9504246]
11. S. Goswami, Phys. Rev. D **55**, 2931–2949 (1997). [hep-ph/9507212]
12. J. Barry, W. Rodejohann, H. Zhang, JHEP **1107**, 091 (2011). [arXiv:1105.3911]
13. L. Lavoura, *J. Phys. G* **42** (2015) 105004, arXiv:1502.0300; S. Choubey, W. Rodejohann, and P. Roy, *Nucl. Phys. B* **808** (2009) 272–291, arXiv:0807.4289; S. Goswami and A. Watanabe, *Phys.Rev. D* **79** (2009) 033004, arXiv:0807.3438
14. M. Ghosh, S. Goswami and S. Gupta; JHEP 1304 (2013) 103 [hep-ph]; arXiv:1211.0118

Chapter 71
Consequences of a CP-Transformed $\mu\tau$-Flavored Friedberg-Lee Symmetry in a Neutrino Mass Model

Roopam Sinha, Sukannya Bhattacharya, and Rome Samanta

Abstract We propose a neutrino mass model with a generalized $\mu\tau$-flavored CP symmetry assuming the light Majorana neutrino mass term enjoys an invariance under a generalized Friedberg-Lee (FL) transformation. While both Normal Ordering (NO) as well as the Inverted Ordering (IO) are allowed, the absolute scale of neutrino masses is dictated as a consequence of FL symmetry. For both NO and IO, we show that the atmospheric mixing angle $\theta_{23} \neq \pi/4$. The Dirac CP phase $\delta = \pi/2, 3\pi/2$ for IO and nearly maximal for NO due to $\cos \delta \propto \sin \theta_{13}$. For the NO, very tiny CP violation might arise through one of the Majorana phases, namely, β. Beside fitting the neutrino oscillation data, we present a study of $\nu_\mu \to \nu_e$ oscillation which is expected to reveal Dirac CP violation in long-baseline experiments. We also calculate the Ultra High Energy (UHE) neutrino flavor flux ratios at neutrino telescopes, from which statements regarding the octant of θ_{23} have been made in our model.

71.1 Introduction

Various neutrino oscillation experiments have determined the three mixing angles and the two independent mass-squared differences to decent accuracy. However, the octant of the atmospheric mixing angle θ_{23} is yet undetermined though the current

R. Sinha (✉) · S. Bhattacharya
Saha Institute of Nuclear Physics, HBNI 1/AF Bidhannagar, 700064 Kolkata, India
e-mail: roopamsinha123@gmail.com

S. Bhattacharya
e-mail: sukannya.bh@gmail.com

S. Bhattacharya
Theoretical Physics Division, Physical Research Laboratory, Navrangpura, Ahmedabad 380009, India

R. Samanta
Physics and Astronomy, University of Southampton, Southampton SO17 1BJ, UK
e-mail: romesamanta@gmail.com

© Springer Nature Singapore Pte Ltd. 2021
P. K. Behera et al. (eds.), *XXIII DAE High Energy Physics Symposium*,
Springer Proceedings in Physics 261,
https://doi.org/10.1007/978-981-33-4408-2_71

best-fit values are 47.2° for NO and 48.1° for IO [1]. Therefore, a precise prediction of θ_{23} can be used to exclude and discriminate models in the light of forthcoming precision measurements. On the other hand, the current best-fit values of the Dirac CP phase δ are close to 234° for NO and 278° for IO. The possibility of CP conservation ($\sin \delta = 0$) is allowed at slightly above 1σ, and one of the CP violating value $\delta = \pi/2$ is disfavored at 99% CL. Thus, the other CP violating value $\delta = 3\pi/2$ and deviations around it remains a viable possibility. A specific generalization of $\mu\tau$ symmetry combined with a nonstandard CP transformation [2] (CP$^{\mu\tau\theta}$) is implemented in the neutrino Majorana mass term via the field transformation

$$\nu_{Ll} \to i G^{\theta}_{lm} \gamma^0 \nu^C_{Lm}. \tag{71.1}$$

In the neutrino flavor space, $G^{\mu\tau\theta}$ has the generic form

$$G^{\mu\tau\theta} = \begin{pmatrix} -1 & 0 & 0 \\ 0 & -\cos\theta & \sin\theta \\ 0 & \sin\theta & \cos\theta \end{pmatrix}, \tag{71.2}$$

with 'θ' being an arbitrary mixing angle that mixes the $\nu_{L\mu}$ and $\nu_{L\tau}$ flavor fields. In this work, to have testable predictions for masses as well as mixing, we consider a Friedberg-Lee (FL) transformation [3–6] in combination with (71.1),

$$\nu_{Ll} \to i G^{\mu\tau\theta}_{lm} \gamma^0 \nu^C_{Lm} + \eta_l \xi. \tag{71.3}$$

This leads to

$$M^\nu \eta = 0, \quad \text{and} \quad (G^{\mu\tau\theta})^T M_\nu G^{\mu\tau\theta} = M^*_\nu, \tag{71.4}$$

where η_l ($l = e, \mu, \tau$) are three arbitrary complex numbers, $\eta = (\eta_e\ \eta_\mu\ \eta_\tau)^T$ and ξ is a fermionic Grassmann field. Clearly, $G^{\mu\tau\theta}_{lm}$ in (71.2) is a symmetry which mixes ν_μ and ν_τ flavors and reduces to '$\mu\tau$-interchange' symmetry in the limit $\theta \to \pi/2$. The first condition in (71.4) is satisfied for a nontrivial eigenvector η if det $M_\nu = 0$. The latter condition implies that at least one of the light neutrino masses is zero. Thus, by construction, this model predicts the absolute light neutrino mass scale and has been investigated in substantial detail in [7].

71.2 FL Transformed CP$^{\mu\tau\theta}$ Invariance of M_ν

Using (71.4), a 3 × 3 symmetric mass matrix can most generally be parametrized as[1]:

---

[1]In rest of the paper, η_e, η_μ and η_τ are referred to as η_1, η_2 and η_3, respectively.

$$
M_\nu = \begin{pmatrix} -\frac{2a_1}{(1+c_\theta)}\frac{\eta_2}{\eta_1} & a_1 + ia_2 & -a_1 t_\frac{\theta}{2} + ia_2 t_\frac{\theta}{2}^{-1} \\ a_1 + ia_2 & c_1 t_\frac{\theta}{2} - a_1\frac{\eta_1}{\eta_2} - ia_2(1+c_\theta)\frac{\eta_1}{\eta_2} & c_1 - ia_2 t_\theta^{-1} c_\theta\frac{\eta_1}{\eta_2} \\ -a_1 t_\frac{\theta}{2} + ia_2 t_\frac{\theta}{2}^{-1} & c_1 - ia_2 t_\theta^{-1} c_\theta\frac{\eta_1}{\eta_2} & c_1 t_\frac{\theta}{2}^{-1} - a_1\frac{\eta_1}{\eta_2} + ia_2(1+c_\theta)\frac{\eta_1}{\eta_2} \end{pmatrix},
$$

$$(71.5)$$

where $c_\theta \equiv \cos\theta$, $s_\theta \equiv \sin\theta$ and $t_{\theta/2} = \tan\frac{\theta}{2}$. For simplicity, we choose the ratios $\frac{\eta_1}{\eta_1}$, $\frac{\eta_2}{\eta_3}$ and $\frac{\eta_3}{\eta_1}$ to be all real. In (71.5), there are five real free parameters: a_1, a_2, c_1, $\frac{\eta_1}{\eta_2}$ and θ. These can be constrained using the neutrino oscillation global-fit data. The mass matrix M_ν in (71.5) can be diagonalized a unitary matrix U as

$$
U^T M_\nu U = M_\nu^d \equiv \text{diag}(m_1, m_2, m_3), \tag{71.6}
$$

where m_i ($i = 1, 2, 3$) are real and we assume that $m_i \geq 0$. Without any loss of generality, we work in the diagonal basis of the charged lepton so that U can be related to the $PMNS$ mixing matrix U_{PMNS} as

$$
U = P_\phi U_{PMNS} \equiv P_\phi \begin{pmatrix} c_{12}c_{13} & e^{i\frac{\alpha}{2}}s_{12}c_{13} & s_{13}e^{-i(\delta-\frac{\beta}{2})} \\ -s_{12}c_{23} - c_{12}s_{23}s_{13}e^{i\delta} & e^{i\frac{\alpha}{2}}(c_{12}c_{23} - s_{12}s_{13}s_{23}e^{i\delta}) & c_{13}s_{23}e^{i\frac{\beta}{2}} \\ s_{12}s_{23} - c_{12}s_{13}c_{23}e^{i\delta} & e^{i\frac{\alpha}{2}}(-c_{12}s_{23} - s_{12}s_{13}c_{23}e^{i\delta}) & c_{13}c_{23}e^{i\frac{\beta}{2}} \end{pmatrix},
$$

$$(71.7)$$

where $P_\phi = \text{diag}(e^{i\phi_1}, e^{i\phi_2}, e^{i\phi_3})$ is an unphysical diagonal phase matrix and $c_{ij} \equiv \cos\theta_{ij}, s_{ij} \equiv \sin\theta_{ij}$ with the mixing angles $\theta_{ij} \in [0, \pi/2]$. Here, $\alpha, \beta \in [0, 2\pi]$ denote Majorana phases.

71.3 Mass Ordering, Mixing Angles and CP Properties

Equations (71.4) and (71.6) jointly imply

$$
G^\theta U^* = U\tilde{d}. \tag{71.8}
$$

where $\tilde{d} = \text{diag}(\tilde{d}_1, \tilde{d}_2, \tilde{d}_3)$, where each $\tilde{d}_i$ ($i = 1, 2, 3$) is either $+1$ or -1. From (71.8), we find that either $\alpha = 0$ or $\alpha = \pi$, and either $\beta = 2\delta$ or $\beta = 2\delta - \pi$. In addition, we also obtain, $\cot 2\theta_{23} = \cot\theta \cos(\phi_2 - \phi_3)$ and,

$$
\cos^2\delta = \cos^2\theta \sin^2(\phi_2 - \phi_3) = \frac{\cos^2\theta \sin^2 2\theta_{23} - \sin^2\theta \cos^2 2\theta_{23}}{\sin^2 2\theta_{23}}. \tag{71.9}
$$

Note that, (71.9) reduces to the co-bimaximal prediction of $CP^{\mu\tau}$ in the $\theta \to \pi/2$, as expected. Now, due to FL invariance, M_ν has a vanishing eigenvalue with corresponding normalized eigenvector given by

$$
\mathbf{v} = N^{-1} \begin{pmatrix} -\frac{\eta_1}{\eta_2}\cot\frac{\theta}{2} \\ -\cot\frac{\theta}{2} \\ 1 \end{pmatrix} e^{i\gamma}, \quad \text{with } N = \left[\left(1 + \frac{\eta_1^2}{\eta_2^2}\right)\cot^2\frac{\theta}{2} + 1\right]^{1/2}. \tag{71.10}
$$

If the zero eigenvalue is associated with $m_1 = 0$ ($m_3 = 0$), we discover additional consequences for the NO (IO). For NO, $\mathbf{v}$ is associated with the first column of PMNS matrix (71.7) which gives

$$\cos^2 \delta = \frac{\sin^2 2\theta_{12} s_{13}^2 \cos^2 \theta}{\sin^2 2\theta_{12} s_{13}^2 \cos^2 \theta + 4 \left[1 + \left(1 + \frac{\eta_1^2}{\eta_2^2} \right) \cot^2 \frac{\theta}{2} \right]^2 \cot^2 \frac{\theta}{2}} \tag{71.11}$$

and,

$$\cos^2 \theta_{23} = \frac{\left[\left\{ 1 + \left(1 + \frac{\eta_1^2}{\eta_2^2} \right) \cot^2 \frac{\theta}{2} \right\} s_{12}^2 - 1 \right] \cot \theta + \cot \frac{\theta}{2}}{\left(\cot^2 \frac{\theta}{2} - 1 \right) \cot \theta + 2 \cot \frac{\theta}{2}}. \tag{71.12}$$

In general, $\cos \delta \neq 0$ for NO. However, the numerically allowed range of δ is very close to $3\pi/2$: $\delta \in [269.6° - 270.4°]$. For IO, $\mathbf{v}$ is associated with the third column of PMNS, from which we obtain

$$\tan \theta_{23} = \cot \frac{\theta}{2}, \quad (\phi_2 - \phi_3) = \pi. \tag{71.13}$$

Since $(\phi_2 - \phi_3) = \pi$, it follows from (71.9) that the Dirac CP violation is maximal irrespective of the value of θ_{23}, i.e.,

$$\cos \delta = 0. \tag{71.14}$$

Though any significant departure from maximality in δ will exclude $CP^{\mu\tau}$ as well as this model ($CP^{\mu\tau\theta}$ + FL), if the experiments favor nonmaximal θ_{23} and a maximal value of δ, our model will survive while $CP^{\mu\tau}$ will be in tension.

71.4 Numerical Results

We use the (3σ) ranges of the globally fitted neutrino oscillation data [1] together with the upper bound of 0.17 eV on the sum of the light neutrino masses from PLANCK. Next, we discuss the predictions in our model on $0\nu\beta\beta$ decay, CP violation in $\nu_\mu \to \nu_e$ oscillations and flavor flux ratios at neutrino telescopes.

71.4.1 Neutrinoless Double Beta ($0\nu\beta\beta$) Decay Process

According to the PDG parametrization of U_{PMNS}, M_{ee} is given by

$$M_{ee} = c_{12}^2 c_{13}^2 m_1 + s_{12}^2 c_{13}^2 m_2 e^{i\alpha} + s_{13}^2 m_3 e^{i(\beta - 2\delta)}. \tag{71.15}$$

For a NO, δ deviates from $3\pi/2$, and $m_1 = 0$, and for an IO, $\delta = 3\pi/2$, and $m_3 = 0$. With these conditions, and with the four sets of values of α, β (obtained in Sect. 71.3), (71.15) M_{ee} we plot $|M_{ee}|$ against $\sum_i m_i$. It is clear from Fig. 71.1 that $|M_{ee}|$ in each plot of Fig. 71.1 leads to an upper limit which is below the sensitivity reach of the GERDA phase-II data. The upper bounds on $|M^{ee}|$ are expected to improve with upcoming experiments which may probe our model better.

71.4.2 Effect of CP Asymmetry in Neutrino Oscillations

The effect of leptonic Dirac CP violation makes its appearance in neutrino oscillation experiments though the phase δ in the asymmetry parameter $A_{\mu e}$, [8]

$$A_{\mu e} = \frac{P(\nu_\mu \to \nu_e) - P(\bar{\nu}_\mu \to \bar{\nu}_e)}{P(\nu_\mu \to \nu_e) + P(\bar{\nu}_\mu \to \bar{\nu}_e)} = \frac{2\sqrt{P_{atm}}\sqrt{P_{sol}}\sin\Delta_{32}\sin\delta}{P_{atm} + 2\sqrt{P_{atm}}\sqrt{P_{sol}}\cos\Delta_{32}\cos\delta + P_{sol}}$$
(71.16)

where $\Delta_{ij} = \Delta m_{ij}^2 L/4E$ and δ is to be substituted from (71.11) to (71.14) for NO and IO, respectively. The quantities P_{atm}, P_{sol} are given by

$$\sqrt{P_{atm}} = \sin\theta_{23}\sin\theta_{13}\frac{\sin(\Delta_{31} - aL)}{(\Delta_{31} - aL)}\Delta_{31},$$
(71.17)

$$\sqrt{P_{sol}} = \cos\theta_{23}\cos\theta_{13}\sin 2\theta_{12}\frac{\sin aL}{aL}\sin\Delta_{21},$$
(71.18)

where $a = G_F N_e/\sqrt{2}$ and N_e is the electron number density in the medium (Fig. 71.2).

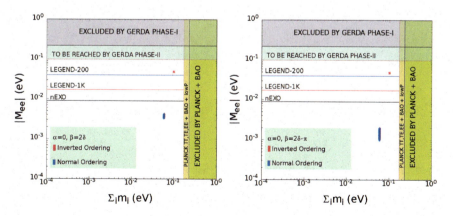

Fig. 71.1 Plots of $|M^{ee}|$ versus m_{min} for both the mass ordering with two of the four possible choices of the Majorana phases α and β

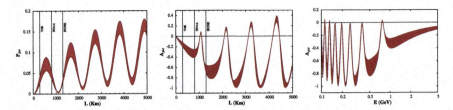

Fig. 71.2 First two figures respectively represent the variation of $P_{\mu e}$ and $A_{\mu e}$ against baseline length L for IO with energy $E = 1$ GeV and $\delta = 3\pi/2$. The bands correspond to 3σ ranges in θ_{12} and θ_{13}. The rightmost figure shows the variation of $A_{\mu e}$ with E for $L = 810$Km (NOνA)

71.4.3 Octant of θ_{23} from Flavor Flux Measurements

The flavor flux ratios R_l at the neutrino telescope are given by

$$R_l \equiv \frac{\phi_l^T}{\sum_m \phi_m^T - \phi_l^T} = \frac{1 + \sum_i |U_{li}|^2(|U_{\mu i}|^2 - |U_{\tau i}|^2)}{2 - \sum_i |U_{li}|^2(|U_{\mu i}|^2 - |U_{\tau i}|^2)}, \quad (l, m = e, \mu, \tau) \quad (71.19)$$

where each R_l depends on all three θ_{ij} and $\cos \delta$. For NO, θ_{23} and $\cos \delta$ are given by (71.12) and (71.11) while for IO those are given by (71.13) and (71.14), respectively. For both types of ordering, we display in Fig. 71.3 the variation of $R_{e,\mu,\tau}$ w.r.t θ.

In case of IO, the expressions for $R_{e,\mu,\tau}$ are relatively simple:

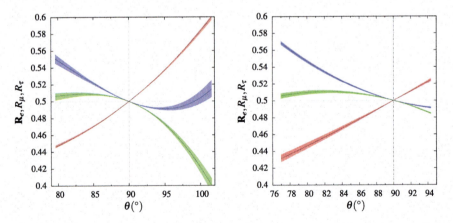

Fig. 71.3 Variation of R_e (red), R_μ (blue) and R_τ (green) with θ for NO (left panel) and IO (right panel). The solid lines correspond to the best-fit values of the mixing angles while the bands are caused by the current 3σ ranges of the mixing angles θ_{12} and θ_{13}

$$R_e = \frac{2 - \sin^2 2\theta_{12}c_\theta}{4 + \sin^2 2\theta_{12}c_\theta}, \quad R_{\mu,\tau} = \frac{1 + \frac{1}{4}\sin^2 2\theta_{12}c_\theta \pm (1 - \frac{1}{4}\sin^2 2\theta_{12})c_\theta^2}{2 - \frac{1}{4}\sin^2 2\theta_{12}c_\theta \mp (1 - \frac{1}{4}\sin^2 2\theta_{12})c_\theta^2}, \quad (71.20)$$

where we ignore terms of $\mathcal{O}(s_{13}^2)$. Clearly, for $R_e < \frac{1}{2}(R_e > \frac{1}{2}), \theta < \frac{\pi}{2}(\theta > \frac{\pi}{2})$. Since (71.13) implies $2\theta_{23} = \pi - \theta$, observational value of R_e will give a definite value of θ_{23}. In particular, $\theta > \frac{\pi}{2}$ implies $\theta_{23} < \frac{\pi}{4}$ and vice versa. Similar statement holds for observational values of R_μ. Conversely, a precision measurement of θ_{23} in long-baseline experiments will uniquely predict the range of R_l for all l.

71.5 Conclusion

We proposed a model where the light neutrino Majorana mass term enjoys a $\mu\tau$-flavored CP symmetry together with a Friedberg-Lee transformation. Both the mass ordering are allowed with vanishing smallest neutrino mass. In general, $\theta_{23} \neq \pi/4$, the phase $\delta = \pi/2$ for IO and nearly maximal for NO. For the latter, the deviation from maximality does not exceed $0.4°$ on either side of $\delta = 3\pi/2$. For the IO, θ_{23} is, in general, nonmaximal but δ is maximal. Evidently, any large departure of δ from $3\pi/2$ will exclude the model. We study the role of Dirac CP violation in different long-baseline $\nu_\mu \to \nu_e$ oscillation experiments. Finally, from the UHE neutrino flavor flux ratios at neutrino telescopes we comment on the predictability of the octant of θ_{23}.

References

1. I. Esteban et al., JHEP **1701**, 087 (2017)
2. R. Sinha, P. Roy, A. Ghosal, Phys. Rev. D **99**(3), 033009 (2019)
3. Z.Z. Xing, H. Zhang, S. Zhou, Phys. Lett. B **641**, 189 (2006)
4. S. Luo, Z.Z. Xing, Phys. Lett. B **646**, 242 (2007)
5. C.S. Huang, T.J. Li, W. Liao, S.H. Zhu, Phys. Rev. D **78**, 013005 (2008)
6. T. Araki, R. Takahashi, Eur. Phys. J. C **63**, 521 (2009)
7. R. Sinha, S. Bhattacharya, R. Samanta, JHEP **1903**, 081 (2019)
8. G.J. Ding, F. Gonzalez-Canales, J.W.F. Valle, Phys. Lett. B **753**, 644 (2016)

Chapter 72
Physics Potential of Long-Baseline Neutrino Oscillation Experiments in Presence of Sterile Neutrino

Rudra Majhi, C. Soumya, and Rukmani Mohanta

Abstract Recent result from Mini Booster Neutrino Experiment (MiniBooNE) agrees well with the Liquid Scintillator Neutrino Detector (LSND) experiment for excess event appearance and is a strong evidence for the existence of eV-scale sterile neutrino. Sterile neutrino can mix with the active neutrinos, which will be an exciting tool for sensitivity analysis of long-baseline neutrino oscillation experiments. Here we explore the effect of active-sterile mixing on the degeneracy resolution capabilities of these experiments. We found that the existence of sterile neutrino can lead to new kind of degeneracies among the oscillation parameters which are not present in the three active neutrino oscillation paradigm.

72.1 Introduction

Since last four decades several neutrino oscillation experiments have been confirmed the existence of non-zero neutrino mixing and masses. As of now, the study of neutrino physics has entered into an era of high precision. However, there exists some unsettled issues in this sector: (a) the value of atmospheric mixing angle (θ_{23}) can be non-maximal, i.e., either $\theta_{23} < 45°$ or $\theta_{23} > 45°$, (b) the sign of atmospheric mass splitting Δm_{31}^2 can be positive or negative, for $\Delta m_{31}^2 > 0$, it is known as normal hierarchy (NH) and inverted hierarchy (IH) for $\Delta m_{31}^2 < 0$, and (c) the value of CP-violating phase δ_{CP} is not yet determined which can tell us whether CP violation exists in lepton sector or not. The long-baseline experiments play a major role in the determination of these unknowns due to the presence of enhanced matter effect [1].

R. Majhi (✉) · R. Mohanta
University of Hyderabad, Hyderabad 500046, India
e-mail: rudra.majhi95@gmail.com

R. Mohanta
e-mail: rukmani98@gmail.com

C. Soumya
Institute of Physics, Sachivalaya Marg, Bhubaneswar 751005, India
e-mail: soumyac20@gmail.com

© Springer Nature Singapore Pte Ltd. 2021
P. K. Behera et al. (eds.), *XXIII DAE High Energy Physics Symposium*,
Springer Proceedings in Physics 261,
https://doi.org/10.1007/978-981-33-4408-2_72

However, the existence of fourfold degeneracies among the oscillation parameters greatly affects the sensitivities of these experiments [2]. Therefore, the resolution of degeneracies among the oscillation parameters is the primary concern in neutrino oscillation studies.

Another unresolved issue is the existence of sterile neutrino. Moreover, latest result from the MiniBooNE is also in good agreement with the excess of events reported by LSND regarding the existence of eV-scale sterile neutrino [3]. Though sterile neutrinos are blind to weak interaction, they can mix with active neutrinos. The mixing of sterile neutrino with active neutrinos can be explained by an effective 4×4 matrix. In addition to standard neutrino oscillation parameters, we need three mixing angles (θ_{14}, θ_{24}, θ_{34}), two phases (δ_{14}, δ_{34}), and one mass-squared difference (Δm_{41}^2) for parametrization of neutrino mixing. The vacuum oscillation probability in 3+1 framework is given by

$$P_{\mu e} \sim 8 \sin\theta_{13} \sin\theta_{12} \cos\theta_{12} \sin\theta_{23} \cos\theta_{23} (\alpha\Delta) \sin\Delta \cos(\Delta \pm \delta_{13})$$
$$+ 4\sin^2\theta_{23} \sin^2\theta_{13} \sin^2\Delta + 4\sin\theta_{14} \sin\theta_{24} \sin\theta_{13} \sin\theta_{23} \sin\Delta \sin(\Delta \pm \delta_{13} \mp \delta_{14}) \,,$$

where $\Delta \equiv \Delta m_{31}^2 L/4E$, $\alpha \equiv \Delta m_{21}^2 / \Delta m_{31}^2$, $\Delta m_{ij}^2 = m_i^2 - m_j^2$, and L, E are, respectively, baseline and energy of the neutrino used in the experiment. Upper (lower) sign is for neutrino (anti-neutrino), in double sign case.

72.2   Simulation Details

As we focus on currently running long-baseline experiments NOνA and T2K, we simulate these experiments using GLoBES software package along with snu plugin [5, 6]. The auxiliary files and experimental specification of these experiments that we use for the analysis are taken from [7–9]. In our analysis, we use the values of oscillation parameters as given in Table 72.1.

72.3   Degeneracies and MH Sensitivity

In order to analyze degeneracies among the oscillation parameters at probability level, in the top panel of Fig. 72.1 we show ν_e ($\bar{\nu}_e$) appearance oscillation probability as a function of δ_{CP}. From the left panel of figure, it can be seen that the bands NH-HO and IH-LO bands are very well separated in neutrino channel, whereas the bands for NH-LO and IH-HO are overlapped with each other which results degeneracies among the oscillation parameters. It can be shown that in the anti-neutrino channel the case is just opposite. Therefore, a combined analysis of neutrino and anti-neutrino data helps in degeneracies resolution and improves the sensitivity of LBL experiment to the unknowns in standard paradigm. From the right panel, it can be

Table 72.1 The values of oscillation parameters that we consider in analysis [4]

Parameters	True values	Test value range
$\sin^2 \theta_{12}$	0.304	NA
$\sin^2 2\theta_{13}$	0.085	NA
$\sin^2 \theta_{23}$	0.5	$0.4 \longrightarrow 0.6$
	(LO 0.44)	$0.4 \longrightarrow 0.5$
	(HO 0.56)	$0.5 \longrightarrow 0.6$
δ_{CP}	$-90°$	$-180° \longrightarrow 180°$
$\frac{\Delta m^2_{12}}{10^{-5}eV^2}$	7.4	NA
$\frac{\Delta m^2_{31}}{10^{-3}eV^2}$	2.5(NH)	$2.36 \longrightarrow 2.64$
	-2.5(IH)	$-2.64 \longrightarrow -2.36$
$\sin^2 \theta_{14}$	0.025	$\theta_{14}(0°, 15°)$
$\sin^2 \theta_{24}$	0.025	$\theta_{14}(0°, 15°)$
$\sin^2 \theta_{34}$	0	NA
$\frac{\Delta m^2_{14}}{1^e V^2}$	1	NA
δ_{14}	$-90°$	$-180° \longrightarrow 180°$
δ_{34}	$0°$	$-180° \longrightarrow 180°$

seen that there emerged new types of degeneracies among the oscillation parameters in presence of sterile neutrino for $\delta_{14} = -90$ which can deteriorate the sensitivity. Another way of representing these degeneracies among oscillation parameters is by using the bi-probability plot as given in bottom panel of Fig. 72.1. The ellipses in the figure correspond to three flavor case, whereas the bands represent the oscillation probabilities in presence of sterile neutrino with all possible values of δ_{14} phase. From the figure, it can be seen that the ellipses for LO and HO are very well separated for both hierarchies, whereas the ellipses for NH and IH for both LO and HO are overlapped each other and give rise to degeneracies. Therefore, NOνA experiment is more sensitive to octant of θ_{23} than that of mass hierarchy. New degeneracies are found in 3+1 paradigm, for all combinations of mass hierarchy and θ_{23}. More degeneracies between lower and higher octant along with standard case, indicate that experiment is loosing its sensitivity in presence of sterile neutrino.

Finally, we discuss how MH sensitivity of NOνA get modify in presence of sterile neutrino. From the Fig. 72.2, one can see that the wrong mass hierarchy can be ruled out significantly above 2 σ in the favorable regions, i.e., lower half plane (upper half plane) for NH (IH) in the standard paradigm. Whereas, in presence of sterile neutrino the δ_{CP} coverage for the mass hierarchy sensitivity is significantly reduced. The synergy of NOνA+T2K in 3+1 case showed a significant increase in the MH sensitivity with the increase of δ_{CP} coverage above 3σ C.L.

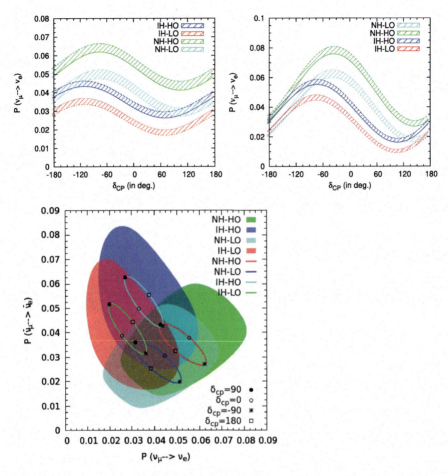

Fig. 72.1 The neutrino oscillation probability as a function of δ_{CP} is in the top panel. The top left panel is for 3 flavor case and top right panel is for 3+1 case with $\delta_{14} = -90°$. [Bottom panel] Bi-probability plots for NovA in 3 years in neutrino and 3 years in anti-neutrino mode

72.4   Effect on Neutrino-Less Double Beta Decay

Neutrino-less double beta decay ($0\nu\beta\beta$) is a lepton number violating process with $\Delta L = 2$. Implication of this decay mode on Neutrino Physics will be that it can confirm the Majorana nature of neutrinos and neutrino masses. In this perspective, this decay process is important and we will study the effect of light sterile neutrino on $0\nu\beta\beta$ decay.

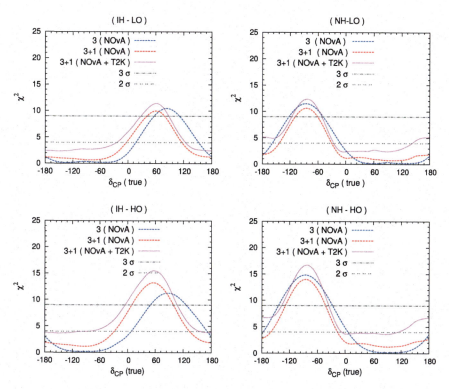

Fig. 72.2 MH sensitivity as a function of true values of δ_{CP}. The left (right) panel is for inverted (normal) hierarchy and the upper (bottom) panel is for LO (HO)

Various experiments like KamLAND-Zen, GERDA, EXO-200, etc. have provide effective Majorana mass parameter $|M_{ee}|$. Current best upper limit on the $|M_{ee}|$ from KamLAND-Zen collaboration is $|M_{ee}| < (0.061 - 0.165)$ eV at 90% C.L. [10].

The effective Majorana mass, which is the key parameter of $0\nu\beta\beta$ decay process, is defined in the standard three neutrino formalism, as

$$|M_{ee}| = \left| U_{e1}^2 m_1 + U_{e2}^2 m_2 e^{i\alpha} + U_{e3}^2 m_3 e^{i\beta} \right|, \qquad (72.1)$$

where U_{ei} are the PMNS matrix elements and α, β are the Majorana phases. In terms of the lightest neutrino mass, atmospheric and solar mass-squared differences, $|M_{ee}|$ can be written for NH and IH case. If sterile neutrino will be exist then analogically one can write the expression as

$$|M_{ee}| = \left| U_{e1}^2 m_1 + U_{e2}^2 m_2 e^{i\alpha} + U_{e3}^2 m_3 e^{i\beta} + U_{e4}^2 m_4 e^{i\gamma} \right|. \qquad (72.2)$$

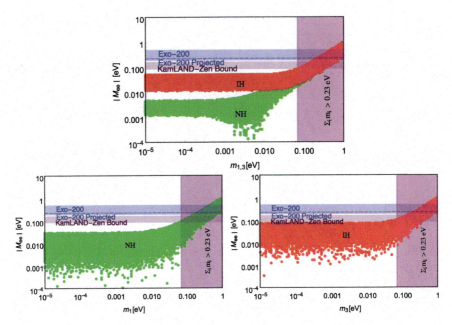

Fig. 72.3 Parameter space between effective Majorana mass parameter $|M_{ee}|$ and lightest neutrino mass

Using PMNS matrix elements, Dirac CP phase, Majorana phases α and β in between $[0, 2\pi]$, we show the variation of $|M_{ee}|$ with lightest neutrino mass for both 3 neutrino and 3+1 model in Fig. 72.3. Top panel is for three neutrino and bottom panel in presence of sterile neutrino. In bottom panel left panel is for NH and right panel for IH case. Horizontal regions show bounds from different $0\nu\beta\beta$ experiments and vertical shaded regions are disfavoured from PLANCK data on sum of light neutrinos. Current bound on the sum of neutrino masses is $\sum_i m_i < 0.23$ eV at 95% C.L. from Planck+WP+highL+BAO data. From the figure it is clear that in presence of sterile neutrino the IH parameter space of $|M_{ee}|$ is within sensitivity region of the KamLAND-Zen experiment. Also some overlap regions are there in between NH and IH case.

72.5 Conclusion

In this paper, we discussed the effect of active-sterile mixing on the degeneracy resolution capability and MH sensitivity of NOνA experiment. We found that introduction of sterile neutrino gives rise to new kind of degeneracies among the oscillation parameters which results in reduction of δ_{CP} coverage for MH sensitivity of NOνA experiment. We also found that addition of T2K data helps in increment of MH sensitivity analysis by increase of δ_{CP} coverage above 3σ C.L. in presence of sterile neutrino. Also we have shown that the inclusion of sterile neutrino enhances the effective parameter space for $|M_{ee}|$ and for IH case it could be observable in KamLAND-Zen experiment.

Acknowledgements We thank INSPIRE fellowship of DST and SERB for financial support.

References

1. M.V. Diwan et al., Long-baseline neutrino experiments. Ann. Rev. Nucl. Part. Sci. **66**, 47–71 (2016). https://doi.org/10.1146/annurev-nucl-102014-021939
2. V. Barger et al., Phys. Rev. D **65**, 073023 (2002)
3. A. Aguilar-Arevalo et al., Observation of a significant excess of electron-like events in the MiniBooNE short-baseline neutrino experiment. FERMILAB-PUB-18-219, LA-UR-18-24586 (2018)
4. D. Dutta et al., Capabilities of long-baseline experiments in the presence of a sterile neutrino. JHEP **1611**, 122 (2016). https://doi.org/10.1007/JHEP11(2016)122
5. P. Huber et al., Simulation of long-baseline neutrino oscillation experiments with GLoBES. Comput. Phys. Commun. **167**, 195 (2005). https://doi.org/10.1016/j.cpc.2005.01.003
6. P. Huber et al., New features in the simulation of neutrino oscillation experiments with GLoBES 3.0. Comput. Phys. Commun. **177**, 432–438 (2007). https://doi.org/10.1016/j.cpc.2007.05.004
7. M.A. Acero et al. (NOvA Collaboration), Phys. Rev. D **98**, 032012
8. K. Abe et al. (T2K Collaboration), Phys. Rev. Lett. **118**, 151801 (2017)
9. C. Soumya et al., A comprehensive study of the discovery potential of NOvA, T2K, and T2HK experiments. Adv. High Energy Phys. (2016). 9139402. https://doi.org/10.1155/2016/9139402
10. A. Gando et al. (KamLAND-Zen Collaboration), Phys. Rev. Lett. **117**(8), 082503 (2016)

Chapter 73
Using A4 to Ameliorate Popular Lepton Mixings: A Model for Realistic Neutrino Masses and Mixing Based on Seesaw

Soumita Pramanick

Abstract A neutrino mass model conserving discrete flavour symmetry $A4$ is proposed. Using an interplay of both Type-I and Type-II seesaw mechanisms the discrepancies present in the popular lepton mixing patterns, viz., Tribimaximal (TBM), Bimaximal (BM), and Golden Ratio (GR) mixings are cured and they were harmonized with the present neutrino oscillation observations. A scenario with no solar mixing (NSM) was also considered. The model has several predictions testable in the light of future oscillation experiments.

73.1 Introduction

This talk is based on [1]. A model with two-component Lagrangian formalism based on Type-I and Type-II seesaw mechanism is proposed for tree-level generation of realistic neutrino masses and mixing using the discrete flavour symmetry $A4$ in [1]. The Type-II seesaw is the dominant contribution, Type-I seesaw is the sub-dominant contribution. Type-II seesaw features vanishing solar splitting ($\Delta m_{solar}^2 = 0$), $\theta_{13} = 0$, $\theta_{23} = \pi/4$ and can have any solar mixing angle. Choices of θ_{12}^0 corresponding to certain mixing patterns, viz., $\theta_{12}^0 = 35.3°$ (tribimaximal), $45.0°$ (bimaximal), $31.7°$ (golden ratio) were considered. Tribimaximal, Bimaximal, Golden Ratio mixing are abbreviated as TBM, BM, and GR mixing, respectively. No solar mixing (NSM), i.e., $\theta_{12}^0 = 0$ was also studied. Including first-order corrections originating from the sub-dominant Type-I seesaw can yield the values of these oscillation parameters into the ranges allowed by the data leading to interesting interrelationships between them that are testable in future oscillation experiments. For example, normal (inverted) ordering gets associated with first (second) octant of θ_{23}. The 3σ neutrino oscillation parameters global fits [2, 3]:

S. Pramanick (✉)
Department of Physics, University of Calcutta, 92 Acharya Prafulla Chandra Road, Kolkata 700009, India
e-mail: soumita509@gmail.com; soumitapramanick5@gmail.com

Harish-Chandra Research Institute, Chhatnag Road, Jhunsi 211019, Allahabad, India

© Springer Nature Singapore Pte Ltd. 2021
P. K. Behera et al. (eds.), *XXIII DAE High Energy Physics Symposium*,
Springer Proceedings in Physics 261,
https://doi.org/10.1007/978-981-33-4408-2_73

$$\Delta m_{21}^2 = (7.02 - 8.08) \times 10^{-5}\,\text{eV}^2, \quad \theta_{12} = (31.52 - 36.18)°,$$
$$|\Delta m_{31}^2| = (2.351 - 2.618) \times 10^{-3}\,\text{eV}^2, \quad \theta_{23} = (38.6 - 53.1)°,$$
$$\theta_{13} = (7.86 - 9.11)°, \quad \delta = (0 - 360)° \ . \tag{73.1}$$

The above numbers are from NuFIT2.1 of 2016 [2]. Models with similar objectives based on $S3$ and $A4$ at tree level were studied in [4]. A model-independent approach of the same can be found in [5].

73.2 The Model

The Type-I and Type-II seesaw contributions of the model arise out of an underlying $A4$ symmetry of the Lagrangian. $A4$ has four irreducible representations, viz., one three-dimensional representation 3 and three one-dimensional representations 1, 1′, and 1″. Note $1' \times 1'' = 1$. Two three-dimensional representations can be combined as

$$3 \otimes 3 = 1 \oplus 1' \oplus 1'' \oplus 3 \oplus 3 \ . \tag{73.2}$$

Thus when two triplets $3_a \equiv a_i$ and $3_b \equiv b_i$, with $i = 1, 2, 3$ combine they yield

$$1 = a_1 b_1 + a_2 b_2 + a_3 b_3 \equiv \rho_{1ij} a_i b_j \ ,$$
$$1' = a_1 b_1 + \omega^2 a_2 b_2 + \omega a_3 b_3 \equiv \rho_{3ij} a_i b_j \ ,$$
$$1'' = a_1 b_1 + \omega a_2 b_2 + \omega^2 a_3 b_3 \equiv \rho_{2ij} a_i b_j \ ,$$
$$c_i = \left(\frac{a_2 b_3 + a_3 b_2}{2}, \frac{a_3 b_1 + a_1 b_3}{2}, \frac{a_1 b_2 + a_2 b_1}{2} \right) \ , \quad \text{or,} \quad c_i \equiv \alpha_{ijk} a_j b_k \ ,$$
$$d_i = \left(\frac{a_2 b_3 - a_3 b_2}{2}, \frac{a_3 b_1 - a_1 b_3}{2}, \frac{a_1 b_2 - a_2 b_1}{2} \right) \ , \quad \text{or,} \quad d_i \equiv \beta_{ijk} a_j b_k \ , \quad (i, j, k, \text{ are cyclic}) \ . \tag{73.3}$$

For detailed discussion of $A4$, see [1, 6, 7]. The particle spectrum of the model with their respective quantum numbers are shown in Tables 73.1 and 73.2. An extensive study of the scalar potential arising out of the scalars in Table 73.2 can be found in the appendix of [1]. The $SU(2)_L \times U(1)_Y$ conserving Lagrangian that preserves $A4$ symmetry is given by

$$\begin{aligned}
\mathcal{L}_{mass} = & \ y_j \rho_{jik} \bar{l}_{Li} l_{Rj} \Phi_k^0 \quad \text{(charged lepton mass)} \\
& + f \rho_{1ik} \bar{\nu}_{Li} N_{Rk} \eta^0 \quad \text{(neutrino Dirac mass)} \\
& + \frac{1}{2} \left(\sum_{n=a,b} \hat{Y}_n^L \alpha_{ijk} \nu_{Li}^T C^{-1} \nu_{Lj} \hat{\Delta}_{nk}^{L0} + Y_\zeta^L \rho_{\zeta ij} \nu_{Li}^T C^{-1} \nu_{Lj} \Delta_\zeta^{L0} \right) \quad \text{(neutrino Type–II see–saw mass)} \\
& + \frac{1}{2} \left(\sum_{p=a,b,c} \hat{Y}_p^R \alpha_{ijk} N_{Ri}^T C^{-1} N_{Rj} \hat{\Delta}_{kp}^{R0} + Y_\gamma^R \rho_{\gamma ij} N_{Ri}^T C^{-1} N_{Rj} \Delta_\gamma^{R0} \right) \quad \text{(rh neutrino mass)} + h.c.
\end{aligned} \tag{73.4}$$

Table 73.1 Leptons with their $A4$ and $SU(2)_L$ properties. Lepton number (L) and hypercharge (Y) are shown

Fields	Notations	A4	$SU(2)_L$ (Y)	L
Left-handed leptons	$(\nu_i, l_i)_L$	3	2 (−1)	1
	l_{1R}	1		
Right-handed charged leptons	l_{2R}	$1'$	1 (−2)	1
	l_{3R}	$1''$		
Right-handed neutrinos	N_{iR}	3	1 (0)	−1

No soft breaking $A4$ terms were allowed. Symmetries were broken spontaneously when the scalars acquire their vacuum expectation values ($vevs$). In the Lagrangian basis this leads to the mass matrices:

$$M_{e\mu\tau} = \frac{v}{\sqrt{3}} \begin{pmatrix} y_1 & y_2 & y_3 \\ y_1 & \omega y_2 & \omega^2 y_3 \\ y_1 & \omega^2 y_2 & \omega y_3 \end{pmatrix} \; , \; M_{\nu L} = \begin{pmatrix} (Y_1^L + 2Y_2^L)u_L & \frac{1}{2}\hat{Y}_b^L v_{Lb} & \frac{1}{2}\hat{Y}_b^L v_{Lb} \\ \frac{1}{2}\hat{Y}_b^L v_{Lb} & (Y_1^L - Y_2^L)u_L & \frac{1}{2}(\hat{Y}_a^L v_{La} + \hat{Y}_b^L v_{Lb}) \\ \frac{1}{2}\hat{Y}_b^L v_{Lb} & \frac{1}{2}(\hat{Y}_a^L v_{La} + \hat{Y}_b^L v_{Lb}) & (Y_1^L - Y_2^L)u_L \end{pmatrix} .$$

$$(73.5)$$

Here $M_{e\mu\tau}$ represents the charged lepton mass matrix and $M_{\nu L}$ is the left-handed Majorana neutrino mass matrix arising out of Type-II seesaw when $Y_2^L = Y_3^L$. From the charged lepton mass matrix we have $y_1 v = m_e$, $y_2 v = m_\mu$, $y_3 v = m_\tau$. The neutrino Dirac mass matrix (M_D) and right-handed Majorana neutrino mass matrix ($M_{\nu R}$) we get that take part in Type-I seesaw are given by

$$M_D = fu \, \mathbb{I} \; , \quad M_{\nu R} = m_R \begin{pmatrix} \chi_1 & \chi_6 & \chi_5 \\ \chi_6 & \chi_2 & \chi_4 \\ \chi_5 & \chi_4 & \chi_3 \end{pmatrix} . \qquad (73.6)$$

If we denote the scale of neutrino Dirac masses by m_D then we can identify $fu = m_D$. The χ_i in (73.6) are given by

$$m_R \chi_1 \equiv (Y_1^R u_{1R} + Y_2^R u_{2R} + Y_3^R u_{3R})$$

$$m_R \chi_2 \equiv (Y_1^R u_{1R} + \omega Y_2^R u_{2R} + \omega^2 Y_3^R u_{3R})$$

$$m_R \chi_3 \equiv (Y_1^R u_{1R} + \omega^2 Y_2^R u_{2R} + \omega Y_3^R u_{3R})$$

$$m_R \chi_4 \equiv \frac{1}{2}(\hat{Y}_a^R v_{Ra} + \hat{Y}_b^R v_{Rb} + \hat{Y}_c^R v_{Rc})$$

$$m_R \chi_5 \equiv \frac{1}{2}(\hat{Y}_a^R v_{Ra} + \omega \hat{Y}_b^R v_{Rb} + \omega^2 \hat{Y}_c^R v_{Rc})$$

$$m_R \chi_6 \equiv \frac{1}{2}(\hat{Y}_a^R v_{Ra} + \omega^2 \hat{Y}_b^R v_{Rb} + \omega \hat{Y}_c^R v_{Rc}). \qquad (73.7)$$

Table 73.2 Scalars in the model with their $A4$ and $SU(2)_L$ behaviours. Their lepton number (L), hypercharge (Y) and vev configurations are also given

Purpose	Notations	A4	$SU(2)_L$ (Y)	L	vev
Charged fermion mass	$\Phi = \begin{pmatrix} \phi_1^+ & \phi_1^0 \\ \phi_2^+ & \phi_2^0 \\ \phi_3^+ & \phi_3^0 \end{pmatrix}$	3	2 (1)	0	$\langle\Phi\rangle = \frac{v}{\sqrt{3}}\begin{pmatrix} 0 & 1 \\ 0 & 1 \\ 0 & 1 \end{pmatrix}$
Neutrino Dirac mass	$\eta = (\eta^0, \eta^-)$	1	2 (−1)	2	$\langle\eta\rangle = \left(u, 0\right)$
Type-II seesaw mass	$\hat{\Delta}_a^L = \begin{pmatrix} \hat{\Delta}_{1a}^{++} & \hat{\Delta}_{1a}^+ & \hat{\Delta}_{1a}^0 \\ \hat{\Delta}_{2a}^{++} & \hat{\Delta}_{2a}^+ & \hat{\Delta}_{2a}^0 \\ \hat{\Delta}_{3a}^{++} & \hat{\Delta}_{3a}^+ & \hat{\Delta}_{3a}^0 \end{pmatrix}^L$	3	3 (2)	−2	$\langle\hat{\Delta}_a^L\rangle =$ $v_{La}\begin{pmatrix} 0 & 0 & 1 \\ 0 & 0 & 0 \\ 0 & 0 & 0 \end{pmatrix}$
Type-II seesaw mass	$\hat{\Delta}_b^L = \begin{pmatrix} \hat{\Delta}_{1b}^{++} & \hat{\Delta}_{1b}^+ & \hat{\Delta}_{1b}^0 \\ \hat{\Delta}_{2b}^{++} & \hat{\Delta}_{2b}^+ & \hat{\Delta}_{2b}^0 \\ \hat{\Delta}_{3b}^{++} & \hat{\Delta}_{3b}^+ & \hat{\Delta}_{3b}^0 \end{pmatrix}^L$	3	3 (2)	−2	$\langle\hat{\Delta}_b^L\rangle =$ $v_{Lb}\begin{pmatrix} 0 & 0 & 1 \\ 0 & 0 & 1 \\ 0 & 0 & 1 \end{pmatrix}$
Type-II seesaw mass	$\Delta_\zeta^L =$ $(\Delta_\zeta^{++}, \Delta_\zeta^+, \Delta_\zeta^0)^L$	1	3 (2)	−2	$\langle\Delta_1^L\rangle = \left(0, 0, u_L\right)$
		$1'$	3 (2)	−2	$\langle\Delta_2^L\rangle = \left(0, 0, u_L\right)$
		$1''$	3 (2)	−2	$\langle\Delta_3^L\rangle = \left(0, 0, u_L\right)$
Right-handed neutrino mass	$\hat{\Delta}_a^R = \begin{pmatrix} \hat{\Delta}_{1a}^0 \\ \hat{\Delta}_{2a}^0 \\ \hat{\Delta}_{3a}^0 \end{pmatrix}^R$	3	1 (0)	2	$\langle\hat{\Delta}_a^R\rangle = v_{Ra}\begin{pmatrix} 1 \\ 1 \\ 1 \end{pmatrix}$
Right-handed neutrino mass	$\hat{\Delta}_b^R = \begin{pmatrix} \hat{\Delta}_{1b}^0 \\ \hat{\Delta}_{2b}^0 \\ \hat{\Delta}_{3b}^0 \end{pmatrix}^R$	3	1 (0)	2	$\langle\hat{\Delta}_b^R\rangle = v_{Rb}\begin{pmatrix} 1 \\ \omega \\ \omega^2 \end{pmatrix}$
Right-handed neutrino mass	$\hat{\Delta}_c^R = \begin{pmatrix} \hat{\Delta}_{1c}^0 \\ \hat{\Delta}_{2c}^0 \\ \hat{\Delta}_{3c}^0 \end{pmatrix}^R$	3	1 (0)	2	$\langle\hat{\Delta}_c^R\rangle = v_{Rc}\begin{pmatrix} 1 \\ \omega^2 \\ \omega \end{pmatrix}$
Right-handed neutrino mass	$\Delta_1^R = (\Delta_1^0)^R$	1	1 (0)	2	$\langle\Delta_1^R\rangle = u_{1R}$
Right-handed neutrino mass	$\Delta_2^R = (\Delta_2^0)^R$	$1'$	1 (0)	2	$\langle\Delta_2^R\rangle = u_{2R}$
Right-handed neutrino mass	$\Delta_3^R = (\Delta_3^0)^R$	$1''$	1 (0)	2	$\langle\Delta_3^R\rangle = u_{3R}$

Needless to mention m_R and χ_i are mass scale of the right-handed Majorana neutrino and $\mathcal{O}(1)$ dimensionless quantities respectively. To diagonalize the charged lepton mass matrix in (73.5) one can appoint a unitary transformation U_L on the left-handed lepton doublets, keeping the right-handed charged leptons unchanged. In order to keep the neutrino Dirac mass matrix still proportional to identity in this changed basis, the unitary transformation V_R is applied to the right-handed neutrino fields.

$$U_L = \frac{1}{\sqrt{3}} \begin{pmatrix} 1 & 1 & 1 \\ 1 & \omega^2 & \omega \\ 1 & \omega & \omega^2 \end{pmatrix} = V_R . \tag{73.8}$$

In this basis the charged lepton mass matrix is given by $diag(m_e, m_\mu, mu_\tau)$ and the entire lepton mixing therefore arises out of the neutrino sector. We call this basis as the flavour basis. For structures of all the mass matrices in flavour basis see [1]. The left-handed Majorana neutrino mass matrix in the flavour basis $M_{\nu L}^{flavour}$ is not diagonal. It has to be diagonalized by another unitary transformation say U^0 of the following form, to go the mass basis of the neutrinos in which the left-handed Majorana neutrino mass matrix is diagonal:

$$U^0 = \begin{pmatrix} \cos\theta_{12}^0 & \sin\theta_{12}^0 & 0 \\ -\frac{\sin\theta_{12}^0}{\sqrt{2}} & \frac{\cos\theta_{12}^0}{\sqrt{2}} & \frac{1}{\sqrt{2}} \\ \frac{\sin\theta_{12}^0}{\sqrt{2}} & -\frac{\cos\theta_{12}^0}{\sqrt{2}} & \frac{1}{\sqrt{2}} \end{pmatrix} . \tag{73.9}$$

In other words, $M^0 = M_{\nu L}^{mass} = U^{0T} M_{\nu L}^{flavour} U^0 = diag(m_1^{(0)}, m_1^{(0)}, m_3^{(0)})$. This M^0 is the dominant Type-II seesaw contribution in the mass basis. Note the first two mass eigenstates are degenerate to ensure the vanishing solar splitting. Thus degenerate perturbation theory has to be considered for the solar sector. From the form of U^0 one can readily infer $\theta_{13} = 0$, $\theta_{23} = \pi/4$ and θ_{12}^0 can be of any value. We consider θ_{12}^0 values corresponding to TBM, BM, GR, and NSM scenarios. Again $M_D = m_D \mathbb{I}$ can be retained by applying the $U^{0\dagger}$ on the right-handed fields. The right-handed neutrino mass matrix so obtained can be found in [1] giving rise to the Type-I seesaw contribution:

$$M' = \left[M_D^T (M_{\nu R})^{-1} M_D \right] = \frac{m_D^2}{m_R} \begin{pmatrix} 0 & y\,e^{i\phi_1} & y\,e^{i\phi_1} \\ y\,e^{i\phi_1} & \frac{x\,e^{i\phi_2}}{\sqrt{2}} & \frac{-x\,e^{i\phi_2}}{\sqrt{2}} \\ y\,e^{i\phi_1} & \frac{-x\,e^{i\phi_2}}{\sqrt{2}} & \frac{x\,e^{i\phi_2}}{\sqrt{2}} \end{pmatrix} . \tag{73.10}$$

Here x and y are dimensionless $\mathcal{O}(1)$ real quantities. The details of the analysis can be found in [1].

73.3 Results

The CP-conserving scenario, i.e., ($\phi_1 = 0$ or π, $\phi_2 = 0$ or π) is well studied in [1]. For the CP-violating case, to keep hermiticity of the theory intact, the combination $(M^{0\dagger} M' + M'^\dagger M^0)$ is treated as the sub-dominant correction to the dominant $M^{0\dagger} M^0$ term. Using degenerate perturbation theory to obtain the first-order corrections one gets

$$\theta_{12} = \theta_{12}^0 + \zeta \ , \quad \tan 2\zeta = 2\sqrt{2}\, \frac{y}{x}\, \frac{\cos\phi_1}{\cos\phi_2} \ , \tag{73.11}$$

where

$$\sin\epsilon = \frac{y\cos\phi_1}{\sqrt{y^2\cos^2\phi_1 + x^2\cos^2\phi_2/2}} \ , \quad \cos\epsilon = \frac{x\cos\phi_2/\sqrt{2}}{\sqrt{y^2\cos^2\phi_1 + x^2\cos^2\phi_2/2}} \ , \quad \tan\epsilon = \frac{1}{2}\tan 2\zeta \ . \tag{73.12}$$

The solar splitting is given by

$$\Delta m_{solar}^2 = \sqrt{2} m_1^{(0)}\, \frac{m_D^2}{m_R}\sqrt{x^2\cos^2\phi_2 + 8y^2\cos^2\phi_1} = \sqrt{2} m_1^{(0)}\, \frac{m_D^2}{m_R}\, \frac{x\cos\phi_2}{\cos 2\zeta} = \sqrt{2} m_1^{(0)}\, \frac{m_D^2}{m_R}\, \frac{2\sqrt{2}y\cos\phi_1}{\sin 2\zeta} \ . \tag{73.13}$$

In Table 73.3 allowed ranges of ζ and ϵ for different mixing patterns are shown. The third first-order corrected ket yields

$$\sin\theta_{13}\cos\delta = \kappa_c\, \sin(\epsilon - \theta_{12}^0) \quad \text{and} \quad \tan(\pi/4 - \theta_{23}) \equiv \tan\omega = \frac{\sin\theta_{13}\cos\delta}{\tan(\epsilon - \theta_{12}^0)}, \tag{73.14}$$

with

$$\kappa_c = \frac{m_D^2}{m_R m^-}\sqrt{y^2\cos^2\phi_1 + x^2\cos^2\phi_2/2} \ , \tag{73.15}$$

where $m^- \equiv m_3^{(0)} - m_1^{(0)}$. Thus m^- is positive for normal ordering and negative for inverted ordering. Hence the sign of κ_c is dictated by the mass ordering. The sign of κ_c along with $(\epsilon - \theta_{12}^0)$ determine the octant of θ_{23} and quadrant of δ. Therefore the neutrino mass ordering gets co-related to the octant of θ_{23} and quadrant of δ as a consequence of this model as shown in Table 73.4 for all the four mixing patterns. For example, normal ordering is found always associated with θ_{23} in first octant. Such inter-relations in Table 73.4 can be tested by oscillation experiments in future

Table 73.3 Values of ζ, ϵ and $(\epsilon - \theta_{12}^0)$ allowed by 3σ oscillation data for different mixings

Model (θ_{12}^0)	TBM (35.3°)	BM (45.0°)	GR (31.7°)	NSM (0.0°)
ζ	$-4.0° \leftrightarrow 0.6°$	$-13.7° \leftrightarrow -9.1°$	$-0.4° \leftrightarrow 4.2°$	$31.3° \leftrightarrow 35.9°$
ϵ	$-4.0° \leftrightarrow 0.6°$	$-14.5° \leftrightarrow -9.3°$	$-0.4° \leftrightarrow 4.2°$	$44.0° \leftrightarrow 56.7°$
$\epsilon - \theta_{12}^0$	$-39.2° \leftrightarrow -34.6°$	$-59.5° \leftrightarrow -54.4°$	$-39.2° \leftrightarrow -30.0°$	$44.0° \leftrightarrow 56.7°$

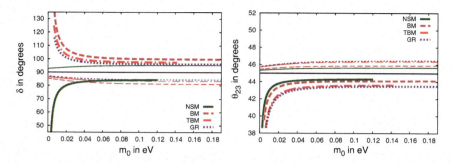

Fig. 73.1 Model predictions for CP-phase δ (θ_{23}) for all four mixing patterns are shown as a function of the lightest neutrino mass m_0 with best-fit values of the oscillation input parameters in left (right) panel. Thick (thin) curves are for Normal Ordering (Inverted Ordering). The green solid, pink dashed, red dot-dashed, and violet dotted curves denote NSM, BM, TBM, and GR mixing, respectively

Table 73.4 The octant of θ_{23} and the quadrant of the CP-phase δ for different mixing patterns for both orderings of neutrino masses are exhibited

Mixing Pattern	Normal ordering		Inverted ordering	
	δ quadrant	θ_{23} octant	δ quadrant	θ_{23} octant
NSM	First/Fourth	First	Second/Third	Second
BM, TBM, GR	Second/Third	First	First/Fourth	Second

and therefore the model has testable predictions. Using (73.14), CP-phase δ and θ_{23} predicted by the model can be plotted for the four mixing patterns as shown in Fig. 73.1. Thus after including the corrections coming from Type-I seesaw into the Type-II seesaw dominant contribution, realistic neutrino oscillation parameters can be obtained. Details of the procedure can be found in [1].

73.4 Conclusion

A seesaw model based on $A4$ for realistic neutrino mass and mixing has been studied. The Lagrangian has a dominant component originating from Type-II seesaw and a sub-dominant contribution arising from Type-I seesaw. The Type-II seesaw component has $\Delta m_{solar}^2 = 0$, $\theta_{13} = 0$, $\theta_{23} = \pi/4$ and solar mixing angle of the TBM, BM, GR, NSM kind. Including corrections offered by the Type-I seesaw sub-dominant contribution can produce the oscillation parameters in the ranges allowed by data. The model [1] predicts inter-relations between neutrino mass ordering, octant of θ_{23} and quadrant of δ that can be tested by oscillation experiments in future.

Acknowledgements I thank the organizers of the XXIII DAE-BRNS High Energy Physics Symposium 2018. I acknowledge support from CSIR (NET) Senior Research Fellowship during initial part of the work done at University of Calcutta. I thank Prof. Amitava Raychaudhuri for valuable discussions.

References

1. S. Pramanick, Phys. Rev. D **98**(7), 075016 (2018). arXiv:1711.03510 [hep-ph]
2. M.C. Gonzalez-Garcia, M. Maltoni, J. Salvado, T. Schwetz, JHEP **1212**, 123 (2012). NuFIT 2.1. arXiv:1209.3023v3 [hep-ph]
3. D.V. Forero, M. Tortola, J.W.F. Valle, Phys. Rev. D **86**, 073012 (2012). arXiv:1205.4018 [hep-ph]
4. S. Pramanick, A. Raychaudhuri, Phys. Rev. D **93**(3), 033007 (2016). arXiv:1508.02330 [hep-ph]; S. Pramanick, A. Raychaudhuri, Phys. Rev. D **94**(11), 115028 (2016). arXiv:1609.06103 [hep-ph]
5. S. Pramanick, A. Raychaudhuri, Phys. Rev. D **88**(9), 093009 (2013). arXiv:1308.1445 [hep-ph]; S. Pramanick, A. Raychaudhuri, Phys. Lett. B **746**, 237 (2015). arXiv:1411.0320 [hep-ph]; S. Pramanick, A. Raychaudhuri, Int. J. Mod. Phys. A **30**(14), 1530036 (2015). arXiv:1504.01555 [hep-ph]
6. E. Ma, G. Rajasekaran, Phys. Rev. D **64**, 113012 (2001). [hep-ph/0106291]; G. Altarelli, F. Feruglio, Nucl. Phys. B **741**, 215 (2006). [hep-ph/0512103]; H. Ishimori, T. Kobayashi, H. Ohki, Y. Shimizu, H. Okada, M. Tanimoto, Prog. Theor. Phys. Suppl. **183**, 1 (2010). arXiv:1003.3552 [hep-th]
7. S. Pramanick, A. Raychaudhuri, JHEP **1801**, 011 (2018). arXiv:1710.04433 [hep-ph]

Chapter 74
Impact of Nuclear Effects on CP Sensitivity at DUNE

Srishti Nagu, Jaydip Singh, and Jyotsna Singh

Abstract The precise measurement of neutrino oscillation parameters is one of the highest priorities in neutrino oscillation physics. To achieve the desired precision, it is necessary to reduce the systematic uncertainties related to neutrino energy reconstruction. An error in energy reconstruction is propagated to all the oscillation parameters; hence, a careful estimation of neutrino energy is required. To increase the statistics, neutrino oscillation experiments use heavy nuclear targets like Argon (Z=18). The use of these nuclear targets introduces nuclear effects that severely impact the neutrino energy reconstruction which in turn poses influence in the determination of neutrino oscillation parameters. In this work we have tried to study the impact of nuclear effects on the determination of CP phase at DUNE using final state interactions.

74.1 Introduction

Neutrino oscillation physics has entered into the era of precision measurement from the past two decades. Significant achievements have been made in the determination of the known neutrino oscillation parameters and continuous attempts are being made to estimate the unknown neutrino oscillation parameters precisely. The neutrino oscillation parameters governing the three flavor neutrino oscillation physics are three mixing angles $\theta_{12}, \theta_{13}, \theta_{23}$, a leptonic CP phase δ_{CP} and two mass-squared differences, Δm_{21}^2 (solar mass splitting) and Δm_{31}^2 (atmospheric mass splitting). The unknown oscillation parameters in the picture are (1) the octant of θ_{23} whether

S. Nagu (✉) · J. Singh · J. Singh
Department of Physics, Lucknow University, Lucknow 226007, India
e-mail: srishtinagu19@gmail.com

J. Singh
e-mail: jaydip.singh@gmail.com

J. Singh
e-mail: singh.jyotsnalu@gmail.com

© Springer Nature Singapore Pte Ltd. 2021
P. K. Behera et al. (eds.), *XXIII DAE High Energy Physics Symposium*,
Springer Proceedings in Physics 261,
https://doi.org/10.1007/978-981-33-4408-2_74

it lies in the lower octant ($\theta_{23} < \pi/4$) or in the higher octant ($\theta_{23} > \pi/4$) (2) the sign of $|\Delta m_{31}^2|$, i.e., neutrino mass eigenstates $m_i (i = 1, 2, 3)$ are arranged in normal order ($m_1 \ll m_2 \ll m_3$) or inverted order ($m_2 \approx m_1 \gg m_3$) (3) the leptonic δ_{CP} phase which can lie in the entire range $-\pi < 0 < +\pi$. Accurate measurement of the leptonic CP phase can lead to further studies on the origin of leptogenesis [1] and baryon asymmetry of the universe [2]. Determination of precise δ_{CP} value is also required for explaining the phenomenon of sterile neutrinos [3]. The global analysis results as indicated in [4] report current bounds on oscillation parameters which have been performed by several experimental groups.

A defining challenge for neutrino experiments is to determine the incoming neutrino energy since the configuration of the outgoing particles and kinematics of the interaction within the nucleus are completely unknown. In collider experiments, the neutrino beams are generated via secondary decay products which assign a broad range of energies to the neutrinos thus causing their energy reconstruction to be difficult. Hence neutrino energy is reconstructed from final state particles. The present-day neutrino oscillation experiments use heavy nuclear targets like argon (Z=18), in order to collect large event statistics. With a nuclear target, where neutrinos interact with fermi moving nucleons, uncertainties in the initial state particles produced at the primary neutrino-nucleon interaction vertex arise. These nuclear effects are capable enough to change the identities, kinematics, and topologies of the outgoing particles via final state interactions (FSI) and thus hiding the information of the particles produced at the initial neutrino-nucleon interaction vertex which gives rise to fake events. Detailed discussion regarding the impact on atmospheric oscillation parameters due to the presence of FSI in the QE interaction process can be found in [5] and due to fake events stemming from QE and RES processes can be found in [6]. The impact of cross-sectional uncertainties on the CP violation sensitivity can be found in [7]. In this work, we attempt to study the impact of nuclear effects imposed by FSI in the QE (Quasi Elastic), resonance (RES), and deep inelastic scattering (DIS) interaction processes. For a detailed discussion one can look into [8]. Understanding nuclear effects will give us a handle to filter out true events from the fake events in a given neutrino-nucleon interaction which will lead to an accurate measurement of neutrino oscillation parameters.

74.2 Neutrino Oscillation Studies with the Long-Baseline Experiment-DUNE

The Deep Underground Neutrino Experiment, DUNE [9], an upcoming long-baseline neutrino oscillation experiment, to be set up in the US is aiming for discovering the unknown oscillation parameters and explore new physics. The 1300-km baseline, stretching from LBNE facility at Fermilab to Sanford Underground Research Facility (SURF) at South Dakota, is ideal for achieving the desired sensitivity for CP violation and mass hierarchy. The far detector will be composed of 40 ktons of liq-

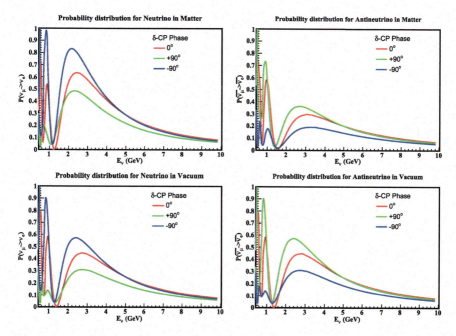

Fig. 74.1 Neutrino oscillation probability for neutrino (left panel) and antineutrino (right panel) with three values of leptonic δ_{CP} phase, i.e., $0°$, $+\pi/2$ and $-\pi/2$ in matter (upper panel) and vacuum (bottom panels) for the $\nu_\mu \to \nu_e$ channel

uid argon (nuclear target with Z=18) as detector material which will provide large event statistics. The DUNE-LBNF flux spreads in the energy range 0.5–10 GeV, with an average energy peaking at 2.5 GeV. It is composed of QE, Resonance, DIS, and Coherent neutrino-nucleon interaction processes with resonance being the dominant interaction process in this energy regime. To evaluate the sensitivity of LBNE and to optimize the experimental design, it is important to accurately predict the neutrino flux. The neutrino oscillation probability is presented in Fig. 74.1 for ν_e appearance channel.

74.3   Simulation and Result

Approximately 2 lakh events are generated using DUNE flux for the muon disappearance channel with the help of GiBUU (Giessen Boltzmann-Uehling-Uhlenbeck) [10]. The migration matrices are obtained using GiBUU and are inserted in the required format into GLoBES [11, 12]. The systematics considered in our work are as follows—signal efficiency is 85%, normalization error and energy calibration error for the signal and background are—5%, 10% and 2% respectively. The running time considered is 10 years in neutrino mode with 35 kton fiducial mass of the detector.

Table 74.1 True Oscillation Parameters considered in our work

θ_{12}	33.58°
θ_{13}	8.48°
θ_{23}	45°
δ_{CP}	180°
Δm_{21}^2	$7.50e^{-5}eV^2$
Δm_{31}^2	$2.40e^{-3}eV^2$

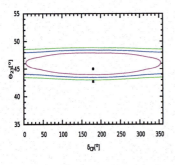

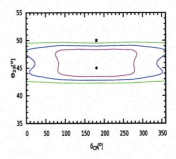

(a) Confidence regions in the $(\theta_{23}, \delta_{CP})$ plane- obtained using the migration matrices of pure QE and QE like events. The * point is the shift due to nuclear effects.

(b) Confidence regions in the $(\theta_{23}, \delta_{CP})$ plane-obtained using the migration matrices of pure QE+RES+DIS and QE+RES+DIS like events. The * point is the shift due to nuclear effects.

Fig. 74.2 Confidence regions in $\theta_{23} - \delta_{CP}$ plane

The values of oscillation parameters used in this work are presented in Table 74.1 and are motivated from [13, 14]. The value of δ_{CP} considered in this work lies within the present set bounds on δ_{CP} [15].

Here, we consider a parameter α, which help us to incorporate nuclear effects in our analysis. It can be considered as a way of including systematic uncertainties, such approach has been considered previously in [5, 6]. We present the position of the best fit corresponding to values of α taken as 0 and 1, by plugging them in equations (74.1) and (74.2)

$$N_i^{test}(\alpha) = \alpha \times N_i^{QE} + (1 - \alpha) \times N_i^{QE-like} \tag{74.1}$$

$$N_i^{test}(\alpha) = \alpha \times N_i^{QE+RES+DIS} + (1 - \alpha) \times N_i^{QE-like+RES-like+DIS-like}, \tag{74.2}$$

where N is the total number of events. Two cases arise—

1. When $\alpha = 1$, the second term in each of the equations (74.1) and (74.2) cancels out which imply that nuclear effects are completely disregarded.
2. When $\alpha = 0$, the presence of fake events are registered which imply that nuclear effects are incorporated.

We notice from Fig. 74.2a that the value of octant of θ_{23} shifts from maximal toward lower octant when QE and QE-like events are examined while from Fig. 74.2b we can see that the θ_{23} shifts to the higher octant when we include contribution of nuclear effects from QE+RES+DIS processes.

74.4 Conclusion

We notice a 3σ shift in the best fit point value for Charged Current QE events in Fig. 74.2a and a shift of more than 3σ for QE+RES+DIS processes in Fig. 74.2b. In a future work we will report the results by considering different values of the parameter α as defined above, since α can take on any value between 0 and 1. In an outlook of the study, we need to perform an extensive authentication of the accuracy of nuclear models employed in data analysis. Employment of nuclear targets in neutrino oscillation experiments aid in boosting the event statistics which reduce the statistical error but we need to pin down the systematic uncertainties arising from the persistent nuclear effects that will bring us a step closer in achieving our goals.

Acknowledgements This work is partially supported by Department of Physics, Lucknow University. Financially it is supported by government of India, DST project no-SR/MF/PS02/2013, Department of Physics, Lucknow.

References

1. G. Engelhard, Y. Grossman, Y. Nir, JHEP **0707**, 029 (2007), arXiv:hep-ph/0702151
2. S. Pascoli, S.T. Petcov, A. Riotto, Phys. Rev. D 75, 083511 (2007); Nucl. Phys. B **774**, 1 (2007)
3. J. Singh, Constraining the effective mass of majorana neutrino with sterile neutrino mass for inverted ordering spectrum. Adv. High Energy Phys. (2019). Article ID 4863620. https://doi.org/10.1155/2019/4863620, arXiv:1902.08575
4. P.F. de Salas et al., Status of neutrino oscillations 2017 (2017), arXiv:1708.01186 [hep-ph]
5. P. Coloma, P. Huber, Impact of nuclear effects on the extraction of neutrino oscillation parameters (2013), arXiv:1307.1243v2 [hep-ph]
6. S. Naaz, A. Yadav, J. Singh, R.B. Singh, Effect of final state interactions on neutrino energy reconstruction at DUNE. https://doi.org/10.1016/j.nuclphysb.2018.05.018
7. S. Nagu, J. Singh, J. Singh, R.B. Singh, Impact of cross-sectional uncertainties on DUNE sensitivity due to nuclear effects, 23 May 2019 (2019), arXiv:1905.13101v1 [hep-ph]
8. S. Nagu, J. Singh, J. Singh, Nuclear effects and CP sensitivity at DUNE, 5 Jun 2019 (2019). arXiv:1906.02190v1 [hep-ph]
9. The DUNE far detector interim design report. Physics, Technology and Strategies, vol. 1, 26 July 2018 (2018), arXiv:1807.10334
10. O. Buss et al., Transport-theoretical description of nuclear reactions. Phys. Rep. **512**, 1 (2012), arXiv:1106.1344
11. P. Huber, M. Lindner, T. Schwetz, W. Winter, First hint for CP violation in neutrino oscillations from upcoming superbeam and reactor experiments. JHEP **0911**, 044 (2009), arXiv:0907.1896
12. P. Huber, M. Lindner, W. Winter, Simulation of long-baseline neutrino oscillation experiments with GLoBES. Comput. Phys. Commun. **167**, 195 (2005). hep-ph/0407333

13. F. Capozzi, E. Lisi, A. Marrone, D. Montanino, A. Palazzo, Nucl. Phys. **B908**, 218 (2016), arXiv:1601.07777
14. I. Esteban, M. C. Gonzalez-Garcia, M. Maltoni, I. MartinezSoler, T. Schwetz, JHEP **01**, 087 (2017), arXiv:1611.01514
15. I. Esteban et al., Global analysis of three flavour neutrino oscillations: synergies and tensions in the determination of θ_{23}, δ_{CP} and the mass ordering, arXiv:1811.05487v1

Chapter 75
Study of Phase Transition in Two-Flavour Quark Matter at Finite Volume

Anirban Lahiri, Tamal K. Mukherjee, and Rajarshi Ray

Abstract We study some aspects of the phase transition of the finite size droplets with u and d quarks. We modelled the system through the Polyakov Nambu Jona-Lasinio Model (PNJL) and employ multiple reflection expansion to introduce the finite size effects. We discuss the qualitative behaviour of the order parameters, pressure and susceptibilities. We find increased fluctuations around the transition temperature region due to finite size effects and the fluctuations are more for small system size.

PACS numbers: 25.75.Nq · 12.39.-x · 11.10.Wx · 64.60.an

75.1 Introduction

For the last few decades, intense efforts have been going on to understand the phase structure of Quantum Chromodynamics (QCD). But due to many complexities involved in the theoretical (as well as experimental) study of the QCD, we are yet to draw a clear definitive picture of the phase diagram amongst the many possibilities. Current understanding, on the basis of majority of the studies (see, for example, [1–13]), indicates that the QCD phase diagram of two light flavour quarks, is a crossover transition from hadron degrees of freedom to quark degrees of freedom

A. Lahiri
Fakultät für Physik, Universität Bielefeld, 33615 Bielefeld, Germany
e-mail: alahiri@physik.uni-bielefeld.de

T. K. Mukherjee (✉)
Department of Physics, School of Sciences, Adamas university, Barasat - Barrackpore Rd 24 Parganas North Jagannathpur, Kolkata 700126, India
e-mail: tamalkumar.mukherjee@adamasuniversity.ac.in; tamal.k.mukherjee@gmail.com

R. Ray
Department of Physics & Center for Astroparticle Physics & Space Science, Bose Institute, EN-80, Sector-5, Bidhan Nagar, Kolkata 700091, India
e-mail: rajarshi@jcbose.ac.in

© Springer Nature Singapore Pte Ltd. 2021
P. K. Behera et al. (eds.), *XXIII DAE High Energy Physics Symposium*,
Springer Proceedings in Physics 261,
https://doi.org/10.1007/978-981-33-4408-2_75

543

at around zero chemical potential. Whereas for high value of chemical potential, it is a first-order phase transition [2, 5–14]. This first-order phase boundary line ends at a second-order phase transition point. This second-order phase transition point is known as Critical End Point (CEP). However, we should keep in mind that there are studies which indicate no first-order phase transition at large chemical potential [14–20] and the CEP does not exist in the QCD phase diagram. The only definitive answer regarding the existence and location of the critical end point should come from the experiments. The affirmative answer would also indirectly supports the existence of first-order phase transition line at large chemical potential and crossover line at the small chemical potential region. Various heavy-ion collision experiment programmes, the Beam Energy Scan (BES) programme at RHIC, the FAIR at GSI, and the NICA at DUBNA all are devoted to find an answer to the structure of the QCD phase diagram and problem of locating the CEP.

However, the deconfined matter of quarks and gluons created in relativistic heavy-ion collisions has finite size. Whereas most of the theoretical studies regarding the QCD phase diagram have been carried out in thermodynamic limit. So the pertinent question is that whether the size of the matter created is large enough to be considered to be in the thermodynamic limit or not. If it is not large enough, then we need to take into account the finite size effects as well in the theoretical calculations. This will allow better comparison of the theoretical result with the experimental one.

Various groups tried to include the finite size effects through various methods, some of which can be found here [21–27]. Here, we are going to include the effects of finite size by the modification of density of states. This is achieved following the prescription of Multiple Reflection Expansion (MRE) formalism. In this work, we discuss some of the aspects of the QCD phase transition in finite volume in terms of behaviour of the order parameters and pressure. We also illustrate the behaviour of fluctuations in finite volume with the help of second- and fourth-order quark number susceptibilities. The paper is organised as follows. The model and the MRE formalism are briefly discussed in the next section. Relevant results are presented in the last section along with a outlook.

75.2   Formalism

We start with the thermodynamic potential of two-flavour PNJL model within a finite sphere, given by

$$\Omega' = \mathcal{U}'(\bar{\Phi}, \Phi, T) + \frac{\sigma^2}{2G} - 6N_f \int \frac{d^3 p}{(2\pi)^3} \rho_{\mathrm{MRE}}(p, m, R) E \, \theta \left(\Lambda^2 - |\mathbf{p}|^2\right)$$

$$- 2N_f T \int \frac{d^3 p}{(2\pi)^3} \rho_{\mathrm{MRE}}(p, m, R) \left\{ \ln \left[1 + 3\Phi e^{-(E-\mu)/T} + 3\bar{\Phi} e^{-2(E-\mu)/T} + e^{-3(E-\mu)/T} \right] \right.$$

$$\left. + \ln \left[1 + 3\bar{\Phi} e^{-(E+\mu)/T} + 3\Phi e^{-2(E+\mu)/T} + e^{-3(E+\mu)/T} \right] \right\}, \tag{75.1}$$

where the quark condensate is given by $\sigma = G\langle\bar{\psi}\psi\rangle$. The quasi-particle energy is given by $E = \sqrt{|\mathbf{p}|^2 + m^2}$, where $m = m_0 - \sigma$ is the constituent quark masses. Λ is the three-momentum cutoff as in the NJL model. Free parameters of the NJL part, i.e., m_0, G and Λ are fixed by zero temperature observables [9].

For the Polyakov loop potential, a Landau–Ginzburg type potential with Z(3) symmetry may be written [9, 28] with proper consideration of the Haar measure as [29],

$$\frac{\mathcal{U}'\left(\Phi, \bar{\Phi}, T\right)}{T^4} = -\frac{b_2\left(T\right)}{2} \bar{\Phi}\Phi - \frac{b_3}{6}\left(\Phi^3 + \bar{\Phi}^3\right) + \frac{b_4}{4}\left(\bar{\Phi}\Phi\right)^2 - \kappa \ln[J(\Phi, \bar{\Phi})]$$

$$\equiv \frac{\mathcal{U}\left(\Phi, \bar{\Phi}, T\right)}{T^4} - \kappa \ln[J(\Phi, \bar{\Phi})]$$

(75.2)

with the temperature dependent coefficient of the form, $b_2\left(T\right) = a_0 + a_1\left(\frac{T_0}{T}\right) + a_2\left(\frac{T_0}{T}\right)^2 + a_3\left(\frac{T_0}{T}\right)^3$. The constants could be obtained by fitting Polyakov loop and pressure obtained in pure gauge QCD simulations on the lattice [9, 28, 29]. Here T_0 is the temperature where the above-written Polyakov loop potential shows a transition. Originally it was found to be 270 Mev, later it was tuned to be 190 MeV to bring the pseudo-critical temperature down [9]. We also choose $T_0 = 190$ MeV for the present work.

$J(\Phi, \bar{\Phi})$ is the VdM term [29] which comes as a Jacobian when one writes the Polyakov loop potential in terms of scalar-valued traced Polyakov loop.

Effect of finite size of the system has been incorporated by a modified density of states following MRE formalism [30–34]. For a spherical droplet of radius R, the modified density of states is given by

$$\rho_{\text{MRE}}(p, m, R) = 1 + \frac{6\pi^2}{pR} f_S + \frac{12\pi^2}{(pR)^2} f_C,$$

(75.3)

where the surface term

$$f_S = -\frac{1}{8\pi}\left(1 - \frac{2}{\pi}\arctan\frac{p}{m}\right)$$

(75.4)

was derived within MRE formalism [31] and the curvature term is from an ansatz by Madsen [32]

$$f_C = \frac{1}{12\pi^2}\left[1 - \frac{3p}{2m}\left(\frac{\pi}{2} - \arctan\frac{p}{m}\right)\right]$$

(75.5)

Suppression of MRE density of states is more pronounced in the low momentum regime where it takes unphysical negative values [35]. To overcome this issue an IR cutoff has been introduced in [35], which is defined as the largest zero of ρ_{MRE}. The mean fields are obtained by minimization of thermodynamic potential,

$$\frac{\partial \Omega'}{\partial X} = 0 \quad \text{with} \ X = \Phi, \bar{\Phi}, \sigma. \tag{75.6}$$

During this minimization process we did not take derivatives of the ρ_{MRE} w.r.t. σ or equivalently m. The density of states are evaluated for a fixed mass scale and at that point we do not care about the microscopic mechanism of mass generation. This philosophy was also adapted for MIT bag model [32].

Bulk pressure is given by $P = -\Omega$ where we did not take the VdM term as explained in [29]. Although the relation $P = -\Omega$ is in principle valid only in thermodynamic limit, but it has been shown in [26] that this relation holds very well down to a system of radius 5 fm or less. Susceptibilities of different order is defined as $\chi_n = \frac{\partial^n (P/T^4)}{\partial(\mu/T)^n}\big|_{\mu=0}$ for $n = 2, 4, \dots$ χ_n for odd values of n vanishes at $\mu = 0$ due to CP symmetry.

75.3 Result and Outlook

As mentioned in the previous section, we are including the effects of finite size by modifying the density of states. Now, the modified density of states $\rho_{MRE}(p, m, R)$ is a function of momentum (p), mass (m) and Radius (R). The forms of f_S and f_C were first used in the context of MIT bag model. However, in MIT bag model the mass is not a dynamical quantity, whereas in PNJL model the mass is a dynamical quantity and defined through the dynamical auxiliary field σ. For this reason, whilst using the form of density states ρ_{MRE} derived in the context of MIT bag model, we treat mass in ρ_{MRE} as parameter whose value is fixed from the value of mass at each value of temperature and chemical potential. This is an ad hoc assumption made in this work. The detailed analysis of this assumption is beyond the scope of this work and will be presented in the subsequent work. As a result of this assumption, we do not consider the variation of the density of states w.r.t σ in the minimization process.

Another assumption is the equality $P = -\Omega$ which is valid in the thermodynamic limit. But, we used this equality to find the pressure of a system of finite size, assuming the error is within 10% as illustrated in [26].

We are going to present some of the qualitative features of the phase transition in finite sized strongly interacting matter consisting of two light flavoured quarks u and d at zero chemical potential. In (Fig. 75.1), we have shown the behaviour of the order parameters as a function of temperature. As evident from the figure, the traced Polyakov loop is not much affected by the finite size of the system. This is expected as we have not introduced any size dependence in the Polyakov loop potential. However, they split depending on the system size only during the rapid rise within a narrow temperature band where the crossover transition takes place. On the other hand, the chiral condensate σ is reallty influenced by the finite size effects. The vacuum value of the condensate is greatly reduced as we decrease the system size and transition temperature shifted to the lower temperature as we decrease the system size.

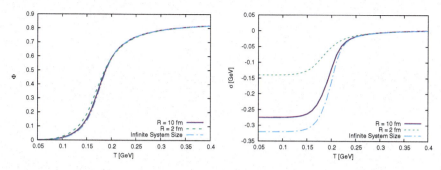

Fig. 75.1 Traced Polyakov loop Φ (left) Chiral Condensate σ (right) as a functions of temperature for different system sizes

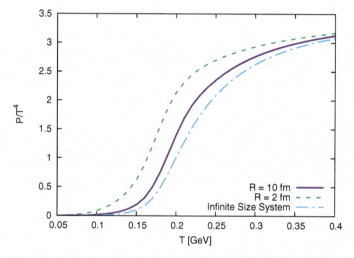

Fig. 75.2 Pressure as a functions of temperature for different system sizes

Variation of pressure with temperature is presented in the (Fig. 75.2). Qualitatively, pressure for different system sizes behave similarly and the transition takes place at a lower temperature for smaller system size. For any particular value of temperature, we find within our formalism, pressure is higher for small system size. We find first and fourth terms (Polyakov loop potential and medium contribution to the pressure) of (75.1) are not affected by the system size effects. Whereas, second and third terms of (75.1) are sensitive to the system size. Their contribution makes the pressure differ for different system sizes. This is expected as chiral condensate changes drastically with system sizes as can be seen from (Fig. 75.1).

To discuss the finite size effects in fluctuations we have plotted second and fourth-order susceptibilities in (Fig. 75.3). As can be seen from the figure, finite size corrections introduce non-trivial effects in the fluctuations and they are qualitatively different from the susceptibilities in the thermodynamic limit. Especially the second-

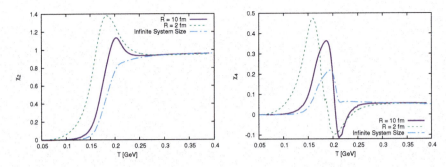

Fig. 75.3 (left) Second-order (χ_2) and (right) Fourth-order (χ_4) susceptibilities as a functions of temperature for different system sizes

order fluctuations do not smoothly approach the saturation value as we increase the temperature. There is a non-trivial increase of fluctuations around the transition temperature region. When we analyse the individual contributions of surface and curvature terms, we find it is the surface term that introduces the non-trivial effects and main contributor of the finite size effects in various thermodynamic quantities.

To summarise, we have tried to study the effects of the finite size of the system of strongly interacting matter created in the heavy-ion collision experiments using the PNJL model with a suitable density of states. We find the effects to be quite significant for the pressure as well as the quark number fluctuations. This implies that a careful analysis of the data from the experiments is necessary taking finite size effects into account. Further progress in the study will be reported elsewhere.

Acknowledgements TKM gratefully acknowledges the financial support from SERB-DST through Ramanujan Fellowship under Project no- SB/S2/RJN-29/2013.

References

1. O. Philipsen, Prog. Part. Nucl. Phys. **70**, 55 (2013)
2. S. Gupta, X. Luo, B. Mohanty, H.G. Ritter, N. Xu, Science **332**, 1525 (2011)
3. Y. Aoki, G. Endrodi, Z. Fodor, S.D. Katz, K.K. Szabo, Nature **443**, 675 (2006)
4. Y. Aoki, S. Borsanyi, S. Durr, Z. Fodor, S.D. Katz, S. Krieg, K.K. Szabo, J. High Energy Phys. **06**, 088 (2009)
5. S. Ejiri, Phys. Rev. D **78**, 074507 (2008); A.Y. Li, A. Alexandru, K.-F. Liu, Phys. Rev. D **84**, 071503 (R) (2011)
6. S.X. Qin, L. Chang, H. Chen, Y.X. Liu, C.D. Roberts, Phys. Rev. Lett. **106**, 172301 (2011)
7. X.Y. Xin, S.X. Qin, Y.X. Liu, Phys. Rev. D **90**, 076006 (2014)
8. C.S. Fischer, J. Luecker, J.A. Mueller, Phys. Lett. B **702**, 438 (2011); C.S. Fischer, J. Luecker, Phys. Lett. B **718**, 1036 (2013); C.S. Fischer, J. Luecker, C.A. Welzbacher, Phys. Rev. D **90**, 034022 (2014)
9. C. Ratti, M.A. Thaler, W. Weise, Phys. Rev. D **73**, 014019 (2006)
10. B.-J. Schaefer, J.M. Pawlowski, J. Wambach, Phys. Rev. D **76**, 074023 (2007)
11. W.J. Fu, Z. Zhang, Y.X. Liu, Phys. Rev. D **77**, 014006 (2008)

12. L.J. Jiang, X.Y. Xin, K.L. Wang, S.X. Qin, Y.X. Liu, Phys. Rev. D **88**, 016008 (2013)
13. X.Y. Xin, S.X. Qin, Y.X. Liu, Phys. Rev. D **89**, 094012 (2014)
14. K. Fukushima, Phys. Rev. D **77**, 114028 (2008)
15. S. Klimt, M.F. Lutz, W. Weise, Phys. Lett. B **249**, 386 (1990)
16. M. Kitazawa, T. Koide, T. Kunihiro, Y. Nemoto, Prog. Theor. Phys. **108**, 929 (2002)
17. C. Sasaki, B. Friman, K. Redlich, Phys. Rev. D **75**, 054026 (2007)
18. K. Fukushima, Phys. Rev. D **78**, 114019 (2008)
19. B.-J. Schaefer, M. Wagner, Phys. Rev. D **79**, 014018 (2009)
20. N.M. Bratovic, T. Hatsuda, W. Weise, Phys. Lett. B **719**, 131 (2013)
21. M. Cristoforetti, T. Hell, B. Klein, W. Weise, Phys. Rev. D **81**, 114017 (2010)
22. R.-A. Tripolt, J. Braun, B. Klein, B.-J. Schaefer, Phys. Rev. D **90**, 054012 (2014)
23. L. Abreu, M. Gomes, A. da Silva, Phys. Lett. B **642**, 551 (2006)
24. S. Yasui, A. Hosaka, Phys. Rev. D **74**, 054036 (2006)
25. L.M. Abreu, A.P.C. Malbouisson, J.M.C. Malbouisson, Phys. Rev. D **83**, 025001 (2011)
26. A. Bhattacharyya, P. Deb, S.K. Ghosh, R. Ray, S. Sur, Phys. Rev. D **87**, 054009 (2013)
27. A. Bhattacharyya, R. Ray, S. Sur, Phys. Rev. D **91**, 051501 (2015)
28. S.K. Ghosh, T.K. Mukherjee, M.G. Mustafa, R. Ray, Phys. Rev. D **73**, 114007 (2006)
29. S.K. Ghosh, T.K. Mukherjee, M.G. Mustafa, R. Ray, Phys. Rev. D **77**, 094024 (2008)
30. R. Balian, C. Bloch, Ann. Phys. **60**, 401 (1970)
31. M.S. Berger, R.L. Jaffe, Phys. Rev. C **35**, 213 (1987); **44**, 566(E) (1991)
32. J. Madsen, Phys. Rev. D **50**, 3328 (1994)
33. O. Kiriyama, A. Hosaka, Phys. Rev. D **67**, 085010 (2003)
34. O. Kiriyama, Phys. Rev. D **72**, 054009 (2005)
35. O. Kiriyama, T. Kodama, T. Koide, arXiv:hep-ph/0602086

Chapter 76
Measurement of Azimuthal Correlations of Heavy-Flavour Hadron Decay Electrons with Charged Particles in P–Pb and Pb–Pb Collisions at $\sqrt{s_{NN}} = 5.02$ TeV with ALICE at the LHC

Bharati Naik

Abstract The ALICE (A Large Ion Collider Experiment) apparatus at the LHC is designed to study the properties of QGP (Quark-Gluon Plasma), a deconfined state of quarks and gluons produced in heavy-ion collisions. The study of angular correlations between the heavy-flavour hadron decay electrons and charged particles can offer information about potential heavy-flavour jet quenching in the QGP. It also provides information about any possible medium-induced modification of heavy-quark fragmentation and hadronization. In these proceeding, the measurement of azimuthal correlations between high-p_T heavy-flavour hadron decay electrons and charged particles in Pb–Pb collisions at $\sqrt{s_{NN}} = 5.02$ TeV will be presented. A comparison of the results measured in p–Pb collisions at $\sqrt{s_{NN}} = 5.02$ TeV using LHC Run 2 data will also be shown.

76.1 Introduction

The ALICE apparatus at the LHC is dedicated to study the properties of the QGP. Due to their heavy masses heavy quarks (charm and beauty) are produced in the early stages of the collision via hard-scattering processes. As a result, they experience the entire evolution of the system formed in such collisions and therefore, are effective probes for investigating the properties of the medium. The angular correlations between the heavy-flavour hadron decay electrons (HFe) and charged particles in p–Pb collisions offer information about the cold nuclear effects on charm jets. The long-range v_2-like structure in high-multiplicity p–Pb collisions and in Pb–Pb

B. Naik (✉)
Indian Institute of Technology Bombay, Mumbai 400076, India
e-mail: bharati.naik@cern.ch

© Springer Nature Singapore Pte Ltd. 2021
P. K. Behera et al. (eds.), *XXIII DAE High Energy Physics Symposium*,
Springer Proceedings in Physics 261,
https://doi.org/10.1007/978-981-33-4408-2_76

collisions, it can provide information about any possible medium-induced modification of heavy-quark fragmentation, hadronization, and in-medium Parton energy loss mechanisms.

76.2 Analysis Method

The main detectors used for this analysis are the: ITS (Inner Tracking System) for tracking and vertexing, TPC (Time Projection Chamber) for tracking and particle identification, TOF (Time Of Flight) for particle identification, EMCal (ElectroMagnetic calorimeter) for electron identification and triggering and the V0 detector for triggering and multiplicity determination. The various sub-systems of the ALICE apparatus and their performance are described in details in [1, 2]. The data sets used for both p–Pb and Pb–Pb analysis were collected in 2016 with an integrated luminosity $L_{int} = 291 \ \mu b^{-1}$ and $L_{int} = 225 \ \mu \ b^{-1}$, respectively.

76.2.1 Electron Identification

The HFe contributed by the semi-leptonic decays of open heavy-flavour hadrons via the following channels: $D \rightarrow e + X$ (BR: 10%), $B \rightarrow e + X$ (BR: 10%). The low p_T ($p_T <$ 4.0 GeV/c) electrons are identified by using information from the TOF and TPC, the high p_T ($p_T \geq$ 4.0 GeV/c) electrons are identified using the E/p ratio of EMCal. The detailed procedure can be found in [3].

76.2.2 Azimuthal Correlation and Corrections

Each selected electron is correlated with charged tracks (hadrons), produced in the same event and the ($\Delta \eta$, $\Delta \varphi$) correlation distribution is built. Here, $\Delta \eta = \eta_{electron} - \eta_{hadrons}$ and $\Delta \varphi = \varphi_{electron} - \varphi_{hadrons}$. The effects on the distribution due to the limited detector acceptance and inhomogeneities are corrected via the event-mixing technique. The background, electrons from non-heavy flavoured hadron decays (non-HFe), can be subtracted from the inclusive electron distribution to obtain a relatively pure sample of HFe. The non-HFe are mainly from photon conversions in the detector material and Dalitz decays of light neutral mesons (π^0, η). They can be removed by an invariant mass (M) technique and by utilizing the photon-electron tagging method. For the latter, the invariant mass of the electron pairs are reconstructed and those satisfying $M_{e^+e^-} <$ 140 MeV/c^2 are tagged as non-HFe. Further details of this procedure are described in [3, 4]. The ($\Delta \eta$, $\Delta \varphi$) distributions are projected onto $\Delta \varphi$ axis and are normalized by the number of trigger particles (HFe).

76.3 Results

The angular correlations between HFe and charged particles are studied in two different event multiplicity classes, i.e., low multiplicity (60–100%) and high multiplicity (0–20%) in p–Pb collisions [7]. The correlation function obtained from events at low multiplicities is subtracted from that measured in high multiplicities to remove the jet contribution, assuming that the jet fragmentation is independent of the event multiplicity. The left panel of Fig. 76.1 shows the azimuthal-correlation distribution between HFe and charged particles, for high-multiplicity p–Pb collisions after subtracting the jet contribution from low-multiplicity collisions. The left panel of Fig. 76.1 demonstrates a v_2-like modulation as observed in di-hadron correlation distribution of light flavours [5]. A Fourier fit is applied to the azimuthal-correlation distribution to extract the HFe v_2. The right panel of Fig. 76.1 shows a positive v_2 of heavy-flavour decay electron with a significance of 5.1σ in the range $2 < p_\mathrm{T}^e < 4$ GeV/c.

Similarly, the azimuthal correlation between HFe and charged particles is obtained for central (0–20%) and semi-central (20–50%) in Pb–Pb collisions. After removal of the pedestal and flow contribution from the correlation distribution, the near-side associated yield is calculated by taking the integral of $\Delta\varphi$ distribution in the range $-1 < \Delta\varphi < 1$. The left panel of Fig. 76.2 shows the near-side associated yield measured in central and semi-central Pb–Pb collisions, along with the results measured in p–Pb collisions. The ratios of the near-side associate yield between Pb–Pb and p–Pb collisions are shown in the right panel. The ratio increases with decreasing associated particle p_T, which indicates a possible medium-induced modification of heavy-quark fragmentation in heavy-ion systems. The ratio also demonstrates a centrality dependence.

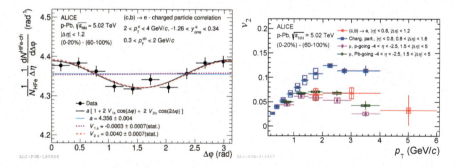

Fig. 76.1 Left panel: azimuthal-correlation distribution between heavy-flavour decay electrons and charged particles, for high-multiplicity p–Pb collisions after subtracting the jet contribution from low-multiplicity collisions. Right panel: heavy-flavour decay electron v_2 as a function of p_T compared to the v_2 of unidentified charged particles [5] and inclusive muons [6]

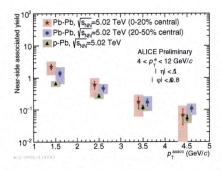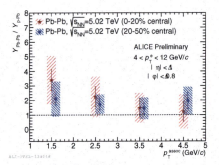

Fig. 76.2 Left panel: near-side associated yield in central and semi-central Pb–Pb and p–Pb collisions. Right panel: ratio of near-side associated yield of Pb–Pb w.r.t p–Pb collisions

76.4 Summary and Outlook

The results of azimuthal correlations between HFe and charged particles in p–Pb and Pb–Pb collisions, extracted in different p_T intervals of HFe and associated charged particles, are presented. A positive v_2 for both light-flavour hadrons and heavy-flavour particles seen in p–Pb collisions. The near-side associated yield in both p–Pb and Pb–Pb collisions and the ratio of near-side associated yield of Pb–Pb with respect to p–Pb collisions are presented. The mild increase in the near-side associated yield in Pb–Pb collisions compared to p–Pb hints at a possible medium-induced modification of the fragmentation of heavy quarks.

References

1. K. Aamodt et al. (ALICE Collaboration), JINST **3** S08002 (2008)
2. B. Abelev et al. (ALICE Collaboration), Int. J. Mod. Phys. A **29**, 1430044 (2014)
3. J. Adam et al. (ALICE Collaboration), Phys. Lett. B **754**, 81 (2016)
4. B. Abelev et al. (ALICE Collaboration), Phys. Lett. B **738**, 97 (2014)
5. B. Abelev et al. (ALICE Collaboration), Phys. Lett. B **726**, 164 (2013)
6. B. Abelev et al. (ALICE Collaboration), Phys. Lett. B **753**, 126 (2016)
7. S. Acharya et al. (ALICE Collaboration), Phys. Rev. Lett. **122**, 072301 (2019)

Chapter 77
Evolution of Quarkonia States in a Rapidly Varying Strong Magnetic Field

Partha Bagchi, Nirupam Dutta, Bhaswar Chatterjee, and Souvik Priyam Adhya

Abstract In a transient magnetic field, heavy quarkonium bound states evolve non-adiabatically. In the presence of a strong magnetic field, J/Ψ and $\Upsilon(1S)$ become more tightly bound than we expected earlier for a pure thermal medium. We have shown that in a time- varying magnetic field, there is a possibility of moderate suppression of J/Ψ through the non-adiabatic transition to continuum whereas the $\Upsilon(1S)$ is so tightly bound that it cannot be dissociated through this process. We have calculated the dissociation probabilities up to the first order in the time-dependent perturbation theory for different values of initial magnetic field intensity.

There are possibilities of the production of a very high intensity magnetic field [1–4] in non- central high-energy nucleus-nucleus collisions. The produced magnetic field can modify heavy quarkonia suppression in the deconfined Quark Gluon Plasma [5–7]. Another obvious modification is the Zeeman splitting of quarkonium states in a constant magnetic field which essentially creates various quarkonium states [8, 9] differing by their spin degrees of freedom which is very similar to the case of positronium in quantum electrodynamics [10]. Then, there are possibilities for spin mixing in homogeneous [7, 9] and inhomogeneous [11] magnetic field environments. Besides that, the ionisation [12] of bound states due to the tunnelling caused by the magnetic field can lead to the suppression of quarkonium states. Furthermore, the static quark anti-quark potential in a medium can also be modified in the presence of

P. Bagchi (✉) · S. P. Adhya
Variable Energy Cyclotron Centre, Kolkata, India
e-mail: parphy85@gmail.com

S. P. Adhya
e-mail: sp.adhya@vecc.gov.in

N. Dutta
National Institute of Science Education and Research Bhubaneswar, Odisha, India
e-mail: niripamdu@gmail.com

B. Chatterjee
Department of Physics, Indian Institute of Technology Roorkee, Roorkee 247667, India
e-mail: bhaswar.mph2016@iitr.ac.in

© Springer Nature Singapore Pte Ltd. 2021
P. K. Behera et al. (eds.), *XXIII DAE High Energy Physics Symposium*,
Springer Proceedings in Physics 261,
https://doi.org/10.1007/978-981-33-4408-2_77

a magnetic field. Depending on the non-centrality and the energy of the heavy ion, the magnetic field can be as strong as $B \simeq 50\, m_\pi^2$ where $m_\pi^2 = 10^{18}$ Gauss and decays very quickly as the spectator quarks move away from the fireball. It has been estimated that at time $t \simeq 0.4\ fm$, the magnetic field is practically negligible. However, if QGP forms very early in time, then it can trap the magnetic field because of its high electrical conductivity. So the formation of QGP can increase the persistence time [13] of the magnetic field in Relativistic Heavy Ion Collision (RHIC). In spite of that, the field will decay to a few orders of magnitude within a few fm/c time. Hence, the produced magnetic field is time dependent.

In this article, we have calculated the transition from ground states of quarkonia to the continuum states in the presence of the transient magnetic field. This leads to further suppression of quarkonia which is completely different from the ionisation process discussed earlier [12]. In a time-varying magnetic field, quarkonia evolve non-adiabatically because the quark anti-quark potential becomes time dependent and changes very rapidly as the magnetic field does. The non-adiabatic evolution previously has been addressed in the context of evolving QGP [14] and also in the context of rapid thermalisation [15].

In this work, we will restrict ourselves within the strong magnetic field approximation which means that the magnetic field will dominate over all other scales present in the system such as mass and temperature because $\frac{eB}{m^2} >> 1$ and $\frac{eB}{T^2} >> 1$, where m is the mass of the medium particle affected by the magnetic field and T is the temperature of the system. This is obviously above Schwinger's critical limit [16] that makes it possible to have a classical description of the magnetic field. The effects of the magnetic field are incorporated through the light quark propagator. The fermion propagator in the strong field limit is given by

$$S_0(k) = i\,\frac{m + \gamma \cdot k_\parallel}{k_\parallel^2 - m^2}(1 - i\gamma_1\gamma_2)e^{\frac{-k_\perp^2}{|q_f B|}} \tag{77.1}$$

for zero temperature. Here, we have assumed the magnetic field B to be along a fixed direction (let's say z). q_f is the electric charge of the fermion of flavour f, and K is the fermion 4-momentum expressed as $k_\perp^2 = -(k_x^2 + k_y^2)$, $k_\parallel^2 = k_0^2 + k_z^2$ and $\gamma \cdot k_\parallel = \gamma_0 k_0 - \gamma_3 k_z$. The split in the 4-momentum occurs due to the Landau quantization in the plane transverse to the magnetic field as the fermion energy is given by

$$E = \sqrt{m^2 + k_z^2 + 2n|q_f|B} \tag{77.2}$$

with n being the number of Landau levels which is equal to zero in the strong field limit. At finite temperature, the propagator in real time [17] becomes

$$i S_{11}(p) = \left[\frac{1}{p_{\parallel}^2 - m^2 + i\epsilon} + 2\pi i n_p \delta(p_{\parallel}^2 - m^2) \right] (1 + \gamma^0 \gamma^3 \gamma^5)$$

$$\times (\gamma^0 p_0 - \gamma^3 p_z + m) e^{\frac{-p_{\perp}^2}{|q B|}}, \tag{77.3}$$

where the distribution is

$$n_p(p_0) = \frac{1}{e^{\beta |p_0|} + 1}, \tag{77.4}$$

with the Boltzmann factor β. The Debye screening mass (m_D) heavy quark potential in a strong magnetic field can be obtained by taking the static limit ($|\mathbf{p}| = 0$, $p_0 \to 0$) of the longitudinal part of the gluon self-energy $\pi_{m_D \nu}$. If there is no magnetic field in the medium, then m_D can be written for the three-flavour case as $m_D = g T \sqrt{1 + N_f/6}$ [18]. In the presence of a magnetic field, the Debye mass [19] becomes

$$m_D^2 = g'^2 T^2 + \frac{g^2}{4\pi^2 T} \sum_f |q_f B| \int_0^\infty dp_z \frac{e^{\beta \sqrt{p_z^2 + m_f^2}}}{\left(1 + e^{\beta \sqrt{p_z^2 + m_f^2}}\right)^2} \tag{77.5}$$

where the first term is the contribution from the gluon loops which depends on temperature only. The second term, the contribution from the fermion loop, strongly depends on the magnetic field and is not much sensitive to the temperature of the medium. In the first term, $g'^2 = 4\pi \alpha_s'(T)$ where $\alpha_s'(T)$ is the usual temperature-dependent running coupling where the renormalization scale is taken as $2\pi T$. It is given by

$$\alpha_s'(T) = \frac{2\pi}{\left(11 - \frac{2}{3} N_f\right) \ln \left(\frac{\Lambda}{\Lambda_{QCD}}\right)} \tag{77.6}$$

where $\Lambda = 2\pi T$ and $\Lambda_{QCD} \sim 200\,\text{MeV}$.

In the second term, $g^2 = 4\pi \alpha_s^{\parallel}(k_z, q_f B)$, where $\alpha_s^{\parallel}(k_z, q_f B)$ is the magnetic field-dependent coupling and doesn't depend on temperature. This is given by [20, 21]

$$\alpha_s^{\parallel}(k_z, q_f B) = \frac{1}{\alpha_s^0(\mu_0)^{-1} + \frac{11 N_c}{12\pi} \ln \left(\frac{k_z^2 + M_B^2}{\mu_0^2}\right) + \frac{1}{3\pi} \sum_f \frac{q_f B}{\sigma}} \tag{77.7}$$

where

$$\alpha_s^0(\mu_0) = \frac{12\pi}{11 N_c \ln \left(\frac{\mu_0^2 + M_B^2}{\Lambda_V^2}\right)} \tag{77.8}$$

All the parameters are taken as $M_B = 1\,\text{GeV}$, the string tension $\sigma = 0.18\,\text{GeV}^2$, $\mu_0 = 1.1\,\text{GeV}$ and $\Lambda_V = 0.385\,\text{GeV}$.

Now one has to see the nature of the magnetic field which decreases with time and that essentially makes the Debye screening mass a time-dependent quantity. The intensity of the initial magnetic field B_0 is of the order of a few m_π^2 and decays with time in the following way:

$$B = B_0 \frac{1}{1 + at}, \tag{77.9}$$

using the fitting of the result provided in the article by Tuchin [22] with the value of the parameter $a = 5.0$. The heavy quark potential in the medium can be written as

$$V(r) = -\frac{\alpha}{r} exp(-m_D r) + \frac{\sigma}{m_D}(1 - exp(-m_D r)) \tag{77.10}$$

The effect of the temperature and magnetic field is incorporated in the Debye mass given in (77.5) This is obvious that the potential becomes time dependent due to the time dependence of the magnetic field and temperature. We consider that initially there are only ground states of charmonia (J/Ψ) and bottomonia $(\Upsilon(1S))$. These two states evolve in a time-dependent potential which causes the transition to other excited states as well as to the continuum. We would like to calculate the transition probabilities of the ground states to the continuum which gives us the dissociation probabilities of (J/Ψ) and $(\Upsilon(1S))$. We have adopted time-dependent perturbation theory in this context in order to calculate the dissociation probability up to the first order. The perturbation at any instant t is considered to be $H^1(t) = V(r, t) - V(r, t_i)$. We want to calculate the transition probability to the unbound states which are obviously plane wave states given by

$$\Psi_k = \frac{1}{\sqrt{\Omega}} e^{i\mathbf{k}\cdot\mathbf{r}} \tag{77.11}$$

which is box-normalised over a volume Ω and can have all possible values of the momentum $\mathbf{k}$. The first-order contribution to the transition amplitude can be expressed as

$$a_{ik} = \int \frac{d}{dt} \langle \Psi_k | H^1(t) | \Psi_i \rangle \frac{e^{i(E_i - E_k)}}{(E_i - E_k)} dt. \tag{77.12}$$

$|\Psi_i\rangle$, E_i are initial quarkonium state and the corresponding energy eigenstates, respectively, and E_k is the energy of the dissociated state $|\Psi_k\rangle$. The total transition probability to all continuum states is given by

$$= \int_{k=0}^{\infty} |a_{ik}|^2 \frac{\Omega}{(2\pi)^3} k^2 dk, \tag{77.13}$$

where the number of unbound states between the momentum continuum k and $k + dk$ over 4π solid angle is

$$dn = \left(\frac{L}{2\pi}\right)^3 k^2 dk = \frac{\Omega}{(2\pi)^3} k^2 dk \qquad (77.14)$$

We know that J/Ψ and $\Upsilon(1S)$ can survive in the thermal medium (QGP) almost up to $2.2T_c$ and $4T_c$, respectively [23], but in the presence of a magnetic field, the binding energies of these states get modified. The binding energy is given by

$$E_{disso} = E_B - 2m_q - \frac{\sigma}{m_D}, \qquad (77.15)$$

where m_q is the mass of quark and E_B is the energy eigenvalue calculated from the time-independent Schrödinger equation by using the Numerov's method. We have plotted the binding energy of J/Ψ at a temperature $1.7T_c$ and $\Upsilon(1S)$ at a temperature $3T_c$ as a function of the magnetic field intensity in Fig. 77.1. The binding energies do not change much over a span of magnetic field intensity from $1 - 15m_\pi^2$. In other words, these quarkonium states can survive at a higher temperature if there is a magnetic field present in the medium. Within the specified rage of the magnetic field intensity, the dissociation temperature of J/Ψ and $\Upsilon(1S)$ becomes $2.73 - 2.94T_c$ and $8.12 - 8.89T_c$, respectively.

We have employed first-order perturbation theory to evaluate the dissociation probabilities of both the ground states first by considering a purely thermal QGP which cools off to the temperature T_c of the medium and then the same has been calculated by considering the time-dependent magnetic field in the evolving QGP. For our study on J/Ψ, we have considered that the initial temperature of the medium is $1.7T_c$ which eventually decreases according to the power law given by,

Fig. 77.1 Binding energy of J/Ψ and $\Upsilon(1S)$ as a function of the magnetic field intensity

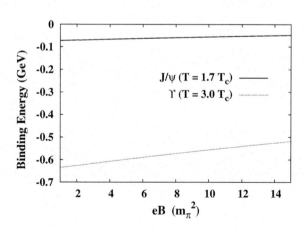

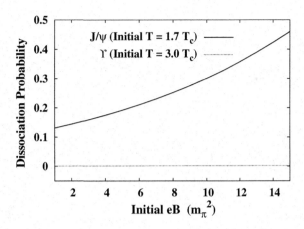

Fig. 77.2 Dissociation probability of J/Ψ and $\Upsilon(1S)$ as a function of the intensity of the initial magnetic field

$$T(t) = T_0 \left(\frac{\tau_0}{\tau_0 + t} \right)^{\frac{1}{3}} , \qquad (77.16)$$

with T_0, the initial temperature and τ_0 be the equilibration time, taken to be approximately $0.5\,fm/c$ for QGP. We have calculated the dissociation probability when the medium temperature falls off to T_c from an initial value in the presence of the time-dependent magnetic field. The initial value of the magnetic field is not known exactly and therefore we have used various initial values of the magnetic field intensity and have shown the dissociation probabilities as a function of the initial magnetic field. The same has been done for the $\Upsilon(1S)$ state by considering the initial temperature around $3T_c$. In Fig. 77.2, the solid black line denotes the dissociation probability of J/Ψ which increases with the initial field intensity. The state J/Ψ can be dissociated from 12 to 50 percent within the range of the field intensity $1-15m_\pi^2$. The dotted blue line shows that the dissociation probability for $\Upsilon(1S)$ is almost zero over the specified span of the field strength.

Summarising the article, we conclude that due to the modification of the heavy quark potential in the presence of a magnetic field, the bound states J/Ψ and $\Upsilon(1S)$ become more strongly bound compared to those in a pure thermal QGP. As a result, the bound states can survive much higher temperatures than we have expected previously. Although J/Ψ can be dissociated by making non-adiabatic transitions to the unbound states, $\Upsilon(1S)$ still remains bound. We have estimated the dissociation probability within the limits of first-order perturbation theory.

Acknowledgements P. B. acknowledges SERB (NPDF Scheme: PDF/2016/003837), Government of India, for the financial assistance and also thanks Jan-e Alam for useful discussion.

References

1. V. Skokov, A.Y. Illarionov, V. Toneev, Int. J. Mod. Phys. A **24**, 5925 (2009)
2. D.E. Kharzeev, L.D. McLerran, H.J. Warringa, Nucl. Phys. A **803**, 227 (2008)
3. N. Mueller, J.A. Bonnet, C.S. Fischer, Phys. Rev. D **89**(9), 094023 (2014)
4. V.A. Miransky, I.A. Shovkovy, Phys. Rep. **576**, 1 (2015)
5. C. Bonati, M. D'Elia, A. Rucci, Phys. Rev. D **92**(5), 054014 (2015)
6. X. Guo, S. Shi, N. Xu, Z. Xu, P. Zhuang, Phys. Lett. B **751**, 215 (2015)
7. D.L. Yang, B. Muller, J. Phys. G **39**, 015007 (2012)
8. P. Filip, PoS CPOD **2013**, 035 (2013)
9. J. Alford, M. Strickland, Phys. Rev. D **88**, 105017 (2013)
10. S.G. Karshenboim, Int. J. Mod. Phys. A **19**, 3879 (2004) [hep-ph/0310099]
11. N. Dutta, S. Mazumder, Eur. Phys. J. C **78**(6), 525 (2018)
12. K. Marasinghe, K. Tuchin, Phys. Rev. C **84**, 044908 (2011)
13. A. Das, S.S. Dave, P.S. Saumia, A.M. Srivastava, Phys. Rev. C **96**(3), 034902 (2017)
14. N. Dutta, N. Borghini, Mod. Phys. Lett. A **30**(37), 1550205 (2015)
15. P. Bagchi, A.M. Srivastava, Mod. Phys. Lett. A **30**(32), 1550162 (2015)
16. J.S. Schwinger, Phys. Rev. **82**, 664 (1951)
17. M. Hasan, B. Chatterjee, B.K. Patra, Eur. Phys. J. C **77**(11), 767 (2017)
18. F. Karsch, H. Satz, Z. Phys. C **51**, 209 (1991)
19. M. Hasan, B.K. Patra, B. Chatterjee, P. Bagchi (2018), arXiv:1802.06874 [hep-ph]
20. M.A. Andreichikov, V.D. Orlovsky, Y.A. Simonov, Phys. Rev. Lett. **110**, 162002 (2013)
21. E.J. Ferrer, V. de la Incera, X.J. Wen, Phys. Rev. D **91**(5), 054006 (2015)
22. K. Tuchin, Phys. Rev. C **93**(1), 014905 (2016)
23. A. Mocsy, P. Petreczky, Phys. Rev. Lett. **99**, 211602 (2007)

Chapter 78
Predictions for Transverse Momentum Spectra and Elliptic Flow of Identified Particles in Xe+Xe Collisions at $\sqrt{s_{NN}} =$ 5.44 TeV Using a Multi-phase Transport Model (AMPT)

Sushanta Tripathy, R. Rath, S. De, M. Younus, and Raghunath Sahoo

Abstract Relativistic collisions of Xe+Xe at Large Hadron Collider give us an opportunity to study the properties of possible state of deconfined quarks and gluons, where the size of the produced system lies between the p+p and Pb+Pb collisions. In the present study, we have incorporated nuclear deformation in a multi-phase transport (AMPT) model to study the identified particle production and elliptic flow in Xe+Xe at $\sqrt{s_{NN}} = 5.44$ TeV. In more comprehensive way, we have studied p_T-differential and p_T-integrated particle ratios to pions and kaons as a function of centrality and the number of constituent quark (n_q) and transverse mass (m_T) scaling of elliptic flow. The effect of deformation on particle production has also been highlighted by comparing with non-deformation case. The p_T-differential particle ratios show a strong dependence on centrality whereas p_T-integrated ratios are almost independent of centrality.

78.1 Introduction

Ultra-relativistic heavy-ion collisions at Relativistic Heavy-Ion Collider (RHIC) and Large Hadron Collider (LHC) give us an opportunity to produce the matter created in early universe, few micro-seconds after the Big-bang, in laboratory and study its properties. In such collisions, the produced system contains deconfined quarks and gluons and because of initial energy density and pressure the system expands and passes into a phase of hadron gas. This transition basically consists of couple of phase boundary, namely, chemical freeze-out, where the inelastic collisions cease and kinetic freeze-out boundary, where the elastic scatterings cease or the particle

S. Tripathy (✉) · R. Rath · S. De · R. Sahoo
Department of Physics, Indian Institute of Technology Indore, Simrol, Indore 453552, India
e-mail: sushanta.tripathy@cern.ch

M. Younus
Department of Physics, Nelson Mandela University, Port Elizabeth 6031, South Africa

© Springer Nature Singapore Pte Ltd. 2021
P. K. Behera et al. (eds.), *XXIII DAE High Energy Physics Symposium*,
Springer Proceedings in Physics 261,
https://doi.org/10.1007/978-981-33-4408-2_78

transverse momentum spectra get fixed. Hence, the hadrons produced in such collisions may carry the information about the space-time evolution of the produced system till the occurrence of the final freeze-out. Thus, the final state hadrons are very important to study the properties of the produced system in heavy-ion collisions. In recent times, symmetrical nuclei like Pb-ions or spherical gold Au-ions have been used in the ultra-relativistic colliders to produce the matter at extreme conditions. However, recent interests have been shown to collide with deformed nuclei like Uranium (U) at RHIC, BNL, and Xenon (^{129}Xe) ions at $\sqrt{s_{NN}} = 5.44$ TeV at LHC to look at the sensitivity of the final state particle production [1] and elliptic flow [2] on initial geometry of the colliding nuclei. Xenon (^{129}Xe) ions at $\sqrt{s_{NN}} = 5.44$ TeV at LHC also bridge the final state multiplicity gap between the larger Pb+Pb systems and smaller systems like p+p and p+Pb.

In this work, the nuclear deformation has been implemented in AMPT model. The effects of nuclear deformation have been investigated by studying the p_T-differential and p_T-integrated particle ratios to pions and kaons as a function of multiplicity. Also we have looked into the p_T-differential ratio of proton and ϕ meson, which have similar mass and studied the hydrodynamical behaviour of the particle production in central collisions and effect of deformation on it. Furthermore, we have studied the scaling behaviour of elliptic flow with respect to constituent quarks as a function of p_T/n_q and $(m_T - m_0)/n_q$. For the incorporation of deformation in AMPT model we refer to [1]. Now lets proceed to next section to discuss the obtained results using AMPT model.

78.2   Results and Discussion

In this works events are generated for Xe+Xe at $\sqrt{s_{NN}} = 5.44$ TeV at mid-rapidity using the AMPT model, which could be compared with upcoming ALICE data once available.

78.2.1   Particle Ratios

Identified particle ratios to pions and kaons for different centrality classes are shown in upper panel of Fig. 78.1 as a function of transverse momentum. These two ratios are independent of centrality classes for the bulk part ($p_T < 1$ GeV/c). Above 1 GeV/c, the ratios are higher for the most central collisions. For p/K ratio, above 1.4 GeV/c the ratio is greater than one, which suggest higher production rate of protons in the intermediate p_T-range, which may be due to contributions from the recombination.

Lower panel of Fig. 78.1 shows the p_T-integrated identified particle ratios to pions and kaons as a function of centrality. These p_T-integrated ratios remain independent of centrality classes whereas we see a strong centrality dependence of p_T-differential particle ratios. This is attributed to the fact that the integrated yield is defined by bulk

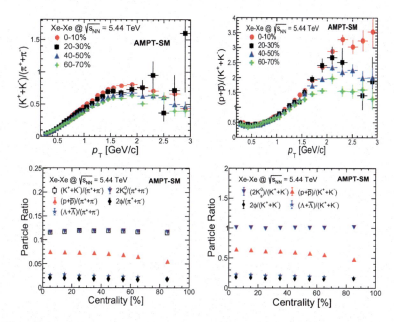

Fig. 78.1 (Color online) (Upper panel) Identified p_T-differential particle ratios to pions and kaons for different centrality classes. (Lower panel) Identified p_T-integrated particle ratios to pions and kaons as a function of centrality [1]

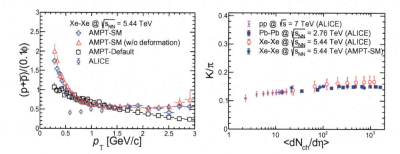

Fig. 78.2 (Color online) (Left plot) p/ϕ ratio as a function of transverse momentum. (Right plot) K/π ratio as function of charged particle multiplicity [1]

part of the system, which are at low-p_T region and from the observations from upper panel, identified particle ratios for the bulk part are independent of centrality. Left plot of Fig. 78.2 shows the p_T-differential p/ϕ ratio for different AMPT settings, i.e., AMPT default, string melting without deformation and string meting with deformation. Here, we have also compared our observations with the ALICE experimental data. From hydrodynamic point of view the p/ϕ ratio in ALICE data is almost constant at low-p_T because of the fact that they have similar masses. From all the AMPT settings, default AMPT with deformation explains the ratio qualitatively. Similarly,

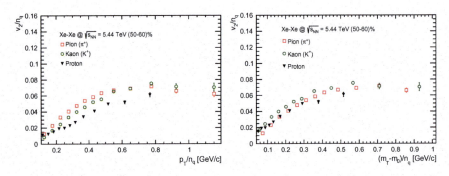

Fig. 78.3 (Color online) **a**: v_2/n_q as a function of p_T/n_q for π, K and p. **b**: v_2/n_q as a function of $(m_T - m_0)/n_q$ for π, K and p [2]

right plot of Fig. 78.2 shows the K/π ratio as a function of charged particle multiplicity. The ratio obtained from Xe+Xe collisions at $\sqrt{s_{NN}} = 5.44$ TeV fill up the gap between p+p and Pb+Pb collisions and obtained values are similar for these three systems for a given charged particle multiplicity.

78.2.2 Elliptic Flow (v_2)

The effect of collision geometry on elliptic flow has been studied using AMPT-SM. Figure 78.3 shows the elliptic flow scaled with the number of constituent quarks for π, K and p as a function of p_T/n_q and $(m_T - m_0)/n_q$ in 50–60% centrality class. We do not observe scaling of v_2 with respect to number of constituent quarks as a function of p_T/n_q. However, a possible n_q-scaling of v_2 as a function of $(m_T - m_0)/n_q$ is evident from Fig. 78.3b.

78.3 Summary and Conclusion

In this work, a detailed study has been performed for identified particle production and their elliptic flow, p_T-differential and p_T-integrated particle ratios to pions and kaons as a function of multiplicity and centrality for Xe+Xe collisions at $\sqrt{s_{NN}} = 5.44$ TeV using AMPT model. The elliptic flow scaled with number of constituent quarks for π, K and p does not show any scaling behaviour as a function of p_T/n_q, however, a possible scaling is observed as a function of $(m_T - m_0)/n_q$. The p_T-differential identified particle ratios show a strong dependence of centrality classes above $p_T >$ 1 GeV/c, however the p_T-integrated particle ratios are independent of centralities. Furthermore, default AMPT with deformation explains p_T-differential p/ϕ ratio qualitatively. Finally, the obtained K/π ratio using AMPT-SM with deformation agrees well with ALICE experimental data as a function of charged particle multiplicity. For details of this work, one may look into [1, 2].

References

1. R. Rath, S. Tripathy, R. Sahoo, S. De, M. Younus, Phys. Rev. C **99**, 064903 (2019). And references therein
2. S. Tripathy, S. De, M. Younus, R. Sahoo, Phys. Rev. C **98**, 064904 (2018). And references therein

Chapter 79
Infrared Effective Dual QCD at Finite Temperature and Densities

H. C. Chandola and H. C. Pandey

Abstract An infrared effective version of dual QCD based on the topological structure of non-Abelian gauge theories has been discussed and its extension to finite temperature and baryon densities has been analyzed to explore the dynamics of quark-hadron phase transition and QGP. The topologically effective magnetic symmetry-based dual QCD has been discussed for its color confining aspects, and its thermal version has been developed using the grand canonical ensemble formalism. The construction of the equation of state within the dual QCD framework has been shown to lead to critical parameters and critical points in the phase diagram for a QGP phase transition. For quark matter at finite baryon densities, various thermodynamical profiles have been shown to indicate a weak first-order phase transition with some critical end points.

79.1 Introduction

It is widely believed that Quantum chromodynamics as a non-Abelian theory of gauge fields can be very well used for the fundamental description of strong interactions [1, 2] between quarks and gluons. However, the low energy region of QCD still remains far from clear inaccessible by first principles due to the highly nonperturbative nature of the hadronic system in the infrared sector of QCD. In the absence of a satisfactory model to explain the low energy behavior (color confinement, hadron mass spectrum, etc.), the dual models for QCD have been put forward by a number of authors [3–6] with their own merits and demerits. The dual QCD model based on the topological structure of non-Abelian gauge theories [6–9] leads to a gauge-independent description of dual superconducting QCD vacuum, and the associated flux-tube structure may suitably be used for analyzing deconfinement/QGP

H. C. Chandola (✉)
Department of Physics (UGC-Centre of Advanced Study), Kumaun University, Nainital, India
e-mail: chandolahc@kunainital.ac.in

H. C. Pandey
Birla Institute of Applied Sciences, Bhimtal, India

© Springer Nature Singapore Pte Ltd. 2021
P. K. Behera et al. (eds.), *XXIII DAE High Energy Physics Symposium*,
Springer Proceedings in Physics 261,
https://doi.org/10.1007/978-981-33-4408-2_79

phase transition [5, 7, 9]. In addition, it has also been realized that there should be a qualitative change in a hadronic system with different thermal and density conditions, and QCD must have a complex phase structure under such extreme environments [10, 11]. This is further supplemented by various recent studies related to heavy-ion collisions to create QGP at RHIC in BNL and LHC at CERN [5, 12]. Such phase transitions of hadronic matter are of extreme importance due to their role not only in QCD but in early stages of the evolutionary universe also [11, 13].

In the present paper, following the statistical approach, the thermodynamical description of topologically effective dual QCD has been presented to analyze the phase structure of QCD in the presence of non-zero bario-chemical potential. Discussing the confining features of dual QCD in Sect. 79.2, its thermal response has been analyzed in Sect. 79.3 using grand canonical ensemble formalism for a hadronic system with a finite chemical potential. Constructing a suitable equation of state for quark matter at finite baryon densities and profiles of various thermodynamical variables have been analyzed for the phase structure of QCD.

79.2   Infrared Effective Dual QCD and Implications

The infrared effective dual QCD is basically based on the gauge-independent Abelian projection defined by the magnetic isometry and has important implications on the confining nature of the resulting dual QCD formulation. It establishes the Abelian dominance by using Abelian isometry to project the Abelian part for the Abelian sub-dynamics. In view of the fact that the non-Abelian gauge theory can be viewed [7, 8] as the Einstein theory of gravitation in a higher dimensional unified space that allows the introduction of some additional internal symmetries, the magnetic symmetry may be introduced as a set of self-consistent Killing vector fields of the internal space which, while keeping the full gauge degrees of freedom intact, restricts and reduces some of the dynamical degrees of freedom of the theory. It in turn may be shown to define the dual dynamics between the color isocharges and the topological charges of the underlying gauge group G of the theory. For the simplest choice of the gauge group $G \equiv SU(2)$ with its little group $H \equiv U(1)$, the gauge covariant magnetic symmetry condition, resulting from the Lie condition $\mathcal{L}_{m_a} g_{AB} = 0$, is expressed in the form,

$$D_\mu \hat{m} = 0, \quad i.e. \quad (\partial_\mu + g \mathbf{W}_\mu \times) \hat{m} = 0 \tag{79.1}$$

where $\hat{m}$ is a scalar multiplet that constitutes the adjoint representation of the gauge group G and $\mathbf{W}_\mu$ is the associated gauge potential of the underlying group G. The condition (79.1) thus implies that the magnetic symmetry imposes a strong constraint on the metric as well as connection and may, therefore, be regarded as the symmetry of the potential. The monopole, therefore, emerges as the topological object associated with the elements of the second homotopic group $\Pi_2(G/H)$. The typical potential satisfying the condition (79.1) is identified as

$$\mathbf{W}_\mu = A_\mu \hat{m} - g^{-1}(\hat{m} \times \partial_\mu \hat{m}), \tag{79.2}$$

where $\hat{m} \cdot \mathbf{W}_\mu \equiv A_\mu$ is the color electric potential unrestricted by magnetic symmetry, while the second term is completely determined by magnetic symmetry and is topological in origin. Thus, the virtue of the magnetic symmetry is that it can be used to describe the topological structure of gauge symmetry in such a way that the multiplet $\hat{m}$ may be viewed to define the homotopy of the mapping $\Pi_2(S^2)$ on $\hat{m} : S_R^2 \to S^2 = SU(2)/U(1)$. It clearly shows that the imposition of magnetic symmetry on the potential brings the topological structure into the dynamics explicitly. The associated field strength corresponding to the potential (79.2) is then given by

$$\mathbf{G}_{\mu\nu} = \mathbf{W}_{\nu,\mu} - \mathbf{W}_{\mu,\nu} + g\mathbf{W}_\mu \times \mathbf{W}_\nu \equiv (F_{\mu\nu} + B_{\mu\nu}^{(d)})\hat{m} \tag{79.3}$$

where, $F_{\mu\nu} = A_{\nu,\mu} - A_{\mu,\nu}$ and $B_{\mu\nu}^{(d)} = B_{\nu,\mu} - B_{\mu,\nu} = g^{-1}\hat{m}.(\partial_\mu \hat{m} \times \partial_\nu \hat{m})$. In order to explain the dynamics of the resulting dual QCD vacuum and its implications on confinement mechanism, we start with the SU(2) Lagrangian with a quark doublet source $\psi(x)$ as given by

$$\mathcal{L} = -\frac{1}{4}\mathbf{G}_{\mu\nu}{}^2 + \bar{\psi}(x)i\gamma^\mu D_\mu \psi(x) - m_0\bar{\psi}(x)\psi(x) \tag{79.4}$$

However, in order to avoid the problems due to the point-like structure and the singular behavior of the potential associated with monopoles, we use the dual magnetic potential $B_\mu^{(d)}$ coupled to a complex scalar field $\phi(x)$. Taking these considerations into account, the modified form of the dual QCD Lagrangian (79.4) in quenched approximation is given as follows:

$$\mathcal{L}_m^{(d)} = -\frac{1}{4}B_{\mu\nu}^2 + \left|\left[\partial_\mu + i\frac{4\pi}{g}B_\mu^{(d)}\right]\phi\right|^2 - V(\phi^*\phi) \tag{79.5}$$

where the effective potential $V(\phi^*\phi)$ appropriate for inducing the dynamical breaking of magnetic symmetry in the near-infrared region of QCD is the quartic potential of the following form:

$$V(\phi^*\phi) = 3\lambda\alpha_s^{-2}(\phi^*\phi - \phi_0^2)^2. \tag{79.6}$$

In order to analyze the nature of magnetically condensed vacuum and the associated flux-tube structure, responsible for its non-perturbative behavior, we use the Neilsen and Olesen [14] interpretation of vortex-like solutions of the field equations associated with Lagrangian (79.6) which leads to the possibility of the existence of the monopole pairs inside the superconducting vacuum in the form of thin flux tubes responsible for the confinement of any colored flux. Under cylindrical symmetry (ρ, φ, z) and the field ansatz given by, $B_\varphi^{(d)}(x) = B(\rho)$, $B_0^{(d)} = B_\rho^{(d)} = B_z^{(d)} = 0$ $\ and \ \ \phi(x) = \chi(\rho)exp(in\varphi)$ $(n = 0, \pm1, \pm2, ----)$ along with a representation with dimensionless parameters as $r = 2\sqrt{3\lambda}\alpha_s^{-1}\phi_0\rho$, $F(r) = (4\pi\alpha_s^{-1})^{\frac{1}{2}}\rho B(\rho)$, $H(r) = \phi_0^{-1}\chi(\rho)$ the associated field equations acquire the following form:

$$H'' + \frac{1}{r}H' + \frac{1}{r^2}(n + F)^2 + \frac{1}{2}H(H^2 - 1) = 0, \quad F'' - \frac{1}{r}F' - \alpha(n + F)H^2 = 0 \tag{79.7}$$

where $\alpha = 2\pi\alpha_s/3\lambda$ and the prime stands for the derivative with respect to r. Using the asymptotic boundary conditions given by $F \to -n$, $H \to 1$ as $r \to \infty$, the asymptotic solution for the function F is obtained in the following form:

$$F(\rho) = -n + C\rho^{\frac{1}{2}}exp(-m_B\rho) \tag{79.8}$$

where $C = 2\pi B(3\sqrt{2}\lambda g^{-3}\phi_0)^{\frac{1}{2}}$ and $m_B = (8\pi\alpha_s^{-1})^{\frac{1}{2}}\phi_0$ is the mass of the magnetic glueballs which appear as vector mode of the magnetically condensed QCD vacuum. Since the function $F(\rho)$ is associated with the color electric field through gauge potential $B(\rho)$ as $E_m(\rho) = -\rho^{-1}\partial_\rho(\rho B(\rho))$, it indicates the emergence of the dual Meissner effect leading to the confinement of the color isocharges in the magnetically condensed dual QCD vacuum. Utilizing the asymptotic solutions of the associated dual QCD fields, the energy per unit length of the resulting flux-tube configuration acquires the following form:

$$k = 2\pi\phi_0^2 \int_0^\infty rdr \left[\frac{6\lambda}{g^2}\frac{(F')^2}{r^2} + \frac{(n + F)^2}{r^2}H^2 + (H')^2 + \frac{(H^2 - 1)^2}{4}\right], \tag{79.9}$$

which, on using its relationship with the Regge slope parameter ($\alpha' = 1/2\pi k = 0.93 GeV^{-2}$), leads to the numerical identification of unknown ϕ_0 and, hence, in turn, m_B for different strong couplings in full infrared sector of QCD [8] as (2.11, 1.51, 1.21 and 0.929) GeV for $\alpha_s = 0.12, 0.22, 0.47, 0.96$, respectively. The unique confining multi-flux-tube configuration of dual QCD leads to a typical phase structure to QCD which becomes more evident for the thermalized hadronic system as discussed in the next section.

79.3 Thermal Dynamics and QGP

The phase structure of hadronic matter in QCD and the associated QGP phase of matter has been a subject of immense importance especially during post LHC/RHIC era with state-of-the-art facilities for high-energy heavy-ion collisions. Hence, the investigations for the thermodynamical evolution of dual QCD are greatly desired to describe the dynamics of QCD matter in different phases including the QGP phase. Using equilibrium thermodynamics, statistical concepts may then be used to study the properties of hadronic matter with a large number of particles under extreme conditions of temperature/density for exploiting to determine the critical parameters of phase transitions. With these considerations, using grand canonical ensemble formalism, the partition function for a thermodynamical system in thermal and chemical equilibria may be given by

$$Z = Tr\left[exp\left(-\frac{1}{T}(\hat{H} - \mu\hat{N})\right)\right], \tag{79.10}$$

where $\hat{H}$ and $\hat{N}$ are the Hamiltonian and particle number operators, respectively, and μ is the associated chemical potential. It, in turn, leads to thermodynamic quantities of physical interest such as energy density (ε), particle number density (n), pressure (P), entropy (S), baryon number (n_B) and internal energy (E), as the appropriate first derivatives expressed in the following form:

$$\varepsilon = \frac{T^2}{V}\frac{\partial}{\partial T}(lnZ) + \mu n, \quad n = \frac{T}{V}\frac{\partial}{\partial \mu}logZ, \quad P = \frac{\partial(TlnZ)}{\partial V} \quad S = \frac{\partial(TlnZ)}{\partial T},$$
$$n_B = \frac{1}{3}\frac{\partial(TlnZ)}{\partial \mu_q}, E = TS - PV + \mu n_B, \tag{79.11}$$

where $\mu = 3\mu_q$ in terms of quark chemical potential with $\mu_u = \mu_d = \mu_q$. As a measure of thermal fluctuations, the second derivative parameters such as specific heat at constant volume and the square of speed of sound may also be defined in the form, $C_V = \left(\frac{\partial \varepsilon}{\partial T}\right)_V$, $c_s^2 = \frac{dP}{d\varepsilon} = \frac{s(T)}{C_V(T)}$ where $s(T) = \frac{S}{V}$ is the entropy density providing a measure of the deviation of equation of state from the conformal behavior. The associated non-conformal behavior of a special QCD medium is further related with the parameters of trace anomaly and conformal measure expressed in the form, $\Delta(\tau) = \frac{\varepsilon - 3P}{T^4}$, $\zeta = (\varepsilon - 3P)/\varepsilon$.

Using Eq. (79.10), the logarithm of partition function in particle number representation for a system of particles in the large volume limit for the gases of massless fermions (f) and massless bosons (b) is obtained as

$$(TlnZ)_b = \frac{g_b V}{90}\pi^2 T^4, \quad (TlnZ)_f = \frac{g_f V}{24}\left(\frac{7\pi^2 T^4}{30} + \mu^2 T^2 + \frac{1}{2\pi^2}\mu^4\right), \tag{79.12}$$

which for the system of free quarks, antiquarks and gluons,

$$(TlnZ)_{q,g} = \frac{2}{3}\left(\frac{\pi^2}{37}T^4 + \mu_q^2 T^2 + \frac{1}{2\pi^2}\mu_q^4\right)V. \tag{79.13}$$

On the other hand, for addressing the QGP phase transition in QCD, the thermal evolution of a quark-gluon system may be analyzed in terms of the infrared effective model using magnetic symmetry-based dual QCD [7, 8], where the multi-flux-tube system as a periodic system on a S^2-sphere on energetic grounds leads to the bag constant given as

$$B^{1/4} = \left(\frac{12}{\pi^2}\right)^{1/4} \frac{m_B}{8}. \tag{79.14}$$

Assuming that the whole quark matter is enclosed inside a big bag leading to the shift in ground state from physical vacuum into the QCD vacuum achieved by including the bag energy term (BV) to (79.13), the grand canonical partition function for QGP may be evaluated as

$$(T \ln Z)_p = \frac{2}{3} V \left(\frac{\pi^2 T^4}{3} + \mu_q^2 T^2 + \frac{1}{2\pi^2} \mu_q^4\right) - BV. \tag{79.15}$$

The resulting expressions for the energy density, pressure and entropy density for the quark-gluon plasma phase may be evaluated as

$$\varepsilon_p = \frac{2}{3}\pi^2 T^4 + 2\mu_q^2 T^2 + \frac{\mu_q^4}{\pi^2} + B, \quad P_p = \frac{2}{9}\pi^2 T^4 + \frac{2}{3}T^2\mu_q^2 + \frac{\mu_q^4}{3\pi^2} - B \text{ and}$$

$$s_p = \frac{8}{9}\pi^2 T^3 + \frac{4}{3}\mu_q^2 T. \tag{79.16}$$

For the study of equilibrium phase transition, we consider baryonic matter composed of nucleons with non-vanishing baryo-chemical potential and neglect the inter-quark interactions so that the associated phase transition is dominated by entropy considerations. Hence, using the expression (79.12) with the degeneracy factor $g_f = 2 \times 2$ for nucleons, the pressure and energy density for hadronic matter may be expressed in the following form:

$$P_h = \frac{7}{180}\pi^2 T^4 + \frac{1}{6}\mu_q^2 T^2 + \frac{1}{12\pi^2}\mu_q^4, \quad \varepsilon_h = \frac{7}{60}\pi^2 T^4 + \frac{1}{2}\mu_q^2 T^2 + \frac{1}{4\pi^2}\mu_q^4 \text{ and}$$

$$s_h = \frac{7}{45}\pi^2 T^3 + \frac{1}{3}\mu_q^2 T. \tag{79.17}$$

In the domain of non-zero quark chemical potential, the coexistence of matter in hadronic (h) and plasma (p) phases as per Gibbs criteria given by $P_h = P_p = P_c$, $T_h = T_p = T_c$, $\mu = 3\mu_q = \mu_c$ leads to the critical point of phase transition from hadron to the QGP phase at a typical temperature and chemical potential (subscript c refers to the critical point of the QGP phase transition). As such, using the phase equilibrium conditions for the hadron pressure and plasma pressure given by equations (79.16) and (79.17), respectively, along with the critical point as per Gibbs free energy minimization $\mu_q = \frac{1}{3}\mu_c$, $T = T_c$ results in an identity given as

$$\frac{11}{60}\pi^2 T_c^4 + \frac{1}{18}T_c^2\mu_c^2 + \frac{\mu_c^4}{324\pi^2} = B. \tag{79.18}$$

The relations given by Eqs. (79.16)–(79.18) play an important role in the phase structure of QCD and are useful to compute various critical and thermodynamical properties of QGP as discussed below.

Analyzing the equation of state for the QGP phase transition for non-vanishing bario-chemical potential as presented in $(T - \mu)$ plane in Fig. 79.1 using the normalized pressure (P/T^4) for hadron and QGP phases given by equations (79.17) and (79.16), respectively, leads to the critical temperatures of the QGP phase transition which for the case of the optimal value of the strong coupling, α_s=0.12, in the infrared sector of QCD, yielding the values of (μ_c, T_c^μ) as $(0.66, 0.071)$ GeV. The corresponding plots for vanishing chemical potential lead to the point of the QGP phase transition at the critical temperatures of (T_c^0) of 0.239 GeV for the case of $\alpha_s = 0.12$ [8]. It, therefore, indicates a considerable reduction in the critical temperature of the QGP phase transition for a non-vanishing bario-chemical potential. In addition, the thermal variation of the bario-chemical potential as per equation (79.18) depicted in Fig. 79.2 indicates the existence of a critical end point where a weak first-order QCD phase transition goes into a possible crossover.

A similar indication for a possible crossover beyond the critical end point is obtained by the quantities derived from pressure derivatives like quark number density and susceptibility and scale pressure difference which indeed show growing fluctuations around the critical end point.

Similarly, in view of the rapid rise in the degree of freedom around the critical point, thermodynamic variables like energy density, entropy density, specific heat and sound speed become typical indicators of phase transition in QCD and need to be investigated for the case of finite bario-chemical potentials. In the presence of chemical potential, the differences in energy densities of plasma and hadronic phases

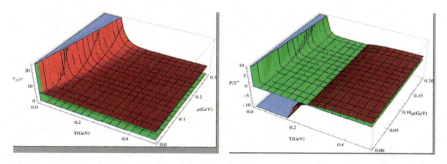

Fig. 79.1 (Color online) Variation of energy density (left) and pressure (right) with T and μ

Fig. 79.2 (Color online)
$T - \mu$ phase diagram

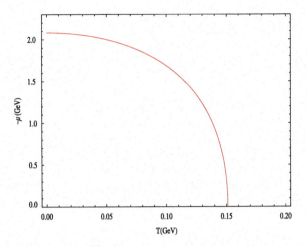

appearing as a measure of non-zero finite latent heat and the associated entropy differences may be evaluated in the following form:

$$\Delta \varepsilon = \varepsilon_p(T_c^\mu) - \varepsilon_h(T_c^\mu) = \frac{33\pi^2 T_c^{\mu 4}}{60} + \frac{3}{2}\mu_q^2 T_c^{\mu 2} + \frac{3}{4\pi^2}\mu_q^4 + B,$$

$$\Delta s = s_p(T_c^\mu) - s_h(T_c^\mu) = \frac{11}{15}\pi^2 T_c^{\mu 3} + \frac{2}{3}\mu_q^2 T_c^\mu. \qquad (79.19)$$

Equations (79.16) and (79.18) have been depicted in $(T - \mu)$ plane in Fig. 79.2 for strong coupling in the near-infrared sector of QCD. The slight enhancement in said quantities for finite chemical potential is characterized by the process of hadrons slowly losing their identities and quarks and gluon gradually becoming the fundamental degrees of freedom in the QGP phase.

Furthermore, the squared sound speed corresponding to the softening of the equation of state and as a sensitive indicator of the critical behavior in the QGP phase may be expressed using Eq. (79.16) in the following form:

$$c_s^2 = \frac{(2\pi^2 T^4/9 + 2T^2\mu_q^2/3 + \mu_q^4/3\pi^2 - B)}{(2\pi^2 T^4/3 + 2\mu_q^2 T^2 + \mu_q^4/\pi^2 + B)}. \qquad (79.20)$$

Figure 79.3 represents the plot for the square of sound speed where in the QGP phase, near the transition region, c_s^2 drops to its minimum and then rises to the value $c_s^2 = 0.33$.

In addition, the dynamics of the associated phase transitions as determined by analyzing the derivatives of the energy density with respect to temperature, i.e. specific heat also plays an important role to determine the nature of phase transition associated with the change in degrees of freedom in the medium. Using Eqs. (79.16)

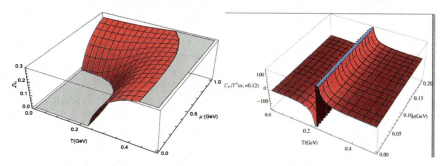

Fig. 79.3 (Color online) Variation of squared sound speed (left) and normalized specific heat (right) with T and μ

and (79.19), the normalized specific heat for the QGP phase may be expressed in the following form:

$$\frac{C_V}{T^3} = (8\pi^2/9 + 4\mu_q^2/3T^2)\frac{(2\pi^2T^4/3 + 2\mu_q^2T^2 + \mu_q^4/\pi^2 + B)}{(2\pi^2T^4/9 + 2T^2\mu_q^2/3 + \mu_q^4/3\pi^2 - B)}. \quad (79.21)$$

The normalized specific heat profile in Fig. 79.3, in fact, corresponds to the lambda transition similar to that in liquid helium, which clearly reflects a very sharp rise associated with large energy fluctuations around the critical point followed by a subsequent large reduction at high temperatures. A huge enhancement in the values of the normalized value of specific heat and squared speed of sound for critical values of (μ_c, T_c^{μ}) has been found in comparison to those for vanishing chemical potential near seemingly crossover region.

79.4 Conclusions

The infrared effective dual QCD based on gauge-independent Abelian projection establishes dual dynamics between color isocharges and topological magnetic charges which ensures the confinement of colored sources as a result of magnetic condensation of QCD vacuum due to magnetic symmetry breaking in a dynamical way. Extending the analysis for thermal domains and finite baryon density environments through a statistical approach, the equation of state for the quark-hadron system with magnetic glueball mass-dependent bag pressure leads to various thermodynamical parameters vital for analyzing the phase structure of QCD. The profiles of various thermodynamical functions under variations of temperature and chemical potential has been shown to lead to a complementary behavior of thermal and density parameters for a QGP phase transition. The use of the Gibbs criteria has been shown to lead to a range of temperatures for the QGP phase transition and indicates a weak first-order phase transition possibly reconciling into an rapid analytic crossover. As a measure of thermodynamical fluctuations and critical behavior, the squared speed of sound and specific heat have also been computed which for a non-zero chemical potential case leads to the deviation from conformal symmetry.

Acknowledgements The authors are thankful to the organizers of XXIII DAE-BRNS High Energy Physics symposium 2018 for their hospitality during the course of the symposium.

References

1. D. Gross, F. Wilczek, Phys. Rev. D **8**, 3497 (1973)
2. H.D. Politzer, Phys. Rev. Lett. **26**, 1346 (1973)
3. Y. Nambu, Phys. Rev. **D10**, 4262 (1974); S. Mandelstam, Phys. Rep. **C67**, 109 (1980); ibid **23**, 245 (1976); G. 't Hooft, Nucl. Phys. **B138**, 1 (1978); Nucl. Phys. **B190**, 455 (1981)
4. M. Baker, J.S. Ball, F. Zachariasen, Phys. Rep. **209**, 73 (1991); M. Baker, Phys. Rev. **D78**, 014009 (2008); H. Suganuma, S. Sasaki, H. Toki, Nucl. Phys. **B435**, 207 (1995)
5. H. Ichie, H. Suganuma, H. Toki, Phys. Rev. **D54**, 3382 (1996); ibid **52**, 2944 (1995); T. Suzuki et al., Phys. Rev. **D80**, 054504 (2009); M.A. Lampar, B. Svetitsky, Phys. Rev. **D61**, 04011 (2000)
6. Y.M. Cho, Phys. Rev. **D21**, 1080 (1980); ibid **D23**, 2415 (1981); Int. J. Mod. Phys. **A29**, 1450013 (2014); Y.M. Cho, X.Y. Pham et al., Phys. Rev. **D91**, 114020 (2015)
7. H.C. Panndey, H.C. Chandola, Phys. Lett. **B476**, 193 (2000); H.C. Chandola, H.C. Pandey, Mod. Phys. Lett. **A17**, 599 (2002); H.C. Chandola, D. Yadav, H.C. Pandey, H. Dehnen, Int. J. Mod. Phys. **A20**, 2743 (2005); H.C. Chandola, D.S. Rawat, H.C. Pandey, D. Yadav, H. Dehnen (2019), arXiv:1904.11714 [hep-th]
8. H.C. Chandola, D. Yadav, Nucl. Phys. **A829**, 151 (2009); H.C. Chandola, G. Punetha, H. Dehnen, Nucl. Phys. **A945**, 226 (2016)
9. N. Cundy, Y.M. Cho, W. Lee, J. Leem, Phys. Lett. **B729**, 192 (2014); Nucl. Phys. **B895**, 64 (2015)
10. K. Adeox, S.S. Adler et al., Nucl. Phys. **A757**, 184 (2005); K. Aamodt, A.A. Quintana et al., Phys. Rev. Lett. 3032301 (2011); Phys. Lett. **B696** (2011); B. Muller, J. Schukraft, B. Wyslouch, Ann. Rev. Nucl. Part. Sci. **62**, 361 (2012)
11. S. Ejiri, Phys. Rev. **D78**, 074507 (2008); H.C.M. Lizardo, E.V.L. Mello, Physica **A305**, 340 (2002)
12. V. Riabov, EPJ Web Conf. **204**, 01017 (2019); P. Foka, J. Phys. **455** (2013)
13. H. Toki, H. Suganuma, Prog. Part. Nucl. Phys. **45**, 5397 (2000)
14. H. Neilson, B. Olesen, Nucl. Phys. B **61**, 45 (1973)

Chapter 80
Spin Alignment Measurements of K^{*0} Vector Mesons with ALICE Detector at the LHC

Sourav Kundu

Abstract We present recent results on K^{*0} polarization from the ALICE experiment at mid-rapidity ($|y|$ <0.5) in Pb–Pb collisions at $\sqrt{s_{NN}} = 2.76\,\text{TeV}$ and $5.02\,\text{TeV}$ and in pp collisions at $\sqrt{s} = 13\,\text{TeV}$. The polarization of the K^{*0} vector meson (spin = 1) is studied with respect to the production plane and second-order event plane by measuring the zeroth element of spin density matrix ρ_{00}. In pp collisions, ρ_{00} is consistent with 1/3 in all measured p_T region ranges from $0.0 < p_T < 10$ GeV/c. However, in Pb–Pb collisions the ρ_{00} values are below 1/3 for $p_T < 2$ GeV/c and consistent with 1/3 for $p_T > 2$ GeV/c, with respect to both the production and event planes. ρ_{00} values are also measured for K^0_S (spin = 0) in Pb–Pb collisions at $\sqrt{s_{NN}} = 2.76\,\text{TeV}$ in the 20–40% centrality as a null test and the measurements are consistent with 1/3 in $0.0 < p_T < 5$ GeV/c. We have also observed a centrality dependence of measured ρ_{00} values for K^{*0}, with the maximum deviation from 1/3 occurring in mid-central collisions.

80.1 Introduction

In a non-central heavy-ion collision, the overlapping region retains the large initial angular momentum [1] in the direction perpendicular to the reaction plane. In the presence of this initial angular momentum, vector mesons (spin = 1) can be polarized due to the spin angular momentum interaction. The presence of this initial state effect can be probed by measuring the angular distribution of the decay daughters of vector mesons [2–4], in the rest frame of the vector mesons with respect to a quantization axis. The quantization axis can be perpendicular to the production plane (defined by the momentum direction of the vector meson and the beam direction) or

Sourav Kundu—For the ALICE collaboration.

S. Kundu (✉)
School of physical sciences, national institute of science education and research, hbni,
jatni 752050, india
e-mail: sourav.kundu@cern.ch

© Springer Nature Singapore Pte Ltd. 2021
P. K. Behera et al. (eds.), *XXIII DAE High Energy Physics Symposium*,
Springer Proceedings in Physics 261,
https://doi.org/10.1007/978-981-33-4408-2_80

perpendicular to the reaction plane (defined by the impact parameter direction and the beam direction) of the system. In the experiment, the event plane is used as a proxy for the reaction plane. The angular distribution of the decay daughters of the vector meson is expressed as [5]

$$\frac{dN}{d\cos\theta^*} = N_0\left[1 - \rho_{00} + \frac{1}{R}\cos^2\theta^*(3\rho_{00} - 1)\right]. \tag{80.1}$$

Here, N_0, R and ρ_{00} are the normalization constant, second-order event plane resolution and zeroth element of the spin density matrix, respectively. The $1/R$ coefficient becomes 1 in the production plane analysis. The angle θ^* is the angle made by any of the decay daughters with the quantization axis in the rest frame of the vector mesons. In the absence of polarization, $\rho_{00} = 1/3$, which leads to a uniform distribution of $\cos\theta^*$ whereas, in the presence of polarization, ρ_{00} deviates from 1/3, leading to a non-uniform angular distribution. In this work, we present recent results on K^{*0} polarization from the ALICE experiment [6] at mid-rapidity ($|y|<0.5$) in Pb–Pb collisions at $\sqrt{s_{NN}} = 2.76$ TeV and 5.02 TeV and in pp collisions at $\sqrt{s} = 13$ TeV. The value of ρ_{00} is measured as a function of p_T and centrality with respect to both the production and event planes.

80.2 Analysis Details

We have used a data sample of 14 M minimum bias events in Pb–Pb collisions at $\sqrt{s_{NN}} = 2.76$ TeV and 30 M minimum bias events in Pb–Pb collisions at $\sqrt{s_{NN}} = 5.02$ TeV to measure the polarization of K^{*0} in Pb–Pb collisions. In addition, 43 M minimum bias pp collisions at $\sqrt{s} = 13$ TeV are also used to measure the polarization of K^{*0} in pp collisions. The ρ_{00} values for K^{*0} are extracted at mid-rapidity ($-0.5 < y < 0.5$) in different p_T and centrality regions. In addition, a null hypothesis test is performed by measuring the ρ_{00} value for spin zero hadron K_S^0 in 20–40% central Pb–Pb collisions at $\sqrt{s_{NN}} = 2.76$ TeV. Since K^{*0} is a resonance particle (lifetime $\sim 10^{-23}$ s), it cannot be detected directly by the detector. Therefore, it is reconstructed with the invariant mass technique by identifying oppositely charged K and π decay daughters, as discussed in [7]. The K_S^0 is reconstructed via the identification of pairs of pion daughters with opposite charges by applying selection cuts on the V0 decay topology, as reported in [8]. Charged tracks are selected using a set of standard track-quality criteria, described in detail in [7]. The daughter kaons and pions are identified by using the specific ionization energy loss (dE/dx) measured in the Time Projection Chamber (TPC) [6] and the flight time measured in the Time Of Flight (TOF) [6] detector. Charge pions for K_S^0 reconstruction are identified by using only TPC. Triggering and centrality determination are provided by the V0 detectors [6]. These are the two plastic scintillator detectors, V0A and V0C, placed in the pseudorapidity ranges $2.8 < \eta < 5.1$ and $-3.7 < \eta < -1.7$, respectively, covering the full azimuthal angle. The V0 detectors are also used for the estimation of the 2nd order event plane.

The invariant mass distribution of unlike charge $K\pi$ pairs from the same event contains the K*0 signal along with a large combinatorial background. This combinatorial background is reconstructed by adopting mixed event techniques [7] and subtracted from the $K\pi$ invariant mass distribution to extract the K*0 signal. After the background subtraction, the K*0 peak is visible on top of a residual background, which is due to the production of correlated $K\pi$ pairs from the decays of other hadrons and from jets. In order to extract the K*0 yield, the mixed event background subtracted unlike charge $K\pi$ invariant mass distribution from the same event is fitted with a Breit–Wigner function added to a second-order polynomial in $M_{K\pi}$. The Breit–Wigner function represents the K*0 signal and the second-order polynomial function describes the residual background. The area under the Breit–Wigner function corresponds to the K*0 yield. The left panel of Fig. 80.1 shows the invariant mass distributions of unlike charged $K\pi$ pairs from the same event along with the normalized mixed event background. The right panel of Fig. 80.1 shows the invariant mass distribution of unlike charge $K\pi$ pairs after mixed event background subtraction. K*0 yields are extracted in each p_T and $\cos\theta^*$ bin and then corrected for the detector acceptance and efficiency. The left panel of Fig. 80.2 shows the corrected $\cos\theta^*$ distribution at mid-rapidity in 10–30% central Pb–Pb collisions at $\sqrt{s_{NN}} = 5.02$ TeV for $0.8 \leq p_T < 1.2$ GeV/c using the production plane, and the right panel shows the corrected $\cos\theta^*$ distribution at mid-rapidity in 10–30% central Pb–Pb collisions at

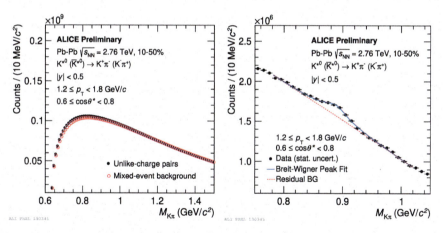

Fig. 80.1 (Color online) Left Panel: Invariant mass distribution of unlike charge $K\pi$ pairs from the same event along with the normalized mixed event background in Pb–Pb collisions at $\sqrt{s_{NN}} = 2.76$ TeV for $1.2 < p_T < 1.8$ GeV/c and $0.6 < \cos\theta^* < 0.8$, w.r.t. the production plane. Right panel: Mixed event background subtracted invariant mass distribution of unlike charged $K\pi$ pairs, fitted with a Breit–Wigner function + second-order polynomial function in $M_{K\pi}$

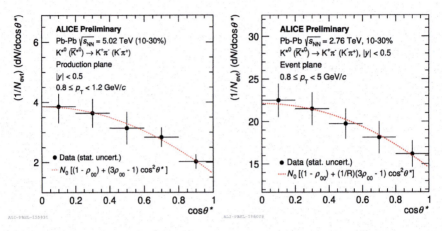

Fig. 80.2 (Color online) Left Panel: $dN/d\cos\theta^*$ versus $\cos\theta^*$ distribution at $|y| < 0.5$ and $0.8 \leq p_T < 1.2$ GeV/c in 10–30% central Pb–Pb collisions at $\sqrt{s_{NN}} = 5.02$ TeV using the production plane. Right Panel: $dN/d\cos\theta^*$ versus $\cos\theta^*$ distribution at $|y| < 0.5$ and $0.8 \leq p_T < 1.2$ GeV/c in 10–30% central Pb–Pb collisions at $\sqrt{s_{NN}} = 2.76$ TeV using the event plane

$\sqrt{s_{NN}} = 2.76$ TeV for $0.8 \leq p_T < 5.0$ GeV/c using the event plane. The corrected $\cos\theta^*$ distributions are fitted with Eq.(80.1) to extract ρ_{00} values in each p_T bin and centrality class.

80.3 Results

Figure 80.3 shows the transverse momentum dependence of ρ_{00} values for K^{*0} with respect to the production plane in pp collisions at $\sqrt{s} = 13$ TeV and in 10–50% central Pb–Pb collisions at $\sqrt{s_{NN}} = 2.76$ TeV and 5.02 TeV, along with the measurements for K_S^0 in 20–40% central Pb–Pb collisions at $\sqrt{s_{NN}} = 2.76$ TeV. The measured ρ_{00} values for K^{*0} in pp collisions and for K_S^0 in Pb–Pb collisions are consistent with 1/3 in all measured p_T regions whereas, in Pb–Pb collisions the ρ_{00} values for K^{*0} are consistent with 1/3 for $p_T > 2$ GeV/c and a deviation from 1/3 is observed for $p_T < 2$ GeV/c. The measured ρ_{00} values for K^{*0} in Pb–Pb collisions at $\sqrt{s_{NN}} = 2.76$ TeV and 5.02 TeV are consistent with each other. Figure 80.4 shows a comparison of ρ_{00} values for K^{*0} using the production and event planes. Measurements w.r.t. the two different planes are consistent with each other within the uncertainties. In Pb–Pb collisions at $\sqrt{s_{NN}} = 2.76$ TeV, the measured ρ_{00} values for K^{*0} vector meson in the lowest p_T bin are 2.5 σ and 1.7 σ below 1/3 in the production and event plane analyses, respectively.

Fig. 80.3 (Color online) ρ_{00} for K*0 as a function of p_T with respect to the production plane in pp collisions at $\sqrt{s} = 13$ TeV and in 10–50% central Pb–Pb collisions at $\sqrt{s_{NN}} = 2.76$ TeV and 5.02 TeV, along with the measurements for K_S^0 in 20–40% central Pb–Pb collisions at $\sqrt{s_{NN}} = 2.76$ TeV. Statistical and systematic uncertainties on ρ_{00} are shown by bars and boxes, respectively

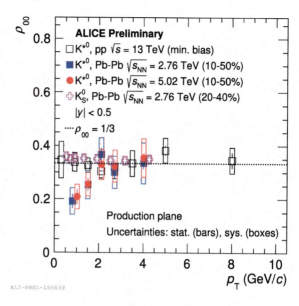

Fig. 80.4 (Color online) Comparison of ρ_{00} versus p_T between the production and event plane analyses in 10–50% central Pb–Pb collisions at $\sqrt{s_{NN}} = 2.76$ TeV. Statistical and systematic uncertainties on ρ_{00} are shown by bars and boxes, respectively

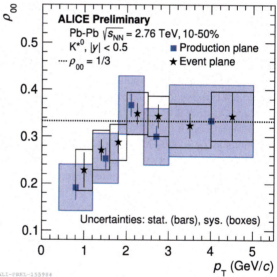

The ρ_{00} values as a function of $\langle N_{part} \rangle$ in Pb–Pb collisions for the lowest p_T bin are shown in Fig. 80.5. A clear centrality dependence of ρ_{00} for K*0 is observed. The maximum deviation of ρ_{00} from 1/3 occurs in mid-central collisions where the angular momentum is expected to be large whereas, in central and peripheral collisions, the ρ_{00} values are close to 1/3.

Fig. 80.5 (Color online) ρ_{00} as a function of $\langle N_{part} \rangle$ at mid-rapidity for lowest p_T bin in Pb–Pb collisions at $\sqrt{s_{NN}} = 2.76$ TeV and 5.02 TeV with respect to both the production and event planes. Statistical and systematic uncertainties on ρ_{00} are represented by bars and boxes, respectively

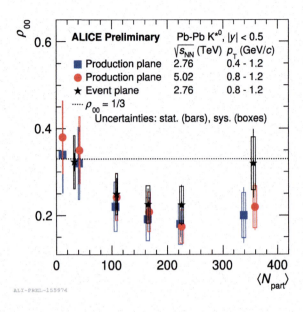

ALI-PREL-155974

80.4 Summary

Recent spin alignment measurements for K^{*0} vector mesons in pp collisions at $\sqrt{s} = 13$ TeV and in Pb–Pb collisions at $\sqrt{s_{NN}} = 2.76$ TeV and 5.02 TeV are presented. We do not observe any polarization for K^{*0} vector meson in pp collisions, in Pb–Pb collisions at $p_T > 2.0$ GeV/c and for the spin zero K_S^0 in Pb–Pb collisions. However, a deviation of ρ_{00} for the K^{*0} vector meson from 1/3 is observed in mid-central Pb–Pb collisions at $p_T < 2.0$ GeV/c with respect to both the production and event planes. We observe a centrality dependence of ρ_{00}, with the maximum deviation from 1/3 occurring in mid-central collisions, where the overlapping region retains the large initial angular momentum.

Spin alignment measurements for ϕ mesons are ongoing. Measurements with new Pb–Pb data set at $\sqrt{s_{NN}} = 5.02$ TeV is underway for better statistical precision.

References

1. F. Becattini, F. Piccinini, J. Rizzo, Phys. Rev. C **77**, 024906 (2008)
2. B.I. Abelev et al. (STAR Collaboration), Phys. Rev. C **77**, 061902 (2008)
3. B. Mohanty (for the ALICE Collaboration), EPJ. Web. Conf. **171**, 16008 (2018)
4. R. Singh (for the ALICE Collaboration), Nucl. Phys. A **982**, 515 (2019)
5. K. Schilling, P. Seyboth, G.E. Wolf, Nucl. Phys. B **15**, 397 (1970)
6. B. Abelev et al. (ALICE Collaboration), Int. J. Mod. Phys. A **29**, 1430044 (2014)
7. J. Adam et al. (ALICE Collaboration), Phys. Rev. C **95**, 064606 (2017)
8. B. Abelev et al. (ALICE Collaboration), Phys. Rev. Lett. **111**, 222301 (2013)

Chapter 81
Intriguing Similarities Between High-p_T Particle Production in pp and A-A Collisions

Aditya Nath Mishra

Abstract In this paper, we study the particle production at high transverse momentum ($p_T > 8\,\text{GeV}/c$) within $|\eta| < 0.8$ in both pp and Pb-Pb collisions at LHC energies. The characterization of the spectra is done using a power-law function and the resulting power-law exponent (n) is studied as a function of x_T for minimum-bias pp collisions at different $\sqrt{s}$. The functional form of n as a function of x_T exhibits an approximate universal behavior. PYTHIA 8.212 reproduces the scaling properties, and therefore, it is used to study the multiplicity-dependent particle production. Going from low to high multiplicities, the power-law exponent decreases. A similar behavior is also observed in heavy-ion collisions when one studies the centrality-dependent particle production. The interpretation of heavy-ion results requires the quantification of the impact of this correlation (multiplicity and high-p_T) on jet-quenching observables.

81.1 Introduction

The similarity between analogous observables in large (A-A) and small (pp and p-A) collision systems has been extensively studied by the heavy-ion community [1–4]. A vast number of quantities as a function of the charged-particle density ($dN_{ch}/d\eta$) in small systems have been documented in recent works [5]. These observables (azimuthal anisotropies, radial flow, and strangeness enhancement [6–8]) have been measured in the low- and intermediate-transverse momentum regimes ($p_T < 8\,\text{GeV}/c$). For higher transverse momenta, the traditional treatments intend to isolate the QGP effects using reference data where the formation of a partonic medium is not expected. Minimum-bias proton–proton collisions have been used for this purpose. However, this assumption is questionable [9, 10].

A. N. Mishra (✉)
Instituto de Ciencias Nucleares, UNAM, Apartado Postal 70-543, CDMX 04510, Mexico
e-mail: Aditya.Nath.Mishra@cern.ch

© Springer Nature Singapore Pte Ltd. 2021
P. K. Behera et al. (eds.), *XXIII DAE High Energy Physics Symposium*,
Springer Proceedings in Physics 261,
https://doi.org/10.1007/978-981-33-4408-2_81

In the present work, we study the multiplicity dependence of particle production at high transverse momenta ($p_T > 8\,\text{GeV}/c$) within $|\eta| < 0.8$ in pp collisions at LHC energies. The message of the present paper is that the **shape** of R_{AA} for high-p_T particles may not fully attributed to the parton energy loss; since as we will demonstrate, a similar shape is observed for the analogous ratios in pp collisions, i.e., high-multiplicity p_T spectra normalized to that for minimum-bias events.

81.2 Particle Production at Large Transverse Momenta

In heavy-ion collisions, particle production at high p_T is commonly used to study the opacity of the medium to jets. Experimentally, the medium effects are extracted by means of the nuclear modification factor, R_{AA}, which is defined as

$$R_{AA} = \frac{\mathrm{d}^2 N_{AA}/\mathrm{d}y\,\mathrm{d}p_T}{\langle N_{\text{coll}}\rangle \mathrm{d}^2 N_{pp}/\mathrm{d}y\,\mathrm{d}p_T} \tag{81.1}$$

where $\mathrm{d}^2 N_{AA}/\mathrm{d}y\,\mathrm{d}p_T$ and $\mathrm{d}^2 N_{pp}/\mathrm{d}y\,\mathrm{d}p_T$ are the invariant yields measured in A-A and minimum-bias pp collisions, respectively. The ratio is scaled by the average number of binary nucleon–nucleon collisions (N_{coll}) occurring within the same A–A interaction, which is usually obtained using Glauber simulations. The resulting ratio is supposed to account (at least from 8 GeV/c onward) for the so-called jet quenching whereby the high-momentum partons would be "quenched" in the hot system created in the collision of nuclei. For instance, in the 0–5% Pb-Pb collisions at the LHC energies the suppression is about 70–80% for p_T of around 6–7 GeV/c [14]. For higher p_T, R_{AA} exhibits a continuous rise and approaches unity. As suggested by the p_T-differential baryon-to-meson ratio, for p_T larger than 8 GeV/c radial flow effects are negligible, and therefore, the shape of R_{AA} is expected to be dominated by parton energy loss.

In the present paper, we study the shape of the p_T spectra of charged particles measured in heavy-ion and pp collisions separately within the ALICE acceptance ($|\eta| < 0.8$). The aim is to discuss the origin of the rise of the R_{AA} for $p_T > 6\,\text{GeV}/c$. Since PYTHIA 8.212 (tune Monash 2013) reproduces rather well many features of LHC data, we base our studies on PYTHIA 8.212 simulations of pp collisions for different multiplicity classes. In this paper, we use the event multiplicity classes used by ALICE based on the number of tracklets ($N_{\text{SPDtracklets}}$) within $|\eta| < 0.8$ for 13 TeV analysis [11].

81.3 Results and Discussion

ALICE recent result for the multiplicity-dependent p_T spectra for pp collisions at $\sqrt{s} = 13$ TeV shows that for $p_T > 8$ GeV/c the spectra become harder with increasing multiplicity [11]. It also shows the ratios of the p_T spectra for the different multiplicity classes divided by that for minimum-bias pp collisions which exhibit an important increase with p_T, similar to the one observed in the R_{AA} measured in Pb-Pb collisions [14]. To characterize the changes with multiplicity we fitted a power-law function ($\propto p_T^{-n}$) to the p_T spectrum of a specific colliding system and for a given multiplicity class. The power-law exponents allow us to investigate in a bias-free manner various systems, multiplicities, and energies.

Figure 81.1 displays the multiplicity-dependent p_T-spectra and their corresponding fitted power-law functions for pp collisions at $\sqrt{s} = 13$ TeV. We observed that it is not possible to describe the full p_T (8–200 GeV/c) interval assuming the same power-law exponent. For instance, for p_T larger than 20 GeV/c the ratios go beyond 20%. In order to check this, we have performed the fit considering sub-intervals of p_T. This allows the extraction of local power-law exponents for different p_T sub-intervals. The results indicate that the exponent has an important dependence on p_T. This is shown in the right-hand side of Fig. 81.1, where the multiplicity dependence of n as a function of p_T is shown. The exponents have a very specific behavior with multiplicity. At low multiplicities the exponents rise more rapidly than the minimum-bias ones. We observe that for all multiplicity classes there is a trend to have smaller

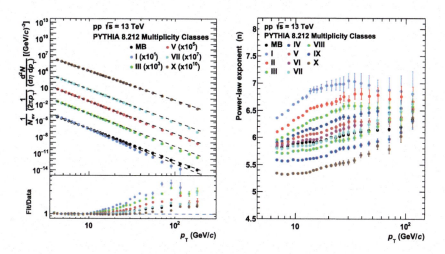

Fig. 81.1 Left: Transverse momentum distributions of charged particles for minimum-bias (MB) and different multiplicity classes in pp collisions at $\sqrt{s} = 13$ TeV simulated with PYTHIA 8.212: fitted with power-law functions (dashed lines) for $p_T > 8$ GeV/c. The ratios of fit and data are shown in the lower panel. Right: The power-law exponents extracted from the fits, considering sub-intervals of p_T, are plotted as a function of transverse momentum

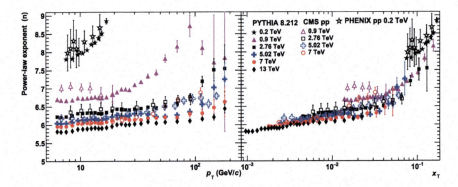

Fig. 81.2 Power-law exponent as a function of transverse momentum (right) and x_T (left) for minimum-bias pp collisions at different energies. The data have been taken from [14, 17–21]. Results are compared with PYTHIA 8.212 predictions

exponents (softening of the spectra) at higher momenta; the tendency getting smaller for high multiplicities. Theoretically p_T can range from 0 to half of the center-of-mass energy, $\sqrt{s}/2$, of the collision. Therefore, the distribution can also be presented as a function of the dimensionless variable $x_T = 2p_T/\sqrt{s}$ [15], which varies between 0 and 1.

In Fig. 81.2, we show n as a function of p_T for minimum-bias pp data at different $\sqrt{s}$ (0.2, 0.9, 2.76, 5.02, 7 and 13 TeV). The results are compared with PYTHIA 8.212 (tune Monash 2013) simulations. Going from low to high energies the power-law exponent decreases in both data and PYTHIA 8.212. This is expected since at higher energies the production cross sections of hard processes increase resulting in a change in the slope of the spectra at large transverse momenta. A different representation is shown in the right-hand-side plot, where the power-law exponent is presented as a function of x_T. Within 10% the data, that were before distinctly different, fall now approximately on an universal curve. Prominent in this respect is the case of the $\sqrt{s} = 0.2$ TeV data. The approximate scaling property is well reproduced by PYTHIA 8.212.

Applying now the same treatment to the Pb-Pb data we observe that the exponents as a function of x_T and centrality behave very similar to those for pp collisions simulated with PYTHIA 8.212. The comparison is shown in Fig. 81.3 for pp and Pb-Pb collisions at $\sqrt{s_{NN}} = 2.76$ TeV. Going from 70 to 90% peripheral to 0–5% central Pb-Pb collisions the exponent exhibits an overall decrease for x_T below 0.02. This is consistent with the hardening of the p_T spectra going from peripheral to central Pb-Pb collisions, a feature that still does not have a coherent physical explanation.[1] For higher x_T (>0.02), the exponents gradually rise toward the minimum-bias value at $x_T \approx 0.04$. On the other hand, flattening of the R_{AA} at high-p_T is natural since the corresponding pp minimum-bias spectra and the central Pb-Pb ones have the same exponents. The multiplicity dependence of n vs x_T in pp collisions simulated with

---

[1]For a recent effort to elucidate the phenomenon see Ref. [16].

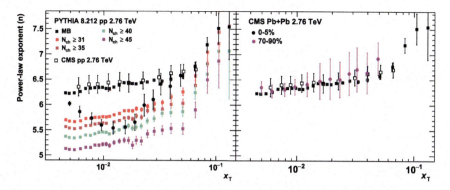

Fig. 81.3 Power-law exponent as a function of x_T for central (left) and peripheral (right) Pb-Pb collisions at $\sqrt{s_{NN}} = 2.76$ TeV. Heavy-ion data are compared with minimum-bias (MB) CMS data for pp collisions and high-multiplicity ($|\eta| < 0.8$) pp collisions at $\sqrt{s} = 2.76$ TeV simulated with PYTHIA 8.212

PYTHIA 8.212 is qualitatively similar to that observed in heavy-ion data. The same behavior is also observed at higher energies; in particular for Pb-Pb collisions at $\sqrt{s_{NN}} = 5.02$ TeV [22]. The observed behavior invites one interesting consequence. The accepted view which entirely attributes the rise in R_{AA} to the decrease of the parton energy loss should be revised. It is well-known that the mean p_T continues rising with multiplicity both in pp and in heavy-ion collisions, implying that high multiplicity, which is proportional to the energy density, is correlated with the high-momentum particle production.

81.4 Conclusion

We have studied the high-p_T ($p_T > 8$ GeV/c) charged-particle production in both pp and Pb-Pb collisions. Considering different p_T sub-intervals, power-law functions were fitted to the transverse momentum distributions of minimum-bias pp collisions measured by experiments at the RHIC and LHC. The local exponents of the power-law fits were compared to those obtained from Pb-Pb data. With respect to minimum-bias pp collisions, we have determined the following:

- The high-p_T part of the p_T spectra cannot be described by a single power-law function (same exponent value) within a wide p_T interval (8–100 GeV/c).
- The minimum-bias p_T spectra, when represented in terms of the local exponent as a function of the Bjorken variable x_T, obey an approximate scaling behavior over a wide range of center-of-mass energy, $\sqrt{s} = 0.2$ to 13 TeV.
- For heavy-ion collisions the evolution of the local exponent as a function of x_T and collision centrality is qualitatively similar to that for pp collisions.

It would be very important to produce experimental results on high-multiplicity pp collisions over a wide p_T interval in order to be able to assess in details the source of the apparent similarity between pp and A-A data.

Acknowledgements A.M. acknowledges the post-doctoral fellowship of DGAPA UNAM.

References

1. E.K.G. Sarkisyan, A.S. Sakharov, Relating multihadron production in hadronic and nuclear collisions. Eur. Phys. J. C **70**, 533–541 (2010)
2. A.N. Mishra, R. Sahoo, E.K.G. Sarkisyan, A.S. Sakharov, Effective-energy budget in multi-particle production in nuclear collisions. Eur. Phys. J. C **74**, 3147 (2014)
3. E.K.G. Sarkisyan, A.N. Mishra, R. Sahoo, A.S. Sakharov, Multihadron production dynamics exploring the energy balance in hadronic and nuclear collisions. Phys. Rev. D **93**, 054046 (2016)
4. E.K.G. Sarkisyan, A.N. Mishra, R. Sahoo, A.S. Sakharov, Centrality dependence of midrapidity density from GeV to TeV heavy-ion collisions in the effective-energy universality picture of hadroproduction. Phys. Rev. D **94**, 011501 (2016)
5. C. Loizides, Experimental overview on small collision systems at the LHC. Nucl. Phys. A **956**, 200–207 (2016)
6. Abelev, B. *et al.*: Multiplicity Dependence of Pion, Kaon, Proton and Lambda Production in p-Pb Collisions at $\sqrt{s_{NN}}$ = 5.02 TeV, (ALICE Collaboration), Phys. Lett. B728, 25-38 (2014)
7. V. Khachatryan et al., Evidence for collectivity in pp collisions at the LHC, (CMS Collaboration). Phys. Lett. B **765**, 193–220 (2017)
8. J. Adam et al., Enhanced production of multi-strange hadrons in high-multiplicity proton-proton collisions, (ALICE Collaboration). Nature Phys. **13**, 535–539 (2017)
9. B.G. Zakharov, Parton energy loss in the mini quark-gluon plasma and jet quenching in proton-proton collisions. J. Phys. **G41**, 075008 (2014)
10. Mangano, M. L. and Nachman, B.: Observables for possible QGP signatures in central pp collisions, arXiv:1708.08369 (2017)
11. Acharya, S. *et al.*: Charged-particle production as a function of multiplicity and transverse spherocity in pp collisions at $\sqrt{s}$ = 5.02 and 13 TeV, (ALICE Collaboration), arXiv:1905.07208
12. Adare, A. *et al.*: Scaling properties of fractional momentum loss of high-p_T hadrons in nucleus-nucleus collisions at $\sqrt{s_{NN}}$ from 62.4 GeV to 2.76 TeV, (PHENIX Collaboration), Phys. Rev. C93, 024911 (2016)
13. Ortiz, A. and Vázquez, O.: Energy density and path-length dependence of the fractional momentum loss in heavy-ion collisions at $\sqrt{s_{NN}}$ from 62.4 to 5020 GeV, Phys. Rev. C97, 014910 (2018) (ALICE Collaboration), arXiv:1802.09145 (2018)
14. Khachatryan, V. *et al.*: Charged-particle nuclear modification factors in PbPb and pPb collisions at $\sqrt{s_{NN}}$ = 5.02 TeV, (CMS Collaboration), JHEP 04, 039 (2017)
15. D. Sivers, S.J. Brodsky, R. Blankenbecler, Large transverse momentum processes. Phys. Reports **23**, 1–121 (1976)
16. Mishra, A. N. and Paic, G.: Did we miss the "melting" of partons in pp collisions?, arXiv:1905.06918 (2019)
17. Khachatryan, V. *et al.*: Transverse momentum and pseudorapidity distributions of charged hadrons in pp collisions at $\sqrt{s}$ = 0.9 and 2.36 TeV, (CMS Collaboration), JHEP 02, 041 (2010)
18. Chatrchyan, S. *et al.*:Study of high-pT charged particle suppression in PbPb compared to *pp* collisions at $\sqrt{s_{NN}}$ = 2.76 TeV, (CMS Collaboration), Eur. Phys. J. C72, 1945 (2012)
19. V. Khachatryan et al., Transverse-momentum and pseudorapidity distributions of charged hadrons in *pp* collisions at $\sqrt{s}$ = 7 TeV, (CMS Collaboration). Phys. Rev. Lett. **105**, 022002 (2010)

20. B. Abelev, Energy Dependence of the Transverse Momentum Distributions of Charged Particles in pp Collisions Measured by ALICE, (ALICE Collaboration). Eur. Phys. J. C **73**, 2662 (2013)
21. A. Adare, Inclusive cross section and double helicity asymmetry for π^0 production in pp collisions at $\sqrt{s} = 200$ GeV: Implications for the polarized gluon distribution in the proton, (PHENIX Collaboration). Phys. Rev. D **76**, 051106(R) (2007)
22. A.N. Mishra, A. Ortiz, G. Paic, Intriguing similarities of high-p_T particle production between pp and AA collisions. Phys. Rev. C **99**, 034911 (2019)

Chapter 82
Chiral Symmetry Breaking, Color Superconductivity, and Equation of State for Magnetized Strange Quark Matter

Aman Abhishek and Hiranmaya Mishra

Abstract We investigate the vacuum structure of dense quark matter in strong magnetic fields at finite temperature and densities in a three-flavor Nambu Jona Lasinio (NJL) model including the Kobayashi–Maskawa–t'Hooft (KMT) determinant term using a variational method. The method uses an explicit structure for the 'ground' state in terms of quark–antiquark condensates as well as diquark condensates. The mass gap equations and the superconducting gap equations are solved self-consistently and are used to compute the thermodynamic potential along with charge neutrality conditions. We also derive the equation of state for charge neutral strange quark matter in the presence of strong magnetic fields which could be relevant for neutron stars.

82.1 Introduction

Study of the ground state in quantum chromodynamics under extreme conditions of temperature and density is an important theoretical and experimental challenge [2]. Non-perturbative aspects of QCD such as chiral symmetry breaking, confinement, and asymptotic freedom are expected to play an important role in such conditions. Especially important is the study of quark gluon plasma, which is a new form of matter. It is expected to form at sufficient high temperature and densities. It has been observed in heavy ion collisions at high temperatures. At high densities, a few times of the nuclear matter density, also one may expect the formation of quark gluon plasma. Such densities are reached in the interior of neutron stars. However, QCD at such densities is non-perturbative. Therefore, one cannot make first principle calculation except at very high densities where the coupling is weak due to asymptotic freedom and perturbative QCD can be employed. Such analysis predicts a Color Flavor Locked

A. Abhishek (✉) · H. Mishra
Physical Research Laboratory, Navrangpura, Ahmedabad 380009, India
e-mail: aman@prl.res.in

H. Mishra
e-mail: hm@prl.res.in

© Springer Nature Singapore Pte Ltd. 2021
P. K. Behera et al. (eds.), *XXIII DAE High Energy Physics Symposium*,
Springer Proceedings in Physics 261,
https://doi.org/10.1007/978-981-33-4408-2_82

(CFL) phase to be the ground state at asymptotic densities. In this state, the color and flavor degrees of freedom of three light quark are correlated to form cooper pairs. At intermediate densities the situation is less clear. There may be several possible phases such as 2SC, quarkyonic and LOFF. However, studies based on effective models such as Nambu Jona Lasinio (NJL) model suggest a phase transition from hadronic to quark gluon phase at few times the nuclear matter density. Other than extreme temperatures and densities, the effect of strong magnetic fields [3, 4] on strongly interacting matter has been of recent interest. It has been suggested that very high magnetic fields might be produced in ultra relativistic heavy ion collisions. Other than heavy ion collision, such high magnetic fields may be present in the interior of neutron stars. Magnetars are known to have very high surface magnetic fields of $10^{15}G$ [5–7]. In the interior of the star, the magnetic field could be much higher. The effect of such high magnetic fields has been studied in effective models and lattice QCD [8–10]. From lattice studies, it is known that magnetic field enhances the value of chiral condensate at low temperatures, and at high temperatures it weakens the chiral condensate. These two effects are known as magnetic catalysis and inverse magnetic catalysis. Effective models also suggest similar effect of magnetic fields. Hence it is important to study the effect of magnetic field on matter as it affects the equation of state which is important for the modeling of neutron stars.

82.2 Formalism

To study the superconductivity in three-flavor quark matter, we [1] adopt a variational approach within the framework of Nambu Jona Lasinio model with determinant interaction. We assume an ansatz for the ground state with both the chiral condensate and superconducting cooper pairs at finite quark chemical potential μ but zero temperature. The two light flavors u and d are assumed to form cooper pairs in the color anti-symmetric anti-triplet channel. Furthermore, we need to study the system at finite temperature and chemical potential. This is accomplished by three successive Bogoliubov transformations as follows :

$$|\Omega\rangle_\chi = U_\chi|0\rangle \equiv \exp\sum_{flav}(B_i^\dagger - B_i)|0\rangle. \qquad (82.1)$$

The above transformation populates the vacuum annihilated by free field operators by quark–antiquark pairs which form the chiral condensate and are characterized by a function ϕ_i, where i stands for the flavor.

After assuming an ansatz with chiral condensate, we now apply a Bogoliubov transformation to include diquark condensate which is characterized by functions $f(n, p_z)$ and $f_1(n, p_z)$:

$$|\Omega\rangle_{d,\chi} = U_d|\Omega\rangle_\chi \equiv \exp(B_d^\dagger - B_d)|\Omega\rangle_\chi. \qquad (82.2)$$

To include the effects of temperature and density, we next write down the state at finite temperature and density $|\Omega(\beta, \mu)\rangle$ through a thermal Bogoliubov transformation over the state $|\Omega\rangle$ using the thermofield dynamics (TFD) method [? ? ?]. We write the thermal state as

$$|\Omega(\beta, \mu)\rangle_{d,\chi} = \mathcal{U}_{\beta,\mu}|\Omega\rangle_{d,\chi} \equiv e^{\mathcal{B}^{\dagger}(\beta,\mu) - \mathcal{B}(\beta,\mu)}|\Omega\rangle_{d,\chi}. \tag{82.3}$$

Using the above ansatz, we calculate the free energy of the trial ground state and minimize it with respect to functions ϕ_i, $f(n, p_z)$, $f_1(n, p_z)$ and $\theta^{ia}_{\pm}(n, k_z, \beta, \mu)$. We get the following relations :

$$\cot \phi^{ia} = \frac{(m_i - 4G_s I_s^i + K \epsilon^{ijk} I_s^j I_s^k + K/4 I_D^2 \delta_{i3})}{|p_{ia}|} \equiv \frac{M_i}{|p_{ia}|} \tag{82.4}$$

$$\tan 2f(k) = \frac{2(G_D - \frac{K}{4} I_s^{(3)}) I_D}{\bar{\epsilon}_n - \bar{\mu}} \cos(\frac{\phi_1 - \phi_2}{2}) \equiv \frac{\Delta}{\bar{\epsilon}_n - \bar{\mu}} \cos(\frac{\phi_1 - \phi_2}{2}) \tag{82.5}$$

$$\tan 2f_1(k) = \frac{\Delta}{\bar{\epsilon}_n + \bar{\mu}} \cos(\frac{\phi_1 - \phi_2}{2}), \tag{82.6}$$

where we have defined the superconducting gap Δ as

$$\Delta = 2\left(G_D - \frac{K}{4} I_s^{(3)}\right) I_D, \tag{82.7}$$

and $\bar{\epsilon} = (\epsilon_n^u + \epsilon_n^d)/2$, $\bar{\mu} = (\mu^{ur} + \mu^{dg})/2 = \mu + 1/6\mu_E + 1/\sqrt{3}\mu_8$, where we have used Eq. (??) for the chemical potentials. Further, ϵ_n^i is the nth Landau level energy for the ith flavor with constituent quark mass M_i given as $\epsilon_n^i = \sqrt{p_z^2 + 2n|q_i|B + M_i^2}$.

$$\sin^2 \theta^{ia}_{\pm} = \frac{1}{\exp(\beta(\omega_{i,a} \pm \mu_{ia})) + 1}, \tag{82.8}$$

Various ω^{ia}'s $(i, a \equiv$ flavor, color) are explicitly given as

$$\omega^{11}_{n\pm} = \omega^{12}_{n\pm} = \bar{\omega}_{n\pm} + \delta\epsilon_n \pm \delta_\mu \equiv \omega^u_{n\pm} \tag{82.9}$$

$$\omega^{21}_{n\pm} = \omega^{22}_{n\pm} = \bar{\omega}_{n\pm} - \delta\epsilon_n \mp \delta_\mu \equiv \omega^d_{n\pm} \tag{82.10}$$

$$\bar{\omega}_{n\pm} = \sqrt{(\bar{\epsilon}_n \pm \bar{\mu})^2 + \Delta^2 \cos^2(\phi_1 - \phi_2)/2} \tag{82.11}$$

for the quarks participating in condensation. $\delta\epsilon_n = (\epsilon_n^u - \epsilon_n^d)/2$ is half the energy difference between the quarks which condense in a given Landau level and $\delta\mu = (\mu_{ur} - \mu_{dg})/2 = \mu_E/2$ is half the difference between the chemical potentials of the two condensing quarks. For the charged quarks which do not participate in the superconductivity $\omega_{n\pm}^{ia} = \epsilon_n^i \pm \mu^{ia}$. In the above, the upper sign corresponds to antiparticle excitation energies while the lower sign corresponds to the particle excitation energies.

82.2.1 Charge Neutrality

In the context of neutron star matter, the quark phase that could be present in the interior consists of the u,d,s quarks as well as electrons, in weak equilibrium

$$d \rightarrow u + e^- + \bar{\nu}_{e^-}, \tag{82.12}$$

$$s \rightarrow u + e^- + \bar{\nu}_{e^-}, \tag{82.13}$$

and,

$$s + u \rightarrow d + u, \tag{82.14}$$

leading to the relations between the chemical potentials μ_u, μ_d, μ_s, μ_E as

$$\mu_s = \mu_d = \mu_u + \mu_E. \tag{82.15}$$

The neutrino chemical potentials are taken to be zero as they can diffuse out of the star. So there are *two* independent chemical potentials needed to describe the matter in the neutron star interior which we take to be the quark chemical potential μ_q and the electric charge chemical potential, μ_e in terms of which the chemical potentials are given by $\mu_s = \mu_q - \frac{1}{3}\mu_e = \mu_d$, $\mu_u = \mu_q + \frac{2}{3}\mu_e$ and $\mu_E = -\mu_e$. In addition, for description of the charge neutral matter, there is a further constraint for the chemical potentials through the following relation for the particle densities given by

$$Q_E = \frac{2}{3}\rho_u - \frac{1}{3}\rho_d - \frac{1}{3}\rho_s - \rho_E = 0. \tag{82.16}$$

The color neutrality condition corresponds to

$$Q_8 = \frac{1}{\sqrt{3}} \sum_{i=u,d,s} \left(\rho^{i1} + \rho^{i2} - 2\rho^{i3}\right) = 0 \tag{82.17}$$

In the above, ρ^{ia} is the number density for quarks of flavor i and color a.

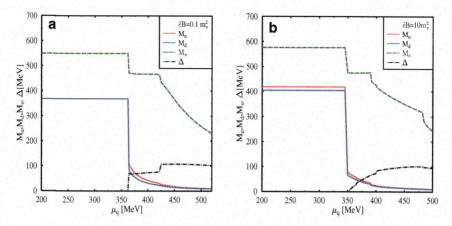

Fig. 82.1 Constituent quark masses and superconducting gap when charge neutrality conditions are imposed. Figure 82.1a shows the masses and superconducting gap at zero temperature as a function of quark chemical potential for magnetic field $\tilde{e}B = 0.1m_\pi^2$ Fig. 81.1b shows the same for $\tilde{e}B = 10m_\pi^2$

82.3 Results

Solving the gap equations self-consistently along with charge neutrality conditions imposed we get

82.3.1 Gapless Modes

Under conditions of charge neutrality, the dispersion relation of superconducting quarks may admit nodes in the spectrum. These are known as gapless modes in analogy with energy spectrum of a massless particle. These have been studied in absence of magnetic field [15, 16]. From dispersion relations given in Eqs. (82.9) and (82.11), we find that it is possible to have zero modes depending upon the values of $\delta\mu$ and $\delta\epsilon_n$. For charge neutral matter, the d quark number density is larger so that $\delta\mu = \mu_E/2$ is negative. This renders $\omega_n^u(p_z) > 0$ for any value of momentum p_z; however, ω_n^d can vanish for some values of p_z. This can be seen in Fig. 82.2-a. The lower plot is for ω_n^d which vanishes for two values of p_z. These are known as gapless modes in analogy with spectrum of a massless particle.

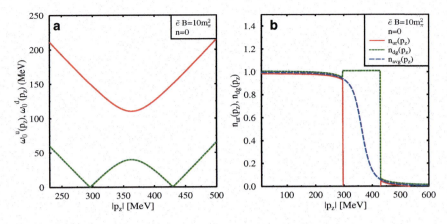

Fig. 82.2 Dispersion relation and the occupation number for condensing quarks at T = 0,μ_q=340 MeV. Figure82.2a shows the dispersion relation for the condensing quarks for zeroth Landau level. The upper curve is for u quark and the lower curve corresponds to d quark dispersion relation. Figure 82.2-b shows the occupation number as a function of momentum for $\tilde{e}B = 10m_\pi^2$

References

1. Aman Abhishek, Hiranmaya Mishra, Phys. Rev. D **99**, 054016 (2019)
2. For reviews see K. Rajagopal and F. Wilczek,arXiv:hep-ph/0011333; D.K. Hong, Acta Phys. Polon. B32,1253 (2001);M.G. Alford, Ann. Rev. Nucl. Part. Sci 51, 131 (2001); G. Nardulli,Riv. Nuovo Cim. 25N3, 1 (2002); S. Reddy, Acta Phys Polon,B33, 4101(2002);T. Schaefer arXiv:hep-ph/0304281; D.H. Rischke, Prog. Part. Nucl. Phys. 52,197 (2004); H.C. Ren, arXiv:hep-ph/0404074; M. Huang, arXiv: hep-ph/0409167;I. Shovkovy, arXiv:nucl-th/0410191
3. D. Kharzeev, L. McLerran, H. Warringa, Nucl. Phys. A **803**, 227 (2008)
4. V. Skokov, A. Illarionov, V. Toneev, Int. j. Mod. Phys. A **24**, 5925 (2009)
5. R.C. Duncan, C. Thompson, Astrophys. J. **392**, L9 (1992)
6. C. Thompson, R.C. Duncan, Astrophys. J. **408**, 194 (1993)
7. C. Thompson, R.C. Duncan, Mon. Not. R. Astron. Soc. **275**, 255 (1995)
8. D. Kharzeev, Ann. of Physics, K. Fukushima, M. Ruggieri and R. Gatto , Phys. Rev. D 81, 114031 (2010)
9. M. D'Elia, S. Mukherjee, F. Sanflippo, Phys. Rev. D **82**, 051501 (2010)
10. A.J. Mizher, M.N. Chenodub and E. Fraga,arXiv:1004.2712[hep-ph]
11. M. Alford, K. Rajagopal, F. Wilczek, Phys. Lett. B **422**, 247 (1998)
12. M. Alford, C. Kouvaris, K. Rajagopal, Phys. Rev. Lett. **92**, 222001 (2004). arXiv:hep-ph/0406137
13. K. Rajagopal, A. Schimitt, Phys. Rev. D **73**, 045003 (2006)
14. M. Buballa, Phys. Rep. **407**, 205 (2005)
15. Mei Huang, Igor Shovkovy, Nucl. Phys. A **729**, 835 (2003)
16. A. Mishra, H. Mishra, Phys. Rev. D **71**, 074023 (2005)

Chapter 83
J/ψ Production as a Function of Charged-Particle Multiplicity in pp Collisions at $\sqrt{s}$ = 5.02 TeV with ALICE

Anisa Khatun

Abstract The relative J/ψ yields as a function of relative charged-particle multiplicity in pp collisions at $\sqrt{s}$ = 5.02 TeV, measured at forward rapidity, are investigated for the first time. A linear increase of the relative J/ψ yield with respect to multiplicity is observed. A comparison of the findings of the present work with the available ALICE measurements obtained in pp collisions at $\sqrt{s}$ = 13 TeV at forward and mid-rapidity indicates that the increase of J/ψ with multiplicity is independent of energy but exhibits a strong dependence on the rapidity gap between the J/ψ and multiplicity measurements. The results are compared with theoretical model calculations.

83.1 Introduction

The event by event multiplicity of charged-particles (N_{ch}) produced in high energy collisions is taken as a simple yet important variable for understanding the collision dynamics. The multiplicity measurement is useful for studying the general properties of particle production. The quarkonium production as a function of charged-particle multiplicity ($dN_{ch}/d\eta$) in proton–proton (pp) and proton–nucleus (p–Pb) collisions is considered as an interesting observable to understand multi-parton interactions (MPI) and to explore the presence of collective behavior in small systems. Such studies can play an important role in understanding the production mechanism of heavy quarks from hard processes, and its relation with soft scale processes [1]. The multiplicity dependence of J/ψ production has been investigated in pp collisions at $\sqrt{s}$ = 7 and 13 TeV and p–Pb collisions at $\sqrt{s_{NN}}$ = 5.02 TeV at forward and mid-rapidity using the ALICE detector [2–4]. A similar study has also been carried out for D mesons [5]. In all cases, an increase of the relative particle yields as a function of the relative charged-particle multiplicity ($dN_{ch}/d\eta/ < dN_{ch}/d\eta >$) has been reported.

(A. Khatun for the ALICE collaboration).

A. Khatun (✉)
Department of Physics, Aligarh Muslim University, Aligarh, India
e-mail: anisa.khatun@cern.ch

© Springer Nature Singapore Pte Ltd. 2021
P. K. Behera et al. (eds.), *XXIII DAE High Energy Physics Symposium*,
Springer Proceedings in Physics 261,
https://doi.org/10.1007/978-981-33-4408-2_83

In these proceedings, we report on the study of the multiplicity dependent relative J/ψ yield (d$N_{J/\psi}$/dη/ < d$N_{J/\psi}$/dη >) at forward rapidity in pp collisions at $\sqrt{s}$ = 5.02 TeV, which has not been measured until now. The findings are compared to the results obtained in pp collisions at $\sqrt{s}$ = 13 TeV to explore the energy and rapidity dependences of the correlation between soft and hard physics processes. The results are also compared with theoretical model calculations.

83.2 Experimental Setup and Analysis Strategy

In ALICE [6], the charmonia are measured via their dilepton decay channels. The Muon Spectrometer is used for the reconstruction of muons coming from J/ψ decays at forward pseudorapidity ($-4 < \eta < -2.5$), while the central barrel detector covers the mid-pseudorapidity range ($|\eta| < 0.9$) and measures J/ψ in the di-electron decay channel. The central barrel detector system includes the Inner Tracking System (ITS) and the Time Projection Chamber. The ITS consists of six layers of silicon detectors. The Muon Spectrometer consists of Muon Tracking Chambers, used for tracking the muons and the Muon Trigger system that allows the selection of collisions that contain opposite sign dimuon pairs. Multiplicity is described in terms of tracklets, which are reconstructed by pairs of hits in the two innermost ITS layers, the Silicon Pixel Detector (SPD). The analysis is restricted to the event class INEL > 0 defined by requiring at least one charged-particle produced in $|\eta| < 1$. Several selection criteria are applied to make sure that both the SPD vertex position and the event charge-particle multiplicity are properly determined. Pile-up events are removed [2]. The relative charged-particle multiplicity in the ith multiplicity range, estimated at mid-rapidity ($|\eta| < 1$) and for INEL > 0 events, is calculated using the following formula:

$$\frac{\langle dN_{ch}/d\eta \rangle^i}{\langle dN_{ch}/d\eta \rangle} = \frac{f(\langle N_{trk}^{corr} \rangle^i)}{\Delta\eta \cdot \langle dN_{ch}/d\eta \rangle_{INEL>0}}, \tag{83.1}$$

where $\langle N_{trk}^{corr} \rangle^i$ is the average number of SPD tracklets, corrected by acceptance and efficiency in the multiplicity bin i. The correlation function f is evaluated from Monte Carlo simulations to convert the corrected number of tracklets into a number of primary charged-particles produced within $|\eta| < 1$. The relative J/ψ yield is estimated in the ith multiplicity bin by using the following equation:

$$\frac{dN_{J/\psi}^i/dy}{\langle dN_{J/\psi}/dy \rangle} = \frac{N_{J/\psi}^i}{N_{J/\psi}} \times \frac{N_{MB}}{N_{MB}^i} \times \epsilon, \tag{83.2}$$

where $N_{J/\psi}$ and N_{MB} are the number of J/ψ and MB events, respectively. The factor ϵ is a combination of several corrections that account for trigger selection, event selection, SPD vertex quality assurance and pile-up rejection. Details of the analysis procedure can be found in Ref. [4].

83.3   Results and Discussions

The multiplicity dependence of the J/ψ yield is displayed for pp collisions at $\sqrt{s} =$ 5.02 TeV in Fig. 83.1. The correlation between multiplicity and J/ψ is compared to a linear function with a slope of 1 (y = x). The ratio of the relative J/ψ yield to this diagonal as a function of multiplicity is displayed in the bottom panel of Fig. 83.1. It is interesting to note that the ratio is consistent with unity over the full multiplicity range. This, in turn, implies that the production of J/ψ scales linearly with the underlying event activity when a gap is present between the rapidities at which $dN_{ch}/d\eta$ and J/ψ are measured. The J/ψ yield observed at $\sqrt{s} = 5.02$ TeV is compared with those reported for forward and mid-rapidity ALICE measurements in pp collisions at $\sqrt{s}$ = 13 TeV and the results are displayed in Fig. 83.2. It is observed that the relative J/ψ yield estimated for different multiplicity bins are practically independent of the beam energy in the forward rapidity region. However, in the mid-rapidity region for $\sqrt{s} = 13$ TeV data, the measured relative J/ψ yield increases faster than linearly with increasing multiplicity. Such a dependence may be attributed to the presence of auto-correlations between the J/ψ and the multiplicity measurements at mid-rapidity, for instance, because of a possible jet bias [7]. In Fig. 83.3, the relative J/ψ yields as a function of multiplicity are compared to predictions from two theoretical models proposed by Kopeliovich et al. [8] and Ferreiro et al. (the percolation model) [9]. The percolation string model shows a linear increase at low density and quadratic increase at higher density and describes the data well at low multiplicity. The Kopeliovich

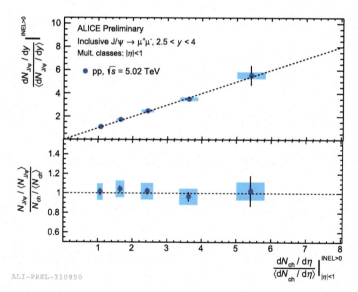

Fig. 83.1   The relative J/ψ yield as a function of the relative charged-particle density measured at forward rapidity in pp collisions at $\sqrt{s} = 5.02$ TeV. Bottom panel: ratio of the relative J/ψ yield to the relative charged-particle density as a function of multiplicity

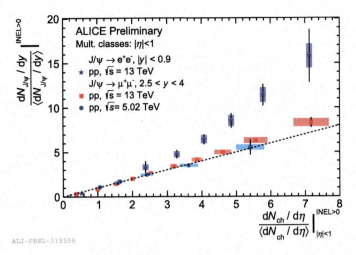

Fig. 83.2 Comparison of the relative J/ψ yields as a function of the relative charged-particle multiplicity at forward rapidity in pp collisions at $\sqrt{s}$ = 5.02 TeV and 13 TeV and at mid-rapidity at $\sqrt{s}$ = 13 TeV

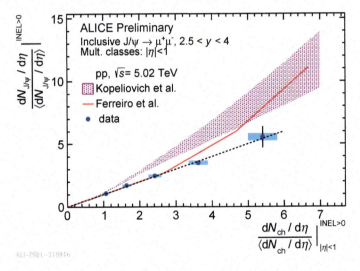

Fig. 83.3 Theoretical models compared to the relative J/ψ yield at $\sqrt{s}$ = 5.02 TeV

model takes into consideration the contributions of higher Fock states to reach high multiplicities in pp collisions. As a result of a higher number of gluons, the J/ψ production rate is also enhanced. A stronger than linear increase with multiplicity is observed.

83.4 Conclusions and Outlook

The multiplicity dependence of J/ψ has been studied in pp collisions at $\sqrt{s} = 5.02$ TeV. A linear increase of the relative J/ψ yield is observed as a function of multiplicity at forward rapidity. The multiplicity dependence is independent of $\sqrt{s}$ at LHC energies. The increase of J/ψ production seems to depend on the rapidity gap between the J/ψ and the multiplicity measurement. The finding reveals too that the percolation model describes the data well at forward-rapidity for relative multiplicities < 3. The multiplicity dependent study of various quarkonium states will shed more light on this topic.

Acknowledgements We are thankful to UGC and DST for financial support.

References

1. T. Sjöstrand, M. van Zijl, Phys. Rev. D **36**, 2019 (1987)
2. B. Abelev et al., ALICE Collaboration. Phys. Lett. B **712**, 165 (2012)
3. S. G. Weber (ALICE Collaboration), (2017) [arXiv:1704.04735 [hep-ex]]
4. Adamová, D. and others (ALICE Collaboration), Phys. Lett. **B776**, 91-104 (2018)
5. J. Adam et al., ALICE Collaboration. JHEP **1509**, 148 (2015)
6. K. Aamodt et al., ALICE Collaboration. JINST **3**, S08002 (2008)
7. S. G. Weber, A. Dubla, A. Andronic and A. Morsch, (2018) arXiv:1811.07744 [nucl-th]
8. B.Z. Kopeliovich et al., Phys. Rev. D **88**, 116002 (2013). [arXiv:1308.3638 [hep-ph]]
9. E.G. Ferreiro, C. Pajares, Phys. Rev. C **86**, 034903 (2012). [arXiv:1203.5936 [hep-ph]]

Chapter 84
Thermodynamics of a Gas of Hadrons with Interaction Using S-Matrix Formalism

Ashutosh Dash, Subhasis Samanta, and Bedangadas Mohanty

Abstract Ideal hadron resonance gas (HRG) is a popular model to study the late stage of QCD matter formed in heavy-ion collisions at finite temperature and chemical potential. An extension of the HRG model is constructed to include attractive interactions using the relativistic virial expansion of partition function. The virial coefficients are related to the phase shifts which are calculated using K-matrix formalism. We calculate various thermodynamics quantities like pressure, energy density, and entropy density of the system. A comparison of thermodynamic quantities with non-interacting HRG model, calculated using the same number of hadrons, shows that the values of thermodynamic quantities from the above formalism are larger. A good agreement between equation of state calculated in K-matrix formalism and lattice QCD simulations is observed. Further, we report the effect of including repulsive interactions on various thermodynamic observables calculated using a S-matrix based HRG model to already available corresponding results with only attractive interactions.

84.1 Introduction

Relativistic heavy ion collisions have contributed immensely to our understanding of strongly interacting matter at finite temperature (T) and baryon chemical potential (μ_B). Lattice quantum chromodynamics (LQCD) provides a first principle approach to study strongly interacting matter at zero baryon chemical potential (μ_B) and finite

A. Dash (✉) · S. Samanta · B. Mohanty
School of Physical Sciences, National Institute of Science Education and Research,
HBNI, Jatni 752050, India
e-mail: ashutosh.dash@niser.ac.in

S. Samanta
e-mail: subhasis.samant@gmail.com

B. Mohanty
e-mail: bedanga@niser.ac.in

© Springer Nature Singapore Pte Ltd. 2021
P. K. Behera et al. (eds.), *XXIII DAE High Energy Physics Symposium*,
Springer Proceedings in Physics 261,
https://doi.org/10.1007/978-981-33-4408-2_84

temperature (T) which indicates a smooth crossover transition from hadronic to a quark-gluon plasma (QGP) phase. On the other hand, at high baryon chemical potential, the nuclear matter is expected to have a first-order phase transition which ends at a critical point. One of the approach to study the properties of hadronic phase formed by hadronization of the QGP is through a statistical model of a gas of hadrons called the hadron resonance gas model (HRG). The hadron resonance gas (HRG) [1–6] models have successfully described the hadron multiplicities produced in relativistic nuclear collisions over a wide range of energies. The result of such an investigation was the observation of rise in the extracted chemical freeze-out temperature values from lower energies to almost a constant value of temperature $T \simeq 155-165$ MeV at higher energies, accompanied with the decrease of the baryon chemical potential (μ_B) with increasing energy [13].

The primary assumption of HRG model is that, the partition function contains all relevant degress of freedom of a confined, strongly interacting medium by treating the resonances as point-like particles and neglecting their mutual interaction. However, relaxing the above assumptions by including overlapping resonances and resonances of finite widths, it has been seen that the variation of thermodynamic variables with temperature changes substantially. Also, it can be argued that such interactions contribute only to the attractive part of partition function and the inclusion of a repulsive part could partially negate the effect of the attractive part. One of such method is the virial expansion of the partion function. In the virial expansion approach, dynamical information obtained either from theoretical or from empirical two-body scattering phase shifts is used to compute the thermodynamic observables for an interacting gas of hadrons at zero or non-zero chemical potential. The S-matrix formulation of statistical mechanics- proposed by Dashen, Ma and Bernstein [20] is an extension of the usual virial expansion to the relativistic case. Since the virial expansion can be expressed as an expansion in powers of the density, the results of this approach will very likely contain the physics for dilute systems. Theoretically, K-matrix formalism can be used to calculate the phase shifts of the resonance spectral function in contrast to the popular Breit-Wigner parametrization. It has been argued previously that the K-matrix formalism preserves the unitarity of the scattering matrix (S-matrix) and neatly handles multiple resonances [14–16]. However, the formalism fails to handle any repulsive channel in the scattering matrix. Therefore,[22] we include the repulsive part by fitting to experimental phase shifts that encodes the information about the nature of interaction. We use the phase shifts data from Scattering Analysis Interactive Database (SAID) partial wave analysis for nucleon-nucleon (NN), pion-nucleon (πN) and kaon-nucleon (KN) interaction in their respective isospin channels [17–19].

84.2   Formalism

The most natural way to incorporate interaction among a gas of hadrons is to use relativistic virial expansion as given in Ref. [20]. In this formalism, the logarithm of the partition function can be written as the sum of non-interacting (ideal) and interacting parts i.e.,

$$\ln Z = \ln Z_0 + \sum_{i_1, i_2} z_1^{i_1} z_2^{i_2} b(i_1, i_2), \tag{84.1}$$

where z_1 and z_2 are fugacities of two species and $z = e^{\beta \mu}$. The chemical potential of jth particle is defined as $\mu_j = B_j \mu_B + S_j \mu_S + Q_j \mu_Q$ where B_j, S_j, Q_j are baryon number, strangeness and electric charge and μ's are the respective chemical potentials. The virial coefficients $b(i_1, i_2)$ are written as

$$b(i_1, i_2) = \frac{V}{4\pi i} \int \frac{d^3 p}{(2\pi)^3} \int d\varepsilon \exp\left(-\beta(p^2 + \varepsilon^2)^{1/2}\right) \left[A\left\{S^{-1} \frac{\partial S}{\partial \varepsilon} - \frac{\partial S^{-1}}{\partial \varepsilon} S\right\}\right]_c . \tag{84.2}$$

In the above expression, the inverse temperature is denoted by β. V, p and ε stand for the volume, the total center of mass momentum and energy respectively. The labels i_1 and i_2 refer to channel of the S-matrix which has initial state containing $i_1 + i_2$ particles. The symbol A denotes the symmetrization (anti-symmetrization) operator for a system of bosons (fermions). We consider baryon and meson octets as the stable hadrons. Non interacting stable hadrons contribute to the ideal part of the pressure whereas two body elastic scattering between any two stable hadrons gives the interacting part of the pressure. The relevant expressions of thermodynamics variables can be computed from the partition function Eq. 84.1 and the explicit form can be found in Ref. [16].

The S-matrix can be expressed in terms of phase shifts δ_l^I as [21]

$$S(\varepsilon) = \sum_{l, I} (2l + 1)(2I + 1) \exp(2i\delta_l^I), \tag{84.3}$$

where l and I denote angular momentum and isospin, respectively. A theoretical way of calculating the attractive phase shifts is to use the K-matrix formalism. The K-matrix formalism elegantly expresses the unitarity of the S-matrix for the processes of type $ab \to cd$, where a, b and c, d are hadrons. Details of the K-matrix formalism can be found in Refs. [14–16]. The K-matrix formalism is applicable for attractive interaction but not for repulsive interactions. Similarly, the formalism is applicable when the information about resonance mass and width is available but not non-resonant interaction for e.g.., in NN interaction such formalism is not applicable. In such cases, one has to fit experimental phase shifts to get the dynamical information about interaction. In Ref [22], information about experimental phase shifts [17–19, 23] for NN interaction was used for $l \leq 7$ in both isospin channel $I = 0$ and $I = 1$ to calculate the second virial coefficient.

84.3 Results

In Fig. 84.1, the results of K-matrix formalism (attractive), 'Total' (attractive+ repulsive) is compared with the IDHRG model (IDHRG-1) with same number of hadrons and resonances and lattice data [16, 22, 26–28]. The results of K-matrix formalism for all thermodynamic observables are larger compared to the ideal HRG values. A similar comparison of thermodynamic observable with the inclusion of repulsive contribution along with the attractive contribution, with K-matrix formal-

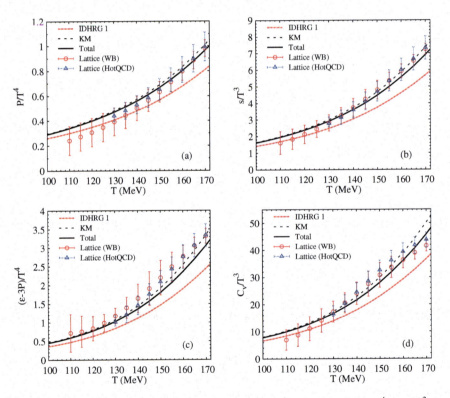

Fig. 84.1 Temperature dependence of various thermodynamic quantities (**a** P/T^4, **b** s/T^3, **c** $(\varepsilon - 3P)/T^4$ **d** C_v/T^3,) at zero chemical potential. Total contains both the attractive and repulsive interaction whereas KM contains only the attractive part. IDHRG-1 corresponds to results of ideal HRG model with same number of hadrons and resonances as in KM. Results are compared with lattice QCD data of Refs. [24] (WB) and [25] (HotQCD)

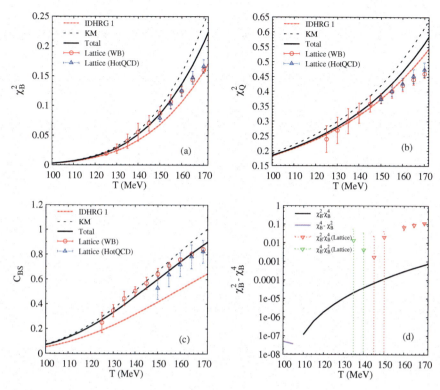

Fig. 84.2 Temperature dependence of second order susceptibilities (**a** χ_B^2, **b** χ_Q^2, **c** C_{BS} and **d** $\chi_B^2 - \chi_B^4$) at zero chemical potential. Total contains both the attractive and repulsive interaction whereas KM contains only the attractive part. IDHRG-1 corresponds to results of ideal HRG model with same number of hadrons and resonances as in KM. Results are compared with lattice QCD data of Refs. [26] (WB), [27] (HotQCD) and [28] (Lattice)

ism shows a reduction in all such observables. The difference between 'KM'and 'Total' is more towards the higher temperature regime and negligible in lower temperature.

A similar comparison of second order diagonal and off diagonal susceptibilities is shown in Fig. 84.2. It is seen that the K-matrix formalism shows better agreement with LQCD data than IDHRG almost across all observables, notably in χ_{BS}^{11}. With the inclusion of repulsion which is almost in the baryonic sector (πN, KN and NN), the results of susceptibilities like χ_B^2 and χ_Q^2 show a lot of improvement than with only attraction. Results of two important observables C_{BS} and $\chi_B^2 - \chi_B^4$ as a function of temperature are also shown Fig. 84.2. The contribution strength from different channels to the repulsive part of the second virial coefficient is in the order such that $\pi N > KN > NN$.

To summarize, the findings suggest that the isospin-weighted sum of higher order attractive and repulsive phase-shifts is non-zero which is reflected across all thermodynamic variables of Figs. 84.1 and 84.2, which with regard to complete cancellation would have coincided with the IDHRG results. Similarly, we could ascertain that in IDHRG increasing the degeneracies just by adding additional resonances can also explain lattice data which is in contrast to genuine interaction that is present in K-matrix/S-matrix formalism.

Acknowledgements BM acknowledges financial support from J C Bose National Fellowship of DST, Government of India. AD and SS acknowledges financial support from DAE, Government of India.

References

1. P. Braun-Munzinger, J. Stachel, J. P. Wessels and N. Xu, Phys. Lett. B **344**, 43 (1995) [nucl-th/9410026]
2. J. Cleymans, D. Elliott, H. Satz and R. L. Thews, Z. Phys. C **74**, 319 (1997) [nucl-th/9603004]
3. G. D. Yen and M. I. Gorenstein, Phys. Rev. C **59**, 2788 (1999) [nucl-th/9808012]
4. P. Braun-Munzinger, I. Heppe and J. Stachel, Phys. Lett. B **465**, 15 (1999) [nucl-th/9903010]
5. J. Cleymans and K. Redlich, Phys. Rev. C **60**, 054908 (1999) [nucl-th/9903063]
6. F. Becattini, J. Cleymans, A. Keranen, E. Suhonen, K. Redlich, Phys. Rev. C **64**, 024901 (2001). [hep-ph/0002267]
7. P. Braun-Munzinger, D. Magestro, K. Redlich, J. Stachel, Phys. Lett. B **518**, 41 (2001). [hep-ph/0105229]
8. F. Becattini, J. Manninen, M. Gazdzicki, Phys. Rev. C **73**, 044905 (2006). [hep-ph/0511092]
9. A. Andronic, P. Braun-Munzinger and J. Stachel, Nucl. Phys. A **772**, 167 (2006) [nucl-th/0511071]
10. A. Andronic, P. Braun-Munzinger and J. Stachel, Phys. Lett. B **673**, 142 (2009) Erratum: [Phys. Lett. B **678**, 516 (2009)]
11. S. Das, D. Mishra, S. Chatterjee, B. Mohanty, Phys. Rev. C **95**(1), 014912 (2017)
12. A. Andronic, P. Braun-Munzinger, K. Redlich, J. Stachel, Nature **561**(7723), 321 (2018)
13. P. Braun-Munzinger, J. Stachel, Nature **448**, 302 (2007)
14. S.U. Chung, J. Brose, R. Hackmann, E. Klempt, S. Spanier, C. Strassburger, Annalen Phys. **4**, 404 (1995)
15. A. Wiranata, V. Koch, M. Prakash, X.N. Wang, Phys. Rev. C **88**(4), 044917 (2013)
16. A. Dash, S. Samanta, B. Mohanty, Phys. Rev. C **97**(5), 055208 (2018)
17. R.L. Workman, W.J. Briscoe, I.I. Strakovsky, Phys. Rev. C **94**(6), 065203 (2016)
18. R.L. Workman, R.A. Arndt, W.J. Briscoe, M.W. Paris, I.I. Strakovsky, Phys. Rev. C **86**, 035202 (2012)
19. J.S. Hyslop, R.A. Arndt, L.D. Roper, R.L. Workman, Phys. Rev. D **46**, 961 (1992)
20. R. Dashen, S.K. Ma, H.J. Bernstein, Phys. Rev. **187**, 345 (1969)
21. J. J. Sakurai and J. Napolitano,
22. A. Dash, S. Samanta, B. Mohanty, Phys. Rev. C **99**(4), 044919 (2019)
23. R. Garcia-Martin, R. Kaminski, J.R. Pelaez, J. Ruiz de Elvira, F.J. Yndurain, Phys. Rev. D **83**, 074004 (2011)
24. S. Borsanyi, Z. Fodor, C. Hoelbling, S.D. Katz, S. Krieg, K.K. Szabo, Phys. Lett. B **730**, 99 (2014)
25. A. Bazavov et al., HotQCD Collaboration. Phys. Rev. D **90**, 094503 (2014)
26. S. Borsanyi, Z. Fodor, S.D. Katz, S. Krieg, C. Ratti, K. Szabo, JHEP **1201**, 138 (2012)
27. A. Bazavov et al., HotQCD Collaboration. Phys. Rev. D **86**, 034509 (2012)
28. R. Bellwied, S. Borsanyi, Z. Fodor, S.D. Katz, A. Pasztor, C. Ratti, K.K. Szabo, Phys. Rev. D **92**(11), 114505 (2015)

Chapter 85
Effects of Baryon-Anti-baryon Annihilation on the Anti-hyperon to Anti-proton Ratio in Relativistic Nucleus-Nucleus Collisions

Ekata Nandy and Subhasis Chattopadhyay

Abstract RHIC's Beam Energy Scan Program at lower energy and future facilities at FAIR offer a unique opportunity to study Quark-Gluon Plasma (QGP) at high baryon density. The formation of QGP is often characterized by the enhanced production of strange over non-strange particles in central heavy-ion collisions relative to peripheral or proton-proton collisions at the same energy . Previous measurements at RHICs' AGS and CERN SPS have presented evidence of strangeness enhancement in the light of non-monotonic variation of kaon-to-pion ratio as a function of collision energy. A similar signature was also observed in the baryon sector, where an enhancement in $\bar{\Lambda}$ to $\bar{p}$ was reported . However, it still remains unclear whether $\bar{\Lambda}$ to $\bar{p}$ enhancements can be uniquely attributed to the strangeness enhancement owing to the deconfinement phase transition. In this article, we will demonstrate that processes like baryon-anti-baryon annihilation in the late hadronic stage may also account for the apparent enhancement in $\bar{\Lambda}$ to $\bar{p}$ ratio.

85.1 Introduction

Quantum Chromodynamics (QCD) calculations have predicted that the collisions of heavy nuclei (A+A) at relativistic energies are likely to produce high-density matter [1] of deconfined quarks and gluons, the Quark Gluon Plasma (QGP). In the coalescence picture of hadronization, it is argued that the strangeness production from a deconfined partonic matter is energetically more efficient than in a hadron gas. As a result, one may expect a relative enhancement of strange-to-non-strange particles in central A+A collisions with respect to peripheral A+A or proton-proton(p+p)

E. Nandy (✉) · S. Chattopadhyay
Variable Energy Cyclotron Centre, HBNI, 1/AF, Bidhan Nagar, Kolkata 700064, India
e-mail: ekatanandy@gmail.com

S. Chattopadhyay
e-mail: sub@vecc.gov.in

E. Nandy
Homi Bhabha National Institute, Mumbai, India

© Springer Nature Singapore Pte Ltd. 2021
P. K. Behera et al. (eds.), *XXIII DAE High Energy Physics Symposium*,
Springer Proceedings in Physics 261,
https://doi.org/10.1007/978-981-33-4408-2_85

interactions at the same collision energy [2]. This makes strangeness enhancement a diagnostic probe of the properties for the partonic matter and hence considered among the potential signatures of the QGP formation [3] .

Strangeness enhancement has been extensively studied at RHIC and SPS. A non-monotonic variation in K/π ratio as a function of beam energy, popular as horn structure, was reported unanimously from both the experimental facilities [4]. This non-monotonic increase in the K/π ratio was considered as the first confirmatory evidence of strangeness enhancement in relativistic heavy-ion collisions. At low-energy collisions, as in AGS, a baryon-rich QGP is expected to be produced. In a baryon-dense QGP, since the production of light anti-quarks are also suppressed, a large enhancement in strange anti-baryons over the ordinary anti-baryons is anticipated. A similar signature was also observed in the baryon sector, where an enhancement in $\bar{\Lambda}$ to $\bar{p}$ was reported. Indeed, a large enhancement $\bar{\Lambda}$ over $\bar{p}$ ratio ($\sim$3.5) was reported by the E917 experiment at AGS [5]. Also, a significant increase from 0.25 in p+p collisions to 1.5 for A+A collisions was published by the NA49 experiment at CERN SPS. However, what has remained unclear is whether this enhancement is uniquely associated with the strangeness enhancement in QGP [6]. The primary reason for this ambiguity is the modification of particle yields in the later stages of a collision, mostly during the hadronic rescattering phase, prior to freeze-out. Based on the hadronic transport model calculations, it was inferred that $\bar{\Lambda}$ and $\bar{p}$ have different baryon-anti-baryon absorption cross-sections, thereby modifying the spectral shapes in a way leading to an apparent increase in $\bar{\Lambda}/\bar{p}$ ratio.

In this work, our goal is to quantitatively demonstrate the sensitivity of $\bar{\Lambda}$ to $\bar{p}$ ratio to the $B\bar{B}$ annihilation processes by analyzing inclusive cross-sections of $\bar{\Lambda}$ and $\bar{p}$ in 0–7% central Au+Au(Pb+Pb) collisions at collision energies, corresponding to beam energies of AGS (SPS), using a hadronic version of A Multi Phase Transport (AMPT) model and Ultra-Relativistic Quantum Molecular Dynamics (UrQMD) by switching off and on $B\bar{B}$ annihilation processes within the model.

85.2   A Multi Phase Transport (AMPT) Model

AMPT is a hybrid Monte Carlo model comprising three main parts: a HIJING-based initial condition, a partonic transport followed by hadronization by quark coalescence or the Lund string fragmentation and a relativistic hadronic transport [7]. In partonic transport, partons are scattered elastically and the scattering cross section is calculated perturbatively from the strong coupling constant (α_s) and gluon screening mass (μ). These two are the main input parameters to the model and have been constrained with a wide range of experimental data. For this work, we have set the scattering cross-section to 10 mb, with the following choice of input parameters : $\alpha = 0.47$ and $\mu = 1.8$ fm.

85.3   Ultra-Relativistic Quantum Molecular Dynamics (UrQMD)

UrQMD is a microscopic transport model where the space-time evolution of the fireball is described in terms of fragmentation of excited color strings followed by the covariant propagation of hadrons and resonances which eventually scatter and decay to the final state stable hadrons. In UrQMD, baryonic and mesonic interactions are modeled by the additive quark model (AQM), considering the number and flavor of the valance quark content of a given hadron [8].

85.4   Results and Discussions

$\bar{\Lambda}/\bar{p}$ ratio has been measured in RHIC AGS ($\sqrt{s_{NN}}$ =4.9 GeV Au+Au collision) energy as well as CERN SPS energy ($\sqrt{s_{NN}}$ = {6.27 GeV,7.62 GeV, 8.77 GeV, 12.3 GeV,17.3 GeV Pb-Pb collisions}). It was shown that the ratio increases with the decrease of beam energy, i.e., with the increase in baryon density. Here, we have calculated yields of $\bar{\Lambda}$ and $\bar{p}$ at mid-rapidity ($|y|$ <0.4) for most central collisions (b<3.5 fm). As final yields of $\bar{\Lambda}$ & $\bar{p}$ are sensitive to hadronic interactions, mainly the baryon-anti-baryon annihilation ,therefore, we have calculated these yields from UrQMD with baryon-anti-baryon (B$\bar{B}$) annihilation on and off conditions. In Figs. 85.1 and 85.2, we have compared the individual yields of $\bar{\Lambda}$ and $\bar{p}$ as a function of $\sqrt{s_{NN}}$ from data and UrQMD with and without B$\bar{B}$ annihilation. The effects of annihilation are seen to be significant on the final yields. Without B$\bar{B}$ annihilation, model calculation overestimates the data for both $\bar{\Lambda}$ and $\bar{p}$. However, with B$\bar{B}$ annihilation, $\bar{\Lambda}$ underestimate data but $\bar{p}$ has a good agreement.

Next we have calculated the $\bar{\Lambda}/\bar{p}$ ratio at different $\sqrt{s_{NN}}$ with and without B$\bar{B}$ annihilation from UrQMD as shown in Fig. 85.3.

We observe, with B$\bar{B}$ annihilation, the ratio is sensitive to the choice of pt range and achieves maximum in the lowest p_T-range. However, it fails to describe the

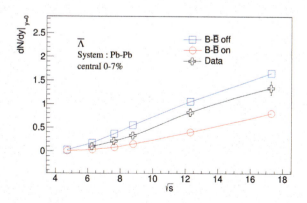

Fig. 85.1 Figure shows the yields at mid-rapidity of $\bar{\Lambda}$ wrt $\sqrt{s_{NN}}$ with B$\bar{B}$ annihilation on and off conditions in UrQMD and that compared with data. Yields are calculated in the Pb-Pb system with 0–7% centrality except for the first point (Au-Au system)

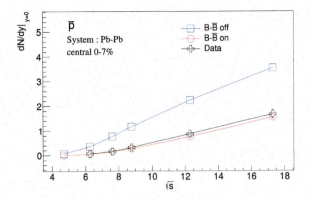

Fig. 85.2 Figure shows the yields at mid-rapidity of p̄ wrt √s_{NN} with BB̄ annihilation on and off conditions in UrQMD and compared with data. Yields are calculated in the Pb-Pb system with 0-7% centrality except for the first point (Au-Au system)

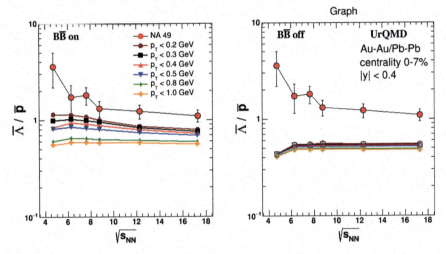

Fig. 85.3 Figure shows the Λ̄/p̄ ratio at different √s_{NN} with and without BB̄ annihilation from UrQMD at mid-rapidity |y|<0.4 and at different p_T and it has been compared with the experimental data. Ratios are calculated in the Pb-Pb system with 0-7% centrality except the first point (Au-Au system)

data quantitatively. Without BB̄ annihilation, UrQMD fails to describe the data both qualitatively and quantitatively and has no sensitivity to the choice of p_T-range. We also studied the ratio within a multi phase transport (AMPT) model that implements a spatial coalescence scheme for hadronization. Figure 85.4 shows the ratio with √s_{NN} from UrQMD, AMPT and experimental data values at mid-rapidity |y|<0.4 and p_T <0.5. AMPT has higher values of the ratios compared to UrQMD. However, AMPT does not include the annihilation of Λ̄.

Fig. 85.4 Figure shows the $\bar{\Lambda}/\bar{p}$ ratio at different $\sqrt{s_{NN}}$ with B$\bar{B}$ annihilation from UrQMD and AMPT at mid-rapidity |y|<0.4 and p_T <0.5 GeV/c and it has been compared with the experimental data. Ratios are calculated in the Pb-Pb system with 0-7% centrality except the first point (Au-Au system)

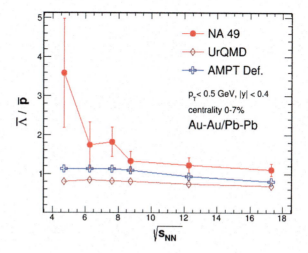

Nevertheless, both the model calculations reproduce the general trend in the $\bar{\Lambda}/\bar{p}$ ratio in data somewhat well. This essentially suggests, with proper tuning of the annihilation cross-sections, models without an explicit implementation of the partonic degrees of freedom may be sufficient for the quantitative description of $\bar{\Lambda}/\bar{p}$ ratio in data, implying enhancement in $\bar{\Lambda}/\bar{p}$ ratio may not be an unambiguous evidence of medium formation with partonic degrees of freedom.

References

1. CBM Collaboration (T. Ablyazimov (Dubna, JINR) et al.), 11 pp, Eur. Phys. J. A53 no.3, 60 (2017)
2. Johann Rafelski1 and Jean Letessier, Journal of Physics G : Nuclear and Particle Physics, Volume 30, Number 1 (2003)
3. Paolo Castorina, Salvatore Plumari, Helmut Satz , 7 pp. Int. J. Mod. Phys. E26 no.12, 1750081 (2017)
4. C. Alt et al., NA49 Collaboration. Phys. Rev. C **77**, 024903 (2008)
5. E917 Collaboration (B. B. Back (Argonne) et al.), 5 pp, Published in Phys.Rev.Lett. 87 242301 (2001)
6. NA49 Collaboration (C. Alt (Frankfurt U.) et al.). 11 pp, e-Print: nucl-ex/0512033, (2005)
7. Zi-Wei Lin, Che Ming Ko, Bao-An Li, Bin Zhang, Subrata Pal, 28 pages. Phys. Rev. C **72**, 064901 (2005)
8. M. Bleicher (Frankfurt U.), E. Zabrodin , J.Phys. G25 1859-1896 (1999)

Chapter 86
Deconfinement to Confinement as PT Phase Transition

Haresh Raval and Bhabani Prasad Mandal

Abstract We consider $SU(N)$ QCD in a new quadratic gauge which highlights a certain characteristic of the theory in the non-perturbative sector. By considering the natural hermiticity property of the ghost fields, we cast this model as non-Hermitian but symmetric under combined Parity (P) and Time reversal (T) transformations. We explicitly study the PT phase transition in this model. This is the very first such study in the non-Hermitian gauge theory. The ghost fields condensate which give rise to spontaneous breaking of PT symmetry. This leads to realize the transition from deconfined phase to confined phase as a PT phase transition in this system. The hidden C-symmetry in this system is identified as inner automorphism in this theory. Explicit representation is constructed for the C-symmetry.

86.1 Introduction

Symmetries and their spontaneous breakdown played a crucial role in the understanding of physics from time to time. About two decades ago, the formulation of usual quantum mechanics where all physical observables are represented by self-adjoint operators, has been extended to include non-self-adjoint operators for their observable. Two important discrete symmetries, namely Parity (P) and Time Reversal (T) are instrumental in such formulation as first shown in Ref. [1, 2]. Consistent formulation with real energy eigenvalues, unitary time evolution and probabilistic interpretation for unbroken PT symmetric non-Hermitian quantum systems have been formulated in a different Hilbert space equipped with positive definite CPT inner product. Such non-Hermitian PT symmetric systems generally exhibit a phase transition (or more specifically a PT breaking transition) that separates two parametric regions: (i) region

H. Raval · B. Prasad Mandal
Department of Physics, Institute of Science, Banaras Hindu University, Varanasi 221005, India
e-mail: bhabani.mandal@gmail.com

H. Raval (✉)
Department of Physics, Indian Institute of Technology Delhi, New Delhi 11016, India
e-mail: haresh@phy.iitb.ac.in

© Springer Nature Singapore Pte Ltd. 2021
P. K. Behera et al. (eds.), *XXIII DAE High Energy Physics Symposium*,
Springer Proceedings in Physics 261,
https://doi.org/10.1007/978-981-33-4408-2_86

of the unbroken PT symmetry in which the entire spectrum is real and eigenstates of the systems respect PT symmetry and (ii) a region of the broken PT symmetry in which the whole spectrum (or a part of it) appears as complex conjugate pairs and the eigenstates of the systems do not respect PT symmetry. The physics at this transition point is extremely rich in nature and the typical characteristics of the non-Hermitian system are reflected by the behavior of the system at the transition point. Thus, the PT phase transition and its realization being extremely important in theories with non-Hermitian systems have been studied frequently [3–14]. Even though self-adjointness of quantum observables have never been challenged, this formulation of complex quantum mechanics created remarkable interest in several fronts of physics including open quantum systems [15], scattering theory [16–24], optics, etc. Particularly the theory of optics, where several physical processes are known to follow Schrödinger like equations, provides a fertile ground to verify the implications of such formulations experimentally. PT symmetric complex potentials are realized through complex refractive index in the optical media and the consequence of PT phase transition has been experimentally realized in optics [14, 16, 25–27]. Therefore, the applicability of this path-breaking formulation of complex quantum theories relies on observing PT phase transition in various physical systems.

In the present work, we for the first time demonstrate the PT phase transition in a gauge theory. We consider $SU(N)$ QCD in the newly found quadratic gauge [28], which is shown recently to have substantial implications in the non-perturbative sector of the theory [28, 29], to study the PT phase transition. This is a novel study of PT phase transition in a gauge theory. Although the non-Hermitian extension of a gauge theory has been explored [30, 31], this particular subject has never been touched upon. The gauge has a few following unusual features. (1) The gauge is not one of Abelian projection gauges [32] and has quark confinement signatures contrary to common studies of the confinement which have been done in Abelian projection. (2) It is the covariant algebraic gauge. In general, algebraic gauges are not covariant. (3) It removes the Gribov ambiguity on the compact manifold contrary to the case of usual gauges. [29]. This theory has two distinct phases, one is normal phase or deconfined phase and in the other ghost fields condensate leading to the confinement phase [28]. The Lagrangian density which represents the deconfinement phase of the theory is shown to be non-Hermitian by adopting the natural but unconventional property of hermiticity for ghosts [33]. However, the theory is invariant under PT transformations of the gluon and ghost fields. We explicitly show that the appearance of the ghost condensed state is the cause of spontaneous break down of the PT symmetry. At this transition point, the theory passes from deconfined phase to confined phase. We further identify the inner automorphism in this system as the C-symmetry which is inherent in all PT symmetric systems and connects the negative PT norm states to positive PT norm states and vice-versa. This C-symmetry is useful to define the non-Hermitian theory in a fully consistent manner in the modified Hilbert space endowed with CPT inner product. We explicitly construct the representation of the C-symmetry for the present non-Abelian theory. This provides us the first example of the explicit representation of the C-symmetry in any gauge theory. Hence, the present theory can be viewed as a consistent non-Hermitian gauge theory.

Now we present the plan of the paper. In the next section, we consider a charged scalar theory with a non-Hermitian mass matrix to set the mathematical preliminaries for the later non-Abelian–non-Hermitian model. Two phases of the QCD in the newly found quadratic gauge have been elaborated in Sec. III. In Sec. IV, non-hermiticity and PT symmetry properties of two phases have been discussed. The transition from deconfined phase to confined phase has also been identified as PT phase transition in this section. Explicit representation of the C-symmetry is constructed. Last Sec. is kept for results and discussions.

86.2   The Toy Model of Non-Hermitian Complex Scalars

In what follows from the next section has a close analogy with simple complex scalar non-Hermitian model discussed in Refs. [31, 34]. Hence, we first study the non-Hermitian theory of charged scalars. This theory is described by the following Lagrangian:

$$L = \partial_\mu \phi_1^* \partial_\mu \phi_1 + \partial_\mu \phi_2^* \partial_\mu \phi_2 + [\phi_1^* \ \phi_2^*] M^2 \begin{bmatrix} \phi_1 \\ \phi_2 \end{bmatrix} \tag{86.1}$$

where

$$M^2 = \begin{bmatrix} m_1^2 & \mu^2 \\ -\mu^2 & m_2^2 \end{bmatrix} \tag{86.2}$$

We will be interested only in cases for which $m_1^2, m_2^2, \mu^2 \geq 0$. We see that the mass matrix M^2 is not Hermitian. Discussion on discrete symmetries become easier when the doublet of two fields is defined as

$$\Phi = \begin{bmatrix} \phi_1 \\ \phi_2 \end{bmatrix}. \tag{86.3}$$

Then, the parity and time reversal, respectively, are defined on the doublet as follows:

$$\Phi \xrightarrow{P} P\Phi \tag{86.4}$$

$$\Phi \xrightarrow{T} T\Phi^* \tag{86.5}$$

where P and T now are 2×2 matrices and complex conjugation in time reversal is due to anti-linearity. We can make a clear guess for the choice of P by the analogy of the parity transformation in $\mathbb{R}^2$, where $x \to x$ and $y \to -y$. The parity in $\mathbb{R}^2$ suggests that the field ϕ_1 transforms as a scalar and the other, ϕ_2 transforms as a pseudoscalar. Therefore, the P has the following matrix form:

$$P = \begin{bmatrix} 1 & 0 \\ 0 & -1. \end{bmatrix}$$ (86.6)

This leaves us with the only choice for the time reversal T under which the Lagrangian is PT-invariant. We must choose $T = \mathbf{1}_2$ [31, 34]. One can in principle swap the roles of P and T however in order to interpret this PT symmetric theory in terms of a coupled system with gain and loss, one should take $T = \mathbf{1}_2$ [31, 34]. The theory remains in the unbroken PT-symmetric state as long as the eigenvalues of the mass matrix given as below remain real

$$M_{\pm}^2 = \frac{1}{2}(m_1^2 + m_2^2) \pm \sqrt{(m_1^2 - m_2^2)^2 - 4\mu^4}$$ (86.7)

So, for $|m_1^2 - m_2^2| \geq 2\mu^2$, we are in the phase of unbroken PT symmetry. When $|m_1^2 - m_2^2| < 2\mu^2$ happens, we step into the region of broken PT symmetry as eigenvalues turn complex and $PT\psi_{\pm} = \pm\psi_{\pm}$ is no longer valid, where $\psi_{\pm}$ are eigenfunctions of the mass matrix M^2 corresponding to eigenvalues $M_{\pm}^2$. We shall encounter similar non-Hermitian mass matrix for gluons in our non-Abelian model to be discussed.

Since the eigenvalues in Eq. (86.7) do not change under $\mu^2 \rightarrow -\mu^2$, there still exists the charge conjugation symmetry under which the theory is CPT-invariant in both PT broken and unbroken phases. The charge conjugation is defined as follows:

$$\Phi \xrightarrow{C} C\Phi^*$$ (86.8)

with C=P [31, 34]. The theory in the region $|m_1^2 - m_2^2| < 2\mu^2$ violates CP also but preserves CT symmetry. Such charge conjugation symmetry exists in the non-Abelian model also as we will see later.

86.3 SU(N) QCD in the Quadratic Gauge

Here, we discuss a model in which we intend to establish a PT phase transition. The model relies on the new type of quadratic gauge fixing of Yang-Mills action as follows [28]:

$$H^a[A^{\mu}(x)] = A_{\mu}^a(x)A^{\mu a}(x) = f^a(x); \quad \text{for each } a$$ (86.9)

where $f^a(x)$ is an arbitrary function of x. The Faddeev–Popov determinant in this gauge is given by

$$\det\left(\frac{\delta(A_\mu^{a\epsilon} A^{\mu a\epsilon})}{\delta\epsilon^b}\right) = \det\left(2A_\mu^a(\partial^\mu \delta^{ab} - gf^{acb} A^{\mu c})\right), \tag{86.10}$$

Therefore, the resulting effective Lagrangian density is given as follows:

$$\mathcal{L}_Q = -\frac{1}{4}F_{\mu\nu}^a F^{\mu\nu a} - \frac{1}{2\zeta}(A_\mu^a A^{\mu a})^2 - 2\overline{c^a} A^{\mu a}(D_\mu c)^a \tag{86.11}$$

where ζ is an arbitrary gauge fixing parameter, the field strength $F_{\mu\nu}^a = \partial_\mu A_\nu^a(x) - \partial_\nu A_\mu^a(x) - gf^{abc} A_\mu^b(x)A_\nu^c(x)$ and $(D_\mu c)^a = \partial_\mu c^a - gf^{abc} A_\mu^b c^c$. The summation over an index a is understood when it appears repeatedly, including when occurred thrice in the ghost term. In particular, it is important for the present paper to understand the structure of the ghost Lagrangian,

$$-\overline{c^a} A^{\mu a}(D_\mu c)^a = -\overline{c^a} A^{\mu a} \partial_\mu c^a + gf^{abc}\overline{c^a} c^c A^{\mu a} A_\mu^b$$

where the summation over indices a, b and c each runs independently over 1 to $N^2 - 1$. As shown in [29], The resulting Lagrangian is BRST invariant[35, 36] which is essential for the ghost independence of the green functions and unitarity of the S-matrix. The substantial non-perturbative implications of this gauge have been studied in Refs. [28, 29]. In a recent in- teresting work, the FFBRST technique itself has been extended to connect non-perturbative sector implied by this gauge to perturbative sector signified by the Lorenz gauge [37].

86.3.1   Phases of the Theory in the Quadratic Gauge

This theory has two different phases [28]: the normal or deconfined phase and the ghost condensed phase showing the confinement. The Lagrangian in normal phase is given by Eq. (86.11) itself. We should note that the ghost Lagrangian does not have kinetic terms. They act like auxiliary fields in the normal phase but play an important role in the IR regime as we discuss now.

Ghost condensation

To demonstrate the significance of ghosts in terms of their condensates in the IR limit and citing their value for the present purpose also, we elaborate the ghost condensation and its implication thoroughly. The ghost condensation as a concept was introduced independently in Refs. [38–40]. The second term in the ghost Lagrangian contains ghost bilinears multiplying terms quadratic in gauge fields. Hence, if the ghosts freeze they amount to a non-zero mass matrix for the gluons as follows:

$$(M^2)^{ab}_{\text{dyn}} = 2g \sum_{c=1}^{N^2-1} f^{abc} \langle \overline{c^a} c^c \rangle. \tag{86.12}$$

We would get masses of gluons by diagonalizing the matrix and finding its eigenvalues. In an $SU(N)$ symmetric state, where all ghost-anti-ghost condensates are identical as given in Eq. (86.13) below the mass matrix becomes peculiar,

$$\langle \overline{c^1} c^1 \rangle = ... = \langle \overline{c^1} c^{N^2-1} \rangle = ... = \langle \overline{c^{N^2-1}} c^1 \rangle = ... = \langle \overline{c^{N^2-1}} c^{N^2-1} \rangle \equiv K. \tag{86.13}$$

The physical relevance of the theory is lent strength once a physical mechanism is laid out within which the state in the above equation can occur consistently. This objective was achieved by introducing a Lorenz gauge fixing term for one of the diagonal gluons, in addition to the purely quadratic terms of Eq. (86.9). This gauge fixing gives the propagator to the corresponding ghost field. Using this ghost propagator, one can give nontrivial vacuum values to bilinears $\overline{c^a} c^c$ within the framework Coleman- -Weinberg mechanism as described in [28]. This mechanism naturally gives the K to be real. Thus, in the state of condensates given by Eq. (86.13), the mass matrix becomes

$$(M^2)^{ab}_{\text{dyn}} = 2gK \sum_{c=1}^{N^2-1} f^{abc} \tag{86.14}$$

which is an antisymmetric matrix i.e., non Hermitian,

$$(M^2)^\dagger \neq M^2 \tag{86.15}$$

due to the antisymmetry of the structure constants. The matrix in Eq. (86.14) is unique, it has $N(N-1)$ non-zero eigenvalues only and thus has nullity $N-1$ which implies that $N(N-1)$ off-diagonal gluons obtain masses and the $N-1$ diagonal gluons remain massless. Because of the antisymmetry, eigenvalues that occur are purely imaginary and in conjugate pairs. The massive off-diagonal gluons are inferred as evidence of Abelian dominance, which is one of signatures of quark confinement. Further, mass squared of the off-diagonal gluon is purely imaginary, hence, the pole of the off-diagonal gluon propagator is on imaginary p^2 axis which is another important signature of color confinement [41]. The mass for gluons generated through a given dynamical mechanism breaks the gauge symmetry as usual. We note that there exist other interesting mechanisms where the mass can consistently be given to gluons in a gauge-invariant manner, a thorough overview of such mechanisms is found in Refs. [42, 43] and refs. therein. Thus, we see that the ghost condensation acts as the QCD vacuum. Therefore, in the ghost condensed phase the Lagrangian can effectively be given as follows

$$\mathcal{L}_{GC} = -\frac{1}{4}F^a_{\mu\nu}F^{\mu\nu a} - \frac{1}{2\zeta}(A^a_\mu A^{\mu a})^2 + M^2_a A^a_\mu A^{\mu a} \qquad (86.16)$$

Here, for the diagonal gluons $M^2_a = 0$, e.g., for $SU(3)$, $M^2_3 = M^2_8 = 0$. While for the off-diagonal gluons, $M^2_1 = +im^2_1$, $M^2_2 = -im^2_1$, $M^2_4 = +im^2_2$, $M^2_5 = -im^2_2$, $M^2_6 = +im^2_3$, $M^2_7 = -im^2_3$ (m^2_1, m^2_2, m^2_3 are positive real). So the gluons 1 and 2 can be considered as conjugate of each other. The same is true for other pairs. Hence for $SU(3)$, the last term of the effective Lagrangian in Eq. (86.16) would be

$$M^2_a A^a_\mu A^{\mu a} = +im^2_1 A^1_\mu A^{\mu 1} - im^2_1 A^2_\mu A^{\mu 2} + im^2_2 A^4_\mu A^{\mu 4} - im^2_2 A^5_\mu A^{\mu 5}$$
$$+ im^2_3 A^6_\mu A^{\mu 6} - im^2_3 A^7_\mu A^{\mu 7} \qquad (86.17)$$

Thus, we end our discussion on the quadratic gauge model. Having reviewed all the prerequisites, we are now in a position to move on to the outlined objective of the work.

86.4 PT Phase Transition in the Gauge Theory

There have been studies on the non-Hermitian extension of a gauge theory. However, the subject of PT phase transition has not been explored in gauge theories. Here, we show that the non-Abelian gauge theory of interest exhibits the PT phase transition. Since the discussion on PT symmetry becomes meaningful only in non-Hermitian systems, we first discuss the hermiticity property of two mentioned phases and demonstrate that they both are non-Hermitian.

86.4.1 Hermiticity of the Theory

The effective Lagrangian in the normal phase is given in Eq. (86.11)

$$\mathcal{L}_{\text{eff}} = -\frac{1}{4}F^a_{\mu\nu}F^{\mu\nu a} - \frac{1}{2\zeta}(A^a_\mu A^{\mu a})^2 - 2\overline{c^a}A^{\mu a}(D_\mu c)^a \qquad (86.18)$$

Now the hermiticity property of fields A^a_μ is well defined since they describe real degrees of freedom. Fields must be Hermitian in order to define the real degrees of freedom, i.e.,

$$A^{a\dagger}_\mu = A^a_\mu \qquad (86.19)$$

However, such is not the case for ghosts. Their Hermiticity remains unclear. Based on the following heuristic argument, we shall define this property for ghosts under which the present theory is cast as non-Hermitian model. As the operation of conjugation in principle transforms particle to its anti particle, the following is the natural choice of Hermiticity property for ghosts[1] [33]

$$c^{a\dagger} = \overline{c^a}$$
$$\overline{c^a}^\dagger = c^a \tag{86.20}$$

Under Eqs. (86.19),(86.20), the Lagrangian in the normal phase in Eq. (86.11) is not Hermitian since the ghost Lagrangian is not Hermitian as shown below

$$
\begin{aligned}
(\overline{c^a} A^{\mu a} (D_\mu c)^a)^\dagger &= (\overline{c^a} A^{\mu a} \partial_\mu c^a - g f^{abc} \overline{c^a} c^c A^{\mu a} A^b_\mu)^\dagger \\
&= -c^a A^{\mu a} \partial_\mu \overline{c^a} + g f^{abc} c^a \overline{c^c} A^{\mu a} A^b_\mu \\
&= -c^a A^{\mu a} (D_\mu \overline{c^a}) \neq \overline{c^a} A^{\mu a} (D_\mu c)^a.
\end{aligned}
\tag{86.21}
$$

The effective Lagrangian in the ghost condensed (confinement) phase (86.16) is also not Hermitian as the mass term for gluons is purely imaginary as explained. Important point here is to note that non hermiticity of the Lagrangian in this ghost condensed phase is free of the hermiticity convention for ghosts as they do not appear in this phase and thus the non hermiticity of the ghost condensed phase is profound. The Lagrangian (86.16) obeys the extended hermiticity [28], i.e., when the following inner automorphisms is applied hermiticity gets restored, viz. $\mathfrak{T}\mathcal{L}^\dagger_{GC}\mathfrak{T}^\dagger = \mathcal{L}_{GC}$,

$$
\begin{array}{llll}
\mathfrak{T}L_1\mathfrak{T}^\dagger = L_2 & \mathfrak{T}L_4\mathfrak{T}^\dagger = L_5 & \mathfrak{T}L_6\mathfrak{T}^\dagger = L_7 & \mathfrak{T}L_3\mathfrak{T}^\dagger = L_8 \\
\mathfrak{T}L_2\mathfrak{T}^\dagger = L_1 & \mathfrak{T}L_5\mathfrak{T}^\dagger = L_4 & \mathfrak{T}L_7\mathfrak{T}^\dagger = L_6 & \mathfrak{T}L_8\mathfrak{T}^\dagger = L_3
\end{array}
\tag{86.22}
$$

with the property

$$\mathfrak{T}^2 = \mathfrak{T}^{\dagger 2} = 1 \tag{86.23}$$

where L_i refers to any individual Lagrangian term such as $-\frac{1}{4}F^i_{\mu\nu}F^{\mu\nu i}$, $-\frac{1}{2\zeta}(A^i_\mu A^{\mu i})^2$, $im^2 A^i_\mu A^{\mu i}$ appearing in Eq. (86.16). The inner automorphism amounts to exchanging group indices as $1 \leftrightarrow 2, 4 \leftrightarrow 5, 6 \leftrightarrow 7, 3 \leftrightarrow 8$. In the adjoint representation it is given by

---

[1] In literature, at times unconventional Hermiticity property of ghost fields is invoked under which the effective theory in a given usual gauge can be reinterpreted as Hermitian theory. However, this is not natural and methodological way of treatment. (See Ref. [33]). The discussion on Hermiticity of Eq. (86.11) with natural Hermiticity Eq. (86.20) is general to all usual gauges such as Lorenz gauge.

$$\begin{bmatrix} 0 & 1 & 0 & 0 & 0 & 0 & 0 & 0 \\ 1 & 0 & 0 & 0 & 0 & 0 & 0 & 0 \\ 0 & 0 & 0 & 0 & 0 & 0 & 0 & 1 \\ 0 & 0 & 0 & 0 & 1 & 0 & 0 & 0 \\ 0 & 0 & 0 & 1 & 0 & 0 & 0 & 0 \\ 0 & 0 & 0 & 0 & 0 & 0 & 1 & 0 \\ 0 & 0 & 0 & 0 & 0 & 1 & 0 & 0 \\ 0 & 0 & 1 & 0 & 0 & 0 & 0 & 0 \end{bmatrix}$$

We have thus shown that both deconfined and confined phases are non-Hermitian, later being profoundly. There is no spontaneous breaking of Hermiticity and the system is consistently non Hermitian. Hence, it becomes interesting to discuss state of PT symmetry in this theory which we shall commence now.

86.4.2 PT Symmetry of the Theory

As in the case of the hermiticity, parity and time reversal properties of the gluons are well defined but not for ghosts. For gluons, parity is given as

$$A_i^a(\mathbf{x}, t) \xrightarrow{\text{P}} -A_i^a(-\mathbf{x}, t)$$
$$A_0^a(\mathbf{x}, t) \xrightarrow{\text{P}} A_0^a(-\mathbf{x}, t). \tag{86.24}$$

The rule for parity is same for all gluons as it is a linear operator. It is easy to see that Lagrangian in the normal phase (86.11) is invariant under parity if we choose ghosts to be scalars,

$$c^a(\mathbf{x}, t) \xrightarrow{\text{P}} c^a(-\mathbf{x}, t)$$
$$\overline{c^a}(\mathbf{x}, t) \xrightarrow{\text{P}} \overline{c^a}(-\mathbf{x}, t). \tag{86.25}$$

The ghosts being scalars under parity are consistent with the BRST transformations of fields under which the $\mathcal{L}_Q$ is invariant. Such convention is chosen in Ref. [44].

The case of the time reversal is not straight forward unlike parity as the time reversal is an anti-linear operation. Since some of the generators of $SU(N)$ are purely imaginary, the time reversal property is not same for all gluons. We shall explain it using $SU(3)$ group, further generalization to $SU(N)$ is obvious. In $SU(3)$, three generators namely, second, fifth, and seventh are purely imaginary. Therefore, time reversal for gluons is given by

$$A_i^P(\mathbf{x}, t) \xrightarrow{\text{T}} -A_i^P(\mathbf{x}, -t)$$
$$A_0^P(\mathbf{x}, t) \xrightarrow{\text{T}} A_0^P(\mathbf{x}, -t), \tag{86.26}$$

where index p is $1, 3, 4, 6, 8$ and,

$$A_i^q(\mathbf{x}, t) \xrightarrow{\text{T}} A_i^q(\mathbf{x}, -t)$$
$$A_0^q(\mathbf{x}, t) \xrightarrow{\text{T}} -A_0^q(\mathbf{x}, -t), \tag{86.27}$$

where index q is $2, 5, 7$. Therefore, the field strength with any spacetime and group indices can utmost change up to overall negative sign, i.e.,

$$F_{\mu\nu}^a \xrightarrow{\text{T}} \pm F_{\mu\nu}^a. \tag{86.28}$$

Thus, the action in the normal phase (86.11) is invariant under time reversal given that the time reversal property for ghosts is defined in the following manner:

$$c^p(\mathbf{x}, t) \xrightarrow{\text{T}} -c^p(\mathbf{x}, -t)$$
$$\overline{c^p}(\mathbf{x}, t) \xrightarrow{\text{T}} \overline{c^p}(\mathbf{x}, -t) \tag{86.29}$$

and,

$$c^q(\mathbf{x}, t) \xrightarrow{\text{T}} c^q(\mathbf{x}, -t)$$
$$\overline{c^q}(\mathbf{x}, t) \xrightarrow{\text{T}} \overline{c^q}(\mathbf{x}, -t) \tag{86.30}$$

where the description of indices p and q are as above. Anti-linearity makes two sets of ghosts transform in a completely different manner. Thus, the theory in normal phase is individually both parity and time reversal invariant. It can be shown that these time reversal conventions are the only choice which are consistent with BRST transformations. This PT symmetry breaks down spontaneously in the confined phase as we explain now.

The theory in the confined phase is given by Eq. (86.16),

$$\mathcal{L}_{GC} = -\frac{1}{4}F_{\mu\nu}^a F^{\mu\nu a} - \frac{1}{2\zeta}(A_\mu^a A^{\mu a})^2 + M_a^2 A_\mu^a A^{\mu a}$$

It is easy to check that parity (86.24) is still a symmetry. However, the time reversal is broken due to pure complex nature of the mass term,

$$\begin{aligned} M_a^2 A_\mu^a A^{\mu a} &= +im_1^2 A_\mu^1 A^{\mu 1} - im_1^2 A_\mu^2 A^{\mu 2} + im_2^2 A_\mu^4 A^{\mu 4} - im_2^2 A_\mu^5 A^{\mu 5} \\ &\quad + im_3^2 A_\mu^6 A^{\mu 6} - im_3^2 A_\mu^7 A^{\mu 7} \xrightarrow{\text{T}} \\ &\quad - im_1^2 A_\mu^1 A^{\mu 1} + im_1^2 A_\mu^2 A^{\mu 2} - im_2^2 A_\mu^4 A^{\mu 4} + im_2^2 A_\mu^5 A^{\mu 5} \\ &\quad - im_3^2 A_\mu^6 A^{\mu 6} + im_3^2 A_\mu^7 A^{\mu 7} \\ &= -M_a^2 A_\mu^a A^{\mu a} \end{aligned} \tag{86.31} \tag{86.32}$$

and also $PT\psi \neq \pm\psi$, where ψs are eigenfunctions of the mass matrix (86.14). The first two terms of $\mathcal{L}_{GC}$ remain unaffected by the time-reversal. Thus, PT symmetry is violated in this phase. We can see that the anti-symmetric nature of structure constant appearing in the mass matrix has led to this breaking. An important point again here is to note that the PT symmetry violation in the confined phase is profound as it is independent of the convention for ghosts. Therefore, the transition from the normal phase to the confinement phase with $SU(N)$ symmetric ghost condensates can be identified as PT phase transition from unbroken to broken PT phase. We note here that interestingly association between color confinement and spontaneous PT breaking is model and mechanism independent even though in this model the link is through ghost condensation since one prime signature of quark confinement, the pole of the propagator on purely imaginary p^2 axis, inevitably breaks PT symmetry. The usefulness of a consistent model such as one in this paper lies in that it gives valuable insight into a process through which the link can take place.

Conventionally, the order parameter is the one whose value tuning separates two phases, e.g.., in the toy model of Sec. II, tuning of $\eta \equiv \frac{2\mu^2}{|m_1^2 - m_2^2|}$ from less than 1 to greater than 1 separates PT unbroken and broken phase. We note the following regarding an order parameter in the present model. For the phase transition from deconfined phase which is PT unbroken to confined phase which PT broken, different ghost bilinears $\overline{c}^a c^c$ (a and c runs over 1 to $N^2 - 1$ independently) need to condense first and that also to a stated $SU(N)$ symmetric vacuum given by Eq. (86.13). This mechanism is quite similar to the Higgs mechanism in the electroweak theory where ground state of theory parameterized by the expectation value of the Higgs field spontaneously breaks the electroweak symmetry and thus in this sense, the expectation value of the Higgs field acts as the order parameter. In the same way, the present theory has the ground state parameterized by K of Eq. (86.13) which breaks the PT symmetry. Therefore, K provides the order parameter of the PT transition in this non-Abelian model. Thus, we have provided a gauge theory in which PT phase transition is explicitly shown for the first time.

86.4.3 C-Symmetry

In the PT symmetric non-Hermitian quantum mechanics, a C-symmetry (not the charge conjugation) is defined to improve the probabilistic interpretation of the PT-inner product and is inherent in all PT symmetric systems hence it becomes essential to find C-symmetry in the given model. We show that in the setup of quantum field theory in which we are working the inner automorphism provides the representation of this C-symmetry. So far, no explicit representation of the C-symmetry is known in the framework of gauge theories. This symmetry in quantum mechanics must satisfy the following three conditions:

$$[H, C]\psi = 0, \ [PT, C]\psi = 0, \ C^2 = \mathbf{1} \tag{86.33}$$

The inner automorphism satisfies QFT analogue of the conditions (86.33) as we explain now.

(1) The inner automorphism exchanges group indices i.e., $1 \leftrightarrow 2$, $4 \leftrightarrow 5$, $6 \leftrightarrow 7$, $3 \leftrightarrow 8$ and the Lagrangian of the initial unbroken PT theory in the normal phase contains sum over group index a. Hence, the Lagrangian and therefore Hamiltonian in this phase remain invariant under the inner automorphism. Thus, QFT analogue of the first of conditions (86.33) is obeyed.

(2) PT is a space-time symmetry and the inner automorphism is the operation in the group space. Therefore, it is easy to check that changing the order of inner automorphism and PT operations on Lagrangians of both the phases in Eqs. (86.11) and (86.16) does not alter the final result. In other words, they commute. This proves the QFT analogue of the second condition in Eq. (86.33).

(3) The third of Eq. (86.33) has already been shown. Therefore, we see that the inner automorphism forms an explicit representation of the C-symmetry, which in adjoint representation is given by the matrix (86.24).

It is clear that the theory in both the phases is invariant under CPT. In the broken PT phase, the theory also violates CP symmetry but preserves the CT, in complete analogy with the scalar model described in sec. II.

86.5   Conclusion

The main features of the non-self adjoint theories are encoded in the rich characteristics of the PT phase transition, hence it is extremely important to study the PT phase in PT symmetric non-self adjoint theories. Even though non-Hermitian extension of various models in quantum field theory has been studied, PT phase transition was not realized in the framework of a gauge theory. In the present work, we have demonstrated the PT phase transition by constructing an appropriate non-Hermitian but PT symmetric model of QCD in a recently introduced quadratic gauge which throws light on certain typical characteristics in non-perturbative sector. In this particular gauge, we have ghost fields condensation leading to confinement phase. We have shown the transition from deconfinement phase to confinement phase is a PT phase transition in this model of QCD. Ghost condensates result in PT symmetry breakdown. To have a fully consistent non-Hermitian quantum theory, it is important to explicitly find the C-symmetry which is inherent in all PT symmetric non-Hermitian systems. We have found the C-symmetry with its explicit representation in this model which is nothing but the inner automorphism. Thus, we give a new example where representation of C-symmetry in a gauge theory is constructed. Importantly, we note that there is a direct association between quark confinement and PT breaking which we bring to light through this model for the first time. It would be interesting to further study the relevance that the implications of the PT phase transition in this model may hold for the other areas of research.

Acknowledgements This work is partially supported by the Department of Science and Technology, Govt. of India under National Postdoctoral Fellowship scheme with File no. 'PDF/2017/000066'.

References

1. C. M. Bender and Boettcher S, Phys. Rev. Lett. 80, 5243 (1998)
2. C. M. Bender, Repts. Prog. Phys. 70, 947 (2007) and Refs. therein.;A. Mostafazadeh, Int. J. Geom. Meth. Mod. Phys. 7, 1191(2010) and references therein
3. M. Znojil J. Phys. A 36 (2003) 7825
4. B.P. Mandal, B.K. Mourya, K. Ali, A. Ghatak, Annals of Physics **363**, 185–193 (2015)
5. C. M. Bender, S. Boettcher and P. N. Meisinger : J. Math. Phys. 40 2201 (1999)
6. A. Khare, B.P. Mandal, Phys. Lett A **272**, 53 (2000)
7. B.P. Mandal, Mod. Phys. Lett. A **20**, 655 (2005)
8. B.P. Mandal, A. Ghatak, J. Phys. A: Math. Theor. **45**, 444022 (2012)
9. C.T. West, T. Kottos, T. Prosen, Phys. Rev. Lett. **104**, 054102 (2010)
10. A. Nanayakkara, Phys. Lett. A **304**, 67 (2002)
11. C.M. Bender, G.V. Dunne, P.N. Meisinger, M. Simsek. Phys. Lett. A **281**, 311–316 (2001)
12. B.P. Mandal, B.K. Mourya, R.K. Yadav, Physics Letters A **377**, 1043 (2013)
13. G. Levai, J. Phys. A **41**, 244015 (2008)
14. C. E. Ruter, K. G. Makris, R. El-Ganainy, D. N. Christodulides, M. Segev, and D. Kip: Nature (London) Phys. 6, 192 (2010)
15. I. Rotter and J.P. Bird, Rep. Prog. Phys. 78, 114001 (2015) and Refs. therein
16. W. Wan, Y. Chong, L. Ge, H. Noh, A.D. Stone, H. Cao, Science **331**, 889 (2011)
17. N. Liu, M. Mesch, T. Weiss, M. Hentschel, H. Giessen, Nano Lett. **10**, 2342 (2010)
18. H. Noh, Y. Chong, A. Douglas Stone, and Hui Cao, Phys. Rev. Lett. 108, 6805 (2011)
19. C.F. Gmachl, Nature **467**, 37 (2010)
20. A. Ghatak, R. D. Ray Mandal, B. P. Mandal, Ann. of Phys. 336, 540 (2013)
21. A. Ghatak, Md Hasan, B.P. Mandal, Phys. Lett. A **379**, 1326 (2015)
22. S. Longhi, J. Phys. A: Math. Theor. **44**, 485302 (2011)
23. R.K. Yadav, A. Khare, B. Bagchi, N. Kumari, B.P. Mandal, Journal of Mathematical Physics **57**, 062106 (2016)
24. A. Mostafazadeh, Phys. Rev. Lett. **102**, 220402 (2009)
25. Z.H. Musslimani, K.G. Makris, R. El-Ganainy, D.N. Christodoulides, Phys. Rev. Lett. **100**, 030402 (2008)
26. R. El-Ganainy, K.G. Makris, D.N. Christodoulides, Z.H. Musslimani, Opt. Lett. **32**, 2632 (2007)
27. A. Guo et al., Phys. Rev. Lett. **103**, 093902 (2009)
28. H. Raval and U. A. Yajnik, Phys. Rev. D 91, no. 8, 085028 (2015). H. Raval and U. A. Yajnik, Springer Proc. Phys. 174, 55 (2016)
29. Haresh Raval, Eur. Phys. J. C **76**, 243 (2016)
30. Jean Alexandre, Carl M. Bender, Peter Millington, JHEP **11**, 111 (2015)
31. Jean Alexandre, Peter Millington, Dries Seynaeve, Phys. Rev. D **96**, 065027 (2017)
32. G.'t Hooft, Nucl. Phys. B190, 455 (1981)
33. T. Kugo, I. Ojima, Prog. Theor. Phys. Suppl. **66**, 1 (1979)
34. Jean Alexandre, Peter Millington, Dries Seynaeve, arXiv:1710.01076v1 [hep-th]
35. C. Becchi, A. Rouet, R. Stora, Commun. Math. Phys. **42**, 127 (1975)
36. C. Becchi, A. Rouet, R. Stora, Annals Phys. **98**, 287 (1976)
37. B.P. Haresh Raval, Mandal. Eur. Phys. J. C **78**, 416 (2018)
38. Kei-Ichi Kondo, Toru Shinohara, Phys. Lett. B **491**, 263 (2000)
39. D. Dudal, H. Verschelde, J. Phys. A **36**, 8507–8516 (2003)

40. D. Dudal, J.A. Gracey, V.E.R. Lemes, M.S. Sarandy, R.F. Sobreiro, S.P. Sorella, H. Verschelde, Phys. Rev. D **70**, 114038 (2004)
41. C.D. Roberts, A.G. Williams, G. Krein, Int. J. Mod. Phys. A **07**, 5607 (1992)
42. A.C. Aguilar, D. Binosi, J. Papavassiliou,. arXiv:1511.08361 [hep-ph]
43. Arlene C. Aguilar, Joannis Papavassiliou, JHEP **12**, 012 (2006)
44. O.M. Cima, Phys. Lett. B **750**, 1–5 (2015)

Chapter 87
Bottomonium Suppression in p−Pb and Pb−Pb Collisions at the CERN Large Hadron Collider: Centrality and Transverse Momentum Dependence

Captain R. Singh, S. Ganesh, and M. Mishra

Abstract The deconfined state of QCD matter in heavy-ion collisions has been a topic of paramount interest for many years. Quarkonium suppression in heavy-ion collisions at the Relativistic Heavy-Ion Collider (RHIC) and Large Hadron Collider (LHC) experiment indicates the formation of quark-gluon plasma (QGP) in heavy-ion collision experiments. Experiments at LHC have given indications of deconfined QCD matter effect in asymmetric p−Pb nuclear collisions. Here, we use a theoretical model to investigate the bottomonium suppression in Pb−Pb, 2.76 TeV, 5.02 TeV and in p−Pb at 5.02 TeV center-of-mass energies under a QGP formation scenario. Our current formulation is based on a unified model consisting of suppression due to color screening, gluonic dissociation along with collisional damping. Regeneration due to correlated quark and antiquark pairs has also been taken into account in the current work. We obtain here the net bottomonium suppression in terms of survival probability under the combined effect of suppression plus regeneration in the deconfined QGP medium. We concentrate here on the centrality, number of participants and transverse momentum, and p_T dependence of bottomonium 1S and 2S state suppression in Pb−Pb and p−Pb collisions at mid-rapidity. We compare our model predictions for bottomonium 1S and 2S suppression to the corresponding experimental data obtained from the LHC. We find that the experimental observations on transverse momentum p_T- and centrality N_{PART}-dependent suppression agree reasonably well with our model predictions.

87.1 Introduction

There are various QGP effects which affect the quarkonium yields [1–3]. Until the mid-2000s, color screening (also known as Debye color screening) was thought to be the only possible suppression mechanism for quarkonia in a QGP medium. However,

C. R. Singh (✉) · S. Ganesh · M. Mishra
Department of Physics, Birla Institute of Technology and Science, Pilani 333031, India
e-mail: captainriturajsingh@gmail.com

M. Mishra
e-mail: madhukar.12@gmail.com

© Springer Nature Singapore Pte Ltd. 2021
P. K. Behera et al. (eds.), *XXIII DAE High Energy Physics Symposium*,
Springer Proceedings in Physics 261,
https://doi.org/10.1007/978-981-33-4408-2_87

experimental results involve some puzzling features which defy explanations based on color screening alone. The other suppression mechanisms like, gluonic dissociation and collisional damping may breakup the bound state of quarkonium states. However, there are some other mechanisms which can mimic the QGP effects up to a certain extent without the actual QGP effects. To separate out these effects from the QGP, it is required to study the smaller collision systems like p-p, p-A, d-A, etc. Thus a comprehensive study of QGP formation in A-A collisions includes p-p and p-A collision systems at the same or nearby energies. In such small collision systems, the formation of QGP was considered very unlikely. But after recent observation on collectivity in small systems and a few other results, it has now become the most debated/discussed topic of current interest. If we assume that there is no QGP in small collision systems, then what are the factors which suppress the quarkonia production? This modification in the yield is explained as the initially produced quarkonia that get strongly affected by the surrounding nuclear environment. This environment is considered as a hadronic fireball which contains an admixture of hadrons, like nucleons and mesons. The environmental effects on the particles productions in hard processes or soft processes are described in terms of the "cold nuclear matter (CNM) effects". The most significant CNM effect is the shadowing effect which comes into the picture due to the modification of parton distribution functions in the nuclear medium. Since a nucleus may contain some other hadrons along with protons and neutrons, the deviation in the nucleus isospin due to the presence of these hadrons affects all the perturbative QCD calculations. In the present work, we considered the combined effect of QGP and non-QGP effects to obtain the bottomonium suppression in p-A at $\sqrt{s_{NN}} = 5.02$ TeV and A-A collisions at $\sqrt{s_{NN}} = 2.76, 5.02$ TeV collision energies. We named this framework as Unified Model of Quarkonia Suppression, "UMQS".More details about the model can be found in Refs. [4, 5].

87.2 UMQS Model Formulation

In our model, we start with realistic in medium lattice−based potential extractions from dynamical lattice QCD. Using this potential, we investigated the bottomonium decay width in the QGP medium. This static potential between two heavy quarks placed in a QCD medium consists of two parts. The first part represents the standard time-independent Debye screened potential. This is used to obtain the bottomonium suppression due to color screening.

87.2.1 Shadowing Effect and Color Screening

Color Screening: In color screening, free flowing partons in the QGP medium screen the color charges present between the $q\bar{q}$ bound states which leads to the dissociation of bound states, or prevents to form bound states. As mentioned earlier, this screening of color charges in QGP is analogous to the screening of electric charges in the ordinary QED plasma. The color screening mechanism strongly depends on the dissociation temperature (T_D) of the quarkonia. For $\Upsilon(1S)$, $T_D \approx 700$ MeV while for $\Upsilon(2S)$, it is ≈ 220 MeV. So if the temperature of the QGP medium is the less than the dissociation temperature ($T < T_D$) of any quarkonium species, then that state will not be dissociated due to the color screening effect. The Survival of bottomonium around screening radius (r) is obtained in the form of survival probability:

$$S_c(p_T, N_{part}) = \frac{2(\alpha + 1)}{\pi R_T^2} \int_0^{R_T} dr\, r\, \phi_{max}(r) \left\{ 1 - \frac{r^2}{R_T^2} \right\}^{\alpha}, \qquad (87.1)$$

where $\alpha = 0.5$ and R_T is the transverse radius of the QGP. In our calculation, we have found that the color screening effect for $\Upsilon(1S)$ state is negligible at $\sqrt{s_{NN}} = 2.76, \& 5.02$ TeV collision energies, due to its high dissociation temperature (T_D), while a significant color screening effect on $\Upsilon(2S)$ can be seen in all the above mentioned collision energies, as shown in Fig. 87.1a.

Shadowing Effect: Being a cold nuclear matter effect, it suppresses the initial bottomonium production in heavy-ion collisions. It firmly depends on the center of mass collision energy of the colliding ions [6]. The shadowing effect arises due to the difference in the parton distribution between the nucleon present in the nucleus and out of the nucleus. Suppression due to shadowing is plotted in Fig. 87.2b for the various collision systems at different center-of-mass collision energies ($\sqrt{s_{NN}}$). Figure 87.2b clearly depicts the dominance of $\sqrt{s_{NN}}$ in the shadowing effect, as for higher values of $\sqrt{s_{NN}}$ shadowing increases the suppression.

We use the EPS09 parametrization to obtain the shadowing $S^i(A, x, \mu)$ for nucleus with mass A, momentum fraction x and scale μ. Thus, the shadowing factor is

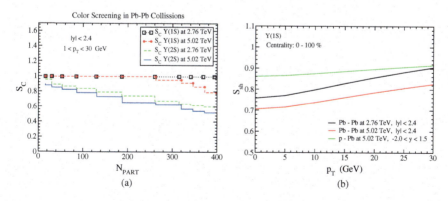

(a)   (b)

Fig. 87.1 **a** Color screening for $\Upsilon(1S)$ and $\Upsilon(2S)$ versus N_{PART} in Pb−Pb Collisions at LHC energies, $\sqrt{s_{NN}} = 2.76, 5.02$ TeV. **b** Variation of $\Upsilon(1S)$ shadowing factor ($S_s h$) versus transverse momentum (p_T) plotted corresponding to central rapidity region for Pb−Pb at $\sqrt{s_{NN}} = 2.76, 5.02$ TeV and p−Pb at $\sqrt{s_{NN}} = 5.02$

obtained as follows:

$$S_p^i(A, x, \mu, \mathbf{r}) = 1 + N_\rho(S^i(A, x, \mu) - 1)\frac{\int dz\rho_A(\mathbf{r}, z)}{\int dz\rho_A(0, z)} \tag{87.2}$$

where N_ρ is determined by the following normalization condition:
$\frac{1}{A}\int d^2r dz\rho_A(s)S_p^i(A, x, \mu, \mathbf{r}) = S^i(A, x, \mu)$. The suppression due to the shadowing effect is thus determined by

$$S_{sh} = \frac{dN_{AB}/dy}{T_{AB}(b)d\sigma_{pp}/dy}. \tag{87.3}$$

87.2.2 Collisional Damping, Gluonic Excitation and Gluonic De-Excitation

Collisional Damping: Collisional damping is the dissociation of the bound state of quark and antiquark pairs due to the exchange of low frequency fields between $q - \bar{q}$ pairs. Hence, we find the associated decay width, which is obtained by taking the expectation value of the imaginary part of effective quark-antiquark potential in QGP [7]:

$$\Gamma_{damp,nl} = \int g_{nl}(r)^\dagger Im(V)g_{nl}(r)dr, \tag{87.4}$$

where $g_{nl}(r)$ is the singlet wave function.

Gluonic Excitation: The gluons absorbed by singlet state trigger a transition from singlet to octet state. As the octet state is a colorful state, it represent a dissociated state of $b - \bar{b}$ pair. So the basic principle behind the suppression of a meson due to gluonic dissociation is that, there is an excitation of the singlet state to the octet state. The cross section and other details about this process are given in Ref. [5]. We have taken the thermal average of gluonic cross section over a modified Bose–Einstein distribution to obtain the decay width $\Gamma_{gd,nl}$ corresponding to the gluonic dissociation, given as

$$\Gamma_{gd,nl}(\tau, p_T, b) = \frac{g_d}{4\pi^2}\int_0^\infty \int_0^\pi \frac{dp_g \, d\theta \, \sin\theta \, p_g^2\sigma_{d,nl}(E_g)}{e^{\{\frac{\gamma E_g}{T_{eff}}(1+v_T\cos\theta)\}} - 1}, \quad ; \quad g_d = 16 \tag{87.5}$$

where $\sigma_{diss,nl}$ is the gluonic dissociation cross section.

Finally, we calculated the combined effect of $\Gamma_{gd,nl}$ and $\Gamma_{damp,nl}$ by unifying them in a single entity Γ_D, given as $\Gamma_D = \Gamma_{damp} + \Gamma_{gd,nl}$.

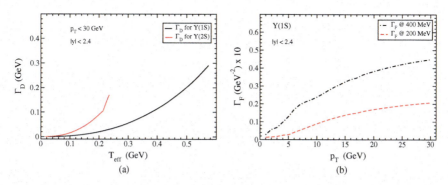

Fig. 87.2 **a** Variation of $\Upsilon(1S)$ and $\Upsilon(2S)$ total decay width along with its components, i.e., gluonic dissociation and collisional damping versus effective temperature. **b** Variation of $\Upsilon(1S)$ recombination factor (Γ_F) versus transverse momentum (p_T) plotted for $T_{eff} = 200$ MeV and 400 MeV

In Fig. 87.2a, the net effect of gluonic dissociation and collisional damping is plotted for $\Upsilon(1S)$ and $\Upsilon(2S)$ states. The total decay width for $\Upsilon(1S)$ is a monotonically increasing function of effective temperature as shown in Fig. 87.2a, but a non-monotonic behavior is observed for $\Upsilon(2S)$ as shown in the same figure. For $\Upsilon(2S)$, boost in Γ_D around $T_{eff} \approx 200$ MeV is due to the Debye mass (M_D) which is also a function of T_{eff}. The Debye mass initiates the sequential melting of $\Upsilon(2S)$ near its dissociation temperature and dissociates it completely at $T_{eff} > T_D$.

Gluonic De-excitation: There is another process which works against the suppression mechanisms. We named it regeneration due to correlated $b - \bar{b}$ pairs. In this mechanism, the particle which is already in the octet state can make the transition to the color singlet state via emitting a gluon. Therefore, We have considered the possibility of regeneration of bottomonia due to de-excitation of octet to singlet state. We calculated the recombination cross section using the detailed balance from gluonic dissociation cross section, given as

$$\sigma_{f,nl} = \frac{48}{36}\sigma_{diss,nl} \frac{(s - M_{nl}^2)^2}{s(s - 4\,m_b^2)} \tag{87.6}$$

where s is the Mandelstam variable, M_{nl}, and m_b are the masses of bottomonium states and bottom quark, respectively. Now define the recombination factor $\Gamma_{F,nl}$ as the thermal average of the product of the above cross section and relative velocity between $b - \bar{b}$ [5].

87.2.3 Survival Probability

The net production of Υs includes hot and cold nuclear matter effects. We combined all the effects in the UMQS model and finally obtained the survival probability. The survival probability due to shadowing, gluonic dissociation along with collisional damping and recombination is defined by $S_{gd}^{\Upsilon}(p_T, b)$. So far color screening is not included because it works independent of all other mechanisms. So, the net yield is obtained after including the survival probability (S_c) of bottomonium due to color screening:

$$S_P(p_T, b) = S_{gd}^{\Upsilon}(p_T, b)\, S_c^{\Upsilon}(p_T, b). \qquad (87.7)$$

Feed-down: The higher resonances of bottomonia may decay into their respective lower states, so the feed-down of the higher resonances into lower states becomes important to include in the final calculation of survival probability.

Feed-down for $\Upsilon(1S)$ is obtained by considering that $\sim 68\%$ of $\Upsilon(1S)$ come up by direct production whereas $\sim 17\%$ is from the decay of $\chi_b(1P)$ and $\sim 9\%$ is from the decay of $\Upsilon(2S)$. The feed-down of $\chi_b(2P)$ and $\Upsilon(3S)$ into $\Upsilon(1S)$ is taken as $\sim 5\%$ and $\sim 1\%$, respectively. Similarly, the feed-down of $\chi_b(2P)$ and $\Upsilon(3S)$ into $\Upsilon(2S)$ effectively suppress its production. From the feed-down fractions for $\Upsilon(2S)$, we have considered that $\sim 65\%$ of $\Upsilon(2S)$ come up by direct production whereas $\sim 30\%$ from the decay of $\chi_b(2P)$ and $\sim 5\%$ is from the decay of $\Upsilon(3S)$.

87.3 Results and Discussion

In this section, we have compared our model predictions for bottomonium suppression with the corresponding experimental results obtained at mid-rapidity at LHC energies. Our UMQS model determines the p_T and centrality-dependent survival probability of quarkonium states at mid-rapidity in Pb–Pb collisions at $\sqrt{s_{NN}} = 2.76$ and $5.02\ TeV$ and in p–Pb collision at $\sqrt{s_{NN}} = 5.02$ TeV [8–11].

87.3.1 Transverse Momentum-Dependent Suppression

The bottomonium transverse momentum (p_T)-dependent nuclear modification factor R_{AA} data sets are available corresponding to minimum bias ($0 - 100\%$ centrality). Therefore, we have calculated the p_T-dependent survival probability (S_P) at minimum bias via taking the weighted average over all centrality bins and compared with the corresponding R_{AA} data.

Figure 87.3a suggests that $\Upsilon(1S)$ suppression is a slowly varying function of transverse momentum p_T in comparison with $\Upsilon(2S)$ in the QGP medium. Figure 87.3b

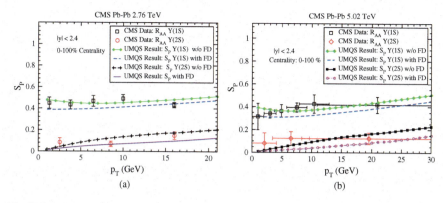

Fig. 87.3 **a** Survival probability of $\Upsilon(nS)$ versus p_T is compared with $\Upsilon(nS)$ nuclear modification factor R_{AA} [9] in Pb−Pb collisions at $\sqrt{s_{NN}} = 2.76$ TeV. **b** Survival probability of $\Upsilon(nS)$ versus p_T is compared with $\Upsilon(nS)$ nuclear modification factor R_{AA} [10] in Pb−Pb collisions at $\sqrt{s_{NN}} = 5.02$ TeV

depicts the suppression for Pb−Pb collision at $\sqrt{s_{NN}} = 5.02$ TeV else it is very similar to what is shown in Fig. 87.3a.

Double ratio plotted in Fig. 87.4a represents the production of $\Upsilon(2S)$ over $\Upsilon(1S)$ and quantifies the medium effects since the shadowing effect is almost the same for all bottomonium states. Thus suppression in yield ratio is purely due to the QGP medium effect. It is clear from Fig. 87.4a that except for the first data point, our calculated p_T variation agrees well with the measured double ratio of bottomonium states. In Fig. 87.4b, we have plotted our model predictions in terms of survival probability of $\Upsilon(1S)$ and $\Upsilon(2S)$ versus p_T along with a small suppression in $\Upsilon(1S)$ at low p_T and a bit enhancement or almost no suppression at high p_T observed in the central rapidity region in $p - Pb$ collision at 5.02 TeV energy. Our model calculation showing small suppression of $\Upsilon(1S)$ at low p_T which decreases at high p_T is consistent with the observed suppression data. The indirect $\Upsilon(2S)$ suppression in terms of the double ratio is plotted in Fig. 87.5b. The comparison of calculated yield ratio and the measured double ratio in Fig. 87.5b clearly supports our prediction of $\Upsilon(2S)$ suppression in p−Pb collisions at $\sqrt{s_{NN}} = 5.02$ TeV.

87.3.2 Centrality-Dependent Suppression

In our model calculations, we have used a number of participants N_{PART} to relate the centrality of collisions to the measured relative yield in terms of R_{AA}. We obtained the centrality (N_{PART})-dependent survival probability for $\Upsilon(1S)$ and $\Upsilon(2S)$ by averaging over p_T.

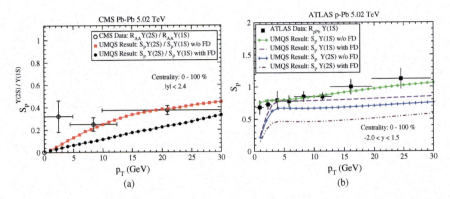

Fig. 87.4 a The predicted yield ratio of $\Upsilon(2S)$ to $\Upsilon(1S)$ is compared with the observed double ratio, $\Upsilon(2S)$ to $\Upsilon(1S)$ in $Pb - Pb$ collision [10] at 5.02 TeV LHC energy. **b** Survival probability of $\Upsilon(1S)$ versus p_T is compared with $\Upsilon(1S)$ nuclear modification factor R_{AA} [12] in p$-$Pb collisions at $\sqrt{s_{NN}} = 5.02$ TeV. S_P of $\Upsilon(2S)$ is predicted for the same collision system

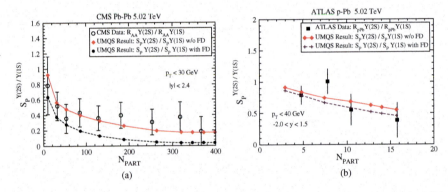

Fig. 87.5 a The centrality variation of our calculated yield ratio of $\Upsilon(2S)$ to $\Upsilon(1S)$ is compared with the measured double ratio as a function of centrality in Pb$-$Pb collisions obtained from the CMS experiment at $\sqrt{s_{NN}} = 5.02$ TeV [11]. **b** The centrality variation of our calculated yield ratio of $\Upsilon(2S)$ to $\Upsilon(1S)$ is compared with the measured double ratio as a function of centrality in p$-$Pb collisions obtained from the ATLAS experiment at $\sqrt{s_{NN}} = 5.02$ TeV [12]

The yield ratio of $\Upsilon(2S)$ to $\Upsilon(1S)$ is compared with double ratio as plotted in Fig. 87.5a; it is consistent with our model prediction for $\Upsilon(1S)$ and $\Upsilon(2S)$ suppression in Pb$-$Pb collision at 5.02 TeV LHC energy. A significant effect of feed-down is seen at $\sqrt{s_{NN}} = 5.02$ TeV energy over the most peripheral to the most central collision. After taking the feed-down, our predicted results for $\Upsilon(1S)$ yield is showing good agreement with the data, while it predicts oversuppression for $\Upsilon(2S)$ at the mid-central region.

87.4 Conclusion

Outcomes of our model show that bottomonium suppression is the combined effect of hot and cold nuclear matter. The color screening effect is almost insignificant to suppress the $\Upsilon(1S)$ production while it significantly suppressed $\Upsilon(2S)$ production in Pb−Pb and p−Pb collisions at all the LHC energies. The gluonic dissociation along with the collisional damping mechanisms plays an important role in $\Upsilon(1S)$ and $\Upsilon(2S)$ in Pb−Pb and p−Pb collisions. Our model suggests an effective regeneration of $\Upsilon(1S)$ in Pb−Pb at $\sqrt{s_{NN}} = 2.76$ & 5.02 TeV. QGP formation in p−Pb collision clearly explained by $\Upsilon(1S)$ suppression is around unity with large uncertainty, and no direct experimental results are available for $\Upsilon(2S)$ suppression. However, indirect experimental information of $\Upsilon(2S)$ suppression is available in the form of a double ratio. Our model predicted the $\Upsilon(2S)$ suppression in p−Pb collisions.

Acknowledgements M. Mishra is grateful to the Department of Science and Technology (DST), New Delhi, for the financial assistance. Captain R. Singh acknowledges BITS−Pilani, Pilani, for the financial support.

References

1. T. Matsui, H. Satz, Physics Letters B **178**, 416 (1986)
2. F. Nendzig, G. Wolschin, Phys. Rev. C **87**, 024911 (2013)
3. N. Brambilla, M.A. Escobedo, J. Soto, A. Vairo, Physical Review D **96**, 034021 (2017)
4. C.R. Singh, P. Srivastava, S. Ganesh, M. Mishra, Physical Review C **92**, 034916 (2015)
5. C.R. Singh, S. Ganesh, M. Mishra, The European Physical Journal C **79**, 147 (2019)
6. R. Vogt, Phys. Rev. C **81**, 044903 (2010)
7. M. Laine, O. Philipsen, M. Tassler, P. Romatschke, Journal of High Energy Physics **2007**, 054 (2007)
8. C.M.S. Collaboration, S. Chatrchyan et al., Phys. Rev. Lett. **109**, 222301 (2012)
9. C.M.S. Collaboration, V. Khachatryan et al., Physics Letters B **770**, 357 (2017)
10. C.M.S. Collaboration, A.M. Sirunyan et al., Physics Letters B **790**, 270 (2019)
11. C.M.S. Collaboration, A.M. Sirunyan et al., Phys. Rev. Lett. **120**, 142301 (2018)
12. M. Aaboud et al., The European Physical Journal C **78**, 171 (2018)

Chapter 88
Nonlinear Effects in Singlet Quark Distribution Predicted by GLR-MQ Evolution Equation

Mayuri Devee

88.1 Introduction

The DGLAP evolution equation at the twist-2 level predicts a sharp growth of the gluon densities as x grows smaller which is clearly observed in the DIS experiments at HERA as well. It is quite obvious from the perturbative QCD calculations that the sea quark distributions in a hadron evolve rapidly with $\ln(1/x)$ at fixed Q^2 in the same manner as the gluon distribution. However, in the region of very small x, the individual gluons necessarily start to overlap or recombine with each other. Therefore the sharp growth of the sea quark distribution is expected to slow down in order to restore the Froissart bound [1] on physical cross sections. Accordingly, the DGLAP equation was modified by incorporating correlations among the initial partons by Gribov-Levin-Ryskin (GLR) and Mueller-Qiu (MQ) in their pioneering works [2, 3] at the twist-4 level.

The solution of the GLR-MQ equation is important to examine the effect of non-linear corrections due to the gluon-gluon recombination at small x. In our previous works [4, 5], (related works: Refs. [6, 7]), we solved the GLR-MQ equation to study the Q^2 and x dependence of the gluon distribution function using Regge-like behaviour of gluon distribution. In the present work, we have solved the GLR-MQ equation for sea quark distribution using the same method in leading twist approximation and investigate the effect of nonlinear corrections on the small x and Q^2 behaviour of singlet structure function, $F_2^S(x, Q^2)$. Our semi-analytical results are compared with NMC [8] and E665 [9] experimental data as well as with the NNPDF collaboration [10]. Moreover, we perform a comparative analysis of our predictions of $F_2^S(x, Q^2)$ obtained from the nonlinear GLR-MQ equation with those obtained from the linear DGLAP equation.

M. Devee (✉)

Department of Physics, University of Science and Technology Meghalaya, Ri-Bhoi, Baridua 793101, India
e-mail: deveemayuri@gmail.com

© Springer Nature Singapore Pte Ltd. 2021
P. K. Behera et al. (eds.), *XXIII DAE High Energy Physics Symposium*,
Springer Proceedings in Physics 261,
https://doi.org/10.1007/978-981-33-4408-2_88

88.2 Formalism

88.2.1 General Framework

The nonlinear corrections arising from the recombination of two gluon ladders into
one gluon or a $q\bar{q}$ pair modify the evolution equations of singlet structure function
as [11]

$$\frac{\partial F_2^S(x, Q^2)}{\partial \ln Q^2} = \frac{5}{18} \frac{\partial F_2^S(x, Q^2)}{\partial \ln Q^2}\bigg|_{DGLAP} - \frac{5}{18} \frac{27}{160} \frac{\alpha_s^2(Q^2)}{R^2 Q^2} G^2(x, Q^2), \quad (88.1)$$

which is known as the GLR-MQ equation. The first term on the right-hand side is
given by the standard linear DGLAP equation whereas the term quadratic in G is the
result of gluon recombination. The negative sign in front of the nonlinear term tames
the strong growth of singlet quark distribution generated by the linear term at very
small x. The size of the nonlinear term crucially depends on the correlation radius
R between two interacting gluons.

The first term of Eq. (88.1) in the leading twist approximation is given by [12]

$$\frac{\partial F_2^S(x, Q^2)}{\partial \ln Q^2}\bigg|_{DGLAP} = \frac{\alpha_s(Q^2)}{2\pi}\left[\frac{2}{3}\Big(3 + 4\ln(1-x)\Big)F_2^S(x, Q^2)\right.$$

$$+ \frac{4}{3}\int_x^1 \frac{dw}{1-w}\left\{(1+w^2)F_2^S\Big(\frac{x}{w}, Q^2\Big) - 2F_2^S(x, Q^2)\right\}$$

$$\left. + N_F \int_x^1 \Big(w^2 + (1-w)^2\Big)G\Big(\frac{x}{w}, Q^2\Big)dw\right]. \quad (88.2)$$

Here, $\alpha_s(Q^2) = \frac{4\pi}{\beta_0 \ln(Q^2/\Lambda^2)}$ with $\beta_0 = 11 - \frac{2}{3}N_f$ and N_f being the number of active
quark flavours.

88.2.2 Solution of GLR-MQ Equation for Singlet Structure Function and Effects of Gluon Recombination

To solve the GLR-MQ equation, we have taken into account a simple form of Regge-
like behaviour of singlet structure function as

$$F_2^S(x, Q^2) = f(Q^2)x^{-\lambda_S}, \quad (88.3)$$

where $f(Q^2)$ is a function of Q^2 and λ_S is the Regge intercept for the singlet structure
function.

Now by employing the Regge ansatz of Eq. (88.3) in Eq. (88.1), we arrive at

$$\frac{\partial F_2^S(x, Q^2)}{\partial Q^2} = A_1(x)\frac{F_2^S(x, Q^2)}{\ln(Q^2/\Lambda^2)} - A_2(x)\frac{[F_2^S(x, Q^2)]^2}{Q^2\ln(Q^2/\Lambda^2)}, \tag{88.4}$$

where $A_1(x)$ and $A_2(x)$ are some functions of x. Here, we have assumed the relation $G(x, Q^2) = k(x)F_2^S(x, Q^2)$, with the ad hoc parameter $k(x)$ to be determined from phenomenological analysis. This assumption is justifiable as the evolution equations of gluon and singlet structure function are in the same forms of the derivative with respect to Q^2.

Equation (88.4) is a partial differential equation for the singlet structure function $F_2^S(x, Q^2)$ with respect to the variables x and Q^2. It can be solved to obtain the Q^2 and x evolutions of the nonlinear singlet structure function as

$$F_2^S(x, t) = \frac{t^{A_1(x)} F_2^S(x, t_0)}{t_0^{A_1(x)} + A_2(x)\left[\int t^{A_1(x)-2}e^{-t}dt - \int t_0^{A_1(x)-2}e^{-t_0}dt_0\right]F_2^S(x, t_0)} \tag{88.5}$$

and

$$F_2^S(x, t) = \frac{t^{A_1(x)} F_2^S(x_0, t)}{t^{A_1(x_0)} + \left[A_2(x)\int t^{A_1(x)-2}e^{-t}dt - A_2(x_0)\int t^{A_1(x_0)-2}e^{-t}dt\right]F_2^S(x_0, t)}, \tag{88.6}$$

respectively. Here, we have used the variable $t = \ln(\frac{Q^2}{\Lambda^2})$. The input functions $F_2^S(x, t_0)$ and $F_2^S(x_0, t)$ are defined as $F_2^S(x, t) = F_2^S(x, t_0)$ at $t = t_0$ where $t_0 = \ln\left(\frac{Q_0^2}{\Lambda^2}\right)$ for some lower value of $Q^2 = Q_0^2$ and $F_2^S(x, t) = F_2^S(x_0, t)$, at some high $x = x_0$.

Again the linear DGLAP equation at the leading order defined by Eq. (88.2) can be solved using the Regge ansatz defined by Eq. (88.3) as

$$F_2^S(x, Q^2) = f_{10}t^{A_1(x)-A_1(x_0)} \tag{88.7}$$

for fixed Q^2. The input distributions f_{10} have to be chosen from the initial boundary conditions.

Now to estimate the effect of nonlinear corrections as a consequence of gluon recombination in our predictions of the singlet structure function at small x, we calculate the ratio of Eqs. (88.6) and (88.7) given by

$$R_{F_2^S} = \frac{F_2^{S\,GLR-MQ}(x, t)}{F_2^{S\,DGLAP}(x, t)}, \tag{88.8}$$

as a function of variable x for different values of Q^2.

88.3 Result and Discussion

The effects of nonlinear corrections to the evolution of the singlet structure function, $F_2^S(x, Q^2)$, are examined at small x by solving the GLR-MQ evolution equation.

We have performed our analysis in the kinematic region $0.6 \leq Q^2 \leq 25$ GeV^2 and $10^{-4} \leq x \leq 10^{-1}$ where the suggested solution is found to be legitimate. In Fig. 88.1, we plot the Q^2 dependence of $F_2^S(x, Q^2)$ computed from Eq.(88.5) for fixed $x = 0.008$, whereas Fig. 88.2 represents the small-x behaviour of $F_2^S(x, Q^2)$ computed from Eq.(88.6) at fixed values of $Q^2 = 1.25$, 1.094 and 4.03 GeV^2 for both $R = 2$ GeV^{-1} and $R = 5$ GeV^{-1}, respectively. The consistency of our results are examined with the NMC, E665 and NNPDF data. The vertical error bars represent the total combined statistical and systematic uncertainties of the experimental data. It is observed that the obtained results of $F_2^S(x, Q^2)$ increase with increasing Q^2

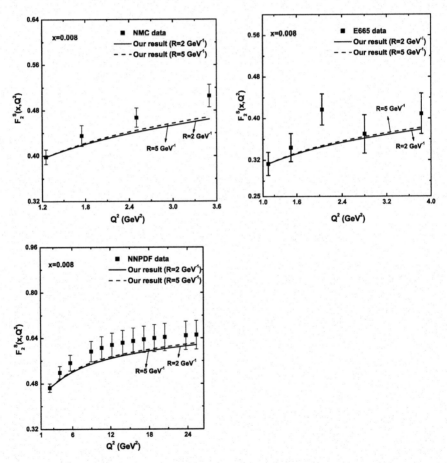

Fig. 88.1 Q^2 dependence of singlet structure function with nonlinear corrections

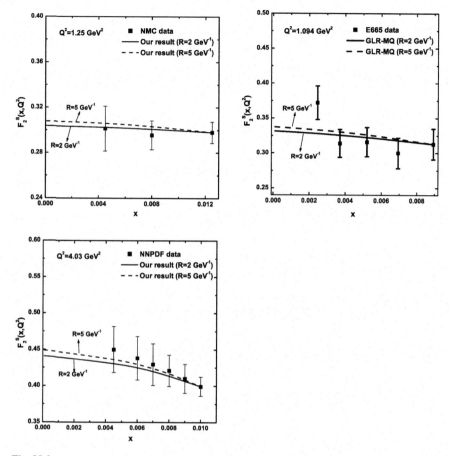

Fig. 88.2 x dependence of singlet structure function with nonlinear corrections

and decreasing x, but this attitude is tamed with respect to the nonlinear terms in the GLR-MQ equation. The effect of nonlinearity is observed to be more at $R = 2$ GeV^{-1} than at $R = 5$ GeV^{-1}.

In Fig. 88.3, the ratio $R_{F_2^S}$ defined in Eq.(88.8) is plotted against the variable x in the range $10^{-4} \leq x \leq 10^{-2}$ for three representative values Q^2. It is observed that, towards smaller values of x and Q^2, the GLR-MQ/DGLAP ratio for $F_2^S(x, Q^2)$ decreases which implies that the effect of nonlinearity increases towards small x.

Fig. 88.3 The
GLR-MQ/DGLAP ratio for
singlet structure function

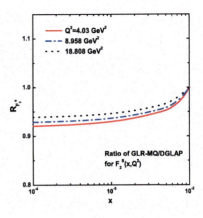

88.4 Summary

To summarize, the nonlinear GLR-MQ equation for $F_2^S(x, Q^2)$ is solved in the leading order and the effect of nonlinear corrections due to the gluon recombination
processes has been investigated at small x and Q^2. The suggested solution is found
to be valid in the kinematic domain $10^{-4} \leq x \leq 10^{-1}$ and $0.6 \leq Q^2 \leq 25$ GeV^2,
where the gluon recombination processes play an important role on the QCD evolution of $F_2^S(x, Q^2)$. Our results reveal the general trend of experimental data and
parametrization, however, with the inclusion of the nonlinear terms, this behaviour
of the singlet structure function is slowed down towards small x. The effect of nonlinear corrections become significant at the hot spot with $R = 2$ GeV^{-1} when the
gluons and the sea quarks are assumed to be condensed in a small region within the
proton. Moreover, the predictions of the GLR-MQ/DGLAP ratio also indicate that
nonlinearity increases towards smaller values of x and Q^2.

References

1. M. Froissart, Asymptotic behavior and subtractions in the Mandelstam representation. Phys.
   Rev. **123**, 1053–1057 (1961)
2. L.N. Gribov, E.M. Levin, M.G. Ryskin, Semihard processes in QCD. Phys. Rep. **100**, 1–150
   (1983)
3. A.H. Mueller, Small-x behavior and parton saturation: A QCD model. Nucl. Phys. B **335**,
   115–137 (1990)
4. M. Devee, J.K. Sarma, Nonlinear GLR-MQ evolution equation and Q^2-evolution of gluon
   distribution function. Eur. Phys. J. C **74**, 2751 (2014)
5. M. Devee, J.K. Sarma, Analysis of the small-x behavior of gluon distribution and a search for
   gluon recombination. Nucl. Phys. B **885**, 571–582 (2014)
6. P. Phukan, M. Lalung, J.K. Sarma, NNLO solution of nonlinear GLR-MQ evolution equation
   to determine gluon distribution function using Regge like ansatz. Nucl. Phys. A **968**, 275–286
   (2017)

7. M. Lalung, P. Phukan, J.K. Sarma, On phenomenological study of the solution of nonlinear GLR-MQ evolution equation beyond leading order. Nucl. Phys. A **984**, 29–43 (2019)

8. M. Arneodo et al., Measurement of the proton and deuteron structure functions F_2^p and F_2^d, and of the ratio $\sigma_L \sigma_T$. Nucl. Phys. B **483**, 3–43 (1997)

9. M.R. Adams et al., Proton and deuteron structure functions in muon scattering at 470 GeV. Phys. Rev. D **54**, 3006–3056 (1996)

10. S. Forte et al., Neural network parametrization of deep inelastic structure functions. JHEP **0205**, 062 (2002)

11. K. Prytz, Signals of gluon recombination in deep inelastic scattering. Eur. Phys. J. C **22**, 317–321 (2001)

12. L.F. Abbott, W.B. Atwood, R. Michael Barnett, Quantum-chromodynamic analysis of eN deep-inelastic scattering data. Phys. Rev. D **22**, 582–594 (1980)

Chapter 89
Possible Restoration of Z_3 Symmetry in the Presence of Fundamental Higgs

M. Biswal, S. Digal, and P. S. Saumia

Abstract We study the Z_3 symmetry in SU(3) Higgs theory using Monte Carlo simulation methods. We focus mainly on the distribution of the Polyakov loop and related parameters to study the Z_3 symmetry. We show that the Z_3 symmetry is explicitly broken when Higgs condesate acquires a nonzero value. However, we also show that there is a possibility of the Z_3 symmetry in the Higgs symmetric phase where Higgs condensate vanishes.

89.1 Introduction

It is expected that at high enough temperatures, hadrons melt into quark-gluon plasma. These conditions existed in the early universe. Recently, heavy-ion collision experiments are able to reach such extreme conditions. Theoretical studies in Quantum Chromodynamics (QCD) show that the melting proceeds via a transition known as confinement-deconfinement (CD) transition [1]. The CD transition is present in all SU(N) [$N \geq 2$] gauge theories like QCD and Electroweak theory. In pure SU(N) gauge theories, the CD transition is described by an order parameter, the average of Polyakov loop ($\langle L \rangle$) and the Z_N symmetry. In QCD, at '$\mu = 0$' the presence of dynamical quarks (in the fundamental representation) breaks this symmetry explicitly. The CD transition is a cross-over for realistic quark masses. In Electroweak theory, the presence of Higgs and other matter fields breaks this symmetry explicitly. In SU(N) Higgs theory, there are very few non-perturbative studies on the explicit breaking of the Z_N symmetry. And also it is important to understand the similarities (differences) between bosonic and fermionic matter as to how they affect the Z_N

M. Biswal (✉)
Institute of Physics, Sachivalaya Marg, Bhubaneswar 751005, India
e-mail: minati.b@iopb.res.in

S. Digal
The Institute of Mathematical Sciences, Chennai 600113, India

P. S. Saumia
Bogoliubov Laboratory of Theoretical Physics, JINR, 141980 Dubna, Russia

© Springer Nature Singapore Pte Ltd. 2021
P. K. Behera et al. (eds.), *XXIII DAE High Energy Physics Symposium*,
Springer Proceedings in Physics 261,
https://doi.org/10.1007/978-981-33-4408-2_89

symmetry. So in this work, we study the Z_3 symmetry in SU(3) Higgs theory by Monte Carlo simulations of the CD transition. Our results suggest that the explicit symmetry breaking is vanishingly small in part of the Higgs symmetric phase. Preliminary results indicate that the explicit symmetry breaking vanishes in the entire Higgs symmetric phase. This work is a follow up to our previous work [2].

89.2 Pure Gauge Theory and Z_N Symmetry

The partition function of a pure SU(N) gauge theory at high temperature $(T = \frac{1}{\beta})$ is

$$\mathcal{Z} = \mathrm{Tr} e^{-\beta H} = \int dA \langle A | e^{-\beta H} | A \rangle = \int_{bc} DA e^{-S(A)} \tag{89.1}$$

Here, S(A) is the Euclidean gauge action given by

$$S(A) = \int_0^\beta d\tau \int_V d^3x \left\{ \frac{1}{2} \mathrm{Tr} \left(F^{\mu\nu} F_{\mu\nu} \right) \right\} \tag{89.2}$$

where $F_{\mu\nu} = \partial_\mu A_\nu - \partial_\nu A_\mu + ig[A_\mu, A_\nu]$, the allowed A's in the path integral are periodic in β,

$$A_\mu(\vec{x}, 0) = A_\mu(\vec{x}, \beta) \tag{89.3}$$

$S(A)$ and $\mathcal{Z}$ are invariant under the gauge transformation $V(\vec{x}, \tau)$, A_μ transforms

$$A_\mu \longrightarrow V A_\mu V^{-1} - \frac{i}{g} \left(\partial_\mu V \right) V^{-1} \tag{89.4}$$

$V(\vec{x}, \tau)$ need not be periodic, as long as it satisfies the following equation:

$$V(\vec{x}, \tau = 0) = z V(\vec{x}, \tau = \beta) \tag{89.5}$$

where $z \in Z_N$, with, $z = \mathbb{1} \, exp(\frac{2\pi i n}{N})$, $n = 0, 1, 2...N - 1$. Therefore, all the allowed gauge transformations at finite temperature are classified by the Z_N group. Z_N is a symmetry of $\mathcal{Z}$.

89.3 Z_3 Symmetry (with Fundamental Higgs)

The action in the presence of a fundamental Higgs field is given by

$$S_E = \int_0^\beta d\tau \int_V d^3x \left[\frac{1}{2}\text{Tr}\left(F^{\mu\nu}F_{\mu\nu}\right) + \frac{1}{2}|D_\mu\Phi|^2 + \frac{m^2}{2}\Phi^\dagger\Phi + \frac{\lambda}{4!}(\Phi^\dagger\Phi)^2 \right]$$

$$(89.6)$$

Being a bosonic field, $\Phi(\vec{x}, 0) = \Phi(\vec{x}, \beta)$. Under the above non-periodic gauge transformations, $\Phi'(0) \neq \Phi'(\beta)$ (when $z \neq \mathbb{1}$). It is not clear how this Z_3 explicit breaking will affect the CD transition. Fluctuations of the gauge and Higgs fields need to be considered. For simulations, we discretize the action on a 4D Euclidean space, $\Phi(x) \to \Phi_n$, $e^{iagA_{n,\mu}} \to U_{n,\mu}$. Further, we scale Φ, $\bar{\lambda}$ and m as

$$\Phi(x) \to \frac{\sqrt{\kappa}\Phi_n}{a}, \bar{\lambda} \to \frac{\lambda}{\kappa^2}, m^2 \to \frac{(1 - 2\lambda - 8\kappa)}{\kappa a^2}$$

The discretized action is given by

$$S(U, \Phi) = \beta_g \sum_p \text{Tr}(1 - \frac{1}{2N}(U_p + U_p^\dagger)) - \kappa \sum_{\mu,n} \text{Re}\left[\text{Tr}(\Phi_{n+\mu}^\dagger U_{n,\mu}\Phi_n)\right]$$

$$+ \sum_n \left[\frac{1}{2}\text{Tr}\left(\Phi_n^\dagger\Phi_n\right) + \lambda\left(\frac{1}{2}\text{Tr}\left(\Phi_n^\dagger\Phi_n\right) - 1\right)^2\right].$$

$$(89.7)$$

Here $\beta_g = \frac{6}{g^2}$. Plaquette U_p is the product of links around an elementary square 'p' $(U_p = U_{n,\mu}U_{n+\mu,\nu}U_{n+\nu,\mu}^\dagger U_{n,\nu}^\dagger)$. In the Monte Carlo simulations, an initial configuration of Φ_n and $U_{\mu,n}$ is repeatedly updated to generate a Monte Carlo history. In an update, a new configuration is generated from an old one according to the Boltzmann probability factor e^{-S} taking care of the principle of detailed balance. The Boltzmann factor and the principle of detailed balance are implemented using the pseudo-heat-bath algorithm [3],[4] for the Φ field and the standard heat-bath algorithm for the link variables U_μ's [5]. To reduce auto-correlation between consecutive configurations, we use the over-relaxation method.

In Fig. 89.1(a), there is no Z_3 symmetry in the distribution of the Polyakov loop for SU(3). Here, the Z_3 symmetry is explicitly broken. The largest distributed peak corresponds to the stable state and others correspond to meta-stable states. In Fig. 89.1(b), the distribution of the Polyakov loop for SU(3) has the Z_3 symmetry. Similar behavior has been seen for the SU(2) case in Ref [2]. So away from the Higgs transition line in the Higgs symmetric phase within the numerical accuracy, there is a possibility of the Z_3 symmetry, though a detailed study on this symmetry restoration needs to be done.

89.4   Conclusion

Our results suggest that the Higgs condensate plays a role of the symmetry breaking field like an external field in the Ising model. Under Z_3, $A \to A'$. But $\Phi \to \Phi' = V\Phi$ is not considered as Φ' is not periodic, so $S(A, \Phi) \neq S(A', \Phi)$:

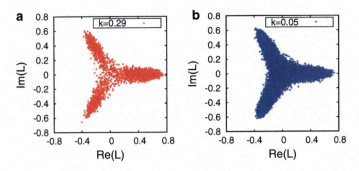

Fig. 89.1 (a) Distribution of Polyakov loop real versus imaginary part for SU(3), at $\kappa = 0.29$, (b) at $\kappa = 0.05$

$$S(A, \Phi) = S(A) + S(\Phi) + S_I(A, \Phi) \qquad (89.8)$$

It seems with an increase in N_τ, Φ' becomes available so that $S(A, \Phi) = S(A', \Phi')$ giving rise to restoration of the Z_3 symmetry. We believe increase in phase space of the Higgs field is responsible for the Z_3 restoration.

References

1. F. Karsch, Lect. Notes Phys. **583**, 209–249 (2002). (arXiv:hep-lat/0106019)
2. M. Biswal, S. Digal, P.S. Saumia, Nucl. Phys. B **910**, 30 (2016)
3. B. Bunk, Nucl. Phys. B **42**, 556 (1995)
4. A.D. Kennedy et al., PLB **156**, 393 (1985)
5. M. Creutz, Phys. Rev. D **29**, 306 (1984)

Chapter 90
Electrical Conductivity of Hot QCD Matter: A Color String Percolation Approach

Pragati Sahoo, Swatantra Kumar Tiwari, and Raghunath Sahoo

Abstract The study of the transport coefficients like shear viscosity, electrical conductivity, etc. of strongly interacting matter produced in heavy-ion collisions has gained interest in the current scenario. We have attempted to study some of these coefficients, and this work comprises the study of the normalized electrical conductivity (σ_{el}/T) of hot QCD matter as a function of temperature (T) using the Color String Percolation Model (CSPM). We also investigate the temperature dependence of shear viscosity and its ratio with electrical conductivity for the QCD matter. The σ_{el}/T in CSPM is showing a very weak dependence on the temperature. We compare CSPM results with those obtained in the Boltzmann Approach to Multi-Parton Scatterings (BAMPS) model, and a good agreement is found between CSPM results and predictions of BAMPS with a fixed strong coupling constant.

90.1 Introduction

The transport properties play an important role to study and characterize the properties of matter produced at extreme conditions of temperature and energy densities. These are mainly the theoretical inputs to the hydrodynamical calculations and affect various observables. A very small shear viscosity to entropy density ratio explains the elliptic flow of identified hadrons produced at RHIC and LHC energies [1] and suggests a nearly perfect fluid nature of the QCD matter produced. Various methods are used to estimate the shear viscosity (η) and Electrical conductivity (σ_{el}), which is another key transport coefficient in order to understand the behavior and properties of the hot QCD matter produced. Experimentally, it has been observed that very strong electric and magnetic fields are created in the early stages (1-2 fm/c) of non-central

P. Sahoo (✉) · S. K. Tiwari · R. Sahoo
Discipline of Physics, School of Basic Sciences, Indian Institute of Technology Indore, Indore 453552, India
e-mail: pragati.sahoo@cern.ch

S. K. Tiwari
e-mail: sktiwari4bhu@gmail.com

© Springer Nature Singapore Pte Ltd. 2021
P. K. Behera et al. (eds.), *XXIII DAE High Energy Physics Symposium*,
Springer Proceedings in Physics 261,
https://doi.org/10.1007/978-981-33-4408-2_90

653

collisions of nuclei at RHIC and LHC [2, 3]. The large electrical field produces an electric current in the early stage of the heavy-ion collision, which depends on the electrical conductivity (σ_{el}). The high interaction rates of the produced QCD matter lower the shear viscosity to entropy density (η/s) ratio; this also balances the dominance of the associated electric charges of the quarks of the σ_{el}. In view of this, a detailed study of electrical conductivity in the strongly interacting QCD matter is inevitable. It is impossible to experimentally measure the electrical conductivity (σ_{el}) of the matter, and its information can be extracted from flow parameters measured in heavy-ion collision experiments [2].

Color String Percolation Model is a QCD-inspired model [4–8], which can be used as an alternative approach to Color Glass Condensate (CGC). In CSPM, the color flux tubes are stretched between the colliding partons in terms of the color field. The strings produce $q\bar{q}$ pair in finite space filled similarly as in the Schwinger mechanism of pair creation in a constant electric field covering all the space [9]. With the growing energy and the number of nucleons of participating nuclei, the number of strings grows. The number of strings grows and starts to overlap and interact to form clusters as the energy and size of the colliding nuclei grow. After a critical string density is reached, a macroscopic cluster appears that marks the percolation phase transition which spans the transverse nuclear interaction area. When the initial density of interacting colored strings (ξ) exceeds the 2D percolation threshold (ξ_c), i.e. $\xi > \xi_c$, a macroscopic cluster appears, which defines the onset of color deconfinement. The critical density of percolation is related to the effective critical temperature and thus percolation may be a possible way to achieve deconfinement in ultrarelativistic heavy-ion collisions [10] and in high multiplicity pp collisions [11, 12]. Recently, we have performed collision centrality, energies and species-dependent studies of the deconfinement phase transition at RHIC Beam Energy Scan (BES) energies using the color string percolation model [13]. We have also studied various thermodynamical and transport properties at RHIC BES energies in this approach [14]. And a detailed investigation can be found in Ref. [15]. In Sect. 90.2, we give the detailed formulation for the calculation of electrical conductivity and shear viscosity in CSPM. The results and discussions are presented in Sect. 90.3.

90.2 Electrical Conductivity and Shear Viscosity

We develop the formulation for evaluating the electrical conductivity of strongly interacting matter using the color string percolation approach. The percolation density parameter, ξ for central Au+Au collisions at RHIC energies, is calculated by using the parameterization of pp collisions at $\sqrt{s} = 200$ GeV as discussed below. In CSPM one obtains

$$\frac{dN_{ch}}{dp_T^2} = \frac{a}{(p_0 + p_T)^\alpha},\tag{90.1}$$

where a is the normalization factor and p_0, α are fitting parameters given as $p_0 = 1.982$ and $\alpha = 12.877$ [16]. Due to the low string overlap probability in pp collisions, the fit parameters are then used to evaluate the interactions of the strings in Au+Au collisions as

$$p_0 \rightarrow p_0 \left(\frac{\langle n S_1/S_n \rangle_{\text{Au+Au}}}{\langle n S_1/S_n \rangle_{\text{pp}}} \right)^{1/4}. \tag{90.2}$$

Here, S_n corresponds to the area occupied by n overlapping strings. Now,

$$\langle \frac{n S_1}{S_n} \rangle = \frac{1}{F^2(\xi)}, \tag{90.3}$$

where $F(\xi)$ is the color suppression factor. To calculate the electrical conductivity of strongly interacting matter, we proceed as follows. The mean free path, which describes the relaxation of the system far from equilibrium, can be written in terms of number density and cross-section as

$$\lambda_{\text{mfp}} = \frac{1}{n\sigma_{\text{tr}}}, \tag{90.4}$$

where n is the number density of an ideal gas of quarks and gluons and σ_{tr} is the transport cross-section. Using the expressions for number density in CSPM, the λ_{mfp} can be written in terms of ξ as follows:

$$\lambda_{\text{mfp}} = \frac{L}{(1 - e^{-\xi})}, \tag{90.5}$$

where σ_{tr} is the transverse area of the effective strings equal to $S_1 F(\xi)$. And L is the longitudinal extension of the string ~ 1 fm.

Now we derive the formula for electrical conductivity by using the Anderson–Witting model [17] and solving the Boltzmann transport equation, in the relaxation time approximation. And this is given by

$$\sigma_{\text{el}} = \frac{1}{3T} \sum_{k=1}^{M} q_k^2 n_k \lambda_{\text{mfp}}. \tag{90.6}$$

Here, q_k and n_k are the charge and number density of the quarks and gluon species. Putting Eq. (90.5) in Eq. (90.6) and considering the density of up quark(u) and its antiquark($\bar{u}$) in the calculation, we get the expression for σ_{el} as

$$\sigma_{\text{el}} = \frac{1}{3T} \frac{4}{9} e^2 n_q(T) \frac{L}{(1 - e^{-\xi})}. \tag{90.7}$$

Here, the pre-factor 4/9 reflects the fractional quark charge squared $(\sum_f q_f^2)$ and n_q denotes the total density of quarks or antiquarks. Here, e^2 in the natural unit is taken as $4\pi\alpha$, where $\alpha = 1/137$.

In the framework of a relativistic kinetic theory, the shear viscosity over entropy density ratio, η/s, is given by [18–20]

$$\eta/s \simeq \frac{T\lambda_{mfp}}{5}. \qquad (90.8)$$

In the context of CSPM, the above equation can be reduced using Eq.(90.5) as

$$\eta/s \simeq \frac{TL}{5(1 - e^{-\xi})}. \qquad (90.9)$$

90.3 Results and Discussions

The results obtained in CSPM along with those obtained in various approaches will be discussed in this section. In Fig. 90.1, we show σ_{el}/T as a function of temperature. The details can be found in Ref. [15]. The green and brown dotted lines are the result of the microscopic transport model BAMPS [21], in which the relativistic (3+1)-dimensional Boltzmann equation is solved numerically to extract the electric conductivity for a dilute gas of massless and classical particles described by the

Fig. 90.1 (color online) σ_{el}/T versus T plot. The black solid line is the result obtained in CSPM and various results from other model calculations are also shown in the figure

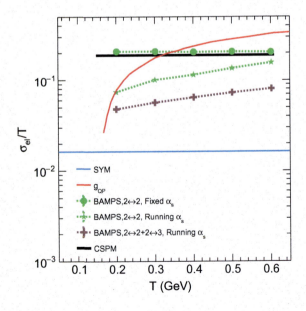

Fig. 90.2 (color online) The ratio η/s as a function of T/T_c. Other estimations with the CSPM result are shown in the black solid line

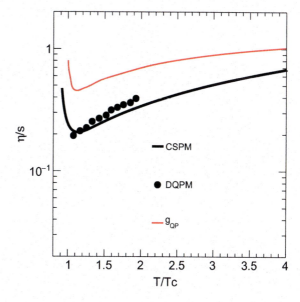

Fig. 90.3 (color online) The ratio η/s and σ_{el}/T with respect to T/T_c. The black solid line is the CSPM result. The DQPM and QP results are shown by the black circles and red solid line, respectively [23]

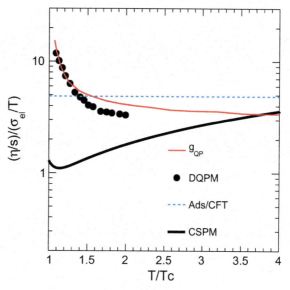

relativistic Boltzmann equation. The BAMPS results show a slower increase of σ_{el}/T with temperature for both the cases of elastic and inelastic processes with running α_s as the effective cross-section changes with the temperature, while σ_{el}/T remains almost independent of temperature for the case of constant α_s. The solid black line shows our results of CSPM for u-quark and antiquark calculated using Eq. (90.7). We observe that σ_{el}/T is almost independent of temperature and matches with the results of BAMPS with constant α_s, which may be due to the similar basic ingredients and procedure for the estimation of σ_{el}/T. Due to an almost similar approach of both the models, BAMPS and CSPM for the calculation of σ_{el}/T, the observations are in agreement.

Figure 90.2 shows the variation of η/s as a function of T/T_c. Here, T_c is the critical temperature which is different in different model calculations and for CSPM $T_c = 167.7$ MeV. The black solid line is the CSPM result. The black circles are the estimations from the dynamical quasi-particle model (DQPM) [22]. In CSPM, η/s first decreases and after reaching a minimum value, it starts increasing with temperature. Thus, it forms a dip which occurs at $T/T_c = 1$. We notice that CSPM results are close to the DQPM predictions. Since we know that η/s is affected by the gluon-gluon and quark-quark scatterings while σ_{el} is only affected by the quark-quark scatterings [23], the ratio between them is important to quantify the contributions from quarks and gluons in various temperature regions. In Fig. 90.3, we show the ratio of η/s and σ_{el}/T versus T/T_c for CSPM and various other measurements. It is observed that this ratio behaves in a similar fashion as η/s.

90.4 Summary and Outlook

In summary, we have estimated the electric conductivity of strongly interacting matter using the color string percolation approach. We use the well-known Drude formula for the estimation of electrical conductivity, which can be obtained after solving the Boltzmann transport equation in relaxation time approximation assuming tiny electric fields and no cross effects between heat and electrical conductivity. We see that the CSPM results for the conductivity stay almost constant with increasing temperature in a similar fashion as shown by BAMPS data and match the results obtained in BAMPS with the fixed strong coupling constant considering elastic cross-section only. We have shown η/s as a function of T/T_c and compared our results with DQPM and QP model results. CSPM results go in line with that obtained in DQPM. We have also plotted the ratio, $(\eta/s)/(\sigma_{el}/T)$ as a function of T, which behaves in a similar manner as η/s varies. We have confronted CSPM results with the results obtained from DQPM and QP models and also compared them with various other model predictions.

References

1. J. Adams et al., STAR Collaboration. Nucl. Phys. A **757**, 102 (2005)
2. Y. Hirono, M. Hongo, T. Hirano, Phys. Rev. C **90**, 021903 (2014)
3. K. Tuchin, Adv. High Energy Phys. **2013**, 490495 (2013)
4. N. Armesto, M.A. Braun, E.G. Ferreiro, C. Pajares, Phys. Rev. Lett. **77**, 3736 (1996)
5. M. Nardi, H. Satz, Phys. Lett. B **442**, 14 (1998)
6. M.A. Braun, C. Pajares, Eur. Phys. J. C **16**, 349 (2000)
7. M.A. Braun, C. Pajares, J. Ranft, Int. J. Mod. Phys. A **14**, 2689 (1999)
8. M.A. Braun, C. Pajares, Phys. Rev. Lett. **85**, 4864 (2000)
9. M.A. Braun et al., Phys. Reports. **509**, 1 (2015)
10. J. Dias de Deus, C. Pajares, Phys. Lett. B **642**, 455 (2006)
11. L.J. Gutay et al., Int. J. Mod. Phys. E **24**, 1550101 (2015)
12. A.S. Hirsch, C. Pajares, R.P. Scharenberg, B.K. Srivastava,. arXiv:1803.02301 [hep-ph]
13. P. Sahoo, S. De, S.K. Tiwari, R. Sahoo, Eur. Phys. J. A **54**(8), 136 (2018)
14. P. Sahoo, S.K. Tiwari, S. De, R. Sahoo, R.P. Scharenberg, B.K. Srivastava, Mod. Phys. Lett. A **34**, 1950034 (2019)
15. P. Sahoo, S.K. Tiwari, R. Sahoo, Phys. Rev. D **98**, 054005 (2018)
16. M.A. Braun, J. Dias de Deus, A.S. Hirsch, C. Pajares, R.P. Scharenberg, B.K. Srivastava, Phys. Rept. **599**, 1 (2015)
17. J. Anderson, H. Witting, Physica (Utrecht) **74**, 466 (1974)
18. J. Dias de Deus et al., Eur. Phys. J. C **72**, 2123 (2012)
19. P. Danielewicz, M. Gyulassy, Phys. Rev. D **31**, 53 (1985)
20. T. Hirano, M. Gyulassy, Nucl. Phys. A **769**, 71 (2006)
21. M. Greif, I. Bouras, C. Greiner, Z. Xu, Phys. Rev. D **90**, 094014 (2014)
22. R. Marty, E. Bratkovskaya, W. Cassing, J. Aichelin, H. Berrehrah, Phys. Rev. C **88**, 045204 (2013)
23. A. Puglisi, S. Plumari, V. Greco, Phys. Lett. B **751**, 326 (2015)

Chapter 91
Effect of Inverse Magnetic Catalysis on Conserved Charge Fluctuations in the Hadron Resonance Gas Model

Ranjita K. Mohapatra

Abstract We discuss the inverse magnetic catalysis effect on conserved charge fluctuations and correlations along the chemical freeze-out curve in the hadron resonance gas model. We have compared fluctuations and correlations with and without charge conservation. Charge conservation plays an important role in the calculation of fluctuations at nonzero magnetic field and for fluctuations in strange charge at zero magnetic field. Charge conservation diminishes the correlation χ_{BS} and χ_{QB}, but enhances the correlation χ_{QS}. We point out that baryonic fluctuations (second order) at $B = 0.25\ GeV^2$ increase more than two times compared to $B = 0$ at higher μ_B.

PACS numbers: 25.75.-q · 12.38.Mh

91.1 Introduction

The basic features of the physical system created at the time of chemical freeze-out in heavy-ion collisions are well described in terms of the hadron resonance gas (HRG) model [1, 2]. There is an excellent agreement between experimental data on particle ratios in heavy-ion collisions with corresponding thermal abundances calculated in the HRG model at an appropriately chosen temperature and baryon chemical potential with different conserved charges taken into account [3]. The universal chemical freeze-out curve in the $T - \mu_B$ plane is determined by the condition $E/N = \epsilon/n \simeq$ 1 GeV, where $E(\epsilon)$ is the internal energy (density) and $N(n)$ is the particle number (density) [4]. It has been already proposed that event-by-event fluctuations of conserved quantities such as net baryon number, net electric charge, and net strangeness

R. K. Mohapatra (✉)
Department of Physics, Indian Institute of Technology Bombay, Mumbai 400076, India
e-mail: ranjita@iitb.ac.in

Institute of Physics, Bhubaneswar 751005, India

Homi Bhabha National Institute, Training School Complex, Anushakti Nagar, Mumbai 400085, India

© Springer Nature Singapore Pte Ltd. 2021
P. K. Behera et al. (eds.), *XXIII DAE High Energy Physics Symposium*,
Springer Proceedings in Physics 261,
https://doi.org/10.1007/978-981-33-4408-2_91

as a possible signal of the QGP formation and quark-hadron phase transition [5, 6]. The fluctuations in net electric charge are suppressed in the QGP phase compared to the hadronic gas phase due to the fractional charge carriers in the QGP phase compared to unit charge carriers in the hadronic gas phase.

Moreover, higher order moments of conserved charge fluctuations are more sensitive to the large correlation lengths in the QGP phase and relax slowly to their equilibrium values at the freeze-out [7]. The divergence of correlation length or higher order fluctuations will hint toward the existence of a critical point in the QCD phase diagram. So, higher moments, different ratios, and skewness and kurtosis of conserved charges have been measured experimentally and compared with the HRG model predictions along the freeze-out curve [8]. The deviation of experimental results from the HRG model predictions may conclude the presence of non-hadronic constituents or non-thermal physics in the primordial medium [9]. Ratios of susceptibilities in lattice QCD have been shown to be consistent with the HRG model predictions with near-zero chemical potential [10, 11].

However, it is very important to study these fluctuations in the presence of a magnetic field because of the huge magnetic field produced in non-central relativistic heavy-ion collision due to the valence charges of colliding nuclei [12]. The effect of the magnetic field on the conserved charge fluctuations has been studied in the HRG model along the universal freeze-out curve and compared with the available experimental data [13].

However, it has been shown that the inverse magnetic catalysis (IMC) effect arises in the presence of an external magnetic field in lattice QCD, in which the chiral transition temperature decreases [14]. But the system exhibits magnetic catalysis at zero temperature where the chiral condensate increases in an external magnetic field. Since the chiral transition temperature decreases in the presence of the magnetic field, the freeze-out curve in the $T - \mu_B$ plane will correspond to a lower temperature [15] in the HRG model. It has been shown that electric charge conservation and strangeness conservation play an important role at higher baryon chemical potential in a nonzero magnetic field. Electric charge susceptibility along the freeze-out curve is very large without charge conservation at higher μ_B, but it decreases significantly when the charge conservation is taken into account [15]. So, it is very important to consider electric charge conservation and strangeness conservation at higher μ_B in the presence of the magnetic field.

This paper is organized as follows. We discuss the essential aspects of the HRG model in the presence of an external magnetic field in section II. Section III describes different universal freeze-out conditions. Section IV describes the conserved charge densities and fluctuations along the freeze-out curve.

91.2 HRG Model in the Presence of the Magnetic Field

The HRG model in the presence of the magnetic field has been studied and thermodynamic quantities like pressure, energy density, entropy density, magnetization, and the speed of sound are presented as functions of the temperature and the magnetic field [16]. The basic quantity in the HRG model is the grand partition function defined for each hadron species i as

$$\ln Z_i = \pm V g_i \int \frac{d^3 p}{(2\pi)^3} \ln \left[1 \pm e^{-(E_i - \mu_i)/T}\right] \qquad (91.1)$$

For a constant magnetic field along the Z-axis, the well-known phenomena of Landau quantization of energy levels for a charged particles takes place along the plane perpendicular to the magnetic field [18]. The single-particle energy for a charged particle in the presence of the magnetic field is given by $E = \sqrt{p_z^2 + m^2 + 2|qB|(n + 1/2 - s_z)}$. Here, n is the Landau level and s_z is the z component of the spin of the hadron.

The grand partition function in the presence of a magnetic field is given by

$$\ln Z_i = \pm V \sum_{s_z=-s}^{+s} \sum_{n=0}^{\infty} \frac{|qB|}{2\pi} \int \frac{dp_z}{2\pi} \ln \left[1 \pm e^{-(E_i - \mu_i)/T}\right] \qquad (91.2)$$

The vacuum part is in general divergent and it needs to be regularized and renormalized [16].

91.3 Universal Freeze-Out Curve

The first unified condition for chemical freeze-out is given by $E/N = \epsilon/n \simeq 1$ GeV. This makes sense because the freeze-out occurs when the average energy per particle $\sim m + \frac{3}{2}T$ crosses 1 GeV for non-relativistic particles. The second condition for unified chemical freeze-out is given by a fixed value for the sum of baryon and anti-baryon densities, i.e. $n_B + n_{\bar{B}} \simeq 0.12$ fm^{-3} [19], and the third unified chemical freeze-out is given by a fixed value of entropy density, i.e. $s/T^3 \simeq 7$ [20].

The magnetic field produced in the relativistic heavy-ion collisions at LHC energies can reach the order of 0.25 GeV^2. Keeping this in mind, we have shown the effect of inverse magnetic catalysis on the freeze-out curve in the HRG model at $B = 0.25$ GeV^2 in Fig. 1. We have compared the freeze-out curve determined by the condition $E/N = \epsilon/n \simeq 1$ GeV for zero and nonzero magnetic fields with charge conservation and without charge conservation in Fig. 1a. A similar plot is already shown in [15]. We can clearly see the freeze-out temperature is lowered at nonzero B due to the effect of inverse magnetic catalysis with charge conservation. However, the freeze-out temperature at nonzero B increases at higher μ_B without charge conservation. This is due to the fact that at higher μ_B, there are more baryons, particularly more protons at nonzero B. If there is no charge conservation; then the number density increases due to more protons produced at nonzero B. So, the freeze-out curve determined by constant $E/N \simeq 1$ GeV should be pushed to higher temperature [15].

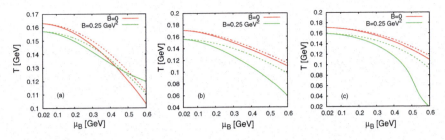

Fig. 91.1 Chemical freeze-out curve determined by **a** $E/N \simeq 1$ GeV, **b** $n_B + n_{\bar{B}} \simeq 0.12$ fm^{-3}, and **c** $s/T^3 \simeq 7$ with (dashed line) and without (solid line) charge conservation

Figure 91.1b represents the freeze-out curve determined from the condition $n_B + n_{\bar{B}} \simeq 0.12$ fm^{-3}. The freeze-out temperature is lowered in nonzero magnetic field with and without charge conservation. However, the freeze-out temperature decreases sufficiently at nonzero B without charge conservation at higher μ_B. This is due to the fact that the proton density increases sufficiently at nonzero B and higher μ_B when there is no charge conservation. So, to keep $n_B + n_{\bar{B}} \simeq 0.12$ fm^{-3} fixed, the temperature will be pushed downwards.

Figure 91.1c represents the freeze-out curve determined from the condition $s/T^3 \simeq 7$. The freeze-out temperature is also lowered at nonzero B with this freeze-out condition. However, the freeze-out temperature is very low (~ 20 MeV) at nonzero B without charge conservation at higher $\mu_B = 600$ MeV. At higher μ_B, the entropy density decreases sufficiently due to the increase in baryon density (proton density) at nonzero B. To keep the ratio $s/T^3 \simeq 7$ fixed, the temperature should decrease sufficiently. The chemical freeze-out parameters determined from different universal conditions are almost the same at zero B. But, they are very different at nonzero B at higher μ_B without charge conservation. From all these plots, it is clear that it is very important to use charge conservation at nonzero B to determine the chemical freeze-out parameters from different freeze-out conditions.

91.4 Results and Discussion

91.4.1 Quantities Related to Conserved Charges Along the Freeze-Out Curve

Figure 91.2 represents μ_S and μ_Q as a function μ_B along the freeze-out curve described in Fig. 91.1a. The solid line in Fig. 91.2 shows the variation of μ_S and dashed line shows the variation of μ_Q along the freeze-out curve. μ_S at nonzero $B = 0.25$ GeV^2 is always larger than μ_S without a magnetic field. At higher μ_B, there are more baryons (protons and neutrons) in the system. Imposing charge conservation, i.e. $B/Q \simeq 2.52$, one needs negative μ_Q. μ_Q is of the order 0.1 GeV

Fig. 91.2 Strangeness and electric charge chemical potential along the freeze-out curve $E/N \simeq 1$ GeV

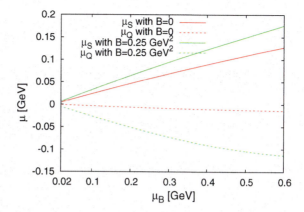

at higher μ_B at $B = 0.25\ GeV^2$, and it is of the order 0.01 GeV at zero magnetic field.

The chemical freeze-out temperature decreases due to the IMC effect at nonzero B. However, if one assumes the freeze-out parameters at nonzero B are same as at $B = 0$, then one could use the fitted freeze-out parameters of T, μ_B, μ_S and μ_Q obtained in Eq. 91.3 and Eq. 91.4 as given below [8] at nonzero B.

$$T = a - b\mu_B^2 - c\mu_B^4 \tag{91.3}$$

where $a = (0.166 \pm 0.002)$ GeV, $b = (0.139 \pm 0.016)$ GeV^{-1}, and $c = 0.053 \pm 0.021)$ GeV^{-3}

$$\mu_X(\sqrt{s_{NN}}) = \frac{d}{1 + e\sqrt{s_{NN}}} \tag{91.4}$$

Here, X is the chemical potential for different conserved charges and the corresponding values of d and e are given in Table 91.1.

Table 91.1 Parameterization of chemical potential μ_X along the freeze-out curve

X	d[GeV]	e[GeV^{-1}]
B	1.308(28)	0.273(8)
S	0.214	0.161
Q	0.0211	0.106

91.4.2 Fluctuations and Correlations Along the Freeze-Out Curve

The fluctuations and correlations are given by the diagonal and off-diagonal components of susceptibility. These are defined by

$$\chi_{xy}^{ij} = \frac{\partial^{i+j}\left(\sum_k P_k/T^4\right)}{\partial(\frac{\mu_x}{T})^i \partial(\frac{\mu_y}{T})^j} \tag{91.5}$$

It also has been pointed out that the correlation between baryon number and strangeness is stronger in the QGP phase compared to the hadronic gas phase since the strange quarks have nonzero baryon number in the QGP phase [21]. But, in the hadronic gas phase, this correlation decreases since strange mesons don't carry baryonic charge.

The most dominant contribution toward strange susceptibility comes from the lowest mass strange particle, i.e. kaons. At $B = 0$, χ_S^2 is large without charge conservation (solid line) compared with charge conservation (dash-dotted line) along the freeze-out curve (Fig. 91.3a). This is also true for nonzero B. So, charge conservation diminishes strangeness fluctuations along the freeze-out curve for zero and nonzero B. We can see χ_S^2 with charge conservation at nonzero B is slightly below the curve at zero B because the effective mass of kaon increases at nonzero B. χ_S^2 at $B = 0.25$ GeV^2 from the fitted parameters (dotted line) is very different than the results at nonzero B with charge conservation.

At $B = 0$, χ_B^2 is almost the same with (dash-dotted line) and without charge conservation (solid line) as shown in Fig. 91.3b. At $B = 0.25$ GeV^2, the second-order baryon susceptibility (baryon fluctuations) is very large when there is no charge conservation (due to more proton production). However, when charge conservation

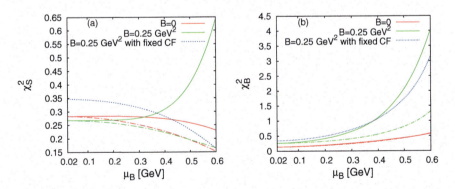

Fig. 91.3 **a** χ_S^2 and **b** χ_B^2 along the freeze-out curve at $B = 0$, $B = 0.25$ GeV^2, and $B = 0.25$ GeV^2 with fixed chemical freeze-out (CF) parameters obtained at zero B (dotted line). Here, the solid line corresponds to the charge susceptibility without charge conservation and the dash-dotted line corresponds to the charge susceptibility with charge conservation

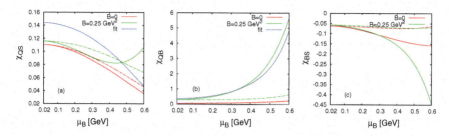

Fig. 91.4 **a** χ_{QS}, **b** χ_{QB}, and **c** χ_{BS} along the freeze-out curve with and without B. Here, the solid line corresponds to the conserved charge correlations without charge conservation and the dash-dotted line corresponds to the conserved charge correlations with charge conservation

is taken into account, the baryon fluctuations decrease at nonzero B. χ_B^2 at $B = 0.25$ GeV^2 from the fitted parameters (dotted line) is very different from the results at nonzero B with charge conservation. From the above discussions, it is clear that if the freeze-out parameters at nonzero B are the same as at zero B (i.e. using zero B fitted parameters at nonzero B), the fluctuations in conserved charges are different compared to the fluctuations at nonzero B with charge conservation. So it is always important to fix strange and electric charge chemical potentials using conservation laws at nonzero B, rather than using zero B fitted parameters at nonzero B. This is also true for all higher order fluctuations. From here onwards, we will not compare the results from fitted parameters.

We have shown conserved charge correlations along the freeze-out curve in Fig. 91.4. Figure 91.4a presents χ_{QS} as a function of μ_B along the freeze-out curve. The dominant contribution toward strangeness and electric charge correlation comes from kaons (same nature of the corresponding charge). At zero B, the correlation decreases toward higher μ_B, because higher μ_B corresponds to lower freeze-out temperature and the number of kaons decreases due to a decrease in temperature. The most important thing here to note is that the correlation increases with charge conservation compared to no charge conservation at zero B. This is due to the fact that kaon density increases with charge conservation due to the increase in chemical freeze-out temperature compared to no charge conservation. The correlation increases at nonzero B compared to zero B due to more production of electric charges with nonzero strangeness (kaons) in a nonzero magnetic field. The correlation also increases with charge conservation at nonzero B due to the increase in chemical freeze-out temperature. At nonzero B and higher $\mu_B = 500$ MeV, we can see the correlation increases without charge conservation due to the increase of chemical freeze-out temperature as shown in Fig. 91.1a.

We have shown the variation χ_{QB} as a function of μ_B along the freeze-out curve in Fig. 91.4b. At zero B, the electric and baryonic charge correlation is the same with and without charge conservation. The correlation increases at nonzero B due to more production of protons and Δ^{++} without charge conservation. This correlation decreases when charge conservation is taken into account.

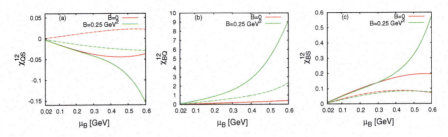

Fig. 91.5 **a** χ_{QS}^{12}, **b** χ_{BQ}^{12}, and **c** χ_{BS}^{12} along the freeze-out curve with and without B. Here, the solid line corresponds to the conserved charge correlations without charge conservation and the dash-dotted line corresponds to the conserved charge correlations with charge conservation

Figure 91.4c presents the variation of χ_{BS} along the freeze-out curve. The correlation is always negative because of the opposite nature of the corresponding charges. The dominant contribution comes from Σ baryons. The correlation (modulus value) is larger without charge conservation than with charge conservation at zero and nonzero B. This is mainly due to more baryon production with nonzero strangeness when there is no charge conservation.

Higher order mixed correlations are presented in Fig. 91.5. Charge conservation always diminishes the absolute value of the correlations. Higher order mixed correlations (absolute value) are always larger compared to lower order mixed correlations.

91.5 Conclusions

We have studied the IMC effect on conserved charge fluctuations and correlations along the freeze-out curve in the HRG model. We have obtained the fluctuations and correlations with and without charge conservation. At $B = 0$, charge conservation does not play a role in the fluctuations (second and higher order) along the freeze-out curve for the conserved charges of electric charge and baryon number. But charge conservation plays an important role for strange charge at $B = 0$. Charge conservation diminishes the fluctuations in strange charge at $B = 0$ compared to the fluctuations without charge conservation. For nonzero B, charge conservation plays a very important role. If there is no charge conservation at nonzero B, then the fluctuations increase by a huge amount compared to zero B at higher μ_B.

Acknowledgements We would like to thank Ajay K. Dash and Rajarshi Ray for very useful discussions and suggestions.

References

1. P. Braun-Munzinger, K. Redlich, J. Stachel, nucl-th/0304013
2. A. Andronic, P. Braun-Munzinger, J. Stachel, Nucl. Phys. A **772**, 167 (2006)
3. P. Braun-Munzinger, D. Magestro, K. Redlich, J. Stachel, Phys. Lett. **B 518**, 41 (2001), J. Cleymans, K. Redlich, Phys. Rev. **C 60**, 054908 (1999), F. Becattini, et al., Phys. Rev. **C 64**, 024901 (2001), J. Cleymans, B. Kampfer, M. Kaneta, S. Wheaton, N. Xu, Phys. Rev. **C 71**, 054901 (2005), A. Andronic, P. Braun-Munzinger, J. Stachel, Phys. Lett. **B 673**, 14 (2009)
4. J. Cleymans, K. Redlich, Phys. Rev. Lett. **81**, 5284 (1998)
5. M. Asakawa, U.W. Heinz, B. Muller, Phys. Rev. Lett. **85**, 2072 (2000)
6. S. Jeon, V. Koch, Phys. Rev. Lett. **83**, 5435 (1999)
7. M.A. Stephanov, Phys. Rev. Lett. **102**, 032301 (2009)
8. F. Karsch, K. Redlich, Phys. Lett. B **695**, 136 (2011)
9. B. Abelev et al., ALICE Collaboration. Phys. Rev. Lett. **110**, 152301 (2013)
10. R.V. Gavai, S. Gupta, Phys. Lett. B **696**, 459 (2011)
11. M. Cheng et al., Phys. Rev. D **79**, 074505 (2009)
12. V. Skokov, A. Illarionov, and V. Toneev, Int. J. Mod. Phys. **A 24**, 5925 (2009), W.-T. Deng and X.-G. Huang, Phys. Rev. **C 85**, 044907 (2012)
13. A. Bhattacharyya, S.K. Ghosh, R. Ray, S. Samanta, Europhys. Lett. **115**, 62003 (2016)
14. G. Bali, F. Bruckmann, G. Endrodi, Z. Fodor, S. Katz, S. Krieg, A. Schäfer, and K. K. Szabó, J. High Energy Phys. **02**, 044 (2012), G. S. Bali, F. Bruckmann, G. Endrodi, Z. Fodor, S. D. Katz, and A. Schafer, Phys. Rev. **D 86**, 071502 (2012), G. S. Bali, F. Bruckmann, G. Endrodi, S. D. Katz, and A. Schafer, J. High Energy Phys. **08**, 177 (2014)
15. K. Fukushima, Y. Hidaka, Phys. Rev. Lett. **117**, 102301 (2016)
16. G. Endrodi, J. High Energy Phys. **04**, 023 (2013)
17. Particle Data Group, J. Beringer et al., Phys. Rev. **D 86**, 010001 (2012)
18. L. Landau and E. Lifshits, Quantum Mechanics: Non-Relativistic Theory. Course of theoretical physics (Vol 3.) (Landau, L. D, 1908-1968). Pergamon Press, 1977
19. P. Braun-Munzinger, J. Stachel, J. Phys. G: Nucl. Part. Phys. **28**, 1971 (2002)
20. J. Cleymans, H. Oeschler, K. Redlich, S. Wheaton, Phys. Lett. B **615**, 50 (2005)
21. V. Koch, A. Majumder, J. Randrup, Phys. Rev. Lett. **95**, 182301 (2005)

Chapter 92
Study of the Compatibility of Parton Distribution Functions Using High-X Ep Collision Data from ZEUS

Ritu Aggarwal

Abstract For Bjorken-x values larger than 0.6, the Parton Distribution Functions (PDFs) have large uncertainties. A part of the problem is that knowledge of the proton structure at these high-x values relies mainly on the data collected by different fixed-target experiments. However, there is clean DIS data collected by ZEUS available in this intense region but it has not been included in any of the PDF fits. Transfer Matrix has been developed for the high-x ZEUS data, using which an expectation for the ZEUS data can be obtained from different PDFs. A comparison of the agreement of various modern PDFs with ZEUS data has been done using the Transfer Matrix, and p-values are calculated for every PDF set. A wide variation in the p-values are observed for the different PDF sets and are reported here.

92.1 Introduction

The composition of matter and its fundamental particles has always been an area of interest for physicists. Many big particle colliders have been built which study the collisions of different particles at various energies and increasing luminosities. The present era's biggest particle collider is located at CERN [1] which collides two proton beams at energies and luminosities than was never achieved before. To be able to analyze the information available from collisions of the protons, the knowledge of the distribution of partons inside the proton is of utmost importance [2, 3]. However, our understanding of the proton structure at the kinematic region of Bjorken-$x > 0.65$ is limited to the data available from the fixed-target experiments [4, 5]. The HERA Neutral Current (NC) *ep* scattering data is one of the cleanest data available (as it is free from nuclear corrections that are involved in the fixed-target experiments),

---

For the ZEUS collaboration.

R. Aggarwal (✉)
Savitribai Phule Pune University, Pune 411007, India
e-mail: ritu.aggarwal1@gmail.com

Department of Science and Technology, New Delhi, India

© Springer Nature Singapore Pte Ltd. 2021
P. K. Behera et al. (eds.), *XXIII DAE High Energy Physics Symposium*,
Springer Proceedings in Physics 261,
https://doi.org/10.1007/978-981-33-4408-2_92

but HERA data included in the Parton Distribution Function (PDF) extractions by different theory groups is only up to 0.65 in Bjorken-x [6–8]. And this is the region where different PDFs start exhibiting large uncertainties. The ZEUS experiment at HERA has measured the ep cross sections up to a value of 1 in Bjorken-x [9]. This data is unique as it is the only DIS data available which measures cross sections at Bjorken-x up to a value of 1. As some of the bins at high-x have low statistics and Poisson errors are quoted this data could not be included in the PDF extraction by the different theory groups. This paper demonstrates the use of Transfer Matrix developed for the ZEUS high-x data to predict cross sections from different commonly used PDFs and comparison to ZEUS high-x data.

92.2   Transfer Matrix for ZEUS High-X Data

The events generated with given true x-Q^2 co-ordinates could end up in a different x-Q^2 co-ordinate when QED radiative corrections are included. Their value further gets smeared at the reconstructed level by the detector effects (finite resolution, limited acceptance, selection criterion, etc.).

Figure 92.1 (left) shows simulated data events reconstructed in the bins where cross sections were reported for e^+p scattering [9]. These events when plotted in the true $(x$-$Q^2)$ co-ordinate phase space have a distribution as shown in Fig. 92.1 (right). The binning here at the generated level is refined to contain each event in the true co-ordinates and is also made finer than the cross section bins depending upon the bin statistics. It is to be noted that the events at the generated level have radiative corrections applied which were calculated using HERACLES [10]. Building a Transfer Matrix for the ZEUS high-x data facilitates us with a probability that an event which is generated in the ith bin gets reconstructed with certain x-Q^2 co-ordinates in the jth bin. Each element a_{ij} in the Transfer Matrix $\mathbf{T}$ is given as

$$a_{ij} = \frac{\sum_{m=1}^{M_i} \omega_m I(m \in j)}{\sum_{m=1}^{M_i} \omega_m^{MC}} \tag{92.1}$$

where ω_m is the weight given to the mth event, I is 1 if $(m \in j)$, else it is zero. Using Transfer Matix $\mathbf{T}$ and the generated distribution of events $\mathbf{M}$ in the true x-Q^2 phase space, the expectation for events reconstructed in the cross section bins $\mathbf{N}$ can be calculated as

$$\mathbf{N} = \mathbf{TM}$$

where

$$\mathbf{M} = \mathbf{KS}$$

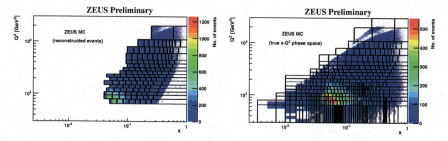

Fig. 92.1 Distribution of event weights in the data (left) in the cross section bins used in [9] after all selection cuts are applied, (right) generated bins used to calulate elements in the **M** vector

Here, **K** represents the radiative corrections which are applied as the scaling factor to the integrated Born cross sections in the true x-Q^2 phase space (represented as vector **S**).

The expectation **N** from any other PDF set (say k) can be obtained by using the following equation:

$$\mathbf{N}_k = \mathbf{TKS}_\mathbf{k}$$

Here, the Transfer Matrix **T** and the radiative corrections **K** are independent of the PDF set whereas the vectors $\mathbf{N_k}$, $\mathbf{M_K}$ and $\mathbf{S_k}$ are PDF dependent.

Figure 92.2 shows the comparison of different PDFs, namely from MMHT [11, 11], CT10 [12, 13], ABM [14, 15] and NNPDF [16, 17] theory groups, at the generated level (vectors $\mathbf{M_k}$) with respect to HERAPDF2.0. It is observed that there is a difference in the different PDFs when compared to HERAPDF2.0 and this difference increases to a level of 10% at Bjorken-x of 0.6. The difference between the different PDFs is more than the PDF uncertainties shown as the shaded region for HERAPDF2.0 in Fig. 92.2.

92.3   Comparison to the High-x Data

Figure 92.3 shows the comparison of reconstructed events from HERAPDF 2.0 (NLO) with ZEUS NC high-x data for $e^- p$ scattering (in left) and for $e^+ p$ scattering (in right). The bands in green, yellow and green show the 68%, 95% and 99.9% probability regions (calculated as explained in [18]) for the HERAPDF2.0 expectations given the ZEUS high-x data. The comparison shows that ZEUS high-x data agrees with HERAPDF2.0 mostly within 68% of the region.

The general agreement of expectations from a given PDF to the ZEUS high-x data can also be estimated by calculating the p-values [19]. Table 92.1 shows the p-values for different PDFs (at NLO) given the ZEUS high-x data. It is observed that for e+p ZEUS high-x data, most of the PDFs give a good p-value, whereas for ZEUS $e^- p$ high-x data the p-values are generally lower. For the ZEUS $e^- p$ high-x

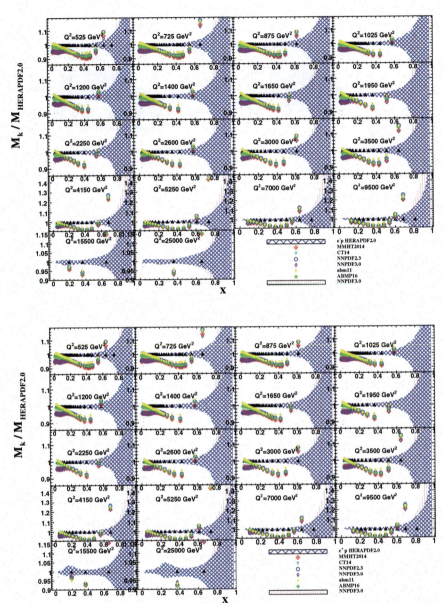

Fig. 92.2 Ratios of total generated events (**M** vectors) for given PDF sets to those calculated using HERAPDF2.0 as functions of x in different Q^2 for e^-p data (top) and e^+p data (bottom). The shaded band represents the uncertainty quoted by HERAPDF2.0 and from NNPDF3.0

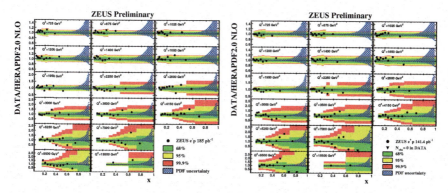

Fig. 92.3 The ratio of the number of observed events to the expectations from the HERAPDF2.0 for e^-p (left) and e^+p (right) data, respectively. The green, yellow and red bands give the 68, 95, 99.9 % probability intervals calculated using Poisson statistics

Table 92.1 p-values for full Bjorken-x range for different PDFs (at NLO)

PDF	e^-p	e^+p
$HERAPDF2.0$	0.05	0.5
$CT14$	0.002	0.8
$MMHT2014$	0.002	0.8
$NNPDF2.3$	0.00007	0.6
$NNPDF3.0$	0.0002	0.7
$ABMP16$	0.01	0.8
$ABM11$	0.001	0.6

data, HERAPDF2.0 and ABMP 16 show the best results and other PDFs lie in the tail of the p-value distribution. The comparison of different PDF sets is done in two x ranges (x greater and lower than 0.6, respectively) and the results are shown in Table 92.2. It is observed that there are large differences in the behavior of $e^- p$ and $e^+ p$ data sets as well as in the two different x ranges. The disagreement among the PDFs is largest at low x for the $e^- p$ data.

All the uncorrelated and correlated sources of systematic uncertainty were checked (a new transfer matrix was produced and a new expectation was calculated for each source of uncertainty and it was then compared to the data). The effect of the statistical and all sources of systematic uncertainties was found to be negligible. The only and most significant source of systematic uncertainty was found to be the normalization uncertainty which is taken as 1.8%. It scales the M distribution (and hence the N distribution) up and down systematically. It is observed that p-values from different PDFs increase, favoring different directions in the change of normalization.

Table 92.2 p-values for two different x ranges

	e⁻p		e⁺p	
PDF	$x < 0.6$	$x \geq 0.6$	$x < 0.6$	$x \geq 0.6$
HERAPDF2.0	0.06	0.2	0.6	0.1
CT14	0.0008	0.2	0.7	0.6
MMHT2014	0.00003	0.1	0.6	0.6
NNPDF2.3	0.00007	0.2	0.6	0.6
NNPDF3.0	0.00003	0.2	0.6	0.6
ABMP16	0.01	0.2	0.8	0.5
ABM11	0.03	0.3	0.7	0.4

92.4 Summary

In this paper, different PDF sets are compared at the generated level and with ZEUS high-x data at the reconstructed level. While this data set has some overlap with data used in other ZEUS publications [7], this high-x data has not been previously used in the extraction of parton densities. The predictions for the expected number of events from the different PDF sets are compared to the number observed by the ZEUS Collaboration using Poisson statistics. Despite the fact that the event numbers are small, the data set contains significant information on the behavior of the parton densities at the highest values of x.

References

1. The Large Hadron Collider, CERN https://home.cern/science/accelerators/large-hadron-collider
2. Juan Rojo et al., J. Phys. G: Nucl. Part. Phys. **42**, 103103 (2015)
3. Jon Butterworth et al., J. Phys. G: Nucl. Part. Phys. **43**, 023001 (2016)
4. BCDMS Coll., A.C. Benvenuti et al., Phys. Lett. **B 223**, 485 (1989)
5. L.W. Whitlow et al., Phys. Lett. B **282**, 475 (1992)
6. ZEUS Coll., H. Abramowicz et al., Eur. Phys. **J. C 75**, 580 (2015)
7. ZEUS Coll., H. Abramowicz et al., Phys. Rev. **D 87**, 052014 (2013)
8. H1 Coll., F. D. Aaron et al., JHEP **1209**, 061 (2012)
9. ZEUS Coll., H. Abramowicz et al., Phys. Rev. **D 89**, 072007 (2014)
10. A. Kwiatkowski, H. Spiesberger and H.-J. Möhring, Comp. Phys. Comm. **69**, 155 (1992). Also in *Proc. Workshop Physics at HERA*. eds. W. Buchmüller and G. Ingelman, (DESY, Hamburg, 1991)
11. A.D. Martin, W.J. Stirling, R.S. Thorne, G. Watt, Eur. Phys. J C **63**, 189 (2009). [arXiv:0901.0002]. L. Harland-Lang, A. D. Martin, P. Motylinski and R. Thorne, Eur. Phys. J. C 75 204 (2015)
12. S. Dulat et al., Phys. Rev. D **93**, 033006 (2016)
13. M. Guzzi et al., Phys. Rev. **D 82**, 074024 (2010), [axXiv:1007.2241]
14. S. Alekhin et al., Phys. Rev. **D 89**, 054028 (2014). S. Alekhin et al., Phys. Lett. **B 672**, 166 (2009), [axXiv:0811.1412]

15. S. Alekhin et al., Phys. Rev. **D 86**, 054009 (2012), [axXiv:1202.2281]
16. R.D. Ball et al., NNPDF Collaboration. Nucl. Phys. B **809**, 1 (2009). [arXiv:0808.1231]
17. R.D. Ball et al. [NNPDF Collaboration], JHEP **1004**, 040 (2015), [arXiv:1410.8849v2]
18. R. Aggarwal, A. Caldwell, Eur. Phys. J. Plus **127**, 1–8 (2012)
19. F. Beaujean et al., Phys. Rev. D **83**, 012004 (2011)

Chapter 93
Φ and K*^{0} Production in p–Pb and Pb–Pb Collisions with ALICE at the LHC

Sandeep Dudi

Abstract We report recent measurements of ϕ and K*^{0} production in p–Pb and Pb–Pb collisions at $\sqrt{s_{NN}}$ = 8.16 and 5.02 TeV, respectively. The integrated yield and mean transverse momentum are reported as a function of charged particle multiplicity to explore the particle production mechanism. Particle ratios K*^{0}/K and ϕ/K are studied as a function of charged particle multiplicity to explore rescattering and regeneration effects. The nuclear modification factor (R_{AA}) as a function of p_T is studied to explore parton energy loss.

93.1 Introduction

It has been established that in high-energy heavy-ion collisions, a hot and dense, strongly interacting matter, often described as a strongly coupled quark-gluon plasma (sQGP), is created [1–3]. The K*(892)^0 ($d\bar{s}$) and ϕ(1020) ($s\bar{s}$) particles produced in these collisions contain strange (anti-strange) quarks and they can be used for the systematic studies of the particle-species dependence of partonic energy loss in the medium. The K*^{0} and ϕ have lifetimes in vacuum of 4.16±0.05 fm/c and 46.3±0.4 fm/c, respectively [4]. Due to their short lifetimes, resonance particles can be used to probe the system at different timescales during its evolution. Resonance yields measured by hadronic decay channels can be affected by particle rescattering and regeneration in the hadron gas phase.

A Large Ion Collider Experiment (ALICE) [5] is a multi-purpose heavy-ion detector, which has excellent tracking and particle-identification capabilities [6]. Resonances ϕ and K*^{0} are reconstructed from their hadronic decay daughters ($\phi \rightarrow$ KK and K*$^0 \rightarrow$ Kπ) using an invariant mass technique. The details of the analysis can be found in [7].

For ALICE Collaboration.

S. Dudi (✉)
Department of Physics, Panjab University, Chandigarh 160014, India
e-mail: sandeep.dudi3@gmail.com

© Springer Nature Singapore Pte Ltd. 2021
P. K. Behera et al. (eds.), *XXIII DAE High Energy Physics Symposium*,
Springer Proceedings in Physics 261,
https://doi.org/10.1007/978-981-33-4408-2_93

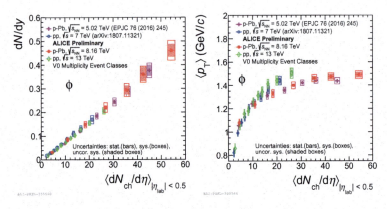

Fig. 93.1 (Left) The integrated yield (dN/dy) and (right) the mean transverse momentum ($\langle p_T \rangle$) of ϕ meson as a function of charged particle multiplicity ($\langle dN_{ch}/d\eta \rangle$) in pp collisions at $\sqrt{s} = 7$ and 13 TeV and in p–Pb collisions at $\sqrt{s_{NN}} = 5.02$ and 8.16 TeV

93.2 Results

Figure 93.2 (left panel) shows the integrated yield of ϕ as a function of the charged particle multiplicity in pp collisions in rapidity ($|y| < 0.5$) at $\sqrt{s} = 7$ and 13 TeV and in p–Pb collisions in rapidity ($-0.5 < y < 0$) at $\sqrt{s_{NN}} = 5.02$ [8] and 8.16 TeV. The integrated yield increases with increasing charged particle multiplicity ($\langle dN_{ch}/d\eta \rangle$) independently of the identities of the colliding ions and their energies. The right panel of Fig. 93.1 shows the mean transverse momentum ($\langle p_T \rangle$) of ϕ as a function of $\langle dN_{ch}/d\eta \rangle$ in pp collisions at $\sqrt{s} = 7$ and 13 TeV and in p–Pb collisions at $\sqrt{s_{NN}} = 5.02$ and 8.16 TeV. The $\langle p_T \rangle$ increases as a function of the charged particle multiplicity and seems to saturate at high multiplicity in p–Pb collisions. A similar trend of $\langle p_T \rangle$ dependence on multiplicity is also observed for other resonances [8]. The new results for $\langle p_T \rangle$ may lead to a better understanding of the mass ordering (particles with similar masses have similar $\langle p_T \rangle$) observed in central and semi-central Pb–Pb collisions, which is consistent with expectations from the hydrodynamic expansion of the system [7]. The mass ordering in $\langle p_T \rangle$ is not observed in small collision systems [9].

The K^{*0}/K and ϕ/K ratios as a function of charged particle multiplicity ($\langle dN_{ch}/d\eta \rangle$) in Pb-Pb collisions at $\sqrt{s_{NN}} = 5.02$ TeV are shown in Fig. 93.2. The results are compared with p–Pb and Pb–Pb collisions at $\sqrt{s_{NN}} = 5.02$ and 2.76 TeV, respectively. The K^{*0}/K ratio decreases as we go from p–Pb and peripheral Pb–Pb collisions to the most central Pb–Pb collisions. This decrease in the K^{*0}/K ratio can be understood as the rescattering of K^{*0} decay daughters in the hadronic medium [7]. The ϕ/K ratio shows no suppression in p–Pb and Pb–Pb collisions and almost no dependence on $\langle dN_{ch}/d\eta \rangle$. The ϕ/K ratio suggests that the ϕ (which has a lifetime of an order of magnitude larger than the K^{*0}) might decay predominantly outside the hadronic medium.

Fig. 93.2 p_T integrated K^{*0}/K and ϕ/K ratios as a function of changed particle multiplicity in p–Pb collisions at $\sqrt{s_{NN}} = 5.02$ TeV and in Pb–Pb collisions at $\sqrt{s_{NN}} = 2.76$ and 5.02 TeV

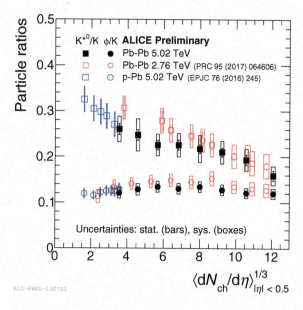

Fig. 93.3 Nuclear modification factor (R_{AA}) of π, K, p, ϕ and K^{*0} as a function of p_T in Pb–Pb collisions at $\sqrt{s_{NN}} = 5.02$ TeV in the 0–10 % multiplicity or centrality class

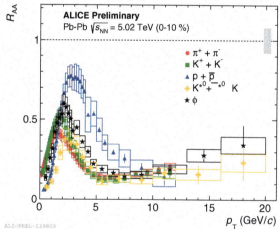

The nuclear modification factor (R_{AA}) of π, K, p, ϕ and K^{*0} as a function of p_T in Pb–Pb collisions at $\sqrt{s_{NN}} = 5.02$ TeV in the 0–10 % collision centrality class is shown in Fig. 93.3. In the intermediate p_T range (2–6 GeV/c), the R_{AA} of ϕ and K^{*0} are similar to that of the charged kaons, whereas protons and ϕ exhibit a different trend despite their similar masses [10]. For $p_T > 8$ GeV/c, all the light flavoured species, π, K, p, ϕ and K^{*0}, show a similar suppression within uncertainties.

93.3 Conclusions

The production of ϕ and K^{*0} mesons in p–Pb and Pb–Pb collisions in various centrality classes at $\sqrt{s_{NN}}$ = 8.16 and 5.02 TeV, respectively, has been measured. The integrated yields (dN/dy) of ϕ and K^{*0} mesons (only the ϕ measurement is shown here) are independent of collision system and energy, indicating that particle production is driven by multiplicity. The $\langle p_T \rangle$ of the ϕ meson saturates at high multiplicity in p–Pb collisions. The suppression in K^{*0}/K ratio indicates that rescattering dominates over regeneration. The R_{AA} value of ϕ mesons and protons differ at intermediate p_T despite similar masses. At high p_T ($p_T >$ 8 GeV/c), all particles show similar suppression within uncertainties.

References

1. J. Adams et al., (STAR collaboration). Nucl. Phys. A **757**, 102 (2005)
2. K. Adcoxet et al., (PHENIX collaboration). Nucl. Phys. A **757**, 184 (2005)
3. K. Aamodtet et al., (ALICE collaboration). Phys. Rev. Lett. **105**, 252302 (2010)
4. K.A. Olive et al., (Particle data group). Rev. Partic. Phys. Chin. Phys. C **38**, 090001 (2014)
5. K. Aamodt et al., ALICE collaboration. J. Inst. **3**, S08002 (2008)
6. J. Adam et al., ALICE collaboration. Eur Phys. J. C **75**, 226 (2015)
7. B. Abelev et al., ALICE collaboration. Phys. Rev. C **91**, 024609 (2015)
8. J. Adam et al., ALICE collaboration. Eur. Phys. J. C **76**, 245 (2016)
9. A.K. Dash (for the ALICE Collaboration), *Proceedings of QM-2018* https://doi.org/10.1016/j.nuclphysa.2018.11.011
10. J. Adam et al., ALICE collaboration. Phys. Rev. C **93**, 034913 (2016)

Chapter 94
An Insight into Strangeness with $\phi(1020)$ Production in Small to Large Collision Systems with ALICE at the LHC

Sushanta Tripathy

Abstract Hadronic resonances are unique tools to investigate the interplay of re-scattering and regeneration effects in the hadronic phase of heavy-ion collisions. As the ϕ meson has a longer lifetime compared to other resonances, it is expected that its production will not be affected by regeneration and re-scattering processes. Measurements in small collision systems such as proton-proton (pp) collisions provide a necessary baseline for heavy-ion data and help to tune pQCD inspired event generators. Given that the ϕ is a bound state of strange-antistrange quark pair (s$\bar{s}$), measurements of its production can contribute to the study of strangeness production. Recent results obtained by using the ALICE detector show that although ϕ has zero net strangeness content, it behaves like a particle with open strangeness in small collision systems and the experimental results agree with thermal model predictions in large systems. The production mechanism of ϕ is yet to be understood. We report on measurements with the ALICE detector at the LHC of ϕ meson production in pp, p–Pb, Xe–Xe and Pb–Pb collisions. These results are reported for minimum bias event samples and as a function of the charged-particle multiplicity or centrality. The results include the transverse momentum (p_T) distributions of ϕ as well as the $\langle p_T \rangle$ and particle yield ratios. The ϕ effective strangeness will be discussed in relation to descriptions of its production mechanism, such as strangeness canonical suppression, non-equilibrium production of strange quarks and thermal models.

94.1 Introduction

Resonances are ideal candidates to probe the hadronic phase formed in heavy-ion collisions due to their short lifetimes. The lifetime of ϕ (46.3 fm/c) is longer compared to that of other hadronic resonances as well as the lifetime of the fireball produced in

For ALICE collaboration.

S. Tripathy (✉)
Department of Physics, Indian Institute of Technology Indore,
Simrol 453552, India
e-mail: Sushanta.Tripathy@cern.ch

683

heavy-ion collisions [1]. Thus, it is expected that a ϕ meson will not be affected by the re-scattering and regeneration processes during the hadronic phase [2]. ϕ being a bound state of a strange-antistrange quark pair (s$\bar{s}$), a measurement of its production can help shed light on strangeness production mechanisms. Also, the study of ϕ in small colliding systems helps in the search for the onset of collectivity and provides a necessary baseline for heavy-ion collisions.

This article focuses on measurements of ϕ production with the ALICE detector at the LHC in pp collisions at $\sqrt{s}$ = 0.9, 2.76, 5.02, 7, 8 and 13 TeV, p–Pb collisions at 5.02 and 8.16 TeV, Xe–Xe collisions at $\sqrt{s_{NN}}$ = 5.44 TeV and Pb–Pb collisions at $\sqrt{s_{NN}}$ = 2.76 and 5.02 TeV. In particular, p_T spectra at different energies and colliding systems as well as p_T-integrated particle ratios to long-lived hadrons are compared for minimum bias collisions and as a function of the charged-particle multiplicity ($\langle dN_{ch}/d\eta \rangle$). In this paper, we aim at addressing one of the major questions, namely whether ϕ behaves like a non-strange or strange particle. The strangeness of ϕ will be discussed in relation to its production mechanism, such as strangeness canonical suppression, non-equilibrium production of strange quarks and thermal models.

94.2 ϕ Meson Reconstruction and p_T Spectra

The $\phi(1020)$ is reconstructed at mid-rapidity ($|y| < 0.5$) through an invariant-mass analysis via its hadronic decay channel [3, 4] into K^+K^- (branching ratio: 49.2%) [5]. Figure 94.1 shows the invariant-mass distribution for the ϕ in pp collisions at $\sqrt{s}$ = 5.02 TeV in the p_T range $0.5 < p_T < 0.7$ GeV/c in V0M Multiplicity class VII. The left plot of Fig. 94.1 shows the unlike-charge invariant-mass distribution with a combinatorial background. The event mixing and like-sign techniques are used to estimate the combinatorial background and after combinatorial background subtraction, a residual background remains as shown in the right plot of Fig. 94.1, together with a fit used to describe the peak of ϕ and the residual background. The latter is mainly due to mis-identified particle decay products or from other sources of correlated pairs (e.g. mini-jets). The $\phi(1020)$ peak is fitted with a Voigtian function, a convolution of Breit–Wigner and Gaussian functions [3, 4]. For some cases, the ϕ peak is fitted without any combinatorial background subtraction when the combinatorial background shows large statistical fluctuation.

In each p_T intervals, raw yields are obtained from the fit to the signal peak and then corrected for the detector efficiency × acceptance and the branching ratio to determine the final p_T spectrum. Figure 94.2 shows the p_T spectra of ϕ mesons in pp collisions at $\sqrt{s}$ = 13 TeV (left) and p–Pb collisions at $\sqrt{s_{NN}}$ = 8.16 TeV (right) in different V0 multiplicity classes. The lower panels show the ratio of the p_T spectra to the 0–100% p_T spectrum. The evolution of p_T spectra is observed at low p_T. For high p_T, the slopes of the spectra in different multiplicity classes seem to be similar to those observed in minimum bias pp collisions. Figure 94.3 shows the p_T spectra of ϕ mesons in Xe–Xe collisions at $\sqrt{s_{NN}}$ = 5.44 TeV (left) and Pb–Pb collisions at $\sqrt{s_{NN}}$ = 5.02 TeV (right) for different centrality classes.

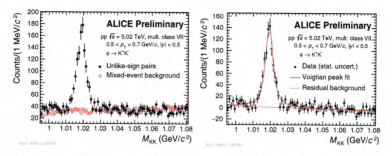

Fig. 94.1 Invariant-mass distribution for the ϕ in pp collisions at $\sqrt{s} = 5.02$ TeV in one of the measured p_T ranges in V0M Multiplicity class VII. Left: the unlike-charge invariant-mass distribution with mixed-event backgrounds. Right: Invariant-mass distribution after subtraction of the mixed-event background with a Voigtian fit to describe the peak of the ϕ and the residual background

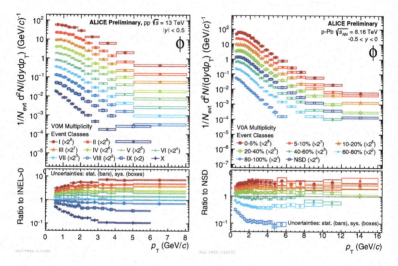

Fig. 94.2 p_T spectra of ϕ mesons in pp collisions at $\sqrt{s} = 13$ TeV (left) and p–Pb collisions at $\sqrt{s_{NN}} = 8.16$ TeV (right) in different multiplicity classes. In the bottom panels of the figure, the ratios of the p_T spectra to the multiplicity-integrated p_T spectrum are reported

94.3   Results and Discussion

Figure 94.4 shows the ratios of p_T spectra of ϕ in inelastic pp collisions at various center-of-mass energies to the spectrum obtained in pp collisions at $\sqrt{s} = 2.76$ TeV. These ratios indicate that in the range 1–2 GeV/c, the yields increase as a function of collision energy, but the production at low p_T does not strongly depend on collision energy.

The left panel of Fig. 94.5 shows the integrated yield of ϕ in pp collisions at $\sqrt{s} = 7$ and 13 TeV and p–Pb collisions at $\sqrt{s_{NN}} = 5.02$ and 8.16 TeV. The integrated

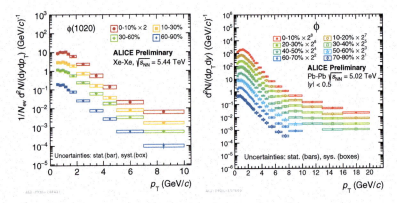

Fig. 94.3 p_T spectra of ϕ mesons in Xe–Xe collisions at $\sqrt{s_{NN}}$ = 5.44 TeV (left) and in Pb–Pb collisions at $\sqrt{s_{NN}}$ = 5.02 TeV (right) in different centrality classes

Fig. 94.4 Ratios of p_T spectra of ϕ in inelastic pp collisions at various center-of-mass energies to the spectrum obtained in pp collisions at $\sqrt{s}$ = 2.76 TeV. Statistical uncertainties are represented by bars and systematic uncertainties are represented by boxes

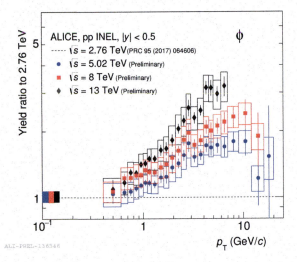

yield shows a linear increase as a function of charged-particle multiplicity for both pp and p–Pb collisions. The right panel of Fig. 94.5 shows the ϕ yield normalized by the $\langle dN_{ch}/d\eta \rangle$ value as a function of average charged-particle multiplicity in pp collisions at $\sqrt{s}$ = 13 TeV and in p–Pb collisions at $\sqrt{s_{NN}}$ = 5.02 and 8.16 TeV. The ratio is independent of collision energy, which suggests that the event multiplicity drives the particle production, irrespective of collision system type and energy.

The top left panel of Fig. 94.6 shows the yield ratios of ϕ and K^{*0} to kaons as a function of charged-particle multiplicity for different colliding systems at different collision energies. As the lifetime of K^{*0} is almost 10 times shorter compared to ϕ, it is expected that K^{*0} is affected by the regeneration and/or re-scattering processes in a long-lasting hadronic phase of the expanding system. A decreasing trend in the K^{*0}/K ratio is observed, suggesting that the re-scattering mechanism dominates

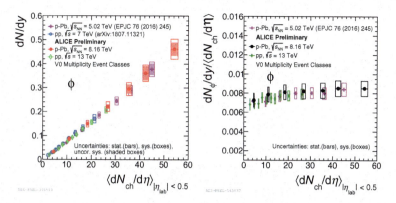

Fig. 94.5 Left: dN/dy of ϕ as a function of charged-particle multiplicity in pp collisions at $\sqrt{s}$ = 7 and 13 TeV and p–Pb collisions at $\sqrt{s_{NN}}$ = 5.02 and 8.16 TeV. Right: $(dN/dy)/\langle dN_{ch}/d\eta \rangle$ for ϕ as a function of average charged-particle multiplicity in pp collisions at $\sqrt{s}$ = 13 TeV and p–Pb collisions at $\sqrt{s_{NN}}$ = 5.02 and 8.16 TeV. Statistical uncertainties are represented by bars and systematic uncertainties are represented by boxes

over-regeneration. As expected, the ϕ/K ratio remains fairly flat, which indicates that either the regeneration and re-scattering are balanced or ϕ decays after the hadronic phase without being affected by these processes. The top right panel of Fig. 94.6 shows the ϕ/π ratio as a function of $\langle dN_{ch}/d\eta \rangle$. The production of ϕ in Pb–Pb and Xe–Xe collisions is well described by a grand-canonical thermal model (GSI-Heidelberg) [6], while for small systems (pp and p–Pb collisions) the increase of the ϕ/π ratio with multiplicity is in contrast to the expectation from strangeness canonical suppression [7]. This behavior favors the non-equilibrium production of ϕ and/or strange particles. The bottom panel of Fig. 94.6 shows the Ξ/ϕ ratio as a function of $\langle dN_{ch}/d\eta \rangle$. The Ξ/ϕ ratio remains fairly flat or slightly increases across a wide multiplicity range. In addition, a multiplicity dependence of the ratio is observed, particularly at low multiplicities. Comparing the ϕ with particles with strange quark content 1 or 2, we observe that ϕ behaves like a particle with open strangeness [8].

94.4 Summary

ALICE has studied ϕ production as a function of collision energy and charged-particle multiplicity in different colliding systems. The event multiplicity seems to drive the production of hadrons, including ϕ production, irrespective of collision energy for pp and p–Pb collisions at the LHC. The ϕ/K ratio remains rather flat across a wide range of multiplicity and across colliding systems, which indicates that either regeneration and re-scattering are balanced or that the ϕ decays after the hadronic phase in Pb–Pb collisions and is not affected by re-scattering and regeneration. The latter seems to

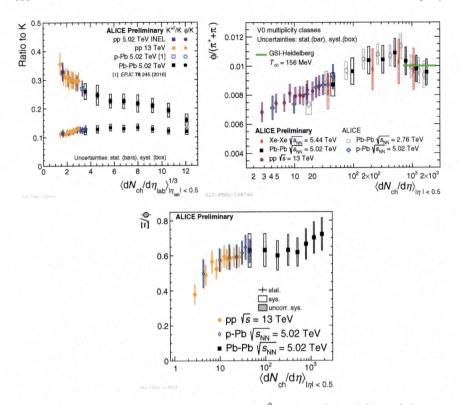

Fig. 94.6 Ratios of the p_T-integrated yield of ϕ and K*^{0} relative to K (top left), ϕ relative to π (top right) and Ξ relative to ϕ (bottom) as a function of the charged-particle multiplicity in different collision systems for different center-of-mass energies

be the likely scenario as ϕ has an almost 10 times longer lifetime than K*^{0}. Looking at the ϕ/π, ϕ/K and Ξ/ϕ ratios, the ϕ meson seems to show similar behavior to that of particles with open strangeness.

Acknowledgements ST acknowledges the financial support by the DST-INSPIRE program of the Government of India.

References

1. C. Markert, R. Bellwied, I. Vitev, Phys. Lett. B **669**, 92 (2008)
2. M. Bleicher, J. Aichelin, Phys. Lett. B **530**, 81 (2002)
3. J. Adam et al., [ALICE collaboration], Phys. Rev. C **95**, no. 6, 064606 (2017)
4. B. Abelev et al., [ALICE collaboration], Eur. Phys. J. C **72**, 2183 (2012)
5. M. Tanabashi et al., [Particle data group]. Rev. Partic. Phys., Phys. Rev. D **98**, 030001 (2018)

6. J. Stachel, A. Andronic, P. Braun-Munzinger, K. Redlich, J. Phys: Conf. Ser. **509**, 012019 (2014)
7. S. Acharya et al., [ALICE collaboration], Phys. Rev. C **99**, no. 2, 024906 (2019)
8. S. Tripathy, [ALICE collaboration], Nucl. Phys. A **982**, 180 (2019)

Chapter 95
Study of Charge Separation Effect in Pb-Pb Collisions Using AMPT

Sonia Parmar, Anjali Sharma, and Madan M. Aggarwal

Abstract The strong magnetic field created by the spectator protons in non-central heavy-ion collision causes the separation of oppositely charged particles along the magnetic field direction which is an important consequence of the Chiral Magnetic Effect. The charge-dependent multiparticle azimuthal correlations are studied as a function of $Db_{\pm}^{max}$ bins obtained using the Sliding Dumbbell Method (SDM). The results reported in this article are performed on AMPT generated Pb-Pb collision events at $\sqrt{s_{NN}} = 2.76$ TeV.

Keywords Chiral magnetic effect · Multiparticle azimuthal correlations · Sliding dumbbell method

95.1 Introduction

The interaction of two high-energy colliding nuclei produces deconfined overlap region and the highly energetic spectator protons generate a strong magnetic field which induces the electric current. This electric current results in the motion of more positively charged particles in one direction and more negatively charged particles in the opposite direction along the system orbital angular momentum direction, a phenomenon known as Chiral Magnetic Effect (CME) [1]. The two main ingredients required for the search of CME signal are the strong magnetic field and the non-zero axial charge density. The fast-moving spectators have a highly positive charge (Pb^{82+}) and create the strongest magnetic field during and after the impact of two heavy ions for a very short interval of time, and a non-zero charge density is also created in the hot dense matter produced during the collision. In order to hunt the CME experimentally, Voloshin [2] introduced the multiparticle correlator defined as

$$\gamma_{ab} = \langle cos(\phi_a + \phi_b - 2\phi_c)\rangle / v_{2,c} \qquad (95.1)$$

S. Parmar (✉) · A. Sharma · M. M. Aggarwal
Panjab University, Chandigarh 160014, India
e-mail: sonia.parmar@cern.ch

© Springer Nature Singapore Pte Ltd. 2021
P. K. Behera et al. (eds.), *XXIII DAE High Energy Physics Symposium*,
Springer Proceedings in Physics 261,
https://doi.org/10.1007/978-981-33-4408-2_95

where ϕ_a, ϕ_b and ϕ_c are the azimuthal angles of charged particles. The indices a, b and c represent the charges of the particles and could be $+$ or $-$. The $v_{2,c}$ is the elliptic flow of particle c. The evidences for the charge separation effect have been reported by the ALICE collaboration at the Large Hadron Collider (LHC) for Pb-Pb data at $\sqrt{s_{NN}}$ = 2.76 TeV [3, 4] and STAR collaboration at Relativistic Heavy Ion Collider (RHIC) in Au+Au collisions for center-of-mass energies ranging 7.7–200 GeV [5]. The charge-dependent three-particle correlator has been measured for different charge combinations, viz., the same sign (SS) ($+$ $+$, $-$ $-$) and the opposite sign (OS). The results obtained are qualitatively in good agreement with the expectations for the CME. CMS collaboration also reported the results on three-particle correlator relative to the reaction plane at $\sqrt{s_{NN}}$ = 5.02 TeV [6].

95.2 Monte Carlo Event Generator

A Multi Phase Transport (AMPT) model, a heavy-ion event generator, is an important tool to study the theoretical predictions for distinct physics observables. AMPT comprises various stages of collision, viz., initial conditions, partonic interactions, conversion from partonic to hadronic matter and hadronic interactions. A detailed information about the AMPT model can be found in Ref. [7]. One million events are generated for the Pb-Pb collisions at $\sqrt{s_{NN}}$ = 2.76 TeV with the string melting on configuration and the preliminary results are presented here. The charged particles in the transverse momentum (p_T) range $0.2 < p_T < 5.0$ GeV and pseudorapidity (η) interval $-0.8 < \eta < 0.8$ are selected for the study.

95.3 Methodology

The Sliding Dumbbell Method (SDM) [8] is used in this analysis for the observation of charge separation effect. This method scans the full $\eta - \phi$ phase space at the microscopic level and locates the events exhibiting charge separation. Events with an excess of positively charged particles on one side of the dumbbell and negatively charged particles on the other side are obtained by a sliding dumbbell of 60° in steps of one degree to get a maximum value of $Db_\pm$ in each event. We obtained the distributions for a maximum value of $Db_\pm$ in each centrality interval for AMPT generated events. Further, the $Db_\pm^{max}$ distributions in each centrality are sliced into ten bins from 0–10% to 90–100% corresponding to highest and lowest $Db_\pm^{max}$ values, respectively. For the analysis, the three-particle correlator defined in Eq. 95.1 is calculated using the Q-cumulants method [9] and studied as a function of $Db_\pm^{max}$ bins. The background correlations due to elliptic flow, resonance decays, etc., can also produce a signal similar to CME. Thus, it is necessary to suppress the background contributions to locate the CME signal events. The background estimation is performed by randomly reshuffling the charges of the particles over the azimuthal

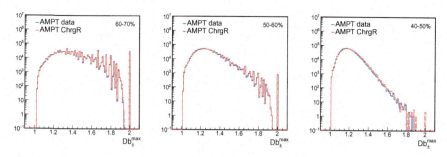

Fig. 95.1 $Db_{\pm}^{max}$ distributions obtained using the sliding dumbbell method for AMPT generated and charge reshuffled events in different centrality intervals

plane keeping the η and ϕ the same. The results from AMPT and charge reshuffle (ChrgR) events are compared to estimate the CME-type effects.

95.4 Results and Discussion

Figure 95.1 displays the $Db_{\pm}^{max}$ distributions obtained for AMPT events and charge reshuffled events in different centralities. Both the distributions seem similar and exhibit a shift toward higher $Db_{\pm}^{max}$ values with decreasing centrality.

For further understanding, the usefulness of SDM and the $Db_{\pm}^{max}$ distributions in each centrality are divided into ten bins and the three-particle correlator (γ_{ab}) are measured in each $Db_{\pm}^{max}$ bin for SS and OS. The difference of opposite and same sign pairs correlator is a natural choice to study the CME contribution since it suppresses the charge-dependent background correlations. The centrality dependence of OS (γ_{opp}) and SS (γ_{same}) charge pairs for the AMPT events and charge reshuffled events are displayed in Fig. 95.2. For higher $Db_{\pm}^{max}$ bins, the OS charge pairs exhibit positive correlation whereas the pairs of SS have negative values showing strong correlation among themselves. A similar trend is observed for AMPT and charge reshuffled events.

Figure 95.3 presents the centrality dependence of difference between opposite sign and same sign correlators ($\gamma_{opp} - \gamma_{same}$). The correlation values are positive for higher $Db_{\pm}^{max}$ bins in each centrality. Reasonably good agreement is seen between the AMPT and charge reshuffle events.

To isolate the events with large charge separation, ten data points shown in Fig. 95.3 are grouped into two bins where the first bin contains the events with top 30% $Db_{\pm}^{max}$ values (i.e., large charge separation) while the second bin corresponds to 30–100% $Db_{\pm}^{max}$ values. Again the correlation for the OS-SS charge pairs is studied for the two bins and it is found that the top 30% $Db_{\pm}^{max}$ have positive correlation whereas the rest of the sample events exhibit almost zero values except for the 50–60% centrality class. The average values for $\gamma_{opp} - \gamma_{same}$ correlator are negligible in comparison to 0–30% $Db_{\pm}^{max}$ bins. Figure 95.4 also shows that though

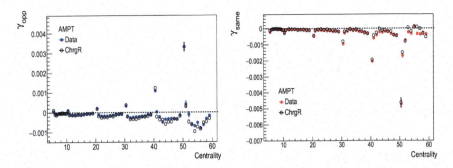

Fig. 95.2 Centrality dependence of three-particle correlator for the opposite sign (left) and same sign charge pairs (right) obtained in different $Db_{\pm}^{max}$ bins for AMPT data and AMPT ChrgR

Fig. 95.3 Centrality dependence of three-particle correlator for the difference of opposite sign and same sign charge pair correlators in different $Db_{\pm}^{max}$ bins for AMPT data and AMPT ChrgR

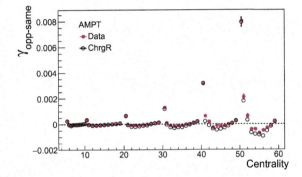

Fig. 95.4 Comparison of $\gamma_{opp} - \gamma_{same}$ correlator for top 30% $Db_{\pm}^{max}$ values and for rest of the sample for AMPT and charge reshuffled events. The average values in each centrality are also displayed

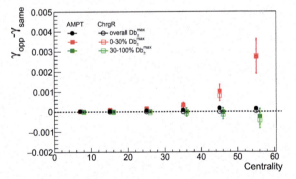

the top 30% $Db_{\pm}^{max}$ bins have more positive values than the other bins, the values for AMPT events match with those of charge reshuffled points within the statistical uncertainties which is expected since there is no CME signal in AMPT events.

Thus, SDM isolates the events showing charge separation effect from the given sample of events and is a powerful tool to observe the CME-type effects in the real data. Also, the results from charge reshuffle agree with the normal AMPT events which lead to the conclusion that the charge reshuffle method can be used to calculate the background correlations while analyzing the experimental data.

Acknowledgements The financial assistances from the Department of Science and Technology and University Grants Commission of the Government of India are gratefully acknowledged.

References

1. K. Fukushima et al., Phys. Rev. D **78**, 074033 (2006)
2. S. Voloshin, Phys. Rev. C **70**, 057901 (2004)
3. B. Abelev et al., ALICE Collaboration. Phys. Rev. Lett. **110**, 012301 (2013)
4. S. Acharya et al., ALICE Collaboration. Phys. Lett. B **777**, 151 (2018)
5. L. Adamczyk et al., STAR Collaboration. Phys. Rev. Lett. **113**, 052302 (2014)
6. V. Khachatryan et al., CMS Collaboration. Phys. Rev. Lett. **118**, 122301 (2017)
7. Zi-Wei Lin et al., Phys. Rev. C **72**, 064901 (2005)
8. Sonia Parmar (for the ALICE Collaboration), Proc. of DAE-BRNS Symp. Nucl. Phys. **61** (2016)
9. A. Bilandzic et al., Phys. Rev. C **83**, 044913 (2011)

Chapter 96
A Portable Cosmic Muon Tracker (CMT) Using Resistive Plate Chambers (RPCs) an Outreach Perspective

B. Satyanarayana, R. R. Shinde, Honey, E. Yuvaraj, Pathaleswar, M. N. Saraf, L. Umesh, and S. Rajkumarbharathi

Abstract Cosmic Muon Tracker (CMT) is a portable charged particle (Muons) tracker, which can be used to track and acquire live Muon events. This compact detector module is expected to serve as an excellent outreach tool for a wide range of students—schools to university. Eight layers of Resistive Plate Chamber (RPC) gaseous detectors are used to track Muons by ionization. Each Muon event from the tracker is acquired by the FPGA-based Data Acquisition system (RPC-DAQ) and generates events based on coincidence. Also, events are displayed in the form of LEDs in a real-time basis. Apart from outreach, CMT can be used to plot the angular distribution of muons, counting rates of strips and their dependence on ambient parameters such as temperature, relative humidity and barometric pressure, which are recorded by the DAQ module. In this paper, the Construction and operating details of this detector will be presented.

Keywords Charged particle · RPC detector · Track · Event acquisition · LED display

96.1 Introduction

Cosmic rays are high-energy particles—comprising mainly nuclei of hydrogen (protons), helium and other heavier elements, arriving from outer space. When they arrive toward Earth, they collide with the atoms in the upper atmosphere, creating more particles, mainly Pions. The charged Pions decay very quickly, producing particles called muons. Unlike Pions, muons do not interact strongly with matter, therefore can travel through the atmosphere to penetrate below ground. At sea level, the average flux of muons is about 1 per square centimeter per minute. Cosmic ray muons

B. Satyanarayana (✉) · R. R. Shinde · Honey · E. Yuvaraj · Pathaleswar · M. N. Saraf · L. Umesh · S. Rajkumarbharathi
Tata Institute of Fundamental Research, Mumbai, India
e-mail: bsn@tifr.res.in

© Springer Nature Singapore Pte Ltd. 2021
P. K. Behera et al. (eds.), *XXIII DAE High Energy Physics Symposium*,
Springer Proceedings in Physics 261,
https://doi.org/10.1007/978-981-33-4408-2_96

697

Fig. 96.1 Cosmic Muon Tracker

can be detected and tracked using stacks of particle detectors such as scintillators or Resistive Plate Chambers (RPCs).

A cosmic muon tracker comprising a vertical stack of eight RPCs—each of 27 cm × 27 cm in area—is built mainly for use during outreach campaigns, Fig. 96.1. The RPCs are operated in gas-sealed mode, thus don't require gas cylinders/system to be carried along. Each RPC provides signals on eight pickup strips (3 cm wide) each on X- and Y-planes. Typically, cosmic ray muons, passing through the stack, produce tracks in all the eight RPCs and induce signals on the strips, which are in the path of the tracks. LEDs—one per strip, mounted on the front-end boards of X- and Y-planes of each RPC in the stack—display the tracks of these cosmic ray muons in real time. The hit strips' information is also stored in an FPGA-based digital Data Acquisition (DAQ) module, which is connected to a computer via an Ethernet port. The DAQ module also produces trigger signals to acquire the data, based on the pattern of hit strips. The data acquired is transferred to the computer and is also displayed as a 3D image of the event in the tracker. The live display of events can also be ported onto the Internet for viewing from anywhere. The low voltages as well as high voltage required to bias the RPCs are all locally generated from mains power, thus making the tracker highly portable.

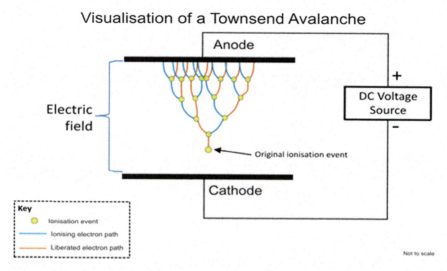

Fig. 96.2 Townsend Avalanche inside RPC

96.2   RPC Detector and Detection Principle

Resistive Plate Chamber (RPC) detectors are gaseous detectors. The Chambers are made by sandwiching two glass electrode plates using poly carbonate spacers. Sealing all sides of the two glass plates forms a chamber. Glass electrodes are painted with a conductive coating whose surface resistance is around 1 Mega Ohm. Pickup panels are made up of Honeycomb material. Copper strips are pasted on one side are used to picking induced signals on the glass. On the other side, Aluminum acts as the ground. Both sides of the chamber pickup panels are kept orthogonal. Ionizing charged particles traversing the gap initiates a streamer in the gas volume. That results in a local discharge of the electrodes. This discharge is limited to a tiny area of about 0.1 cm^2 due to the high resistivity of the glass electrodes and the quenching characteristics of the gas. The discharge induces an electrical signal on external pickup strips on both sides orthogonal to each other, which can be used to record the location and time of ionization (Figs. 96.2 and 96.3).

96.3   Front-end Board

Each layer of RPC will have 8 X side signals and 8 Y side signals. It is necessary to process these signals near the detector. So a Front-End board is connected to each layer on both X and Y sides as shown in Fig. 96.4. These FE boards receive raw RPC signals 8 from each side of the RPC. MAX9108 IC discriminates these signals with a common threshold. A TTL buffer is used to convert the discriminated logic signals

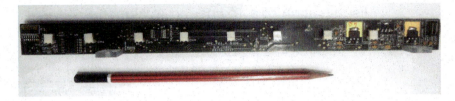

Fig. 96.3 Fabricated Front-end Board

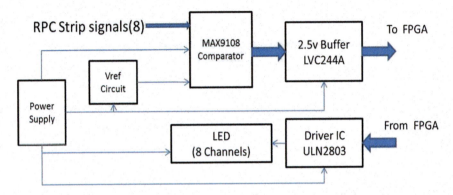

Fig. 96.4 Front-end Block diagram

to standard 2.5V TTL signals. These TTL signals from each FE board are sent to the RPC-DAQ module using high-density cables. Each FE board consist of 8 LEDs mounted accordingly to the strip width of the RPC. These LEDs are driven by the RPC-DAQ via an LED driver IC ULN2803. A detailed block diagram of FE boards is shown in Fig. 96.5.

96.4 Data Acquisition

The data acquisition module RPC-DAQ consists of a Cyclone 4 FPGA. It contains 256 TTL compatible IO channels. Out of 256 IOs, 128 channels are used as output for driving LED corresponding to each RPC strip. The remaining 128 channels are used as inputs which receive all the discriminated outputs from the FE boards of all layers both X and Y. These signals are then processed and used for Trigger formation. The same signals are given to a High Performance Time to Digital Converter (HPTDC) for time stamping input signals with 100 picoseconds resolution. A 100 nanosecond Real-Time Clock timestamps every event. On successful chance coincidence, a Trigger will be generated. The trigger criteria are programmable. The same set of

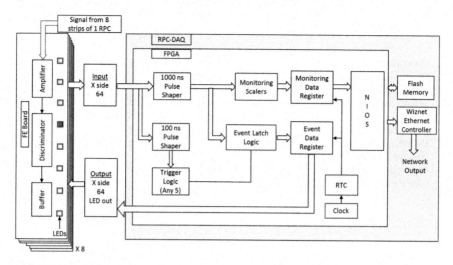

Fig. 96.5 Data Acquisition Module

input signals are also stretched to 1us, to keep the signal available until the trigger forms. On trigger, these 1us stretched input signals are latched. This latch will have binary high for strips that are responsible for trigger formation. This latch is directly connected to the FE board LEDs. So the LEDs which are in front of the strips which are responsible for the Muon trigger will be high, others low.

Until the next trigger forms, the previous latch state is maintained in LEDs of all layers. Each event consists of position, timing and RTC information. A Softcore Processor NIOS is instantiated in the same FPGA to handle data collection and transportation. Wiznet W5300 Module is interfaced with the FPGA which is an Ethernet offload engine. Using Wiznet, data can be transferred using standard Ethernet protocols like UDP and TCP. Server access RPC-DAQ using a dedicated IP address. Figure 6 shows the logic inside the DAQ FPGA.

96.5 Conclusion

Presently, only hit and count rates are being collected by the Back-end server. TDC and TPH interfaces will be added to the Firmware. Autonomous mode and normal modes of operation will be implemented. Also, the possibility of SD card and wireless interfaces with RPC-DAQ will be studied. An Android App will be published.

References

1. INO-Collaboration, India-based neutrino observatory, Tech. Rep. INO/2006/01, Tata Institute of Fundamental Research (January 2006). http://www.ino.tifr.res.in/ino/OpenReports/INOReport.pdf
2. Soft-Core Processor Based Data Acquisition Module for ICAL RPCs with Network Interface Chapter 86, XXI DAE-BRNS High Energy Physics Symposium, Proceedings, Guwahati, India, December 8–12, 2014
3. INO Prototype Detector and Data Acquisition System. http://www.ino.tifr.res.in/ino-old/Pubs/2009/ELE_PUB_2009_01.pdf
4. A compact muon tracking system for didactic and outreach activities. https://www.sciencedirect.com/science/article/pii/S0168900215014813

Chapter 97
Aging Study for Resistive Plate Chambers (RPC) of the CMS Muon Detector for HL-LHC

Priyanka Kumari, Vipin Bhatnagar, and J. B. Singh

Abstract During the phase II of the LHC physics program, called High Luminosity LHC (HL-LHC), the accelerator will increase the instantaneous luminosity up to $5 \times 10^{34} \, \mathrm{cm}^{-2} \, \mathrm{s}^{-1}$. At HL-LHC, the CMS Resistive Plate Chambers (RPC) system will be subjected to high background radiations which could induce non-recoverable aging effects and can alter the detector properties. A new longevity test is then needed to estimate the impact of HL-LHC conditions up to an integrated charge equivalent to the integrated luminosity of $3000 \, \mathrm{fb}^{-1}$, to confirm that the RPC system will survive the harsher background conditions. A dedicated consolidation program is ongoing at the CERN Gamma Irradiation Facility (GIF++), where a few RPC detectors are exposed to intense gamma radiation. The main detector parameters (currents, rate, and resistivity) are under monitoring as a function of the integrated charge and the performance studied with a muon beam. After having collected a significant amount of the total irradiation, preliminary results will be presented.

97.1 Introduction

At present, the Compact Muon Solenoid (CMS) muon system consists of three types of gas-ionization detectors : Drift Tube Chambers (DTC), Cathode Strip Chambers (CSC), and Resistive Plate Chambers (RPC). The RPC system [1] covers both Barrel and Endcap regions of CMS contributing to the trigger, reconstruction, and identification of muons. It consists of 1056 RPCs, organized in 4 stations called RB1 to RB4 in the Barrel region, and RE1 to RE4 in the Endcap region.

The RPC system was designed to provide muon identification, excellent triggering, timing, and momentum measurements at the Large Hadron Collider (LHC) at

On behalf of the CMS Collaboration.

P. Kumari (✉) · V. Bhatnagar · J. B. Singh
Panjab University, Chandigarh, India
e-mail: priyanka.kumari@cern.ch

© Springer Nature Singapore Pte Ltd. 2021
P. K. Behera et al. (eds.), *XXIII DAE High Energy Physics Symposium*,
Springer Proceedings in Physics 261,
https://doi.org/10.1007/978-981-33-4408-2_97

the nominal luminosity of $1 \times 10^{34}\,\mathrm{cm^{-2}\,s^{-1}}$. During the LHC Run1 and Run2 data taking, the performance of the muon systems was outstanding [2].

In the second phase of the LHC physics program, HL-LHC, the instantaneous luminosity will reach $5 \times 10^{34}\,\mathrm{cm^{-2}\,s^{-1}}$ (factor five more then the nominal LHC luminosity), providing to CMS an additional integrated luminosity of about 3000 fb^{-1} over 10 years of operation, starting in 2026. The expected conditions in terms of background, pile-up, and the probable aging of the present detectors will make the muon identification and correct p_T assignment a challenge for the muon system. In order to ensure redundancy also under the HL-LHC conditions, two upgrades [3] are planned on the RPC system: the consolidation of the present system and the extension of the muon coverage at the $|\eta| < 2.4$.

97.2 RPC Aging Studies for the Present System

The present RPC system has been certified for 10 LHC years of operation, at a maximum background rate of 300 Hz/cm^2 and a total integrated charge of 50 mC/cm^2. Based on Run2 data and assuming a linear dependence of the background rates as a function of the instantaneous luminosity, the expected background rates and integrated charge at HL-LHC will be, respectively, $\approx$600 Hz/cm^2 and $\approx$ 840 mC/cm^2, (including a safety factor of three) [3]. HL-LHC will therefore be a challenge for the RPC system since the new operating conditions are much higher with respect to those for which the detectors had been designed, and could induce non-recoverable aging effects that can alter the detector properties and performance.

A new longevity test is then needed to estimate the impact of the HL-LHC conditions up to an integrated charge equivalent to the integrated luminosity of 3000 fb^{-1}, in order to confirm that the RPC system will survive the expected HL-LHC conditions. Longevity studies will identify possible aging effects by monitoring the main detector parameters and performance as a function of the integrated charge.

97.2.1 Setup and Test Procedure

A dedicated longevity study was set up at the CERN Gamma Irradiation Facility (GIF++) where it is possible to test real-size detectors. The facility is equipped with a 13 TBq Cs-137 gamma source, and a system of movable filters allows to variate the gamma irradiation conditions, providing a fairly realistic simulation of the HL-LHC background conditions. A 100 GeV muon beam, providing excellent probes for detector performance studies, complements the source [4]. Since the maximum background rate is expected in the endcap region, in July 2016, the irradiation test was started at GIF++ using four spare endcap chambers: two RE2/2 and two RE4/2 types. Two different types of chambers have been used for this test because the endcap RPC production has been performed in two periods: in 2005 for all RPCs

Fig. 97.1 Integrated charge versus time, accumulated during the longevity test at GIF++ for RE2/2 (red) and RE4/2 (blue) chambers. The RE4/2 chamber has been turned on a few months later because of total gas flow limitations. Different slopes account for different irradiation conditions during data taking

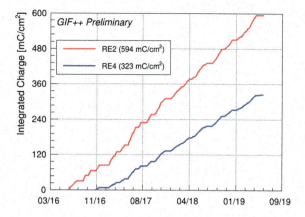

in the endcap system, except the RE4/2 and RE4/3 chambers, which were made in 2013.

In order to study the detectors' longevity, two chambers out of the four (one RE2/2 and one RE4/2) are continuously operated under gamma irradiation, while the remaining two chambers are turned on only from time to time and used as a reference. The main detector parameters are monitored and periodically compared with those of the reference chambers (currents and counting rates at several background conditions, noise and dark current, etc.). Moreover, when the muon beam at GIF++ is available, the detector performance is studied at different irradiation conditions. The method for the data analysis is described in Ref. [5]. All measurements are performed under controlled environmental and gas conditions. The detectors are operating with two gas volume change per hour for the irradiated chambers and one for the reference chambers, and with a relative gas humidity of $\sim$30–40%. The integrated charge versus time is shown in Fig. 97.1. At present, about 592 and 322 mC/cm^2 have been integrated into the RE2/2 and RE4/2 irradiated detectors, which correspond respectively at around 70 and 38% of the expected integrated charge.

97.2.2 Detector Parameter Monitoring

In order to spot possible degradations of the electrode surface due to the irradiation, the detector noise rate and dark current are periodically measured. Figure 97.2 shows the currents (left) and noise rate (right) at the working point as a function of the integrated charge for the RE2/2 irradiated and reference chambers. No significant variations have been observed so far.

The variation of current and rate with background radiation is periodically measured as well. To exclude the dependence on the external parameters, the ratios of the irradiated and the reference chambers are measured as a function of the integrated charge. Figure 97.3 left shows the RE2/2 current and rate ratio.

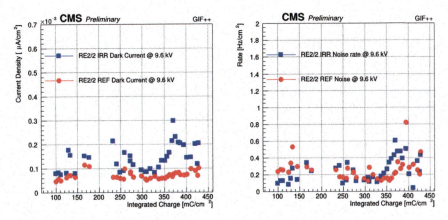

Fig. 97.2 Dark current (left) and noise rate (right) versus the integrated charge, for RE2/2 irradiated (blues) and reference (red) chamber, at the working point voltage

The measurements show a decreasing trend at the beginning of the irradiation period, up to $\approx 300\,mC/cm^2$, when the operating conditions, in terms of gas flow rate and relative gas humidity, were too low with respect to the high gamma background rate. These operating conditions lead to the electrode resistivity increase, which caused the observed rate and current decrease.

The electrodes resistivity increase is confirmed by the measurements performed running the RPCs filled with pure Argon gas. Figure 97.3 (right) shows the coherent correlation between the RE2/2 currents ratio (red) and the resistivity variation (blue). The resistivity variation allow us to cancel out the dependence on the environmental conditions, and is defined as

$$\rho_{var} = \frac{\rho_{irr} - \rho_{ref}}{\rho_{irr}} \tag{97.1}$$

These plots show also that the resistivity increase is a recoverable effect, in fact, the resistivity starts to decrease, and the current and rate increases when the gas flow and the gas relative humidity have been increased.

97.2.3 Detector Performance Monitoring

The detector performance has been tested using the muon beam, before starting the longevity test, and repeated after different irradiation periods at GIF++, up to 51% of the expected integrated charge. Figure 97.4 shows the RE2/2 irradiated chamber efficiency as a function of the effective High Voltage (voltage normalized at the standard temperature and pressure), without irradiation (left), and with a gamma background rate of $\approx 600\,Hz/cm^2$ (right).

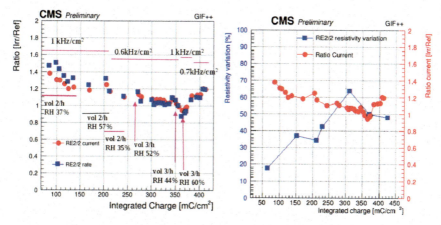

Fig. 97.3 Left: RE2/2 current (red) and rate (blues) ratio between irradiated and reference chamber as a function of the integrated charge. Right: RE2/2 current ratio (red) and resistivity variation (blue)

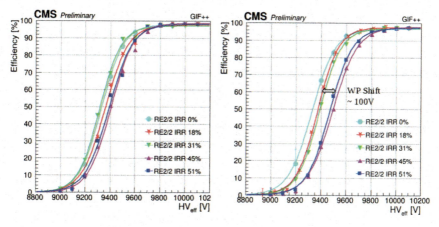

Fig. 97.4 RE2/2 irradiated chamber efficiency as a function of the effective HV, taken with no irradiation (left) and under a gamma background rate of about 600 Hz/cm^2 (right). The efficiency is measured during different Test Beams (TB) corresponding to different fractions of the target charge to integrate

The performance without background rate is stable in time and we do not observe any efficiency degradation or working point shift. With a background rate of $\approx$600 Hz/cm^2, the efficiency remains stable at a working point, but we observe a 100 V shift, starting from 45% of the expected integrated charge. The working point shift is related to the electrode resistivity increase, since the effective voltage applied to the electrodes (HV) is reduced by the voltage drop across the electrodes, which is proportional to the current (I) produced by the ionizing particles and to the bakelite resistance (R) [6]. The effective voltage applied to the gas (HV_{gas}) is therefore defined as

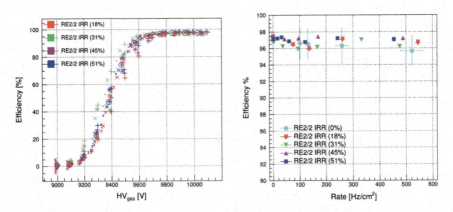

Fig. 97.5 Left: RE2/2 irradiated chamber efficiency as a function of the HV gas, at different background irradiations and at different integrated charge values. Right: RE2/2 irradiated chamber efficiency at the working point as a function of the background rate at different integrated charge values

$$HV_{gas} = HV - RI \tag{97.2}$$

The detector operation regime is invariant with respect to HV_{gas}, therefore the efficiency as a function of HV_{gas} does not depend any longer on the background radiation and on the bakelite resistance. Figure 97.5 (left) represents the RE2/2 irradiated chamber efficiency curves at different background radiation (up to $\approx 600\,\text{Hz/cm}^2$) and at different integrated charge. All the efficiency curves overlap and we do not observe anymore the working point shift, since the electrode resistivity increase effect has been removed. The RE2/2 irradiated chamber efficiency at the working point as a function of the background rate is shown in Fig. 97.5 right. The efficiency is stable in time with a 2% decrease at the highest expected background rate, $600\,\text{Hz/cm}^2$.

97.3   Results and Conclusion

From preliminary results, no evidence of any aging effect has been observed so far in the RPC detectors. The main detector parameters are stable, other than minor variations like the electrode resistivity, that did not affect the performance which remains stable up to 51% of the expected integrated charge. Further investigations are needed to get closer to the final integrated charge of $840\,\text{mC/cm}^2$.

References

1. G. Pugliese [CMS Muon Collaboration], *The RPC System for the CMS Experiment*, 2006 IEEE NSS Conference Record N24-3, January 2007
2. A.M. Sirunyan et al., [CMS Collaboration], *Performance of the CMS Muon Detector and Muon Reconstruction with Proton-proton Collisions* $\sqrt{s} = 13$ TeV, [arXiv:1804.04528 [physics.ins-det]]
3. CMS Collaboration, *The Phase-2 Upgrade of the CMS Muon Detectors*, CERN, Geneva Switzerland, LHC Experiments Committee (2017) [CMS-TDR-016]
4. R. Guida [EN and EP and AIDA GIF++ Collaborations], *GIF++: A new CERN Irradiation Facility to test large-area detectors for the HL-LHC program*, PoS ICHEP **2016** (2016) 260
5. M. Abbrescia et al., Cosmic ray tests of double-gap resistive plate chambers for the CMS experiment. Nucl. Instrum. Meth. A **550**, 116 (2005)
6. G. Pugliese et al., Aging study for resistive plate chambers of the CMS muon trigger detector. Nucl. Instrum. Methods Phys. Res. A **515**, (2003)

Chapter 98
Testing Real-Size Triple GEM Chambers with Pb+Pb Collision at CERN SPS

Ajit Kumar, A. K. Dubey, Jogender Saini, V. Singhal, V. Negi, Ekata Nandy, S. K. Prasad, C. Ghosh, and Subhasis Chattopadhyay

Abstract We discuss the fabrication, assembly and beam tests of two large-size triple Gas Electron Multiplier (GEM) detector prototypes, of CBM Muon Chamber (MuCh), with Pb+Pb collisions at CERN SPS. This was the first prototype test in a multiparticle environment, wherein realistic CBM DAQ collected data from a large number of channels in a free streaming mode, using several nXYTER-based [1] Front End Boards (FEBs) and recording simultaneous hits from over a wide area of the GEM prototypes. The response of the detector and issues related to DAQ, electronics, cooling, etc. have been reported.

A. Kumar (✉)
Homi Bhabha National Institute, Mumbai, India
e-mail: akmaurya@vecc.gov.in

A. K. Dubey · J. Saini · V. Singhal · V. Negi · E. Nandy · C. Ghosh · S. Chattopadhyay
Variable Energy Cyclotron Centre, Kolkata, India
e-mail: anand@vecc.gov.in

J. Saini
e-mail: jsaini@vecc.gov.in

V. Singhal
e-mail: vikas@vecc.gov.in

V. Negi
e-mail: vnegi@vecc.gov.in

E. Nandy
e-mail: ekata@vecc.gov.in

C. Ghosh
e-mail: c.ghosh@vecc.gov.in

S. Chattopadhyay
e-mail: sub@vecc.gov.in

S. K. Prasad
Bose Institute, Kolkata, India
e-mail: sprasad@jcbose.ac.in

© Springer Nature Singapore Pte Ltd. 2021   711
P. K. Behera et al. (eds.), *XXIII DAE High Energy Physics Symposium*,
Springer Proceedings in Physics 261,
https://doi.org/10.1007/978-981-33-4408-2_98

98.1 Introduction

The CBM [2] experiment is a fixed-target heavy-ion experiment at the upcoming facility called FAIR [3] at Darmstadt, Germany. By colliding heavy ions in the energy range of 2–35 AGeV, the goal of CBM is to explore the properties of nuclear matter at high net baryon densities. The physics goals include the study of the fundamental aspects of quantum-chromo-dynamics (QCD) and astrophysics. The CBM experiment will operate at a high interaction rate of $\sim$10 MHz. The challenge thus is to measure rare probes like dilepton pairs decaying from light vector mesons (ρ, ω and ϕ) and charmonium. The MuCh system with its novel arrangement of the alternating absorber and detector stations [4] would detect dimuon signals in a broad momentum range. For the first two stations which face high particle rates and harsh radiation environment, trapezoidal-shaped gas detector modules based on Gas Electron Multiplier (GEM) technology would be employed. Tests of such a large-size ($\sim$80 cm $\times$ $\sim$40 cm) "Mv1" prototype and small ones (10 cm $\times$ 10 cm) with single-particle beams have been reported in [5, 6]. In this contribution, we present the performance of the large-size prototype in Pb+Pb collisions at the H4 beamline of CERN SPS, wherein, a simultaneous response from the full active area of the detector due to a spray of high-energy particles originating from the nucleus-nucleus collision at 150 AGeV/c has been studied for the first time. The new CBM DAQ took data in free streaming mode, involving a large number of FEBs (Fig. 98.2b) for the first time. It was also the prototype tests employing a water-based cooling system.

98.2 Fabrication of Large-Size Chamber

Two real-size detector triple GEM modules were fabricated for the tests. Figure 98.1 shows the fabrication process in a CPDA lab (class 100,1000), clean room at VECC, Kolkata. The assembly procedure is similar to that described in [5, 7]. Large size single-mask GEM foils having 24 segments were used. A translucent glow-box was locally designed and built for the visual inspection of the foils, as shown in Fig. 98.1a. Each foil segment showed a leakage current of $\sim$3–5 nA (at $\sim$50% humidity and $\sim$23° temperature) at 550 V; dimensions of all the components were measured and ascertained to be within specified tolerance levels. After all the satisfactory quality checks, the foils were stacked together in a 3/1/1/1.5 mm gap configuration (Fig. 98.1b) and stretched using glueless "ns2"technique [8]. The stretching of GEM foil is demonstrated in Fig. 98.1c. The picture of the complete assembled module is shown in Fig. 98.1d. The GEM foils were powered using a resistive chain as mentioned in [7].

Each of the trapezoidal GEM modules consisted of about $\sim$1900 readout pads having progressively increasing pad sizes, as guided by the design simulation of the CBM-MUCH. A total of 15 FEBs (each having 128 channels) per module was used. Data were taken with almost full FEB coverage. An elaborate cooling [9, 10] arrangement involving a 10 mm Aluminum plate with controlled water flow either

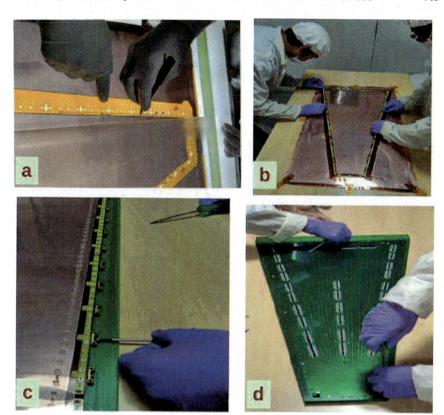

Fig. 98.1 Fabrication of large-size triple GEM detector in CPDA lab, VECC. **a** Inspection of GEM foil. **b** Placing GEM foils on top of each other with a gap of 3 mm / 1 mm / 1 mm / 1.5 mm. **c** Mechanical stretching of the foils. **d** Complete chamber

through grooved channels or through 6 mm Al pipe winding inside the plate volume was used for the first time. Figure 98.2b shows the picture of the cooling plate with all the FEBs mounted on one side. Small, flat copper pieces of 3 mm thickness were thermal-glued below the FEBs, providing the metal contact with the plate. The detector is mounted on the other side of this plate. 10 mm wide slots were machined at 15 readout connector positions. The FEBs were connected with these connectors via 10 cm long flexible Kapton cables each carrying signals from 64 channels.

98.3 Experimental Setup

The schematic of the experimental setup is shown in Fig. 98.2a. Pb beams of 150 GeV/c collided on a 3 mm thick Pb target having an area of 9 cm $\times$ 9 cm. Additionally, 2 blocks of Fe of thicknesses 4 cm and 6 cm were placed downstream of

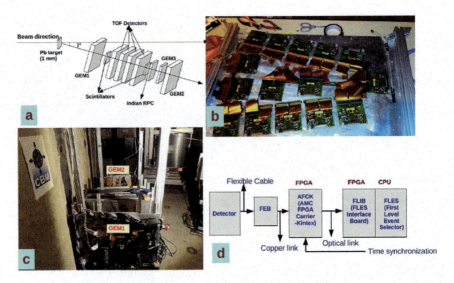

Fig. 98.2 Schematic and picture of test setup and DAQ. **a** Block diagram of experimental setup. **b** FEB boards mounted on Al plate. c)Picture of the experimental setup inside a cave of the H4 beamline. d) Block diagram of DAQ setup

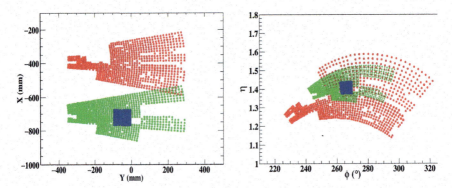

Fig. 98.3 **a** Global X-Y distribution of detector hits. **b** η-ϕ distribution of detector hits (Red-GEM1, Green-GEM2 and Blue-GEM3)

from the Pb target at $\sim$2 cm and $\sim$17 cm, respectively, for increasing interaction rate. A diamond detector was placed just before the target to provide beam information. The prototype detectors were placed on a common mounting frame, tilted at about $\sim$7° from the beam axis, considering the target positioned at (0,0,0). Two "Mv1" modules, one in the front and the other at the back of the setup, were placed as shown in the figure. The first GEM module (GEM-1) was placed at $\sim$3 m distance from the target and its upper boundary about $\sim$20 cm below the beam axis. The second GEM module (GEM-2) was placed at $\sim$6 m from the target position. A third triple GEM chamber of size 10 cm $\times$ 10 cm, GEM-3 (built at GSI), was also put in at a later stage

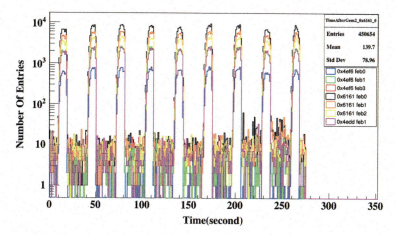

Fig. 98.4 Spill structure seen on GEM2 plane

to provide an additional hit-point for tracking. The global X (mm)-Y (mm) distribution of pad coordinates for the three GEM detectors is shown in Fig. 98.3 (left panel) and the corresponding η-ϕ coordinates are shown in Fig. 98.3 (right panel). A picture of the final setup in the cave of the H4 area of CERN SPS is shown in Fig. 98.2c. A premixed gas mixture containing Ar/CO_2 in a ratio of 70/30 was used as the fill gas for all three GEM detectors. The block diagram of the data acquisition system is shown in Fig. 98.2d. Data from different detectors (TOF, MuCh) were processed by FPGA-based Data Processing Board (DPB) [11]. These DPBs were configured on an AFCK board, and it was mounted on a μTCA crate. Twisted-pair LVDS flat-ribbon cables, ~6 m in length, were used as signal cables from the back-end of FEBs to the front-end of AFCK boards. An optical cable of ~50 m in length was used from the back-end of the AFCK to the FLIB (FLES Interface Board) board which was mounted on FLIB-PC. Time synchronization for combining the detector hits in the two subsystems was carried out via two dedicated AFCK (master and slave) boards placed in the same crate.

98.4   Results and Discussion

Pb-Pb collision data taken for 150 GeV/c have been analyzed. Individual channel hits in the detector prototypes due to particle hits were recorded along with the corresponding time-stamps and amplitude. This free-streaming data was stored in time slices of 10 ms interval. Data were taken for different GEM voltages and also for different noise thresholds. The unsorted raw data was sorted within every time slice during unpacking and later corrected for offsets.

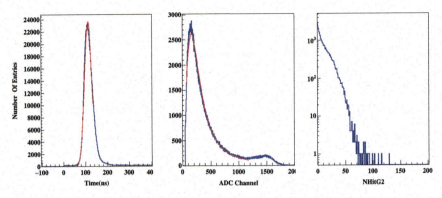

Fig. 98.5 Left: Time difference spectra between GEM and diamond detector. Middle: Pulse height distribution with $3\times\sigma$ window of time correlation. Right: Hit multiplicity distribution

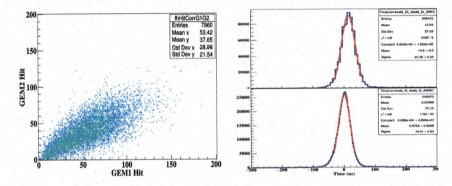

Fig. 98.6 Left: Hit multiplicity correlation between GEM1 and GEM2. Middle: Time difference spectra between one FEB with another FEB of the same GEM plane before time-walk correction. Right: after time-walk correction

The incident Pb beam for collisions was delivered in spills. Figure 98.4 shows the spill structure as observed from the hits in the GEM2 detector. A spill frequency of ~ 30 s and spill width of ~ 9 s can be noted. The various colors in the figure represent different FEBs of GEM2 connected to pads in different regions. Constructing single events from the collection of hits with time-stamp becomes a challenging task. The diamond detector (placed just before the target) was selected as a reference detector for building an event. A simple algorithm wherein all the GEM hits which lie between two consecutive diamond hits was taken to be one event. The time difference spectra of hits in GEM2 and Diamond, event by event, is shown in Fig. 98.5(left). The σ of this spectra is of the order of ~ 14ns, which is the measure of the time resolution of the detector. The pulse height spectra fitted to the Landau distribution for GEM2 within 3σ window of time correlation spectra are shown in Fig. 98.5(middle). The distribution of number hits per event (hit multiplicity) for GEM2 detector and its

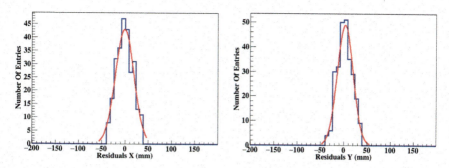

Fig. 98.7 Residuals in X (mm) and Y (mm) at GEM1 plane

correlation with those in GEM1, event-by-event, is shown in Figs. 98.5(right) and 98.6 (left), respectively.

Figure 98.6 (top-bottom) shows the time correlation spectra between hits of one FEB with that of another FEB in the GEM2 plane. The spectra peaks at ~14 ns with σ of ~23 ns and after walk correction, a reduction of ~3ns was observed, as shown in Fig. 98.6 (right-bottom). Detector time resolutions add up in quadrature. The sigma value for the combined FEBs, after walk correction is found to be consistent with this theoretical deduction considering a value of 14 ns from Fig. 98.5.

A track-fit was carried out using hit coordinates in different detector planes. Considering the origin $(0, 0, 0)$ and two GEM hits in detectors GEM2 and GEM3 lying in a common window of $1.39 < \eta < 1.42$ and $262 < \phi < 264$, a straight-line fit was carried out, and the extrapolated hit coordinates in GEM1 plane were calculated based on the fit parameters. The distribution of the residual thus obtained for GEM1 hits in X and Y is shown in Fig. 98.7.

In summary, we have tested for the first time large-size GEM detectors in a multi-particle environment coupled to self-triggered electronics and studied its response in terms of time correlation spectra, hit correlation, and event reconstruction. A straight-line fit was carried out using the hits in the three GEM planes and the diamond. The corresponding residual distributions were also studied.

Acknowledgements AK acknowledges the receipt of the DAE-HBNI fellowship. The authors would like to thank GSI colleagues and CBM-TOF team for their help and support. Also, we would like to thank Jörg Hehner for the 10 cm × 10 cm triple GEM chamber. We also like to thank Nuclear Physics Division (NPD), VECC, for providing the clean room.

References

1. https://cbm-wiki.gsi.de/foswiki/pub/Public/PublicNxyter/nXYTER.pdf
2. CBM, GSI. https://www.gsi.de/work/forschung/cbmnqm/cbm.htm
3. FAIR, GSI. https://fair-center.eu/

4. Muon Chamber, Technical Design Report. http://repository.gsi.de/record/161297/files/much-tdr-final-for-gsi-report.pdf
5. R.P. Adak et al., Nucl. Instr. Meth. A **846**, 29–35 (2017)
6. A.K. Dubey et al., Nucl. Instr. Meth. A **755**, 62–68 (2014)
7. A.K. Dubey et al., A real size prototype for CBM MUCH. In: CBM Progress Report (2015), pp. 56 (2015)
8. L. Franconi, et al., Status of no-stretch no-spacer GEM assembly, the NS2 technique method and experiment result. https://indico.cern.ch/event/176664/contributions/1442160/attachments/229650/321300/The_NS2_Technology.pdf
9. D. Nag, et.al., Study and design of a water based cooling system for MuCh CBM. In: CBM Progress Report (2015), pp. 90 (2016)
10. https://www.sympnp.org/proceedings/61/G76.pdf
11. W. Zabolotny, et.al., Towards the Data Processing Boards for the CBM experiment. In: CBM Progress Report (2015), pp. 97 (2014)

Chapter 99
HARDROC2B: A Readout ASIC for INO-ICAL RPCs

Aman Phogat, Moh. Rafik, Ashok Kumar, and Md. Naimuddin

Abstract The India-based neutrino observatory (INO) is an upcoming multi-purpose underground experiment in the Theni district of Tamil Nadu, with the main aim to determine the neutrino mass hierarchy. In order to satisfy the experimental requirements, the ICAL detector at INO will use 29000 Resistive Plate Chambers (RPCs) as the active detection element, which in turn requires millions of channels to be read out. In this paper, we report on the test results and integration of a new Front-end Application-Specific Integrated Circuit (ASIC), called HARDROC, with an RPC detector. The ASIC comprises 64 readout channels with a variable gain current preamplifiers followed by one slow and three fast shapers and a 10 bit-DACs in each channel. The present study includes the optimisation of various parameters such as preamplifier gain and DAC values for threshold settings. We will also present the important RPC performance results like hit registration efficiency and count rate variation with DAC values.

99.1 Introduction

The proposed 50 kton magnetised Iron Calorimeter (ICAL) detector at the India-based Neutrino Observatory (INO) aims to investigate atmospheric neutrino oscillations. Low cost, good timing capabilities, long time stability and high efficiency are among the main factors which make Resistive Plate Chamber (RPC) [1, 2] favourable as the active detector element. ICAL will employ about 29000 glass Resistive Plate Chambers (RPCs) for charge detection and requires millions of electronic channels to be read out. Therefore, a dedicated multichannel front-end readout application-specific integrated circuit (ASIC) named HARDROC is under consideration as a possible option to fully exploit the advantageous features of INO-ICAL RPC detectors. In fact, due to a large number of electronic channels to be read out, only an ASIC-based readout can assure satisfactory results when specifications in

A. Phogat (✉) · Moh. Rafik · A. Kumar · Md. Naimuddin
Department of Physics and Astrophysics, University of Delhi, Delhi, India
e-mail: amanphogat.phogat@gmail.com

© Springer Nature Singapore Pte Ltd. 2021
P. K. Behera et al. (eds.), *XXIII DAE High Energy Physics Symposium*,
Springer Proceedings in Physics 261,
https://doi.org/10.1007/978-981-33-4408-2_99

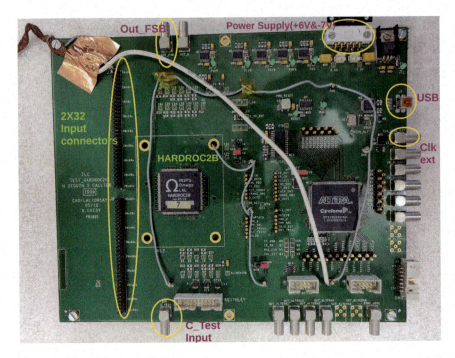

Fig. 99.1 HARDROC2 ASIC Board

terms of compactness, detector-embeddedness, reliability and power consumption for the detection system have to be fulfilled. HARDROC ASIC [3, 4] is a 64 channel analog-digital front-end chip which can read negative fast and short input signals. It is being developed at the Austrian Micro System (AMS) laboratory and fabricated in the 350 nm Silicon-Germanium technology. Version 2 of HARDROC was tested with prototype glass RPC detectors and its performance is reported.

99.2 Description of the HARDROC2 ASIC

HARDROC (HAdronic Rpc Detector ReadOut Chip) is a 64-channel front-end ASIC designed primarily for the readout of a gaseous detector like RPC. A view of the HARDROC2 ASIC board is shown in Fig. 99.1. Each of the 64 channels of HARDROC2 is made of a fast low-input impedance current sensitive preamplifier with 8-bit variable gain (analog gain is between 0 to 2). The received RPC signals are amplified and are shaped further. A variable slow shaping filter with peaking time 50ns to 150ns is followed by a Track and Hold buffer to provide a multiplexed analog charge output up to 10pC charge. The preamplifier is also coupled to 3 variable gain CR-RC bipolar fast shapers (FSBs), with peaking time 20–25 ns. The bipolar fast

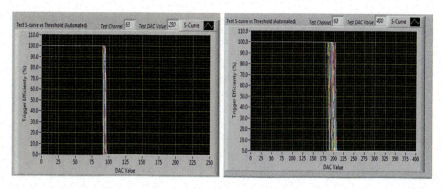

Fig. 99.2 50% efficiency curve of (left) pedestal and (right) with 100 fC charge input for all the 64 channels of the HARDROC2 ASIC

shaper FSB0 is dedicated for input charges from 10fC up to 100fC, FSB1 for input charges from 100fC up to 1pC and FSB2 for input charges from 1pC up to 10pC. This triple-branch shaper stage is followed by 3 low-offset discriminators. Three internal 10-bit Digital to Analog Converters (DAC) are used to adjust the global thresholds. The output signal of the shaper is compared to a programmable threshold by the course of a fast voltage comparator, which generates a trigger signal as soon as a valid event is detected. Three discriminators are cascaded to a 3 input to 2 output encoder. A 127 deep digital memory is used to store the 2-bit encoded outputs [5].

99.3 S-Curve Method

The performance of the HARDROC2 ASIC has been gauged with tests performed for the better apprehension of the different operational parameters and their impact on the readout behaviour of the chip. The s-curve test allows channel response efficiency to be estimated in terms of the applied threshold and noise measurement [6, 7]. The s-curve procedure consists of injecting a given charge on each of the 64 channels of HARDROC2 through the inbuilt input capacitor. In our case, an input charge of 100 fC was injected into the HARDROC test board through an arbitrary waveform generator. The discriminator threshold scan has been performed over the whole dynamic range (0–1023) in steps of 1 DAC units. The declension point on the efficiency curve for each channel for a fixed preamplifier gain is determined. Figure 99.2 (left) shows the s-curve for the 64 channels without giving any input charge, i.e. the pedestal value, and the results (right) show the s-curve for a 100 fC reference charge. The pedestal average was found to be around 94 DAC units. Pedestal subtraction is usually the first step for a given readout system and a cutoff value is then applied to the data. The measured performance is adequate for our application.

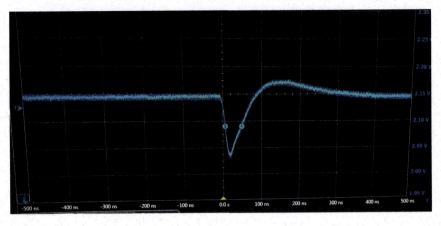

Fig. 99.3 HARDROC response to RPC pulse

99.4 RPC Pulse

The behaviour of the analog channel has been investigated by means of the injection raw avalanche RPC signal on the HARDROC2 front-end board. Figure 99.3 shows the output pulse of the shaper in response to an injected signal, setting preamp gain= 28. The resulting peaking time is about 20 ns.

99.5 Experimental Setup

A prototype 30 cm × 30 cm glass-based Resistive Plate Chamber (RPC) is integrated and commissioned with the HARDROC2 ASIC and tested for cosmic ray muons. RPC is made of two parallel 3mm thin glass plates. A uniform 2 mm gas gap is maintained between the glass plates by polycarbonate button spacers. A thin layer of semi-resistive graphite coating is applied on the outer side of the glass plates to provide the necessary high voltage. When a charged particle crosses the gas gap, it ionises the gas and produces electron-ion clusters. Due to high voltage, an avalanche is produced and the electrons are attracted towards the anode side of the glass plate. They will induce a signal on the copper readout strips which are separated from graphite paint via insulating a mylar sheet. A gas mixture of 94.5% tetrafluoroethane (R134a), 5% iso-butane and 0.5 %sulphur hexafluoride was used to operate the RPC in avalanche mode of operation [8].

A trigger system of three scintillators each coupled to a photomultiplier tube (PMT) was used to carry out efficiency performance measurement when cosmic ray muons cross the detector. Two scintillators of dimension 15 cm × 60 cm and 2.8 cm × 28 cm were placed above the RPC strip under consideration and one large scintillator of dimension 15 cm × 60 cm was placed below the chamber. The coincidence signal

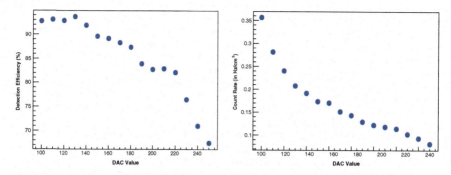

Fig. 99.4 Efficiency (left) and Count Rate (right) variation with DAC threshold value

of these three scintillators has been used for triggering and was chosen for 3 fold counts. The efficiency is obtained by evaluating the ratio between the number of counts in which the RPC strip under consideration has fired to the total number of triggered events in the time window of 200 ns.

99.6   Results

A detailed study of the efficiency dependence upon the DAC threshold was done at the 10.4 kV working point. The trigger threshold was optimised to maximise efficiency while minimising the noise contribution. A common preamplifier gain of 28 was used. Since the pedestal average was found to be around 94 DAC units, the threshold is scanned above the pedestal value. The efficiency and count rate were scanned in 5 DAC unit steps. Figure 99.4 shows the efficiency and count rate variation as a function of the threshold. The efficiency decreases down to 70% at 240 DAC value (see Fig. 99.4 (left)) and count rate moves as expected, i.e. lower as the threshold increases (see Fig. 99.4 (right)).

99.7   Conclusion

We have successfully commissioned and integrated the 30 cm × 30 cm single gap RPCs with the HARDROC front-end ASIC. The variation of the efficiency and count rate with DAC value is reported. The efficiency of RPC at plateau working voltage reached 94% at 140 DAC threshold. The count rate was measured to be less than 0.4 Hz/cm^2. The signal contamination is thus negligible.

References

1. R. Santonico, R. Cardarelli, Development of resistive plate counters. Nucl. Instrum. Methods A. **187**, 377–380 (1981)
2. P. Fonte, Applications and new developments in resistive plate chambers. IEEE Trans. Nucl. Sci. **49**, 881–887 (2002)
3. Moh Rafik, A. Phogat, A. Gaur, A.Kumar, Md. Naimuddin, Front-End Readout for INO-ICAL GRPC, Springer Proc. Phys. **203**, 805–807 (2018)
4. S. Callier, J.B. Cizel, F. Dulucq, C. De La Taille, G. Martin-Chassard, N. Seguin-Moreau, ROC chips for imaging calorimetry at the International Linear Collider **9**, C02022 (2014)
5. A.Kumar, A. Gaur, A. Phogat, Md Rafik, Md. Naimuddin, Development and commissioning of the HARDROC based readout for the INO-ICAL experiment, JINST (2016) https://doi.org/10.1088/1748-0221/11/10/C10004
6. A. Phogat, A. Gaur, Moh Rafik, A.Kumar, Md. Naimuddin, New front-end electronics for INO-ICAL experiment, Nucl. Instrum. Meth. A. **905**, 193–198 (2018)
7. M. Bedjidian et al., Performance of Glass Resistive Plate Chambers for a high-granularity semi-digital calorimeter. JINST **6**, P02001 (2011)
8. A. Phogat, A. Gaur, A.Kumar, Moh Rafik, Md. Naimuddin, Performance Study of Large Size RPC Detector for INO-ICAL Experiment, Springer Proc. Phys. **203**, 755–757 (2018)

Chapter 100
Effect of Variation of Surface Resistivity of Graphite Layer in RPC

Anil Kumar, V. Kumar, Supratik Mukhopadhyay, Sandip Sarkar, and Nayana Majumdar

Abstract The non-uniform surface and bulk resistivity of the Graphite layer present in a Resistive Plate Chamber (RPC) may affect detector response and dead time. In this paper, we present the initial results of a study that is oriented toward investigating the effects of resistivity of different materials on RPC signal generation.

100.1 Introduction

Resistive Plate Chamber (RPC) [1] is an active detector in the Iron CALorimeter (ICAL) detector at the India-based Neutrino Observatory (INO). The INO Project is a multi-institutional effort aimed at building a world-class underground laboratory with a rock cover of approximately 1200 m for non-accelerator-based high energy and nuclear physics research in India as explained in the Physics White Paper of the ICAL (INO) Collaboration [2].

Components of the INO Project

- Construction of an underground laboratory and associated surface facilities at Pottipuram in Bodi West Hills of Theni District of Tamil Nadu, India.

A. Kumar (✉) · V. Kumar · S. Mukhopadhyay · S. Sarkar · N. Majumdar
Saha Institute of Nuclear Physics, Bidhannager, Kolkata, India
e-mail: anil.kumar@saha.ac.in

V. Kumar
e-mail: vishal.kumar@saha.ac.in

S. Mukhopadhyay
e-mail: supratik.mukhopadhyay@saha.ac.in

S. Sarkar
e-mail: sandip.sarkar@saha.ac.in

N. Majumdar
e-mail: nayana.majumdar@saha.ac.in

Homi Bhabha National Institute, Trombay, Mumbai, India

© Springer Nature Singapore Pte Ltd. 2021
P. K. Behera et al. (eds.), *XXIII DAE High Energy Physics Symposium*,
Springer Proceedings in Physics 261,
https://doi.org/10.1007/978-981-33-4408-2_100

- Construction of an Iron Calorimeter (ICAL) detector consisting of 50000 tons of magnetized iron plates arranged in stacks.
- 29000 RPCs of size 2 m × 2 m would be inserted as active detectors in the gap between the iron layers.

There are several reasons that can lead to the non-uniform surface and bulk resistivity which, in turn, can affect RPC detector response and dead time. In this paper, we present the initial results of a study that is oriented toward investigating the effects of resistivity of different materials on RPC signal generation during the ICAL experiment and their analysis. We have simulated potential buildup across the graphite layer for uniform as well as non-uniform surface resistivities and studied their timing behavior.

100.2 Resistive Plate Chamber

A Resistive Plate Chamber is a gaseous detector where the gas is confined between two parallel resistive plates made up of glass or bakelite (Fig. 100.1). The electric field is present inside the gas gap to collect primary ions produced by passing charged particles and also produce secondary electron-ion pairs. The motion of electrons inside the gas gap induces a signal voltage on the pickup strip made up of copper. The resistive plates have very high resistivity which makes them transparent for signal induction but the high voltage cannot be applied to them. The resistive plates have a coating of graphite layer from outside for making electrical contact.

The resistivity of the graphite layer should be low enough to allow the uniform potential buildup and high enough to not disturb the signal induction. The surface resistivity of the graphite layer in the RPC should be around 1 MΩ/$\square$. The surface resistivity depends on the thickness of the layer and a non-uniform surface resistivity may result due to the non-uniform thickness of the layer. We are investigating the effect of non-uniform surface resistivity on the potential distribution and charge transport on the resistive layer.

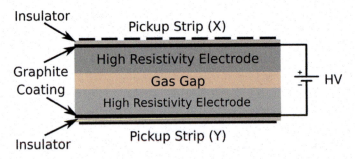

Fig. 100.1 Schematic diagram of RPC

100.3 Charge Transport Simulation

We use a mathematical framework as broadly described in [3], where the simulation of charge transport is carried out by the solution of the Poisson equation through the surface of interest at each time step and calculation of the currents between small subcells of the surface.

$$\nabla^2 V = -\frac{\sigma}{\epsilon}\delta(z) \tag{100.1}$$

Here, "method of moments" is used to solve the corresponding Poisson equation.

The initial conditions are given in terms of the charge on the subcells. The potential, as well as the charge, at a later time is calculated by this program for uniform surface conductivity. In order to make the code more realistic, we have introduced a non-uniform conductivity matrix for various subcells and used it to study the effect of non-uniformity in surface resistivity. The input resistivity of these subcells has been obtained from the experiment as described in the next section.

100.4 Surface Resistivity Measurement

The resistivity of the graphite layer has been obtained by measuring the resistance using a square zig. The resistance of a square region is equal to the surface resistivity. We have used a square probe of size $1\,cm^2$ attached to the AEROTECH PRO165 linear stage programmed in AEROBASIC programming language to traverse the surface of the Graphite layer with an accuracy of $10\,\mu m$. The resistance is measured by a pico-ammeter which can supply voltage and measure current with high accuracy. The data acquisition from the KEITHLEY 6487 pico-ammeter is done through a Python program using a GPIB interface. We have obtained a resistivity map by synchronizing the pico-ammeter as the linear stage traverses the XY plane of the graphite layer. The area of $10 \times 10\,cm^2$ has been divided into 100 cells of size $1\,cm^2$ each as shown in Fig. 100.2b.

100.5 Simulation of Potential Buildup for Uniform Resistivity

A potential of 5000 V is applied at the left side of the resistive layer of size $10 \times 10\,cm^2$ having a uniform resistivity of $2 \times 10^5\,M\Omega$. The relative permittivity of the graphite layer is around 10–15, hence, that of the resistive layer is taken as 10 in the current simulation. The potential and charge at each cell is zero at time $t = 0$. The simulation calculates the charge and potential at each cell after time $t = 40\,\mu s$.

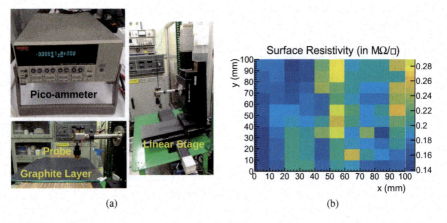

(a)   (b)

Fig. 100.2 **a** Experimental setup with square zig probe to measure surface resistivity. **b** Experimentally measured surface resistivity of graphite layer

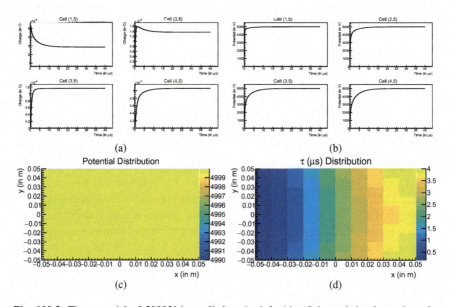

(a)   (b)

(c)   (d)

Fig. 100.3 The potential of 5000 V is applied at the left side of the resistive layer through a conducting contact. **a** Charge buildup as a function of time in a cell (1,5) to (4,5). **b** Potential buildup as a function of time in cells (1,5) to (4,5). **c** Potential distribution for uniform resistivity after 40 μs. **d** Distribution of time required to reach a fixed voltage of $(1 - 1/e) \times 5000$ V for uniform surface resistivity

Figure 100.3a shows that the charge buildup as a function of time in cells (1,5), (2,5), (3,5), and (4,5). A charge pile up can be seen in the cells (1,5) and (2,5) which can be due to a sudden increase in resistivity at the contact point where voltage is applied. The charge increases gradually and then saturates for the cells far away from the applied voltage point.

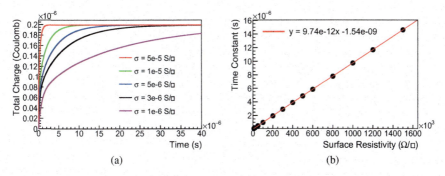

Fig. 100.4   a Comparison of charging behavior for various surface conductivities. **b** Time constant as a function of surface resistivity variation

Figure 100.3b shows that the potential increases with time and then saturates. The rate of increase of the potential slows as we move away from the applied voltage. Figure 100.3c shows nearly constant potential distribution for each cell at time $t = 40$ μs. Although all cells have reached the same potential, it takes different time for cells to saturate. We have defined the time constant τ as the time required to reach a fixed voltage $(1 - 1/e) \times 5000$ V. Figure 100.3d shows the distribution of time constant τ which increases as we move away from the point of application of voltage.

Figure 100.4a shows the total charge of the layer as a function of time for different surface conductivities. It can be observed that the charging becomes faster for higher surface conductivity (or lower surface resistivity) and the total charge at saturation is independent of surface conductivity. Figure 100.4b shows time constant as a linear function of surface resistivity where the time constant is defined as the time required by the total charge to reach $(1 - 1/e) \times Q_{Total}$ at Saturation.

100.6   Simulation of Potential Buildup Across Non-uniform Resistivity

In this section, we describe the simulation of potential buildup across experimentally measured non-uniform surface resistivity as mentioned in Sect. 100.4. Figure 100.5a shows the simulated potential distribution for experimentally measured non-uniform resistivity after 40 μs. It can be observed that the potential distribution is constant but each cell requires a different time to reach the final potential. Figure 100.5b shows the distribution of time required to reach a fixed potential of $(1 - 1/e) \times 5000$ V for experimentally measured non-uniform resistivity. The non-uniformity of surface resistivity results in a slight difference in τ distribution compared to that in Fig. 100.3d.

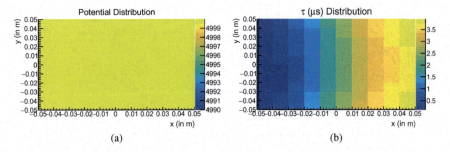

Fig. 100.5 **a** Simulated potential distribution for experimentally measured non-uniform resistivity after 40 μs. **b** Simulated distribution of time required to reach a fixed voltage of $(1 - 1/e) \times 5000$ V for experimentally measured non-uniform resistivity

100.7   Summary

The potential buildup across the graphite layer is simulated for uniform and non-uniform surface resistivities on applying a potential. The experimental setup is developed to measure the two-dimensional distribution of surface resistivity which is given as input to the simulation. The simulation shows that the final potential distribution is constant for uniform as well as non-uniform resistivities. The non-uniformity in surface resistivity disturbs the distribution of time required to reach the final potential.

References

1. R. Santonico, R. Cardarelli, Nuc. Instrum. Methods Phys. Res. **187**(2), 377 (1981)
2. A. Kumar et al., Pramana **88**(5), 79 (2017)
3. N.B. Budanur, Simulation Studies of Charge Transport on Resistive Structures in Gaseous Ionization Detectors. Master's thesis, Istanbul, Tech. U. (2012)